POLYMER CHEMISTRY

POLYMER CHEMISTRY

Bruno Vollmert

Translated from the German by
Edmund H. Immergut

Springer-Verlag
Berlin • Heidelberg • New York
1973

Bruno Vollmert
Director
Polymer Institute
University of Karlsruhe

Edmund H. Immergut
Adjunct Professor
Institute of Polymer Research
Polytechnic Institute of Brooklyn

ISBN 978-3-642-65295-0   ISBN 978-3-642-65293-6 (eBook)
DOI 10.1007/978-3-642-65293-6

FOREWORD

There is, at present, no scarcity of polymer textbooks in the English language. Some of them attempt to cover the entire field, others focus their attention on certain parts of it, e.g., organic chemistry, physical chemistry, solid state physics, etc. This situation must necessarily raise the question, "Why publish another book?" and, even more, "Why translate a book which exists already in German?" and is to a lesser or greater extent legible and comprehensible to many English-speaking scientists.

It appears that a justification can be found in the special character of its content and presentation. As far as content is concerned, Vollmert's book is more encompassing than most existing treatises and, in this sense, almost represents a hybrid between a "textbook" and a "handbook." Numerous figures and tables convey directly a wealth of data. On the other hand, the text is designed to be educational and, in many instances, goes a long way to explain why certain properties are observed and why certain processes take place. These excursions into the intellectual clarification of somewhat complicated phenomena are a refreshing and unusual interruption of the main stream which presents synthesis, characterization and properties of polymeric systems in the classical way.

In the second chapter emphasis is placed on the systematic and complete description of the different types of polymerization processes and of the corresponding techniques with a very welcome chapter on enzymatic and protein syntheses, which can usually not be found in textbooks of polymer chemistry.

In Chapter 3 the properties of the individual macromolecule are presented in much detail. These details are important but are usually treated in only a superficial manner; here experimentation and theory are receiving equal consideration.

In the last chapter — states of molecular aggregation — one finds an up-to-date exposition of the recent important developments in the field of polymer crystallization and deformation.

The writer of this blend of preface and review hopes that this book's special features which have just been enumerated will provide for it a useful and enduring place in the literature.

H. F. MARK
Institute of Polymer Research
Polytechnic Institute of Brooklyn

AUTHOR'S PREFACE

Macromolecular chemistry, or polymer chemistry, founded by the scientific work of H. Staudinger, W. H. Carothers, W. Kuhn and H. F. Mark, can be called "Variations on a Theme of 'Chains.' "

The first part of this book, after a brief survey of different chain structures (Chapter 1), demonstrates the types of synthesis through which long, chain-like macromolecules are formed from small monomer molecules (Chapter 2). Through the nature of the monomer and through the nature and conditions of the polymer synthesis, the primary structure, i.e. the type and arrangement of structural units in the chain, is defined. The primary structure may be changed, intentionally or unintentionally, by chemical reactions of the polymer with low molecular weight reagents or with other polymers. All matters concerning the primary structure have been dealt with in Chapter 2.

The third chapter is devoted to the discussion of factors which do not concern the arrangement of the single links in the chain, but the chain as a whole, thus, the chain length (molecular weight and molecular weight distribution) and the three-dimensional arrangement of the chain, i.e. the secondary structure. If a macromolecule is not influenced by forces of interaction, the long chain exists as a loosely and irregularly folded coil – a "random coil." Random coils exist in dilute polymer solutions as rubberlike, swollen gel particles, which differ from usual material particles in a characteristic manner in that their average density depends on their mass in accordance with theoretical principles, a phenomenon otherwise only found with material "particles" of the magnitude of fixed stars (the reasons are very different indeed).

While the third chapter deals with the properties of the single or free-moving macromolecule, the fourth and last chapter is devoted to the discussion of the different forms of the aggregated state of polymers: the solution, the gel, the rubber, the glass and the fiber.

In the aggregated state, the random coil remains the preferred element of structure. The characteristic property of the single coil, its entropy-elasticity, is carried over into the macroscopic polymer sample where it appears as rubber elasticity. The mobility of the chain defines the temperature at which the rubber freezes to a glass. The degree of regularity of the primary structure decides whether or not, and to what extent, a polymer crystallizes. Within the sphere of crystallinity, the random coil changes into other secondary structures – the helix, double-helix, folded chain-bundle or chain rope. Different secondary structures can exist together in an equilibrium state.

The author of a textbook is obliged to select scientific topics or problems for discussion, and there is no general formula which enables him to appraise the degree of importance of any particular topic. It is not possible, therefore, to write a book in which everybody will find every topic which he considers important enough to warrant representation. Furthermore, in some cases, different opinions or theories are held on the same problem. The reader will find all this in monographs and review articles.

All chapters of this edition are revised, enlarged, and in parts newly written.

The question has often been posed as to whether macromolecular chemistry ought to be regarded as an autonomous discipline of the chemical sciences, in the same way as inorganic, physical or organic chemistry, or whether it could be taught better as a section of organic and physical chemistry. This is not a question of a basic nature, but one of expediency, and capacity of the specialists. If at a university, the professors of organic and inorganic chemistry could decide upon which sections of physical chemistry they would teach in their lectures, the chair of physical chemistry would be superfluous. This situation, of course, assumes that the two scientists, the inorganic and organic chemist, are not only willing to teach physical chemistry, but also able to do so. This could happen. I feel we have a similar situation with regard to polymer science.

During the past few years, I have often been asked for the translation of the Aristotle quotation on the first page of the German edition. Here is the translation:

"The sciences, pursued for their own sake, that is, for the sake of knowledge, deserve to be called wise, to a greater extent than the sciences pursued for the sake of commercial profit. All of these may be more necessary, but none is more worthy."

This quotation was inserted as a dedication to, and in memory of a teacher whom I particularly respect. I have to admit that in our world, worthiness is seldom preferred to benefit and advantage. However, it is perhaps a good thing that we remember from time to time that at all times, men have lived, who did not only claim it but have also done it. The full meaning of the quotation becomes clear only when we take into consideration Aristotle's own interpretation: "Of dignified and divine things we have only little knowledge. But although we may hardly reach up to these higher regions, this kind of knowledge, because of its greater dignity, is more desirable to us than all things of our own world — just as we are more blest when we obtain only a brief glance of a beloved being than when we behold many other and even important things with great clarity."

It was a long time ago when Aristotle wrote these words but their message makes us a little more thoughtful in view of the enormous efforts we expand in exploring "all the things of our own world" — in view also of the efforts which were necessary to write and edit this book. I would very much like to thank everybody who has helped me, above all, my wife, who typed my first and hardly to be deciphered manuscript. I would like to thank Dr. Immergut for great interest in and perseverance with the translation and Mrs. Immergut for the typing of the English manuscript. I also wish to thank all colleagues who have helped me by means of valuable advice and discussion. I particularly wish to thank Professor H. Mark for his kind willingness to write a foreword to this book.

I can only hope that the reader will find something of interest in these pages.

Karlsruhe, May 1973 B. VOLLMERT

TRANSLATOR'S NOTE

This book employs an unusual but highly convenient numbering system for all tables and figures. Thus each table and figure is numbered after the page number on which it appears. For example: Figure 485 appears on page 485 and Table 508 appears on page 508. This makes it very easy to find any figure or table immediately without the need, as with other books, of thumbing through a number of pages before locating it. We hope that the reader will quickly become used to this system and appreciate its simplicity.

New York, May 1973 E. H. IMMERGUT

TABLE OF CONTENTS

POLYMER CHEMISTRY

INTRODUCTION

1. GENERAL CHARACTERIZATION

As the term implies, macromolecular chemistry deals with compounds whose characteristic properties depend mainly on the extraordinarily large size of the molecules: the molecular weight of most natural and synthetic macromolecular compounds lies between 10^4 and 10^7. The compounds are glassy if they are amorphous; they have a fiber structure, especially when they are crystalline; they are rubber-elastic if they have the ability to return spontaneously to the amorphous state after release of the tension which keeps them in a crystalline, or at least oriented, state. The state of aggregation (glassy, fibrous, rubber-elastic) of a macromolecular compound depends on the structure of the molecules and on the temperature. Macromolecular compounds give rise to colloidal solutions in which, in general, the colloidal particles are identical with the macromolecules. (Colloidal solutions or suspensions are those in which the size of the dissolved or suspended particles lies between 50 A and 200 A. Typical examples are milk and rubber latex.) The macromolecules themselves can be compared to long thin threads or strings of pearls. They form more or less dense, tangled coils (random coils) which can imbibe solvent, first forming a gel and finally forming a highly viscous solution whose viscosity increases with increasing molecular weight.

Table 1 presents a classification of natural and synthetic macromolecular compounds. (For their structural formulas, see Table 12.)

TABLE 1. Classification of Macromolecular Compounds

NATURAL MACROMOLECULES

RUBBER, GUTTAPERCHA

POLYSACCHARIDES: cellulose, (cotton, wood), starch, glycogen, pectin, chitin, heparin

NUCLEIC ACIDS: ribonucleic acid (RNA), deoxyribonucleic acid (DNA)

PROTEINS: silk, keratin (wool, hair, feathers), collagen and gelatin (connective tissue), myosin (muscle), albumins (serum, egg), globulins (blood, semen), casein (milk), virus proteins, hormone proteins, toxins.

1

SYNTHETIC MACROMOLECULES

POLYMERS WITH C − C CHAINS:

polyethylene, polypropylene, polyisobutylene, polybutadiene, polystyrene, poly-p-xylylene

polyvinylchloride, polyvinylidenechloride, polyvinylfluoride, polytetrafluoroethylene (Teflon)

polyacrylic acid and its esters, polymethacrylic acid and its esters (Lucite, Plexiglas), polyacrylonitrile (Orlon, Acrilan, PAN),

polyvinylesters (polyvinylacetate)

polyvinylalcohol, polyvinylacetals

polyvinylethers, polyvinylpyrrolidone, polyvinylcarbazole

Copolymers, Terpolymers, Graft- and Block-Copolymers

POLYMERS WITH HETERO-ATOMS IN THE CHAIN:

polyesters (Dacron, Terylene, Trevira, Mylar)

polycarbonates (Lexan, Makrolon)

polyamides (Nylon, Perlon, Rilsan, Caprolan)

polyurethanes, polyureas (Lycra, Moltoprene-foams)

phenol-formaldehyde resins, melamine-formaldehyde resins, urea-formaldehyde resins

polyethers, polyformaldehyde (Delrin, Celcon), polyphenyleneethers (PPO)

polysiloxanes (silicones), polysilicates (inorganic glasses)

polyimides (Kapton)

2. THE MACROMOLECULAR CONCEPT

In macromolecular chemistry, as in general organic chemistry, the molecule (macromolecule) is defined as the smallest part in which the atoms are held together by covalent bonds. However, this definition does not always coincide with the physical definition of a molecule as an aggregate of atoms which in the gaseous state or in solution always remains together. Thus, in macromolecular solutions one sometimes finds aggregates of macromolecules, that is, so-called association occurs and the size of these particles depends on the solvent, the concentration, and the temperature. The formation of molecular aggregates occurs not only with macromolecular compounds but also with low molecular weight compounds, such as soaps and certain dyestuffs. In some solvents, these compounds form particles of colloidal size because of polyaggregation. Such solutions of micelles or molecular aggregates are often highly viscous and form gels, similar to solutions of macromolecular compounds, so that in such cases one speaks justifiably of secondary valence macromolecules. This, of course, implies that the physical definition of the molecule is correct. However, because of the lower stability of secondary valence forces, such polyaggregates are stable only over very narrow temperature and pH regions, and one therefore does not count such micellar colloids among macromolecular compounds. A special type of secondary valence force is the hydrogen bond, whose strength lies between that of the normal secondary valence bonds and the homopolar main valence bonds. As a result, the molecular aggregates formed by means of hydrogen bonding are highly stable and a sharp differentiation from the covalent macromolecules is, therefore, no longer possible. Hydrogen bonding plays an important role, especially with polyamides—both synthetic polyamides, such as nylon, as well as natural polyamides, i.e., the proteins. In protein chemistry, therefore, the phenomena of "macromolecules" and "molecular aggregates" overlap in many respects.

TABLE 3. Macromolecular Compounds of Different Chemical Composition but Similar Physical Properties.

Name of Polymer (or Trademark)	Type of Chemical Bonds and (monomer)	Physical Properties
Cellulose	Polyacetal chain (glucose)	hard, fiber-forming
Asbestos	− Si − O − Si − O − Si − chain (silicic acid)	
Nylon-66	Polyamide-chain (hexamethylene-diamine/adipic acid) − CONH −	
Nylon-6 (Caprolan, Perlon)	Polyamide-chain − CONH − (ε-caprolactam)	
Dacron, Mylar	Polyester-chain − COOR − (ethylene glycol/tere-phthalic acid)	
PAN, Orlon	− C − C − chain (acrylonitrile)	soft, rubberelastic
Saran	− C − C − chain (vinylidene-chloride-copolymer)	
Synthetic Rubber (SBR) or Coral, Ameripol-SN, etc.	− C − C = C − C − chain (styrene/butadiene) or (cis- 1,4-polyisoprene, cis-1,4-polybutadiene)	
Nitrile Rubber (NBR, Buna N)	− C − C = C − C − chain (butadiene/acrylonitrile)	
Butyl Rubber	− C − C − chain (isobutylene)	
Nordel (EPR)	− C − C − chain (ethylene/ propylene)	
Hypalon	− C − C − chain (sufochlorinated polyethylene)	
Vulcollan	polyester-chain with some urethane groups	
Lycra, Vyrene	polyether − or polyester chain with some urethane and urea groups (a block-copolymer)	
Silicone Rubber	− Si − O − Si − O − Si − chain (dialkyldichlorosilane)	
Polyphosphonitril-licchloride	− P = N − P = N − P − chain	

To speak of molecules, however, is meaningful only for such compounds whose molecular weight can be determined, i.e., which can be dissolved. We cannot use the molecular concept when all the atoms of the material are held together by main valence forces as is the case with compounds having a lattice structure, such as diamond or quartz. Similarly, such thermosetting resins as phenolformaldehyde resins, polyester resins, and epoxy resins do not belong under this classification. If these resins are still treated as macromolecular compounds, it is because they are formed from macromolecules by cross-linking through covalent bonds where the cross-linking reaction can occur together with the polymerization process or can be generally combined with the synthesis of the macromolecule.

The most important structural entity in macromolecular chemistry is the chain molecule, which consists of a continuous, more or less regular, sequence of the same or different chain elements or structural units held together by covalent bonds. Cellulose, for example, consists of chain-like macromolecules in which several hundred or thousand anhydroglucose residues are bound together through acetal linkages. The chain molecules do not have to be linear; there are molecular compounds (both natural and synthetic) which consist of more or less strongly branched molecules. Starch, for example, contains a branched polysaccharide. The chain-like structure of these molecules rather than their chemical composition is primarily responsible for such typical properties of macromolecular compounds as film formation, fiber formation, and rubber elasticity. See Table 3.

Although such chemically different compounds as the many different types of elastomers with polyester chains or saturated and unsaturated carbon-carbon chains can show very similar physical properties, polymers which belong to the same chemical compound type (for example, polyesters) can show completely different physical properties—such as, the synthetic fiber Dacron (polyethylene-glycol-terephthalate) and the polyester-based polyurethane elastomers (Vulcollans). Whether a macromolecular material is elastic or hard and glassy depends—at a given temperature—on the internal mobility (flexibility) of the chain and its regular or irregular structure. (A more detailed treatment is given in Chapter 4 which deals with macromolecular aggregation phenomena).

3. MOLECULAR WEIGHT AND POLYMOLECULARITY

It is a part of the concept of a "pure compound" that such a material has a definite molecular weight which is the same for every molecule of that compound. Furthermore, the molecular weight should be measurable not only through direct determination, but also, much more accurately, through the sum of the atomic weights of the atoms contained in the molecule. However, one cannot assign a definite molecular weight to a macromolecular compound. Thus, one cannot say "cellulose has a molecular weight of 500,000," or "polystyrene has a molecular weight of 200,000." Instead, there are many different polystyrenes—some with molecular weights of 10,000, some with molecular weights of 1,000,000, and an infinite number whose molecular weights lie in between, or over, or under these figures. All these compounds, however, are called polystyrene. Furthermore, in a more general sense, one cannot assign a definite uniform molecular weight to macromolecular compounds because all macromolecular compounds (with the exception of a few natural products and certain polymers prepared by special processes) are polymolecular, that is, even the individual molecules of a certain macromolecular compound are of different sizes. The molecular weight of a macromolecular substance, therefore, represents an average value, which may be the result of different types of distribution curves. In very inhomogeneous materials, the entire scale of molecules may be present, from the smallest (the monomer) to the very large molecules with molecular weights of several million, and all of these may be more or less evenly represented, resulting in a flat distribution curve. On the other hand, more homogeneous materials will contain mainly molecules of a certain average size resulting in a distribution curve with a steep maximum (sharp peak). Therefore, for any molecular weight data to be complete, the distribution curve from which the molecular weight average was derived should also be given.

4. CLASSIFICATION ACCORDING TO MOLECULAR SIZE

From the previous discussion it is apparent that one cannot draw a sharp dividing line between low molecular and macromolecular substances. Macromolecular compounds may be built up to increasing sizes from the monomer and may also be broken down again to the monomer. This was shown for the first time by H. Staudinger for the case of polyoxymethylene (from formaldehyde). Staudinger called the resulting series of polymers, having the same chemical structure and increasing molecular weight, a polymer-homologous series. One can say that macromolecular compounds, with the exception of a few natural products and ionically prepared polymers, are mixtures of polymer homologs or polymer homologous mixtures. This molecular heterogeneity or polymolecularity results in the fact that even well-defined low molecular weight compounds—such as styrene, distyrene, or oligosaccharides containing from four to 10 monosaccharide units—can be considered parts of a macromolecular compound and are in fact often found in them. In practice, one usually considers these low-molecular weight components as impurities and tries as much as possible to remove them. Even then, the borderline where a low molecular weight polymer (oligomer) stops being an impurity and starts being a component of the macromolecular material can only be defined through separation by vacuum treatment, solvent extraction, etc. The transition from the low molecular oligomers to the high molecular weight polymer occurs without any recognizable steps. Such physical properties as viscosity of solutions, softening range, and mechanical properties change in a continuous way up to a certain molecular weight, usually in the range of 10,000-100,000. From this point on, some physical properties change only slowly, or not at all, whereas others change continuously. Among those properties whose curve shows a leveling-off after a certain molecular weight range has been reached are the tensile strength and the softening point (softening range). Among the properties which do not show a leveling-off are the viscosities of solutions and melts. This does not mean, however, that the increase in these properties is always linear or remains so.

For polymers with C-C chains (polystyrene, polyvinylchloride, etc.) the leveling-off point lies in the molecular weight range of 150,000-200,000. With polymers containing hetero-atoms in the chain (polyamides, polyesters, etc.) the leveling-off occurs in a much lower molecular weight range, i.e., 15,000-20,000.

Even though the leveling-off molecular weight differs with different polymers, it does present a certain point of reference, perhaps the only one, for a meaningful subdivision of the entire field on a molecular weight basis.

In the technical production and processing of macromolecular materials, a high viscosity of the solution or the melt creates great difficulties; one, therefore, tries to keep the molecular weight as low as possible. This may be accomplished by carrying the molecular weight to the point where the curve of the important technical properties versus molecular weight levels off. The curves in Figure 6 illustrate this for polystyrene. As industrially important properties we selected the stiffness and the solution viscosity on the one hand and the impact resistance and the resistance to deformation on heating (Vicat Number) on the other. The last two properties hardly change after a degree of polymerization of 1,000 is reached, in contrast to the viscosity, which increases continuously without any leveling-off.

5. INORGANIC AND ORGANIC POLYMERS

Even though there are also purely inorganic macromolecular compounds, such as polymeric sulfur, or polyphosphonitrilicchloride, one usually considers macro-molecular chemistry a part of organic chemistry. This is because most

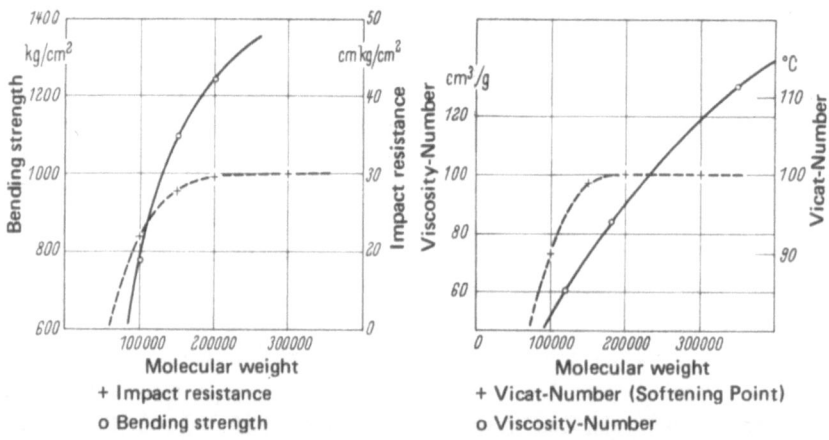

FIG. 6 – Physical properties of polystyrene as a function of the molecular weight

macromolecular compounds, especially the most important natural and synthetic macromolecules such as cellulose, rubber, starch, proteins and nearly all plastics and other synthetic polymers, are based on well-known organic compounds such as glucose, amino acids, butadiene, and styrene and can be prepared from them. Furthermore, the laboratory procedures used in the experimental investigation and synthesis of macromolecular compounds correspond essentially to those used in the organic laboratory. Of course, one also uses physical methods of characterization, which were taken over in great part from colloid chemistry. In spite of the great preponderance of organic macromolecular compounds, however, it would be wrong to consider macromolecular chemistry as simply and only a part of organic chemistry. Neither organic nor inorganic chemistry, macromolecular chemistry is the chemistry of compounds consisting of giant chain-molecules whose chain structure is responsible for the typical physical properties of macromolecular compounds such as rubber elasticity, high viscosity of solutions, and the formation of gels, rather than the chemical nature of the atoms forming the macromolecule. The ability to form long chain-like molecules is not dependent on the presence of a certain element (see Table 3). Only because the carbon atom, due to its strong tendency to combine with itself, is specially suitable for the formation of polymer chains, do the carbon-containing polymers take a dominating position in macromolecular chemistry. All other macromolecular compounds with chains not consisting of carbon atoms take a back seat.

6. NATURAL AND SYNTHETIC MACROMOLECULES

Whether or not macromolecular compounds should be subdivided into natural and synthetic materials (compare Table 1) is more a question of usefulness than principle.

Many natural products, such as proteins, native cellulose, and other polysaccharides, possess special structural characteristics. These are their uniform length (e.g., all molecules of insulin have the same molecular weight or chain length) and their completely identical chemical structure·in which each molecule has exactly the same sequence of different monomer units (e.g., there may be 20 different amino acids in a protein molecule, but they follow each other in the same sequence in every single molecule). Another striking example of natural macromolecules whose structure plays an important role are the nucleic acids. Thus, the sequence of the purine and pyrimidine bases along the chain of a deoxyribonucleic acid molecule (DNA) constitutes the "genetic code" by which information concerning the synthesis of the protein molecules, which make up every living matter or being, is passed along from molecule to molecule.

Another characteristic of natural macromolecules is the special significance of their secondary and tertiary structure. Each molecule of hemoglobin, for example, assumes exactly the same shape to the smallest detail.

The distinction between natural and synthetic macromolecules based on the above considerations loses its significance, of course, once it becomes possible to carry out the synthesis of natural rubber, polysaccharides, proteins or nucleic acids. Whereas the synthesis of *cis*−1,4−polyisoprene (equivalent to natural rubber) was accomplished in the 1950's by means of special catalyst systems, the synthesis of proteins and nucleic acids is much more difficult because of the unique way in which the chain units follow each other. Thus, in the synthesis of sheep insulin (Zahn/Katsoyannis and Dixon), 224 individual reaction steps had to be carried out to produce the insulin molecule (molecular weight = 5734), which is still a relatively small macromolecule. Other natural polymers such as the nucleic acids have molecular weights of $> 10^9$. Knowing that there are thousands of different proteins, nucleic acids, and polysaccharides, it becomes obvious that we still have a long way to go before we can synthesize a macromolecular system able to propagate, or construct a living cell.

In comparison with this laborious, stepwise synthesis of proteins, the preparation of a synthetic polymer in the laboratory or industrial plant and the biosynthesis of a macromolecule in the living cell can be achieved with the greatest of ease and elegance.

It, therefore, seems justified to treat the natural macromolecules as a special group in the field of polymers. One is even more inclined to give them this position, if one reserves a special place for anything associated with "life" in our world.

1. STRUCTURAL PRINCIPLES

11. CHAIN STRUCTURE, DEGREE OF POLYMERIZATION

The simplest form a macromolecule[1] can have is that of an unbranched chain. One speaks of a "chain-like" macromolecule or a "chain" molecule, because macromolecules consist of a large number of links, i.e., they have a chain structure. However, in a real chain the links are not held together in an identical manner from link to link. Furthermore the links can move within each other. It is therefore better to think of a pearl necklace in which the stiffness of the chain is determined by the type of pearls and the material used to string the pearls together. This does not exclude the fact that linear macromolecules in solution have a coiled-up structure similar to a randomly tangled ball of yarn or wire. Because of thermal motion these coils are constantly changing their form. The speed of these changes depends on the temperature. To what extent a macromolecule is stretched or coiled depends on the nature of the chain elements, the type of chain bonding, and the solvent.[2]

In the simplest case, the chain-like macromolecule consists of atoms of a single element, as, for example, with polymeric sulfur. Equally simple is a linear polymethylene chain such as is obtained through the decomposition of diazomethane or through the polymerization of ethylene.[3]

The $-CH_2$-group ($-CH_2-CH_2-$ in polyethylene) is called the structural element, the repeat unit, or the monomer unit of the macromolecule. The number of structural elements (including the one at the beginning and the end of the chain of the molecule) is known as the degree of polymerization, P, which, as was already

---

[1]The term "macromolecule" or "macromolecular compound" was first suggested by Staudinger and is now an internationally recognized official term. In the English literature, one usually finds the word "polymer." In the German literature, one often finds "polymer molecule" instead of "macromolecule," especially if one wants to point out the contrast with the monomer. Instead of "macromolecular compound" one often says "the polymer" or even "the macropolymer." Polymers with only a few structural units (from 5 to 10) are usually called "oligomers." These names are derived from the Greek: "polys" means many, "meros" means part; "oligos" means few.

[2]Further details can be found in Chapter 3.2, "The Shape of the Molecule," and Chapter 4.2, "Macromolecules in Solution."

[3]During the polymerization of ethylene one can obtain either linear macromolecules (low pressure process, Ziegler or Phillips) or molecules with different degrees of branching (mostly C_4) by high pressure polymerization.

pointed out earlier, is in most cases an *average* degree of polymerization ($\overline{P}$). A detailed discussion of $\overline{P}$ can be found in Section 3.13 "Molecular Weight Distribution."

$\overline{P}$ is related to the molecular weight of the polymer, $\overline{M}$, in the following way:

$$\overline{M} = \overline{P} \cdot M_{mon}$$

where $\overline{M}$ = (average) molecular weight of the polymer, $\overline{P}$ = (average) degree of polymerization = (average) number of structural units per polymer chain, and M_{mon} = molecular weight of the structural unit (monomer unit).

The units at the beginning and at the end of the chain must, of course, be chemically different from the structural units in the rest of the chain molecule. The identity of the units at the end and beginning of the chain can often be obtained from a knowledge of the method of preparation of the polymer. Thus, one can expect to find a polyester with COOH or OH groups at the chain ends, depending on whether the condensation was carried out with an excess of glycol or of dicarboxylic acid. In an addition (free radical) polymerization, the two chain ends consist of initiator fragments unless chain termination occurs by disproportionation (see Section 21 13) or by an added terminating agent.

The number of structural units, expressed by the degree of polymerization, is usually not equal to the number of carbon atoms in the chain molecule. Thus, it is usual to express the degree of polymerization not in terms of the number of atoms in the chain, but in terms of the number of monomer molecules or base molecules which form the polymer chain. Thus, polystyrene has only a single structural unit $- CH_2 - CH -$ rather than two units, $- CH_2 -$ and $- CH -$, and the degree of

polymerization refers to this single repeat unit.

With many polymers the molecular weight of the structural unit is equal to the molecular weight of the monomer, e.g., with all vinyl monomers. However, with polymers prepared by condensation reactions under elimination of water or other small molecules, this is obviously not the case (polyamides, polyesters, etc.). Table 12 lists the most important macromolecular compounds with their structural repeat units and the monomer from which they are formed.

Since in most cases the beginning and the end of the chain are not identical, and in many cases are not even exactly known, and, furthermore, since the nature of the chain ends influences the properties of the polymers only slightly because of the great length of the chains, it is correct and customary to express the structure of the polymers only in terms of a chain fragment as shown in Table 12.[4] Often one writes simply the structural unit, and by means of a bracket around it, indicates that one is considering a polymer. For example, for polyvinylchloride one can write:

$$\left[-CH_2 - \underset{\underset{Cl}{|}}{CH} - \right]_{\overline{P} = 1000} \quad \text{or} \quad \left[-CH_2 - \underset{\underset{Cl}{|}}{CH} - \right]_n$$

If one knows the end groups, one can write:

$$H \left[CH_2 - \underset{\underset{CH_3}{|}}{\overset{\overset{CH_3}{|}}{C}} \right] OH \quad \text{or} \quad \left\langle \bigcirc \right\rangle \left[CH_2 \quad CH \right] H$$

$$\overline{P} = n \qquad\qquad\qquad\qquad \overline{P} = n$$

Polyisobutylene

Initiated by $H[BF_3OH]$

Terminated by H_2O or NH_3

Polystyrene

Initiated by C_6H_5 Li

Terminated by H_2O or acid

Comparison of the macromolecular compounds shown in Table 12 according to their chemical composition reveals two types: those in which the macromolecule is a continuous chain of C atoms and those in which the C-C chain is interrupted regularly through other atoms (heteroatoms) or through heteroatom-containing groups (functional groups) at smaller or larger intervals. Within these two large subdivisions, polymers can be grouped according to the different compound groups of organic chemistry, such as esters, ethers, amides, urethanes, ureas, anhydrides, alcohols, amines, and carboxylic acids. In the first case the characteristic functional group is contained *in* the chain and plays a role in the synthesis of the chain. With the other type the characteristic group is a part of the *side* chain (or substituent). This is indicated by calling those with the functional group in the chain poly*ethers*, poly*amides*, poly*esters*, etc., and the other type (with substituent), poly*vinyl*ethers, poly*vinylidene*chloride, poly*acrylic* esters, etc.

Since the polymers with a pure C-C chain have been formed in general through radical or ionic chain polymerization, whereas the polymers with functional groups in the chain are usually formed through reactions in which water is split out (condensation reactions), it has become common to speak in the first case of addition polymers and in the second case of polycondensates or condensation polymers. However, polymers with C-C chains can sometimes be formed through condensation reactions, and polymers with heteroatoms in the chain, through addition reactions; therefore classification of macromolecules in terms of their

---

[4]A significant influence of the end groups of high molecular weight polymers can only be found on those properties which can be measured with extremely sensitive techniques, such as the dielectric properties. (This is of great importance when plastics are used as insulators, especially for high frequency work.)

TABLE 12. Structural Formulas of Macromolecular Compounds

Name	Chain Structure	Monomers
1. Polymers with C—C-Chains		
Polyethylene (Alathon, Hostalen, Marlex)	$\sim\!\sim\!CH_2\!-\!CH_2\!-\!CH_2\!-\!CH_2\!-\!CH_2\!-\!CH_2\!-\!CH_2\!\sim\!\sim$	$CH_2\!=\!CH_2$
Polypropylene	$\sim\!\sim\!CH_2\!-\!CH\!-\!CH_2\!-\!CH\!-\!CH_2\!-\!CH\!\sim\!\sim$ (—CH_3 branches)	$CH_2\!=\!CH$ —CH_3
Polyisobutylene (Vistanex)	$\sim\!\sim\!CH_2\!-\!C(CH_3)_2\!-\!CH_2\!-\!C(CH_3)_2\!-\!CH_2\!-\!C(CH_3)_2\!\sim\!\sim$	$CH_2\!=\!C(CH_3)_2$
Poly-cis-1,4-butadiene	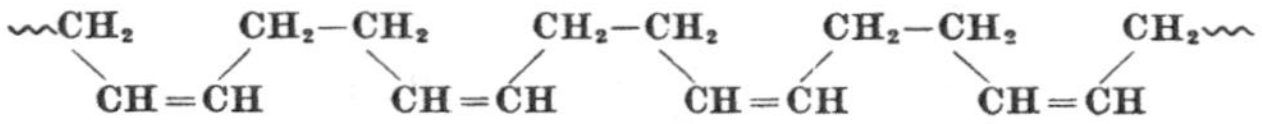 $\sim\!\sim\!CH_2$—CH_2—CH_2—$CH_2\sim\!\sim$ (—$CH\!=\!CH$— cis)	$CH_2\!=\!CH\!-\!CH\!=\!CH_2$
Poly-cis-1,4-isoprene (Natural Rubber)	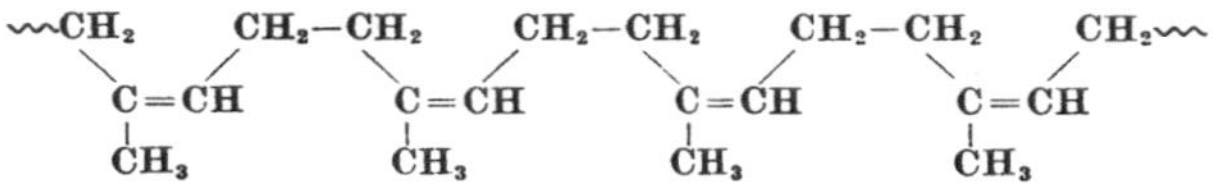 $\sim\!\sim\!CH_2$—CH_2—CH_2—$CH_2\sim\!\sim$ (—$C(CH_3)\!=\!CH$— cis)	$CH_2\!=\!C(CH_3)\!-\!CH\!=\!CH_2$
Poly-trans-1,4-isoprene (Guttapercha, Balata)	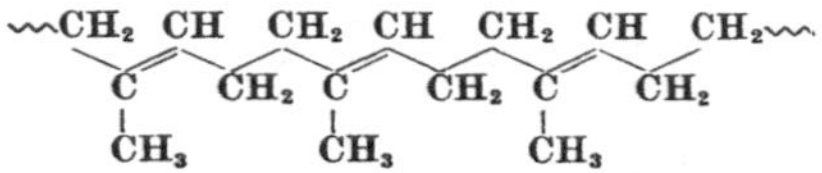 $\sim\!\sim\!CH_2$—CH_2—CH_2—$CH_2\sim\!\sim$ (—$C(CH_3)\!=\!CH$— trans)	$CH_2\!=\!C(CH_3)\!-\!CH\!=\!CH_2$
Poly-chlorobutadiene (Polychloroprene, Neoprene)	$\sim\!\sim\!CH_2\!-\!C(Cl)\!=\!CH\!-\!CH_2\!-\!CH_2\!-\!C(Cl)\!=\!CH\!-\!CH_2\!\sim\!\sim$	$CH_2\!=\!C(Cl)\!-\!CH\!=\!CH_2$ Chloroprene
Polystyrene	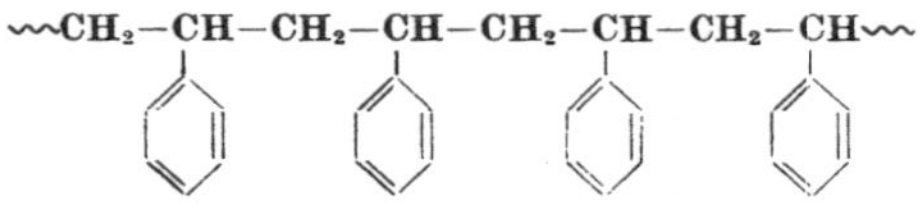 $\sim\!\sim\!CH_2\!-\!CH\!-\!CH_2\!-\!CH\!-\!CH_2\!-\!CH\!-\!CH_2\!-\!CH\!\sim\!\sim$ (—C_6H_5 branches)	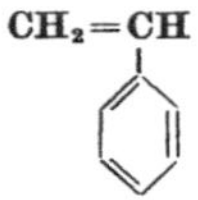 $CH_2\!=\!CH$ —C_6H_5

Name	Polymer structure	Monomer
Poly-α-methylstyrene	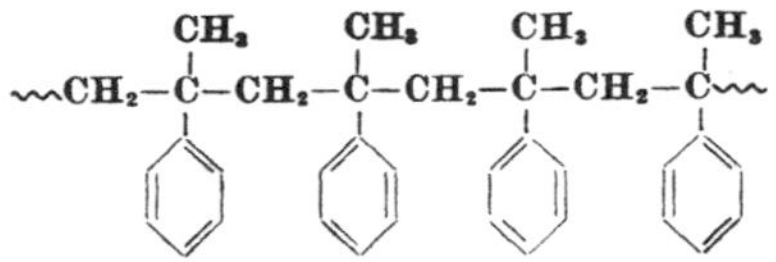 $\sim CH_2-C(CH_3)(C_6H_5)-CH_2-C(CH_3)(C_6H_5)-CH_2-C(CH_3)(C_6H_5)-CH_2-C(CH_3)(C_6H_5)\sim$	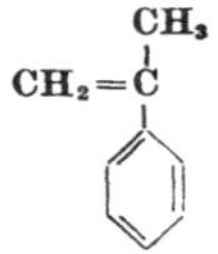 $CH_2=C(CH_3)(C_6H_5)$
Poly-p-xylene	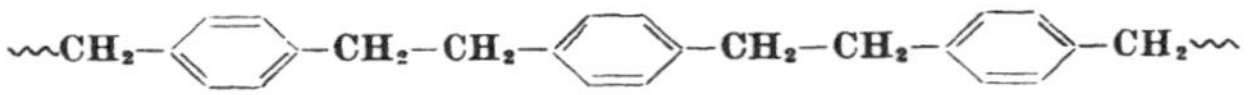 $\sim CH_2-C_6H_4-CH_2-C_6H_4-CH_2-C_6H_4-CH_2\sim$	 $[CH_2=C_6H_4=CH_2]$ Xylene
Polyvinylchloride (PVC)	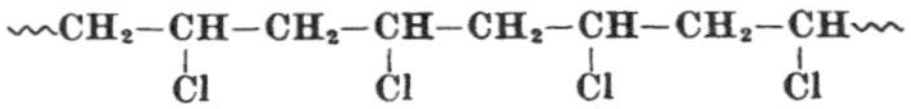 $\sim CH_2-CHCl-CH_2-CHCl-CH_2-CHCl\sim$	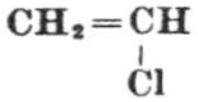 $CH_2=CHCl$
Polyvinylidenechloride (Saran)	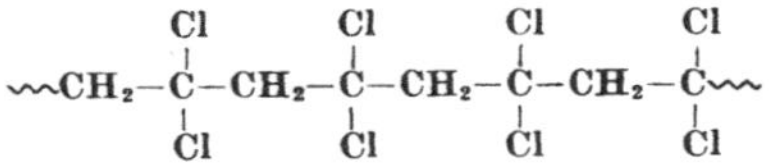 $\sim CH_2-CCl_2-CH_2-CCl_2-CH_2-CCl_2\sim$	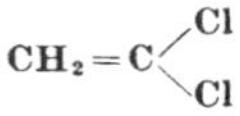 $CH_2=CCl_2$
Polyvinylfluoride (Tedlar)	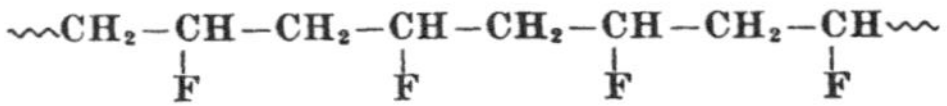 $\sim CH_2-CHF-CH_2-CHF-CH_2-CHF\sim$	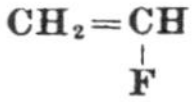 $CH_2=CHF$
Polytetra-fluoroethylene (Teflon)	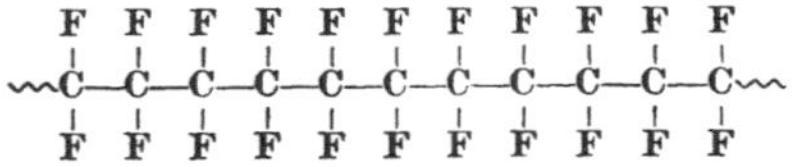 $\sim CF_2-CF_2-CF_2-CF_2-CF_2-CF_2\sim$	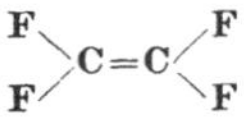 $CF_2=CF_2$
Polyacrylonitrile (PAN, Dralon, Orlon)	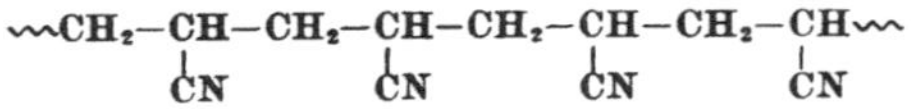 $\sim CH_2-CH(CN)-CH_2-CH(CN)-CH_2-CH(CN)\sim$	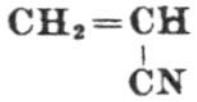 $CH_2=CH(CN)$
Polyvinylidenecyanide (Darvan [USA], Furlon [Jap.])	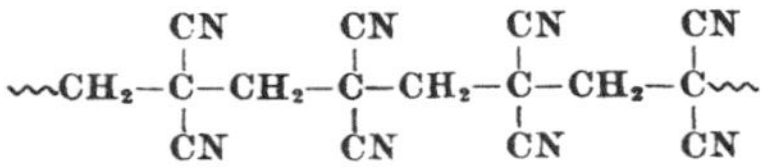 $\sim CH_2-C(CN)_2-CH_2-C(CN)_2-CH_2-C(CN)_2\sim$	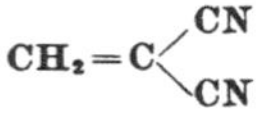 $CH_2=C(CN)_2$

TABLE 12. (Cont'd)

Name	Chain Structure	Monomers
Polyvinylalcohol	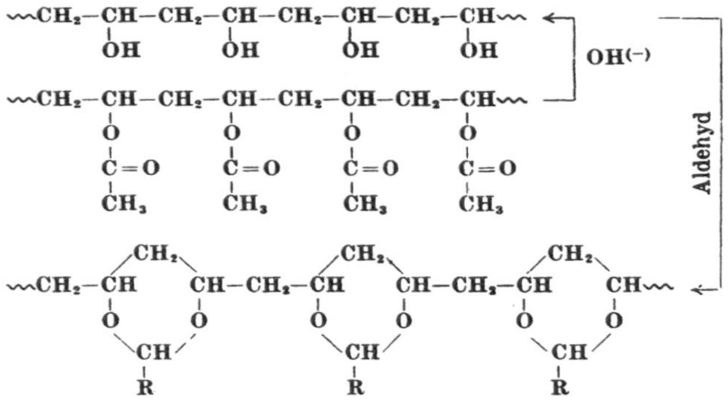 ∼CH₂–CH–CH₂–CH–CH₂–CH–CH₂–CH∼ with OH groups	not bystanding
Polyvinylacetate	∼CH₂–CH₂–CH–CH₂–CH–CH₂–CH∼ with O–C=O, CH₃ groups	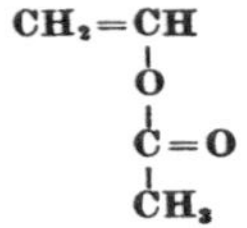 CH₂=CH–O–C=O–CH₃
Polyvinylacetal	chain structure with CH, O–CH–R rings	
Polyvinylenecarbonate	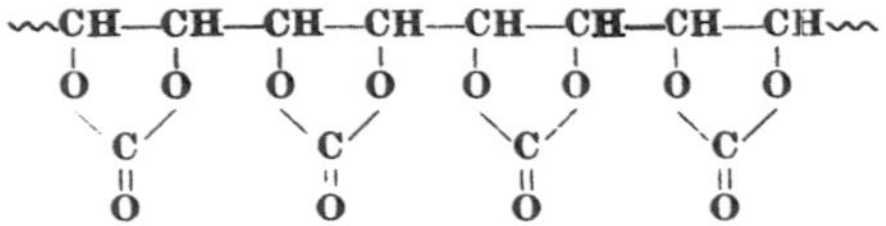 ∼CH–CH–CH–CH–CH–CH∼ with O–C=O groups	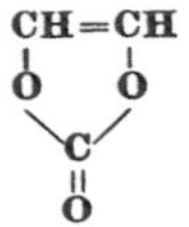 CH=CH, O, O, C=O
Polyacrylicanhydride	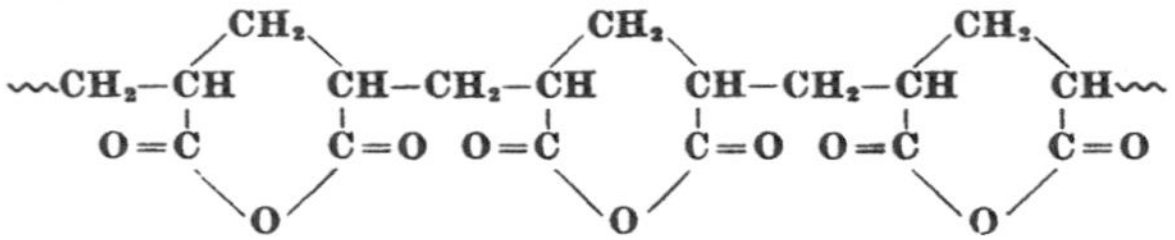 chain with CH₂, C=O, O rings	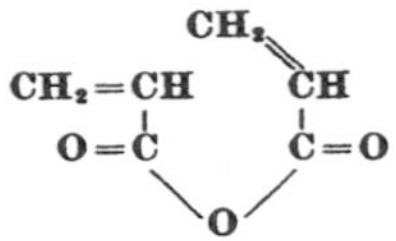 CH₂=CH, CH₂=CH, O=C, O=C, O ; CH₂=CH–CH–C=O, O
Polyacrylicacid	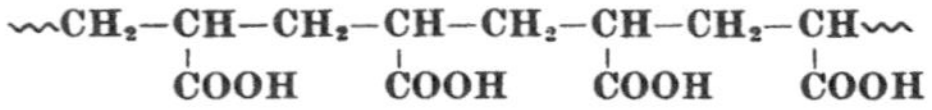 ∼CH₂–CH–CH₂–CH–CH–CH₂–CH–CH₂–CH∼ COOH COOH COOH	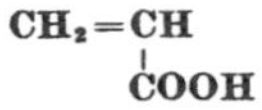 CH₂=CH–COOH
Polyacrylates	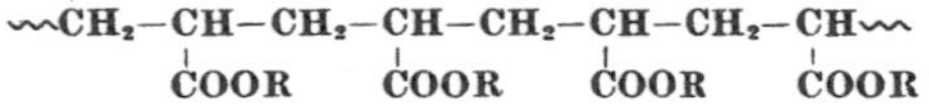 ∼CH₂–CH–CH₂–CH–CH₂–CH–CH₂–CH∼ COOR COOR COOR	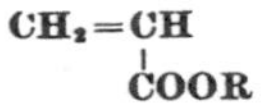 CH₂=CH–COOR

(Aldehyd, OH⁻ notation shown with Polyvinylalcohol/acetal conversion)

Name	Chain structure	Monomer
Poly-α-chloracrylate	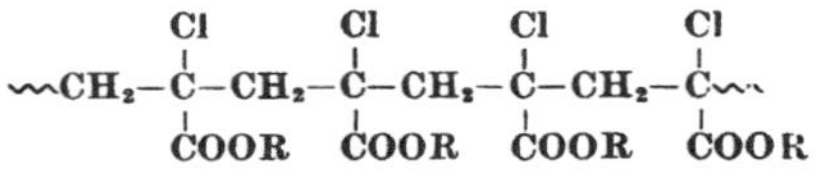 $\sim CH_2-\underset{COOR}{\overset{Cl}{C}}-CH_2-\underset{COOR}{\overset{Cl}{C}}-CH_2-\underset{COOR}{\overset{Cl}{C}}-CH_2-\underset{COOR}{\overset{Cl}{C}}\sim$	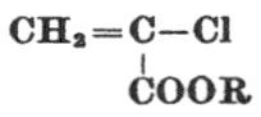 $CH_2=\underset{COOR}{\overset{Cl}{C}}$
Poly-α-cyanoacrylate	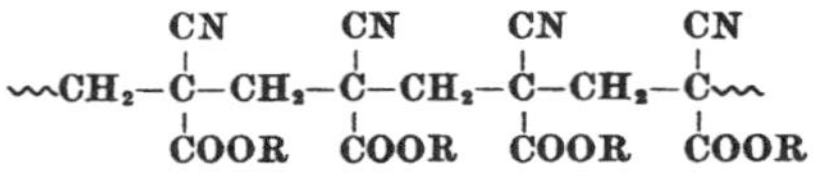 $\sim CH_2-\underset{COOR}{\overset{CN}{C}}-CH_2-\underset{COOR}{\overset{CN}{C}}-CH_2-\underset{COOR}{\overset{CN}{C}}-CH_2-\underset{COOR}{\overset{CN}{C}}\sim$	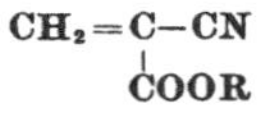 $CH_2=\underset{COOR}{\overset{CN}{C}}$
Polacrylamide	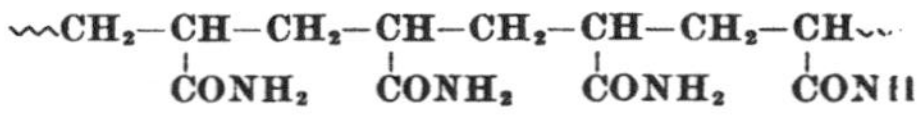 $\sim CH_2-\underset{CONH_2}{CH}-CH_2-\underset{CONH_2}{CH}-CH_2-\underset{CONH_2}{CH}-CH_2-\underset{CONH}{CH}\sim$	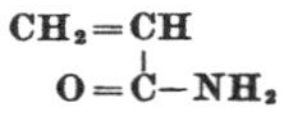 $CH_2=CH$ $O=\overset{}{C}-NH_2$
Polymethylmethacrylate (Plexiglas, Lucite)	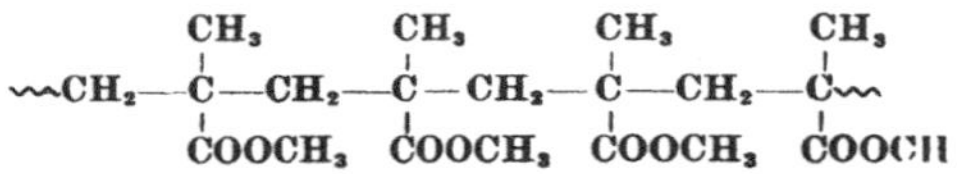 $\sim CH_2-\underset{COOCH_3}{\overset{CH_3}{C}}-CH_2-\underset{COOCH_3}{\overset{CH_3}{C}}-CH_2-\underset{COOCH}{\overset{CH_3}{C}}\sim$	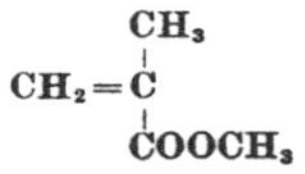 $\overset{CH_3}{\underset{COOCH_3}{CH_2=C}}$
Polymethacrylonitrile	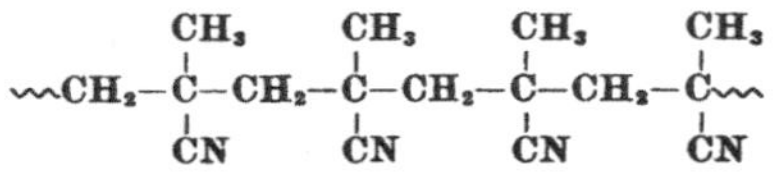 $\sim CH_2-\underset{CN}{\overset{CH_3}{C}}-CH_2-\underset{CN}{\overset{CH_3}{C}}-CH_2-\underset{CN}{\overset{CH_3}{C}}\sim$	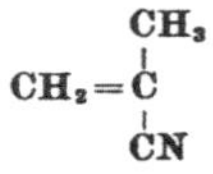 $\overset{CH_3}{\underset{CN}{CH_2=C}}$
Polymethylenemalonate	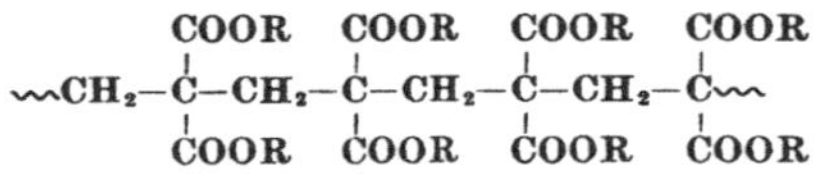 $\sim CH_2-\underset{COOR}{\overset{COOR}{C}}-CH_2-\underset{COOR}{\overset{COOR}{C}}-CH_2-\underset{COOR}{\overset{COOR}{C}}\sim$	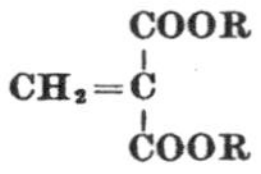 $\overset{COOR}{\underset{COOR}{CH_2=C}}$
Polyvinylethers	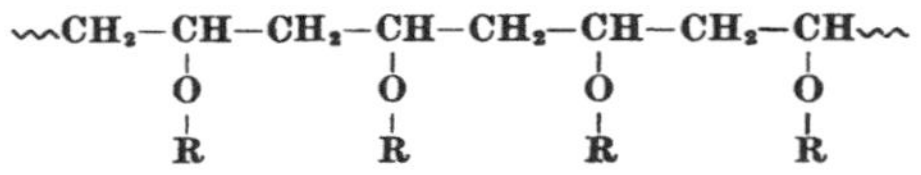 $\sim CH_2-\underset{O-R}{CH}-CH_2-\underset{O-R}{CH}-CH_2-\underset{O-R}{CH}\sim$	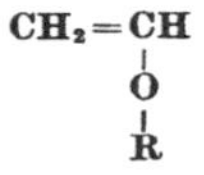 $CH_2=CH$ $O-R$
Polyvinylmethylketone	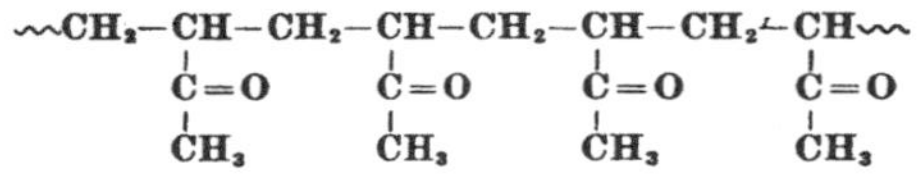 $\sim CH_2-\underset{\overset{C=O}{CH_3}}{CH}-CH_2-\underset{\overset{C=O}{CH_3}}{CH}-CH_2-\underset{\overset{C=O}{CH_3}}{CH}\sim$	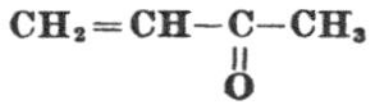 $CH_2=CH-\underset{O}{\overset{}{C}}-CH_3$

TABLE 12. (Cont'd)

Name	Chain Structure	Monomers
Polyvinyl-pyrrolidone	~~CH₂—CH—CH₂—CH—CH₂—CH—CH₂—CH~~ with pyrrolidone N—C=O, CH₂—CH₂, CH₂—CH₂ side groups	CH₂=CH, N, C=O, CH₂—CH₂, CH₂—CH₂
Polyvinyl-carbazole	~~CH₂—CH—CH₂—CH—CH₂—CH~~ with carbazole N rings	CH₂=CH, carbazole
2. Polymers with heteroatoms in the chain		
Polyformaldehyde or Poly-oxy-methylene (Delrin, Celcon)	~~CH₂—O—CH₂—O—CH₂—O—CH₂—O—CH₂—O—CH₂—O—CH₂—O~~	CH₂=O / Trioxane structure
Polyethylene oxide	~~O—CH₂—CH₂—O—CH₂—CH₂—O—CH₂—CH₂—O—CH₂—CH₂—O—CH₂—CH₂~~	CH₂—CH₂ with O bridge
Polypropylene-oxide	~~CH₂—CH—O—CH₂—CH—O—CH₂—CH—O—CH₂—CH—O~~ with CH₃ groups	CH₃—CH—CH₂ with O bridge / Propyleneoxide
Poly-di-chloro-methyloxa-cyclobutane (Penton)	~~CH₂—C—CH₂—O—CH₂—C—CH₂—O—CH₂—C—CH₂—O~~ with CH₂Cl, CH₂Cl groups	Cl—CH₂—C—CH₂—O with CH₂Cl, CH₃ groups / Bis-chloromethyl-oxacyclobutane

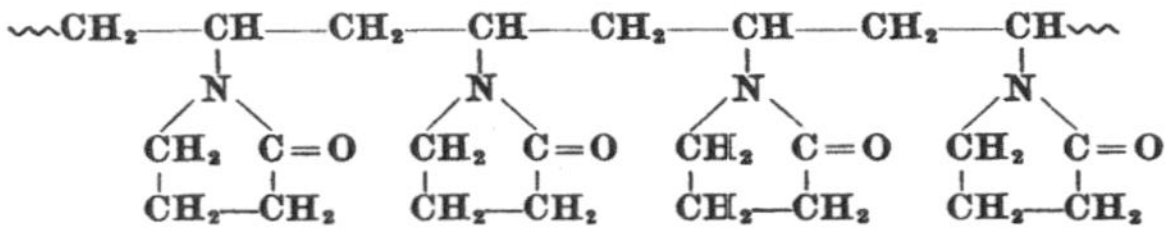

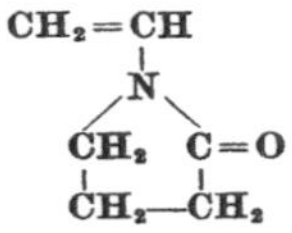

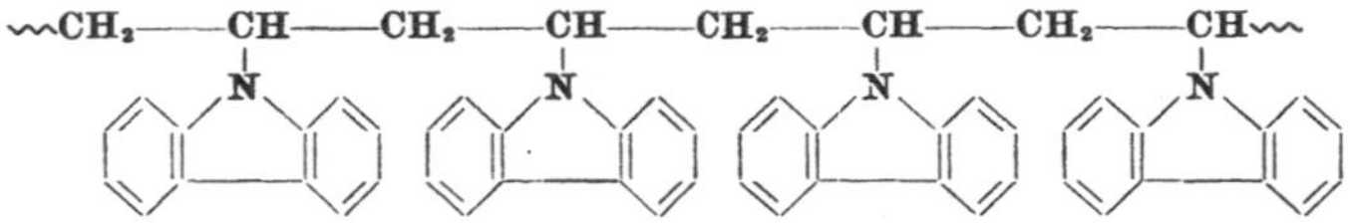

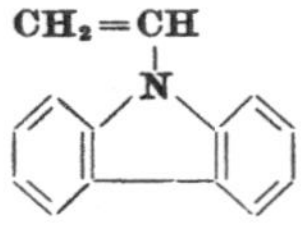

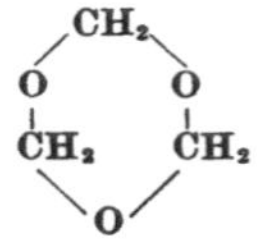

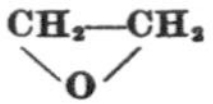

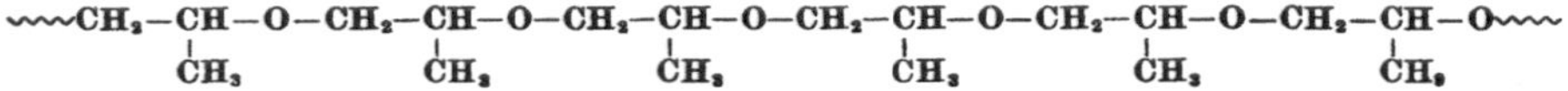

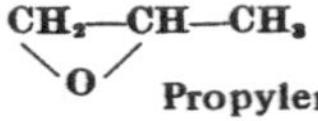

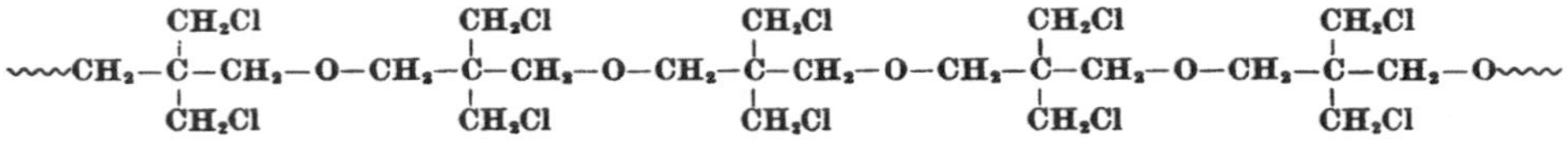

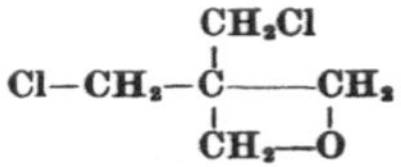

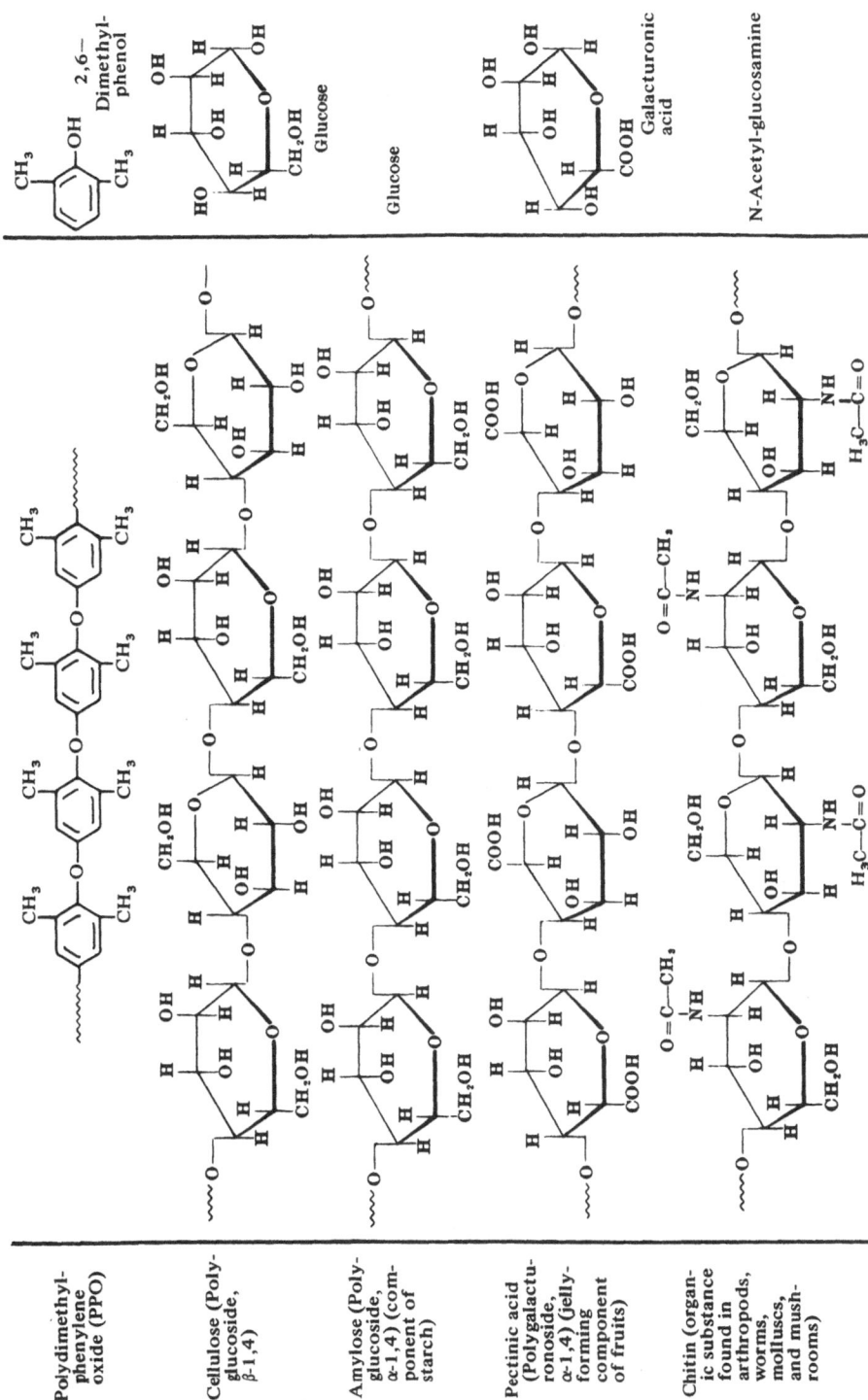

Polydimethyl-phenylene oxide (PPO)

Cellulose (Poly-glucoside, β-1,4)

Amylose (Poly-glucoside, α-1,4) (component of starch)

Pectinic acid (Polygalactu-ronoside, α-1,4) (jelly-forming component of fruits)

Chitin (organic substance found in arthropods, worms, molluscs, and mushrooms)

TABLE 12. (Cont'd)

Name	Chain Structure	Monomers
Polyethylene-glycol-tere-phthalate (Dacron, Terylene, Trevira, Mylar)	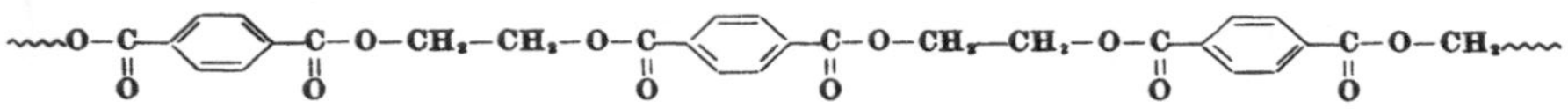 [chain structure]	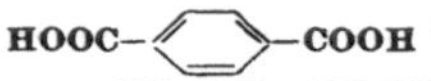 HOOC—⟨ ⟩—COOH Terephthalic acid HO—CH₂—CH₂OH Glycol
Poly-dioxy-diphenylpro-pane-carbonate (Polycar-bonate of Bisphenol-A) (Lexan, Makrolon)	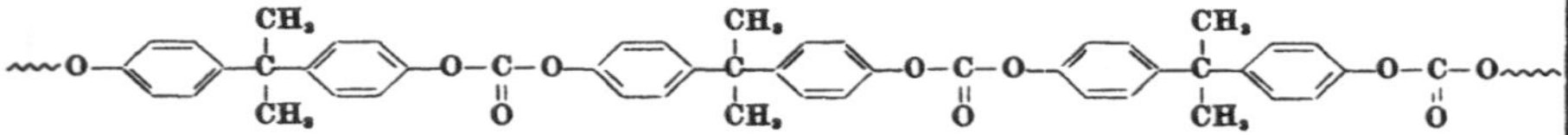 [chain structure]	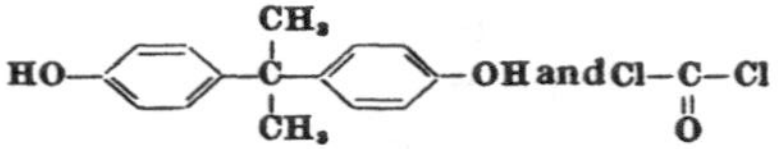 [structure] —OH and Cl—C—Cl ‖O Phosgene 4,4′Dioxy-2,2-diphenylpropane (Bisphenol-A)
Poly-adenyl-ribose-phos-phate (with RNA, in ad-dition to ad-enine, there are 3 other bases: cyto-sine, guanine and uracil; with DNA, thymine re-places uracil among the 4 bases and the OH-group on the C₂ of the ribose is missing)	[chain structure]	[structure] Adenylribose and HO—P—OH ‖O ⟍OH Phosphoric acid
Polyurethane from hexa-methylene-diisocyanate and butane-diol	[chain structure]	O=C=N—(CH₂)₆—N=C=O Hexamethylenediisocyanate and HO—(CH₂)₄—OH

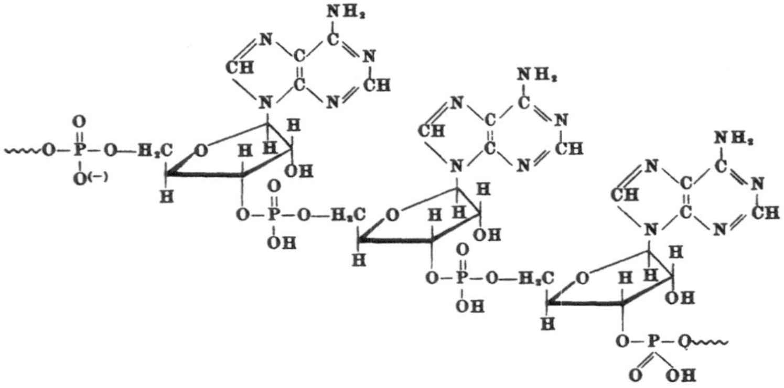

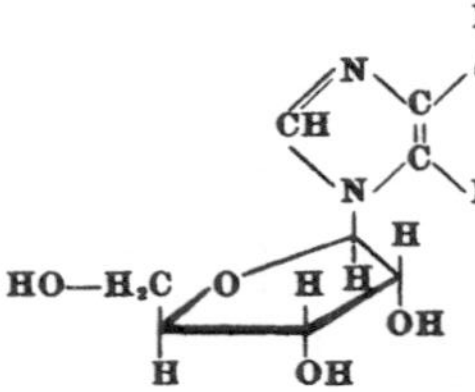

$O{=}C{=}N{-}(CH_2)_6{-}N{=}C{=}O$ Hexamethylenediisocyanate
and $HO{-}(CH_2)_4{-}OH$

	Chain Structure	Components
Polycaprolactam = Nylon 6 (Perlon, Caprolan)	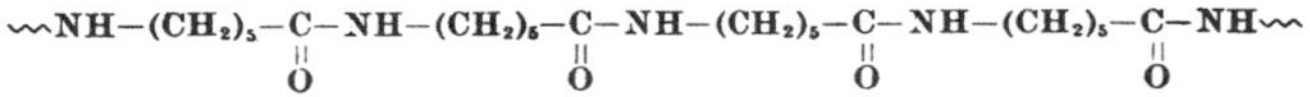 ∼NH—(CH₂)₅—C—NH—(CH₂)₅—C—NH—(CH₂)₅—C—NH∼ (C=O)	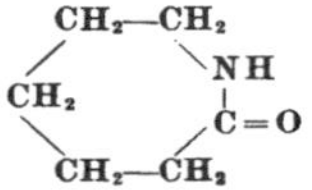 ε-Caprolactam $$\begin{array}{c} CH_2-CH_2 \\ CH_2 \quad NH \\ CH_2-CH_2 \quad C=O \end{array}$$
Nylon-6,6	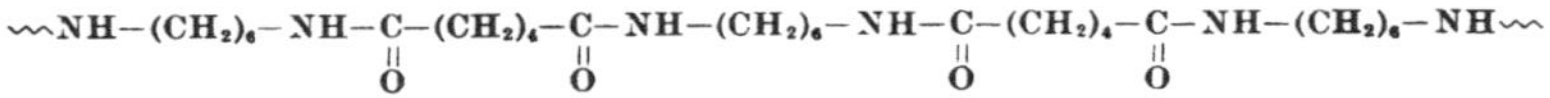 ∼NH—(CH₂)₆—NH—C—(CH₂)₄—C—NH—(CH₂)₆—NH—C—(CH₂)₄—C—NH∼ (C=O)	Hexamethylenediamine and Adipic acid
Polybenzyl-glutamate	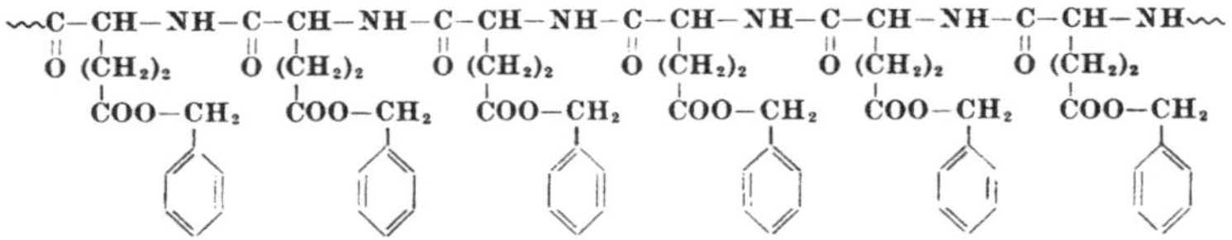 ∼C—CH—NH—C—CH—NH—C—CH—NH—C—CH—NH∼ with O=C (CH₂)₂ / COO—CH₂—C₆H₅ side groups	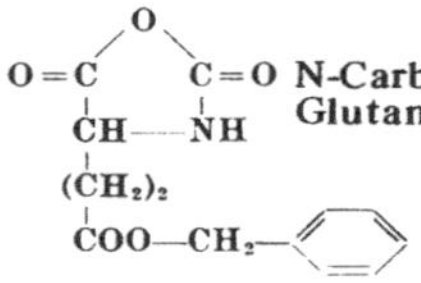 N-Carboxyanhydride of Glutamic acid-benzylester $$O=C \quad C=O$$ CH—NH / (CH₂)₂ / COO—CH₂—C₆H₅
Polyurea	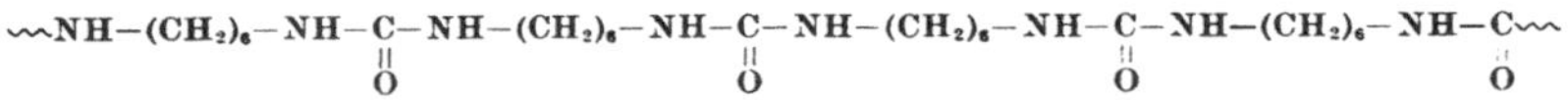 ∼NH—(CH₂)₄—NH—C—NH—(CH₂)₄—NH—C—NH—(CH₂)₄—NH—C∼ (C=O)	Hexamethylenediisocyanate and Hexamethylenediamine
Aromatic Polyimide (Kapton-Film, Pyre-ML-Coating)	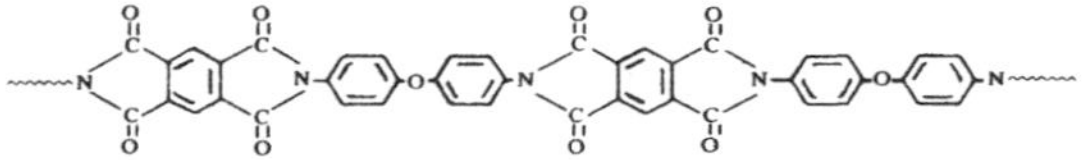 (polyimide chain structure with pyromellitimide and diphenyl ether units)	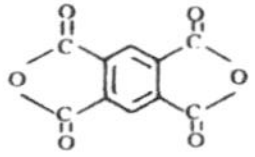 Pyromellitic anhydride (PMDA); 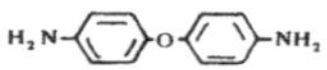 4,4'-Diaminodiphenylether
Polybenzimidazole (Imidite)	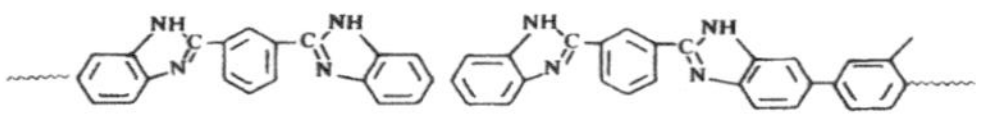 (polybenzimidazole chain structure)	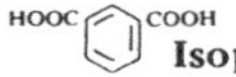 Isophthalic acid; 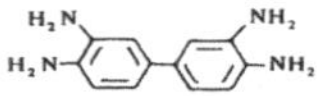 3-3'-Diaminobenzidin

TABLE 12. (Cont'd)

Name	Chain Structure	Monomers
Polybenzothiazole		Phthalamide 3,3′-Dimercaptobenzidine
Polyimidazopyrrolone (Pyrrone)		Pyromellitic-anhydride 3,3′-Diaminobenzidine
Polysulfone		Bisphenol-A 4,4′-Dichlordiphenylsultone
Polydimethylsiloxane (Silicone)	$\sim O-Si(CH_3)_2-O-Si(CH_3)_2-O-Si(CH_3)_2-O-Si(CH_3)_2-O-Si(CH_3)_2-O-Si(CH_3)_2-O-Si(CH_3)_2-O-Si(CH_3)_2-O-Si(CH_3)_2\sim$	$Cl-Si(CH_3)_2-Cl + 2\,H_2O \rightarrow [HO-Si(CH_3)_2-OH] + 2\,HCl$ Dimethyldichlorosilane (or diphenyl-, etc.)
Polysiloxane ladder-polymer		Phenyltrichlorosilane
Thiokol	$\sim CH_2-CH_2-S-S-S-S-CH_2-CH_2-S-S-S-S-CH_2-CH_2-S-S-S-S-CH_2\sim$	Ethylenechloride and Sodiumtetrasulfide

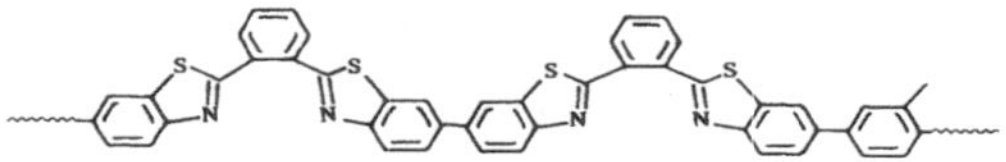

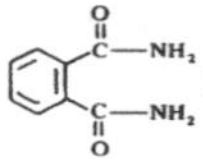

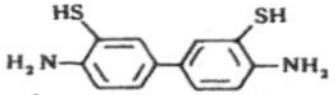

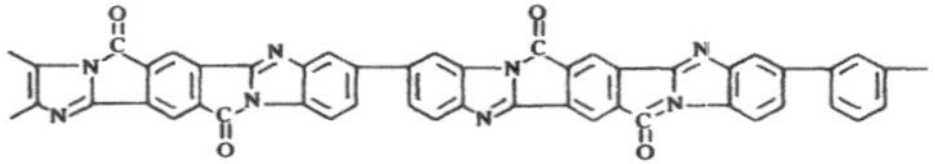

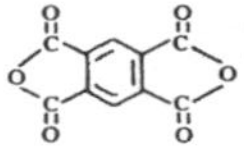

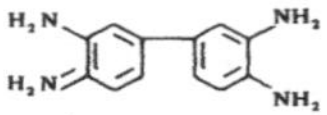

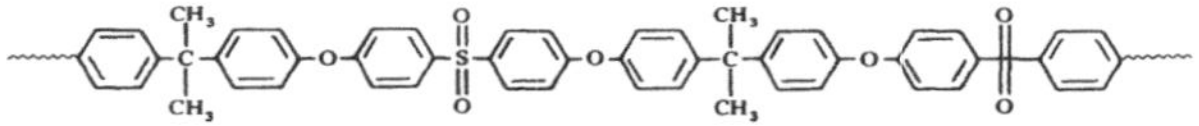

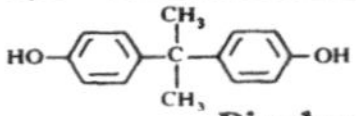

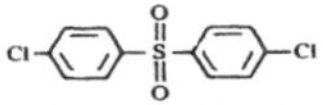

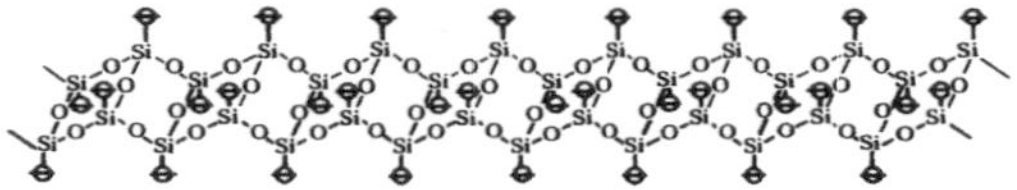

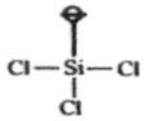

synthesis is not very satisfactory. This is especially so because more and more syntheses have been developed which cannot be easily grouped into one or the other of these two types. So it is better to call only those macromolecular compounds addition polymers which are really formed through addition polymerization and polycondensates those which are really formed through polycondensation. For the classification of macromolecular compounds, it is better to use the chemical structure of the chain itself.

One must, however, distinguish between the structural formulas of macromolecular compounds and the structural formulas of low molecular organic compounds. It is impossible to give a more or less exact structure for a macromolecular compound except in a very small number of cases, because the synthesis of the macromolecules hardly ever occurs quantitatively in the sense of a single reaction equation. The reaction is always accompanied by side reactions which, in the case of macromolecular compounds, do not lead to different separable compounds but only to irregularities in the chain structure. For example, in the polymerization of conjugated dienes, one obtains a mixture of 1,4—and

$$CH=CH \quad CH=CH \quad CH=CH \quad CH=CH$$
$$\sim\sim CH_2 \quad CH_2-CH_2 \quad CH_2-CH_2 \quad CH_2-CH_2 \quad CH_2\sim\sim \qquad \textit{cis-}1,4\text{-Polybutadiene}[5]$$

$$\qquad\quad CH_2-CH_2 \qquad\qquad\qquad CH_2-CH_2$$
$$CH=CH \quad CH=CH \quad CH=CH \quad CH=CH \qquad \textit{trans-}1,4\text{-Polybutadiene}[5]$$
$$\sim\sim CH_2 \qquad\qquad CH_2-CH_2 \qquad\qquad CH_2-CH_2\sim\sim$$

$$\sim\sim CH_2-CH-CH_2-CH-CH_2-CH-CH_2-CH\sim\sim \qquad \text{1,2-Polybutadiene}$$
$$\qquad\quad CH \qquad\quad CH \qquad\quad CH \qquad\quad CH$$
$$\qquad\quad \| \qquad\qquad \| \qquad\qquad \| \qquad\qquad \|$$
$$\qquad\quad CH_2 \qquad\quad CH_2 \qquad\quad CH_2 \qquad\quad CH_2$$

$$\sim\sim CH_2-CH-CH_2-CH-CH_2-CH-CH_2-CH\sim\sim$$
$$\qquad\quad R \qquad\quad R \qquad\quad R \qquad\quad R$$

Head-to-tail structure (1,3-position of the substituents R)

$$\sim\sim CH_2-CH-CH-CH_2-CH_2-CH-CH-CH_2-CH_2-CH-CH-CH_2\sim\sim$$
$$\qquad\quad R \quad R \qquad\qquad R \quad R \qquad\qquad R \quad R$$

Head-to-head structure (1,2-position of the substituents R)

---

[5]These formulas are projection formulas, which show only the position of the carbon atoms relative to the double bond. They can not be used to describe the real structure of the chain. Because of the nearly free rotation around the carbon-carbon single bond, it is not possible to assign to a chain a certain absolute configuration. Use of atomic models to clarify the actual structure and avoid wrong impressions is strongly recommended.

1,2-addition as well as the formation of *cis-* and *trans-*1,4-polymers. With vinyl compounds one can obtain pure head-to-tail polymerization, but also a mixture of head-to-tail, head-to-head polymerization. (Compare pp. 55-57.)

With vinyl polymers one usually finds 1,3-addition of the substituents. However, it is very difficult to prove that the regularity of the structure is not interrupted every now and then by a different form of addition.

When the carbon atoms in the chain are asymmetric (pseudo-asymmetric) there can be an additional heterogeneity of the chain structure because of the steric position of the substituents along the chain. This has been shown mainly by Natta and co-workers.

$$\sim CH_2-\underset{\underset{R}{|}}{\overset{\overset{H}{|}}{C}}-CH_2-\underset{\underset{R}{|}}{\overset{\overset{H}{|}}{C}}-CH_2-\underset{\underset{R}{|}}{\overset{\overset{H}{|}}{C}}-CH_2-\underset{\underset{R}{|}}{\overset{\overset{H}{|}}{C}}\sim \quad \text{Isotactic chain structure}$$

$$\sim CH_2-\underset{\underset{R}{|}}{\overset{\overset{H}{|}}{C}}-CH_2-\underset{\underset{H}{|}}{\overset{\overset{R}{|}}{C}}-CH_2-\underset{\underset{R}{|}}{\overset{\overset{H}{|}}{C}}-CH_2-\underset{\underset{H}{|}}{\overset{\overset{R}{|}}{C}}\sim \quad \text{Syndiotactic chain structure}$$

The statistically irregular steric structure of most technical polymeric materials is described as *atactic*. The *isotactic* polymers, owing to their regular structure, demonstrate a great tendency to crystallize. When dissolved in "good solvents," however, there are no significant differences between the isotactic and atactic polymers. The best way to represent the isotactic structure is with molecular models (see p. 185-190).

The synthesis of isotactic polymers is usually only possible by means of special stereospecific (stereoregulating) initiators (Ziegler-catalysts). Syndiotactic polymers may also be synthesized through ionic or radical polymerization at low temperatures. In most cases the isotactic and syndiotactic polymers still contain more or less large amounts of atactic structures. Even within a single chain one can have atactic and iso- or syndiotactic segments following one another.

In addition to the irregularity in the chain itself, side chains and chain-branching also strongly influence the properties of the polymers.

12. COPOLYMERS

If only one monomer is involved in the synthesis of a macromolecule, it is called a homopolymer. However, it is equally possible to synthesize macromolecules from two or more different monomers, and these are called copolymers. Table 12 shows a number of macromolecules consisting of two different structural units. These are, however, only of the type in which the different units alternate in a completely regular fashion. In such cases it is possible to combine different structural units into a single repeat unit -AB-. This alternating sequence is, however, only a special case

TABLE 23. Types of Copolymers

1. ⟿A—B—B—A—B—A—A—A—B—A—A—B—A—B—B—B—B—A⟿
(statistical (or random)
sequence)

2. ⟿A—B—A—B—A—B—A—B—A—B—A—B⟿
   or   ⟿AB—AB—AB—AB—AB—AB—AB⟿
(alternating sequence)

3. a) ⟿AB—AB'—AB—AB—AB'—AB—AB—AB'—AB'⟿
(statistical distribution)

   b) ⟿AB—A'B—A'B—AB—A'B—AB—AB—A'B—AB⟿
(statistical distribution)

   c) ⟿AB—A'B—AB'—A'B'—A'B—AB—AB'—A'B'—AB'⟿
(statistical distribution)

4. ⟿A—C—B—C—A—B—C—C—A—B—B—B—A—A—C—B—C—A⟿
(statistical distribution)

5. ⟿A—A—A—A ⋯ A—A—A—A—B—B—B⟍B ⋯ B—B—B—A—A—A⟿
(segment polymers = block-
copolymers)

of a series of possibilities as they are shown schematically in Table 23.

Structure 1 shows the most common case of a two-component copolymer in which the two components are distributed randomly along the chain in a certain ratio, which does not always correspond to that of the two monomers in the copolymerizing monomer mixture. This is the chain structure of the majority of the commercially produced copolymers. (For example, those from butadiene and styrene, or butadiene and acrylonitrile, vinylidene chloride and acrylic acid esters, or those from two different acrylic acid esters.) In the same way it is possible to copolymerize three or more monomers together (see structure 4). For example, different acrylic acid esters alone or with styrene, vinyl chloride, acrylonitrile, acrylic acid esters, vinyl pyrrolidone, or also different vinyl ethers by themselves or with isobutylene, acrylic acid esters, or vinyl chloride. Structure 2 corresponds to the normal polyesters and polyamides made from glycols, i.e., diamines, and dicarboxylic acids. There are also combinations of unsaturated compounds, which form copolymers with alternating sequence. To these belong, for example, the following systems: styrene/maleic anhydride, and styrene/acrylonitrile.

Structure 3 allows several possibilities for preparing mixed polyesters or mixed polyamides: (a) for example, from a glycol A and two dicarboxylic acids B and B', (b) from two different glycols A and A' and a dicarboxylic acid B, and (c) from two different glycols A and A' and two dicarboxylic acids B and B'.

There are also many copolymers in nature and, in fact, one can say that it is through copolymerization that nature is able to produce all the many different types of plants and animals. Thus, there is an infinity of different sugars, pentoses and hexoses, and such derivatives as uronic acids and aminosugars from which, by combination in different ways, a large number of polysaccharides are formed. In these different polysaccharides the sequence of the different structural units is by no means always statistically random. There are some with a certain regular sequence which will then affect the properties of the particular polysaccharide. This is even more characteristic of the proteins, whose chains consist of amino acid residues strung together in a very definite sequence. With certain proteins, for example, insulin or the protein component of tobacco mosaic virus, the sequence of amino acids has now been completely established.

Of some interest is the possibility of combining two different homopolymers into a long chain, as shown in structure 5. Such macromolecules are called segment polymers or block copolymers. They can be prepared in different ways, for example, through the reaction of different polyesters having OH end groups with diisocyanates or through the anionic polymerization of certain monomers with sodium- or lithium alkyls (see pages 164-173).

The common copolymers of unsaturated compounds (structure 1) are of much greater commercial importance. The possibilities of forming copolymers and terpolymers are limited because not every vinyl compound is capable of reacting with every other one to form a copolymer. Styrene, for example, shows very little inclination to react with vinyl chloride or vinyl acetate to form copolymers. In such cases, if one goes to high conversions, the monomers polymerize essentially individually so that a mixture of homopolymers results. As most polymers are not compatible with each other, the mixtures obtained in such cases are not transparent, but opaque white or translucent.

13. BRANCHED AND CROSS-LINKED POLYMERS

In principle, the synthesis of chain-like macromolecules always occurs through the reaction of bifunctional components with each other. This does not mean that the formation of macromolecules requires the monomers to have two preformed functional groups, such as OH groups, COOH groups, NH_2 groups, or O=C=N groups. Often the bifunctional character of the monomers arises only with the addition of an initiator, as for example, with vinyl compounds, which, in the presence of unstable radicals, add to each other as if they were biradicals ("opening of the double bond") or, if ions are added, react like polar compounds.

Another way compounds can attain a bifunctional character through an initiation reaction is the ring-opening of such heterocyclic compounds as cyclic esters (lactones), amides (lactams), and ethers.

The presence of compounds with more than two functional groups leads to the formation of branched and cross-linked macromolecules, as shown in Figures 25 and 26.

Whether a macromolecular compound should be classified as branched or cross-linked is a question of the length ($\overline{P}$) of the linear chains and of the degree of cross-linking. A branched macromolecule, however, should not be considered equivalent to a partly cross-linked one. As long as the compound is still soluble, it is considered branched, and one speaks of branches or side chains. If, however, the

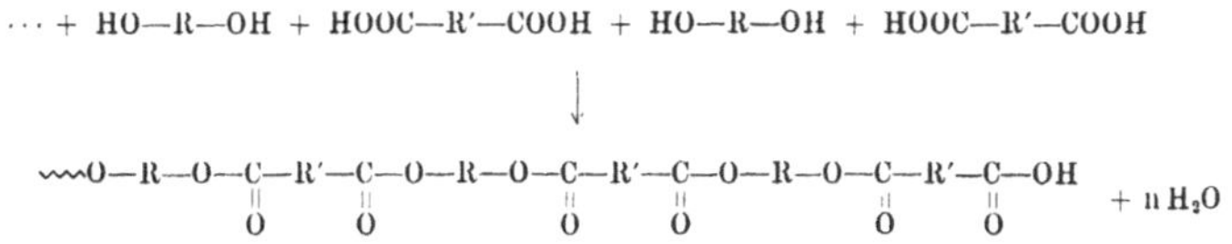

$$\cdots + \text{HO—R—OH} + \text{HOOC—R}'\text{—COOH} + \text{HO—R—OH} + \text{HOOC—R}'\text{—COOH}$$

$$\downarrow$$

$$\text{\large\textbackslash\textbackslash\textbackslash\textbackslash O—R—O—C—R}'\text{—C—O—R—O—C—R}'\text{—C—O—R—O—C—R}'\text{—C—OH} \quad + \text{n H}_2\text{O}$$

Linear Polyester

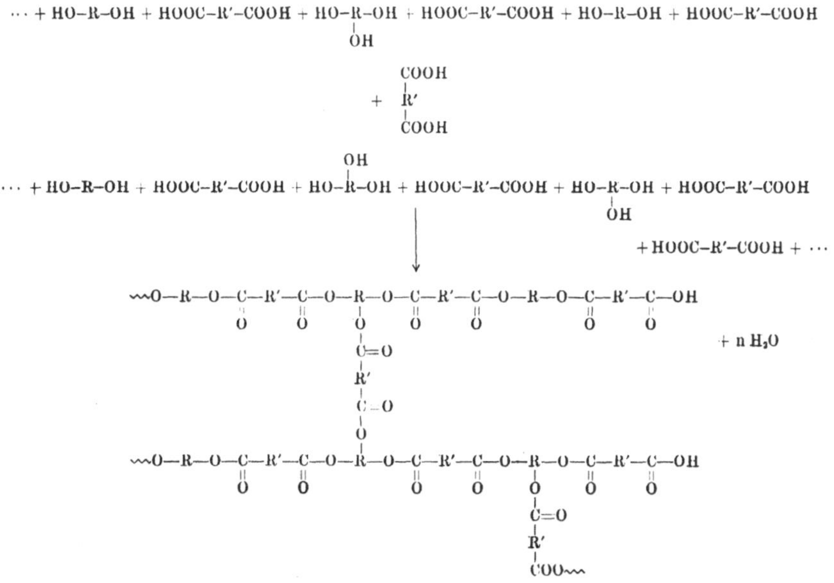

$$\cdots + \text{HO—R—OH} + \text{HOOC—R}'\text{—COOH} + \text{HO—R—OH} + \text{HOOC—R}'\text{—COOH} + \text{HO—R—OH} + \text{HOOC—R}'\text{—COOH}$$

$$\text{OH}$$

$$\text{COOH}$$
$$+ \quad \text{R}'$$
$$\text{COOH}$$

$$\text{OH}$$
$$\cdots + \text{HO—R—OH} + \text{HOOC—R}'\text{—COOH} + \text{HO—R—OH} + \text{HOOC—R}'\text{—COOH} + \text{HO—R—OH} + \text{HOOC—R}'\text{—COOH}$$

$$\text{OH}$$
$$+ \text{HOOC—R}'\text{—COOH} + \cdots$$

$$\downarrow$$

$$\text{\large\textbackslash\textbackslash\textbackslash O—R—O—C—R}'\text{—C—O—R—O—C—R}'\text{—C—O—R—O—C—R}'\text{—C—OH}$$

$$+ \text{n H}_2\text{O}$$

FIG. 25 – Cross-linked (branched) Polyester

compound is insoluble (limited swellability) in all solvents, one speaks of cross-linked polymer systems. In the latter case, all chains of the polymer are tied together by covalent bonds, and one can therefore no longer consider the molecules as the smallest part of the material. Figure 28/1 shows in a schematic way the transition from branched macromolecules to a three-dimensional cross-linked material, first through increase in the degree of polymerization of the linear chains and second through the increase in the number of branching and cross-linking points.

There is at a given concentration of cross-linking agent, a critical degree of polymerization at which a branched polymer becomes a completely .cross-linked material. The greater the concentration of the cross-linking agent, the lower the critical degree of polymerization. Similarly, for a certain degree of polymerization, one can speak of a critical concentration of cross-linking agent. With solution polymerizations it is particularly easy to observe the attainment and surpassing of the critical degree of polymerization, because often in the space of a few minutes there is a transition from a viscous solution to a more or less solid cross-linked gel.

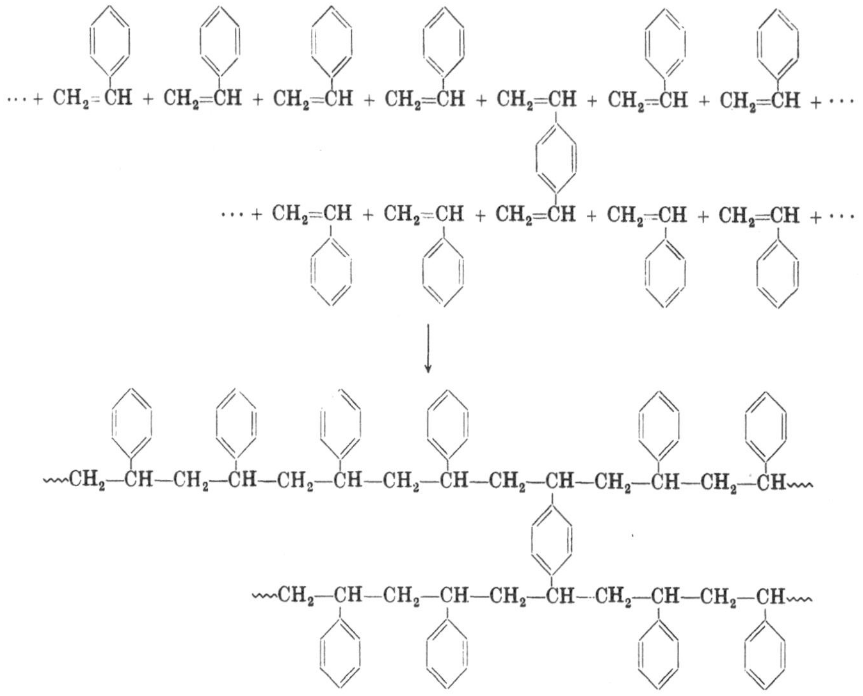

FIG. 26 — Origin of branching (cross-linking) in the polymerization of styrene in the presence of divinylbenzene.

Just as it is possible to convert branched polymers to cross-linked ones by increasing the degree of polymerization, it is possible to convert cross-linked polymers into branched polymers through chain scission. Such a degradation can be brought about in a simple way if one exposes the polymer to strong shearing forces (mechanical degradation). Most macromolecular compounds are thermoplastic, except for some which decompose below their melting point. This means that there is a temperature range in which they show a transition from a hard glassy material into a soft plastic one. This transition usually occurs over a more or less definite stage of rubber elasticity. Thermoplasticity is lost through cross-linking. At low degrees of cross-linking, i.e., large distances between cross-linking points, the cross-linked material remains rubber-elastic above the softening point without being able to flow, whereas strong cross-linking makes the material hard and brittle.

Examples of commercially important cross-linked polymers are vulcanized rubber and the polyester elastomers (for example, Vulcollan). Very tightly cross-linked materials are represented by unsaturated polyester casting resins, ion exchange resins, epoxy resins, and resins based on phenol formaldehyde (Bakelite), melamine-formaldehyde (melamine resins), and phthalic anhydride glycerol (Alkyd Resins). Such resins are infusible, and are therefore called thermosetting resins in contrast to thermoplastic materials.

From a formal point of view one could call all materials with a main valence lattice, such as diamond and quartz, three-dimensional cross-linked macromolecules. Materials with layer lattices, such as mica, talcum, and graphite, could be called two-dimensional cross-linked macromolecular materials.

Cross-linking does not always occur during the polymerization or during the formation of the macromolecular material. It is also possible to first prepare linear polymers with reactive groups or reactive residues, which are then reacted with bifunctional compounds:

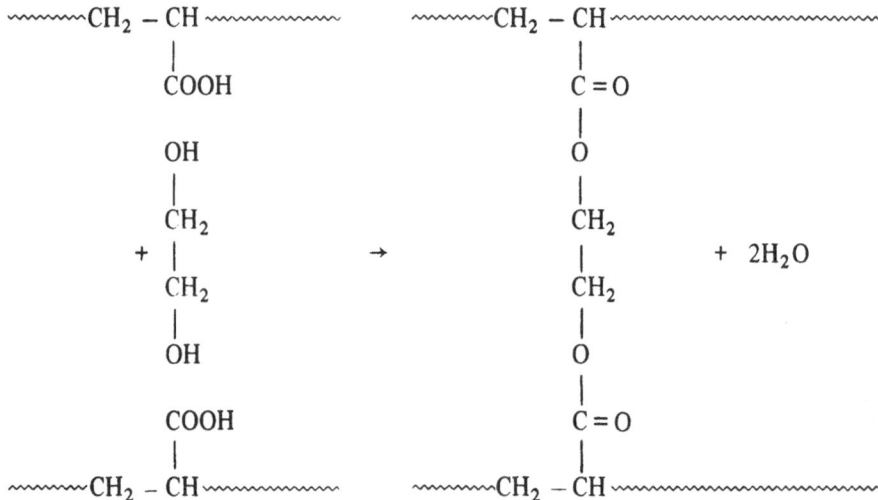

The most important industrial example of this kind of cross-linking is the vulcanization of rubber with sulfur or organic sulfur compounds.

Cross-linked polymers are insoluble in all solvents. However, according to the degree of cross-linking, they can be swollen to a smaller or larger extent. The degree of swelling, that is, the maximum amount of solvent which 1 g of a cross-linked polymer can absorb, can be taken as a measure of the degree of cross-linking (Flory-Rehner-equation). Swollen macromolecules are called gels. Each macromolecular compound exhibits the phenomenon of swelling. The uncross-linked polymers show a gradual transition from swelling to the formation of a viscous solution. However, with cross-linked polymers, swelling is limited (see Chapter 4).

Cross-linking can also come about through secondary valence forces. Examples of macromolecular compounds cross-linked through secondary valence forces are the different polysaccharides and protein gels which play an important role in nature and in living organisms. Depending on temperature and pH, they can occur as solutions or gels. All dissolution of macromolecular compounds occurs through an intermediate stage of the secondary valence gel.

Of particular interest are cross-linked polymers in which the side chains occur at more or less regular intervals along the main chain (Christmas tree structure: Figure 28/2. It is rather difficult to form structures that are completely free of

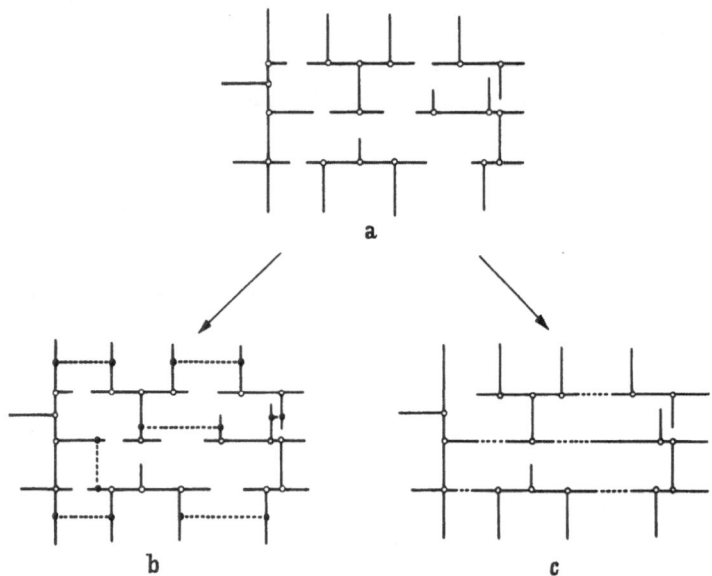

a) branched macromolecules
b) total crosslinking by increasing the number of crosslinking points
c) total crosslinking by increasing the degree of polymerization of the linear chains

FIG. 28/1 — Molecular branching and crosslinking.

continuous cross-linking (structure a and b). In most cases one obtains systems that correspond to structure c. With all these structures the chains are not really rods, but have a coil structure. This is true not only for the main chains but also for the

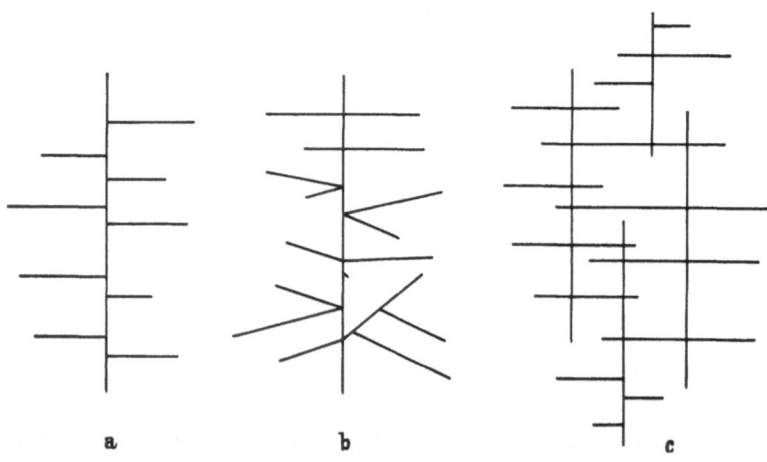

FIG. 28/2 — Branching and crosslinking possibilities

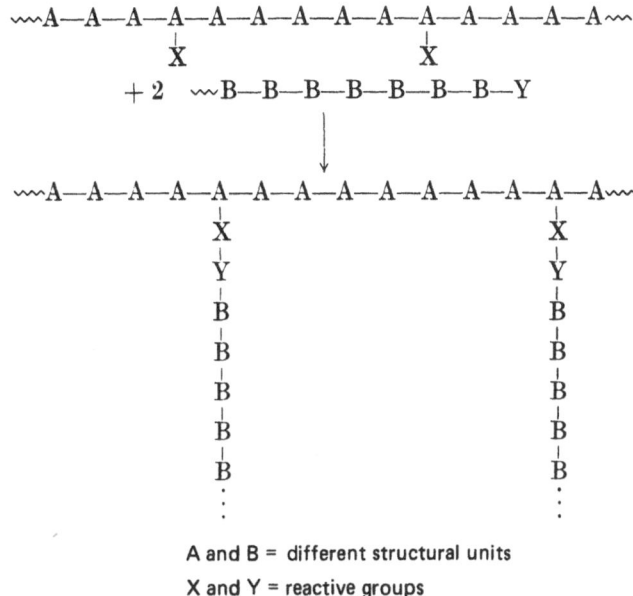

A and B = different structural units

X and Y = reactive groups

FIG. 29 — Graft-copolymers obtained by reaction of two polymers.

side chains. (Because of the coil structure of polymer molecules one always has to distinguish between *inter-* and *intra*molecular cross-linking. In general, both types of cross-links are present. For a more detailed treatment of these phenomena, see Section 2723).

Polymers with structures similar to those shown in Figure 28/1 are called graft-copolymers if the side chains have a different structural unit from the main chain. There are a number of ways such graft-copolymers can be prepared (Chapter 2, "Synthesis of Macromolecular Compounds").

There are essentially two ways of synthesizing graft-copolymers:

(1) The attachment of side chains which already exist as polymers containing the structural unit B to a polymer with the structural unit A through functional groups X and Y, which must have the property that only X can react with Y, but not X with X and Y with Y (see Figure 29).

(2) Polymerization of a monomer B on to a polymer $- A - A - A - A -$ with the aid of initiator molecules R_I- which are somehow attached to the main chain or built into it (see Figure 30).

Both block-copolymers and graft-copolymers have properties which are different from those of normal copolymers and from those of the two corresponding homopolymers. Because there preparation is difficult, their industrial importance, especially that of uncross-linked graft-copolymers, is still rather limited. Cross-linked graft-copolymers, however, (see Figure 28/2, structure c) are prepared in large quantities as polyester casting resins.

PRIMARY, SECONDARY, AND TERTIARY STRUCTURE

A polymer chain consists of a large number of structural units which are to a greater or lesser extent mobile with respect to each other. They can therefore arrange themselves in different ways. All the processes and phenomena which can be observed in the chemistry of low molecular compounds only in a large ensemble of single molecules and which can be described generally with the aid of thermodynamics are found, in macromolecular chemistry, within the region of a single large molecule. Thus one can correctly attribute a state of aggregation to the single macromolecule and, within this meaning, speak of an *intra*molecular or *micro*-Brownian motion. In addition to this motion of single sections or segments of a chain, which may be more or less strong depending on the temperature and the chemical constitution of the chain, the entire macromolecule itself moves (*inter*molecular or *macro*-Brownian motion). The behavior of macromolecular systems is, therefore, especially complicated, not only in respect to thermodynamics (e.g., phase transitions), but also in respect to chemical reactions.

The classification of primary, secondary, and tertiary structure is to be understood as follows. The *primary* structure refers to the kind of structural units a chain consists of and to the sterical arrangement of neighboring structural units (*cis, trans,* isotactic, etc.). The primary structure of a polymer chain means the same as its chemical constitution and configuration.

Table 12 shows the primary structure of the most important macromolecular

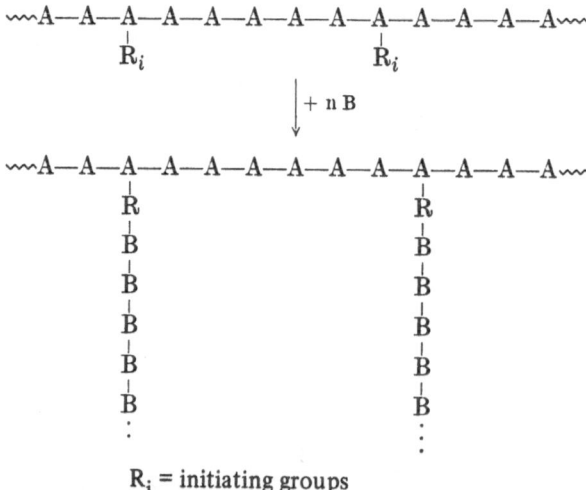

R_i = initiating groups

FIG. 30 — Graft-copolymers obtained by polymerizing a monomer in the presence of a polymer.

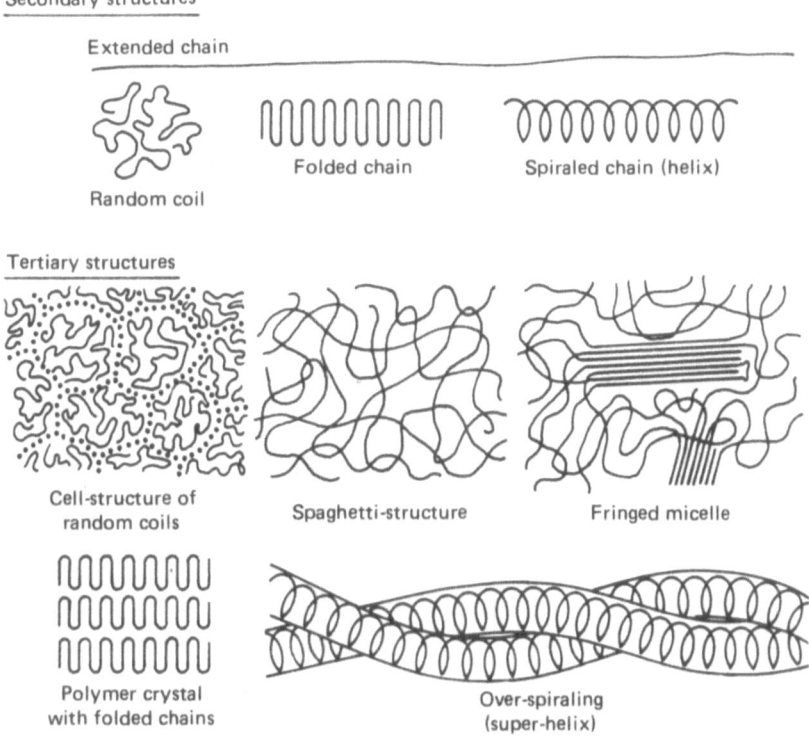

Secondary structures

Extended chain

Random coil

Folded chain

Spiraled chain (helix)

Tertiary structures

Cell-structure of
random coils

Spaghetti-structure

Fringed micelle

Polymer crystal
with folded chains

Over-spiraling
(super-helix)

FIG. 31 — Schematic representation of secondary and tertiary structures
of macromolecules.

compounds. The primary structure affects some characteristics directly—for
example, solubility and chemical behavior. It affects most properties, however, only
indirectly by influencing the degree of flexibility or mobility of the chain and thus
the secondary structure. The secondary structure, in turn, determines and modifies
the tertiary structure and thereby the physical behavior of the material.

The *secondary* structure relates to the arrangement of the polymer chain within
the range of a single macromolecule, i.e., conformation or the type of aggregation
state of an individual macromolecule.

A polymer chain can exist in a number of different shapes: as a completely
extended chain, in the form of a random coil, or in the form of a periodically
regular arrangement of chain segments (compare Figure 31).

The random coil is, at least with synthetic polymers and with polymer solutions,
the predominating type of secondary structure. Its characteristic attribute is the
average coil density, which in solution is exactly proportional to the viscosity
number and to the molecular weight.

The helix plays a most important role with proteins and nucleic acids. The special stability of these helices, even in aqueous gels and in solutions, is caused by hydrogen bonds between adjacent turns in the helix, or, in the DNA double helix, between certain substituents.

The *tertiary* structure, which is strongly influenced by the secondary structure, is concerned with the arrangement of the macromolecules to form more complex aggregates. As Figure 31 shows, a large variety of tertiary structures is possible. The tertiary structure can extend uniformly throughout the whole polymeric material; however, association of two or more tertiary aggregates can lead to *quaternary* structures, which occur quite often in nature. Hemoglobin, for example, the component of the red blood cells responsible for the transportation of oxygen, consists of particles with four subunits, which are formed in turn by a particular folding of helices.

2. SYNTHESIS AND REACTIONS OF MACROMOLECULAR COMPOUNDS

The work, ΔF, which has to be performed in the change of a system according to the reaction $A + B \rightarrow C$, or which has to be performed to obtain the reverse reaction, is a measure of the driving force of this reaction. A system can perform work either because its energy becomes smaller through the heat of reaction ΔH (enthalpy), or because its entropy increases by $T \Delta S$ because it goes from a less probable to a more probable state i.e., from an ordered to a disordered (or less ordered) state. If only the internal energy changes, then $\Delta F = \Delta H$; if only the entropy changes, $\Delta F = -T \cdot \Delta S$. In a reaction if the internal energy and also the enthropy change, ΔF is equal to the sum of both, i.e., $\Delta F = \Delta H - T \cdot \Delta S$.

Accordingly, a reaction can occur (ΔF negative) if (1) both the internal energy decreases (ΔH negative) and the entropy increases (ΔS positive); (2) if the internal energy increases (positive ΔH = endothermic reaction), but the entropy increase (positive ΔS) is so large that this effect outweighs the increase in energy; or (3) if there is an entropy decrease (negative ΔS) which is outweighed by simultaneous decrease of the internal energy (negative ΔH = exothermic reaction).

The formation of macromolecules from small molecules (monomers) belongs to this last type of reaction. For example:

$$n\ CH_2 = CH_2 \rightarrow [-CH_2-CH_2-]_n.$$

Since the polymeric state, in which the monomer residues are tied together into long chains, is always less probable than the state of a completely disordered mixture of the monomer molecules, ΔS must be negative. The formation reaction is, therefore, possible only because it corresponds to the transition of an energy-rich to an energy-poor state (for example, olefin $\rightarrow$ paraffin) and because all formation reactions of macromolecules are exothermal (negative ΔH).

There are a few exceptions, such as the polymerization of sulfur during which S_8 – rings are converted into linear sulfur chains:

$$\overline{S_4\ S_4} \rightarrow \ \text{\small\textasciitilde\textasciitilde}S-S-S-S-S-\text{\small\textasciitilde\textasciitilde}$$

The magnitude of the heat of reaction can be quite different with the different reactions which lead to macromolecules. For example, the heat of polymerization of such unsaturated compounds as ethylene, propylene, isobutylene, styrene, vinylchloride, acrylic acid esters, vinyl ethers, and vinyl esters is in the order of 20 kcal/mol. The ΔH of the polymerization of ethylene oxide, formaldehyde, and the polyaddition of diisocyanates, lies in approximately the same range (30 kcal). In comparison, such polycondensation reactions as occur in the formation of polyamides and polyesters from dicarboxylic acids and amines, i.e., alcohols have a smaller ΔH (about 4 kcal). On the other hand, we also know of polymerizations that have only a very small ΔH—for example, the polymerization of α–methylstyrene. In such a case, the magnitudes of ΔH and $T \cdot \Delta S$ at 20° C are approximately of the same order, so that the system monomer $\rightleftharpoons$ polymer is in a state of reversible equilibrium.

$$n \quad CH_2 = \underset{\underset{\text{(phenyl)}}{\overset{\displaystyle CH_3}{|}}}{C} \quad \rightleftharpoons \quad \left[-CH_2 - \underset{\underset{\text{(phenyl)}}{\overset{\displaystyle CH_3}{|}}}{C} - \right]_n$$

The higher the temperature, the larger the entropy part $T \cdot \Delta S$ and the greater the concentration of monomer, which is in an equilibrium state. The lower the temperature, the smaller $T \cdot \Delta S$; therefore, with low temperatures ΔF becomes negative, and the concentration of the polymer in the equilibrium increases. With the selection of a suitable initiator (lithium alkyls, lithium aryls, and sodium naphthalene in tetrahydrofuran), it is possible through repeated cooling to $-70°C$ and heating to 40°C to bring about reversible polymerization and depolymerization. This is similar to altering the temperature of a saturated solution: crystals form on cooling and dissolve on heating.

In theory all reactions that lead to the formation of macromolecules are equilibrium reactions, but usually the ΔH values are so large that at ordinary temperatures there is practically no monomer in the system. Only at high temperatures (200°C-300°C) does the entropy term $T \cdot \Delta S$ become so large that depolymerization begins and the monomers distill off—for example, if one heats polymethylmethacrylate (Plexiglas). In other cases (for example, polyacrylonitrile, polyvinylalcohol, or cellulose), certain decomposition reactions begin even below the temperature at which depolymerization can be carried out in a preparative manner. Such decomposition reactions are often combined with discoloration and carbonization.

The temperature at which for a certain monomer $\Delta H = T \cdot \Delta S$ is called the ceiling temperature (T_c). At this temperature $\Delta F = 0$, and no further polymerization occurs. In certain cases one can determine T_c by polymerizing at a number of temperatures and determining the conversion. If one extrapolates the temperature conversion curve to a conversion of O, one obtains the ceiling temperature.

If water or other small molecules are formed in the preparation of a polymer, as for example in the formation of polyesters from glycols and dicarboxylic acids, the equilibrium depends not only on the temperature but also on the concentration of water. To obtain polymers of a high molecular weight through polycondensation reactions, one must remove the water carefully from the equilibrium. For example, from 146 g of adipic acid and 100 g of hexamethylenediamine on quantitative conversion one obtains 226 g of a polyamide with a molecular weight of 226 x N_A = 1.361 x 10^{26} (the entire polymeric material is considered a single molecule) and 36 g of water. If one assumes that, on the average, for every 1,000 molecules of adipic acid and diamine (i.e., from 2,000 water molecules) there always remains one molecule of water as NH_2 and COOH end groups, then one obtains a polyamide with an average molecular weight of 226,000. To obtain this molecular weight, one actually has to remove 1,999 of a theoretical 2,000 water molecules, i.e., the conversion has to be 99.95%. In the same way, an excess of acid or amide affects the equilibrium. One obtains macromolecules with two NH_2 or two carboxyl end groups which cannot find a reaction partner, and therefore a further increase in molecular weight becomes impossible.

One can see from this example that it is not very easy to obtain macromolecular compounds with high molecular weights through polycondensation reactions because the removal of the last traces of water in the reaction mixture is coupled with a number of technical difficulties.[1] This is why industrially produced polycondensates such as polyamides and polyesters (for example, nylon and Dacron) have molecular weights of about 20,000, whereas commercial addition polymers (vinyl polymers such as polymethylmethacrylate, polystyrene, and polyvinylchloride) usually have molecular weights on the order of 200,000.

The fact that a polymerization (or a polycondensation, or polyaddition) is thermodynamically possible (negative ΔF) does not mean that it can occur with a useful reaction velocity, even at normal temperatures. In most cases, one has to use higher temperatures or a catalyst to overcome the potential barrier (whose height is given by the magnitude of the activation energy) which prevents the transition from monomer to polymer.

The many reactions used for the preparation of linear macromolecules can be subdivided into two main groups which differ from each other completely in their kinetics:

I. Polymerization of *unsaturated* olefin monomers or *cyclic* monomers by a *chain reaction:*

Initiator $\quad \xrightarrow{\text{Decomposition}} \quad$ 2 Radicals R·

Radical R· + Monomer M $\rightarrow$ R – M·   (Start = Initiation)

R – M· + M $\rightarrow$ R – M – M· $\rightarrow$ Chain Radical R⤳M· (Growth = Propagation)

R⤳M· + ·M⤳R $\rightarrow$ R⤳M – M⤳R   (Termination)

Depending on the type of initiation, one can differentiate between radical, ionic,

---

[1] Another reason might be the inability of the polymer coils to penetrate each other (see Section 431).

or metal-complex polymerizations. Unsaturated compounds of the type of ethylene and its derivatives, but also such cyclic monomers as ethyleneoxide or trioxane, are easily polymerized according to the above scheme. Of the compounds with C=O double bonds, only formaldehyde leads to relatively stable polymers. Other compounds with CO groups (e.g., acetaldehyde, acetone, and carbon monoxide together with ethylene have also been polymerized, but the resulting polymers are too thermally unstable for any important industrial applications.

II. Polymerization by *stepwise reaction* of monomers with *functional groups:*

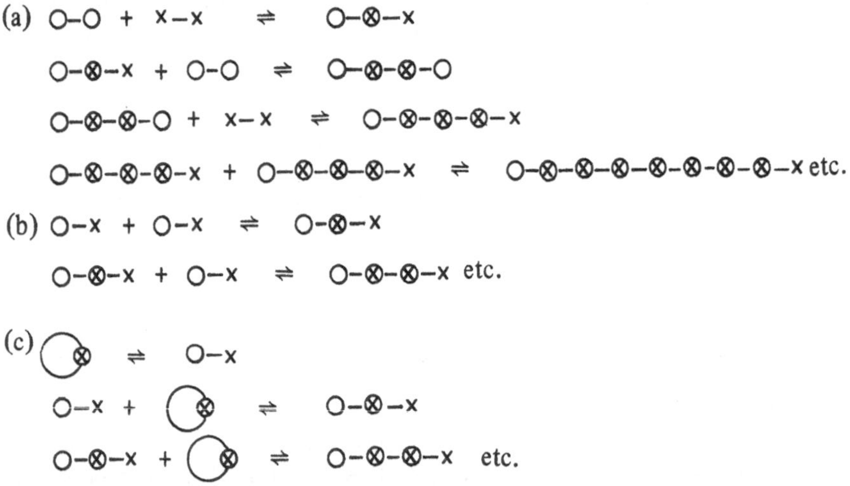

(a) O–O + x–x   ⇌   O–⊗–x

   O–⊗–x + O–O   ⇌   O–⊗–⊗–O

   O–⊗–⊗–O + x–x   ⇌   O–⊗–⊗–⊗–x

   O–⊗–⊗–⊗–x + O–⊗–⊗–⊗–x   ⇌   O–⊗–⊗–⊗–⊗–⊗–⊗–⊗–⊗–x etc.

(b) O–x + O–x   ⇌   O–⊗–x

   O–⊗–x + O–x   ⇌   O–⊗–⊗–x etc.

(c) ⊗   ⇌   O–x

   O–x + ⊗   ⇌   O–⊗–x

   O–⊗–x + ⊗   ⇌   O–⊗–⊗–x   etc.

O and x are any functional groups which can react with each other to form ⊗, a homopolar main-valence bond. For example:

O		x		O		x
– OH	+	– COOH		– NH$_2$	+	– COOH
– OH	+	– COCl		– NH$_2$	+	– N = C = O
– OH	+	– N = C = O		– NH$_2$	+	– COOCH$_3$
– OH	+	– CH – CH$_2$ (O)				

The two different groups, O and x may be, as shown in reaction (a), distributed over two different monomer molecules, (for example, adipic acid and hexamethylenediamine → 6,6–nylon). However, they can also be in one and the same monomer molecule (for example, with ω-amino acids), or they can form a ring in which the two groups are then present within the ring in a condensed form (for example, ε-caprolactam → 6–nylon).

The following are the basic differences between the two types of polymer-forming reactions:

In case I the monomers can react only with the relatively few radicals or ions present but not with each other. By addition of monomers at the few active centers the chain molecules grow very rapidly in a "sudden rush" within a short time span (order of magnitude: 1 second) until they have reached their definitive length by chain termination. Once the growth has been terminated, the formed macromolecules do not participate in the polymerization reaction any more (except where chain transfer reactions occur). This type of chain growth reaction causes the length of the polymer chains to be more or less independent of the degree of conversion (Figure 37a). Just the opposite holds for the other type of reaction, II, which is

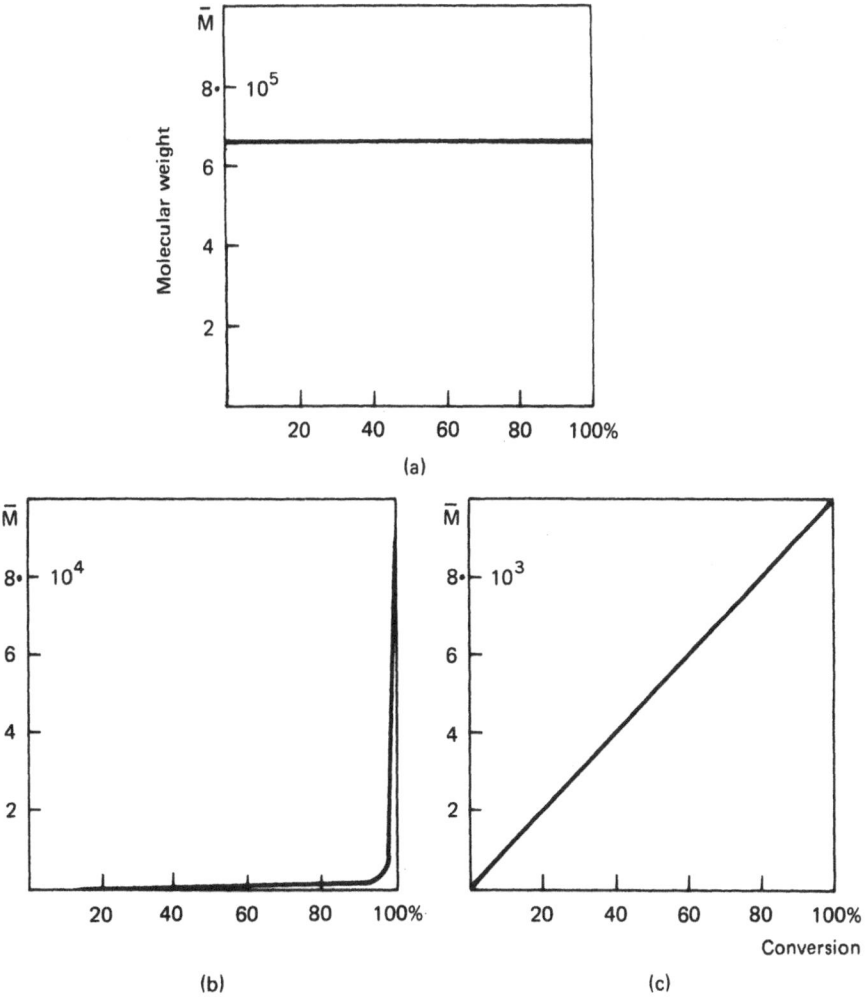

(a)

(b)   (c)

a) with radical polymerizations
b) with polycondensations
c) with living polymerizations and with protein synthesis

FIG. 37 – Dependence of the molecular weight on conversion.

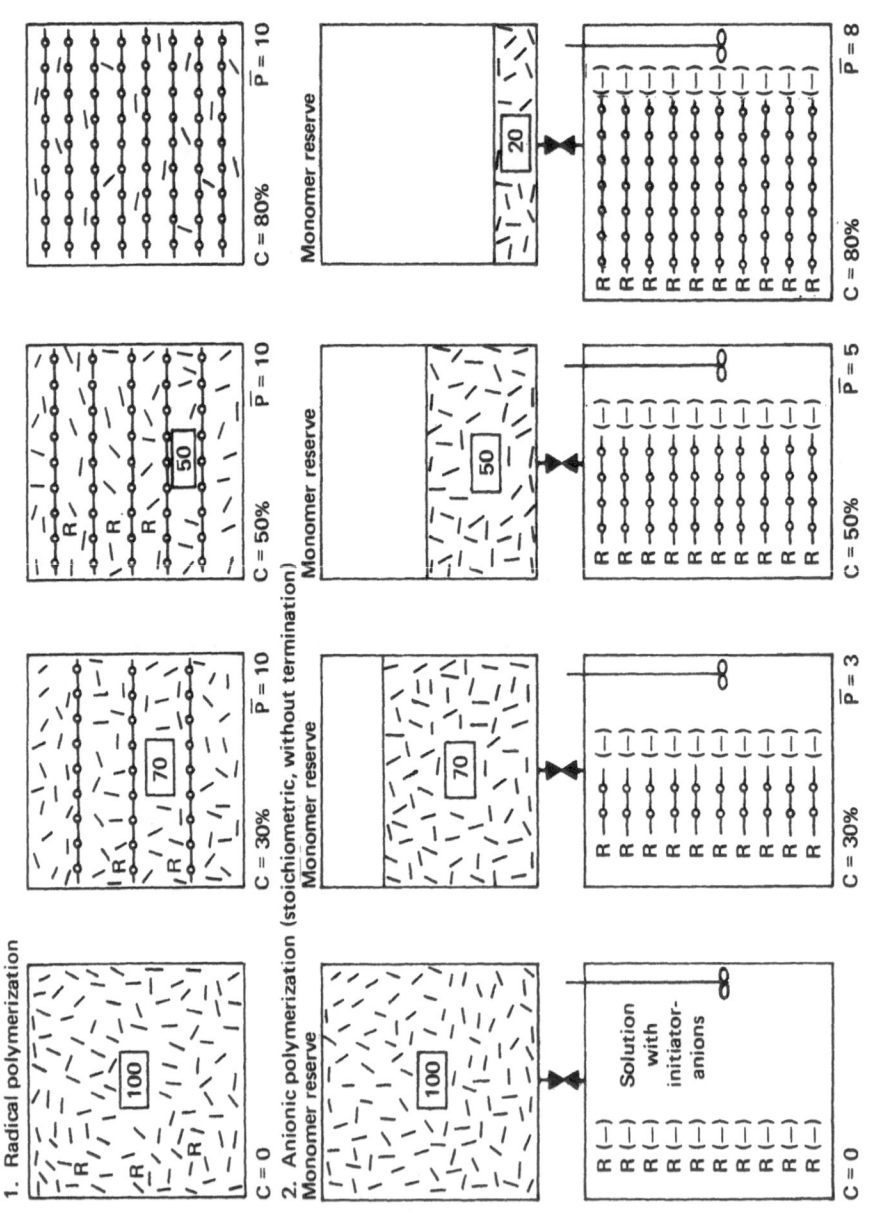

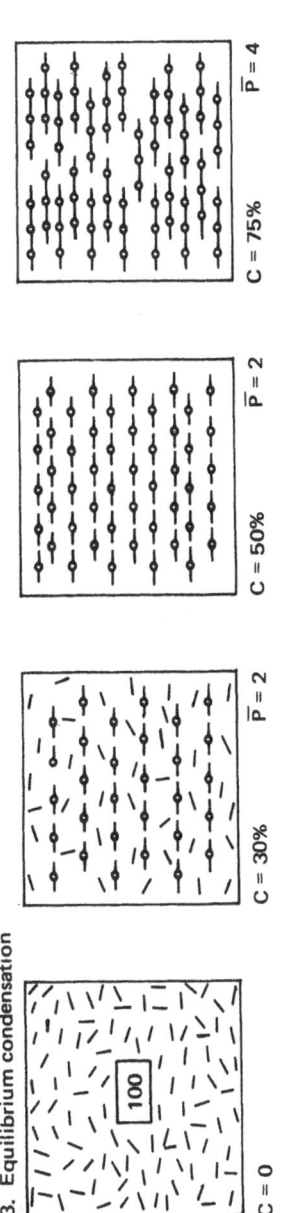

3. Equilibrium condensation

4. Stepwise copolymersynthesis with predetermined monomer sequence (protein synthesis)

For a protein with P = 100:

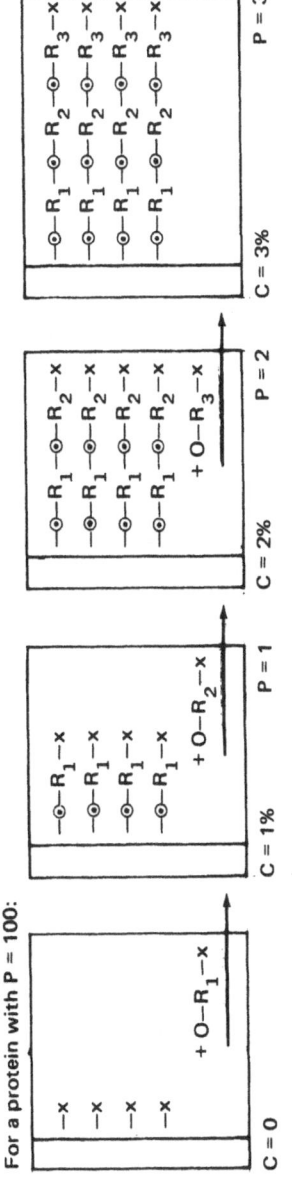

FIG. 38 — Schematic representation of the different possibilities of chain growth with different types of polymer syntheses.

characterized by a very strong dependence of the chain length on the degree of conversion (Figure 37b).

In the case of type II all reaction partners have the same reactivity, and no preferred reaction centers exist. Each monomer molecule can react with each other monomer molecule so that at first (at lower conversions below about 80%) preferably dimers, trimers, and other oligomers are formed. Only at high conversions above 98-99.9%, are long polymer chains formed. Reaction II occurs in a stepwise manner, i.e., it is possible to interrupt the reaction (e.g., by cooling) at any time without affecting the reactivity of the then present chains of oligomers or polymers. The reaction can be continued (e.g., by heating) at any time, and the oligomers and polymers will continue to react with each other.

For both types, the chain-reaction polymerization and the stepwise polycondensation, there is a special case which is characterized by the fact that all chains grow at a constant rate from a constant number of active centers. This leads to a linear relationship between the chain length (i.e., the molecular weight) and the conversion. (Figure 37c).

With the aid of the schematic presentation in Figure 38, it is simple to see the different ways the chain grows in different types of polymer synthesis reactions.

Not all reactions which lead to the formation of macromolecular compounds can be classified without a certain degree of arbitrariness, partly because the reaction mechanism is not completely known in all cases and partly because the reactions of organic synthesis are too numerous to be classified into such a simple four-class scheme. One also should realize that not all reactions which lead to the tying together of molecules are useful for the preparation of macromolecular compounds. Because the purification of polymers, if at all possible, is always difficult and complicated, those reactions which occur without side reactions are preferred, especially by industry, because the removal of side products is always rather expensive. This is also one reason purification of the monomeric starting materials is always carried out with great care.

21. SYNTHESIS OF MACROMOLECULES WITH C-C CHAINS THROUGH POLYMERIZATION OF OLEFINIC UNSATURATED COMPOUNDS

Many olefinic unsaturated compounds are able to form chain-like macromolecules through elimination of the double bond, a phenomenon first recognized by Staudinger. Diolefins polymerize in the same manner, however only one of the two double bonds is eliminated. Such reactions occur through the initial addition of a monomer molecule to an initiator radical or an initiator ion by which the active state is transferred from the initiator to the added monomer. In the same way, by means of a chain reaction, one monomer molecule after the other is added (2,000-20,000 monomers per second) until the active state is terminated through a different type of reaction. The polymerization is a chain reaction in two ways: because of the reaction kinetics and because as a reaction product one obtains a chain molecule. The length of the chain molecule is proportional to the kinetic chain length.

One can summarize the process as follows (R· is equal to the initiator radical):

$$R \cdot + CH_2 = CH + CH_2 = CH + CH_2 = CH + \cdots \cdots \rightarrow$$
$$\qquad\quad | \qquad\qquad | \qquad\qquad |$$
$$\qquad\quad Cl \qquad\qquad Cl \qquad\qquad Cl$$

$$R - CH_2 - CH - CH_2 - CH - CH_2 - CH\sim\sim\sim\sim\sim$$
$$\qquad\qquad | \qquad\qquad | \qquad\qquad |$$
$$\qquad\qquad Cl \qquad\qquad Cl \qquad\qquad Cl$$

One thus obtains polyvinylcholoride from vinylchloride, or polystyrene from styrene, or polyethylene from ethylene, etc.

The length of the chain molecules, measured by means of the degree of polymerization (definition see p. 9), can be varied over a large range through selection of suitable reaction conditions. Usually, with commercially prepared and utilized polymers, the degree of polymerization lies in the range of 1,000 to 5,000, but in many cases it can be below 500 and over 10,000. This should not be interpreted to mean that all molecules of a certain polymeric material consist of chains of 500, or 1,000, or 5,000 monomer units. In almost all cases, the polymeric material consists of a mixture of polymer molecules of different degrees of polymerization. Through such processes, as fractional precipitation and chromatographic methods a polymeric material can be separated into more homogeneous fractions. The greater the number of fractions the more homogeneous they can be made. Often one also specifies a degree of polymerization even for an unfractionated polymer. In such cases, the degree of polymerization is an average, which is defined differently depending on the method used for the determination of the molecular weight. These averages are written $\bar{P}$ or $\bar{M}$ and are usually called $\bar{P}_n, \bar{P}_w$ or $\bar{M}_n, \bar{M}_w$ (compare Section 3135).

Not all olefinic unsaturated compounds can be polymerized. One basic requirement is a favorable potential difference from the monomer to the polymer (negative ΔH), which can be recognized by the exothermic behavior of the polymerization. The example of α-methylstyrene shows that not even this basic assumption is true for all olefins. In addition, there are some olefins which fulfill this basic condition (i.e., where the polymer presents a thermodynamically more stable condition than the monomer) but which do not polymerize. Thus, polymerization can be prevented by too high an activation energy, if suitable intermediate reactions are absent through which it might be possible to overcome the potential barrier, which surrounds the monomer like a wall. For example, all efforts to polymerize propylene to a real macromolecule were unsuccessful for a long time until metal complexes, introduced by K. Ziegler were used as initiators [in this case $TiCl_3/Al(C_2H_5)_3$]. Furthermore, steric reasons or the peculiar case of autoinhibition (see p. 71) may be responsible for the reluctance of an unsaturated compound to polymerize.

If we arrange the polymerizable olefins on a linear scale, we can put at one end of the scale those olefins which are difficult to polymerize (or not at all polymerizable). At the other end will then be those which have such a great tendency to polymerize that it is very difficult to prepare them in a monomerically pure state—they are stable only in the dark under refrigeration or by addition of stabilizers. In between these extremes lie those monomers which polymerize rapidly only upon warming or upon addition of suitable initiators (the term "monomer" is used for all polymerizable unsaturated compounds and also for any bifunctional compounds used in the synthesis of macromolecules). Table 42 gives some examples of these three groups of unsaturated compounds.

TABLE 42. Examples of unsaturated olefins and vinyl compounds with different tendency for polymerization

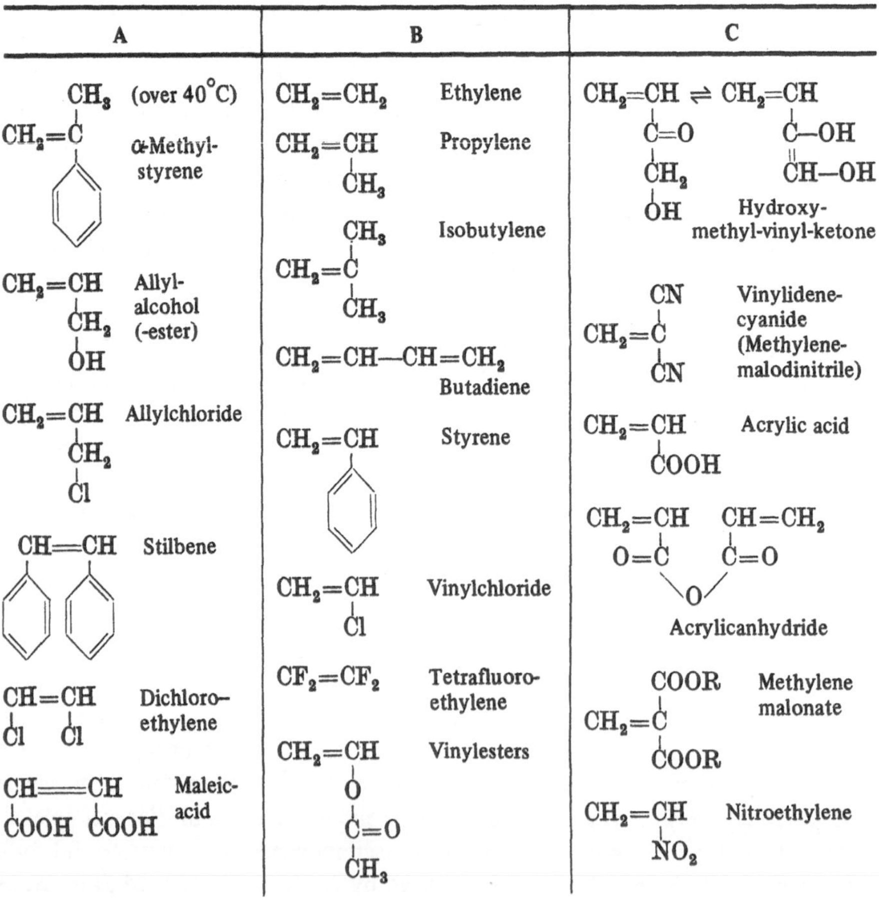

A	B	C
$CH_2=C$ — CH_3 (over 40°C) α-Methyl-styrene	$CH_2=CH_2$ Ethylene	$CH_2=CH \rightleftharpoons CH_2=CH$ $C=O$ $C-OH$ CH_2 $CH-OH$ OH Hydroxy-methyl-vinyl-ketone
$CH_2=CH$ — CH_2 — OH Allyl-alcohol (-ester)	$CH_2=CH$ — CH_3 Propylene	
	$CH_2=C$ — CH_3 Isobutylene	$CH_2=C$ — CN / CN Vinylidene-cyanide (Methylene-malodinitrile)
$CH_2=CH$ — CH_2 — Cl Allylchloride	$CH_2=CH—CH=CH_2$ Butadiene	
	$CH_2=CH$ Styrene	$CH_2=CH$ — $COOH$ Acrylic acid
$CH=CH$ Stilbene		$CH_2=CH$ $CH=CH_2$ $O=C$ $C=O$ $\backslash O \diagup$ Acrylicanhydride
	$CH_2=CH$ — Cl Vinylchloride	
$CH=CH$ — Cl Cl Dichloro-ethylene	$CF_2=CF_2$ Tetrafluoro-ethylene	$CH_2=C$ — $COOR$ / $COOR$ Methylene malonate
	$CH_2=CH$ — O — $C=O$ — CH_3 Vinylesters	
$CH=CH$ — $COOH$ $COOH$ Maleic-acid		$CH_2=CH$ — NO_2 Nitroethylene

TABLE 42. (Cont'd)

A	B	C

Cl   Cl Trichloro-
$\diagdown$C=C$\diagup$ styrene
Cl   ⬡
	$CH_2=CH$ Vinylethers
	    $\mid$
	    O
	    $\mid$
	    R
		CN α-Cyanaocrylate
		$\mid$
		$CH_2=C$
		    $\mid$
		    $COOR$

	$CH_2=CH$ Acrylic acid-
	    $\mid$ esters
	    $COOR$	

	$COOH$
	   $\mid$
	$CH_2=C-CH_2-COOH$
	    Itaconic acid (esters)	

	$CH_2=CH$ Acrylo-
	    $\mid$ nitrile
	    CN	

	$CH_2=CH$ Vinyl-
	    $\mid$ pyrrolidone
	   $\diagup N \diagdown$
	  CH_2 $C=0$
	   $\mid$ $\mid$
	  CH_2-CH_2	

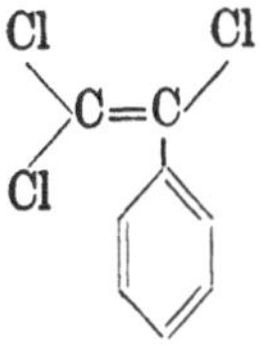

A Unsaturated compounds which do not polymerize by themselves

B Monomers which polymerize under the action of heat, light, or initiators

C Monomers which have a particularly strong tendency to polymerize

Polymerization, a chain reaction, occurs according to the same mechanism as the well-known chlorine-hydrogen reaction and the decomposition of phosgene.

The initiation reaction, which consists of the activation of the double bond, can be brought about by heating, irradiation, ultrasonics, or initiators. The initiation of the chain reaction can be observed most clearly with radical or ionic initiators. These are energy-rich compounds which can add suitable unsaturated compounds (monomers) and maintain the activated radical, or ionic, state so that further monomer molecules can be added in the same manner. For the individual steps of this growth reaction one needs only a relatively small activation energy and therefore through a single activation step (the actual initiation reaction) a large number of olefin molecules are converted, as is implied by the term "chain reaction." Because very small amounts of the initiator bring about the formation of a large amount of polymeric material (1:1,000 to 1:10,000), it is possible to regard polymerization from a superficial point of view as a catalytic reaction. For this reason, the initiators used in polymerization reactions are often designated as polymerization catalysts, even though, in the strictest sense, they are not true catalysts because the polymerization initiator enters into the reaction as a real partner and can be found chemically bound in the reaction product, i.e., the polymer. In addition to the ionic and radical initiators there are now metal complex initiators (which can be obtained, for example, by the reaction of titanium tetrachloride or titanium trichloride with aluminum alkyls), which play an important role in polymerization reactions (Ziegler catalysts). The mechanism of their catalytic action is not yet completely clear.

Table 44 shows the type of initiation by which most monomers are polymerized. When the monomer is also commercially polymerized, the preferred initiation is designated with a circle.

TABLE 44. Types of initiation mechanism by which different monomers polymerize.

Monomer		Polymerization mechanism			
		radical	anionic	cationic	Metal-complex
$CH_2=CH_2$	Ethylene	⊕		+	⊕
$CH_2=CH$ $\|$ CH_3	Propylene			+	⊕
$CH_2=CH-CH_2-CH_3$	Butylene-1				⊕
CH_3 $\|$ $CH_2=C$ $\|$ CH_3	Isobutylene			⊕	+
$CH_2=CH-CH=CH_2$	Butadiene	⊕	+		⊕
$CH_2=C-CH=CH_2$ $\|$ CH_3	Isoprene	+	+		⊕
$CH_2=CH$  ⬡	Styrene	⊕	+	+	+
$CH_2=CH$ $\|$ Cl	Vinylchloride	⊕			+
Cl $\|$ $CH_2=C$ $\|$ Cl	Vinylidenechloride (Dichloroethylene)	⊕			

TABLE 44. (Cont'd)

Monomer		Polymerization mechanism			
		radical	anionic	cationic	Metal-complex
$CH_2=CH$ $\vert$ F	Vinylfluoride	$\oplus$			
$CF_2=CF_2$	Tetrafluoroethylene	$\oplus$			
$CF_2=CF$ $\vert$ CF_3	Perfluoro-propylene	$\oplus$			$+$
$CH_2=CH$ $\vert$ NO_2	Nitroethylene	$+$	$+$		
$CH_2=CH$ $\vert$ O $\vert$ R	Vinylethers			$\oplus$	$+$
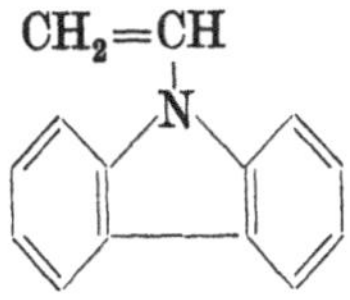 $CH_2=CH$ $\vert$ N (carbazole)	Vinylcarbazole	$\oplus$			$+$
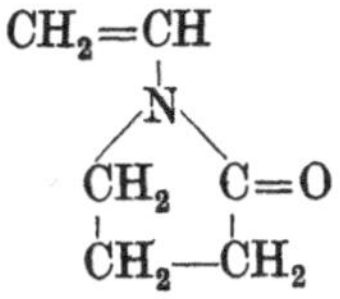 $CH_2=CH$ $\vert$ N CH_2 $C=O$ CH_2-CH_2	Vinylpyrrolidone	$\oplus$		$+$	
$CH_2=CH$ $\vert$ O $\vert$ $C=O$ $\vert$ R	Vinyl esters	$\oplus$			
$CH_2=CH$ $\vert$ $C=O$ $\vert$ O $\vert$ R	Acrylic acid esters (acrylates)	$\oplus$			

TABLE 44. (Cont'd)

Monomer	Polymerization mechanism			
	radical	anionic	cationic	Metal-complex
Methacrylates (R=CH$_3$) $CH_2=C$ with R, C=O, O, R'	⊕	+		+
α-Chloracrylates (R = Cl)	+			
α-Cyanoacrylates (R = CN)	+	+		
Methylenemalonates $CH_2=C$ with COOR, COOR	+	+		
α-Cyanosorbates $CH_3-CH=CH-CH=C$ with CN, COOR	+	+		
Acrylicanhydride $CH_2=CH$ $CH=CH_2$, O=C C=O, O	+			
Itaconicanhydride $CH_2=C$ with C=O, O, CH$_2-C=O$	+			
Oxymethylvinylketone $CH_2=CH$, C=O, CH$_2$OH	+			
Acrylonitrile $CH_2=CH$, CN	⊕	+		

TABLE 44. (Cont'd)

Monomer	Polymerization mechanism			
	radical	anionic	cationic	Metal-complex
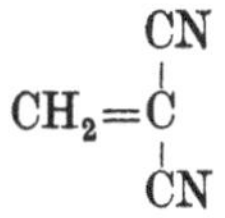 $CH_2=C$ with CN, CN groups — Vinylidenecyanide	$\oplus$	$+$		
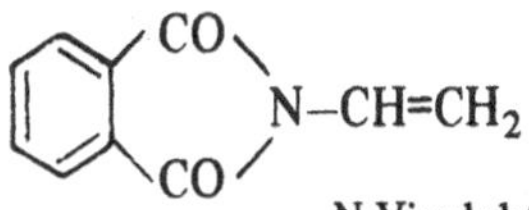 N-Vinylphthalimide (phthalimide with $N-CH=CH_2$)	$+$			
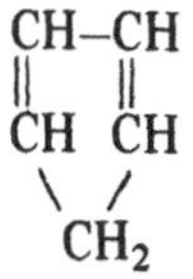 Cyclopentadiene $CH-CH$, $CH\ CH$, CH_2				$+$
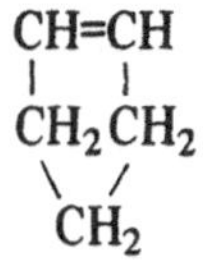 Cyclopentene $CH=CH$, CH_2CH_2, CH_2				$+$
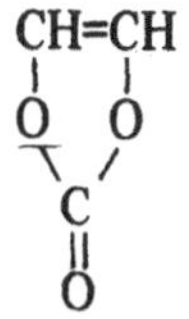 Vinylenecarbonate $CH=CH$, $O\ O$, C, O	$+$			

$\oplus$ indicates an industrial process

211 Radical Polymerization

Compounds which readily decompose into free radicals on heating or under the influence of light are usually used as initiators in radical polymerizations. Peroxides and aliphatic azo-compounds in particular have shown themselves as very suitable, and they are therefore produced commercially for use as polymerization initiators. Table 48 shows some of the most common peroxides and azo-compounds.

The following reactions describe the course of a radical polymerization, using as an example a vinyl compound (for example, where $R = -C_6H_5$

$$-\underset{\underset{O}{\|}}{C}-OR\ ,\ -O-\underset{\underset{O}{\|}}{C}-R,$$

TABLE 48. Initiators for radical polymerization

Name	Formula	Suitable Polymerization Temperature °C
Potassiumpersulfate	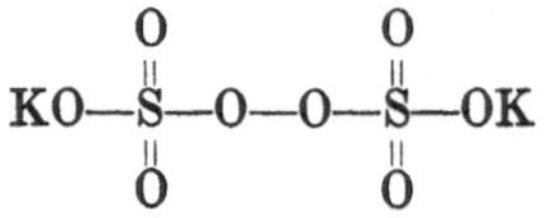 $KO-\overset{\overset{O}{\|\|}}{\underset{\underset{O}{\|\|}}{S}}-O-O-\overset{\overset{O}{\|\|}}{\underset{\underset{O}{\|\|}}{S}}-OK$	40— 80
Dibenzoylperoxyde (Lucidol)	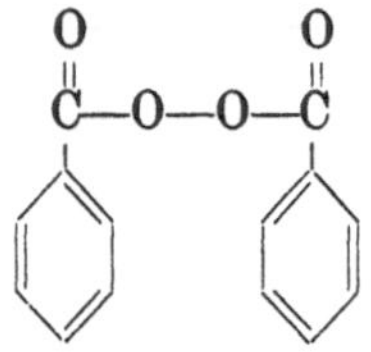 $\overset{\overset{O}{\|\|}}{C}-O-O-\overset{\overset{O}{\|\|}}{C}$ (phenyl groups)	40— 90
Cumenehydroperoxyde	$\overset{\overset{CH_3}{\|}}{\underset{\underset{CH_3}{\|}}{C}}-O-O-H$ (phenyl)	50—100
Cyclohexanoneperoxyde	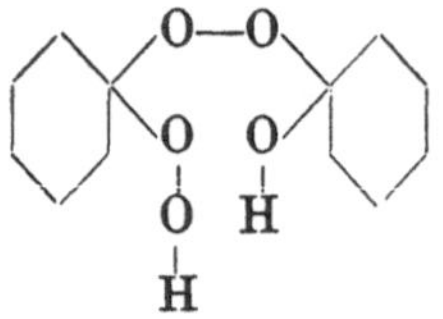 structure with O—O bridge, O O, O H, H	20— 80
Di-t-butylperoxyde	$CH_3-\overset{\overset{CH_3}{\|}}{\underset{\underset{CH_3}{\|}}{C}}-O-O-\overset{\overset{CH_3}{\|}}{\underset{\underset{CH_3}{\|}}{C}}-CH_3$	80—150
Azo-bis-isobutyronitrile (Vazo, AIBN)	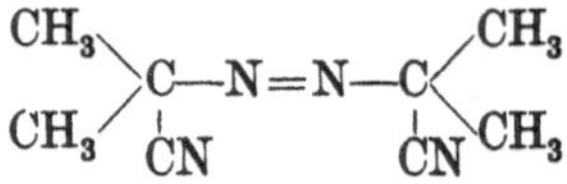 $\overset{CH_3}{\underset{CH_3}{>}}\overset{}{\underset{CN}{C}}-N=N-\overset{}{\underset{CN}{C}}\overset{CH_3}{\underset{CH_3}{<}}$	20—100
Cyclohexylsulfonyl-acetylperoxyde (Percadox ACS)	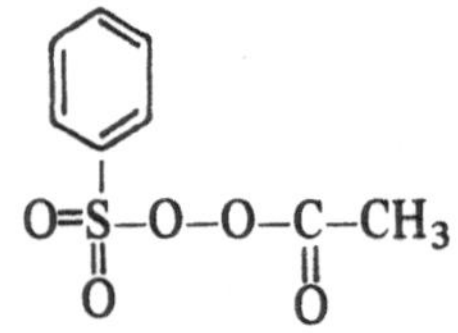 (cyclohexyl) $O=\overset{}{\underset{O}{S}}-O-O-\overset{}{\underset{O}{C}}-CH_3$	0–40
Diisopropylpercarbonate (Percadox JPP)	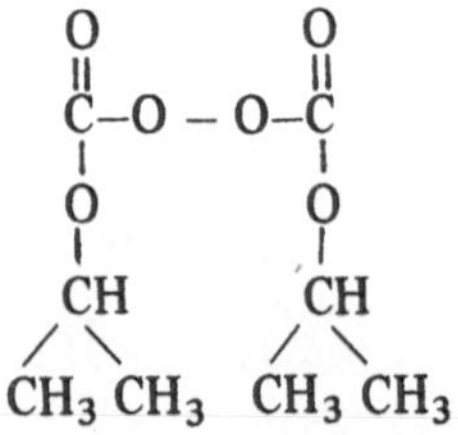 $\overset{\overset{O}{\|\|}}{C}-O-O-\overset{\overset{O}{\|\|}}{C}$, O, O, CH, CH, $CH_3\ CH_3\ CH_3\ CH_3$	40–80

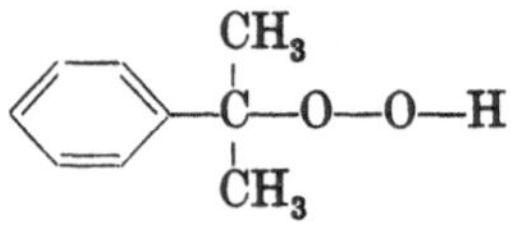

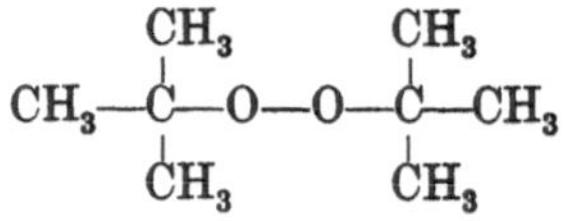

— Cl or — CN) with azobisisobutyronitrile as the initiator. The reactions are valid in principle for all monomers that polymerize by a radical mechanism, even though the individual reaction steps, especially the initiation and termination reactions, may differ in certain cases from the scheme presented here.

Initiation:

$$
\begin{array}{c}
\text{CH}_3 \quad\quad \text{CH}_3 \\
\text{NC}-\overset{|}{\underset{|}{\text{C}}}-\text{N}{=}\text{N}-\overset{|}{\underset{|}{\text{C}}}-\text{CN} \quad \xrightarrow{k_D} \quad 2\,\text{NC}-\overset{\text{CH}_3}{\underset{\text{CH}_3}{\overset{|}{\underset{|}{\text{C}}}}}{\cdot} + \text{N}_2 \\
\text{CH}_3 \quad\quad \text{CH}_3
\end{array}
$$

$$
\text{NC}-\overset{\text{CH}_3}{\underset{\text{CH}_3}{\overset{|}{\underset{|}{\text{C}}}}}{\cdot} + \text{CH}_2{=}\overset{}{\underset{\text{R}}{\text{CH}}} \quad \xrightarrow{k_i} \quad \text{NC}-\overset{\text{CH}_3}{\underset{\text{CH}_3}{\overset{|}{\underset{|}{\text{C}}}}}-\text{CH}_2-\overset{\bullet}{\underset{\text{R}}{\text{CH}}}
$$

Propagation:

$$
\text{NC}-\overset{\text{CH}_3}{\underset{\text{CH}_3}{\overset{|}{\underset{|}{\text{C}}}}}-\text{CH}_2-\overset{\text{H}}{\underset{\text{R}}{\overset{|}{\underset{|}{\text{C}}}}}{\cdot} + \text{CH}_2{=}\overset{}{\underset{\text{R}}{\text{CH}}} \quad \xrightarrow{k_p} \quad \text{NC}-\overset{\text{CH}_3}{\underset{\text{CH}_3}{\overset{|}{\underset{|}{\text{C}}}}}-\text{CH}_2-\overset{}{\underset{\text{R}}{\text{CH}}}-\text{CH}_2-\overset{\text{H}}{\underset{\text{R}}{\overset{|}{\underset{|}{\text{C}}}}}{\cdot}
$$

$$
\Big\downarrow + \text{CH}_2{=}\underset{\text{R}}{\text{CH}}
$$

$$
\text{NC}-\overset{\text{CH}_3}{\underset{\text{CH}_3}{\overset{|}{\underset{|}{\text{C}}}}}-\text{CH}_2-\underset{\text{R}}{\text{CH}}-\text{CH}_2-\underset{\text{R}}{\text{CH}}-\text{CH}_2-\overset{\bullet}{\underset{\text{R}}{\text{CH}}} \text{ etc.}
$$

Termination:

$$
\text{NC}-\overset{\text{CH}_3}{\underset{\text{CH}_3}{\overset{|}{\underset{|}{\text{C}}}}}\Big[\text{CH}_2-\underset{\text{R}}{\text{CH}}\Big]_n\text{CH}_2-\overset{\text{H}}{\underset{\text{R}}{\overset{|}{\underset{|}{\text{C}}}}}{\cdot} + {\cdot}\overset{\text{H}}{\underset{\text{R}}{\overset{|}{\underset{|}{\text{C}}}}}-\text{CH}_2\Big[\underset{\text{R}}{\text{CH}}-\text{CH}_2\Big]_n\overset{\text{CH}_3}{\underset{\text{CH}_3}{\overset{|}{\underset{|}{\text{C}}}}}-\text{CN}
$$

$$
\Big\downarrow k_t \text{ (Combination)}
$$

$$
\text{NC}-\overset{\text{CH}_3}{\underset{\text{CH}_3}{\overset{|}{\underset{|}{\text{C}}}}}\Big[\text{CH}_2-\underset{\text{R}}{\text{CH}}\Big]_n\text{CH}_2-\underset{\text{R}}{\text{CH}}-\underset{\text{R}}{\text{CH}}-\text{CH}_2\Big[\underset{\text{R}}{\text{CH}}-\text{CH}_2\Big]_n\overset{\text{CH}_3}{\underset{\text{CH}_3}{\overset{|}{\underset{|}{\text{C}}}}}-\text{CN}
$$

k_t (Disproportionation)

$$
\text{NC}-\overset{\text{CH}_3}{\underset{\text{CH}_3}{\overset{|}{\underset{|}{\text{C}}}}}\Big[\text{CH}_2-\underset{\text{R}}{\text{CH}}\Big]_n\text{CH}_2-\text{CH}_2 + \underset{\text{R}}{\text{CH}}{=}\text{CH}\Big[\underset{\text{R}}{\text{CH}}-\text{CH}_2\Big]_n\overset{\text{CH}_3}{\underset{\text{CH}_3}{\overset{|}{\underset{|}{\text{C}}}}}-\text{CN}
$$

2111 Chain Initiation

The initiation reaction occurs in two steps: the decomposition reaction of the initiator and the addition of the first monomer molecule to the radical. Not every radical formed through the decomposition reaction must lead to the starting of a chain. A fraction of the radicals can disappear under certain circumstances also through different reactions, e.g., through direct recombination of two radicals or by reaction with atmospheric oxygen or other inhibiting substances, as will be discussed in greater detail later on.

To determine the radical yield, one has to determine the number of radicals formed in a given time and the number of polymer chains formed in the same time. The rate of radical formation can be determined by analytical methods, for example, determination of peroxide. The number of polymer chains can be determined by molecular weight measurements with a method which gives the number-average molecular weight (by end-group determination or osmotic pressure measurements).

The rate of the initiation reaction depends on the rate of the initiator decomposition reaction, i.e., on the half-life of the initiator. This reaction has a high activation energy (with the most commonly used initiators it is on the order of 30 kcal/mole), and it is therefore strongly temperature dependent. The course of the decomposition reaction and, therefore, the constitution of the radicals which are formed, is often dependent on the nature of the solvent in which the decomposition occurs. Thus, dibenzoylperoxide decomposes in inert solvents, such as benzene, into phenyl radicals under formation of CO_2. However, in the presence of styrene, where no CO_2 formation seems to occur, the apparently intermediate benzoyl radicals start the polymerization.

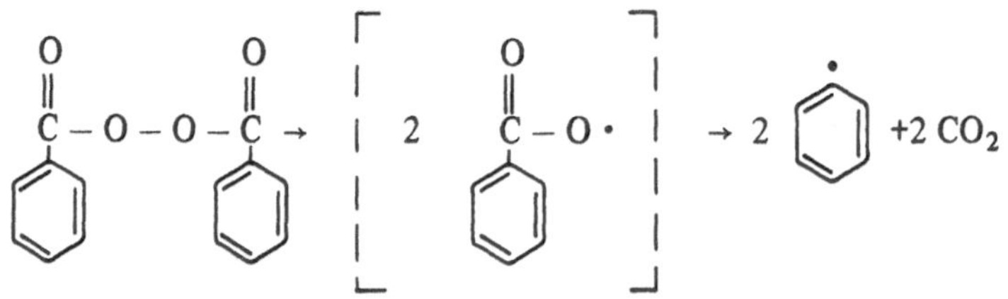

Hydroperoxides may, according to W. Kern, decompose in a bimolecular reaction with the formation of water:

$$R - O - O - H + H - O - O - R \rightarrow R - O \cdot + R - O - O \cdot + H_2O$$

Among the well-known radical formers, azobisisobutyronitrile (available commercially under the name VAZO in the United States), which has found great use as a foaming agent in the manufacture of foam plastics, decomposes especially uniformly and homogeneously to give only a single type of radical. It is therefore very suitable as an initiator for radical polymerizations, both for scientific (kinetic) investigations and for industry. With the aid of azobisisobutyronitrile it is possible to carry out polymerizations already at relatively low temperatures (e.g., 40°C-60°C). Another advantage of azobisisobutyronitrile is that it can be prepared with relatively little danger and can be kept in pure form; because the preparation and use of many peroxides can be very dangerous, one has to keep them either in solution or in suspension. At temperatures of 80°C-90°C, however, even azobisisobutyronitrile decomposes quite violently.

In addition to the rate of decomposition, it is also important to consider the stability of the radicals formed in order to determine the usefulness of a radical-forming compound as a polymerization initiator. Very stable radicals, such as triphenylmethyl and diphenylpicrylhydrazyl, do not show a great tendency to initiate polymerization.

One can reduce the activation energy of the peroxide decomposition reaction and thus increase the rate of decomposition at low temperatures greatly, if one adds reducing agents as "activators"—e.g., sodium hyposulfite, potassium persulfate, amines such as dimethylaniline or hydroxylamine, readily enolizable carbonyl compounds such as ascorbic acid, or metal ions such as Fe^{++} or Cu^+. This is usually done in the case of emulsion polymerizations, as illustrated by the following reactions:

$$H - O - O - H + Fe^{++} \rightarrow HO\cdot + OH^- + Fe^{+++} \quad \text{(Haber-Weiss)}$$

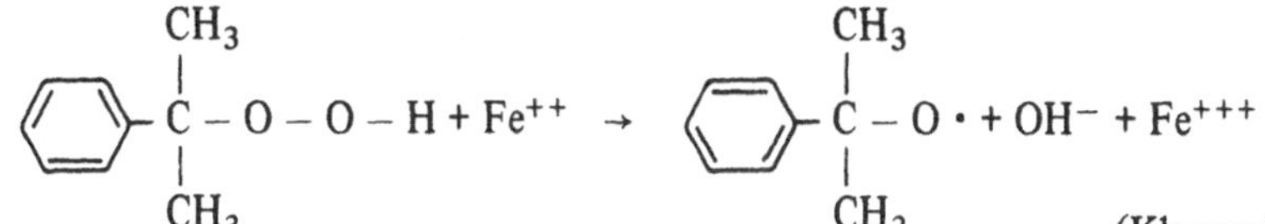

(cumene hydroperoxide)

$$R - O - O - H + Cu^{++} \rightarrow R - O - O\cdot + H^+ + Cu^+ \quad \text{(Kern)}$$

$$R - O - O - H + Cu^+ \rightarrow R - O\cdot + OH^- + Cu^{++} \quad \text{(Kern)}$$

Such redox systems permit radical polymerizations (especially emulsion polymerizations) at 0°C and even lower temperatures. A large-scale industrial application of redox polymerization is the production of synthetic rubber (SBR), which occurs by emulsion polymerization at temperatures down to -5°C.

The emulsion polymerization of vinyl compounds by a radical mechanism can also be initiated in the absence of peroxides by salts of such transition metals as $H_2PtCl_6 \cdot 6H_2O$, $PdCl_2 \cdot 2H_2O$, $RhCl_3 \cdot 3H_2O$, or $TiCl_3$. However, these initiators show a specificity toward different monomers. For example, $TiCl_3$, H_2PtCl_6, and $PdCl_2$ are good initiators for styrene, but not for methylmethacrylate. $RhCl_3$, on the other hand, initiates the polymerization of methylmethacrylate, but not of styrene. Vinylchoride, vinylacetate, and acrylonitrile are not polymerized by these salts. Although the mechanism of initiation is not exactly known, one can assume that a redox polymerization is involved (Berger and Youngman). Just as with the initiation by peroxides, oxygen and quinone act as inhibitors. The composition of the copolymers is the same as with peroxide-initiated polymerization, and the polymers show no stereoregularity (unlike butadiene, which can be polymerized to stereoregular polymers in emulsion by using the same intiators).

In the presence of oxygen or peroxides (e.g., H_2O_2), boron alkyls (for example, triethylborane or tributylborane) also give strongly active initiators for the polymerization of vinyl compounds at low temperatures. Van der Werf explains the initiation in the following manner: the alkyl borane reacts with oxygen to form a boralkylperoxide which, in the presence of a vinyl compound, reacts further to a trialkylboroxine with the formation of alkyl radicals. These alkylradicals then start the polymerization:

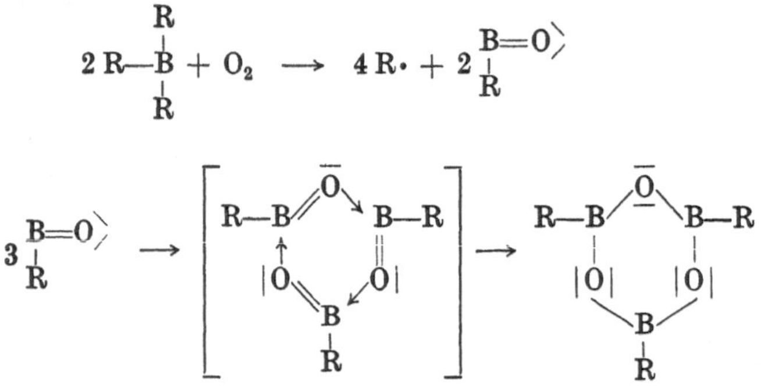

Trialkylboroxine

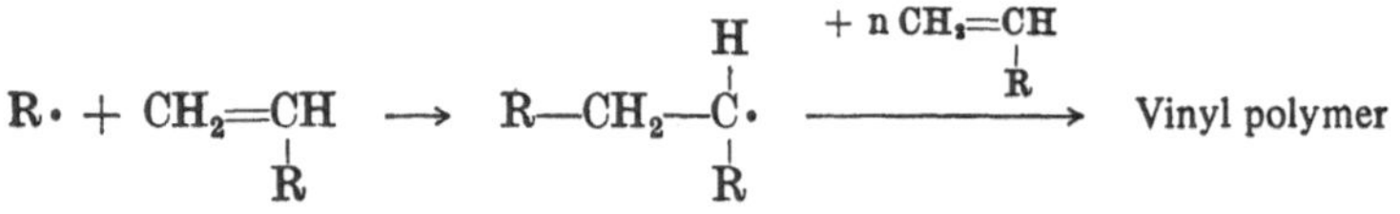

The presence of trialkylboroxine could be demonstrated in the reaction of boronalkyls with oxygen. In the absence of a monomer, one obtains alkyl borates (boric acid esters) as the chief product.

Although oxygen usually acts as an inhibitor with polymerizations initiated by free radicals, in polymerizations initiated by boron alkyls the presence of oxygen is absolutely necessary. It is, therefore, reasonable to assume that oxygen activates the polymerization. There can be no doubt, however, that even with polymerizations initiated by boron alkyls, oxygen acts as an inhibitor if it is present in excess (we assume here that the above mechanism is correct).

In principle, the initiation of polymerization by light, especially UV light, or heat occurs in the same way as with various chemical initiators. However, up to now it has not been possible to demonstrate a plausible mechanism for the initiation reaction occurring in photopolymerization or in a thermal polymerization. It would be simplest to assume that the initiation occurs through biradicals:

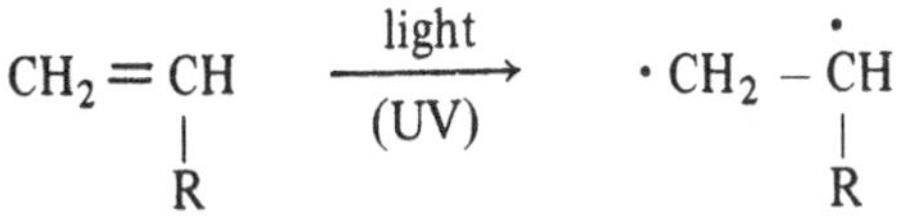

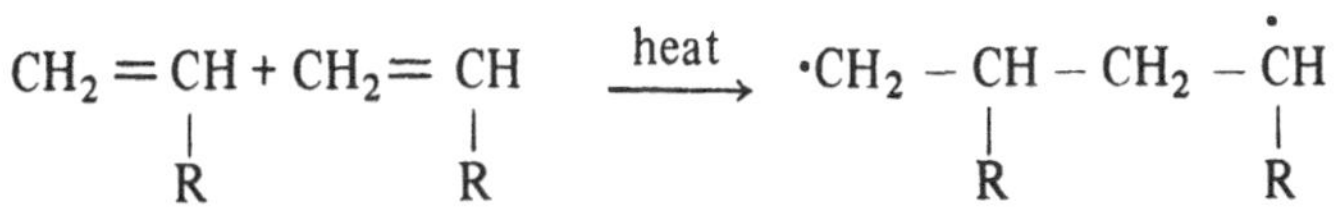

$$CH_2 = CH \quad \xrightarrow[\text{(UV)}]{\text{light}} \quad \cdot CH_2 - \overset{\bullet}{CH}$$
$$\underset{R}{|} \qquad\qquad\qquad \underset{R}{|}$$

$$CH_2 = CH + CH_2 = CH \quad \xrightarrow{\text{heat}} \quad \cdot CH_2 - CH - CH_2 - \overset{\bullet}{CH}$$
$$\underset{R}{|} \qquad \underset{R}{|} \qquad\qquad\qquad \underset{R}{|} \qquad \underset{R}{|}$$

Both of these reactions have a relatively high activation energy. However, the most important argument against such a chain initiation is the following: the tendency of the biradicals for a recombination (i.e., for mutual deactivation) is very high, and therefore there is a strong possibility that at P = 3, six-membered rings are formed:

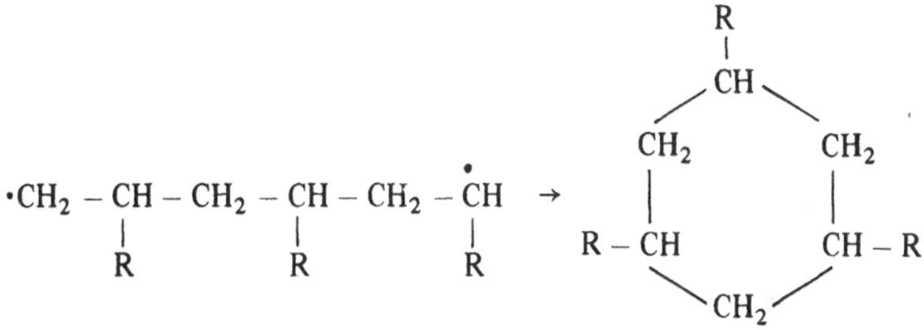

It therefore seems that a biradical chain has almost no possibility to overcome this barrier of 6-ring formation. This has also been experimentally demonstrated: on irradiation of dibutyldisulfide, one obtains butylradicals in addition to sulfur. These butylradicals bring about a rapid polymerization of styrene, whereas the biradicals formed in the photolysis of 1 − oxa − 4,5 − dithiacycloheptane do not bring about any increase in the rate of styrene polymerization:

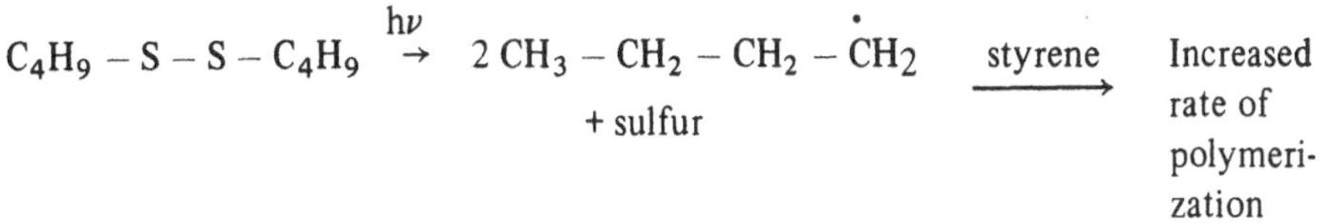

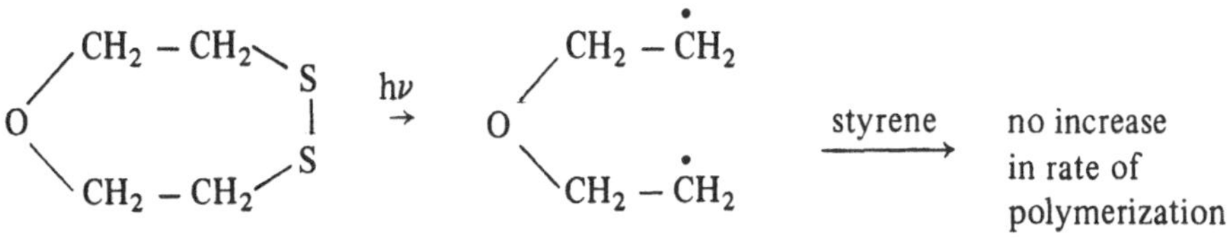

$$C_4H_9 - S - S - C_4H_9 \quad \overset{h\nu}{\rightarrow} \quad 2\,CH_3 - CH_2 - CH_2 - \overset{\bullet}{CH_2} \quad \xrightarrow{\text{styrene}} \quad$$
$$+\text{ sulfur}$$

Increased rate of polymerization

styrene no increase in rate of polymerization

As far as is known today, it is not possible to present a generally valid mechanism for the initiation reaction in photopolymerizations. With certain monomers the molecule is split into many small radicals (e.g., into $\cdot CH_3$, $CH_2 = \overset{\bullet}{C}H$, and $CH_3 - \overset{\bullet}{C} = O$ for the case of methylvinylketone). With other monomers (for example, methylmethacrylate) a condensation to quinoidal systems occurs. With styrene one assumes that a disproportionation reaction between two activated molecules is possible:

$$styrene \overset{h\nu}{\rightarrow} 2 \left[CH_2 = CH \right]^* \longrightarrow CH_3 - \overset{\bullet}{C}H + CH_2 = \overset{\bullet}{C}$$

(Flory)

excited styrene
molecules

Even less is known about the mechanism of a straight thermal initiation reaction, such as the polymerization of styrene, which has been carried out for over 25 years on a large industrial scale without its being possible up until now to elucidate the mechanism completely.

Often one adds certain sensitizers to a photopolymerization. These sensitizers are compounds which, under ultraviolet radiation, decompose into radicals. Commonly used sensitizers are benzoin and its derivatives, condensed aromatic ring systems, sulfur compounds, azo compounds, etc. Among the last is azobisisobutyro-nitrile, which is especially useful. The initiation reaction is the same as without UV irradiation, but the difference is that with UV irradiation the decomposition temperature of the initiator is lowered by approximately 50°C to 80°C. In this way, it is possible to bring about radical polymerizations at −10°C to −30°C.

2112 Chain Propagation

In contrast to the initiation reaction, the growth or propagation reaction requires a much lower activation energy, and its rate (v_p) is therefore also less temperature-dependent than the rate of the initiation reaction. This is important for the dependence of the degree of polymerization on the temperature of polymerization, since the degree of polymerization is proportional to the ratio of the rates of the growth reaction and to the termination reaction (compare pp. 86 and 93).

It is surprising that certain unsaturated compounds polymerize rapidly by a free radical mechanism (for example, acrylic esters, vinyl ketone, styrene, vinyl chloride, and a number of other vinyl compounds), whereas other unsaturated compounds polymerize only very slowly or not at all by a free radical mechanism (for example, allylic esters, allyl chloride, maleic esters, and vinyl ether). This happens not because in one case the propagation rate is much lower than in the other case, but rather because the lifetime of the radicals is very much shorter as a result of chain termination or chain transfer.

The addition of vinyl monomers during chain propagation occurs mainly as 1,2-addition so that the substituents are in 1,3-position:

1,2-addition:

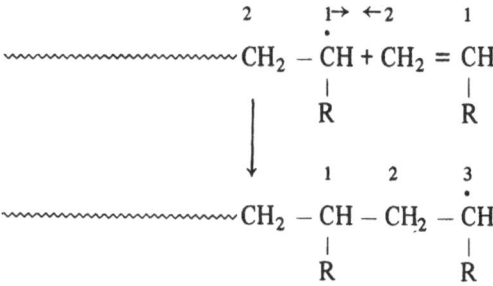

1,1-addition:

$$\overset{2}{CH_2} - \overset{\overset{1\rightarrow \leftarrow 1}{\cdot}}{CH} + \overset{2}{CH} = CH_2$$
$$\qquad\quad\; R \quad R$$

$$\downarrow$$

$$\overset{1}{CH_2} - \overset{2}{CH} - CH - \overset{\cdot}{CH_2}$$
$$\qquad\quad\; R \quad R$$

2,2-addition:

$$\overset{1}{CH} - \overset{\overset{2\rightarrow \leftarrow 2}{\cdot}}{CH_2} + CH_2 = \overset{1}{CH}$$
$$R \qquad\qquad\qquad R$$

$$\downarrow$$

$$\overset{1}{CH} - \overset{2}{CH_2} - \overset{3}{CH_2} - \overset{\overset{4}{\cdot}}{CH}$$
$$R \qquad\qquad\qquad R$$

The fraction of 1,1- and 2,2-addition is usually so small that it cannot be proved experimentally. Only if there are especially sensitive reactions available, such as chain scission of polyvinylalcohol with periodic acid, is it possible to determine the amount of 1,2-glycol groups.

$$\sim CH_2-CH-CH_2-CH-CH-CH_2-CH_2-CH-CH_2-CH\sim$$
$$\qquad\;\; OH \qquad OH \; OH \qquad\qquad OH \qquad\quad OH$$

$$\downarrow HIO_4$$

$$\sim CH_2-CH-CH_2-CHO + OHC-CH_2-CH_2-CH-CH_2-CH\sim$$
$$\qquad\;\; OH \qquad\qquad\qquad\qquad\qquad\qquad OH \qquad\quad OH$$

The striking preference of 1,3-addition with radical polymerization can be explained on the basis of the different stability of the two radicals (I) and (II).

$$\overset{\bullet}{\text{wCH}_2\text{—CH}} \qquad \text{wCH—}\overset{\bullet}{\text{CH}_2}$$
$$\text{(I)}\quad \text{R} \qquad\qquad \text{R}\quad\text{(II)}$$

Since with radical (I), the radical is in most cases stablized by resonance, the formation of this radical, which occurs through 1,2-addition (in the sense of the formula of p. 72), is thermodynamically preferred. With a styrene radical, for example, the following mesomeric forms are possible:

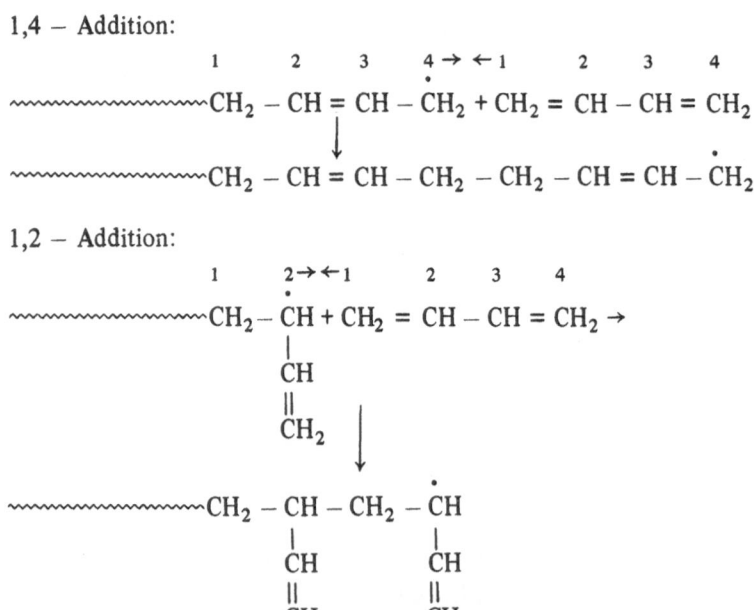

In addition to resonance stabilization, steric hindrance can play an important role in the structure of the polymer. Structures which allow a greater mobility will probably be preferred. It is also possible that polarization effects can have an influence on the position in which a monomer adds to the growing chain.

Of special importance is the question of chain structure with dienes (butadiene and isoprene). Propagation can occur as 1,4- and as 1,2-addition.

1,4 – Addition:

$$\overset{1}{}\quad\overset{2}{}\quad\overset{3}{}\quad\overset{4\to\leftarrow 1}{}\quad\overset{2}{}\quad\overset{3}{}\quad\overset{4}{}$$

$$\text{wwwCH}_2 - \text{CH} = \text{CH} - \overset{\bullet}{\text{CH}_2} + \text{CH}_2 = \text{CH} - \text{CH} = \text{CH}_2$$

$$\downarrow$$

$$\text{wwwCH}_2 - \text{CH} = \text{CH} - \text{CH}_2 - \text{CH}_2 - \text{CH} = \text{CH} - \overset{\bullet}{\text{CH}_2}$$

1,2 – Addition:

$$\overset{1}{}\quad\overset{2\to\leftarrow 1}{}\quad\overset{2}{}\quad\overset{3}{}\quad\overset{4}{}$$

$$\text{wwwCH}_2 - \overset{\bullet}{\text{CH}} + \text{CH}_2 = \text{CH} - \text{CH} = \text{CH}_2 \to$$
$$\qquad\qquad | $$
$$\qquad\qquad \text{CH}$$
$$\qquad\qquad \|$$
$$\qquad\qquad \text{CH}_2 \quad \downarrow$$

$$\text{wwwCH}_2 - \text{CH} - \text{CH}_2 - \overset{\bullet}{\text{CH}}$$
$$\qquad\qquad | \qquad\qquad |$$
$$\qquad\qquad \text{CH} \qquad\qquad \text{CH}$$
$$\qquad\qquad \| \qquad\qquad \|$$
$$\qquad\qquad \text{CH}_2 \qquad\qquad \text{CH}_2$$

With radical polymerization one obtains chains with a mixed structure, but with ionic polymerization one can obtain 1,4- or 1,2-polymers according to the choice of the polymerization conditions. With organo-alkali reagents e.g., phenyllithium in tetrahydrofuran, one obtains mainly, 1,2-polymers, whereas with lithium dispersions or phenyllithium in paraffinic hydrocarbons e.g., heptane as a solvent one obtains mainly 1,4-polymers. The pure crystalline 1,2-polybutadiene which is obtained using an initiator combination of chromium hexacarbonyl and aluminum triethyl turns out to be a hard and tough resin with a crystalline melting point of 131°C. The 1,4-*trans*-polybutadiene is also a hard/tough crystalline compound, whereas the 1,4-*cis*-polybutadiene is a soft and elastomeric material which becomes hard and glassy at temperatures of about − 100°C. This finding shows that, in addition to the chemical constitution, the steric structure of polymers also has a great influence on their properties.

Natural rubber (polyisoprene), as another example, has a pure *cis*-1,4 configuration.

$$\begin{array}{c} \text{CH}_3 \\ | \\ \text{C} = \text{CH} \end{array}$$

~~~~CH$_2$ C = CH   CH$_2$ CH$_2$ C = CH   CH$_2$ CH$_2$~~~~
    \   /   \   / \   /   \   /   \   / \   /
     CH$_2$ CH$_2$ C = CH   CH$_2$ CH$_2$ C = CH
                            | |
                           CH$_3$ CH$_3$

The same polymer with a *trans*-1,4 configuration (guttapercha) is not elastic, but turns out to be a hard and brittle resin.

CH$_3$ CH$_3$ CH$_3$ CH$_3$
   | | | |
~~~~CH$_2$ C   CH$_2$ C   CH$_2$ C   CH$_2$ C   CH$_2$~~~~
    \   / \\   / \   / \\   / \   / \\   / \   / \\ /
     CH$_2$ CH   CH$_2$ CH   CH$_2$ CH   CH$_2$ CH

Synthetic rubber (SBR) is a copolymer of butadiene (75%) and styrene (25%) with a mixed structure of 1,2-, and *cis* and *trans* 1,4-butadiene structural units in the chain. With Ziegler-type catalysts (e.g., TiI$_4$/Al(C$_2$H$_5$)$_3$ or AlR$_2$Cl/CoCl$_2$), or phenyllithium, or lithium in heptane, it is possible to prepare a synthetic rubber from isoprene or from butadiene which has mainly the *cis*-1,4 configuration (see p. 135 ff.). Other examples for different steric structures (isotactic and syndiotactic polymers) will be disucssed in greater detail in the section on polymerization initiated by complex catalysts (section 213).

During the polymerization of vinyl compounds one usually obtains polymers without regular steric structure, i.e., all possible steric configurations are distributed along the polymer chain in a random statistical manner.

A special sort of chain propagation is found with monomers of the type of acrylic anhydride in which two identical double bonds are found in the molecule at such a distance that during the polymerization five-or six-membered rings can be formed.

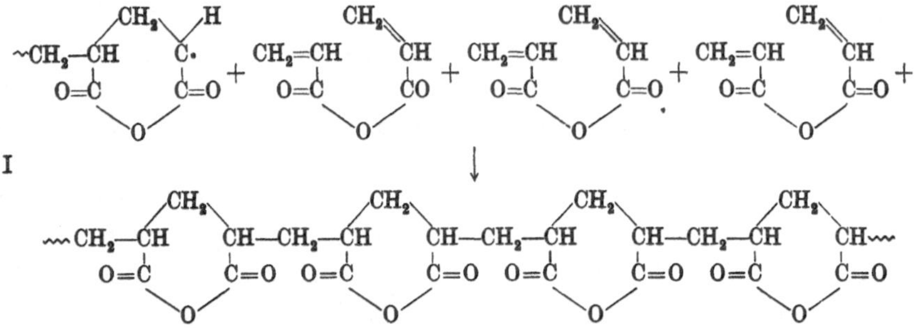

I

$$\sim CH_2\text{-}CH \quad \overset{CH_2}{\overset{|}{C}\cdot} \quad + \quad CH_2=CH \quad \overset{CH_2}{\overset{|}{CH}} \quad + \quad CH_2=CH \quad \overset{CH_2}{\overset{|}{CH}} \quad + \quad CH_2=CH \quad \overset{CH_2}{\overset{|}{CH}} \quad +$$

with carbonyl/anhydride groups $O=C \quad C=O$ bridged by O

↓

$$\sim CH_2\text{-}CH \quad CH\text{-}CH_2\text{-}CH \quad CH\text{-}CH_2\text{-}CH \quad CH\text{-}CH_2\text{-}CH \quad CH\sim$$

with anhydride rings $O=C \quad C=O$ bridged by O

or:

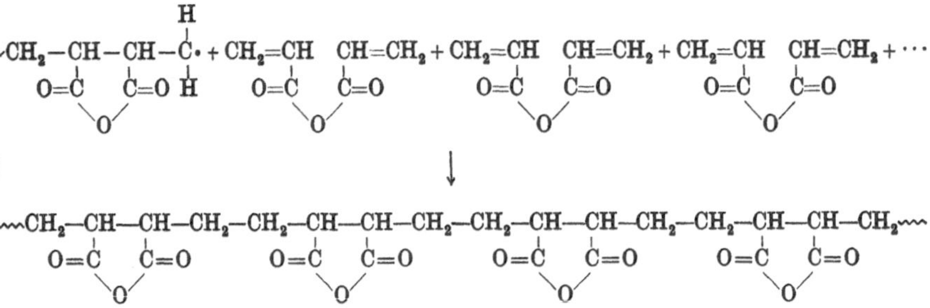

II

$$\sim CH_2-CH-CH-\overset{H}{\overset{|}{C}}\cdot + CH_2=CH \quad CH=CH_2 + CH_2=CH \quad CH=CH_2 + CH_2=CH \quad CH=CH_2 + \cdots$$

with anhydride groups $O=C \quad C=O$ (and H) bridged by O

↓

$$\sim CH_2-CH-CH-CH_2-CH_2-CH-CH-CH_2-CH_2-CH-CH-CH_2-CH_2-CH-CH-CH_2\sim$$

with anhydride groups $O=C \quad C=O$ bridged by O

In reaction (I) chain growth occurs in the manner of a 1,2-addition (see p. 72) and one should therefore assume that this reaction is preferred. However, it was discovered that the infrared spectrum of polyacrylic acid, which was prepared by the hydrolysis of polyacrylic acid anhydride, differs from the normal polyacrylic acid prepared by polymerization of acrylic acid. This finding cannot be understood in terms of structure (I). One therefore has to leave open the question of which of the two chain structures corresponds to the true structure of polyacrylic acid anhydride. It can be said with certainty only that the structure must be one of the two, because a polymerization of the two double bonds of the acrylic acid anhydride in two *different* polymer chains according to the reaction discussed on page 26 for divinylbenzene cannot occur because polyacrylic acid anhydride is a soluble polymer and can therefore not be cross-linked. With chain growth occurring according to reactions I or II, the reaction is called ring-forming polymerization, or cyclopolymerization.

2113 Chain Termination

In contrast to ionic polymerization, which under certain circumstances can occur in such a way that no chain termination takes place, radical polymerization must always involve a chain terminator because the radicals of the growing chain always have a strong tendency to react with each other. Two types of reactions are possible: recombination and disproportionation. The recombination consists of the addition of two growing radical chain ends to form a saturated macromolecule:

$$\text{\small$\sim\!\sim\!$CH}_2\text{—}\overset{\overset{\displaystyle H}{|}}{\underset{\underset{\displaystyle R}{|}}{C}}\cdot \;+\; \cdot\overset{\overset{\displaystyle H}{|}}{\underset{\underset{\displaystyle R}{|}}{C}}\text{—CH}_2\!\sim\!\sim \quad\rightarrow\quad \sim\!\sim\!\text{CH}_2\text{—}\overset{}{\underset{\underset{\displaystyle R}{|}}{CH}}\text{ — }\overset{}{\underset{\underset{\displaystyle R}{|}}{CH}}\text{—CH}_2\!\sim\!\sim$$

In certain cases, by means of osmotic molecular weight determination and quantitative end group determination with polymers which had been prepared using radioactive C^{14}-containing azobisisobutyronitrile, it was demonstrated that each molecule contains two radioactive end groups. The same results were found for emulsion polymers of styrene which had been prepared by initiation with radioactive potassium persulfate (S^{35}). This means that in these quantitatively investigated examples recombination is practically the only termination reaction. The term "recombination" also includes the combination of the growing chain radical with an initiator radical, a reaction which is rather improbable at the beginning of the polymerization as long as there is still a high concentration of monomer but which becomes more and more probable with increasing conversion. (In emulsion polymerization this reaction is the only termination reaction—compare p. 157).

The second reaction which leads to chain termination, called disproportionation, can be described by the following equation:

$$\sim\!\sim\!\sim\text{CH}_2 - \overset{\overset{\displaystyle H}{|}}{\underset{\underset{\displaystyle R}{|}}{C}}\cdot \;+\; \cdot\overset{\overset{\displaystyle H}{|}}{\underset{\underset{\displaystyle R}{|}}{C}} - \text{CH}_2\!\sim\!\sim\!\sim \quad\rightarrow\quad \sim\!\sim\!\sim\text{CH}_2 - \overset{}{\underset{\underset{\displaystyle R}{|}}{CH_2}} \;+\; \overset{}{\underset{\underset{\displaystyle R}{|}}{CH}}\!=\!\text{CH}\!\sim\!\sim\!\sim$$

The reactions for recombination and disproportionation show that recombination leads to macromolecules whose molecular weight is on the average twice that of the growing chain radicals, whereas disproportionation leads to chain molecules of the same average length as that of the chain radicals.

Which of the two termination reactions, recombination or disproportionation, prevails depends both on the type of monomer and on the temperature. Since the recombination reaction, in which the atomic structure of the reaction partners remains unchanged, occurs practically without activation energy, its rate scarcely depends on the temperature. However, in disproportionation a hydrogen atom has to move, and this can not take place without energy. The rate of the disproportionation reaction is therefore temperature dependent. Consequently, disproportionation becomes more important the higher the polymerization temperature. The temperature at which disproportionation becomes significant depends on the individual monomer. With polymerization of styrene, for example, one does not find any disproportionation at 60°C-70°C; with methylmethacrylate, however, disproportionation predominates at this temperature.

In addition to the so-called natural chain termination reactions, recombination and disproportionation, there are a number of additional possibilities for chain termination, which can be brought about more or less at will either through substances which are unintentionally present during the polymerization or through substances which one intentionally adds at a certain stage of the polymerization in order to stop it. In the first case, we are concerned with impurities which are present in the monomer, or in the solvent, or in other additives such as soaps and protective colloids (in emulsion polymerization), or which are present in the reaction vessel (for example, greases or lubricants in technical polymerizations). Atmospheric oxygen must also be included among the impurities which inhibit the polymerization, which is why radical polymerizations are usually carried out under nitrogen. In the second case, the polymerization is intentionally interrupted toward the end of the reaction. Thus while the radical concentration remains nearly constant, only small amounts of monomer remain at the end of the polymerization, and the chain growth is therefore slowed down. However, the termination reactions occur at the same rate as during the beginning of the polymerization, and therefore polymers with relative low molecular weight are preferably formed (see section on kinetics of radical polymerization). These polymers usually have a detrimental influence on the physical properties of the final product. Also, undesirable side reactions, which at high conversion occur more readily, may be a reason for terminating the polymerization at an earlier stage. This is the case during the synthesis of rubber by emulsion polymerization of butadiene and styrene. This reaction is terminated by the addition of stopping agents such as phenyl-β-naphtyla-mine to prevent the formation of larger amounts of cross-linked products.

Inhibitors

In addition to β–naphtylamine there are a large number of other organic compounds suitable for bringing about chain termination in radical polymerization. These compounds usually react with radicals, either giving rise to saturated molecules or to stable radicals too unreactive to start a new chain.

Much more important than bringing about chain termination during polymerization is the use of these compounds during preparation and storage of monomers to prevent polymerization from taking place. One therefore calls chain-terminating substances inhibitors. The following inhibitors, which can be added to a monomer in concentrations of 0.1-0.0001% (0.0001% = 1ppm = 1 part per million) in order to stabilize it, are especially effective and commonly used:

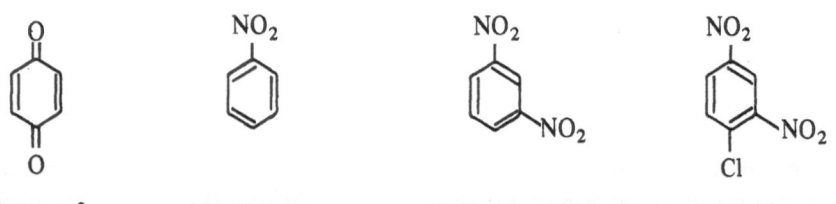

Quinone[2]   Nitrobenzene   Dinitrobenzene   Dinitrochlorobenzene

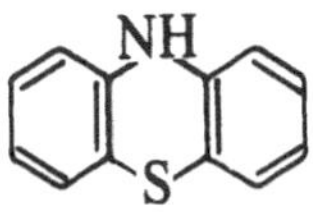

Benzothiazine
(Thiodiphenylamine)

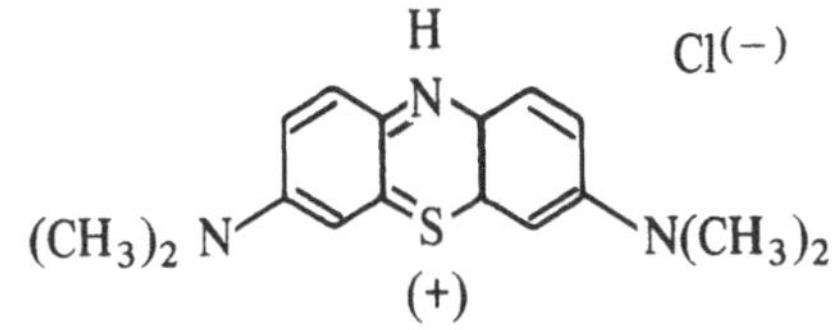

Methylene Blue

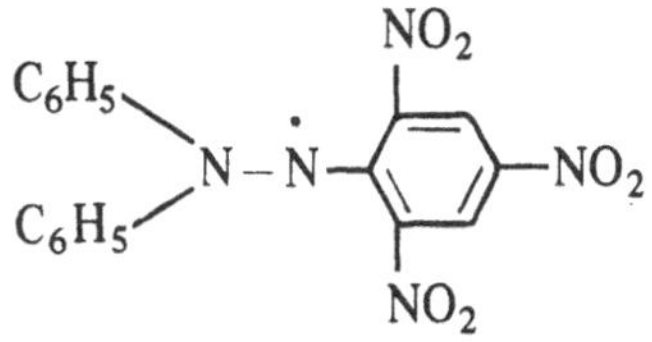

Diphenylpicrylhydrazyl
(DPPH)

In addition to these compounds, the following are also active inhibitors: oxygen, NO (one of the most effective inhibitors), numerous nitroso compounds, sulfur compounds, amines, phenols, aldehydes, and carbamates. Some highly reactive monomers can be distilled only under an atmosphere of NO.

Not all inhibitors are equally useful for all monomers and it is often necessary to select the best inhibitor for a certain monomer by trial and error. The technical importance of inhibitors is just as great as that of polymerization initiators, because without inhibitors the production and storage of many monomers on a technical scale would hardly be possible.

The mechanism of inhibition can be understood most clearly if the inhibitor is a radical such as diphenylpicrylhydrazyl, which is so stable that it does not show any tendency to add to unsaturated compounds in order to start a chain. Such radicals combine with initiator radicals or chain radicals under formation of stable saturated compounds. For example:

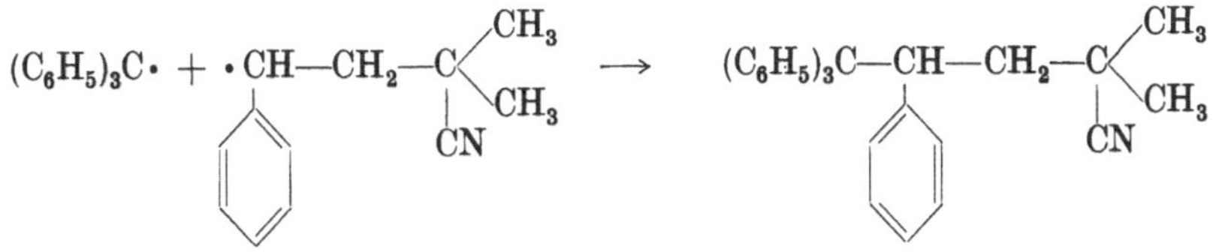

$$(C_6H_5)_3C \cdot + \cdot CH-CH_2-C\underset{\underset{CN}{|}}{\overset{CH_3}{\underset{CH_3}{\diagdown}}} \longrightarrow (C_6H_5)_3C-CH-CH_2-C\underset{\underset{CN}{|}}{\overset{CH_3}{\underset{CH_3}{\diagdown}}}$$

The prerequisite for an inhibitor radical (i.e., no reaction with monomers, but rapid reaction with active radicals) does not apply to many radicals. In many cases, one finds both inhibitor activity and initiator activity, as, for example, with

_____

[2]Hydroquinone is not an inhibitor; however, it acts like one, because it reacts with air to form quinone.

triphenylmethyl which may be used to initiate polymerization as well as to inhibit it. In such cases polymerization is not completely prevented, only slowed down.

In theory this overlapping of inhibition and initiation exists with all inhibitors, because, even with the nonradical inhibitors, the inhibition reaction forms free radicals. During the stabilization with quinone, for example, the following reactions may take place:

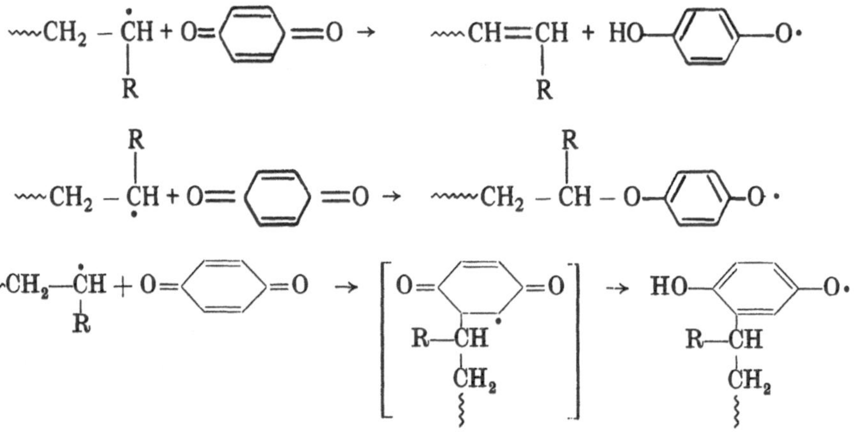

The radicals formed in this way are stabilized by resonance to such an extent that they do not start chains. They disappear partly through disproportionation to quinone and hydroquinone, which explains the formation of hydroquinone observed during the stabilization of styrene with quinone:

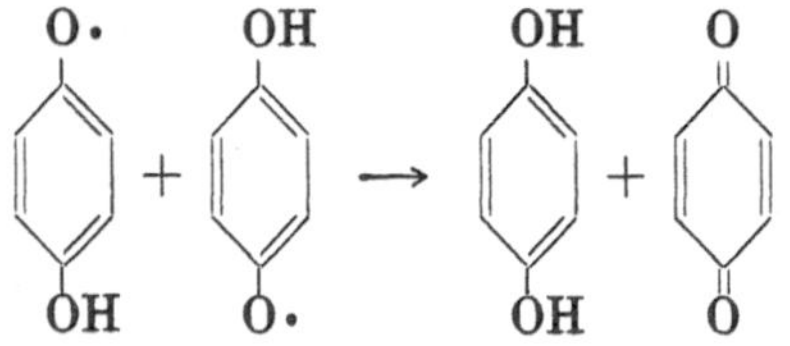

The radicals also combine, both with each other and with new chain radicals, and therefore the radicals formed in the inhibition reaction of quinone are themselves inhibitors, i.e., of the same type as DPPH.

The inhibiting action of oxygen is caused by the biradical character of the oxygen molecule. The first step consists of the addition of oxygen to a chain radical under formation of a peroxide radical:

$$\text{\textasciitilde CH}_2\text{—}\overset{\displaystyle\cdot}{\text{CH}} + \cdot\text{O}{=}\text{O}\cdot \; \rightarrow \; \text{\textasciitilde CH}_2\text{—CH—O—O}\cdot$$
$$\qquad\quad \overset{|}{\text{R}} \qquad\qquad\qquad\quad \overset{|}{\text{R}}$$

The peroxide radical may react further in different ways:

I $\quad$ \text{\textasciitilde CH}_2\text{—CH—O—O}\cdot + \cdot\text{CH—CH}_2\text{\textasciitilde} \; \rightarrow \; \text{\textasciitilde CH}_2\text{—CH—O—O—CH--CH}_2\text{\textasciitilde}
$$\qquad\quad \overset{|}{\text{R}} \qquad\quad \overset{|}{\text{R}} \qquad\qquad\qquad\quad \overset{|}{\text{R}} \qquad\quad \overset{|}{\text{R}}$$

II $\qquad$ $2\ \text{\small\textbackslash\textbackslash\textbackslash}CH_2-CH-O-O\cdot\ \rightarrow\ \text{\small\textbackslash\textbackslash\textbackslash}CH_2-CH-O-O-CH-CH_2\text{\small\textbackslash\textbackslash\textbackslash}+O_2$

$$2\sim\!\!\!\sim\!CH_2-\underset{\underset{R}{|}}{CH}-O-O\cdot\ \rightarrow\ \sim\!\!\!\sim\!CH_2-\underset{\underset{R}{|}}{CH}-O-O-\underset{\underset{R}{|}}{CH}-CH_2\!\sim\!\!\!\sim\ +\ O_2$$

III $\quad\sim\!\!\!\sim\!CH_2-\underset{\underset{R}{|}}{CH}-O-O\cdot\ +\ CH_2\!=\!\underset{\underset{R}{|}}{CH}\ \rightarrow\ \sim\!\!\!\sim\!CH_2-\underset{\underset{R}{|}}{CH}-O-O-CH_2-\overset{\cdot}{\underset{\underset{R}{|}}{CH}}$

In the first two cases an ordinary peroxide is formed; in case III a polymeric peroxide is formed, because the third reaction is not the final one—the chain radical may react further with oxygen (if sufficient oxygen is present) and, therefore, one may think of oxygen as a comonomer:

$$\sim\!\!\!\sim\!CH_2-\underset{\underset{R}{|}}{CH}-O-O-CH_2-\underset{\underset{R}{|}}{CH}-O-O-CH_2-\underset{\underset{R}{|}}{CH}-O-O-CH_2-\underset{\underset{R}{|}}{CH}-O-O\cdot$$

Such polymeric peroxides with a degree of polymerization between 10 and 40 have been isolated. The oxygen radicals as well as the peroxides occurring as the reaction products are evidently fairly stable at low temperatures, otherwise they could not act as inhibitors. At higher temperatures on the other hand, the peroxides thus formed may decompose again and, if no inhibitor is present, bring about a strong polymerization reaction.

Chain Transfer

One must distinguish between the term "chain" used as a designation for a linear macromolecule, and the chain (=reaction chain) as a kinetic concept. Similarly we must distinguish between the chain growth of a macromolecule during polymerization, where the degree of polymerization is determined by the termination of this chain growth and chain growth in a reaction kinetic sense. Thus the termination of the growth of the chain molecule does not always lead to a termination of the kinetic chain. A growing chain radical may abstract an atom (for example, hydrogen or chlorine) from another molecule and thus become saturated, and the molecule from which the atom has been abstracted will then become a radical and start a new chain. The chain reaction therefore continues, even though the chain growth of the first macromolecule has stopped; i.e., it has transferred the reaction chain to another molecule. This reaction is therefore called a transfer reaction, or just chain transfer:

$$\sim\!\!\!\sim\!CH_2-\overset{\cdot}{\underset{\underset{R}{|}}{CH}}+R_1-X\ \rightarrow\ \sim\!\!\!\sim\!CH_2-\overset{\overset{\displaystyle X}{|}}{\underset{\underset{R}{|}}{CH}}+R_1\cdot$$
$$\begin{array}{c}|+CH=CH_2\\ \quad\quad\underset{R}{|}\\ \downarrow\\ R_1-CH_2-\overset{\cdot}{\underset{\underset{R}{|}}{CH}}\ \xrightarrow{\ Polymerization\ }\end{array}$$

Chain transfer of this type may occur with the initiator, with the monomer, with the solvent or other compounds which have been added in order to promote chain transfer (modifiers), and with growing or complete polymer chains.

Among the common initiators, the hydroperoxides (for example, t-butylhydro-peroxide and cumene-hydroperoxide) are the ones with a tendency to transfer reactions, whereas benzoylperoxide, and especially azobisisobutyronitrile, do not take part in transfer reactions. In the transfer reaction, the initiator molecule reacts with a chain radical to terminate it and at the same time gives rise to another radical which then starts a new chain:

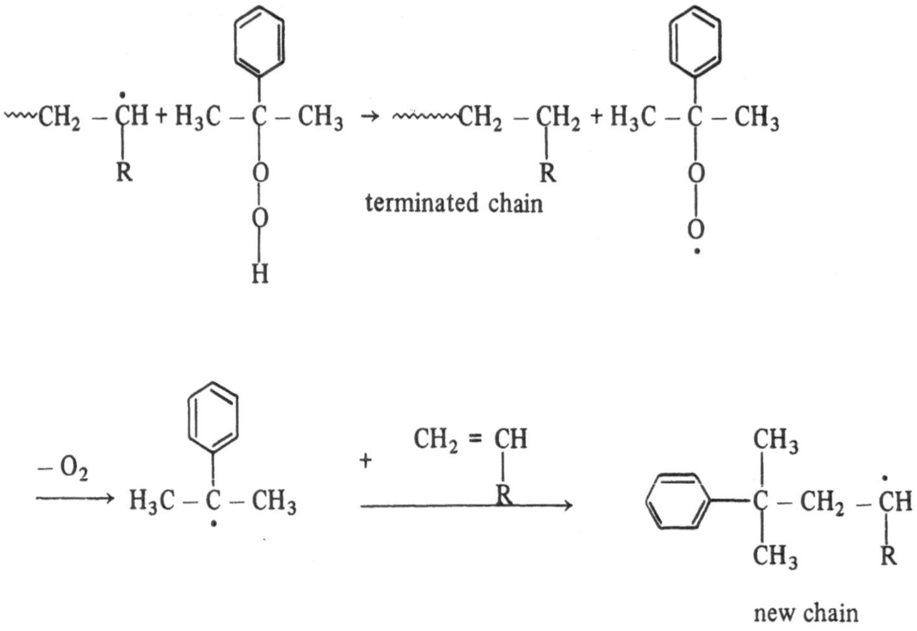

$$\text{~CH}_2 - \overset{\bullet}{\text{CH}} + \text{H}_3\text{C} - \underset{\underset{\underset{\underset{H}{|}}{O}}{\overset{|}{\underset{|}{O}}}}{\overset{C_6H_5}{\underset{|}{C}}} - \text{CH}_3 \rightarrow \text{~CH}_2 - \text{CH}_2 + \text{H}_3\text{C} - \overset{C_6H_5}{\underset{\underset{\underset{\bullet}{|}}{O}}{\underset{|}{C}}} - \text{CH}_3$$

terminated chain

$$\xrightarrow[-O_2]{} \text{H}_3\text{C} - \overset{C_6H_5}{\underset{\bullet}{C}} - \text{CH}_3 \quad + \quad \overset{}{\underset{\underset{R}{|}}{CH_2 = CH}} \xrightarrow{\quad\quad} C_6H_5 - \overset{CH_3}{\underset{\underset{CH_3}{|}}{\overset{|}{C}}} - CH_2 - \overset{\bullet}{\underset{\underset{R}{|}}{CH}}$$

new chain

This type of decomposition of the initiator by a growing chain radical (induced decomposition) must be considered in addition to the spontaneous decomposition. Since such an induced decomposition forms as many new radicals as it consumes, the radical concentration remains unaffected.

Induced decomposition may also result from the reaction of the initiator with solvent molecules. The tendency of different solvents to cause decomposition of the initiator varies greatly. For example, benzoylperoxide decomposes considerably faster in dioxane or acetone than in benzene or toluene.

Usually the initiator concentration is quite small (on the order of 0.1%), and therefore the effect of chain transfer brought about by the initiator molecules is quite small. This could be different with chain transfer to the monomer, because the concentration of the monomer, at least at the beginning of the polymerization, is large. It has been found, however, that most monomers have only a small tendency to chain transfer reactions, probably because hydrogen atoms are tightly held. Important exceptions to this rule will be discussed in the section on autoinhibition (see p. 71) Of practical importance for polymerization reactions are the last two cases: transfer reactions with solvents or modifiers and with polymers.

Chain Transfer Through Solvents and Modifiers

The ability of solvents and modifers to act as chain transfer agents can be measured quantitatively through their chain transfer constant, C_s, which is defined as the ratio of the rate constant of the transfer reaction (k_{tr}) to that of the propagation reaction (k_p):

$$C_s = \frac{k_{tr}}{k_p} \quad .$$

C_s=0.5 therefore means that with equal concentration of solvent and of monomer, the rate of the transfer reaction is only half that of the rate of the growth reaction. With most solvents the chain transfer constant is much smaller than 0.5.

As can be seen from Table 66, the magnitude of the chain transfer constant depends on the constitution of the solvent, the constitution of the chain radical which brings about the transfer reaction, and the temperature. Of interest is the high chain transfer constant of carbon tetrachloride, which is the result of the high mobility of the chlorine in the carbon tetrachloride molecule. In the series benzene, toluene, ethylbenzene, cumene, triphenylmethane, the increase in the mobility of the hydrogen atom makes itself felt in an increase in the chain transfer constant.

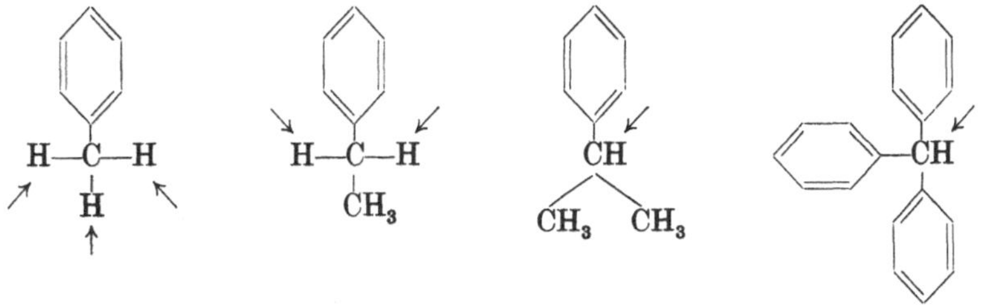

It is evident that in the formulas above, the hydrogen atoms designated by arrows are the ones exchanged during the transfer reaction. It is therefore possible to formulate the chain transfer with ethylbenzene as follows:

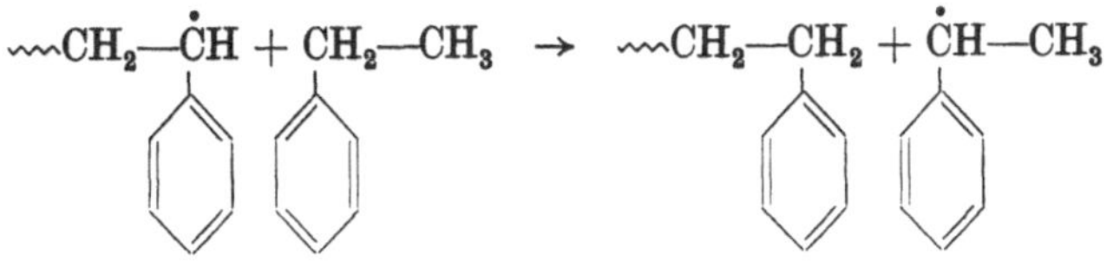

In practice one uses chain transfer to obtain polymers with a definite, not too high, average degree of polymerization in a simple way. A compound added to the polymerization specifically for the purpose of chain transfer is designated as a *modifier,* and its effect may be referred to as *modifying action* (or *modifier action*).

Many sulfur compounds have a much greater modifier action than most solvents. For example, in the synthesis of SBR (styrene-butadiene-rubber) one uses

TABLE 66. Chain-transfer constants of different solvents

Solvent	Transfer constant c_S				
	$\sim CH_2-\overset{\cdot}{C}H$ (phenyl)			$\sim CH_2-\overset{CH_3}{\underset{COOCH_3}{\overset{\mid}{C}}}\cdot$	$\sim CH_2-\overset{\cdot}{C}H$ ($OOC-CH_3$)
	60°	80°	100°		
Benzene	$0.18 \cdot 10^{-5}$	$0.6 \cdot 10^{-5}$	$1.6 \cdot 10^{-5}$	$0.75 \cdot 10^{-5}$	
Cyclohexane	$0.24 \cdot 10^{-5}$	$0.66 \cdot 10^{-5}$	$1.6 \cdot 10^{-5}$	$1.0 \cdot 10^{-5}$	$920 \cdot 10^{-5}$
Toluene	$1.25 \cdot 10^{-5}$	$3.1 \cdot 10^{-5}$	$6.5 \cdot 10^{-5}$	$5.2 \cdot 10^{-5}$	
Ethylbenzene	$6.7 \cdot 10^{-5}$	$10.8 \cdot 10^{-5}$	$16.2 \cdot 10^{-5}$	$13.5 \cdot 10^{-5}$	
Cumene	$8.2 \cdot 10^{-5}$	$13.0 \cdot 10^{-5}$	$20.0 \cdot 10^{-5}$	$19.0 \cdot 10^{-5}$	
Acetic acid esters				$2.4 \cdot 10^{-5}$	
Triphenyl-methane	$35 \cdot 10^{-5}$		$80 \cdot 10^{-5}$		
Butylchloride	$0.4 \cdot 10^{-5}$		$3.7 \cdot 10^{-5}$		
Dichloro-methane (Methylene-chloride).	$1.5 \cdot 10^{-5}$				
Carbontetra-chloride	$900 \cdot 10^{-5}$	$1300 \cdot 10^{-5}$	$1800 \cdot 10^{-5}$	$24 \cdot 10^{-5}$	$1000 \cdot 10^{-5}$

diisopropylxanthate disulfide, which probably reacts as follows (similar to induced initiator decomposition):

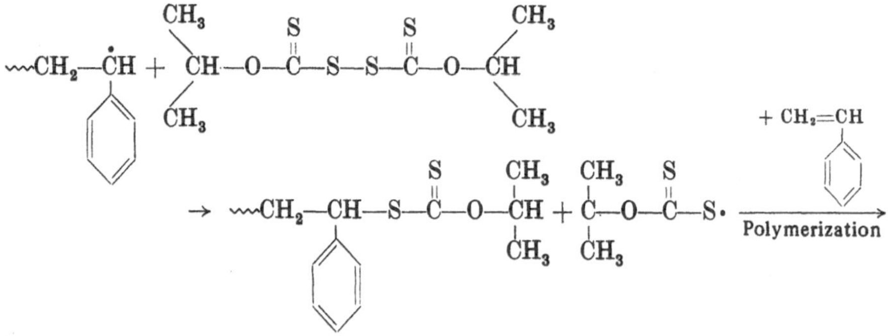

The compounds used most commonly as modifiers are mercaptans (especially dodecylmercaptan). Their chain transfer constants are shown in Table 68. As one can see from these chain transfer constants, the rate constants for the chain transfer reaction are many times considerably larger than the rate constants of the polymerization. One therefore needs only small amounts (on the order of 0.1% based on monomer and even less) to obtain a considerable decrease in the degree of polymerization (see Equation 30, p. 91). The addition of a modifier usually does not have an effect on the rate of polymerization. If one uses peroxide as initiators, mercaptans can at the same time act as redox components, so that in this case the polymerization can be increased considerably (see the section on kinetics of chain transfer, pp. 90-93 for a discussion of the mechanism of modification of the degree of polymerization). The transfer reaction with mercaptans may be formulated as follows:

$$\text{\textasciitilde\textasciitilde CH}_2\text{--}\overset{\overset{\text{H}}{|}}{\underset{|}{\text{C}}}\text{\textbullet} + \text{HS--R} \rightarrow \text{\textasciitilde\textasciitilde CH}_2\text{--}\overset{\overset{\text{H}}{|}}{\underset{|}{\text{C}}}\text{--H} + \text{R--}\overset{\text{\textbullet}}{\text{S}} \xrightarrow[\text{+ CH}_2\text{=CH}]{} \text{R--S--CH}_2\text{--}\overset{\overset{\text{H}}{|}}{\underset{|}{\text{C}}}\text{\textbullet}$$

Chain Transfer Through Polymers

A transfer reaction with a polymer molecule leads to branching unless this transfer occurs at the chain end:

$$\text{\textasciitilde\textasciitilde CH}_2\text{--CH--CH}_2\text{--CH--CH}_2\text{--CH--CH}_2\text{--CH--CH}_2\text{--CH--CH}_2\text{--CH\textasciitilde\textasciitilde}$$
$$\qquad\quad \text{R} \qquad\quad \text{R} \qquad\quad \text{R} \qquad\quad \text{R} \qquad\quad \text{R} \qquad\quad \text{R}$$

$$+ \text{\textasciitilde\textasciitilde CH}_2\text{--}\overset{\text{\textbullet}}{\text{CH}}$$
$$\text{R}$$

$$\downarrow$$

$$\text{\textasciitilde\textasciitilde CH}_2\text{--CH--CH}_2\text{--CH--CH}_2\text{--}\overset{\text{\textbullet}}{\text{C}}\text{--CH}_2\text{--CH--CH}_2\text{--CH--CH}_2\text{--CH\textasciitilde\textasciitilde}$$
$$\qquad\quad \text{R} \qquad\quad \text{R} \qquad\quad \text{R} \qquad\quad \text{R} \qquad\quad \text{R} \qquad\quad \text{R}$$

$$+ \text{\textasciitilde\textasciitilde CH}_2\text{--CH}_2$$
$$\text{R}$$

$$\downarrow + \text{CH}_2\text{=CH}$$
$$\qquad \text{R}$$

$$\text{R}$$
$$\text{\textasciitilde\textasciitilde CH}_2\text{--CH--CH}_2\text{--CH--CH}_2\text{--}\overset{|}{\text{C}}\text{--CH}_2\text{--CH--CH}_2\text{--CH--CH}_2\text{--CH\textasciitilde\textasciitilde}$$
$$\qquad\quad \text{R} \qquad\quad \text{R} \qquad\quad \text{CH}_2 \qquad\quad \text{R} \qquad\quad \text{R} \qquad\quad \text{R}$$
$$\qquad\qquad\qquad\qquad\qquad\qquad \text{R--CH}$$
$$\qquad\qquad\qquad\qquad\qquad\qquad \text{\textbullet}$$

$$\downarrow$$

Branch

This reaction will occur especially often when there is a large concentration of polymer, i.e., in all cases where a polymerization is carried out to a high degree of conversion. It is also possible, before the beginning of the polymerization, to dissolve a polymer in the monomer and thereby favor chain transfer to polymer and the formation of a branched polymer. If one takes a different polymer, for example, if one dissolves polystyrene in the monomer methylmethacrylate, or if one dissolves polybutylacrylate in styrene monomer and then polymerizes, one obtains a branched copolymer whose side chains are different from the primary polymer (the backbone chain). Such branched polymers are called graft-copoly-

TABLE 68. Transfer constants of mercaptans
(according to P.J. FLORY)

Monomer	Mercaptan	c_s
Styrene	n-Butylmercaptan	22
Styrene	t-Butylmercaptan	3.6
Styrene	n-Dodecylmercaptan	19
Methyl-methacrylate	n-Butylmercaptan	0.67
Methyl-acrylate	n-Butylmercaptan	1.7
Vinylacetate	n-Butylmercaptan	48

mers, and the reaction is called a graft-copolymerization. Such graft-copolymers do not occur only through transfer reactions—there are a number of other processes for their formation. In fact, transfer reactions are of no importance in the synthesis of pure graft-copolymers because of the low yields obtained. These will be discussed in a later section (compare page 288).

In a strict sense, every radical polymerization leads to branched polymers because the polymerization obviously occurs in the presence of polymer: the chains which have already been formed. The extent of branching brought about by the chain transfer reaction is determined by the extent to which the polymerization is carried on (degree of conversion) and also by the magnitude of the chain transfer constant of the polymer. This may be quite different depending on the nature of the polymer, just as C_s varies widely with the nature of the solvent. For example, with the acrylates, the transfer constant toward its own chain radicals is relatively high and, therefore, in the bulk polymerization of its monomer, one usually obtains cross-linked, insoluble materials. The cross-linking occurs through a combination of growing side chain radicals:

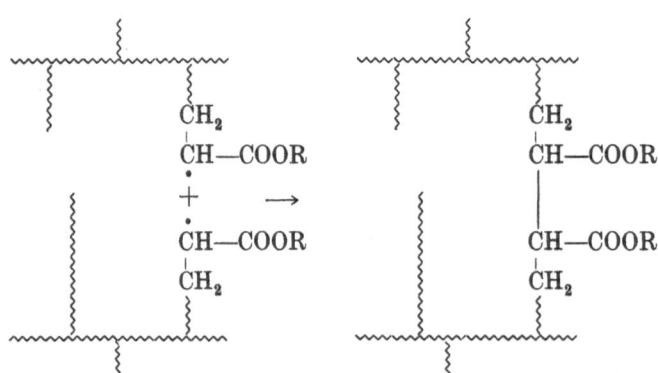

Through the addition of a modifier or through polymerization in solution or emulsion it is possible (as is to be expected) to avoid such cross-linking reactions.

Only in the production of synthetic rubber by polymerization of butadiene-styrene is chain transfer by the polymer chains so effective that cross-linked copolymers are formed at higher conversions even in an emulsion system. The mechanism should be the same as the one shown in the acrylic ester example. The relatively high transfer constant of polybutadiene plays an important role in the production of impact resistant polystyrenes by polymerization of a solution of polybutadiene dissolved in monomeric styrene (see p. 293). With most other polymers the chain transfer constant to polymer is so small that the influence of branching reactions (brought about by chain transfer) on the properties of the polymers is insignificant.

An interesting case of chain transfer to polymer apparently occurs with high pressure polyethylene. The growing chain radicals react every now and then with their own chain in such a way that a six-membered ring intermediate seems to facilitate abstraction of the H atom from the C atom in the fifth position from the chain end:

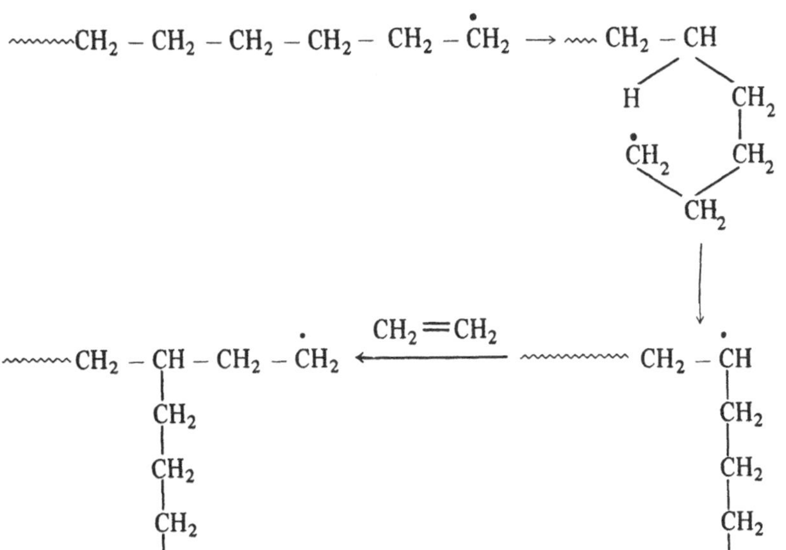

$$\sim\sim\sim CH_2 - CH_2 - CH_2 - CH_2 - CH_2 - \overset{\bullet}{C}H_2 \longrightarrow \sim\sim CH_2 - CH$$

$$\sim\sim\sim CH_2 - \underset{\substack{| \\ CH_2 \\ | \\ CH_2 \\ | \\ CH_2 \\ | \\ CH_3}}{CH} - CH_2 - \overset{\bullet}{C}H_2 \xleftarrow{\quad CH_2 = CH_2 \quad} \sim\sim\sim\sim\sim CH_2 - \underset{\substack{| \\ CH_2 \\ | \\ CH_2 \\ | \\ CH_2 \\ | \\ CH_3}}{\overset{\bullet}{C}H}$$

This reaction accounts for the C_4 side chains, which characterize high pressure polyethylene (short chain branching) and determine its properties to a large extent.

Polyrecombination

A combination of radical transfer and recombination has been employed for the synthesis of macromolecules. For this purpose one allows a compound which contains two groups with the tendency for chain transfer to react with an excess of peroxide at higher temperature. One then obtains through transfer reactions radicals which by recombination can form polymers (V. V. Korshak):

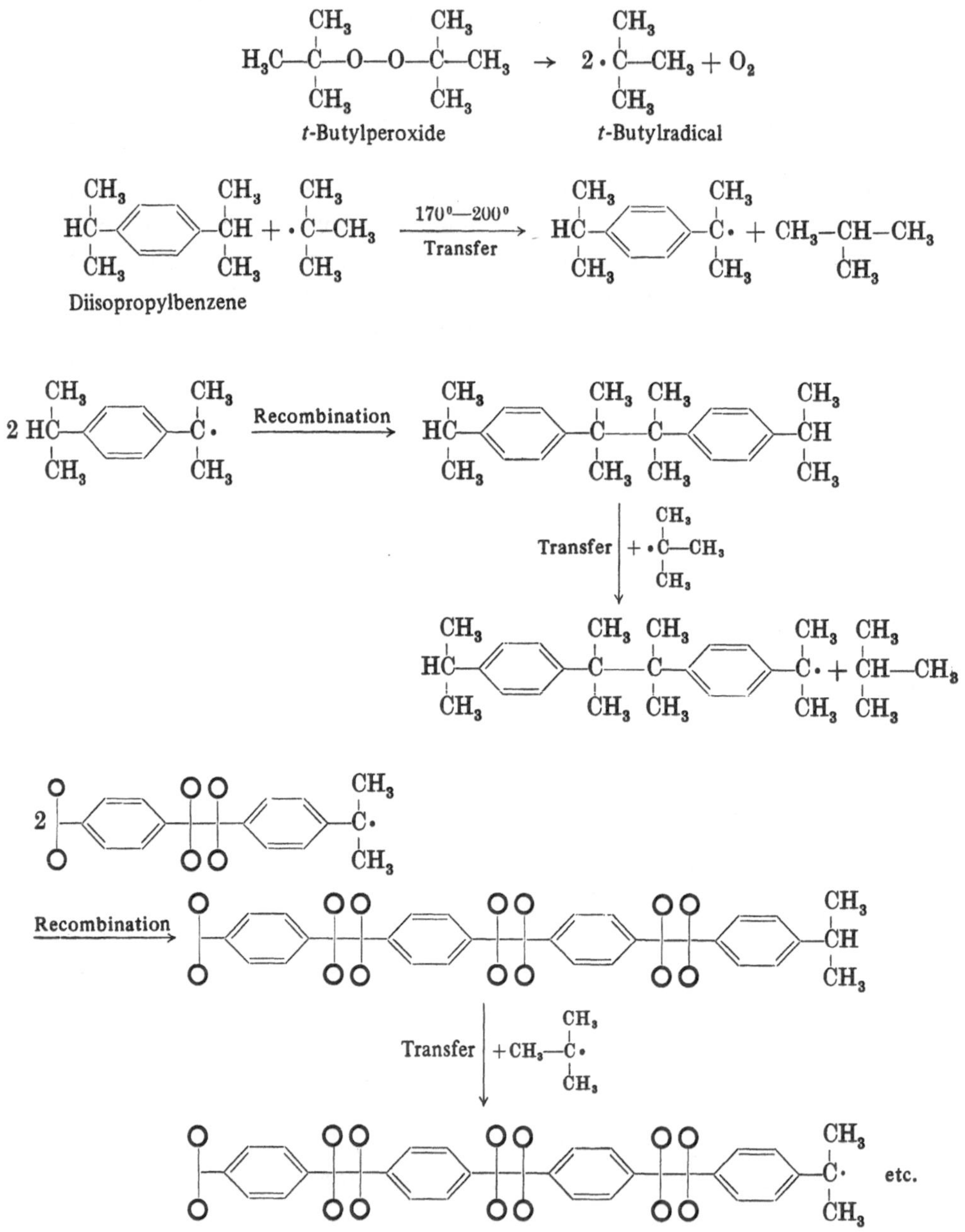

$$H_3C-\underset{\underset{CH_3}{|}}{\overset{\overset{CH_3}{|}}{C}}-O-O-\underset{\underset{CH_3}{|}}{\overset{\overset{CH_3}{|}}{C}}-CH_3 \;\rightarrow\; 2\cdot\underset{\underset{CH_3}{|}}{\overset{\overset{CH_3}{|}}{C}}-CH_3 + O_2$$

t-Butylperoxide t-Butylradical

$$\underset{\underset{CH_3}{|}}{\overset{\overset{CH_3}{|}}{HC}}-\!\!\left\langle\!\!\!\bigcirc\!\!\!\right\rangle\!\!-\underset{\underset{CH_3}{|}}{\overset{\overset{CH_3}{|}}{CH}} + \cdot\underset{\underset{CH_3}{|}}{\overset{\overset{CH_3}{|}}{C}}-CH_3 \xrightarrow[\text{Transfer}]{170^\circ-200^\circ} \underset{\underset{CH_3}{|}}{\overset{\overset{CH_3}{|}}{HC}}-\!\!\left\langle\!\!\!\bigcirc\!\!\!\right\rangle\!\!-\underset{\underset{CH_3}{|}}{\overset{\overset{CH_3}{|}}{C}}\cdot + CH_3-\underset{\underset{CH_3}{|}}{CH}-CH_3$$

Diisopropylbenzene

$$2\; \underset{\underset{CH_3}{|}}{\overset{\overset{CH_3}{|}}{HC}}-\!\!\left\langle\!\!\!\bigcirc\!\!\!\right\rangle\!\!-\underset{\underset{CH_3}{|}}{\overset{\overset{CH_3}{|}}{C}}\cdot \xrightarrow{\text{Recombination}} \text{(dimer)}$$

Transfer $+\cdot\underset{\underset{CH_3}{|}}{\overset{\overset{CH_3}{|}}{C}}-CH_3$

(further recombination and transfer structures)

Such a reaction may be called a polyrecombination. The reaction mechanism is similar to that of a polycondensation of bifunctional compounds.

The mechanism of the reaction between p-diisopropylbenzene and t-butyl-peroxide has not been completely established and must therefore be considered

hypothetical. One should not neglect the possibility of the following disproportionation reaction:

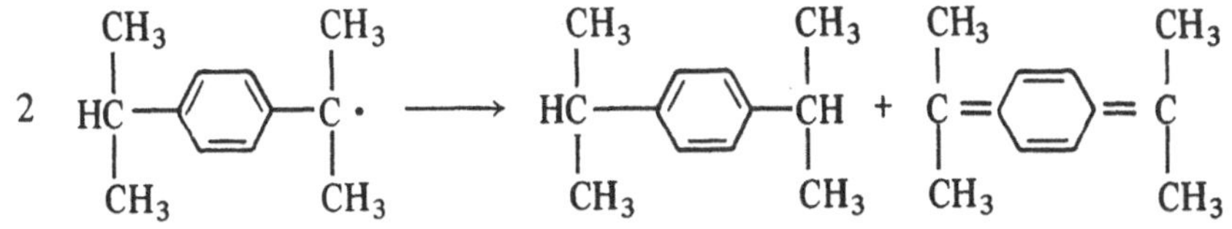

The quinoid intermediate obtained in this manner would lead to the same polymer
resulting from the recombination reaction shown above (see also p. 204).

Polyrecombination reactions may also play a role in the formation of lignin in
plants. (Next to cellulose, lignin is the major constituent of wood. It is a highly
cross-linked material, and therefore its original structure in the plant before its
separation from the other plant components is difficult to elucidate unequivocally.)
According to K. Freudenberg, the following reaction steps may occur:

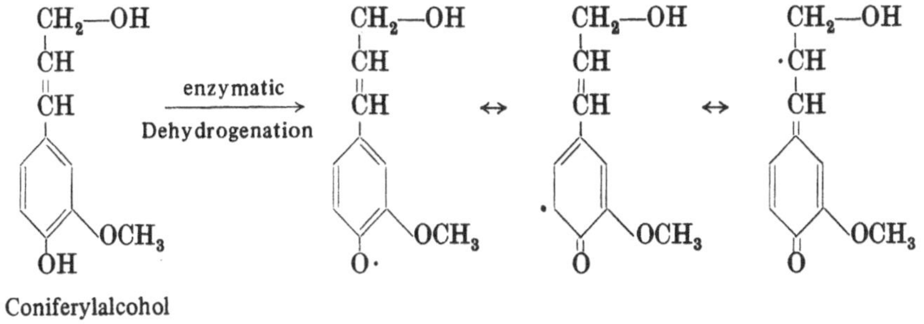

Coniferylalcohol

Through combination of these radicals arise dimeric quinonemethides which
aromatize again partly through proton displacement and partly through the
addition of water. The dimeric phenols formed in this way can again be dehydrated
to form radicals which then further combine, and so on.

The first step of the reaction, enzymatic dehydration, is brought about by
copper-containing enzymes. A synthetic analogy may be the oxidative coupling of
2,6-dimethylphenol in the presence of copper-amine-complexes (see p. 230).

Autoinhibition

Certain unsaturated compounds, which should be able to form stable
macromolecules, are not polymerized at all by free radicals or form only short
chains, and very slowly at that. These are, for example, the allyl compounds,
propylene, isobutylene, vinyl ethers and α-methylstyrene, and probably many other
olefinic unsaturated compounds which have not yet been carefully studied. In the
case of allylchloride, one assumes that through a chain transfer reaction with its
own monomer a resonance-stabilized radical is formed (Bartlett and Altschul):

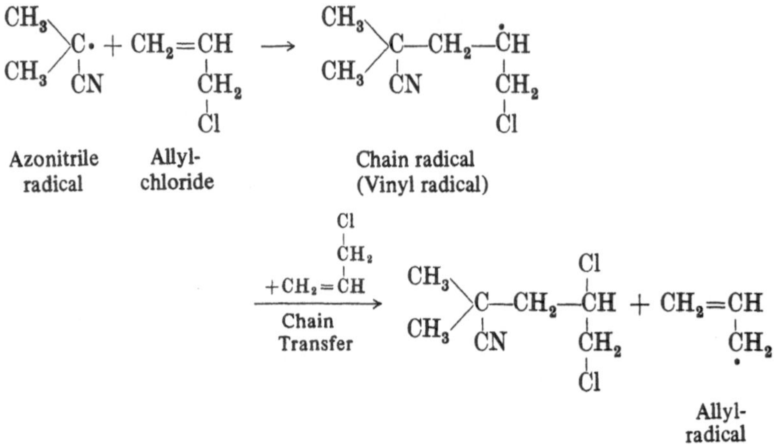

$$\begin{array}{ccc}
\underset{CH_3}{\overset{CH_3}{\diagup}}C\cdot + CH_2{=}CH & \longrightarrow & \underset{CH_3}{\overset{CH_3}{\diagup}}C{-}CH_2{-}\overset{\bullet}{C}H \\
CH_3 \quad CN \qquad\quad CH_2 & & CH_3 \quad CN \qquad CH_2 \\
\qquad\qquad\qquad\quad Cl & & \qquad\qquad\qquad Cl
\end{array}$$

Azonitrile radical	Allyl-chloride	Chain radical (Vinyl radical)

$$\xrightarrow[\text{Transfer}]{\substack{Cl \\ CH_2 \\ +CH_2{=}CH \\ \\ \text{Chain}}} \quad \underset{CH_3}{\overset{CH_3}{\diagup}}C{-}CH_2{-}\overset{\;}{C}H + CH_2{=}CH$$

$$CH_3 \quad CN \qquad CH_2 \qquad\qquad CH_2$$
$$\qquad\qquad\qquad Cl \qquad\qquad \overset{\bullet}{}$$

Allyl-radical

For the allyl radical formed by this chain transfer reaction, one can write two equivalent mesomeric forms:

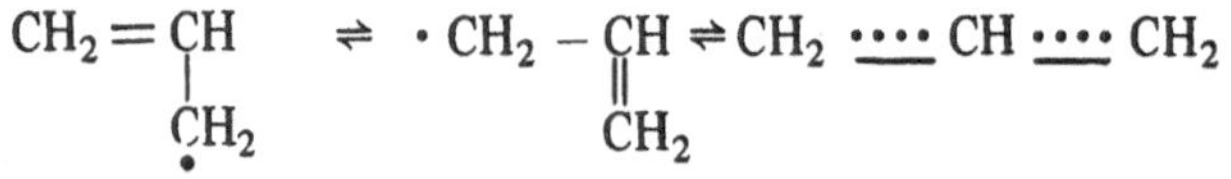

$$CH_2{=}\underset{\underset{\bullet}{CH_2}}{CH} \quad \rightleftharpoons \quad \cdot CH_2 - \underset{\overset{\|}{CH_2}}{CH} \rightleftharpoons CH_2 \cdots CH \cdots CH_2$$

This resonance makes the radical relatively stable and energy poor, so that in this case the transfer reaction with the monomer is thermodynamically favored over the normal chain growth.

Such resonance stabilization in itself is not sufficient reason for the radical not to continue to grow. Thus, the styrene radical (compare p. 56) and many others (acrylic esters, acrylonitrile, butadiene, and vinyl ketone) are stabilized by resonance, and yet the polymerization occurs rapidly and leads to polymers with high molecular weight. Although in these cases the addition of the next monomer always reforms a resonance-stabilized radical, in the case of allylchloride the addition of monomer leads to a reactive, energy-rich radical:

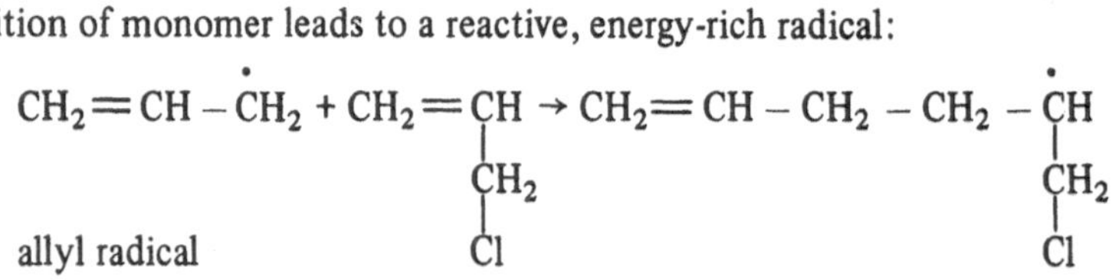

$$CH_2{=}CH - \overset{\bullet}{C}H_2 + CH_2{=}\underset{\underset{Cl}{CH_2}}{CH} \rightarrow CH_2{=}CH - CH_2 - CH_2 - \underset{\underset{Cl}{CH_2}}{\overset{\bullet}{C}H}$$

allyl radical

This is an addition reaction, which for thermodynamic reasons has little probability of occurring at low temperatures. If indeed there is an addition reaction in the sense of a normal polymerization, then this can only lead to very short chains, because

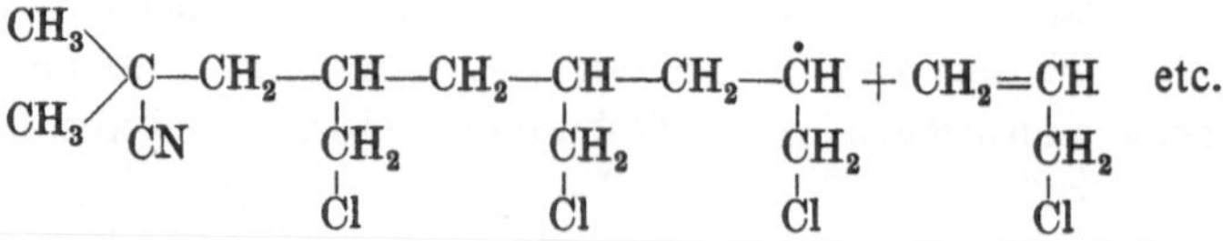

$$\underset{CH_3}{\overset{CH_3}{\diagup}}C{-}CH_2{-}CH{-}CH_2{-}CH{-}CH_2{-}\overset{\bullet}{C}H + CH_2{=}CH \quad \text{etc.}$$

$$CH_3 \quad CN \qquad CH_2 \qquad CH_2 \qquad CH_2 \qquad CH_2$$
$$\qquad\qquad\qquad Cl \qquad\quad Cl \qquad\quad Cl \qquad\quad Cl$$

the growing radical is always exposed to the chain transfer reaction with the monomer, which is favored on thermodynamic grounds. This leads to a resonance-stabilized radical, which has a greater tendency to recombine (thereby terminating a growing chain) than to start a new chain.

Similarly, one can formulate the formation of a resonance-stabilized radical through chain transfer with its own monomer and also with other monomers, however without any experimental evidence (for example, with propylene or isobutylene).

$$\text{\Large\textasciitilde}CH_2\text{---}\overset{\bullet}{C}H + CH_2{=}CH \rightarrow \text{\textasciitilde}CH_2\text{---}CH_2 + CH_2{=}CH \leftrightarrow \overset{\bullet}{C}H_2\text{---}CH$$

$$\underset{CH_3}{|} \qquad \underset{CH_3}{|} \qquad \underset{CH_3}{|} \qquad \underset{\underset{\bullet}{CH_2}}{\|} \qquad \underset{CH_2}{\|}$$

Propylene

$$\underset{CH_3}{\overset{CH_3}{|}} \qquad \underset{CH_3}{\overset{CH_3}{|}} \qquad \underset{CH_3}{\overset{CH_3}{|}} \qquad \underset{CH_2}{\overset{CH_3}{|}} \qquad \underset{CH_2}{\overset{CH_3}{|}}$$

$$\text{\textasciitilde}CH_2\text{---}\overset{|}{C}\cdot + CH_2{=}\overset{|}{C} \rightarrow \text{\textasciitilde}CH_2\text{---}\overset{|}{C}H + CH_2{=}\overset{|}{C} \leftrightarrow \overset{\bullet}{C}H_2\text{---}\overset{|}{C}$$

Isobutylene

Another way of explaining the poor polymerization of these monomers is their position in the Q-e diagram (compare p. 137).

21.24 Kinetics of Radical Polymerization

Determination of the Reaction Rate Constant

Radical polymerization begins with the decomposition of an initiator molecule, I, into two radicals, $R\cdot$, which very soon after their formation add a monomer molecule, M (chain initiation). When the monomer adds, the radical state is transfered from the initiator radical to the added monomer molecule which can then, in the same fashion, add additional monomer molecules (chain growth). This addition continues until two growing chain radicals, $R\text{\textasciitilde}M\cdot$, react with each other, either by recombination or by disproportionation (chain termination).

Initiation: $\quad I \overset{k_d}{\rightarrow} 2\,R\cdot$ (1)

$\quad\quad\quad\quad\quad R\cdot + M \rightarrow R-M\cdot$ $_{+M}$ (2)

Growth: $\quad\quad R-M\cdot + M \rightarrow R-M-M\cdot \overset{+M}{\rightarrow} R-M-M-M\cdot$ etc.

(propagation)

or in general: $\quad R \text{\textasciitilde}M\cdot + M \overset{k_w}{\rightarrow} R\text{\textasciitilde}M-M\cdot$ (3)

Termination: $\quad 2\,R\text{\textasciitilde}M\cdot \overset{k_t}{\rightarrow} R\text{\textasciitilde}M-M\text{\textasciitilde}R$ (recombination)

or

$\quad\quad\quad\quad 2\,R\text{\textasciitilde}M\cdot \overset{k_t}{\rightarrow} R\text{\textasciitilde}MH + R\text{\textasciitilde}M'$ (disporportionation) (4)

At first sight both reactions (1) and (2) could be considered as the initiation reaction. Thus one could say that the reaction chain starts with the formation of the radical $R\cdot$. In that case reaction (2) can be considered the first step of the chain propagation (monomer addition). On the other hand, one can also consider reaction (2) as the real chain starting reaction through which the growth of the polymer chain begins, because it is not certain that all radicals $R\cdot$ lead to the starting of a chain by addition of monomer.

On careful examination, however, this analysis is not correct because reaction (2) has no influence on the kinetics of polymerization, whether or not it enters into the mechanism in a formal way. Thus, if the initiator radicals $R\cdot$ are very active (and only such radicals are normally used for the initiation of polymerization reactions), then reaction (2)—if considered independently of reaction (1)—occurs very rapidly and much more rapidly than reaction (1) $(k_2 \gg k_1)$.

Since the two reactions (1) and (2) are coupled with each other, the effective rate of reaction (2) is determined by the rate of the decomposition reaction (1). In other words, the number of radicals $R\cdot$ which add a monomer molecule according to reaction (2) can be no larger than the number of radicals which are produced by the decomposition of the initiator according to reaction (1). For the simplest case in which all radicals formed by reaction (1) then can react further by reaction (2) with a monomer molecule, this means that the rate of reaction (2), $r_2 = -d(R\cdot)/dt$, is identical to the rate r_1 of the initiator decomposition reaction (1), $r_1 = d(R\cdot)/dt$:

$$r_2 = -d(R\cdot)/dt = r_1 = d(R\cdot)/dt = -2d[I]/dt = r_i, \qquad (5)^3$$

where [I] is the initiator concentration in mol/liter, r is the reaction rate in mol/liter/second, and $d(R\cdot)/dt$ is the number of moles $R\cdot$ per liter formed (or used) in 1 second. As we will see with the discussion of the steady-state equilibrium, $d(R\cdot)/dt$ is not identical with the change in the radical concentration $d[R\cdot]/dt$ (see p. 76).

The number of moles of $R\cdot$ formed per second by decomposition of I according to Equation (1), $r_i = d(R\cdot)/dt$, is twice as large as the number of moles of I disappearing per second according to (1), $r_d = -d(I)/dt$. To avoid the factor 2 in the following equations, the rate of the initiation reaction has been defined by $r_i = d(R\cdot)/dt$ and not by $-d(I)/dt$. The rate of reaction (2), $-d(R\cdot)/dt$, is—in the ideal case—identical with the rate of reaction (1), $d(R\cdot)/dt = -2 d(I)/dt$. It should be noted, however, that the rate constant $k_i = 2 k_d$ is not identical with the rate constant of reaction (2).

The decomposition of the initiator is a reaction of the first order: $-d(I)/dt = k_d[I]$, and one can therefore write for the rate of the initiation reaction:

$$r_i = 2 k_d[I] = k_i[I]. \qquad (6)$$

---

³It should be noted that $r_1 = d(R\cdot)/dt = -2 d[I]/dt = r_i$ is an arbitrary definition. Instead, one could define $r_1 = -d[I]/dt = r_i$ just as well.

If the radical yield is not quantitative [i.e., if not all radicals formed by reaction (1) subsequently react with monomer according to reaction (2), but only a fraction f continues to react], then the rate of the initiation reaction is given by the equation:

$$r_i = -d(R\cdot)/dt = f \cdot k_i[I]. \tag{6a}$$

For initiators, such as peroxides or azo-compounds, one can determine $d(R\cdot)/dt$ by analytical methods: one determines $-d[I]/dt$, and therefore also k_i, from the decrease in the initiator concentration. The determination of the radical yield f further requires molecular weight determinations to obtain $\overline{M}_n$. For many polymerizations $f = 1$ or nearly 1.

For the kinetic treatment of the growth reaction (chain propagation), one can disregard the concentration of the radical $R\text{-}M\cdot$ which is first formed in reaction (2), because reaction (2) is really only one among many thousands of reaction steps which lead to equivalent radicals $R-M\cdot$, $R-M-M\cdot$, $R-M-M-M\cdot$, and so forth. In general $[R\text{\textasciitilde}M\cdot]$ represents the concentration of all the growing radicals present in the reaction mixture at any given time t. One can therefore write:

$$[R\text{\textasciitilde}M\cdot] = [R - M\cdot] + [R - M - M\cdot] +$$

$$[R - M - M - M\cdot] + + \cdots = \sum_{n=1}^{n=P} [R(M)_nM\cdot]$$

If there were no chain termination, the radical concentration would increase, beginning with concentration zero, in a continuous manner according to Equation (6). However, the greater the radical concentration becomes, the greater the possibility that two radicals will meet and disappear through recombination. This means that the rate of the termination reaction increases and finally a radical concentration $[R\text{\textasciitilde}M\cdot]$ will be reached at which, in a given time, just as many radicals will disappear through recombination as there will be new ones formed. Thus the initiation and termination reactions are in a certain phase of dynamic equilibrium (steady state) and the following equation holds:

$$r_i = r_t \tag{7}$$

Under most of the usual polymerization conditions, the equilibrium condition (steady state) is reached in a short time after the beginning of the polymerization (see fig. 77). Therefore the polymerization occurs with a constant radical concentration according to Equation (3). According to the equation for a second order reaction, the polymerization rate (velocity) may be defined as follows:

$$r_P = r_p = -d[M]/dt = k_p \cdot [M] \cdot [R\text{\textasciitilde}M\cdot]. \tag{8}$$

In this equation, k_p is the rate constant of the propagation reaction in $liter \cdot mol^{-1} \cdot sec^{-1}$, $[M]$ is the monomer concentration in moles/liter, and $[R\text{\textasciitilde}M\cdot]$ is the chain radical concentration at the steady state in moles/liter.

The polymerization rate r_p (which is identical with the velocity of the propagation reaction, r_p) is then obtained in terms of $mol \cdot liter^{-1} \cdot sec^{-1}$ (the number of moles of monomer which disappear from one liter per second through addition to the growing chain). r_p may be easily determined experimentally through continous concentration determinations during the polymerization or even more simply by using a dilatometer to measure the increase in density during the polymerization.

During the chain termination reaction two chain radicals [R∽∽M·] combine according to Equation (4), or by disproportionation, to a polymer molecule. If k_t is the rate constant of this reaction, one may write according to a second order equation for the velocity of the termination reaction:

$$r_t = -d(R\text{∽∽}M\cdot)/dt = k_t [R\text{∽∽}M\cdot]^2. \qquad (9)$$

$-d(R\text{∽∽}M\cdot)/dt$ is the rate at which the chain radicals are used up by termination reactions, i.e., $-d (R\text{∽∽}M\cdot)/dt$ = the number of moles $R\text{∽∽}M\cdot$ which react with each other per second by recombination or disproportionation in a volume of 1 liter of the polymerizing system. Just as with R· (see p 74), it is also true for these chain radicals that $d(R\text{∽∽}M\cdot)/dt$ is not identical with the change in the radical concentration $d[R\text{∽∽}M\cdot]/dt$. Whereas according to Equation (9) $-d(R\text{∽∽}M\cdot)/dt = k_t [R\text{∽∽}M\cdot]^2$, the change in the radical concentration $d[R\text{∽∽}M\cdot]/dt$ is always the sum of two rates, i.e., the rate of formation r_i and the rate of destruction r_t of the radicals $R\text{∽∽}M\cdot$:

$$d[R\text{∽∽}M\cdot]/dt = r_i - r_t = d(R\cdot)/dt + d(R\text{∽∽}M\cdot)/dt$$
$$= k_i [I] - k_t [R\text{∽∽}M\cdot]^2. \qquad (9a)$$

r_i normally remains nearly constant in the first hours of a radical polymerization. r_t is at first equal to zero, and then increases with increasing radical concentration during the beginning of the polymerization ($d[R\text{∽∽}M\cdot]/dt$ correspondingly becomes smaller) until at the steady state $r_i = r_t$ and $d[R\text{∽∽}M\cdot]/dt = 0$. The termination rate, $-d(R\text{∽∽}M\cdot)/dt$, on the other hand continues at a constant rate according to Equation (9).

By integration of Equation (9a), one obtains the radical concentration $[R\text{∽∽}M\cdot]_t$ at a time t after the start of the polymerization:

$$[R\text{∽∽}M\cdot]_t = ([I] k_i/k_t)^{1/2} \; \frac{\exp\left[(4k_i k_t [I]^{1/2} \cdot t\right] - 1}{\exp\left[(4k_i k_t [I]^{1/2} \cdot t\right] + 1} \cdot \qquad (9b)$$

Equation (9b) yields the curve shown in Figure 77 for a styrene polymerization. It shows that, after two seconds, the radical concentration has already reached its stationary value $[R\text{∽∽}M\cdot]_s$. One obtains $[R\text{∽∽}M\cdot]_s$ by setting r_i and r_t equal, according to the steady-state requirement or by setting

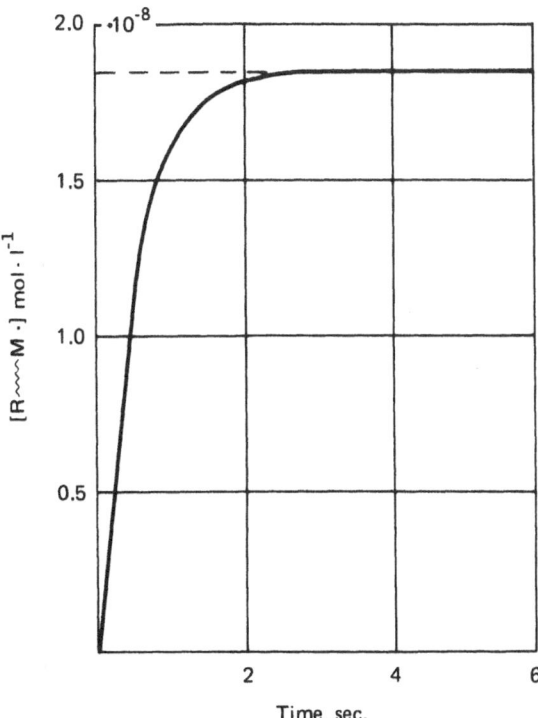

$$[I] = 1 \cdot 10^{-3} \text{ mol} \cdot l^{-1}$$
$$k_i = 1.2 \cdot 10^{-5} \text{ sec}^{-1}$$
$$k_p = 176 \text{ l} \cdot \text{mol}^{-1} \cdot \text{sec}^{-1}$$
$$k_t = 7.2 \cdot 10^7 \text{ l} \cdot \text{mol}^{-1} \cdot \text{sec}^{-1}$$

FIG. 77 – Increase of the radical concentration immediately after the start of a solution-polymerization of styrene in benzene at 60°C (according to D. MARGERISON and G.C. EAST)

Equation (9a)—which is the first derivative of Equation (9b)—equal to zero (since the stationary radical concentration $[R\text{\small$\sim\sim$}M\cdot]_s$ is the maximum value of $[R\text{\small$\sim\sim$}M\cdot]$):

$$d[R\text{\small$\sim\sim$}M\cdot]/dt = r_i - r_t = 0$$

$$k_t \cdot [R\text{\small$\sim\sim$}M\cdot]_s^2 = k_i \cdot [I] \text{ and } [R\text{\small$\sim\sim$}M\cdot]_s = \sqrt{(k_i/k_t) \ [I]}. \tag{10}$$

This calculation shows that $[R\text{\small$\sim\sim$}M\cdot]_s$ remains constant until the initiator concentration has begun to decrease. For the polymerization rate one can then write according to Equation (8):

$$r_p = k_p \, [M] \cdot \sqrt{(k_i/k_t) \cdot [I]} \, . \tag{11}$$

Accordingly, the rate of polymerization depends on the following parameters: 1. The chain-radical concentration $[R\text{\textasciitilde\textasciitilde\textasciitilde}M\cdot]$, which in turn depends on the initiator concentration and the ratio of the reaction rate constants of the initiation and termination reactions k_i/k_t; 2. The monomer concentration $[M]$; and (3) The reaction rate constant of the growth reaction (propagation reaction), k_p.

It may at first sight seem surprising that the rate of the overall polymerization, r_p, depends not only on k_i and $[I]$, but also on k_p and $[M]$, since it was pointed out earlier that the addition of a monomer to a radical (reaction 2)—considered as an independent reaction—occurs much more rapidly than the formation of the radicals by decomposition of the initiator, and that the rate of a multi-step reaction is determined by the slowest step, i.e., the initiator decomposition. This is true only as long as one considers the reaction up to the addition of one monomer molecule (or, in general, for the addition of a specific number of monomer molecules). But this is not true for a description of the overall total polymerization. As with all chain reactions, the growth reaction here also occurs completely independently of the initiation reaction. This means that once a radical $R\cdot$ is formed, the number of monomer molecules which will add to this radical per second depends only on the rate of the growth reaction, i.e. (for a given set of reaction conditions ($[I]$, $[M]$, $[R\text{\textasciitilde\textasciitilde\textasciitilde}M\cdot]$, etc.), it will depend only on k_p. For the polymerization of a monomer with $k_p = 2,000$ liters·mol^{-1}sec^{-1}, therefore 10 times as much polymer will be formed per second and the polymer chains will be 10 times as long, as for a polymerization of a monomer with $k_p = 200$ liters·mol^{-1}·sec^{-1}. It will be recognized that the molecular weight and the growth rate are dependent upon one other [compare p. 86, Equation (25)]. The automatic influence of k_p on the overall rate of polymerization (r_p) therefore holds only for the case where the chains can grow freely and where their length is not limited by a parameter which is independent of k_p. If one could limit the molecular weight for example by the addition of a suitable inhibitor, then the rate of such polymerizations would be determined only by the rate of the initiator decomposition.

The same is true if one limits the discussion to a monomer with a definite reaction constant, k_p. In this case the rate of polymerization is proportional to the square root of the initiator concentration according to Equation (11):

$$r_p = \text{constant} \cdot \sqrt{[I]} \, . \tag{11a}$$

The validity of this relation has been experimentally verified for many monomers (Figure 79). Naturally, this relation holds only for the earlier stages of the polymerization, since the monomer concentration and the initiator concentration changes with continuing polymerization. The linear dependence of the beginning polymerization rate on the monomer concentration, according to Equation (11), has also been experimentally confirmed.

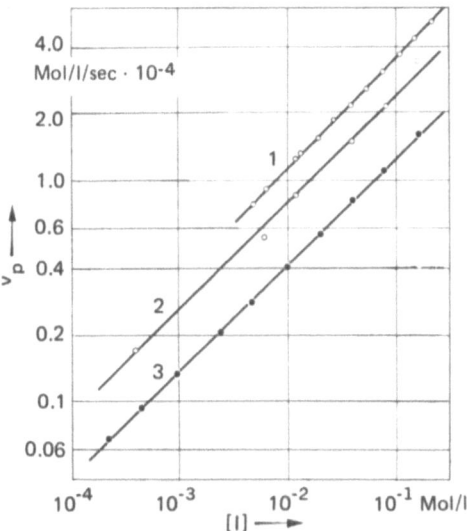

1. Methylmethacrylate with azobisisobutyronitrile at $50°C$;
2. Styrene with benzoyl peroxide at
   $60°$;
3. Methylmethacrylate with benzoyl peroxide at $50°C$.

FIG. 79 – Dependence of the steady state rate of polymerization on the initiator concentration (double logarithmic plot according to P.J. FLORY, based on measurements of SCHULZ and BLASCHKE, ARNETT and MAYO, GREGG and MATHESON).

$$r_P = \text{constant} \cdot [M] . \tag{11b}$$

For the region of constant radical concentration (i.e., with constant initiator concentration), the total polymerization rate r_P behaves like a reaction of first order. If the polymerization is carried out on a continuous basis, one can also hold the monomer concentration constant, and this results in a constant yield at a given time.

Normally, if one does not hold [R〜〜〜M·] and [M] constant by means of special measures, the polymerization rate will decrease with increasing conversion. However, in many cases when a certain conversion has been reached the polymerization rate suddenly increases. This effect, which has been called a gel effect, or an autoacceleration, has been explained in terms of a restriction in the termination reaction caused by the high viscosity of the reaction medium (compare p. 89 ff.).

The Determination of r_i and k_p^2/k_t

The values r_i and r_p can be determined experimentally: r_p by determination of the conversion and r_i by determination of the rate of decomposition of the initiator for example, by determining the rate at which an inhibitor is consumed in the presence of the monomer and initiator. The inhibitor used most is diphenylpicryl-hydrazyl (see also p. 61). From that one can determine the corresponding reaction rate constants, k_i according to Equation (6), and the ratio of k_p^2/k_t according to Equation (11):

$$k_i = \frac{r_i}{[I]} \tag{12}$$

and

$$\frac{k_p}{k_t} = \frac{r_p^2}{[M]^2[I]\,k_i} \quad \text{or} \quad \frac{k_p}{k_t^{1/2}} = \frac{r_p}{[M][I]^{1/2}\,k_i^{1/2}} \; . \tag{13}$$

If in Equation (13) one replaces $k_i[I]$ by v_i according to Equation (6) one obtains:

$$\frac{k_p^2}{k_t} = \frac{r_p^2}{[M]^2 \cdot r_i} \quad \text{or} \quad \frac{k_p}{k_t^{1/2}} = \frac{r_p}{[M] \cdot r_i^{1/2}} \; . \tag{14}$$

The Determination of the Activation Energies

From the temperature dependence of the reaction rate constants one obtains the activation energies according to the Arrhenius equation:

$$k = k_m \cdot e^{-A/RT} \qquad \text{(Arrhenius Equation)}$$

or

$$\ln k = \ln k_m - \frac{A}{RT} = H_n - \frac{A}{RT}$$

and by multiplication with log e:

$$\log k = H_d - \frac{A}{2.3 \cdot R \cdot T}$$

where A = activation energy; k_m = maximum reaction rate constant, i.e., if all collisions would lead to a reaction; and $H = \log k_m$ resp. $\ln k_m$ (for H_d resp. H_n).
If one applies these equations to the ratio $k_p/k_t^{1/2}$ (i.e., if in the expression $\log k_p/k_t^{1/2} = \log k_p - \frac{1}{2} \log k_t$ one replaces the reaction rate constant by the corresponding Arrhenius expressions $\log k_p = H_p - A_p/2.3 \cdot RT$ and $\log k_t = H_t - A_t/2.3 \cdot RT$, one obtains:

$$\log k_p/k_t^{1/2} = (-1/2.3 \cdot RT)(A_p - \frac{1}{2}A_t) + (H_p - \frac{1}{2}H_t). \tag{15}$$

Thus, one can determine $A_p - \frac{1}{2} A_t$ if one plots $\log k_p/k_t^{1/2}$ versus $1/T$. According to Equation (11) one can determine the activation energy of the overall polymerization reaction from this plot. Thus, if in logarithmic Equation (11) one replaces reaction rate constants $\log k_p$, $\log k_i$, and $\log k_t$ by the corresponding Arrhenius expressions $H - A/2.3\,RT$, one obtains:

$$\log r_P/[M] \cdot [I]^{1/2} = H_p + \frac{1}{2}(H_i - H_t) - (A_p - \frac{1}{2}A_t + \frac{1}{2}A_i)/2.3\,RT. \tag{16}$$

According to Equation (11), r_P is equal to $k_p \cdot \sqrt{k_i}/k_t \cdot [M] \cdot \sqrt{[I]}$, and one can therefore consider the expression $k_p \cdot \sqrt{k_i}/k_t$ as a sort of reaction rate constant k_P of the overall polymerization reaction:

$$k_p \sqrt{k_i/k_t} = k_P = r_P/[M] \cdot [I]^{1/2}. \tag{16a}$$

If one now compares Equation (16) with the Arrhenius expression (p. 80), one sees that in Equation (16) the sum of $A_p + \frac{1}{2}(A_i - A_t)$ is the activation energy (corresponding to k_P) of the overall polymerization reaction:

$$A_P = \frac{1}{2}A_i + (A_p - \frac{1}{2}A_t) \tag{17}$$

A_i (which is determined from the temperature dependence of k_i) is approximately 30 kcal/mole for the usual peroxide and azo initiators. For the expression $A_p - \frac{1}{2}A_t$ (determined according to Equation (15)) values between 4 and 7 kcal/mole were found for different monomers so that A_P according to Equation (17) has a value of the order of from 20 to 22 kcal/mole.

The Determination of the Absolute Values of k_p and k_t From the Average Lifetime of the Radicals

It is not as simple to determine the numerical values for the reaction rate constants k_p and k_t. For this purpose it is necessary to determine the average lifetime of the radicals $R\rightsquigarrow M\cdot$ experimentally. This can be done by determining the increase or decrease of the polymerization velocity during the non-steady state (compare p. 76), i.e., during the time in which at the start of the polymerization the radical concentration goes from 0 to $[R\rightsquigarrow M\cdot]_{steady}$ ($= [R\rightsquigarrow M\cdot]_s$) or while it decreases from $[R\rightsquigarrow M\cdot]_s$ to 0 after interruption of UV irradiation. Photopolymerization is especially suitable for investigations of this kind because, unlike decomposition of peroxides, the formation of new radicals can be stopped instantaneously.

According to Equation (9), the decrease in the radical concentration $[R\rightsquigarrow M\cdot]$ as a function of the time t after termination of the UV irradiation is as follows:

$$-d\,[R\rightsquigarrow M\circ]/dt = k_t\,[R\rightsquigarrow M\circ]_t^2$$

As soon as the UV lamp is shut off, the radical formation stops, so that r_i becomes $= o$. Therefore in this special case Equation (9a) is identical with Equation

(9) and integration yields instead of Equation (9b) the simple Equation (18):

$$1/[R\text{\textasciitilde}M\cdot]_t - 1/[R\text{\textasciitilde}M\cdot]_0 = k_t \cdot t \tag{18}$$

assuming that $[R\text{\textasciitilde}M\cdot]_0$ is the radical concentration at the moment of interruption of the UV irradiation. If the UV irradiation has occurred for a sufficient length of time, then $[R\text{\textasciitilde}M\cdot]_0$ equals $[R\text{\textasciitilde}M\cdot]_{\text{steady}}$ $(= [R\text{\textasciitilde}M\cdot]_s)$ If one then multiplies this with Equation (18) one obtains:

$$([R\text{\textasciitilde}M\cdot]_s/[R\text{\textasciitilde}M\cdot]_t)-1 = k_t \cdot [R\text{\textasciitilde}M\cdot]_s \cdot t. \tag{19}$$

The expression $k_t \cdot [R\text{\textasciitilde}M\cdot]_s$ is of course nothing other than the reciprocal average lifetime of the radical. This is because the average lifetime τ of a radical in the steady state is defined as:

$$\frac{\text{Radical concentration}}{\text{Decrease of radical conc.}}$$

The decrease of the radical concentration, $-d[R\text{\textasciitilde}M\cdot]_t/dt$, is given by the rate of the termination reaction (see Equation (9)) and therefore:

$$\tau = [R\text{\textasciitilde}M\cdot]_s/k_t[R\text{\textasciitilde}M\cdot]_s^2 = 1/k_t[R\text{\textasciitilde}M\cdot]_s. \tag{20}$$

If one further considers that the polymerization rate r_P is directly proportional to the corresponding radical concentration (compare Equation 8), then one can replace Equation (19) by the expression:

$$(r_P)_s/(r_P)_t = (1/\tau)\cdot t + 1 \tag{21}$$

Thus if one determines the rate of polymerization during continuous irradiation $(r_p)_s$ and the rate of polymerization as a function of the time t after the termination of the irradiation and plots $(r_P)_s/(r_P)_t$ versus t, one obtains a straight line, the slope of which is equal to $1/\tau$.

A practicable and more exact method for determination of the average lifetime τ of the chain radicals is based on initiation of polymerization by intermittent UV irradiation: A disc rotates through the incident UV beam. Several sectors have been cut out of this disc in such a way for instance that the remaining sectors are as large as those which have been cut out. Therefore, after each period of irradiation, Δt, there follows an equally long period of darkness, and the radical concentration oscillates around a mean value $[R\text{\textasciitilde}M\cdot]_{\text{mean}}$. This mean radical concentration, or rather the ratio of $[R\text{\textasciitilde}M\cdot]_{\text{mean}} /[R\text{\textasciitilde}M\cdot]_s$ depends in a theoretically clear manner on the ratio of the length of the irradiation interval Δt to the mean lifetime of the radicals, τ, according to the following relation: $[R\text{\textasciitilde}M\cdot]_{\text{mean}}/[R\text{\textasciitilde}M\cdot]_s = f (\Delta t/\tau)$. With the aid of Figure 83 one can illustrate this

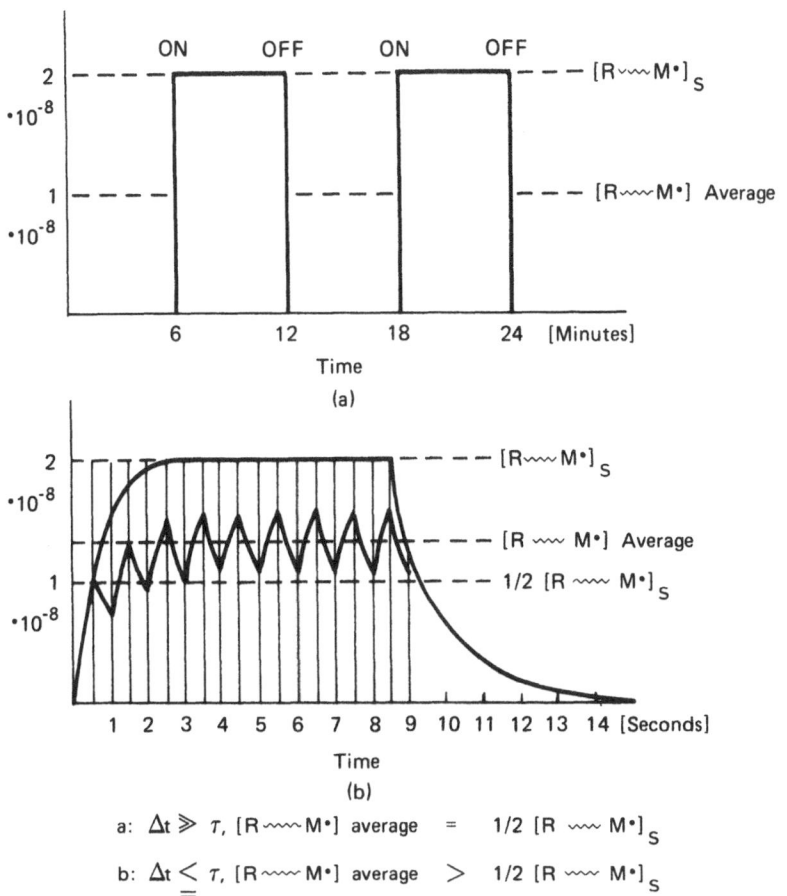

FIG. 83 – Radical concentration with polymerizations initiated by inter-mittent UV-irradiation.

qualitatively for the case of equally long light and dark periods: Part (a) of the figure shows that $[R\text{-----}M\cdot]_{mean} = [R\text{-----}M\cdot]_s/2$ when Δt is much greater than τ, because in that case the times in which $[R\text{-----}M\cdot]$ increases from 0 to $[R\text{-----}M\cdot]_s$ and decreases from $[R\text{-----}M\cdot]_s$ to 0 can be neglected in comparison to Δt. This however, is no longer true when Δt is about equal to τ or when Δt is smaller than τ. Diagram (b) can help explain why in this case the increase in the radical concentration during the irradiation period is initially larger than the decrease in the following dark period. The reason is that at relatively low radical concentrations, one is in a region where the termination reaction still occurs only slowly (i.e., the increase of $[R\text{-----}M\cdot]$, according to Equation (9a), occurs rapidly, but the decrease, according to Equation (9), occurs only slowly). The result is that the increase of the radical concentration only levels off at a medium radical concentration, which is greater than $[R\text{-----}M\cdot]_s/2$. As the quantitative treatment of this process has shown, such a mean radical concentration is reached when the ratio of Δt to τ is approximately equal to 1/10. On further decrease of Δt, this radical concentration does not become any larger, and the ratio of $[\text{R}\sim\sim\text{M}\cdot]_{\text{mean}}$ to $[\text{R}\sim\sim\text{M}\cdot]_s$ at this stage is approximately equal to 0.72. Thus the ratio $[\text{R}\sim\sim\text{M}\cdot]_{\text{mean}}/[\text{R}\sim\sim\text{M}\cdot]_s$ as a function of $\Delta t/\tau$ will have values of between 0.72 and 0.5 if one varies $\Delta t/\tau$ from values around 1/10 to large values of 1,000/1. The quantitative theoretical calculation of this function yields the curve in Figure 84. Since the rate of polymerization r_P is always proportional to the radical concentration $[\text{R}\sim\sim\text{M}\cdot]$, one can replace $[\text{R}\sim\sim\text{M}\cdot]_{\text{mean}}/[\text{R}\sim\sim\text{M}\cdot]_s$ by $r_{P(\text{mean})}/r_{P(s)}$. Then with the aid of the theoretical function $r_{P(\text{mean})}/r_{P(s)} = f(\Delta t/\tau)$, it is now possible to determine τ for any given system if one can determine the function $r_{P(\text{mean})}/r_{P(s)} = f(\Delta t)$ experimentally through the mean rate of polymerization $r_{P(\text{mean})}$ at different rotation

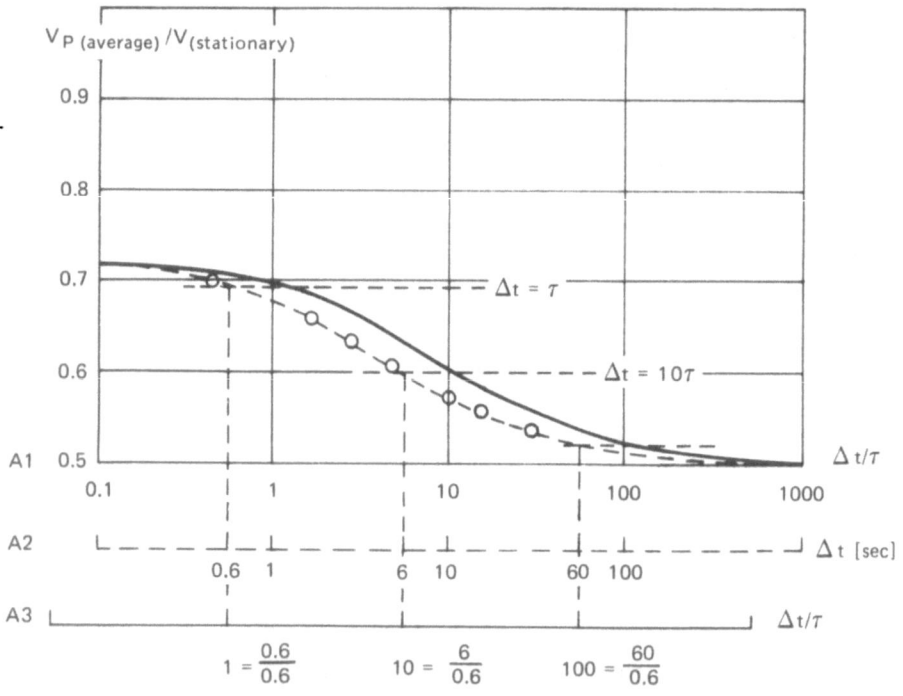

_____ = Theoretical curve (abscissa A1, logarithmic scale)

--O-- --O-- = Experimental curve (abscissa A2, logarithmic scale) arbitrarily chosen

A_3 = Abscissa A1 displaced until both curves coincide

FIG. 84 — Dependence of the average rate of polymerization on the length of the irradiation interval during intermittent photopolymerization (determination of the average life-time of the radicals)

rates of the sector. τ is obtained simply by displacement of the ordinate value (about 0.69) which belongs to the value $\Delta t/\tau = 1$, parallel to the abscissa until it intersects with the experimental curve. To illustrate this clearly, both the theoretical and the experimental curve have been plotted on the same diagram, using the same scale. However, it is not necessary for the determination of τ that both curves be superimposed using the same ordinate scale. Thus for every experimental curve $r_{P(mean)}/r_{P(s)}$ vs. Δt, regardless of the scale, the theoretical result holds true that the Δt value that corresponds to the ordinate value of 0.69 is identical with τ, because only at this point will $\Delta t/\tau = 1$. Similarly, the Δt value corresponding to the ordinate value 0.6 is equal to 10 times the τ value, and so forth.

In this way, it was found that the average lifetime of the radicals is on the order of one to a few seconds for the most important polymerization reactions carried out under the usual conditions. In each individual case the value of τ depends, of course, on the reaction conditions and especially on the concentration of initiator and the temperature.

If one knows the value of τ (i.e., the average radical lifetime), one can then determine the individual values of k_p and k_t. According to the definition of the average lifetime, Equation (20), $\tau = (1/k_t)\,[R\text{\scriptsize$\sim\!\sim\!\sim$}M\boldsymbol{\cdot}]_s$. If one replaces the radical concentration according to Equation (8) by the expression $r_p/k_p\cdot[M]$, one obtains:

$$\tau = \frac{k_p}{k_t}\cdot\frac{[M]}{r_p} \tag{22}$$

Since it is possible to determine the ratio of k_p^2/k_t according to Equation (14), one can determine the individual values of the reaction rate constants with the aid of the numerical values of k_p/k_t [determined according to Equation (22)] and k_p^2/k_t. Thus, if one finds $k_p/k_t = a$ and $k_p^2/k_t = b$, then $k_t = k_p/a$, $k_p = b/a$, and $k_t = b/a^2$. Table 85 gives the absolute values of the kinetic parameters for the photopolymerization of certain vinyl compounds at 25°C and at 60°C.

$d[M]$ is the change in the monomer concentration during the time dt. The monomer concentration is measured in moles/liter, and therefore the polymeriza-

TABLE 85. Absolute values of the rate constants k_p and k_t for the radical polymerization of certain vinyl monomers at 25° and 60°C.

	Vinylacetate Photopolymerization 25°C with different intensities		Styrene 60°	Methylmethacrylate 60°	Methylacrylate 60°	Dimensions
ν_P	0.45 10^{-4}	1.2 10^{-4}				mol l^{-1} sec^{-1}
τ	4	1.5				sec
k_p	940	1010	176	367	2090	l mol^{-1} sec^{-1}
k_t	$5.6\cdot10^7$	$6.1\cdot10^7$	$3.6\cdot10^7$	$0.93\cdot10^7$	$0.47\cdot10^7$	l mol^{-1} sec^{-1}
$[R\text{\scriptsize$\sim\!\sim$}M\boldsymbol{\cdot}]$	$0.44\cdot10^{-8}$	$0.54\cdot10^{-8}$				mol l^{-1}

tion rate r_P equals $-d[M]/dt$ [mol. liter^{-1}·sec^{-1}]. This is the decrease in the amount of monomer per second in a volume of 1 liter caused by the addition of monomer molecules to the chain radicals. The amount of chain-radicals per liter = [R⌇⌇⌇M·] mol/liter, so that one can also define the rate of polymerization as the number of moles of the monomer which adds per second to [R⌇⌇⌇M·] moles of a chain radical. This amount, according to Equation (8), equals k_p ·· [M] [R⌇⌇⌇M·]. This means that to [R⌇⌇⌇M·] moles of a chain radical R⌇⌇⌇M· per second k_p · [M] [R⌇⌇⌇M·] moles of a monomer M will add, so that to 1 mole of chain radicals R⌇⌇⌇M· per second k_p·[M] moles of the monomer will add, or, which is equivalent, to one chain radical R⌇⌇⌇M· per second

$$n_M = k_p \cdot [M] \tag{23}$$

monomer molecules will add.

The molecular weight of many technically important monomers is approximately 100 (for example, styrene 104, methylmethacrylate 100, vinylacetate 86). The monomer concentration [M] of the pure monomer then has an order of magnitude of 8-10 moles/liter. Since k_p (according to Table 85) is approximately equal to 200 to 2,000 liters·mole^{-1}sec^{-1}, we find that 2,000 to 20,000 single monomer molecules add to a single radical per second, depending on the type of monomer and the polymerization conditions.

Kinetic Chain Length and Degree of Polymerization

According to Equation (23), $n_M = k_p$·[M] monomer molecules add to a single radical per second. The number L_{kin} of monomer molecules which are added during the lifetime τ of a radical is therefore equal to:

$$L_{kin} = n_M \cdot \tau = k_p \cdot [M] \cdot \tau. \tag{24}$$

The parameter L_{kin} defined by Equation (24) is called the kinetic chain length. If, according to Equation 20, one replaces the average lifetime τ by the expression [R⌇⌇⌇M·]/k_t[R⌇⌇⌇M·]^2, one obtains:

$$L_{kin} = \frac{k_p[M][R⌇⌇⌇M·]}{k_t \cdot [R⌇⌇⌇M·]^2} = \frac{k_p \cdot [M]}{k_t \cdot [R⌇⌇⌇M·]} = \frac{r_P}{r_t} = \frac{r_P}{r_i} \tag{25}$$

If no chain transfer occurs (i.e., if with the end of molecular growth the radical which carries on the reaction also disappears), then the degree of polymerization is equal to the kinetic chain length, or is proportional to it. If chain termination occurs only by disproportionation, then $\bar{P} = \bar{L}_{kin}$. If chain termination occurs only through recombination then $\bar{P} = 2\bar{L}_{kin}$.

If the proportionality factor is smaller than 2, then this is a sign that chain termination occurs to some extent through disproportionation. If there is no linear dependence between $\bar{P}$ and $\bar{L}_{kin}$ then the size of the molecules is determined not through chain termination but through chain transfer (modifier action).

Propagation reaction rate constants (k_p) of different orders of magnitude are, according to Equation (25), the reason that not all monomers under the same reaction conditions yield polymers with the same molecular weight (i.e., the greater k_p, and therefore r_P, the greater will be the chain length and therefore the higher the molecular weight for corresponding rates of termination. Acrylic esters (k_p = 2,090 liters·mol^{-1}sec^{-1} at 60°C) polymerize very rapidly; styrene (k_p = 176 liters·mol^{-1}sec^{-1}) polymerizes relatively slowly. The molecular weight of the polymers obtained from acrylic esters is correspondingly higher than that obtained from styrene under comparable conditions. On the other hand, one should not conclude from Equation (25) that one can simply increase the degree of polymerization by increasing r_P. Actually, one obtains just the opposite. The kinetic chain length, and therefore the degree of polymerization, is only increased by an increase of r_P if the termination rate r_t remains unchanged. Most conditions under which the rate of polymerization is increased (for example, by increasing the concentration of the initiator or the temperature) give rise to an increase in the radical concentration $[R \sim\sim\sim M \cdot]$, which enters into Equation (25) as a square term, and therefore an increase in r_P caused by a corresponding and much greater increase in r_t leads to a decrease in the ratio r_P/r_t. One sees this immediately if one replaces the ratio in Equation (25) by $k_p \cdot [M]$, and one thus finds–if $[M]$ is constant–that the kinetic chain length is inversely proportional to the rate of polymerization:

$$
L_{kin} = \frac{k_p \, [M] \, [R \sim\sim\sim M \cdot] \, k_p \, [M]}{k_t \, [R \sim\sim\sim M \cdot]^2 \, k_p \, [M]}
$$

$$
= \frac{k_p}{k_t} \cdot \frac{[M]^2}{k_p \, [R \sim\sim\sim M \cdot] \, [M]} = \frac{k_p^2}{k_t} \cdot \frac{[M]^2}{r_P} = const. \, \frac{1}{r_P}
$$

(27)

To represent this function one usually plots $1/P$ against $r_P/[M]^2$. One has to remember however that $\bar{L}_{kin}$ is usually not equal to $\bar{P}$, and one therefore requires a proportionality factor K. Furthermore, it is not always possible to neglect chain transfer. One discovers this immediately by the fact that the lines do not pass through the zero point (see Fig. 88). The intercept at the ordinate gives, in this case, the numerical value of the transfer constant (as long as one only has chain transfer to the monomer), or, a value which is proportional to the transfer constant (for example, $C_{tr} \cdot [\bar{P}]/[M]$, if chain transfer occurs with the polymer of concentration $[\bar{P}]$. If chain transfer occurs with the solvent of concentration $[S]$, then the proportionality factor becomes $C_{tr}[S]/[M]$ (compare pages 63 to 69 and page 91):

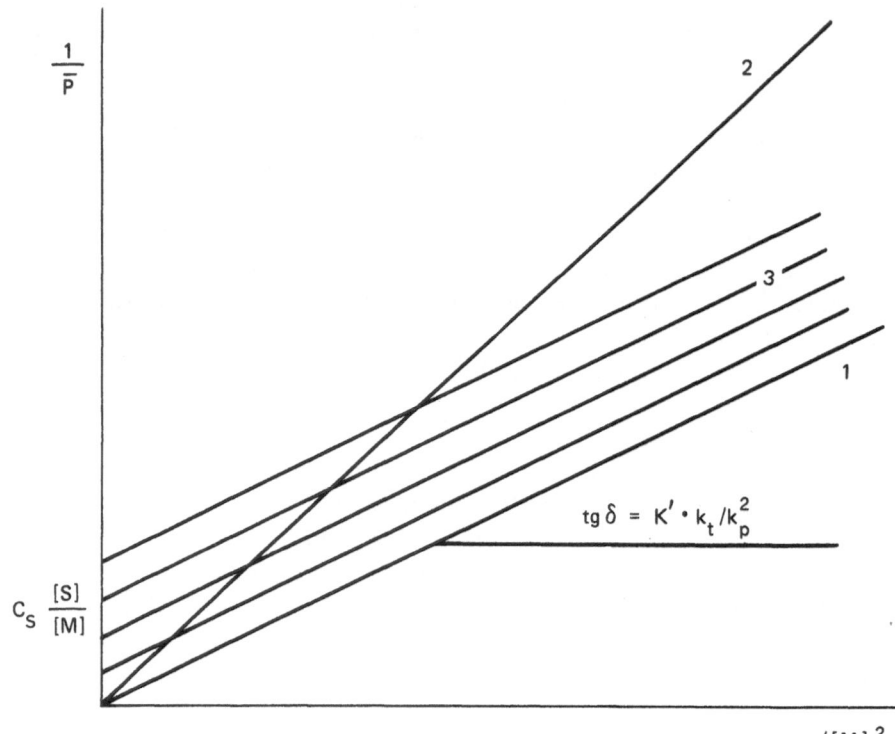

1) Bulk polymerization; termination by disproportionation $(K' \approx 1)$
2) Bulk polymerization; termination by recombination $(K' \approx 2)$
3) Solution polymerization

FIG. 88 — Dependence of the reciprocal degree of polymerization, $1/\overline{P}$, on $v_P/[M]^2$ according to Equ. (27a)

$$\overline{P} = K \cdot \overline{L}_{kin} = K \frac{k_p^2}{k_t} \cdot \frac{[M]^2}{r_P} + \frac{1}{C_s}$$

$$\text{or} \quad \frac{1}{\overline{P}} = K' \frac{k_t}{k_p^2} \cdot \frac{r_P}{[M]^2} + C_s \frac{[S]}{[M]}$$

(27a)

Equation (27a) is identical to Equation (28), which can easily be shown by replacing in both equations r_P and r_t by $k_p \cdot [M] \cdot [R\text{---}M\cdot]$ and $k_t \cdot [R\text{---}M\cdot]^2$, and r_{tr} by $k_{tr} \cdot [S] \cdot [R\text{---}M\cdot]$. If one determines the degree of polymerization $\overline{P}$ at different rates of polymerization r_P, or different monomer concentrations, and plots according to Figure 88, $1/\overline{P}$ vs. $r_P/[M]^2$, then the slopes of the lines permit determination of K, and therefore they also give information about the type of chain termination, if k_p^2/kt is known [compare

Equation (13)], or conversely, they permit a determination of k_p^2/k_t, if the type of chain termination is known. Furthermore, the value of the intercept at the ordinate gives information about the influence of the chain-transfer reactions (compare p. 91). The validity of Equation (27a) is typical for radical polymerizations and can in certain doubtful cases (for example, for initiation with new initiator systems, or by irradiation) be used for the elucidation of the type of polymerization. In order to vary $r_P/[M]^2$, one can carry out the polymerizations at different temperatures or with increasing concentrations of initiator.

Self-Acceleration (Autocatalysis) of the Polymerization (Norrish-Trommsdorf Effect)

The polymerization rate and the degree of polymerization are not always coupled in such a way that with increasing rate of polymerization the degree of polymerization diminishes. For example, the degree of polymerization can increase, if in emulsion polymerization the rate of polymerization is increased by increasing the soap concentration (compare p. 154 ff.). This is also observed during the self-acceleration phase of a radical polymerization at higher conversions. Thus when the conversion with bulk-, or solution-, or pearl-polymerization has reached such a high value that the viscosity of the polymerizing solution has become very high, then the mobility of the polymer molecules is hindered so strongly that a collision between two chain-radicals occurs more seldom.

For the kinetics of the polymerization reaction this phenomenon means that the reaction rate constant of the termination reaction, $k_t(k_t$ = the number of moles per second of chain radicals disappearing through termination when $[R\text{\textasciitilde\textasciitilde} M\cdot] = 1$ mole/1), will decrease after the viscosity has reached a certain relatively high value. The creation of new radicals R· through initiator decomposition continues undisturbed, however, so that the radical concentration increases over that given by the steady state equilibrium. Therefore according to Equation (10):[4]

$$[R\text{\textasciitilde\textasciitilde} M\cdot]_\eta = (\sqrt{k_i \cdot [I]})\,(1/\sqrt{k_{t(\eta)}}) = \text{Constant} \cdot \ 1/\sqrt{k_{t(\eta)}} \qquad (27b)$$

Whereas normally an increase in the radical concentration (for example, by increasing the temperature or by increasing the initiator concentration) always brings about an increase in the termination rate, this is not the case here, where the increase in the radical concentration is effected only by the decrease in the rate constant k_t. This becomes clear, if one substitutes Equation (27b) into Equation (9);

$$-d(R\text{\textasciitilde\textasciitilde} M\cdot)/dt = r_t = k_t \cdot [R\text{\textasciitilde\textasciitilde} M\cdot]^2 = k_t \cdot \text{Constant}\,(1/\sqrt{k_t})^2 = \text{Constant} \qquad (27c)$$

---

[4]One could ask here whether Equation (10), which is based on the assumption that $r_i = r_t$, is still valid. If one remembers that for each new value of $k_{t(\eta)}$ there is again a new stationary equilibrium, with a new radical concentration $[R\text{\textasciitilde\textasciitilde} M\cdot]_\eta$ corresponding to this equilibrium, one sees that the equation is valid at least for a sufficiently short interval of time. [R $\text{\textasciitilde\textasciitilde} M\cdot]_\eta$ is the equilibrium radical concentration corresponding to $k_{t(\eta)}$. When $[R\text{\textasciitilde\textasciitilde} M\cdot]_\eta$ is effectively reached, k_t effective is always somewhat smaller than $k_{t(\eta)}$, but the difference between $k_{t(\eta)}$ and $k_{t(eff)}$ is smaller, the shorter the discussed time interval.

Now we can describe the Norrish-Trommsdorf effect as follows: The slowing down of the reaction rate r_t of the chain termination (caused by increasing viscosity and decreasing k_t) causes an increase of $[R \text{\textasciitilde\textasciitilde} M\cdot]$ according to Equation 27b. This increase of $[R \text{\textasciitilde\textasciitilde} M\cdot]$ would suffice exactly to compensate for the slowing down of r_t, if the viscosity of the polymerizing system (and thus also k_t) could be held constant for a moment. In reality, however, the viscosity continues to rise, effecting again a decrease of r_t and increase of $[R \text{\textasciitilde\textasciitilde} M\cdot]$, etc. Through the compensation mechanism in the sense of Equation (27c), the difference between r_i and r_t remains constant in spite of continuously rising $[R \text{\textasciitilde\textasciitilde} M\cdot]$ –values (if the increase of viscosity runs linearly). By this, the non-steady state in the self-acceleration stage of a radical polymerization is distinguished from the non-steady state in the beginning stage of the polymerization.

The unchecked increase in the radical concentration $[R \text{\textasciitilde\textasciitilde} M\cdot]$ has two important results as far as the polymerization reaction and the formed polymers are concerned:

1. The polymerization rate r_P increases rapidly with increasing viscosity because of the Trommsdorf effect ($r_P = k_p \cdot [M] \cdot [R \text{\textasciitilde\textasciitilde} M\cdot]$).

2. The kinetic chain length, and therefore the degree of polymerization, increases with increasing rate of polymerization and constant termination rate ($\overline{L}_{kin} = r_P/r_t$).

For example, if the termination rate constant decreases because of the high viscosity to one-fourth of its normal value, the radical concentration $[R \text{\textasciitilde\textasciitilde} M\cdot]$ increases according to Equation (27b) to twice its value. The result is that also r_P and L_{kin} increase to twice their normal value.

As the polymerization rate and the degree of polymerization increase, one finds an even more rapid increase in the viscosity. This in turn leads to a further increase in the polymerization rate and so on. The result is often that the polymerization "runs away" unless it is possible to remove the heat of polymerization by certain reaction conditions. Because in a high viscosity polymerizing medium the heat exchange is very much restricted, a bulk polymerization carried out on a technical scale always requires certain experimental procedures to remove the heat.

The increase in the kinetic chain length and the resulting increase in the degree of the polymerization during the stage of autocatalysis can be experimentally proved by means of molecular weight determinations. The cause for the Norrish-Trommsdorf effect is basically the same as that which leads to higher degrees of polymerization and higher polymerization rates with emulsion polymerizations, i.e., the partial or complete isolation of the chain radicals from each other (compare pp. 154-163).

Kinetics of Chain Transfer and Determination of the Transfer Constants

Especially in the presence of solvents or modifiers (compare p. 65) one often finds that the growth of a macromolecule is terminated without the disappearance of the radical which carries on the chain reaction. The radical state is actually transferred from the growing chain end to a solvent or modifier molecule from which a new molecule then begins to grow. The molecular chain is terminated; however, the radical chain is not. Thus in such cases the degree of polymerization is determined not only by the chain termination reaction, but also by the chain transfer reaction. With strongly effective modifiers, the chain transfer reaction is actually the one which alone determines the degree of polymerization. This is why the rate of the transfer reaction r_{tr} plays the same role in determining the degree of polymerization as the rate of the termination reaction plays in determining the kinetic chain length. In analogy to Equation (25), one can write:

$$\bar{P} = \frac{r_P}{K \cdot r_t + r_{tr}} \qquad \text{or} \qquad \frac{1}{\bar{P}} = K' \cdot \frac{r_t}{r_P} + \frac{r_{tr}}{r_P} = \frac{1}{\bar{P}_0} + \frac{r_{tr}}{r_P} \qquad (28)$$

$K \cdot r_t / r_P = 1/\bar{P}_0$ is the reciprocal of the degree of polymerization which would be obtained if the polymerization were carried out under the same conditions but without chain transfer. The chain transfer reaction (similar to the propagation reaction; compare p. 73) is a bimolecular reaction between the chain radical and a solvent or modifier molecule. One therefore obtains (if [S] is the solvent or modifier concentration)

$$r_{tr} = k_{tr} [R \text{~~~} M \cdot] [S] . \qquad (29)$$

If one then replaces r_P by $k_p \cdot [R \text{~~~} M \cdot] [M]$ according to Equation (8), one obtains from Equation (28)

$$\frac{1}{\bar{P}} = \frac{1}{\bar{P}_0} + \frac{k_{tr} [R \text{~~~} M \cdot] [S]}{k_p [R \text{~~~} M \cdot] [M]} = \frac{1}{\bar{P}_0} + \frac{k_{tr} [S]}{k_p [M]}$$

$$(30)$$

$$= \frac{1}{\bar{P}_0} + C_s \frac{[S]}{[M]}$$

This equation permits the calculation of the transfer constant $C_s = k_{tr}/k_t$ of technically important solvents and modifiers by varying the ratio [S]/[M] (i.e., the solvent concentration) and determining the molecular weight obtained for each ratio. If one then plots $1/\bar{P}$ versus [S]/[M], one obtains C_s as the slope of the straight line. In order that $1/\bar{P}_0$ remain constant, one has to select the initiator concentration for each solvent concentration in such a way that r_t/r_P remains constant. This is the case if the initiator concentration is changed according to the relation $[I] = \text{const.} \cdot [M]^2$. Especially simple are the conditions during the thermal polymerization of a monomer (without initiator). In that case the initiation reaction is bimolecular ($r_i = k_i \cdot [M]^2$), and the ratio $r_t/r_p = r_i/r_p$ is therefore independent of the monomer concentration. Figure 92 shows the effect of increasing the concentration of different solvents on the molecular weight obtained during the thermal polymerization of styrene. The slope of the lines yields, according to Equation (30), the transfer constants C_s listed in Table 66.

One can use Equation (30) (if C_s is known) to calculate the decrease of the molecular weight to be expected during radical polymerization on the addition of certain solvents and modifiers.

For a given C_s, the decrease of the degree of polymerization depends only on the ratio $[S]/[M]$. However, one has to take into consideration that this ratio can change significantly if a polymerization is carried out to a high degree of conversion. By addition of monomers during the polymerization, one can counteract such changes occuring in the ratio $[S]/[M]$.

In addition to the above method for the determination of C_s-values, there are still other methods. For example, for high values of C_s, it might be better to determine the decrease in concentration of the modifier with continuing

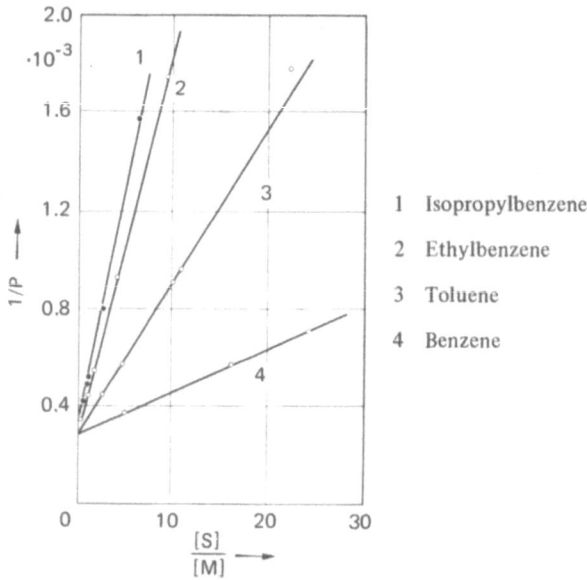

FIG. 92 — Dependence of the reciprocal degree of polymeriza-tion, 1/P, on the solvent concen-tration, i.e. $[S]/[M]$ for the poly-merization of styrene in different solvents at $100°C$ (according to Equ. (30) after GREGG and MAYO).

polymerization. If one then plots log [S] versus log [M] one obtains the transfer constant C_s as the slope of the line log [S] = C_s·log [M] .

Temperature Dependence of the Degree of Polymerization

Experience indicates that when the temperature of polymerization is increased, the degree of polymerization decreases. This is especially important in polymerizations carried out on a technical scale, where one tries to obtain a well-defined degree of polymerization from one batch to another. One therefore has to maintain a definite temperature in order to obtain a constant degree of polymerization. On the other hand, it is possible by variation of the temperature, and/or by addition of modifiers, to predetermine the value of the degree of polymerization in a relatively convenient way.

The negative temperature coefficient of the degree of polymerization (i.e., of the kinetic chain length L_{kin}) results from the value of the activation energy of the different stages of the chain reaction: if one replaces in Equation (25) (L_{kin} = k_p[M]/k_t[R $\sim\sim\sim$ M·]) the radical concentration by the expression $\sqrt{(k_i/k_t)}$· [I] according to Equation (10), one obtains:

$$L_{kin} = \frac{k_p \cdot [M]}{k_t \sqrt{(k_i/k_t) \cdot [I]}} = \frac{k_p \cdot [M]}{\sqrt{k_t \cdot k_i \cdot [I]}} \qquad (31)$$

If one takes the logarithm of Equation (31) and replaces the reaction rate constants k_p, k_t, and k_i by the corresponding Arrhenius expressions ln k = ln k_m − A/RT = H_n − A/RT, one obtains (see page 81):

$$\ln L_{kin} = -\frac{A_p}{RT} + \frac{1}{2}\frac{A_t}{RT} + \frac{1}{2}\frac{A_i}{RT}$$

$$+ H_p - \frac{1}{2} H_t - \frac{1}{2} H_i - \frac{1}{2} \ln [I] + \ln[M]$$

By differentiation with respect to T, one obtains the temperature coefficient of the kinetic chain length:

$$d(\ln L_{kin})/dt = (A_p - \frac{1}{2} A_t - \frac{1}{2} A_i)/RT^2 \qquad (32)$$

Since A_i is approximately 30 kcal, but (A_p − ½A_t) is only 4-7 kcal, d (ln L_{kin})/dT and, therefore, also the temperature coefficient of the degree of polymerization is negative. The decrease in the degree of polymerization with increasing temperature results from the fact that the activation energy of the initiation reaction is relatively high compared to the activation energy of the growth reaction. Although the rate of chain growth (propagation) also increases with increasing temperature (which by itself would lead to an increase in the degree of polymerization), the increase in the rate of initiation is the outweighing factor and therefore according to Equation (25) the degree of polymerization decreases.

For a polymerization where the degree of polymerization is either constant or increases with an increase in temperature, one therefore would need an initiation reaction which has an extremely small activation energy, i.e., one where the rate is independent of the temperature. Such an initiation reaction is possible by initiation through UV irradiation. Thus, with photopolymerization the degree of polymerization increases with increasing temperature, provided that strong chain transfer does not cause a decrease in the degree of polymerization.

2115 Copolymerization

If one polymerizes a mixture of different monomers, one usually obtains macromolecules whose structure contains all the monomers that are present in the reaction mixture.[6] However, one should not expect that these monomers are present in the same ratio in the polymer molecule as in the monomer mixture. In an extreme case, (which, however, can not be realized for theoretical reasons (see p. 107) one might even obtain on polymerizing a mixture of two monomers, M_1 and M_2, polymer molecules which consist exclusively of monomer M_1 and other polymer molecules which consist exclusively of monomer M_2. In such a case one has not obtained a copolymer, but only a mixture of two homopolymers. The other extreme case (and this is encountered quite frequently) occurs if the copolymerization of the monomers is preferred so strongly to the formation of homopolymer that, independent of the concentration of the monomer mixture, only polymer molecules in which the two different monomers alternate regularly are obtained. As soon as the monomer present in the lower concentration is used up, the polymerization stops completely, or the monomer present in excess polymerizes (usually considerably slower) by itself. Between these two extremes lies the ideal case of those copolymers whose macromolecules (with random [statistical] distribution of the structural elements) contain exactly the ratio of the two monomers present in the monomer mixture. Most of the technically important copolymers approach this ideal case to a smaller or larger extent.

The behavior of a monomer mixture is determined to a large extent by the rate (i.e., the ratio of the rates) of the reactions taking part in chain growth. With a copolymerization involving two different monomers, the growth reactions are the following four reactions:

$$1. \ R \rightsquigarrow M_1 \cdot + M_1 \ \xrightarrow{k_{11}} \ R \rightsquigarrow M_1 M_1 \cdot$$

$$2. \ R \rightsquigarrow M_1 \cdot + M_2 \ \xrightarrow{k_{12}} \ R \rightsquigarrow M_1 M_2 \cdot$$

$$3. \ R \rightsquigarrow M_2 \cdot + M_2 \ \xrightarrow{k_{22}} \ R \rightsquigarrow M_2 M_2 \cdot$$

$$4. \ R \rightsquigarrow M_2 \cdot + M_1 \ \xrightarrow{k_{21}} \ R \rightsquigarrow M_2 M_1 \cdot$$

---

[6]For the general structure of copolymers, block-copolymers and graft-copolymers, see pp. 22-30.

If the reaction rate constant k_{11} for the normal chain growth is much greater than k_{12}, and correspondingly k_{22} is very much greater than k_{21}, then reactions 2 and 4, which lead to the formation of copolymers, would be practically insignificant and one should obtain block-copolymers or mixtures of homopolymers. On the other hand, if the constants k_{12} and k_{21} are much greater than k_{11} and k_{22}, chains with $M_1 \cdot$ at the growing chain end add practically only monomer M_2, and $R \sim\!\sim\!\sim M_2 \cdot$ chains add practically only monomer M_1, so that copolymers with an alternating sequence of the structural units are formed, independent of the ratio of the monomers present in the monomer mixture. Finally, if $k_{12} = k_{11}$ and $k_{21} = k_{22}$, the frequency with which an M_1 or an M_2 monomer adds to the chain end is simply determined by their concentration, which leads to the result that the ratio of the two monomers in the monomer mixture is equal to the ratio of the two monomers in the copolymer molecules (ideal copolymerization[7]).

One cannot calculate the rate constants r_{12} and r_{21} from the rate constants r_{11} and r_{22}. Therefore, the behavior of a monomer in homopolymerization does not give any indication of its behavior in copolymerization. If one wants to obtain an idea about the behavior of a certain monomer pair during copolymerization, then one experimentally determines the composition of the copolymers with different $[M_1]/[M_2]$ ratios in the monomer mixture. This composition of the copolymers is usually represented by $d[M_1]/d[M_2]$, because the ratio of the monomers in the copolymer is identical with the ratio with which two monomers disappear from the monomer mixture. Since the ratio of the monomers usually changes continuously during polymerization (except with an ideal copolymerization), only investigations in which the polymerization is stopped at low conversions are meaningful.

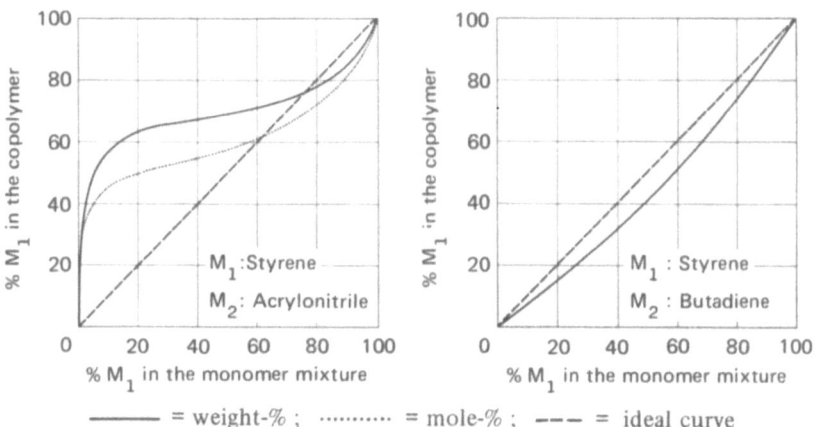

FIG. 95 — Copolymerization diagrams for styrene/acrylonitrile ($r_1 = 0.4$, $r_2 = 0.03$) and styrene/butadiene ($r_1 = 0.7$, $r_2 = 1.4$)

---

[7]The description "ideal copolymerization" is used in the literature mostly for systems where the product of the parameters r_1 and r_2 is equal to 1 (compare p. 112).

Therefore, one interrupts the polymerization with conversions of 5-10% and isolates the copolymer. In many cases one of the monomers contains an atom or functional group which can be determined easily by analytical means. With the aid of this group one can then determine the composition of the copolymer (for example, chlorine with vinyl chloride; or the ester group with acrylic ester or vinyl esters; nitrogen with acrylonitrile or vinyl pyrrolidone; oxygen with vinyl ethers; or a double bond with butadiene). With copolymers of two pure hydrocarbons, such as isobutylene and styrene, or two acrylic esters, the analysis becomes much more complicated. In such cases infrared spectroscopy is useful, or one has to provide one of the monomers with an easily determined atom which, however, does not influence the polymerization behavior of that particular monomer, e.g., C_{14}. If one plots the result of such experiments in the form of the composition of the copolymer $d[M_1]/d[M_2]$ versus $[M_1]/[M_2]$, or better, the weight percent or molar percent of M_1 in the monomer mixture as a function of the percent of M_1 in the copolymer, one obtains characteristic curves whose form is completely analogous to the curves that one obtains for the distillation of binary systems.

Figure 95 shows diagrams for two technically very interesting systems. The dashed curve corresponds to an ideal copolymerization. The curves without inversion correspond to (as we shall see) monomer pairs where one of the monomers reacts more rapidly with its own radical than with the other radical, whereas the second monomer adds more rapidly to the different radical. Inversion curves, such as for the system styrene/acrylonitrile, are typical for monomer pairs which tend to alternating addition. If one obtains curves whose inversions are opposite to those shown in Figure 95, then one has systems where each monomer prefers homopolymerization, and therefore the polymers obtained would be segment- or block-copolymers. (No clearly proven cases are known, see p. 167).

Kinetics of Copolymerization

The interpretation of such curves becomes possible through the following simple considerations. The monomer M_1 is used up by the reactions 1 and 4 (p. 94). According to the equation for bimolecular reactions, one can write for the rate at which monomer M_1 is used up:

$$-d[M_1]/dt = k_{11}[R \mathord{\wedge\mkern-6mu\wedge\mkern-6mu\wedge} M_1 \cdot][M_1] + k_{21}[R \mathord{\wedge\mkern-6mu\wedge\mkern-6mu\wedge} M_2 \cdot][M_1]. \qquad (33)$$

Correspondingly, for the rate of disappearance of monomer M_2 we can combine reactions 2 and 3 to give Equation (34):

$$-d[M_2]/dt = k_{12}[R \mathord{\wedge\mkern-6mu\wedge\mkern-6mu\wedge} M_1 \cdot][M_2] + k_{22}[R \mathord{\wedge\mkern-6mu\wedge\mkern-6mu\wedge} M_2 \cdot][M_2] \qquad (34)$$

As has been explained on page 75, the radical concentration increases at the beginning of the polymerization because of the decomposition of the initiator, until the termination reaction, whose velocity increases with the square of the radical concentration, reaches the same rate as the initiation reaction (i.e., until just as many radicals disappear per unit time as new ones are formed). With copolymerization (we are considering here only binary systems), there are two types of radicals—$R\wwwM_1\cdot$ and $R\wwwM_2\cdot$—such that the formation reaction of the one radical is at the same time the disappearance reaction for the other. $R\wwwM_1\cdot$ is formed through reaction 4 and disappears through reaction 2. Obviously $R\wwwM_1\cdot$ radicals are formed and disappear also through initiation and termination reactions, but the rate of these reactions, compared to the rate of reactions 2 and 4, is so small that it can be disregarded.[8] As is generally true for the relationship between initiation and termination, we can also say for reactions 2 and 4 that the rate of reaction 2 (disappearance of $R\wwwM_1\cdot$ radicals) is determined through 4 and vice versa, that the concentration of $R\wwwM_1\cdot$ radicals grows according to reaction 4 until the rate of reaction 2 has become equal to that of reaction 4 and therefore just as many radicals $R\wwwM_1\cdot$ disappear per unit time as new ones are formed.

$$k_{21}[R\wwwM_2\cdot][M_1] = k_{12}[R\wwwM_1\cdot][M_2]$$

or

$$[R\wwwM_2\cdot] = \frac{k_{12}}{k_{21}} \cdot \frac{[R\wwwM_1\cdot][M_2]}{[M_1]} \qquad (35)$$

If one then substitutes this expression for $[R\wwwM_2\cdot]$ (with $[R\wwwM_1\cdot]$ it would work just as well) in Equation (33) and (34), one obtains:

$$-d[M_1]/dt = k_{11}[R\wwwM_1\cdot][M_1] + k_{12}[R\wwwM_1\cdot][M_2]$$

and

$$-d[M_2]/dt = k_{12}[R\wwwM_1\cdot][M_2] + k_{22}\frac{k_{12}}{k_{21}}\frac{[R\wwwM_1\cdot][M_2]^2}{[M_1]}$$

$$= k_{12}[R\wwwM_1\cdot][M_2]\left(1 + \frac{k_{22}}{k_{21}}\frac{[M_2]}{[M_1]}\right)$$

If one divides the first equation with the second, one obtains:

$$\frac{-d[M_1]/dt}{-d[M_2]/dt} = \frac{d[M_1]}{d[M_2]} = \frac{k_{11}[M_1]}{k_{12}[M_2]} \cdot \frac{1}{1 + \dfrac{k_{22}[M_2]}{k_{21}[M_1]}} + \frac{1}{1 + \dfrac{k_{22}[M_2]}{k_{21}[M_1]}}$$

The ratios k_{11}/k_{12} and k_{22}/k_{21} are called the copolymerization parameters, and the symbols used for them are r_1 and r_2:

$$r_1 = k_{11}/k_{12} \text{ and } r_2 = k_{22}/k_{21}.$$

[8]If k_{21} is extremely small and/or $[M_1]$ is very small, more radicals $(R\wwwM_1\cdot)$ may be formed by decomposition of the initiator than through reaction 4. In that case Equations (33)-(38) no longer apply.

After introduction of these parameters one obtains:

$$\frac{d[M_1]}{d[M_2]} = \frac{r_1[M_1]/[M_2]}{1+(r_2[M_2]/[M_1])} + \frac{1}{1+(r_2[M_2]/[M_1])}$$

$$\frac{d[M_1]}{d[M_2]} = \frac{(r_1[M_1]/[M_2])+1}{(r_2[M_2]/[M_1])+1} \qquad \text{(Mayo and Lewis)}, \qquad (36)$$

where $d[M_1]/d[M_2]$ is the ratio of the rates with which the two monomers $[M_1]$ and $[M_2]$ disappear from the monomer mixture through addition to the growing polymer chains. In the same ratio these monomers will appear as structural units in the chains of the copolymer. Therefore, $d[M_1]/d[M_2]$ is equal to the molar concentration ratio of the structural units in the copolymer. If one designates this ratio with the letter b and the molar concentration ratio of the monomers in the monomer mixture with the letter a, then:

$$a = \frac{[M_1]}{[M_2]} \quad \text{and,} \quad b = \frac{-d[M_1]/dt}{-d[M_2]/dt} = \frac{d[M_1]}{d[M_2]} \quad \text{or} \quad \frac{m_1}{m_2},$$

and one obtains the copolymerization Equation (36) in the following simple form:

$$b = \frac{r_1 a + 1}{(r_2/a)+1} \qquad (36a)$$

or, if one factors out $1/a$, one obtains:

$$b = a\,\frac{r_1 a + 1}{a + r_2} \qquad (36b)$$

If, in this equation a is replaced by $[M_1]/[M_2]$ and in the numerator 1 is replaced by $[M_2]/[M_2]$, and finally $1/[M_2]$ is factored out, one obtains Equation (36) in a form often found in the literature:

$$\frac{d[M_1]}{d[M_2]} = \frac{m_1}{m_2} = \frac{[M_1]}{[M_2]} \cdot \frac{r_1[M_1]+[M_2]}{r_2[M_2]+[M_1]} \qquad (36c)$$

Equation (36) describes the composition of the copolymer as a function of the composition of the monomer mixture. One has to remember that a, and therefore also b, usually changes continuously during a copolymerization (exceptions: ideal copolymerization and those carried out at the azeotrope point, compare p. 104).

In graphical representations of copolymerization reactions one usually plots the concentration of M_1 in the monomer mixture versus the concentration of M_1 in the polymer chain. As a measure of concentration one uses weight percent or mole percent, and thus the sum of the concentrations becomes 100.:

$$[M_1] + [M_2] = 100 \text{ and } d[M_1] + d[M_2] = 100.$$

For b one can write:

$$b = d[M_1]/(100 - d[M_1])$$

This gives the following expression for the molar concentration of structural units of the monomer M_1 in the copolymer:

$$d[M_1] = b(100 - d[M_1]) = 100 \cdot b/(1 + b) \text{ mole } \% \qquad (37)$$

If one substitutes Equation (36a) in Equation (37) one obtains:

$$d[M_1] = (r_1 a + 1)\, 100/(r_2/a + r_1 a + 2). \qquad (38)$$

The copolymerization equation derived in the preceding paragraphs [Equations (36) and (38)] has demonstrated its general validity. (One assumes, however, that the kinetics underlying this derivation hold true, but this may not always be the case. For example, equilibrium polymerization follows different kinetics, such as the anionic polymerization of α-methyl-styrene). The parameters r_1 and r_2 are therefore universally used for the characterization of monomer pairs with regard to their behavior in copolymerization (compare Table 107). Depending upon the magnitude of r_1 and r_2, Equation (38) yields curves similar to those shown in Figures 95a or 95b, which correspond closely to the experimentally determined curves.

Determination of the Parameters r_1 and r_2

The parameters r_1 and r_2 can be estimated fairly accurately after a sufficient amount of practice with copolymerization curves from the shape of the curve. Conversely, from the values of r_1 and r_2 one can sketch the behavior of the copolymerization curve. To determine the values for r_1 and r_2 from the experimentally determined b values, one has to try to obtain the copolymerization equation, Equation (36a), as a linear function. This is done by solving Equation (36a) for r_2 or r_1 (according to Lewis and Mayo) or, by solving Equation (36a) for a-(a/b) (according to Fineman and Ross). The solution for r_2 yields:

$$r_2 = a\,\frac{r_1 a + 1}{b} - a = r_1\,\frac{a^2}{b} + \left(\frac{a}{b} - a\right),$$

If one then substitutes different values determined from the experimental curve for a and b, one obtains a series of straight lines corresponding to different values of r_2

as a function of r_1. By substituting increasing values for r_1 (for example, from 0 to 1)[9] and calculating r_2 according to Equation (39), one obtains a family of straight lines which all intersect at a certain point. At this point all a and b values, taken from the experimental curves have the same r_1 and r_2 values. Therefore, the resulting r_1 and r_2 values are those which one has to substitute in Equations (36), (37), or (38) to obtain curves identical to those determined experimentally. Figure 100 represents such a graphical r_1-r_2 determination for the monomer pair styrene/methylmethacrylate. The solution of Equation (36a), or (39) for a-(a/b) gives:

$$a\text{-}(a/b) = r_1(a^2/b)\text{-}r_2 \qquad (39a)$$

If one now plots the ordinate a-(a/b) against a^2/b as the abscissa, as is shown in Figure 102, one obtains (-r_2) as the intercept at the ordinate and r_1 as tangent δ. Figure 102, shows as an example the determination of the r values for the monomer pair, acrylic acid (1)-acrylonitrile (2). The Fineman-Ross procedure is simpler because one only has to calculate a single straight line by substituting the experimentally determined (a/b)-ratios directly in Equation (39a).

A different way of determination of r_1 and r_2 has been proposed and tested by V. Jaacks. This method uses the integrated copolymerization Equation (36) after reducing it to a very simple form. This is possible if a copolymerization is carried out with a large excess of M_1 so that the ratio $[M_1]/[M_2]$ becomes very large and,

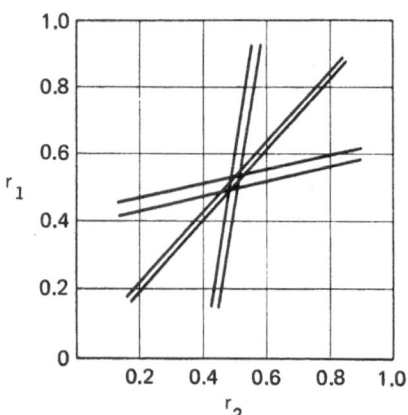

FIG. 100 — Graphical determination of r_1 and r_2 according to Equ. (39) for the system styrene/methylmethacrylate (M_1 = styrene; M_2 = methylmethacrylate) (after LEWIS and MAYO).

[9]From the shape of the copolymerization curves it is possible to select the correct range of r_1 values (compare Figures 103, 104, 106, 112 and 113).

correspondingly, $[M_2]/[M_1]$ becomes very small. In this case we can neglect the 1 in the numerator of Equation (36), and the denominator of Equation (36) becomes equal to 1:

$$\frac{d[M_1]}{d[M_2]} = \frac{r_1\,([M_1]/[M_2]) + 1}{r_2\,([M_2]/[M_1]) + 1} \tag{36}$$

$$[M_1]/[M_2] \gg 1$$

$$[M_2]/[M_1] \ll 1$$

$$\frac{d[M_1]}{d[M_2]} = r_1[M_1]/[M_2] \quad \text{or} \quad \frac{d[M_1]}{[M_1]} = r_1 \cdot d[M_2]/[M_2] \tag{36d}$$

That is, with a sufficient excess of M_1, the composition of the resulting copolymer is determined only by r_1 and is independent of r_2.

By integration of Equation (36d) between the limits $[M_1]_o$ and $[M_2]_o$ at the time before starting the copolymerization, and $[M_1]_t$ and $[M_2]_t$ at the time t after the start of the copolymerization, one obtains:

$$\lg \frac{[M_1]_t}{[M_1]_o} = r_1 \cdot \lg \frac{[M_2]_t}{[M_2]_o} \tag{36e}$$

Equation (36e) describes the course of the whole copolymerization up to high conversions. Therefore, r_1 can be calculated by determination of the monomer ratio $[M_1]_t/[M_1]_o$ at any time after starting the copolymerization; the determination can be done in many cases with sufficient precision by gas chromatography. ($[M_2]_t$ is known when $[M_1]_t$ has been determined, because $[M_1]_t + [M_2]_t = 100$.) For a more precise determination of r_1, the monomer concentration $[M_1]_t$ is determined at different conversions. If $\lg [M_1]_t/[M_1]_o$ is plotted against $\lg [M_2]_t/[M_2]_o$, one obtains r_1 as the slope of the resulting straight line. In the same way, r_2 is determined through a copolymerization using a large excess of M_2. One cannot say that any one of the described methods is the best, but one has to decide which method is most suitable for a particular case. With time it seems that the Fineman-Ross procedure is often preferred. The Jaacks method has the advantage that it is not necessary to isolate and analyze a series of copolymers with different $[M_1]/[M_2]$ ratios. This is especially important, if there is no characteristic group, which can easily be determined quantitatively, in one of

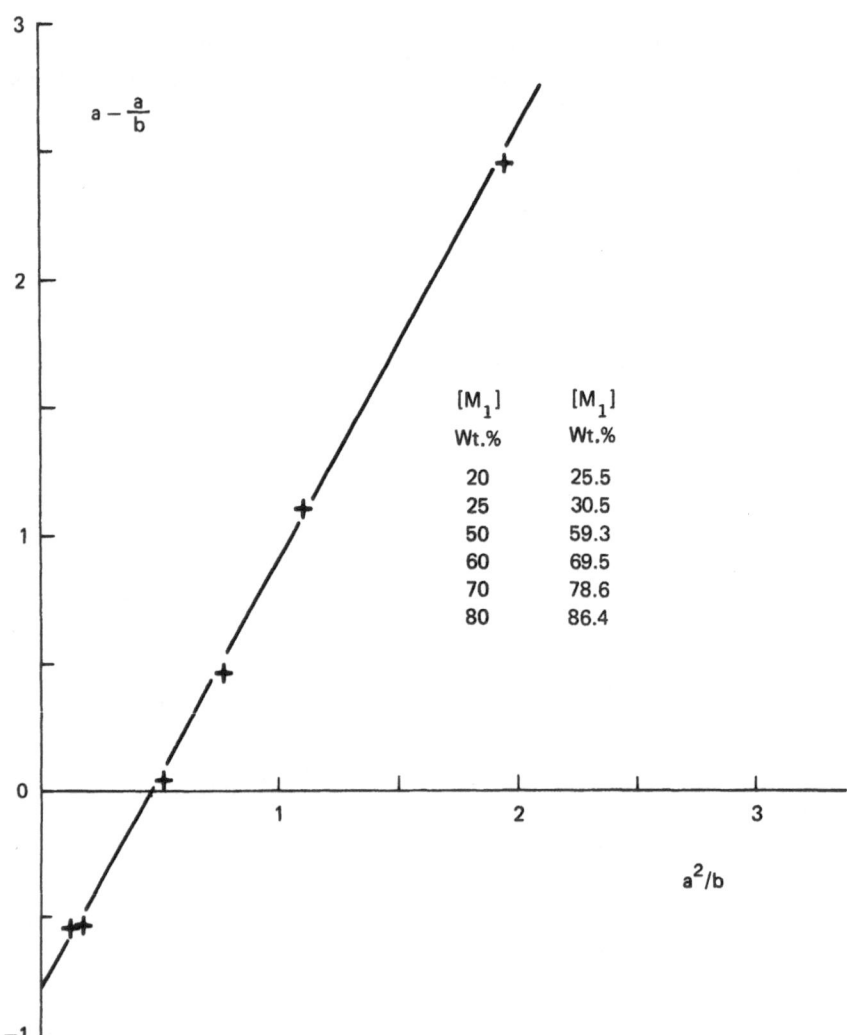

FIG. 102 – Graphical determination of r_1 and r_2 according to FINEMAN and ROSS for the system acrylic acid/acrylonitrile.

the monomers. A further advantage becomes evident with ternary (and higher) copolymerization. Here, it is not necessary to carry out the copolymerizations two by two, but they can be determined by three ternary copolymerizations with a large excess of M_1, or M_2, or M_3.

In the following section we will discuss the most characteristic polymerization diagrams in detail.

Diagrams With Inflection Point Curves

The curves shown in Figure 103 are characteristic for cases where both r_1 and r_2 are smaller than 1. If one remembers that $r_1 = k_{11}/k_{12}$ and $r_2 = k_{22}/k_{21}$, one

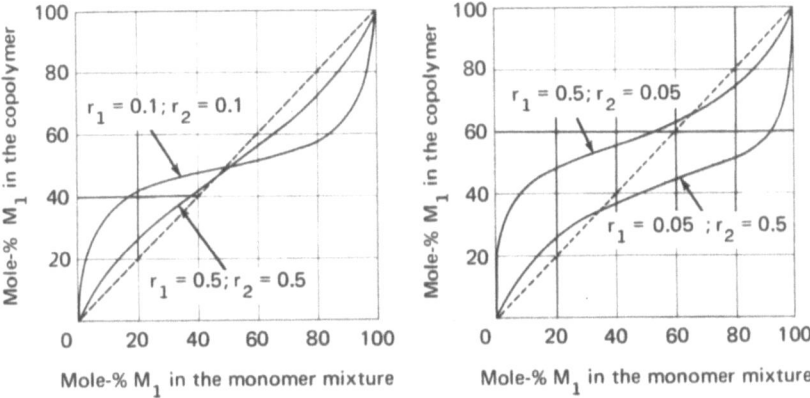

FIG. 103 — Copolymerization curves with inflection-point, $r_1 < 1$ and $r_2 < 1$. - - - - - - $= r_1 = 1$ and $r_2 = 1$.

immediately realizes the kinetic explanation for this behavior, which is that the reaction rate constants k_{12} and k_{21} are larger than k_{11} and k_{22} (i.e., that the addition of the opposite monomer to the radicals $R \leadsto M_1 \cdot$ and $R \leadsto M_2 \cdot$ occurs more rapidly than the addition of its own respective monomer). Furthermore, the aversion of the radicals to add their own monomers is the greater, the smaller the corresponding r_1 or r_2 value. If the parameters r_1 and r_2 are equal and if the molar concentration of the monomers is identical, then the addition of M_2 to $R \leadsto M_1 \cdot$ occurs more frequently than the addition of M_1 to $R \leadsto M_1 \cdot$ by the same amount as the addition of M_1 to $R \leadsto M_2 \cdot$ occurs more frequently than the addition of M_2 to $R \leadsto M_2 \cdot$. The inflection point lies at 50 mol% M_1, and the copolymer molecules contain on the average as many M_1 units as M_2 units. If the parameters r_1 and r_2 are not only equal but also much smaller than 1, then one obtains copolymers with regular alteration of M_1 and M_2 in the chain. The smaller the parameters r_1 and r_2, the more strictly this alternation of M_1 and M_2 in the polymer chain is observed (the larger k_{12} and k_{21} are compared to k_{11} and k_{22}, respectively). On the other hand, the smaller the difference between the parameters r_1 and r_2 and 1, the more frequently one finds that M_1 adds to $\leadsto M_1 \cdot$ and that M_2 adds to $\leadsto M_2 \cdot$ in addition to alternation. If only r_2 is much smaller than 1, but r_1 is close to 1 (for example, styrene-acrylonitrile: $r_1 = k_{11}/k_{12} = 1/2.5$ and $r_2 = k_{22}/k_{21} = 1/33$), then one finds that $\leadsto M_2 \cdot + M_1$-addition and $\leadsto M_1 \cdot + M_2$-addition are preferred to $\leadsto M_2 \cdot + M_2$- and $\leadsto M_1 \cdot + M_1$-addition, respectively. Of these two the $\leadsto M_1 \cdot + M_2$-addition is not as strongly preferred as the $\leadsto M_2 \cdot + M_1$-addition. This means that one will find a number of sections in the copolymer molecule where two or more styrene residues succeed each other. There will be fewer cases where two or more acrylonitrile residues follow each other in the chain. This behavior shows up in the curve in the following way: the inflection point is no longer at 50 mol% but at higher or lower M_1 concentrations (if r_1 is greater than r_2, it is found at higher M_1 concentrations; if r_1 is smaller than r_2 it is found at lower M_1 concentrations).

The inflection point of the curve gives the composition at which the resulting copolymer has the same composition as the monomer mixture. One calls this (just as in the distillation of binary mixtures) an azeotropic composition, or azeotropic concentration, and one can even speak of an azeotropic copolymer.

The azeotropic monomer ratio can be determined immediately by substitution of r_1 and r_2 in the following equation:

$$[M_1]/[M_2] = (r_2 - 1)/(r_1 - 1). \tag{39b}$$

In the azeotrope, namely, the concentration ratio a of the monomers is equal to the concentration ratio b of the monomer units in the copolymer chain. Therefore, if one substitutes $a = b$ in Equation (36b), one obtains:

$$(r_1 a + 1)/(a + r_2) = 1$$

Solving for a, this leads to Equation (39a). For industrial purposes the knowledge of this azeotropic concentration is of great interest, because at this ratio the composition of the monomer mixture and of the copolymer does not change during the polymerization reaction. Therefore one can continue the polymerization to very high conversions (98-99%), without the possibility of the polymer's becoming inhomogeneous. On the other hand, if one polymerizes a monomer mixture whose M_1 concentration is above or below azeotropic composition, then one first obtains a copolymer whose composition is nearer to that of the azeotrope than that of the monomer mixture, and, with strongly curved diagrams (r_1 and r_2 much smaller than 1), one even obtains copolymers that are identical with the azeotropic

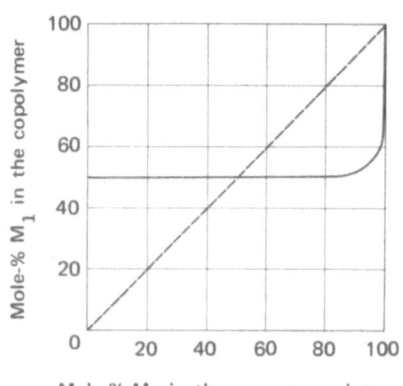

FIG. 104 — Copolymerization diagram for the system styrene/ maleicanhydride ($r_1 \approx 0.0095$, $r_2 \approx 0$)

composition over a wide range of mixtures. As one can easily demonstrate by means of the copolymerization curves, the monomer mixture rapidly becomes poorer in that monomer, which was present in lower concentration, and one soon finds that only the other monomer is left, which then polymerizes, if one continues the polymerization, to a homopolymer. One therefore obtains copolymer that corresponds mainly to the azeotropic composition and, in addition, a homopolymer of M_1 or of M_2. Since with few exceptions different polymers are not miscible, one obtains inhomogeneous mixtures which have poor cohesion and are therefore of little technical importance. Such mixtures result, for example, from the polymerization of styrene-acrylonitrile mixtures which contain less than 65% styrene (azeotropic composition approximately 65:35; compare Figure 95).

Frequently, one of the monomers does not polymerize by itself, or only very slowly, and then r_1 or r_2 becomes O. In such cases the polymerization stops as soon as the other monomer has been used up by azeotropic copolymerization. This is found to be the case with mixtures of fumaric acid esters and isobutylene, or mixtures of styrene and maleic anhydride. Excess isobutylene, in the one case, and maleic anhydride in the other example, remain as unconverted monomers. If more than 50 mol% styrene is present in the monomer mixture, then one obtains homopolystyrene in addition to the $M_1 M_2$ copolymers. Figure 104 shows the copolymerization diagram for the system styrene/maleic anhydride ($r_1 = 0.0095$ and $r_2 \cong 0$), and Table 105 shows the calculation of the curves according to Equation (38).

In this case one finds a degenerate inflection curve where the inflection point is lengthened to an inflexion tangent parallel to the x-axis. The curve shows that over a wide range one obtains copolymer with the same composition of styrene and

TABLE 105. Calculation of the copolymerization curve for the system styrene/maleicanhydride from the parameters $r_1 = 0.0095$, and $r_2 = 0$ according to Equ. (38). (All concentrations in mole-%)

Monomer mixture			Copolymer	
$[M_1]$ Styrene	$[M_2]$ (Maleic- anhydride)	$a = \dfrac{[M_1]}{[M_2]}$	$\dfrac{(r_1 a + 1) \cdot 100}{\dfrac{r_2}{a} + r_1 a + 2}$	$d[M_1]$
1	99	0.01	100.01/2.0001	50
5	95	0.05	100.05/2.0005	50
10	90	0.1	100.1 /2.001	50
25	75	0.3	100.3 /2.003	50
50	50	1	101/2.01	50.3
75	25	3	103/2.03	50.4
90	10	9	109/2.09	52.2
95	5	19	119/2.19	53.4
97	3	32	132/2.32	57.0
99	1	99	199/2.99	66.6

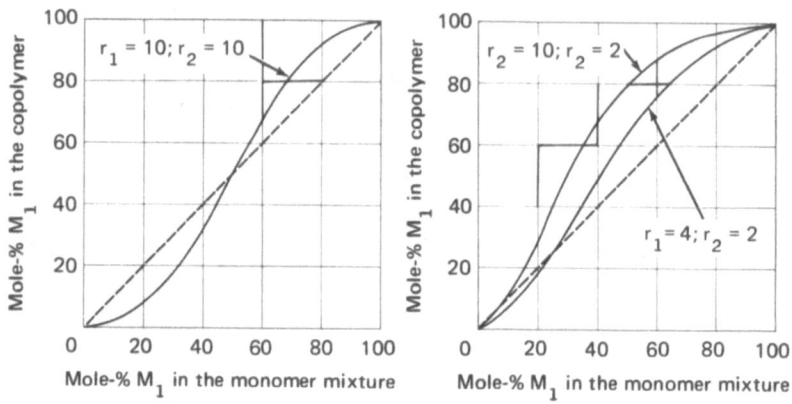

FIG. 106 — Copolymerization diagrams for $r_1 > 1$ and $r_2 > 1$

maleic anhydride independent of the composition of the monomer mixture. This is the result of the strong alternating tendency[10] of the system (k_{12} much greater than k_{11}, and k_{21} much greater than k_{22}).

If *both* parameters are equal to 0, then one obtains a straight line parallel to the x-axis at $d[M_1]$-concentration = 50%.

This is true for monomer pairs where each of the two monomers is unable to polymerize by itself (i.e., where k_{11} and k_{22} are equal to zero). The chains of the copolymers resulting from such systems have a completely regularly alternating sequence of the structural units M_1 and M_2, regardless of the composition of the monomer mixture. The value of systems where $r_1 = 0$ and $r_2 = 0$ for the preparation of copolymers with alternating structure depends entirely on the absolute magnitude of the corresponding reaction rate constants k_{12} and k_{21}. Only if the numerical values of k_{12} and k_{21} are of the order of 100 to 1,000 liters·mol^{-1}sec^{-1} (comparable with the k_p values for styrene, methylmethacrylate, and vinylacetate) is it possible to produce copolymers with high molecular weight. Thus, while listed values of $r_1 = 0$ and $r_2 = 0$ mean that one *cannot* prepare *homo*polymers from these monomers, the values *do not* mean that copolymers *can* be prepared since the values of k_{12} and k_{21} may be too small.

It also happens that one of the monomers (for example, maleic anhydride) is not polymerizable in the usual laboratory experiment, and that k_{11} is not exactly 0, but is very small. In that case r_1 is only approximately 0 if k_{12} is larger by several powers of 10. However, if k_{12} is only slightly larger than k_{11} (10-100 times), then r_1 is not 0, but 0.1 to 0.01. A typical example of this kind is the combination maleicanhydride/stilbene, which has the values $r_1 = 0.03$ and $r_2 = 0.03$. Neither maleicanhydride nor stilbene forms homopolymers with a high molecular weight. In

---

[10]The condition for the formation of an alternating copolymer chain structure, namely, that $r_1 \cong r_2$ is always met with very small r_1 and r_2 values ($r_1 \ll 1$ and $r_2 \ll 1$). Therefore, in the example of styrene/maleic anhydride, where $r_1 = 0.0095$ and $r_2 = 0$, we can write that $r_1 \cong 0$.

spite of that, the r values are (within experimental error) different from zero, even though the copolymerization steps occur quite slowly in contrast to the extremely rapid copolymerization of maleicanhydride-styrene (r_1 = 0 and r_2 = 0.01). The conclusion "r_1 = 0 and r_2 = 0, therefore the monomers of this pair cannot be homopolymerized" is not reversible, i.e., actual values for r_1 and r_2 do not imply that the monomers are capable of homopolymerization. r_1 = 1 and r_2 = 1 simply means that the copolymerization steps do not occur more rapidly than the corresponding homopolymerization steps. With pairs of monomers which do not homopolymerize, or only homopolymerize slowly, the fact that r_1 and r_2 have finite values simply means that these monomers do not homopolymerize or copolymerize with any preparatively useful rate. Of the almost 1,000 monomer pairs whose parameters have been determined, a considerable fraction belong to this category.

As is discussed later on (see p. 111) the behavior of monomer pairs without azeotrope can readily be explained, expecially of those with symmetrical copolymerization diagrams. A satisfactory explanation can also be given for the behavior of monomer pairs which give a copolymerization curve with an inflection point, such as in Figure 103.

The same considerations lead to the conclusion that monomer pairs for which $r_1>1$ and $r_2>1$ do not exist. If one looks over the known r values (for example, see *Polymer Handbook,* Wiley-Interscience, 1966), there are a few monomer pairs, listed at the end of Table 107, which formally belong to this group. However, it is not clear whether this is due to measurement errors, or if one is here dealing with still unknown effects which determine the rate of the different addition steps. It is striking, however, that systems with $r_1>1$ and $r_2>1$ are more common among the ionic copolymerizations. Monomer pairs with high r values, as will be discussed subsequently, have the tendency to form long sequences. Very large values of $r_1>1$ and $r_2>1$ would therefore indicate that such a system (if it exists) would lead to the synthesis of block copolymers or—in extreme cases— that the two homopolymers (poly-M_1 and poly-M_2) would be formed side-by-side.

TABLE 107. Copolymerization Parameters*

M_1	M_2	r_1	r_2	Temp.
Both Parameters = 1	(ideal Monomer pairs)			
Methylmethacrylate	Vinylidenechloride	1.0	1.0	60°
Vinylacetate	Isopropenylacetate	1.0	1.0	75°
Acrylonitrile	Glycidylacrylate	1.0	1.0	60°
Butylacrylate	Acrylonitrile	1.0	0.92	56°E**

* For a comprehensive tabulation see BRANDRUP-IMMERGUT, Polymer Handbook, Interscience Publishers 1966.

** E = Emulsion/polymerization

TABLE 107. (Cont'd)

M_1	M_2	r_1	r_2	Temp.
Ethylene	Vinylacetate	1.07	1.08	90° 1 000 atm
Isoprene	Butadiene	1.06	0.94	-18°E
Styrene	p-Methoxystyrene	1.13	0.93	60°
Acrylonitrile	Methylacrylate	0.84	0.83	65°
Butylacrylate	Vinylidenechloride	0.83	0.88	50°
Chlorotrifluoroethylene	Vinylfluoride	1.20	0.80	80° E
Styrene	p-Methylstyrene	0.83	0.96	63°

$r = 1, r_2 < 1$

M_1	M_2	r_1	r_2	Temp.
Butadiene	Butylacrylate	1.0	0.1	5° E
Butadiene	p-Chlorstyrene	1.0	0.42	50° E
m-Chlorostyrene	Styrene	1.0	0.60	60°
p-Chlorostyrene	Styrene	1.03	0.74	60°
Glycidylmethacrylate	Methylmethacrylate	1.05	0.80	60°
2-Vinylpyridine	Styrene	1.14	0.55	60°
p-Nitrostyrene	Styrene	1.15	0.19	60°
Acrylonitrile	Vinylidenechloride	1.20	0.49	- -
Methylmethacrylate	Acrylonitrile	1.20	0.15	60°
Butylmethacrylate	Vinylacetate	1.20	0.15	60°
Glycidylmethacrylate	Acrylonitrile	1.32	0.14	60°
Acrylic acid	N-Vinylpyrrolidone	1.30	0.15	75°
Butadiene	Styrene	1.40	0.40	5°
Acrylic acid	Methylmethacrylate	1.51	0.48	45°
2-Fluorobutadiene	Styrene	1.55	0.50	50°
Methacrylic acid	Methacrylonitrile	1.60	0.60	65°
Methylvinylketone	Acrylonitrile	1.80	0.60	60°
Vinylchloride	Vinylacetate	1.80	0.60	40°
Butadiene	Vinylidenechloride	1.90	0.05	5° E
Isoprene	Styrene	1.88	0.66	22° (γ-Rays)
Isoprene	Styrene	1.92	0.54	80°
Methacrylic acid	Methacrylamide	2.0	0.3	70°
Styrene	Vinylidenechloride	2.0	0.14	60°
Vinylchloride	Isobutylene	2.05	0.08	60°
Butadiene	Diethylfumarate	2.13	0.25	- -
Acrylic acid	Methylacrylate	2.3	0.35	45°
Methylmethacrylate	Methylacrylate	2.3	0.47	130°
Methacrylonitrile	Acrylonitrile	2.7	0.3	60°
Acrylonitrile	Vinylchloride	2.8	0.04	40°
Acrylonitrile	Vinylformiate	3.0	0.04	60°
2-Chlorobutadiene	Butadiene	3.41	0.06	50°
Methylacrylate	Vinylchloride	4.4	0.12	50°
Butylacrylate	Vinylchloride	4.4	0.07	45°
Vinylidenechloride	Vinylchloride	4.5	0.2	50°
4-Vinylpyridine	Butylacrylate	5.15	0.46	- -
Styrene	N-Vinylcarbazole	5.7	0.035	75°
Acrylonitrile	Vinylacetate	6.0	0.07	70°

TABLE 107. (Cont'd)

M_1	M_2	r_1	r_2	Temp.
Butylacrylate	Allylchloride	6.0	0.1	$60°$
2-Chlorobutadiene	Acrylonitrile	6.9	0.034	$50°$ E
Vinylidenechloride	Vinylacetate	7.0	0.1	$50°$
Methylvinylketone	Vinylacetate	7.0	0.05	$70°$
Vinylchloride	Vinyllaurate	7.4	0.2	- -
Styrene	Vinylisocyanate	8.13	0.08	$60°$
Butadiene	Vinylchloride	8.8	0.035	$50°$
Methylacrylate	Vinylacetate	9.0	0.1	$60°$
Acrylic acid	Vinylacetate	10.0	0.01	$70°$
Methylvinylsulfide	Vinylenecarbonate	10.6	0.5	$60°$
Methacrylonitrile	Vinylacetate	12.0	0.01	$70°$
Styrene	N, N-Divinylaniline	13.0	0.45	- -
Methylmethacrylate	Vinylchloride	15.0	0.02	$45°$
Styrene	N-Vinylpyrrolidone	15.7	0.045	$50°$
Methylmethacrylate	Vinylacetate	20	0.015	$60°$
Methylmethacrylate	Vinylacetate	28	0.035	$30°$
Styrene	Vinyldichloroacetate	20	0.28	$80°$
Styrene	Vinylchloride	35	0.067	$50°$
n-Butylmethacrylate	Acrylonitrile	46	0.058	$60°$
Styrene	Vinylisobutylether	50	0.01	$50°$
2-Chlorobutadiene	Vinylacetate	50	0.01	$65°$
Styrene	Vinylacetate	55	0.01	$60°$
Heptylmethacrylate	Vinylacetate	60	0.27	$60°$
n-Butylvinylether	Vinylacetate	62	0.13	$60°$
n-Butylmethacrylate	Vinylacetate	62	0.12	$60°$
Methylmethacrylate	Vinylenecarbonate	70	0.005	$70°$
2-Chlorobutadiene	Vinylpropionate	70	0.05	$65°$
$r_1 \geqslant 1, r_2 = 0$				
Acrylonitrile	Isobutylene	1.02	0	$60°$
Acrylic acid	2-Chloroallylacetate	1.0	0	$100°$
Vinylchloride	Allylacetate	1.2	0	$40°$
Butadiene	α-Methylstyrene	1.6	0	$12°$
Styrene	Cinnamic acid	1.85	0	- -
2-Chlorobutadiene	Butadiene	2.8	0	$50°$ E
Styrene	2-Chloroallylacetate	4.0	0	$50°$
2-Chlorobutadiene	Styrene	5.2	0	$50°$ E
Styrene	Diethylmaleate	6.52	0	$60°$
Methylmethacrylate	Allylacetate	23	0	$60°$
Methylmethacrylate	Allylchloride	48	0	$60°$
Methylmethacrylate	Triallylcyanurate	50	0	$60°$
Styrene	Triallylcyanurate	80	0	$60°$
Styrene	Ethylvinylether	90	0	$60°$
$r_1 < 1, r_2 < 1$				
Styrene	p-Methylstyrene	0.83	0.96	$63°$
Vinylidenechloride	Butylacrylate	0.88	0.83	$50°$

TABLE 107. (Cont'd)

M_1	M_2	r_1	r_2	Temp.
Butylacrylate	N-Methylolacrylamide	0.87	0.61	- -
Ethylacrylate	Acrylonitrile	0.95	0.44	80°
Styrene	Ethylacrylate	0.80	0.20	70°
Butylmethacrylate	Methacrylonitrile	0.70	0.50	80°
Butylmethacrylate	Styrene	0.64	0.63	50°
Butadiene	Methylacrylate	0.76	0.05	50° E
Styrene	Methylmethacrylate	0.50	0.50	60°
Methacrylonitrile	Methylmethacrylate	0.65	0.67	60°
Styrene	4-Vinylpyridine	0.70	0.54	60°
N-Vinylpyrrolidone	Vinylenecarbonate	0.70	0.40	60°
Styrene	Glycidylacrylate	0.60	0.17	60°
Vinylacetate	Allylacetate	0.60	0.50	60°
Glycidylmethacrylate	Styrene	0.53	0.44	60°
Styrene	Methylacrylate	0.75	0.20	70°
Styrene	Butylacrylate	0.76	0.15	60°
Styrene	Butylacrylate	0.48	0.15	25°
Methacrylic acid	Butadiene	0.53	0.20	50° E
Vinylchloride	N-Vinylpyrrolidone	0.53	0.38	- -
2-Fluorobutadiene	Acrylonitrile	0.60	0.07	50°
Butadiene	Methylmethacrylate	0.53	0.06	5° E
N-Vinylpyrrolidone	Vinylacetate	0.44	0.38	70°
Glycidylmethacrylate	Styrene	0.44	0.35	60°
Diethylfumarate	Vinylchloride	0.47	0.12	60°
Diethylfumarate	Vinylacetate	0.44	0.01	60°
Styrene	Acrylonitrile	0.41	0.03	75°
Methacrylonitrile	α-Methylstyrene	0.35	0.12	80°
Vinylidenechloride	Butylmethacrylate	0.35	0.22	70°
Acrylic acid	Styrene	0.35	0.22	70°
Butadiene	Methacrylonitrile	0.36	0.04	5° E
Butadiene	Acrylonitrile	0.35	0.05	50° E
Styrene	Itaconicanhydride	0.30	0.20	70°
Styrene	Methacrylonitrile	0.25	0.25	80°
Styrene	Diethylfumarate	0.30	0.07	60°
Ethylacrylate	2-Vinylpyridine	0.20	0.20	75°
Isoprene	Acrylonitrile	0.30	0.05	50°
Styrene	Acrylicanhydride	0.17	0.10	35°
Vinylacetate	Maleicanhydride	0.055	0.003	75°
Isopropenylacetate	Maleicanhydride	0.032	0.002	75°
Vinylchloride	Maleicanhydride	0.3	0.008	75°
Stilbene	Maleicanhydride	0.03	0.03	60°

$r_1 < 1, r_2 = 0$

Styrene	β-Nitrostyrene	0.4	0	80°
Styrene	Fumarodinitrile	0.2	0	60°
Vinylacetate	Fumarylchloride	0.14	0	70°
Styrene	Maleianhydride	0.01	0	60°

TABLE 107. (Cont'd)

M_1	M_2	r_1	r_2	Temp.
$r_1 = 0, r_2 = 0$				
Maleicanhydride	2-Chlorallylacetate	0	0	$120°$
Diethylfumarate	Isobutylene	0	0	$70°$
Poly-(ethyleneglycol-fumarate)	n-Amylvinylether	0	0	$60°$
$r_1 > 1, r_2 > 1$				
Acrylonitrile	N-Octadecylacrylamide	1.10 ± 0.035	1.44 ± 0.019	$60°$
N-Methylolacrylamide	Methylacrylate	1.9 ± 0.7	1.3 ± 0.7	- -
Isoprene	Styrene	2.05	1.38	$50°$
Allylbenzoate	Allylchloride	2.5	1.25	$60°$
Acrylonitrile	Dodecylacrylate	3.2 ± 0.5	1.3 ± 0.1	$60°$
Methylvinylketone	α(2-Cyanoethyl)-acrylonitrile	5.05 ± 0.95	1.24 ± 0.75	- -
Ethylacrylate	Na-acrylate	5.7	1.5	$50°$

Ionic and Metal – Complex-Copolymerization

M_1	M_2	r_1	r_2	Temp.	
Isoprene	Butadiene	1.0	1.0	$-20°$	AlR_2Cl/Co-Complex
p-Chlorostyrene	Isobutylene	1.0	1.0	0	$AlBr_3$ in Hexane
p-Chlorostyrene	Isobutylene	1.2	8.6	0	$SnCl_4$ in Hexane
Ethylene	Propylene	15.7	0.11	$75°$	$TiCl_3/AlR_3$
p-Methoxystyrene	Styrene	100	0.01	$0°$	$SnCl_4$
Styrene	p-Chlorostyrene	2.2	0.45	$0°$	$SnCl_4$ in Nitrobenzene
Isobutylene	Styrene	9.0	2.0	$-90°$	$AlCl_3$ in CH_3Cl
2-Chloroethyl-vinylether	Styrene	36	3		cationic
p-Methoxystyrene	2-Chloroethyl-vinylether	5	45		cationic
Isoprene	Styrene	1.92	0.54	$80°$	radical
Isoprene	Styrene	5.9	0.03	$20°$	with Butyllithium (anionic)
Isoprene	Styrene	1.88	0.66	$22°$	initiated with γ-rays

Diagrams Without Azeotrope

With the other large group of monomer pairs, one of the radicals prefers to add the opposite monomer, whereas the second radical prefers to add its own monomer. Thus either $k_{11}/k_{12} = r_1 > 1$ and $k_{22}/k_{21} = r_2 < 1$, or conversely $k_{11}/k_{12} = r_1 < 1$ and $k_{22}/k_{21} = r_2 > 1$.

Figures 112 and 113 show typical diagrams for such cases. The curves are symmetrical (with regard to the 45°-line), if r_2 is smaller than 1 by the same factor by which r_1 is greater than 1 (or vice versa), i.e., if $r_1 \cdot r_2 = 1$. For such systems in the literature the indication "ideal copolymerization" is mostly used. This, however, is not justified, because those systems are not characterized by any particular theoretical or practical condition. The case that the curves in the diagrams for ($r_1 \cdot r_2 = 1$)-systems are symmetrical with regard to the 45°-line is simply one of many imaginable and real ways in which pairs of monomers can behave during copolymerization. But among all these possibilities, there is only a single solution for which the copolymers and the corresponding monomer mixtures have, over the entire range of mixtures, the same composition, and that is the 45°-line with $r_1 = 1$ and $r_2 = 1$. Therefore, only this case can meaningfully be called "ideal copolymerization."

As long as the r_1 and r_2 values are only slightly different from 1, (for example, $r_1 = 0.8$ and $r_2 = 1.2$), the resulting copolymer has nearly the same composition as that of the monomer mixture for all ratios of the two monomers in the monomer mixture. On the other hand, if the deviations from 1 are considerable (curves deviating very much from the diagonal), then during polymerization one of the two monomers is rapidly enriched. For example, if $r_1 = 5$ and $r_2 = 0.2$, then one has the following situation: if with a 50/50 monomer mixture an M_1^* radical is at the end of the chain, then the addition of another M_1 monomer occurs 5 times as rapidly as the addition of a monomer M_2. Then if a monomer M_2 adds, on the average after 5 M_1-additions, the resulting M_2^* radical also adds a monomer M_1 5 times as rapidly as a monomer M_2 (because $k_{22}/k_{21} = 1/5$). Thus, among 6 M_2^* radicals there is on the average only a single one that adds a monomer M_2. Whatever radical is present at the chain end, monomer M_1 will always be preferred in the ratio of 5 to 1. Consequently, of each 6 structural units in the resulting copolymer, on the average 5 will be M_1 units. Thus, the copolymer obtained from a 50/50 monomer mixture

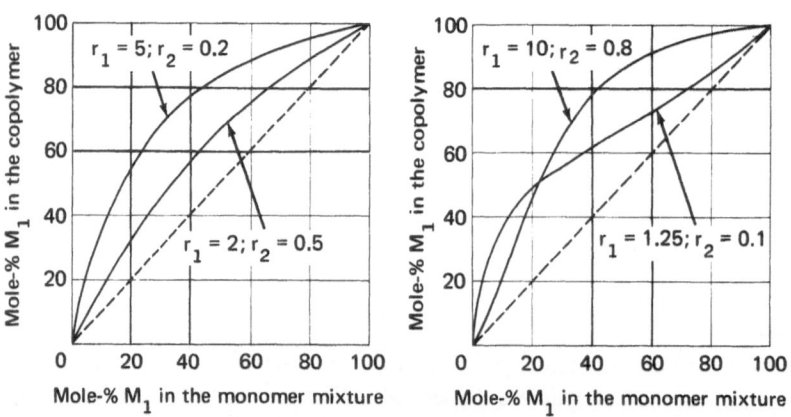

FIG. 112 — Copolymerization diagrams without inflection-point. $r_1 > 1, r_2 < 1$.

with the values $r_1=5$ and $r_2=0.2$ will consist of 83.3% of M_1 units. This can also be demonstrated by observing the resulting copolymerization diagram (see Figure 84/1). The further the parameters deviate from 1, the more the structure of the copolymer approaches that of M_1 homopolymer. With $r_1 = 100$ and $r_2 = 0.01$, a 50/50 monomer mixture would yield a copolymer consisting of 99% of M_1 units. If one allows the polymerization of this mixture to go to high conversion, then the composition of the monomer mixture increases rapidly from 50/50 to 70/30, 90/10, and 95/5 until finally only pure monomer M_2 remains, because monomer M_1 is used up much more rapidly than monomer M_2. Thus, one obtains with increasing M_2 monomer concentration during copolymerization copolymers with increasing M_2 structural units until finally, if the polymerization is continued, a pure homopolymer of M_2 is formed. The result of such a copolymerization is in almost all cases an inhomogeneous mixture of M_1-M_2 copolymer with increasing amounts of M_2 units (from 1-99%) and varying amounts of M_2 homopolymer. In practice, in the case described above, one says that the monomer M_2 does not enter into the chain, and one helps the situation by adding the more slowly used up monomer in high concentration and continuing the addition of the more rapidly used up monomer during copolymerization in such a way that a concentration of the monomer mixture which yields the desired copolymer is maintained. From the diagrams one can in each case determine the experimental conditions required (compare section on "preparative copolymerization", p. 126).

Often one faces the problem of introducing certain functional groups in low concentration into a polymer by means of copolymerization. In such cases it is useful to look at the r_1, r_2 values first and draw the diagram, or, if the r_1, r_2 values are not known, to carry out sufficient copolymerizations to obtain the diagram and to determine the r_1, r_2 values.

Another possibility for copolymerizing two monomers with unsuitable r_1, r_2 values into a copolymer is to use a third monomer. One can usually prepare

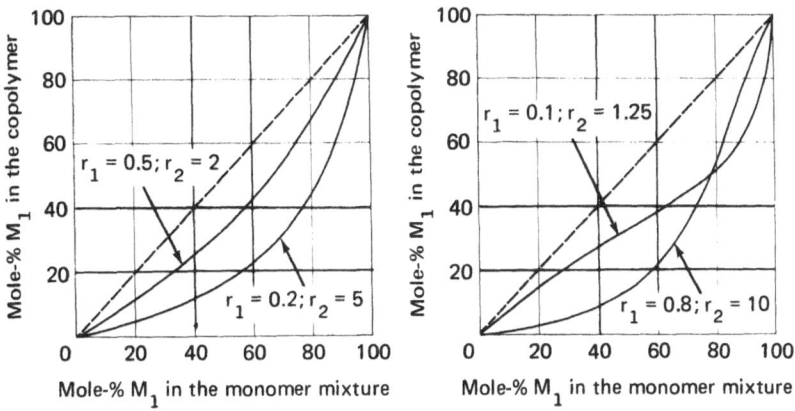

FIG. 113 – Copolymerization diagrams without inflection-point. $r_1 < 1, r_2 > 1$.

copolymers of styrene and vinylcarbazole only with great difficulty. However, by the addition of a small amount of acrylonitrile, it becomes very simple to prepare homogeneous copolymers (terpolymers).

Table 107 shows a number of monomer pairs with their r_1 and r_2 values. They are arranged according to the extent to which they deviate from 1. With the large majority of the monomer pairs the values of r_1 are between 1 and 2 and those of r_2 are between 1 and 0.1. Furthermore, the deviation in general favors values <1. The r_1 and r_2 values in Table 107 are valid for radical polymerization (with the exception of those at the end of the table). Even though the mechanism of ionic polymerization is different from that of radical polymerization, the basic scheme in both cases involves a chain reaction, and ionic copolymerization can be described in the same form and by the same copolymerization equation as radical copolymerization. As the few examples at the end of Table 107 demonstrate, ionic copolymerization parameters usually differ greatly from 1. It is also curious that the combination $r_1>1$ and $r_2>1$ is observed with the cationic rather than the anionic copolymerizations. No explanation for this can be given at this time, however. A series of monomers can be polymerized both by a radical and an ionic mechanism (for example, styrene) or can at least be copolymerized by both mechanisms (for example, vinylethers). The parameters are then different for the same monomer pair, and it is therefore possible to use copolymerization reactions to determine the type of initiation involved (for example, with radiation polymerization).

The accuracy of the numerical r_1 and r_2 values varies greatly depending on the accuracy of the different analytical methods used for the different copolymerizations. (At the end of the book the same monomer pairs are listed once more in alphabetical sequence and with an indication of the error limits. A complete collection of known r_1 and r_2 values can be found in the *Polymer Handbook,* Wiley-Interscience, 1966.)

Copolymerization Equations for Limiting Cases

For the limiting cases discussed previously, the copolymerization equation [Equation (36)] takes on simpler forms which are summarized below:

1. Ideal copolymerization:

Conditions: $r_1 = 1$

  $r_2 = 1$

Equation: $b = \dfrac{r_1 a + 1}{r_2/a + 1} = \dfrac{a + 1}{1/a + 1} = a$

  $m_1 = [M_1]$

2. At the azeotrope point:

   Conditions: $a = b$

   Equation:
   $$b = a = \frac{r_1 a + 1}{(r_2/a) + 1} = \frac{r_1 b + 1}{(r_2/b) + 1}$$

   $$b(r_2/b + 1) = r_1 b + 1$$
   $$r_2 + b = r_1 b + 1$$
   $$r_2 - 1 = r_1 b - b = b(r_1 - 1)$$
   $$b = (r_2 - 1)/(r_1 - 1)$$

3. Strictly alternating copolymerization:

   Conditions: $r_1 = 0$
   $$r_2 = 0$$

   Equation: $b = 1$
   $$m_1 = 50\%$$

4. Symmetrical copolymerization curves
   a) with azeotrope:

   Conditions: $r_1 = r_2 < 1$

   Equation: $$b = \frac{r_1 a + 1}{(r_1/a) + 1}$$

   b) Without azeotrope:

   Conditions: $\dfrac{k_{11}}{k_{12}} = \dfrac{k_{21}}{k_{22}}$ or, $r_1 = \dfrac{1}{r_2}$ or, $r_1 \cdot r_2 = 1$

   Equation:
   $$b = \frac{r_1 a + 1}{(1/r_1 a) + 1} = \frac{r_1 a + 1}{(1/r_1 a)(1 + r_1 a)} = r_1 \cdot a$$

Structure of Copolymers
Mean Sequence Length

The mean number of structural units of M_1 in a sequence is always larger by 1 than the ratio of the number of $(M \cdot_1 + M_1)$-additions to the number of $(M \cdot_1 + M_2)$-additions. It is determined by the ratio between the rates of the two addition steps $(v_{11}/v_{12}) + 1$. For instance, if in the above example v_{11} is 5 times as large as

v_{12}, in an average time interval the two monomers are used up at a rate of 5/1 (with $[M_1]/[M_2] = 1$). Also the monomers are added in that ratio to the growing chain and therefore in a given time t, 5 M_1-monomers and 1 M_2-monomer will add to the radical $\sim\sim\sim M\cdot_1$. This leads to a sequence of $(5/1) + 1 = 6$ M_1 structural units:

$$\sim\sim\sim M_2 - M\cdot_1 + 5\,M_1 + 1\,M_2 \rightarrow$$

$$\sim\sim\sim M_2 - M_1 - M_1 - M_1 - M_1 - M_1 - M_1 - M\cdot_2$$

If $v_{11}/v_{12} = 3/1$, then one obtains sequences of four units:

$$\sim\sim\sim M_2 - M\cdot_1 + 3\,M_1 + 1\,M_2 \rightarrow \sim\sim\sim M_2 - M_1 - M_1 - M_1 - M_1 - M\cdot_2 \quad \text{and if}$$

$v_{11}/v_{12} = 1/1$ one obtains sequences consisting of two units:

$$\sim\sim\sim M_2 - M\cdot_1 + M_1 + M_2 \rightarrow \sim\sim\sim M_2 - M_1 - M_1 - M\cdot_2$$

or in general: the mean sequence length $\bar{l}_1$ (i.e., the average number of M_1 units combined into a sequence) is given by the following equation:

$$\bar{l}_1 = \frac{v_{11}}{v_{12}} + 1 = \frac{k_{11} \cdot [M_1] \cdot [R\sim\sim\sim M\cdot_1]}{k_{12} \cdot [M_2] \cdot [R\sim\sim\sim M\cdot_1]} + 1 = \frac{k_{11}}{k_{12}} \cdot \frac{[M_1]}{[M_2]} + 1 =$$

$$= r_1 \cdot \frac{[M_1]}{[M_2]} + 1 = r_1 a + 1 \tag{39c}$$

Similarly one obtains a relation for the average length of the M_2 sequence:

$$\bar{l}_2 = \frac{v_{22}}{v_{21}} + 1 = \frac{k_{22} \cdot [M_2] \cdot [R\sim\sim\sim M\cdot_2]}{k_{21} \cdot [M_1] \cdot [R\sim\sim\sim M\cdot_2]} + 1 = \frac{k_{22}}{k_{21}} \cdot \frac{[M_2]}{[M_1]} + 1 =$$

$$= r_2 \cdot \frac{[M_2]}{[M_1]} + 1 = (r_2/a) + 1 \tag{39d}$$

As one can immediately see, the average length of the M_1 or M_2 sequences, is equivalent to the composition of the copolymer chain (i.e., the ratio of the structural units in the chain $m_1/m_2 = d[M_1]/d[M_2]$ is identical with the ratio of the mean sequence lengths):

$$\frac{d[M_1]}{d[M_2]} = \frac{\bar{l}_1}{\bar{l}_2} = \frac{r_1[M_1]/[M_2] + 1}{r_2[M_2]/[M_1] + 1} = \frac{r_1 a + 1}{(r_2/a) + 1} \tag{39e}$$

The identity of Equations (36) and (39a) shows that in this way one can derive the copolymerization equation simply and clearly.

Sequence Length Distribution

The copolymerization equation describes the overall composition m_1/m_2 of a copolymer as a function of the monomer mixture composition $[M_1]/[M_2]$ and the parameters r_1 and r_2. There are, of course, many chain structures possible for a given overall composition. These can be regular and irregular, and they differ from each other by the length of the M_1 and M_2 sequences, respectively. However, since the different addition steps occur at random, it is possible with simple statistical considerations to establish a relationship between the r values and the probability with which M_1 or M_2 sequences of increasing length occur in a copolymer molecule:

If the probability that a chain-radical $\sim\sim\sim M_2$-$M_1^•$ [11] adds a monomer molecule M_1 (forming a M_1-M_1 sequence) is given by W_{11}, then the probability that this process immediately occurs again forming a 3 M_1 sequence is given by W_{11}^2 ; that it then occurs a third time (forming a 4 M_1 sequence) is given by W_{11}^3 , and so forth. Therefore the probability for the creation of an M_1-M_1 sequence with n M_1 structural units is equal to $W_{1(n)} = W_{11}^{n-1}$. This sequence is of course an *open* sequence of undetermined length, because the further addition of monomers is not excluded. For a *closed* sequence within the chain, one still always needs the addition of a monomer M_2 given by the probability W_{12}. Therefore, the probability for the existence of a closed sequence with n units of the same type M_1 is given by:

$$W_{1(n)} = W_{11}^{n-1} \cdot W_{12} = W_{11}^{n-1} \cdot (1-W_{11}) \tag{39f}$$

How do we obtain W_{11}, the probability that a chain-radical $\sim\sim\sim M_2$-$M_1^•$ adds a monomer molecule M_1? The probability W for the occurence of an event is defined by the ratio W = number of favorable cases/number of all possible cases. For example, the probability of obtaining a 6 on casting a die is given by 1/6 (among six possible cases, the favorable case exists only once). In the previous case we were concerned with the addition of the monomers M_1 and M_2 to the chain radical $\sim\sim\sim M_2$-$M_1^•$. The number of favorable cases (i.e., the addition of monomer M_1 to $\sim\sim\sim M_2$-$M_1^•$)[12] is proportional to the rate $v_{11} = k_{11} \cdot [R \sim\sim\sim M \cdot_1] \cdot [M_1]$. And the number of all possible cases (i.e., the addition of a monomer M_1 or a monomer M_2 to $\sim\sim\sim M_2$-$M_1^•$), is proportional to the sum $v_{11} + v_{12}$. Therefore for the probability W_{11}, one can write:

---

[11]The probability of throwing a 6 with a die that has six faces is 1/6; the probability of throwing two 6's in a row, is $(1/6)^2$, because the combination 6-6 is one of 36 possible combinations, i.e., 1/36; the possibility of throwing three 6's in a row is $(1/6)^3$, because the combination 6-6-6 is one of 216 possible combinations, etc.

[12]Obviously the probability that other chain radicals $\sim\sim\sim M_i$, such as $\sim\sim\sim M_2-M_1-M_i$ $\sim\sim\sim M_2-M_1-M_i$ add a monomer M_1, is the same in all cases, but we are interested here only in additions to $\sim\sim\sim M_2-M_i$, because only in this case is the number of addition steps indentical with the number of M_1 in the increasing M_1 sequence.

$$W_{11} = \frac{v_{11}}{v_{11} + v_{12}} = \frac{k_{11} [\sim\!\!\sim\!\!\sim M_1 \cdot] [M_1]}{k_{11} [\sim\!\!\sim\!\!\sim M_1 \cdot] [M_1] + k_{12} [\sim\!\!\sim\!\!\sim M_1 \cdot] [M_2]}$$

$$= \frac{k_{11} [M_1]}{k_{11} [M_1] + k_{12} [M_2]} \tag{39g}$$

After division of the numerator and denominator by k_{12} and substitution of r_1 in place of k_{11}/k_{12} one obtains:

$$W_{11} = \frac{r_1 [M_1]}{r_1 [M_1] + [M_2]} \quad \text{and} \quad W_{12} = 1 - W_{11} = \frac{[M_2]}{r_1 [M_1] + [M_2]} \tag{39h}$$

If one then substitutes these calculated values for W_{11} in Equation (39f), one obtains the probability for the creation of an M_1 sequence with $1, 2, 3, 4, 5$, etc., or in general n, structural M_1 units, or, which is the same, the number-fraction of sequences with $1, 2, 3, 4, 5, \ldots \ldots n$ M_1 structural units in a copolymer. This gives a picture of the sequence distribution of the structural units in a copolymer chain. Completely similar equations hold of course also for the M_2 sequences:

$$W_{22} = \frac{r_2 [M_2]}{r_2 [M_2] + [M_1]} \quad \text{and} \quad W_{21} = 1 - W_{22} = \frac{[M_1]}{r_2 [M_2] + [M_1]}$$

$$\text{and} \quad W_{2(n)} = W_{22}^{n-1} \cdot W_{21}$$

The following three examples will show the influence of the r_1-r_2 values and the monomer concentration on the sequence length distribution in a copolymer chain:

(1) $r_1 = 1$, $r_2 = 1$ (ideal copolymerization), $[M_1] = 50$ mol%, $[M_2] = 50$ mol%. On the average the monomer sequence in the chain of such a copolymer represents a chain with alternating $2M_1$-$2M_2$ structure (mean sequence length $\bar{l}_1 = \bar{l}_2 = 2$):

$$\ldots .\, 112211221122112211221122 \ldots .$$

Equation (39f), for $W_{11} = 0.5$ [calculated by means of Equation (39h)] and increasing values for n from 1 to 5, gives the sequence length distribution diagram [Figure 119].

As one can see, of 100 M_1-sequences in the chain molecule of such a polymer, there will be 50 sequences of single units, 25 doubles, 12 triples, and still 6 sequences consisting of 4 M_1-monomer units. The same applies both for monomer M_1 and

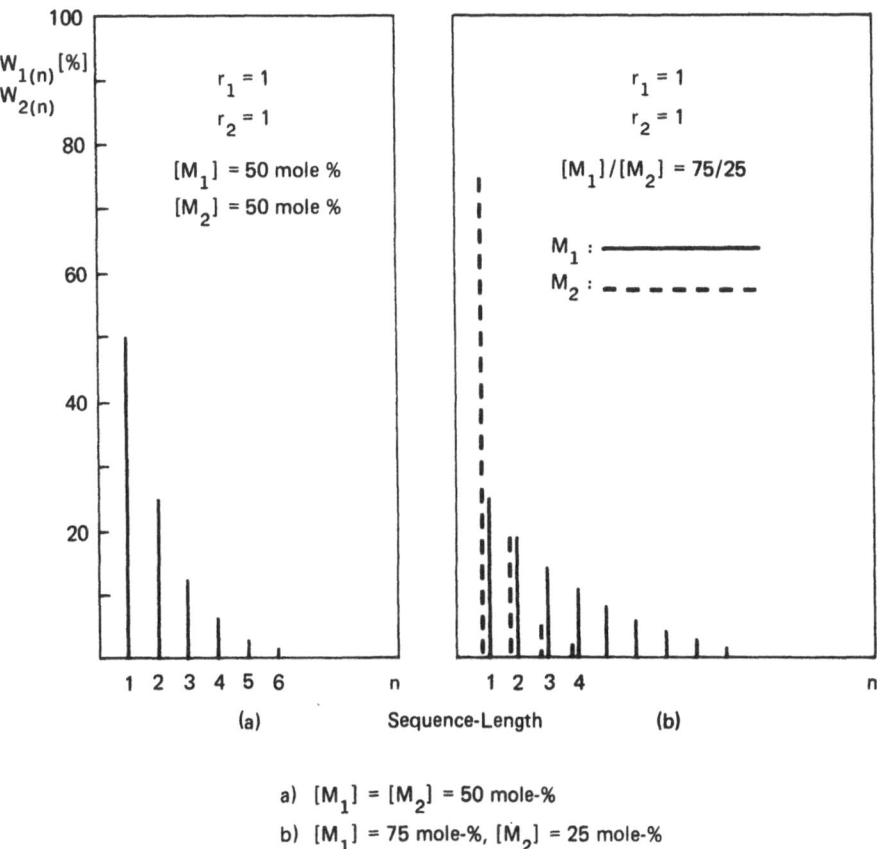

a) $[M_1] = [M_2] = 50$ mole-%

b) $[M_1] = 75$ mole-%, $[M_2] = 25$ mole-%

FIG. 119 — Sequence-length distribution, $W_{1(n)}$, for an ideal copolymerization $(r_1 = 1, r_2 = 1)$

monomer M_2 (for $[M_1]/[M_2] = 50/50$). Although the distribution of the segment length in this case is not very broad, the structure of such an ideal copolymer is significantly different from that of regularly alternating copolymers.

If one starts with unequal monomer concentrations, then the distribution becomes broader for the monomer which is present in higher concentration and becomes narrower for the monomer present in lower concentration [Figure 119b].

The following example shows that the influence of the r values is similar.

(2) $r_1 = 5$, $r_2 = 0.2$, $[M_1] = 50$ mol%, $[M_2] = 50$ mol% (compare Figure 112). The r values $k_{11}/k_{12} = 5/1$ and $k_{22}/k_{21} = 1/5$ show that the addition of the M_1 monomer is preferred by both chain radicals $\sim\sim\sim M_1 \cdot$ and $\sim\sim\sim M_2 \cdot$ in the same ratio of 5/1 (symmetrical copolymerization). This means that on the average after every 5 M_1-additions to $\sim\sim\sim M_2$-$M_1 \cdot$ there occurs an M_2 addition to $\sim\sim\sim M_1 \cdot$, and after every 5 M_1-additions to $\sim\sim\sim M_1$-$M_2 \cdot$ ($\uparrow$) there follows an M_2 addition to

〰〰 M_1-M_2· (↓). The average monomer composition of such a chain is given by the following scheme[13]):

$$
\overset{\downarrow}{}\qquad\qquad\qquad\qquad\qquad\qquad\overset{\downarrow}{}
$$

....1112211111121111112111111211111121111112211111121111111211111
　　↑　　　　　↑　　　　　↑　　　　　↑　　　　↑　　　　　↑　　　　↑　　　　↑

$$
\overset{\downarrow}{}
$$

12111111211111122111....
　↑　　　　↑

Equation (39h) confirms that the probability W_{11} for the addition of a monomer M_1 to 〰〰 M_1· in the above example is 5/6, and W_{12} correspondingly is 1/6. Equation (39f) (using these values and $[M_1] = [M_2] = 50$ mol%) results in the diagram shown in Figure 121).

It is surprising how broad the distribution for M_1 has become and how much the actual chain structure is different from the mean (average) structure. Even sequence lengths of 20 M_1-units can still be encountered, even though there is only about one such sequence among every 200 M_1-sequences. Sequences consisting of 10 M_1-units still amount to over 2% of the total. It is even more surprising that, in spite of the fact that the average sequence length is 6 M_1-units, the isolated single M_1 units represent the largest number fraction (i.e., about 16 out of a 100 sequences), whereas the sequences consisting of 6 units are represented only by 6.7%. On the other hand, the sequences of 5 to 6 units contain the largest number fraction of the M_1 units present in the system.

The sequence length distribution can also be represented in a different manner. Instead of asking after the fraction of sequences, one determines the fraction of the structural units (monomer units) involved in the formation of single, double, and other sequences. As can be seen from Figure 121, with the copolymer under consideration ($r_1 = 5$, $r_2 = 0.2$) one has 16.7 M_1-structural units for single monomer sequences, $(11.6)\cdot(3) = 34.8$ M_1-structural units for triple sequences, and $(8)\cdot(5) = 40$ M_1-structural units for sequences consisting of 5 units, and so forth. If one forms the sum of these fractions over the total sequence length range, one finds that there are 600 M_1-structural units involved in 100 M_1-sequences:

$$
\sum_{n=1}^{n=\infty} n \cdot W_{1(n)} = (1) \cdot (16.7) + (2) \cdot (14) + (3) \cdot (11.6) + (4) \cdot (9.7) + (5) \cdot (8.0) +...... = 600
$$

Thus on the average 6 structural units belong to one sequence:

$$
\overline{l}_{(1)} = \sum_{n=1}^{n=\infty} n \cdot W_{1(n)} \bigg/ \sum_{n=1}^{n=\infty} W_{1(n)} = r_1[M_1]/[M_2] + 1 = (5) \cdot (1) + 1 = 6
$$

---

[13] $l_1 = (r_1[M_1]/[M_2]) + 1 = 5\cdot50/50 + 1 = 6$
$l_2 = (r_2[M_2]/[M_1]) + 1 = 0.2\cdot50/50 + 1 = 1.2$ that means: every 5 M_2-sequences contain 6 monomer units of M_2.

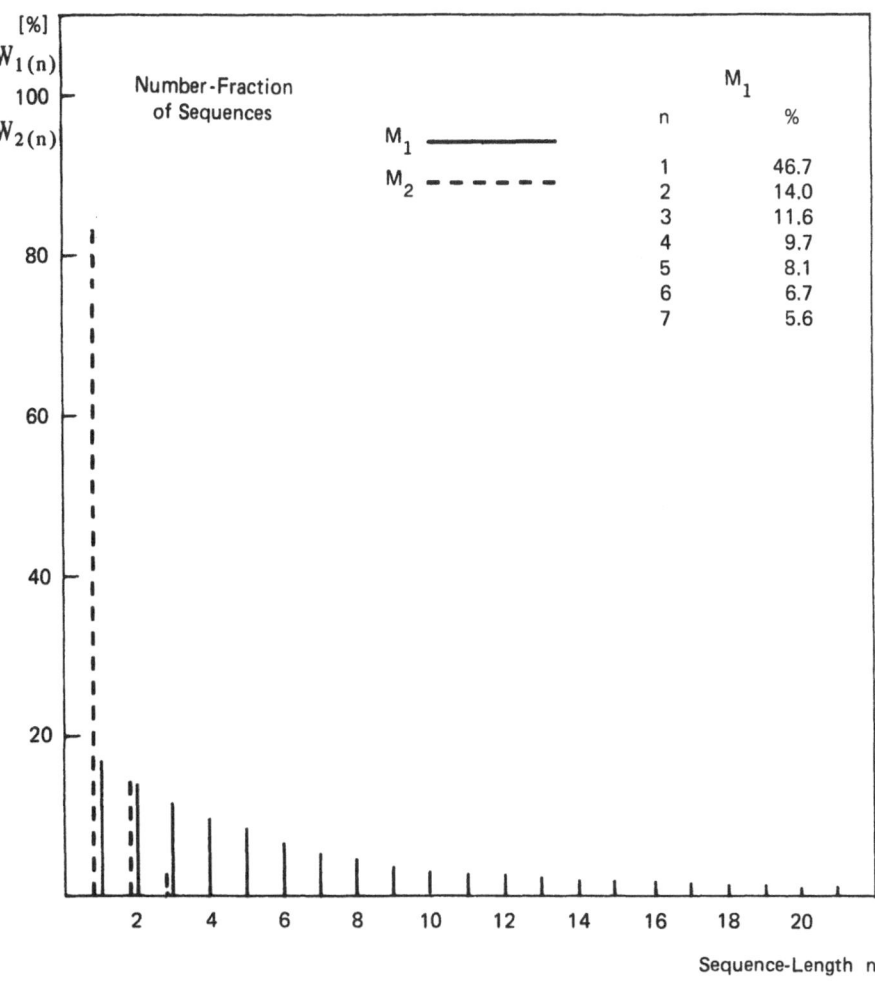

─────── = Number-fraction $W_{1(n)}$ of M_1-sequences of length n among 100 sequences

─ ─ ─ ─ = Number-fraction $W_{2(n)}$ of M_2-sequences of length n among 100 sequences

FIG. 121 — Sequence-length distribution for a 50/50-copolymer with $r_1 = 5$ and $r_2 = 0.2$.

If one now calculates the individual fractions of 600 back to fractions of 100, one obtains the distribution shown in Figure 122, which shows that of 1,000 M_1-structural units in the chain, 67 belong to sequences of 5 units, 47 belong to sequences of 2 units, and 62 belong to sequences of 8 units. etc.

The sequence length distribution for M_2, shown in Figure 121, is much narrower. It is generally true that the sequence length distribution is narrower, i.e., the number of structural units belonging to a sequence has a smaller range, the smaller the r_1 values. Therefore, copolymers of monomer pairs which have a

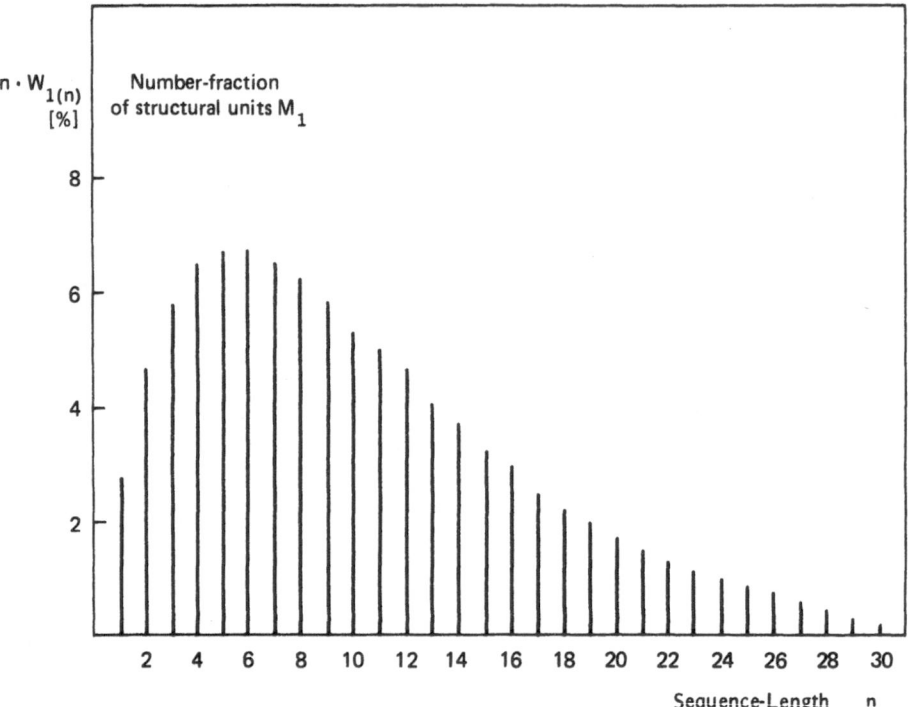

Number-fraction of monomer units M_1 (in %) contained in the sequences of length n.

FIG. 122 — Sequence-length distribution for a 50/50-copolymer with $r_1 = 5$ and $r_2 = 0.2$.

diagram with an inflection point ($r_1 < 1$ and $r_2 < 1$) are always more uniform in structure than copolymers consisting of monomer pairs with $r_1 > 1$ and $r_2 < 1$ as can be seen from the following example:

(3a) $r_1 = 0.5$, $r_2 = 0.5$, $[M_1] = 50$ mol%, $[M_2] = 50$ mol% (compare Figure 103)

$r_1 = 0.1$, $r_2 = 0.1$

$r_1 = 0.05$, $r_2 = 0.05$

Average chain structure (with $r_1 = r_2 = 0.1$):

...21211212121212122121212121211212121212122121212121212121121 ......

(10 sequences contain 11 structural units i.e. $\bar{l}_1 = \bar{l}_2 = 1.1$)

Because of the azeotrope at 50 mol%, the average composition (mol ratio) of the chain in this copolymer does not differ from a copolymer with an ideal diagram. The average chain structure and the sequence length distribution, however, are quite different, as can readily be seen if one substitutes the parameters in Equation (39) and compares the resulting graphical representation of this distribution [Figure 123] with Figure 119. This comparison gives a quantitative confirmation

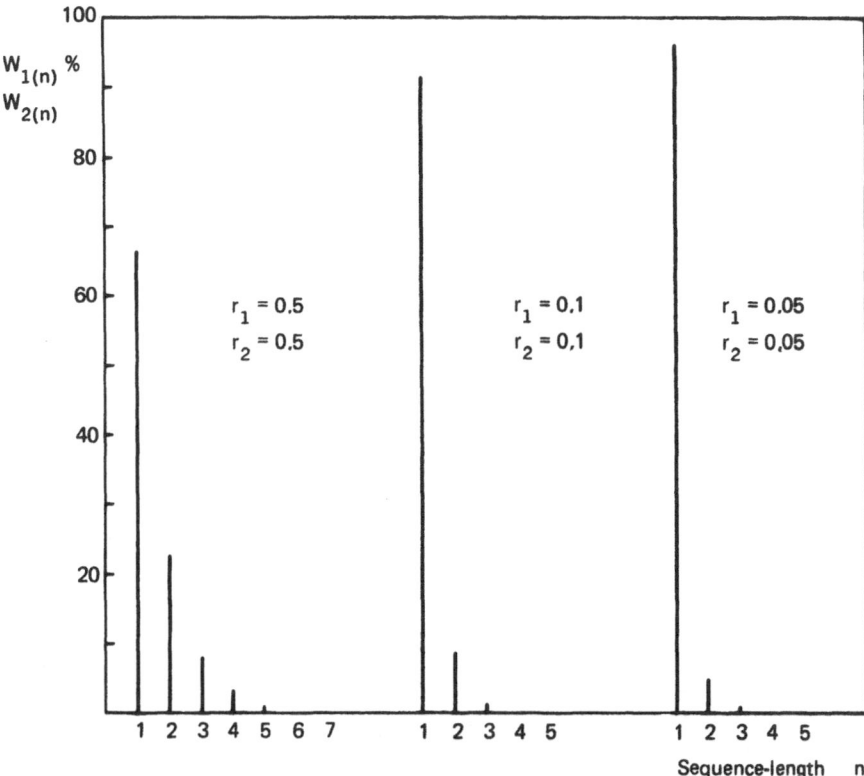

FIG. 123 — Sequence-length distribution with monomer pairs with azeotope.

of the previous qualitative conclusions that monomer pairs of this type tend more to alternation, the smaller the $r_1 \cdot r_2$ values. Monomer pairs with parameters of approximately 0.1 yield copolymers whose structure can be described almost as alternating (about one 1-1-sequence to nine 1-2 sequences).

(3b) $r_1 = 0.1, r_2 = 0.1, [M_1] = 10$ mol%, $[M_2] = 90$ mol%

   $r_1 = 5, r_2 = 0.2, [M_1] = 10$ mol%, $[M_2] = 90$ mol%

In this example, two monomer pairs, one with and the other without an azeotrope, are compared. With a monomer ratio $[M_1]/[M_2] = 10/90$ they have the same copolymer composition, i.e., 35% M_1 and 65% M_2 [compare Figure 124a]. As can be seen from Figure 124, these two copolymers are quite different in their sequence length distribution.

Scattering of the Copolymer Composition from Molecule to Molecule

The formation of a copolymer according to Figure 121 can be simulated by means of a roulette experiment in which of six holes there are always five black ones and one white one (or one can take a corresponding die). If one plots for each black result an M_1 monomer unit and for each white result an M_2 monomer

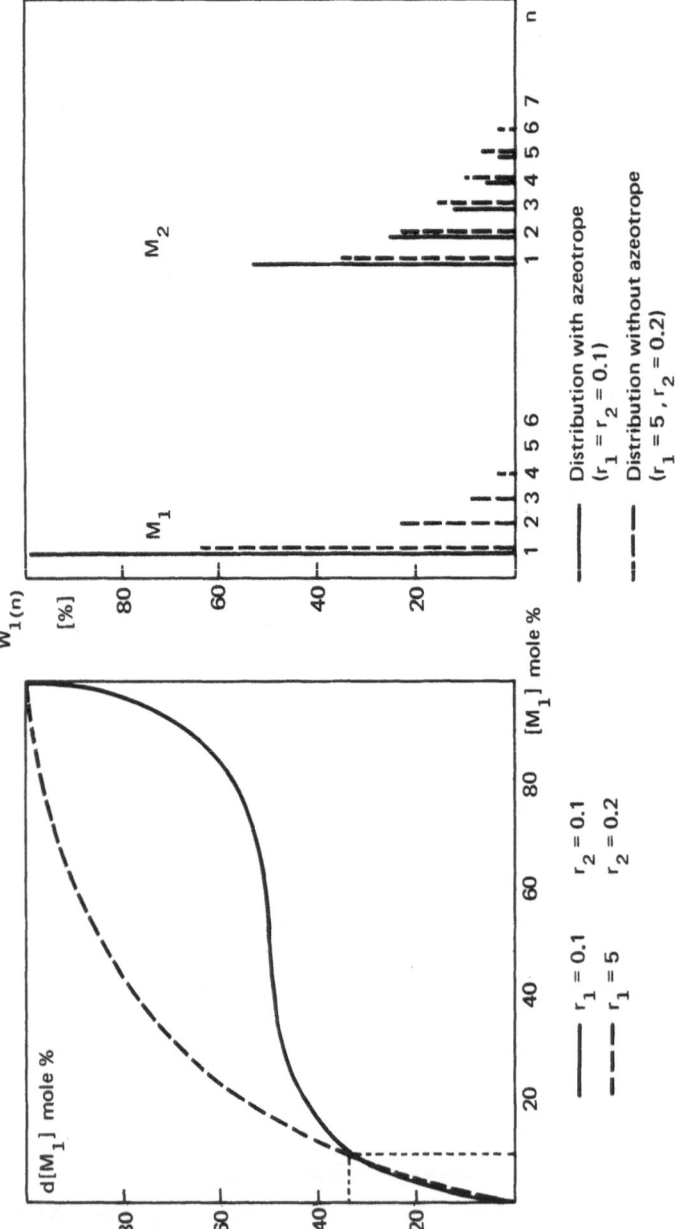

(a) Copolymer diagrams

(b) Sequence-length distribution

FIG. 124 — Comparison of the sequence-length distribution of monomer pairs with and without azeotrope for $[M_1]/[M_2] = 10/90$.

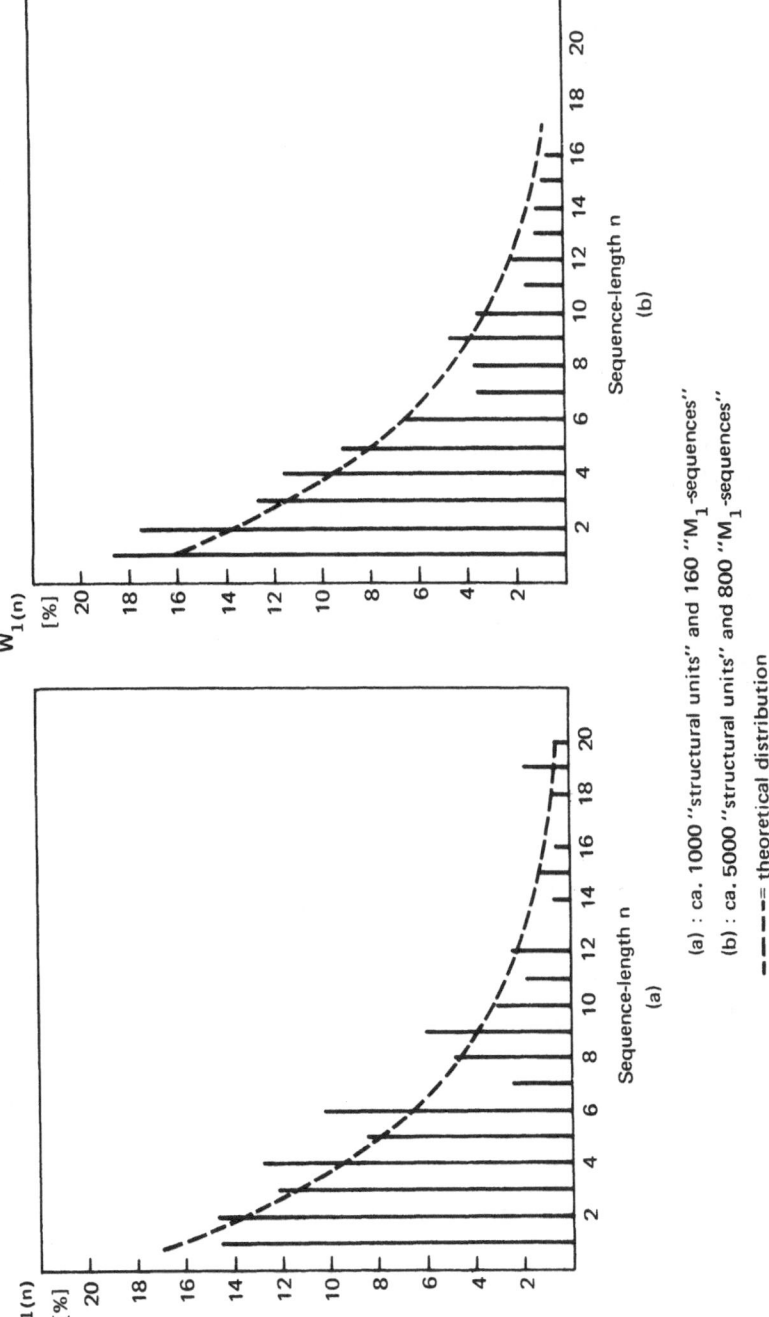

(a) : ca. 1000 "structural units" and 160 "M_1-sequences"

(b) : ca. 5000 "structural units" and 800 "M_1-sequences"

－－－= theoretical distribution

FIG. 125 — "Roulette" — model of the distribution of M_1-sequences for a system where $r_1 = 5$ and $r_2 = 0.2$ ($[M_1]/[M_2] = 50/50$).

unit, one graphs a macromolecule sequence length distribution according to Equation (39f). Figure 125 gives the result after 1,000 and after 5,000 "additions". It can be seen that after 1,000 roulette runs there are still large inhomogeneities in the distribution (gaps at 1 and 7), which are only statistically smoothed out at very much larger numbers. Even after 5,000 experiments, the distribution is still not completely smooth.

This shows that the sequence length distribution is not the same in all the macromolecules. It is more uneven the shorter the chain. Only at very high molecular weights $(P>5,000)$ does the sequence length distribution for all macromolecules of a copolymer become nearly the same.

Not only is the sequence length unevenly distributed over the individual macromolecules (at least for the usual molecular weights from 100,000 to 500,000)—the monomers are also unevenly distributed: if one images that all monomer units of a copolymer are combined in a single macromolecule chain, then this molecule has of course the average monomer ratio. However, if one starts to break up the chain then the probability that one forms parts of the molecule which have more of one monomer and others with more of the other monomer becomes the greater, the smaller the chains. As Stockmayer has calculated, a 1:1 copolymer with a degree of polymerization of $P = 10,000$ still contains 12% of copolymer molecules which have more or less about 1% of one of the monomers.

Preparation of Copolymers

All that has been said up to now about copolymerization theory really applies only to a small range of the copolymerization reaction, because the monomer composition of the polymerizing mixture does not remain constant (if one disregards the monomer pairs where $r_1 = r_2 = 1$ and azeotropic monomer mixtures.)

TABLE 126. Data for a step-wise copolymerization process up to high conversion, $r_1 = 5, r_2 = 0.2; [M_1]/[M_2] = 30/70$

Step Z	Conversion mole %	Starting monomer-mixture $[M_1]/[M_2]$	Copolymer-composition m_1/m_2	Monomer used up at each step M_1 moles	M_2 moles	Residual Monomer M_1 moles	M_2 moles	$\sum_1^z M_1$	$\sum_1^z M_2$	$\dfrac{\sum M_1}{\sum M_2}$ $(M_1 + M_2 = 100)$
1	10	30 / 70	68/ 32	6.8	3.2	23.2	66.8	6.8	3.2	68/32
2	20	26 / 74	62/ 38	6.2	3.8	17	63	13	7	65/35
3	30	21 / 79	55/ 45	5.5	4.5	11.5	58.5	18.5	11.5	62/38
4	40	16 / 84	47/ 53	4.7	5.3	6.8	53.2	23.2	16.8	58/42
5	50	11 / 89	37/ 63	3.7	6.3	3.1	46.9	26.9	23	54/46
6	60	6 / 94	22/ 78	2.2	7.8	0.9	39.1	29.1	31	49/51
7	70	2.3/ 97.7	10/ 90	1.0	9.0	0	30.1	30.1	40	43/57
8	80	0/100	0/100	0	10	0	20	30	50	38/62
9	90	0/100	0/100	0	10	0	10	30	60	33/67
10	100	0/100	0/100	0	10	0	0	30	70	30/70

However, if one continues the polymerization (this is the case with preparative and especially industrial processes) to high conversion, the composition of the product is the sum of all the instantaneous copolymer compositions which have been formed by the different monomer compositions occurring during the polymerization reaction.

Table 126 attempts to show what happens if a monomer pair with $r_1 = 5$ and $r_2 = 0.2$ (compare Figure 112) is polymerized beginning with a molar monomer ratio $[M_1]/[M_2] = 30/70$ and assuming that the copolymer composition within a 10% conversion interval is constant. As the copolymerization equation or the diagram in Figure 112 shows, the copolymer formed in the first interval consists of polymer chains with a monomer unit ratio $m_1/m_2 = 68/32$. This means that during the first step 6.8 moles of M_1 and 3.2 moles of M_2 are used up for the polymerization. A residue of 90 moles of monomer mixture of $[M_1]/[M_2] = 30\text{-}6.8/70\text{-}3.2 = 23.2/66.8 = 26/74$ remains. This mixture will be the starting mixture for the second step. From this mixture, according to Equation (38) and Figure 112, one obtains a copolymer with $m_1/m_2 = 62/38$. Thus for the formation of the copolymer in the second conversion interval, 6.1 moles of M_1 and 3.8 moles of M_2 will be used up, so that there remain 17 and 63 moles of monomer, respectively, etc. After about 70% conversion, monomer M_1 is completely used up, so that from then on only a homopolymer of M_2 is formed. Column 4 of the table shows the composition of the copolymer of each individual step as it would be obtained if one were to separate the fractions after each 10% conversion. Column 2 gives the composition of the "copolymer"[14] up to each individual conversion (integral composition). If one makes the conversion interval smaller and smaller, this corresponds to an integration of the copolymerization equation, which was first done by Mayo, Lewis, and Skeist. The results can be tabulated for different starting monomer mixtures as in Table 126, or they can be shown graphically in groups of curves or in a three-dimensional diagram:

$$m_1 \text{ differential} = f(U), \quad m_1 \text{ integral} = f(U), \text{ or } m_1 \text{ residual} = f(U)$$

Figure 128 shows this function for a molar monomer ratio of 30/70.

More important than the knowledge of how the copolymer composition changes during a copolymerization is how to *prevent* its change, because one wants generally to produce uniform copolymers and not mixtures such as those shown in Table 126 (from $m_1/m_2 = 68/32$ through 62/38, 55/45, 10/90, up to pure m_2).

Sometimes one has the choice between different possible monomer combinations. Then one selects those whose $r_1\text{-}r_2$ values are closest to 1. With monomer pairs where the composition curve has an inflection point, one tries to polymerize at the azeotrope concentration.

---

[14]This copolymer really is a mixture of different copolymers and, at higher conversions, a mixture of these copolymers as well as some M_2 homopolymer.

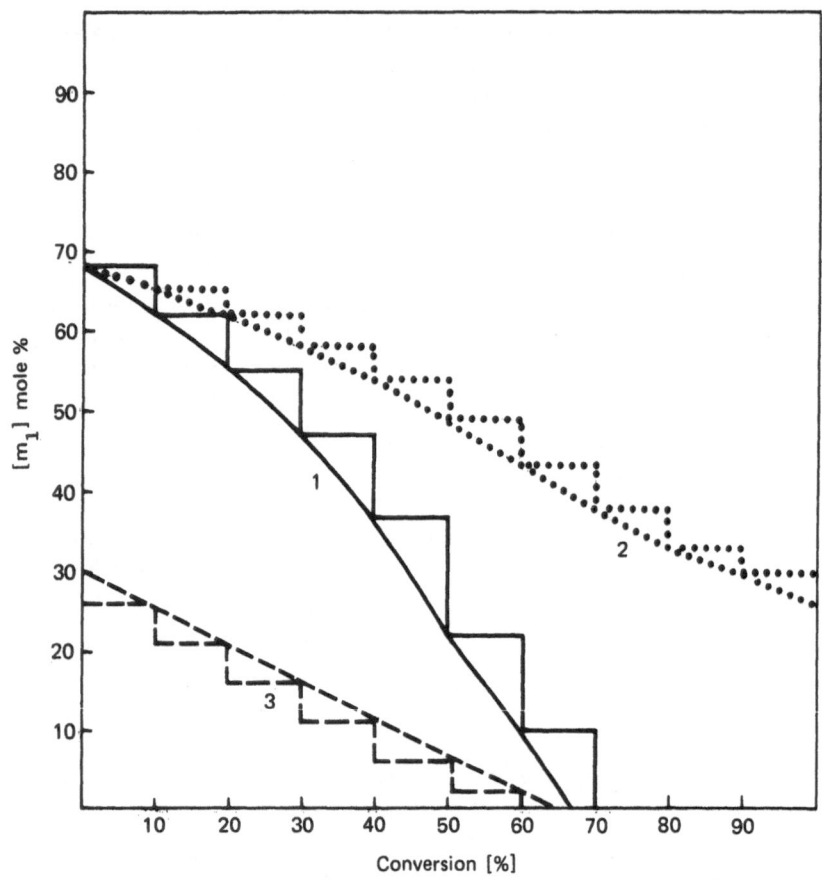

$$[M_1] = 30 \text{ mole} - \%, r_1 = 5, r_2 = 0.2$$

1 ——————　m_1 in the copolymer formed at a particular conversion C (differential composition)

2 ••••••••　m_1 in the total copolymer which has been formed up to the conversion C (integral composition)

3 — — — —　$[M_1]$ in the monomer mixture remaining at conversion C

FIG. 128 — Change in the composition of the copolymer and the monomer mixture during copolymerization. $[M_1]$ = 30 mole-%, r_1 = 5, r_2 = 0.2.

In all other cases one must keep the monomer ratio constant by carefully regulating the monomer input. The monomer concentration one has to maintain for the preparation of a certain copolymer is obtained from the copolymerization diagram. For example, if one has a monomer pair with r_1 = 1.25 and r_2 = 0.1 (diagram in Figure 112), in order to obtain a 50/50 copolymer one must start with a monomer mixture of $[M_1]/[M_2]$ = 20/80 and maintain this monomer concentration during the entire polymerization. This is possible if one adds, at the same rate as the polymer is formed, a monomer mixture with the composition of the copolymer formed, i.e., for this case $[M_1]/[M_2] = 50/50$. Thus the monomers are replenished in the same degree as they are used up, and the monomer ratio in the polymerizing system remains unchanged. Figure 129 shows the course of the copolymer concentration during a polymerization with monomers being continuously added. If one wants to prevent the increase of the concentration over a certain value, for example 50%, then one adds at the same time a calculated amount of an inert solvent. One can see from the curves how the addition of a solvent, together with the monomer, allows one to control the polymer concentration in the mixture at any given value.

In order to prepare such a homogeneous copolymer, it is necessary to have a rapid and accurate control of the rate of polymerization by determining the amount of polymer formed, so that the rate of addition of the monomer mixture can then be regulated accordingly. Since the polymerization rate is constant only

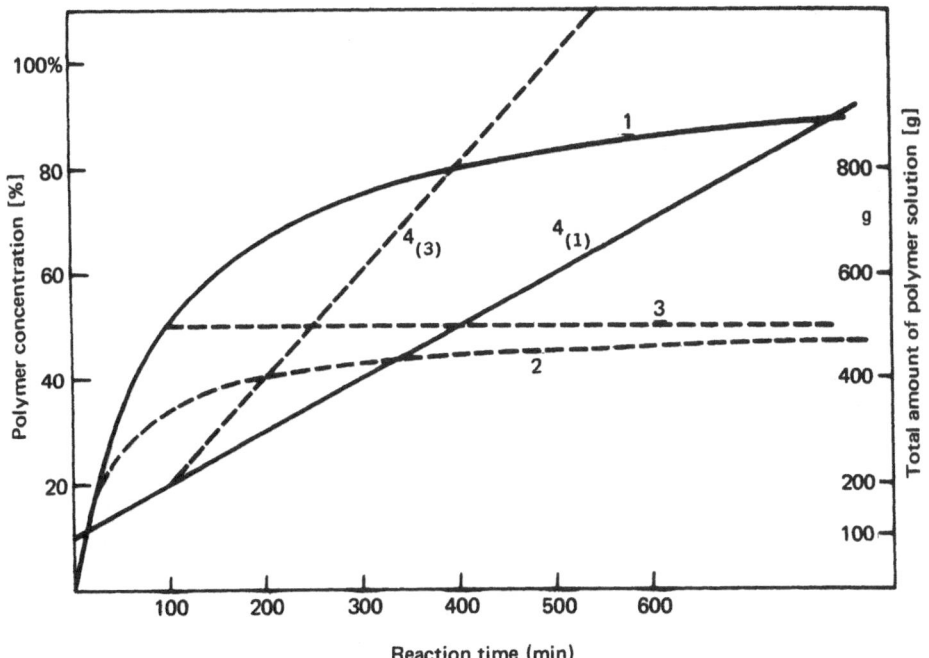

1 : v_p = 1 g polymer/min; input: 1 g monomer/min

2 : v_p = 1 g polymer/min; input: (1 g monomer + 1 g solvent)/min

3 : up to 100 min without solvent (like 1); after 100 min with solvent (like 2)

4 : increase in the total amount of the polymer solution

FIG. 129 — Polymer concentration as a function of reaction time.

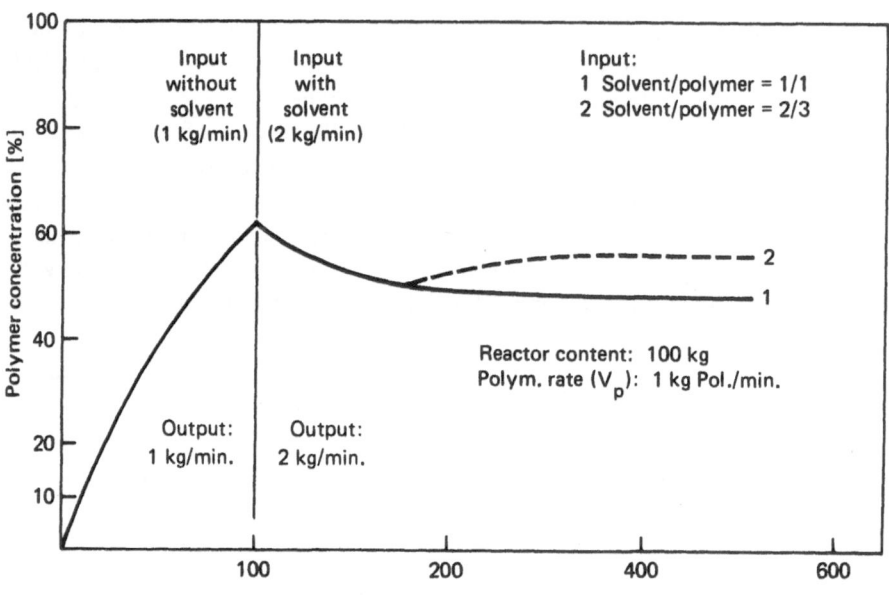

FIG. 130 — Copolymer concentration as a function of reaction time showing the effect of adding monomer and solvent continuously (at the same rate at which polymer is removed).

with a continuous process,[15] one tries to carry out such copolymerizations in a continuous manner by arranging that the same amount of monomer is automatically added in a given time as polymer solution is removed.

Instead of the mixture of M_1 and M_2, one can also add the pure monomer M_1 (or M_2. In every case one has to add the one that is used up more rapidly). The rate of addition in this case has to be somewhat slower than the rate of polymerization, and the control of the processes is therefore somewhat more complicated.

It is remarkable that the properties of the reaction product are changed to such a large extent by changes in purely process parameters (addition rate of the monomer mixture). This is typical for the synthesis of macromolecular compounds, in contrast to the preparation of low molecular weight compounds, where changes in experimental technique may affect the yield, but not the properties of the product.

Copolymerization With Three Monomers (Ternary Copolymer Systems)

In a polymerizing mixture of three monomers, there are three chain radicals with three different possible reaction steps per chain radical. In other words there are nine possible reactions, shown below:

---

[15]The constancy of the rate of polymerization in Figure 130 represents a simplification which is nearly realized only in emulsion polymerization.

$$k_{11}$$
$$\sim\sim M^\cdot{}_1 + M_1 \rightarrow \sim\sim M_1\text{-}M_1^\cdot \quad$$

$$k_{21}$$
$$\sim\sim M_2^\cdot + M_1 \rightarrow \sim\sim M_2\text{-}M_1^\cdot \quad$$

$$k_{31}$$
$$\sim\sim M_3^\cdot + M_1 \rightarrow \sim\sim M_3\text{-}M_1^\cdot$$

$$k_{12}$$
$$\sim\sim M^\cdot{}_1 + M_2 \rightarrow \sim\sim M_1\text{-}M_2^\cdot \quad$$

$$k_{22}$$
$$M_2^\cdot + M_2 \rightarrow \sim\sim M_2\text{-}M_2^\cdot \quad$$

$$k_{32}$$
$$\sim\sim M_3^\cdot + M_2 \rightarrow \sim\sim M_3\text{-}M_2^\cdot$$

$$k_{13}$$
$$\sim\sim M^\cdot{}_1 + M_3 \rightarrow \sim\sim M_1\text{-}M_3^\cdot \quad$$

$$k_{23}$$
$$\sim\sim M_2^\cdot + M_3 \rightarrow \sim\sim M_2\text{-}M_3^\cdot \quad$$

$$k_{33}$$
$$\sim\sim M_3^\cdot + M_3 \rightarrow \sim\sim M_3\text{-}M_3^\cdot$$

In a manner completely analogous to that described for two-component systems, one can also obtain for three- and multi-component systems copolymerization equations which permit calculation in advance of the composition of the terpolymer (using the assumption of a steady state). In order to do this, one needs the six r values of the three monomers combined in pairs (none of the r values must either be zero or ∞):

$$d[M_1] : d[M_2] : d[M_3] = [M_1]\left[\frac{[M_1]}{r_{31}\,r_{21}} + \frac{[M_2]}{r_{21}\,r_{32}} + \frac{[M_3]}{r_{31}\,r_{23}}\right]\left[[M_1] + \frac{[M_2]}{r_{12}} + \frac{[M_3]}{r_{13}}\right]$$

$$: [M_2]\left[\frac{[M_1]}{r_{12}\,r_{31}} + \frac{[M_2]}{r_{12}\,r_{32}} + \frac{[M_3]}{r_{32}\,r_{13}}\right]\left[[M_2] + \frac{[M_1]}{r_{21}} + \frac{[M_3]}{r_{23}}\right]$$

$$: [M_3]\left[\frac{[M_1]}{r_{13}\,r_{21}} + \frac{[M_2]}{r_{23}\,r_{12}} + \frac{[M_3]}{r_{13}\,r_{23}}\right]\left[[M_3] + \frac{[M_1]}{r_{31}} + \frac{[M_2]}{r_{32}}\right]$$

The r values shown in this equation are the r_1-r_2 values of the three monomer pairs M_1-M_2, M_2-M_3, and M_1-M_3. In order to distinguish these better, each r value has now two index numbers:

M_1-M_2 :	M_2-M_3:	M_1-M_3:
$r_1 = k_{11}/k_{12} = r_{12}$	$r_1 = k_{22}/k_{23} = r_{23}$	$r_1 = k_{11}/k_{13} = r_{13}$
$r_2 = k_{22}/k_{21} = r_{21}$	$r_2 = k_{33}/k_{32} = r_{32}$	$r_2 = k_{33}/k_{31} = r_{31}$

Since there is considerable industrial interest in ternary copolymer systems, many such systems have already been calculated with computers. The computer can also carry out the integration according to Table 126, so that one obtains the differential and the integral composition of the copolymer at any desired conversion according to columns 4 and 11 of Table 126. In most cases the calculated composition of the ternary copolymer corresponds very well with the experimentally determined values. The results of a ternary copolymerization are usually plotted in a triangular diagram. Figure 132 shows such a diagram for the

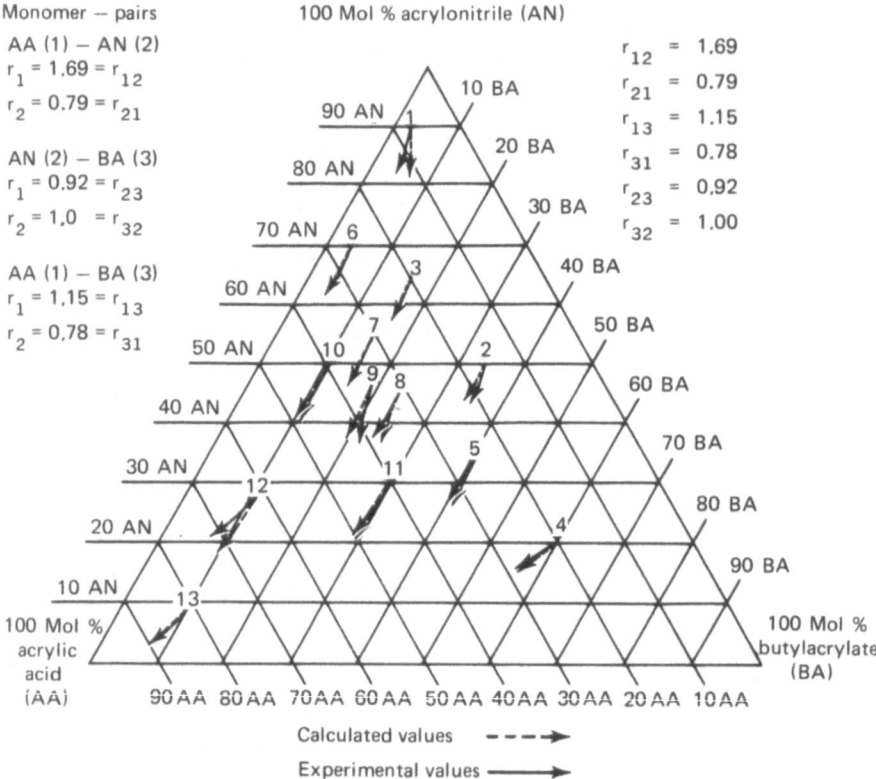

FIG. 132. — Ternary copolymer system acrylic acid/acrylonitrile/butylacrylate

system acrylic acid-acrylonitrile-butylacrylate. As can be seen, the correspondence between the calculated values and the experimentally determined ones is good. The start of the arrow in each case corresponds to the monomer composition at the start of the polymerization, and the head of the arrow shows the composition of the resulting terpolymer. As has been found by P. Wittmer, there are also many azeotropic monomer combinations with ternary systems. [see Figure 133];

Theoretical Interpretation of the r_1 and r_2 Values.

With the aid of the copolymerization parameters r_1 and r_2, it is possible to compare a number of monomer combinations in a reasonable manner. Equations (36-38) do not, however, give any explanation why the different monomers behave so differently in copolymerization, in other words, why the rate of addition of the different monomers to the different radicals is so different.

If one looks at the many binary radical/monomer systems from the point of view of the rate of addition, one discovers that a certain monomer always adds with a high rate if a resonance-stabilized radical is formed from a nonresonance-stabilized

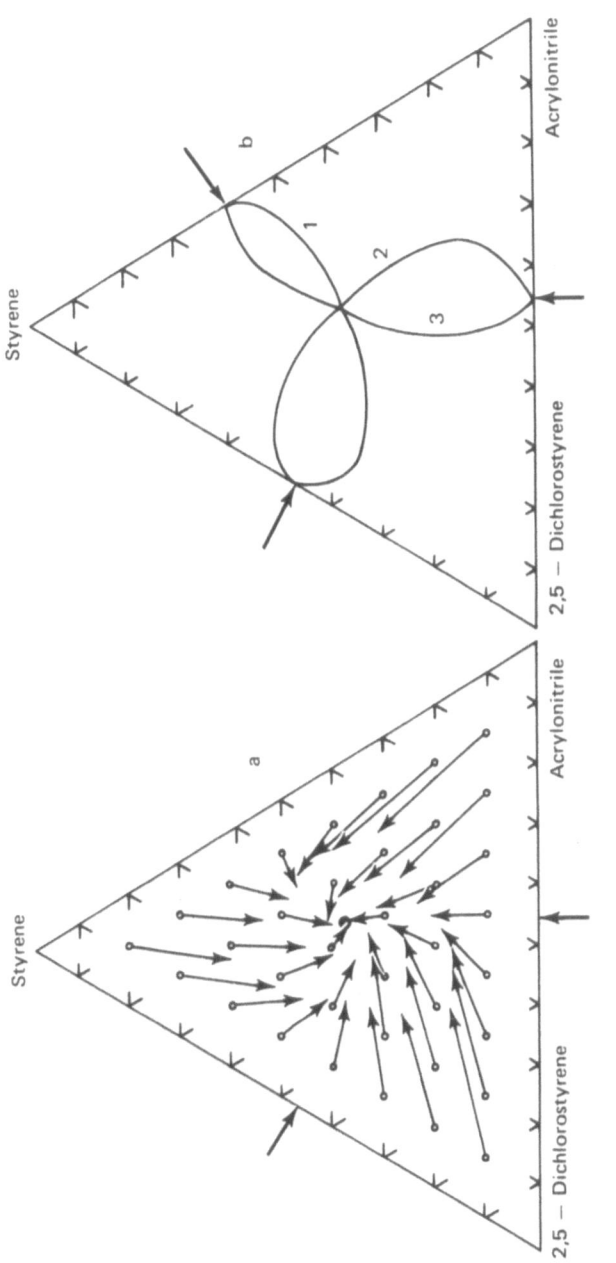

a) Copolymerization as a function of the monomer mixture

b) Partial azeotrope;   1: for styrene
                          2: for 2,5–dichlorostyrene
                          3: for acrylonitrile

(The arrows outside the diagrams indicate the positions of the binary azeotropes)

FIG. 133 – Ternary copolymerization of the system: acrylonitrile/2,5-dichlorostyrene/styrene (according to O. WITTMER).

radical [16]. This is obviously because the reaction has a relatively great potential gradient. For example, if one takes the systems styrene-vinylacetate, the r_1 value ($r_1 = k_{11}/k_{12} = 55$, see Table 107) shows that the reaction:

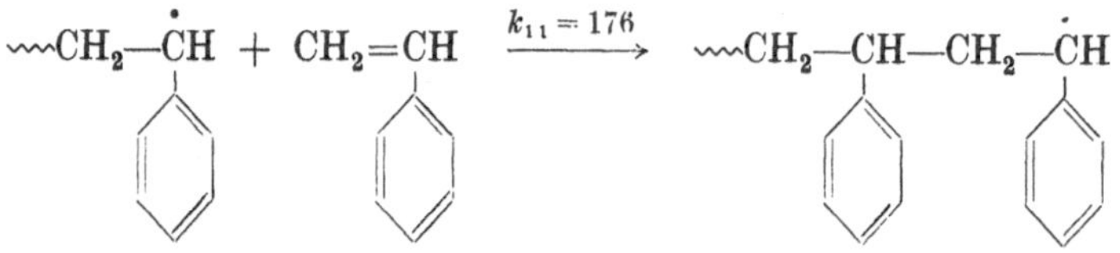

$$\text{ⵡCH}_2\text{—}\overset{\bullet}{\text{CH}} + \text{CH}_2\text{=CH} \xrightarrow{k_{11} = 176} \text{ⵡCH}_2\text{—CH—CH}_2\text{—}\overset{\bullet}{\text{CH}} \qquad 1$$

occurs 55 times more rapidly than the reaction:

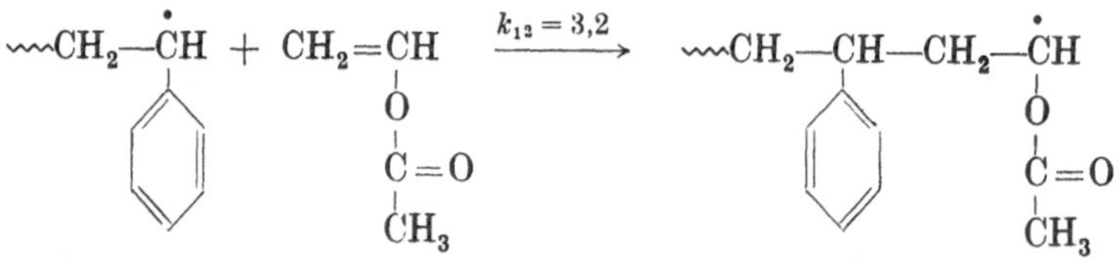

$$\text{ⵡCH}_2\text{—}\overset{\bullet}{\text{CH}} + \text{CH}_2\text{=CH} \xrightarrow{k_{12} = 3,2} \text{ⵡCH}_2\text{—CH—CH}_2\text{—}\overset{\bullet}{\text{CH}} \qquad 2$$

with the side groups O, $C=O$, CH_3

The r_2 value ($r_2 = k_{22}/k_{21} = 0.01$, see Table 107) shows that the reaction:

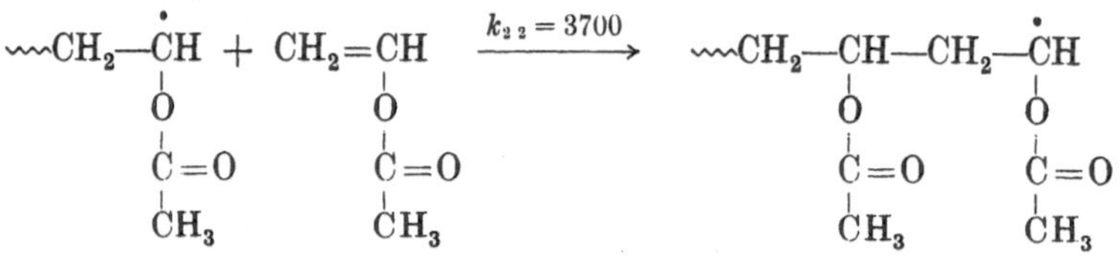

$$\text{ⵡCH}_2\text{—}\overset{\bullet}{\text{CH}} + \text{CH}_2\text{=CH} \xrightarrow{k_{22} = 3700} \text{ⵡCH}_2\text{—CH—CH}_2\text{—}\overset{\bullet}{\text{CH}} \qquad 3$$

with side groups O, $C=O$, CH_3

occurs 100 times more slowly than the addition of styrene to a vinylacetate radical:

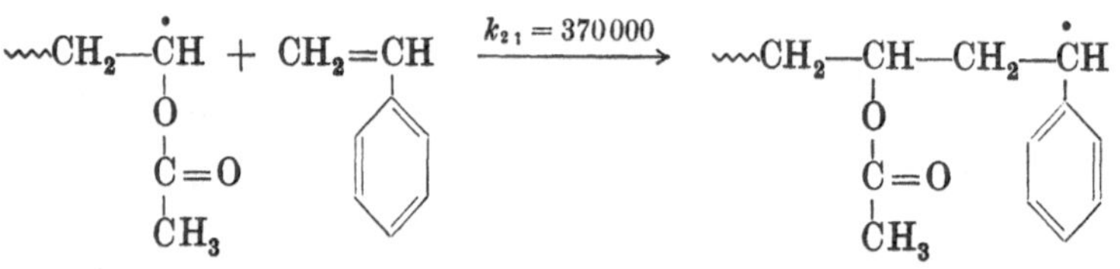

$$\text{ⵡCH}_2\text{—}\overset{\bullet}{\text{CH}} + \text{CH}_2\text{=CH} \xrightarrow{k_{21} = 370000} \text{ⵡCH}_2\text{—CH—CH}_2\text{—}\overset{\bullet}{\text{CH}} \qquad 4$$

with side groups O, $C=O$, CH_3

The styrene radical as a conjugated system has several mesomeric resonance forms (compare p. 56) and is therefore strongly resonance-stabilized. On the other hand, the vinylacetate radical, which is unconjugated, is not resonance-stabilized. This explains why reaction 2, which leads to a relatively energy-rich vinylacetate radical, has the least driving force. The addition of styrene to a styrene radical occurs approximately 50 times faster with the energy level of the radical remaining unchanged. By far the most favored reaction is the addition of a styrene molecule to a vinylacetate radical, because in this case from the relatively energy-rich vinylacetate radical one forms a relatively energy-poor resonance-stabilized styrene radical.

---

[16]For a discussion of resonance stabilization, see p. 56 and p. 72.

Since the absolute values of the reaction rate constants for the homopolymerization of styrene and vinylacetate are known (i.e., for reactions 1 and 3, respectively; according to Table 85 k_{11} = 176 liters/mol·sec and k_{22} = 3,700 liters/mol·sec), it is possible with the aid of the r_1-r_2 values to calculate also the absolute values of the reaction rate constants of reactions 2 and 4:

$$r_1 = \frac{k_{11}}{k_{12}} = 55, \quad \therefore \quad k_{12} = \frac{k_{11}}{r_1} = \frac{176}{55} = 3.2 \; l \cdot mol^{-1} \cdot sec^{-1},$$

$$r_2 = \frac{k_{22}}{k_{21}} = 0.01, \quad \therefore \quad k_{21} = \frac{k_{22}}{r_2} = \frac{3700}{0.01} = 370000 \; l \cdot mol^{-1} \cdot sec^{-1}.$$

Table 135 lists several other examples which show the influence of resonance stabilization on the behavior of the monomer during copolymerization. The values are the reaction rate constants of the addition of the monomers (listed in the first vertical row) to the chain radicals in the uppermost horizontal row. The radicals are listed in such a way that the resonance stabilization decreases from left to right.

Different degrees of resonance stabilization are the explanation for the behavior of monomer pairs without azeotrope. On the other hand, one cannot understand the behavior of monomers with inflection point curves. Thus, if one assumes that the styrene radical is considerably more stable than the acrylonitrile radical, one could explain that the addition of styrene to a chain with an acrylonitrile radical (k_{21}) occurs 33 times as rapidly as the addition of an acrylonitrile monomer (k_{22}) (r_2 = k_{22}/k_{21} = 0.03). However, one would also have to expect that if a styrene radical is at the growing chain end, the addition of styrene again occurs more rapidly than that of acrylonitrile. However, as the r_1 value (r_1 = k_{11}/k_{12} = 0.4) shows, one finds just the opposite: i.e., the addition of acrylonitrile occurs approximately twice as fast as the addition of styrene. Whichever of the two radicals one might find at the growing chain end, the opposite monomer is always

TABLE 135. Absolute rate constants (in l. mole^{-1} . sec^{-1}) for the addition of monomers to different radicals (calculated from the r_1-,r_2 -values and the parameters in Table 85).

	RADICAL: $\sim\!CH_2\!-\!\overset{\bullet}{C}H$	$\sim\!CH_2\!-\!\overset{\overset{\displaystyle CH_3}{\mid}}{\underset{\overset{\mid}{C\!-\!OCH_3}}{\overset{\bullet}{C}}\!\cdot}$	$\sim\!CH_2\!-\!\overset{\overset{\mid}{\underset{\overset{\mid}{C\!-\!OCH_3}}{}}}{\overset{\bullet}{C}H}$	$\sim\!CH_2\!-\!\overset{\overset{\mid}{\underset{\overset{\mid}{O\!-\!C\!-\!CH_3}}{}}}{\overset{\bullet}{C}H}$	$\sim\!CH_2\!-\!\overset{\bullet}{C}H$ $CH_3\!-\!O\!-\!C\!-\!CH$
MONOMER:					
2 Styrene	176	806	11500	370000	185000
3 Methylmeth-acrylate	335	367	232	250000	259000
4 Methyl-acrylate	229	3670	2090	37000	18500
5 Vinylacetate	3.2	18.3	233	3700	
6 Allylacetate	3.5	18.3	230		3700

preferred. This behavior leads to the idea of a positive and a negative charge. Thus, if a particle is positively charged, a negatively charged monomer is attracted, and vice versa.

Whether on the addition of a monomer to a growing chain a positive or a negative chain-end occurs depends on the substituents of the monomer. Electron-attracting substituents (electrophilic substituents) such as $(-C_6H_5)$[17] $<-Cl$ $<-COOR <-NO_2 <-CN$, when present in vinyl compounds, cause the π electrons of the double bond to take part in the electronic system of the substituents. This leads to a certain negative overcharge $\delta-$, which is missing from the double bond and therefore causes the double bond to have a positive overcharge $\delta+$ of the same magnitude. Monomers such as vinylidenechloride, acrylic esters, and acrylonitrile are therefore polarized in such a way that the vinyl groups represent the positive part, and the substituents the negative part of a dipole.

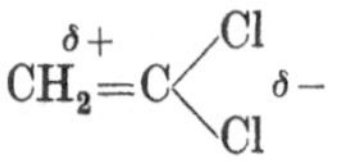

$$CH_2\!\!\overset{\delta+}{=}\!\!C\!\!\begin{array}{l}\diagup Cl \\ \diagdown Cl \end{array}\;\delta-$$ Vinylidenechloride

$$\overset{\delta+}{CH_2}\!\!=\!\!CH\!\!-\!\!\underset{\underset{\delta-}{O}}{\overset{\|}{C}}\!\!-\!\!O\!\!-\!\!CH_3$$ Methylacrylate

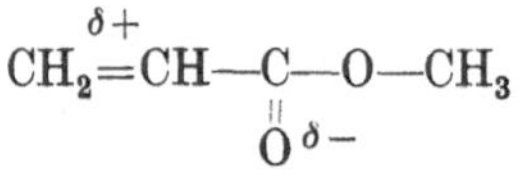

$$\overset{\delta+}{CH_2}\!\!=\!\!CH\!\!-\!\!\overset{\delta-}{C}\!\!\equiv\!\!N$$ Acrylonitrile

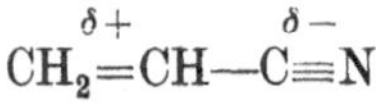

If one assumes that such a polarization is also possible with radicals, one can understand that a chain end at which there happens to be a structural unit with an electron-attracting (accepting) substituent, prefers a monomer with an electron-donating substituent. For example, it prefers to add styrene and not acrylonitrile, if it has the choice between these two monomers:

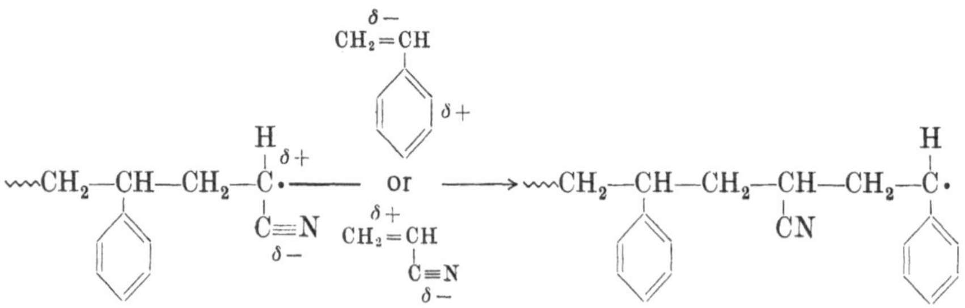

Electron donating groups are: $-C_6H_5 <-CH_3 <-OCH_3 <-N(CH_3)_2$. Monomers such as styrene or vinyl ethers are thus polarized in the opposite sense to vinylidene chloride, acrylic esters, or acrylonitrile. These monomers therefore add in a preferred manner to those chain radicals which, because of polarization, carry a

---

[17]C_6H_5 can be regarded as an electron acceptor as well as an electron donor. It becomes one or the other by inductive effects.

positive overcharge (such as in the example of the styrene-acrylonitrile copolymer shown above). On the other hand, a growing chain end with a structural unit containing an electron-donating substituent (i.e., which, as a result of polarization, carries a negative overcharge) prefers to add a monomer with an electron-attracting substituent:

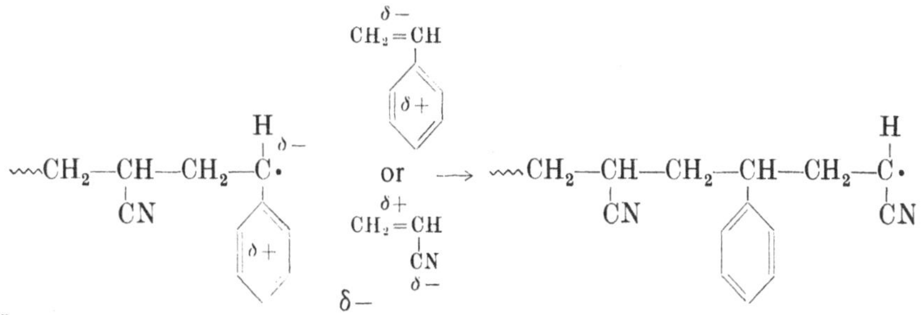

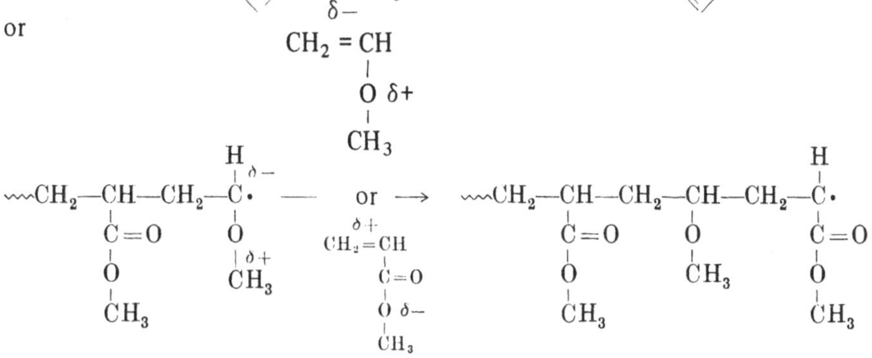

However, the opposite polarization caused by the electron-donating or electron-withdrawing substituents is not solely responsible for the rate of addition of the monomers in a copolymerization. Thus the magnitude of the resonance stabilization is always important too, and the actual behavior of the monomers therefore results from the overlapping of two effects: the desire for the highest possible resonance stabilization and the tendency for charge neutralization.

The Q-e Scheme

This explanation for the behavior of different monomers in radical polymerization is based on the attempt by Alfrey and Price to provide a quantitative description, with the aid of two parameters, for *each* monomer rather than for a monomer *pair*. These parameters are called Q and e. For the rate constant k_{xy} for the addition of a monomer M_y to the radical $\sim\sim\sim M_x^{\cdot}$ according to the reaction:

$$\sim\sim\sim M_x^{\cdot} + M_y \xrightarrow{k_{xy}} \sim\sim\sim M_x - M_y^{\cdot},$$

the following expression can be written:

$$k_{xy} = P_x \cdot Q_y \cdot e^{-e_x \cdot e_y} \qquad (a)$$

P_x is a proportionality factor which is characteristic of the state of the radical $\sim\sim\sim M_x^\bullet$. Q_y is a measure for the reactivity of the monomer M_y and thus a measure for the willingness of the monomer M_y to form the radical $\sim\sim\sim M_x - M_y^\bullet$ by reaction with $\sim\sim\sim M_x^\bullet$. Since this readiness to form a radical ending in M_y is the greater, the greater the resonance stabilization of $\sim\sim\sim M_y^\bullet$, Q_y must be also a measure for the resonance stabilization of the radical $\sim\sim\sim M_y^\bullet$ produced. The quantities e_x and e_y describe the polarization of the two radicals ($\sim\sim\sim M_x^\bullet$ and $\sim\sim\sim M_y^\bullet$) and the two monomers (M_x and M_y) resulting from the substituents (–Cl, –OH, –COOR, –CN, etc.) on the monomers. As a simplification, it is assumed that the corresponding radicals and monomers have the same polarity and therefore the same e value. e_x is therefore the polarity of the radical $\sim\sim\sim M_x^\bullet$ and also of the monomer M_x. The same holds also for e_y, of course. With the aid of Equation (a) one can write the ratios $k_{11}/k_{12} = r_1$ and $k_{22}/k_{21} = r_2$, so that the quantity P_1, which in both cases refers to the radical $\sim\sim\sim M_1 \bullet$, disappears and one obtains Equation (b):

$$r_1 = \frac{k_{11}}{k_{12}} = \frac{P_1 \cdot Q_1 \exp\{-e_1 \cdot e_1\}}{P_1 \cdot Q_2 \exp\{-e_1 \cdot e_2\}} = \frac{Q_1}{Q_2} \exp\{-e_1(e_1 - e_2)\} \tag{b}$$

and correspondingly for r_2:

$$r_2 = \frac{k_{22}}{k_{21}} = \frac{Q_2}{Q_1} \exp\{-e_2(e_2 - e_1)\} . \tag{c}$$

These equations permit us to calculate the Q and e values for single monomers from the $r_1 - r_2$ values, provided we have one monomer for which Q and e have been arbitrarily established. Price chose styrene as the standard monomer with the values Q = 1 and e = –0.8. One can now calculate the Q and e values of any monomer that has been copolymerized with styrene from the r_1 and r_2 values of the copolymerization of styrene with that particular monomer. Knowing the Q and e values of these various monomers, one can then calculate Q and e for any monomer that has been copolymerized with these monomers (i.e., where r_1 and r_2 are given in the literature). Conversely, knowing Q and e for any two monomers, we can calculate the $r_1 - r_2$ values for this monomer pair, whether or not they have ever been copolymerized. This is very useful, since it allows prediction of the copolymerization behavior of any two monomers for which Q and e are known. While the predicted behavior is not always exactly like the experimental result, the Alfrey-Price Q-e scheme nevertheless leads at least to a good approximation.

Table 139 gives a selection of Q and e values for the most common monomers. (A complete tabulation of known Q and e values can be found in *Polymer Handbook*, Interscience Publishers, 1966).

If one constructs a diagram of Q-e values for a number of monomers, such as are shown in Figure 141, one can predict the behavior of any resulting monomer pairs in copolymerization. Such a diagram has been called a "copolymerization map."

TABLE 139. Selected Q and e values*

	e	Q
Isobutylvinylether	− 1.77	0.023
t-Butylvinylether	− 1.58	0.15
p-Dimethylaminostyrene	− 1.37	1.51
α-Methylstyrene	− 1.27	0.98
Isoprene	− 1.22	3.33
n-Butylvinylether	− 1.20	0.087
Na-Methacrylate	− 1.18	1.36
Ethylvinylether	− 1.17	0.032
N-Vinylpyrrolidone	− 1.14	0.14
Allylacetate	− 1.13	0.028
2-Chloroallylacetate	− 1.12	0.53
p-Methoxystyrene	− 1.11	1.36
1,3-Butadiene	− 1.05	2.39
Indene	− 1.03	0.36
p-Methylstyrene	− 0.98	1.27
Isobutylene	− 0.96	0.033
Styrene (reference standard)	− 0.80	1.00
Propylene	− 0.78	0.002
Vinylisocyanate	− 0.70	0.16
Vinylenecarbonate	− 0.65	0.00073
p-Iodostyrene	− 0.40	1.17
m-Chlorostyrene	− 0.36	1.03
p-Chlorostyrene	− 0.33	1.03
p-Styrene sulfonic acid	− 0.26	1.04
n-Butylmethacrylate	− 0.23	0.72
Vinylacetate	− 0.22	0.026
Ethylene	− 0.20	0.015
p-Cyanostyrene	− 0.21	1.86
4-Vinylpyridine	− 0.20	0.82
n-Hexylmethacrylate	− 0.12	0.70
Na-acrylate	− 0.12	0.71
Vinylsulfonic acid	− 0.02	0.093
Glycidylmethacrylate	0.10	0.85
Allylchloride	0.11	0.056
n-Nonylmethacrylate	0.14	0.91
Ethylmethacrylate	0.17	0.56
Vinylchloride	0.20	0.044
Ethylacrylate	0.22	0.52
Allylalcohol	0.29	0.052
Vinylidenechloride	0.36	0.22
N-Methylolacrylamide	0.36	0.43
p-Nitrostyrene	0.39	1.63
Methylmethacrylate	0.40	0.74
Methylcinnamate	0.49	0.12
Methylacrylate	0.60	0.42

*(For a comprehensive tabulation see L.J. YOUNG in "Polymer Handbook", Interscience Publishers, 1966).

TABLE 139. (Cont'd)

	e	Q
Itaconic acid	0.64	0.76
Methacrylic acid	0.65	2.34
Methylvinylketone	0.68	0.69
Acrolein	0.73	0.85
Acrylic acid	0.77	1.15
α-Chloromethylacrylate	0.77	2.02
Methacrylonitrile	0.81	1.12
Itaconicanhydride	0.88	2.50
Glycidylacrylate	0.96	0.55
Acrylyl chloride	1.02	1.78
Butylacrylate	1.06	0.50
n-Octylacrylate	1.07	0.35
n-Octadecylacrylate	1.12	0.42
Acrylonitrile	1.20	0.60
Methacrylamide	1.24	1.46
Diethylfumarate	1.25	0.61
Dimethylmaleate	1.27	0.09
Vinylfluoride	1.28	0.012
Methylvinylsulfone	1.29	0.11
Acrylamide	1.30	1.18
Diethylmaleate	1.49	0.059
Dimethylfumarate	1.49	0.76
N-Butylmaleimide	1.75	3.08
Maleicanhydride	2.25	0.23
Vinylidenecyanide	2.58	20.13

The further on the left of this map the monomer is located, the smaller is the resonance stabilization of its radicals, and therefore the more unstable and energy-rich are the radicals. Such monomers as, for example, vinyl ethers, isobutylene, allylacetate, are therefore trying to escape as quickly as possible from their radical state, if this is possible, into a more strongly resonance-stabilized radical form by reaction with a different monomer, or by chain transfer with its own monomer, in which case further polymerization is prevented (self-inhibition, compare p. 72). In this way one explains the low tendency for polymerization of allyl compounds. If, in addition to the instability of the radicals, there is a relatively strong electron-donating tendency of the substituent, as is the case with vinyl ethers and isobutylene ($e = -1.77$ resp. -0.96), so that the radical and the monomer are polarized in the same sense, then the tendency for addition of the monomer to its own chain radical is decreased. To explain why certain monomers (vinylethers α-methylstyrene, allylacetate, indene, propylene, maleicanhydride, maleic acid esters, fumaric acid esters, allylalcohol, allylchloride, etc.) can *copolymerize* by a radical mechanism, but do not *homo*polymerize by such a mechanism, it would be tempting to blame this on the fact that the carbon atoms involved in the propagation step have the same kind of electrical charge. However, as one can see

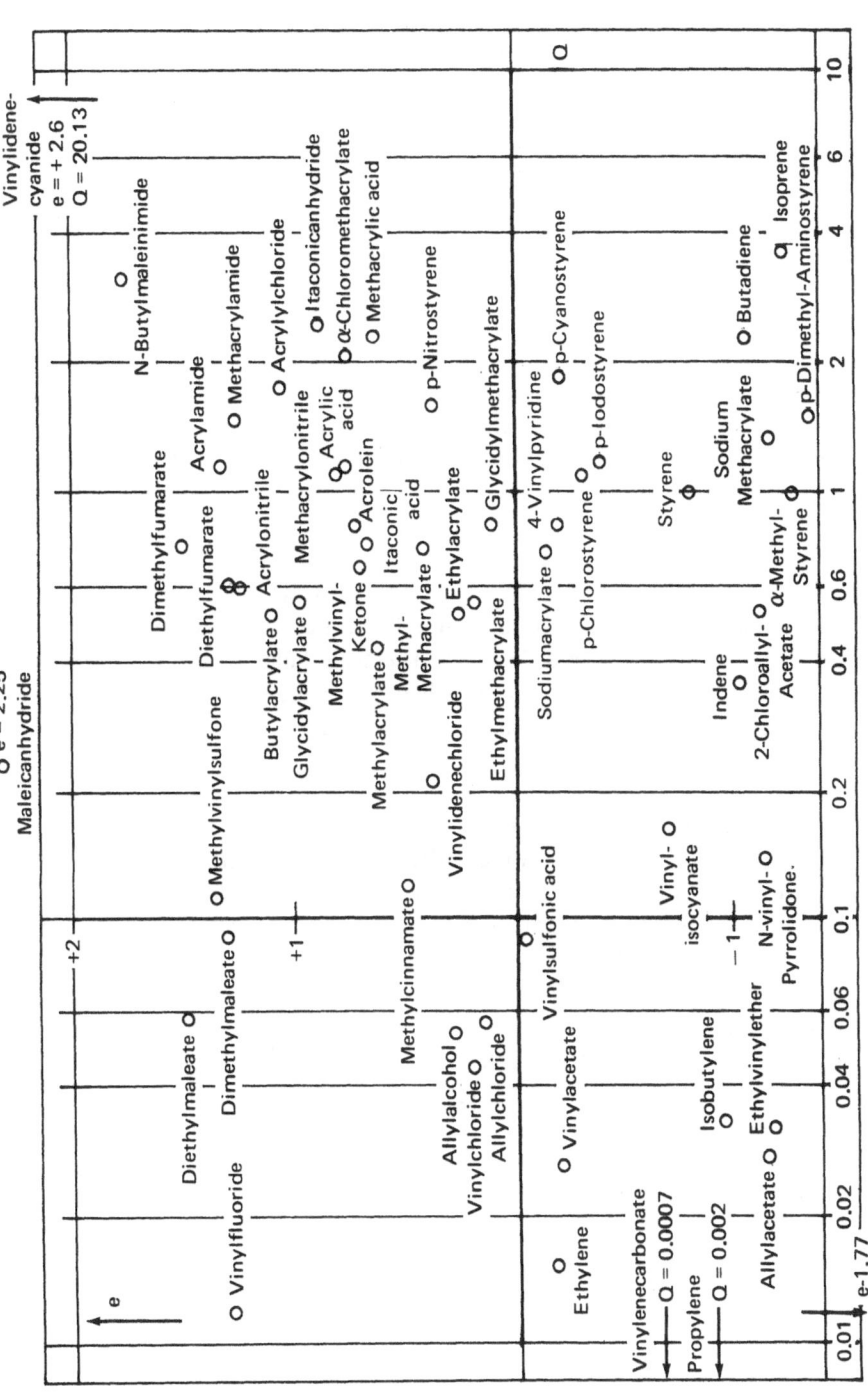

FIG. 141 — Q-e Map.

from the Q-e diagram, this cannot be the case because there is a series of monomers with high e values which still can easily be homopolymerized radically, for example, isoprene, butadiene, sodiummethacrylate, and N-vinylpyrrolidone (e negative), or vinylfluoride, acrylamide, and acrylonitrile (e positive). On the other hand, there is a series of monomers with very small e values which do not homopolymerize by a radical mechanism—such as allylalcohol, allylchloride, methylcinnamate, and propylene (propylene and styrene have the same e-value!). This comparison shows that the e values and the corresponding polarization of the chain radicals and monomers can not be used to explain the remarkable inability of certain monomers to homopolymerize by a radical mechanism.

Let us now consider some examples for an adequate interpretation of the Q-e scheme.

If one is interested in the behavior of vinyl ethers in copolymerization with other monomers, one should probably choose those which have equally low resonance stabilization, but, if possible, no polarization, or only one of opposite sign. The diagram indicates that vinylchloride, vinylacetate, allylchloride, and especially diethylmaleate, are suitable for this purpose. One finds that these monomers (with the exception of allylchloride, which, because of self-inhibition is less suitable) actually are the most used copolymerization components for vinyl ethers. Acrylic esters can also be copolymerized with vinyl ethers, but because of the higher resonance stabilization of their radicals, their addition is clearly preferred and this leads to strongly curved diagrams. The same holds true in even greater degree for the copolymerization of vinyl ethers with styrene, because the styrene radicals are strongly resonance stabilized and furthermore polarized in the same sign.

In general, one finds that the monomers on the right side of the map do not copolymerize well with those on the far left side, because the addition of a monomer with a small Q value to a radical chain end with high resonance stabilization leads from a stable (energy-poor) state to an unstable (energy-rich) state, which on thermodynamic grounds is not very probable. Thus vinylchloride or vinylacetate is not suitable for copolymerization with styrene.

Nearly ideal copolymerization occurs with monomer pairs which are close to each other in the Q-e diagram, i.e., which have similar Q and e values. Examples are styrene-butadiene, vinylchloride-vinylacetate, vinylidenechloride-acrylic esters, different acrylic esters among themselves, acrylic esters-acrylic acid, and acrylonitrile-acrylic esters.

Monomer pairs with opposite polarization (+e and −e) show, as has been discussed before, a tendency to alternating addition (inflection point curves). One expects to find this effect to occur strongly with monomer pairs which have approximately the same Q values and high e values of opposite sign, and which therefore find themselves opposite each other in the two fields of the diagram: for example, styrene-acrylonitrile, styrene-maleic anhydride, styrene-methylmethacrylate, styrene-diethylfumarate, butadiene-fumarates, and dimethylaminostyrene-nitrostyrene. These monomer pairs all have well-defined inflection point curves.

Thus the Q-e diagram presents at least a qualitatively correct picture of the behavior of different monomers in copolymerization. The strongest confirmation of the theory is that the order of the substituents according to increasing tendency to give off electrons (or to take up electrons) in the Q-e diagram (i.e., experimentally determined in copolymerization) corresponds very well with the influence of these same substituents (CH_3, Cl, NH_2, COOR, CN, etc.) in other reactions, for example, the reactions of different substituted benzene derivatives.

It therefore seems reasonable to compare the Q-e equation of Alfrey and Price with the well-known Hammett equation, since in both cases the problem is the same, i.e., to describe the influence of substituents (electron donors and acceptors) on the electron density at the reaction center. One would therefore expect that the e-values of the Alfrey-Price equation are proportional to the sigma-values of the Hammett equation, which describe the influence of the substituents at the benzene ring. This is actually found to be the case. As Taft has shown, this analogy of the problem and the theoretical basis of both equations can be expressed also in a more formal way, if in the Hammett equation the parameter sigma (σ) is divided into two independent variables, one the polar factor $p'\sigma'$, and the other the resonance component R, which, similar to Equation (b) on p. 138, can also be written as ln Q:

Hammett Equation: $\lg(K/K_o) = p \cdot \sigma = p'\sigma' + R = p'\sigma' + \ln Q$ (a)

Where K_o is the dissociation constant of benzoic acid; K the dissociation constant of the substituted benzoic acid; σ (the substituent constant), a measure for the influence of the substituent on the electron density at the reaction center (positive with electron-acceptors, negative with electron-donors); and p (the reaction constant), a measure for the influence of a change in the electron density at the reaction center on the equilibrium.

If instead of the dissociation constant of benzoic acid, one uses the rate constants k_{11} and k_{10} of the growth reactions (and instead of the base 10 log, the natural log) one obtains the following equations:

$$\ln (k_{11}/k_{10}) = p'_1 \sigma'_1 + \ln Q_1 \tag{b}$$
$$\ln (k_{12}/k_{10}) = p'_1 \sigma'_2 + \ln Q_2 \tag{c}$$

Where k_{10} is the reaction rate constant of the copolymerization of monomer M_1 with a reference monomer M_0 (for example, styrene); k_{11}, the reaction rate constant of the homopolymerization of M_1; and k_{12}, the reaction rate constant of the copolymerization of M_1 and M_2.

Subtraction of the two equations (b) and (c) gives:

$$k_{11}/k_{12} = r_1 = Q_1/Q_2 \exp \{p'_1 (\sigma'_1 - \sigma'_2)\} . \tag{d}$$

Equation (d) is nothing else than a transfer of the Hammett equation to the growth step in a copolymerization. The formal correspondence with the Q-e equation [p. 138-(b)] can not be overlooked.

To explain the difference between the r_1-r_2 values calculated from the Q-e values and the experimentally determined r_1-r_2 values, one has to take into account that (a) in the derivation of the copolymerization equation and (b) in the derivation of the Q-e equation certain approximations have been made:

Concerning (a): Only growth steps have been taken into account. The influence of initiation, termination, and transfer reactions was disregarded. The influence of penultimate groups on the addition of the monomer to the growing chain are not considered. All growth steps are considered to be irreversible (contrary to the Alfrey-Price assumption).

[To explain the very different rate of addition of different monomers to a chain radical, one has to take into account the different stabilities of the new radicals formed by the addition step with the different monomers. Additions which lead to resonance-stabilized radicals occur more rapidly than those where very reactive radicals result. One can therefore raise the following question: how does the chain radical $\sim\sim\sim M_1$ know in selecting an M_1 or M_2 monomer, whether after the reaction a stabilized chain radical will result with M_2 and a reactive one with M_1, or vice versa? The explanation is probably that the addition step (at least up to a certain stage) is reversible: thus out of a given number of addition steps, those leading to a stable radical will reverse relatively seldom, whereas those which result in a reactive radical will reverse themselves quite often. Another explanation concerns the different activation energies for the two alternative reaction steps. Thus the reaction leading to a more stable chain radical will always have the lower activation energy (compare, Hammond rule).]

Concerning (b): The steric influence of substituents, which sometimes affects the reactivity of the radical and the monomer in a decisive way, can not be taken into account. It is assumed that the inductive effect of a substituent on the monomer and on its radical is the same.

To these possible sources of errors, one also has to add the considerable difficulty in the analytical determination of the copolymer composition which makes it difficult to give a very narrow limit of error for the r_1 and r_2 values. Therefore give a very narrow limit of error for the r_1 and r_2 values. Therefore a very close correspondence between the experimentally determined r_1 and r_2 values and those calculated from the Q-e values should not really be expected. For a quantitative demonstration of the validity of the Q-e equation see Equation (i).

The "Pattern" Scheme

A relationship with three parameters—for polar effects on the radical (σ), for polar effects on the monomer (α), and for resonance effects on the monomer (β)—was derived by Bamford, Jenkins, and Johnston. It is called the "Pattern" Scheme:

$$\log r_1 = \sigma_1 (\alpha_1 - \alpha_2) + (\beta_1 - \beta_2) \tag{e}$$
$$\log r_2 = \sigma_2 (\alpha_1 - \alpha_2) - (\beta_1 - \beta_2) \tag{f}$$

The quantity σ is the Hammett constant for a substituent R for para-R-benzene derivatives.

These equations are obtained, if one describes the reaction rate constant of any reaction of a monomer with a radical by the semi-empirical equation: $\log k = \log k_{tr(toluene)} + \alpha \cdot \sigma + \beta$ (where k_{tr} is the absolute rate constant for chain transfer of the radical to the solvent toluene at 60°C). With the aid of this equation one can then form the ratio $r_1 = k_{11}/k_{12}$ or its logarithm $\log r_1 = \log k_{11} - \log k_{12}$. The "Pattern" Scheme is only of advantage in cases where the Hammett constant σ can be determined experimentally. Usually it is simply coupled with the α value according to $\alpha = -5.3 \cdot \sigma$, which, however, is rather meaningless.

The q-ε Scheme

Schwan and Price attempted to describe Q and e values theoretically and quantitatively by writing the equations for Q and e in the following manner:

$$Q = \exp\left[-q/RT\right] ; e = \epsilon/(rDRT)^{1/2},$$

where q = resonance energy (cal/mole); ϵ = induced charge (e.s.u.); r = distance between the charge centers; and D = dielectric constant in the charges.

As long as it is not possible to determine the values of Q, ϵ, r, and D experimentally (independent of r_1 and r_2), one is still dependent on the use of styrene as a standard substance ($Q = 1$ and $e = -0.8$). The ϵ and q values therefore did not assume any practical importance.

$r_1 \cdot r_2 \leqslant 1$

Both from the Q-e scheme and also from the "Pattern" Scheme one obtains two interesting conclusions which can be experimentally checked: if one forms with the aid of the Q-e equation [p. 138-(b) and p. 138-(c)] the product $r_1 \cdot r_2$ one obtains:

$$r_1 \cdot r_2 = \frac{Q_1}{Q_2} \cdot \frac{Q_2}{Q_1} \cdot e^{-e_1(e_1 - e_2)} \cdot e^{-e_2(e_2 - e_1)}$$

$$= e^{-(e_1^2 - 2e_1 e_2 + e_2^2)}$$

or

$$r_1 \cdot r_2 = e^{-(e_1 - e_2)^2} \tag{g}$$

In Equation (g) the difference between the e values appears to the square power. The exponent is therefore always negative. The product $r_1 \cdot r_2$ has therefore its maximum value when $(e_1 - e_2) = 0$. Since $e^0 = 1$, one has to conclude:

$$r_1 \cdot r_2 \leqslant 1 \tag{h}$$

One actually finds when looking over the $r_1 \cdot r_2$ values (for example, in Table 107) that there are only a few monomer pairs where the product of the copolymerization parameters is larger than 1. It is difficult to decide if in these cases where $r_1 \cdot r_2 > 1$ there are experimental errors in the determination of r_1 and r_2 which have caused this result, or whether Equation (h), and, therefore, the Q-e relationship is not generally valid.

A similar conclusion from the Q-e scheme (which represents a sharper criticism for its validity), is obtained if one applies the Q-e equation (Equation (a), p. 137), to ternary systems and forms the products $r_{12} \cdot r_{23} \cdot r_{31}$ and $r_{13} \cdot r_{32} \cdot r_{21}$ ($r_{12} = k_{11}/k_{12}, r_{23} = k_{22}/k_{23}$ etc):

$$r_{12} \cdot r_{23} \cdot r_{31} = \frac{Q_1}{Q_2} \cdot \frac{Q_2}{Q_3} \cdot \frac{Q_3}{Q_1} \cdot e^{-e_1(e_1-e_2)} \cdot e^{-e_2(e_2-e_3)} \cdot e^{-e_3(e_3-e_1)}$$

and

$$r_{13} \cdot r_{32} \cdot r_{21} = \frac{Q_1}{Q_3} \cdot \frac{Q_3}{Q_2} \cdot \frac{Q_2}{Q_1} \cdot e^{-e_1(e_1-e_3)} \cdot e^{-e_3(e_3-e_2)} \cdot e^{-e_2(e_2-e_1)}$$

After multiplication, one finds that the exponential terms in the two equations are equal to each other. From this one obtains:

$$r_{12} \cdot r_{23} \cdot r_{31} = r_{13} \cdot r_{32} \cdot r_{21}$$

or,

$$H = \frac{r_{12} \cdot r_{23} \cdot r_{31}}{r_{13} \cdot r_{32} \cdot r_{21}} = 1 \tag{i}$$

In reality the ratio H in Equation (i) is not always equal to 1, and one therefore uses the H value as a measure for the more or less good validity of the Q-e equation, and, since Equation (i) can also be derived from the Bamford Equation [(e) and (f)], for the validity of the "Pattern" Scheme. For ternary systems which consist either of three monomers with conjugated double bonds or three monomers without conjugated double bonds, the H values (calculated from Q and e values) lie between 0.5 and 1. [The H values may be smaller or larger than 1, depending on the numbering of the monomers (M_1, M_2, M_3). One usually chooses the numbering in such a way that H is smaller than 1.] Much larger deviations of H from H = 1 are obtained with ternary systems where some monomers are with, and some without, conjugated double bonds. H values near to 1 are obtained from r values which have been calculated according to the "Pattern" Scheme from experimentally deter-mined σ values, which shows that the influence of a substituent on the polarization of the monomer and the radical is different. (With the Q-e scheme this has been disregarded.)

The π Extension Theory

Fukui has attempted to calculate the copolymerization parameters with the aid of molecular orbital theory (MO theory). He assumes that in the transition state the chain radical and the added monomer molecule form a single extended π orbital, and that the gain in the delocalization energy determines whether M_1 or M_2 is added. Polar effects are taken into account through consideration of the Coulomb integral and the bonding-integral of the heteroatoms in the substituents.

Practical Significance of the Q-e Scheme

The practical significance of all methods for the determination of r values from parameters which have been assigned to individual monomers rather than monomer pairs, becomes smaller the more r values for monomer pairs have been experimentally determined. Even now, almost all even faintly interesting combinations of available monomers have been copolymerized and their r_1 and r_2 values determined. From year to year also the number of calculated ternary systems becomes larger, so that for all practical purposes a method for the characterization of individual monomers becomes unnecessary. The significance of the Q-e values, and similar parameters is therefore mostly of a theoretical nature.

Technical Importance of Copolymerization

Copolymerization plays an especially important role in current technology, because with its aid, it is possible to produce different copolymers whose properties combine and extend the properties of the corresponding homopolymers. Thus it is possible to prepare polymeric materials whose properties can be tailored toward a certain application. Hard polymers, as for example, polyvinylchloride, can be made softer by copolymerization with monomers whose homopolymers are rubber-like, such as with vinyl ethers or acrylic esters (internal plasticization). By copolymerization with acrylonitrile, one obtains polymers characterized by a great solvent resistance, for example, benzene-resistant rubbers from butadiene and acrylonitrile, or benzene-resistant molded products from copolymers of styrene and acrylonitrile. Monomers with a high chlorine or bromine content can be used to reduce the flammability of polymers. Through copolymerization of styrene with unsaturated polyesters (from glycols and maleic anhydride), one obtains polyester casting resins, which, reinforced with glass fibers, are attaining greater and greater importance. Copolymerization of styrene and butadiene leads to SBR synthetic rubber (Styrene-Butadiene Rubber), which may be considered to be the most important of the technical copolymers.

There is a series of monomers, such as maleic anhydride, the fumarates, the maleates, α-methylstyrene, and the vinyl ethers, which by themselves do not polymerize, or only slowly, by a radical mechanism, but which are often used as the components of a copolymerization and then copolymerize by a radical mechanism.

Often one tries to incorporate a small number of functional groups (such as $-OH$, $-COOH$, $-CONH_2$, $-N=C=O$, $-NH_2$, $-CH_2-CH\underset{\underset{O}{\diagdown\diagup}}{-}CH_2$) into a polymer molecule. This can be done in an elegant manner by copolymerization with suitable monomers. Monomers which are often used, and which can be obtained commercially, are, for example, acrylic acid, methacrylic acid, maleic anhydride, acrylamide, methacrylamide, and the following:

Glycidylmethacrylate:

$$CH_2=\overset{\displaystyle CH_3}{\underset{\displaystyle COO-CH_2-\overset{\displaystyle \diagdown O\diagup}{CH-CH_2}}{C}}$$

N-dimethyl-aminoethylacrylate :

$$CH_2=\overset{\displaystyle CH}{\underset{\displaystyle COO-CH_2-CH_2-N\diagup^{CH_3}_{\diagdown CH_3}}{}}$$

Glycolmonoacrylic ester :

$$CH_2=\overset{\displaystyle CH}{\underset{\displaystyle COO-CH_2-CH_2-OH}{}}$$

Isocyanato-ethyl-methacrylic ester: $CH_2=\overset{\displaystyle CH_3}{C}-COO-CH_2-CH_2-N=C=O$

One can utilize such copolymers, with reactive groups on the molecular chain, to prepare graft-copolymers, or to permit cross-linking of these macromolecules by the addition of bifunctional or polyfunctional molecules. This occurs, for example, with the so-called "reactive" coatings, where the copolymer solution is sprayed or painted onto the surface after the addition of the cross-linking agent. On drying or subsequent heating at temperatures from 100°C to 200°C, the functional groups of the copolymer chains react with the groups of the added bifunctional compounds, and cross-linking occurs through which the lacquer becomes insoluble and infusible. It should be remembered that through such cross-linking the coating not only becomes hard and insoluble, but often also brittle, which of course is undesirable. By careful control of the amount of cross-linking agent added, and of the reactive groups in the polymer chain, one can find the optimum conditions for a film or coating which is hard and insoluble without being brittle.

The technical utilization of such copolymers can also be illustrated by poly-acrylonitrile fibers. Pure polyacrylonitrile is very difficult to dye, but copolymers of acrylonitrile with small amounts of acrylic acid, acrylamide, vinylpyridine, or vinylpyrrolidone, on the other hand, are easily dyed. Polypropylene fibers, which also are very difficult to dye, cannot be improved by copolymerization, because acidic or basic groups would destroy the catalyst system which is utilized for the polymerization of propylene (compare pp. 184-201). On the other hand, it is possible to improve the situation by subsequent grafting of acrylic acid, or vinylpyrrolidone, on to the polypropylene chain (for example, by irradiation).

2116   Block- and Graft-Copolymers

In normal copolymerization the different monomers alternate in the chain either in a more or less regular, or in a completely random manner. However in block-copolymerization, the macromolecule consists of *blocks* of considerable length consisting entirely of one type of monomer:

A–A–A ⌁⌁⌁ A–A–A–B–B–B ⌁⌁⌁ B–B–B–C–C–C ⌁⌁⌁ C–C–C

The individual chain segments are usually not equally long. It is possible to prepare copolymers with many different combinations of blocks:

A 〰〰 A–B 〰〰 B–A 〰〰 A–B 〰〰 B–A 〰〰 A

A 〰〰 A–B 〰〰 B–A 〰〰 A

A–B–A–B–A–B 〰〰 A–B–A–A–A–A 〰〰 A–A–A–A–B–A–B–A–B〰〰

The blocks do not always have to be homopolymers, but might in turn be copolymers. (The preparation of block-copolymers by different methods will be described at the end of this chapter, see p. 301.)

A method for the preparation of graft-copolymers was already mentioned earlier on page 67 in discussing chain transfer. However, the methods for the preparation of graft-copolymers are not restricted to those involving polymerization and are therefore discussed in greater detail at the end of this chapter.

2117   Techniques of Radical Polymerization

One of the characteristics of radical polymerization is the fact that it can be carried out also in aqueous medium.[18] This leads to a number of interesting variations in the technical methods used for polymerization. The following techniques have been utilized in industry: bulk polymerization, solution polymerization, precipitation polymerization, suspension polymerization (Pearl polymerization), and emulsion polymerization.

Bulk Polymerization

The polymerization of the pure monomer without diluent is called bulk polymerization or mass polymerization. The monomer (styrene, vinylchloride, vinylacetate, acrylic esters, butadiene, acrylonitrile, vinylpyrrolidone, etc.) is first purified to remove oxygen or other inhibitors (by bubbling nitrogen through it, by distillation, or by evacuation) and then the polymerization is started through heating, UV irradiation, or the addition of initiator (peroxides, azo compounds, $c \cong 0.1\%$). Usually, after a short period of heating, the reaction mixture continues to heat up by itself, and therefore it is necessary to remove the heat by cooling. At the beginning of the polymerization, and with small polymerizations in a test tube, this is still possible. However, with increasing conversion, because of the rapidly increasing viscosity of the polymer-monomer mixture, this becomes more and more difficult. Therefore, with large amounts of monomer, bulk polymerization often takes very turbulent and even explosive form, as a result of the rapidly increasing temperature. The violence of the reaction is even further increased by the increase in the radical concentration which occurs with increasing viscosity (self-acceleration, compare p. 89). The difficulty in heat removal is also the reason why in industry bulk polymerization is only carried out in a few cases. However, in the cases where it is used, it is done on a very large scale, such as in the bulk polymerization of styrene or ethylene (high pressure process) and, to a smaller extent, in the fabrication of Plexiglas (polymethylmethacrylate) and the production of molded products from polyester casting-resin solutions. Because the possibility of chain transfer is relatively small and because of self-acceleration, one often

---

[18]Ionic polymerizations in aqueous solution, on the other hand, are very rare (p. 163).

obtains polymers with a high molecular weight in bulk polymerization. This, depending on the intended application of the polymer, is an advantage or a disadvantage. One of the characteristics of bulk polymerization which is always of technical advantage is the great purity of the polymer resulting from the lack of additives during the polymerization.

Solution Polymerization

If one adds an inert solvent to the monomer, then the polymerization can be controlled much more readily. This is especially the case if the solvent is chosen in such a way that it boils at the desired temperature of polymerization. In that case the heat of polymerization (which is of the order of 20 kcal/mole), is used up for the evaporation of solvent. Furthermore, one can recycle the solvent, letting the cool solvent run back into the polymerization reactor, and, in addition, one can cool from the outside. The concentration of the solvent can be chosen in such a way that the polymerization mixture can still be stirred after complete conversion.

Solution polymerization has been employed almost only in cases where the polymer is then used in the form of solutions (50-60%) for lacquers, adhesives, impregnation materials, etc. To obtain the pure polymer by distilling off the solvent is relatively complicated because the hard polymer cannot be taken out of the vessel after evaporation of the solvent. Only through construction of extruders with vacuum distillation zones, and by using other special evaporaters, has it become possible in an economic manner to separate the polymer from the solvent and to obtain the pure polymer. Therefore, solution polymerization is now becoming increasingly important from a technical point of view. In the choice of a solvent, one has to consider the chain transfer constant, because this influences the molecular weight to a considerable extent. Because of chain transfer with the solvent and because of the lower monomer concentration (compare Equation [25]) the molecular weight of polymers prepared by solution polymerization is usually lower than that of corresponding bulk polymers.

Commercial solution polymerizations are usually not carried out to high conversions (near 100%) but continuously at a constant monomer concentration. The unreacted and evaporated monomer is recycled together with the solvent.

This type of production process has two advantages: The reactor works always in a range of high polymerization rates (high monomer concentration), and the molecular weight distribution curve is not so broad as it is with polymers produced in a discontinuous process with high conversions (compare Equations [27] and [97 p. 420]: varying [M] leads to varying $\bar{P}$ and α values).

Precipitation Polymerization

Not all polymers are soluble in their own monomer. For example, polyvinyl-chloride is insoluble in vinylchloride, and polyacrylonitrile is insoluble in acryloni-trile. These polymers precipitate during polymerization first as gel particles, and these then become a white, more or less fine, powder. Insolubility occurs when one reaches a certain concentration of polymer, so that initially one only observes a slight cloudiness, but as the polymerization continues, the polymer begins to precipitate.

Actually one should call this type of polymerization a bulk-precipitation polymerization, because one can obtain the same effect (i.e., the continuous precipitation of the polymer during the polymerization) if one mixes the monomer with a suitable solvent (for instance, methanol) in which the monomer is soluble in every concentration, but the polymer is insoluble. In this manner (by solution-precipitation polymerization), one can also prepare polymers which are soluble in their own monomer by precipitation polymerization.

Both types of precipitation polymerization have been and are now also employed by industry, one for the production of polyvinylchloride,[19] and the other for the production of acrylonitrile (in water) and styrene-acrylonitrile copolymers (in methanol).

Suspension Polymerization (Pearl Polymerization)

If one takes a water-insoluble monomer, such as styrene or methylmethacrylate, and three to four times the same amount of water, and mixes them intensively, one obtains a system in which the monomer is suspended in the aqueous phase in the form of small spheres of 0.1-1 mm diameter. By the addition of catalysts and by heating, one can start the polymerization. In the course of a few hours (or even faster, depending on the temperature and the catalyst concentration), the monomer droplets become highly viscous and sticky, and the whole mass coagulates into a gel-like precipitate, which, as in bulk polymerization, can continue to polymerize rather violently. This is because the droplets, which collide in short succession with each other, and which agglutinate and melt into each other, can only be brought back into the form of small droplets by stirring, as long as the viscosity is rather low. With increasing viscosity and "stickiness" of the solution, the separation into small particles becomes more and more difficult, so that the system finally cannot be prevented from complete coagulation.

However, coagulation can be prevented by the addition of hydrophilic protective colloids. Even with a low protective colloid concentration ($<0.1\%$), a thin film begins to form at the interface. This results in the highly viscous spheres being repelled by each other, and coagulation is prevented. After completed polymerization one obtains hard, glassy pearls in the form of a polydispersed mixture. By careful choice of the stirring conditions (rotations per second, form of the stirrer, form of the vessel, etc.), and by the type and amount of colloid, one can influence the size of the pearls and their size distribution. With slower stirring one obtains larger pearls, and with faster stirring, smaller pearls.

There are two types of protective colloids useful with pearl polymerization: 1.) water-soluble macromolecular compounds [such as gelatins, agar, pectins, alginates, methylcelluloses, polyvinylalcohols, polyacrylates (or methacrylates), polyvinyl-

---

[19]Not very extensive. More than 95% of polyvinylchloride (PVC) is produced by suspension polymerization.

pyrrolidone, polymethacrylamide, and polymethylvinylethers] and 2.) finely divided water-insoluble minerals (such as clays, bentonites, kieselgur, calcium-phosphates, and barium sulfate). In addition to the protective colloids, one usually also adds water-soluble phosphates (Na-pyrophosphate, Na-tripolyphosphate) in order to lower the surface tension and to stabilize the pH. After the polymerization is completed, the pearls are centrifuged or filtered off, and the protective colloid is removed by intensive washing with water. In view of the fact that on a technical scale it is not very simple to wash the pearls until they are absolutely clean, pearl polymers usually do not have the same high purity as bulk polymers. On a technical scale, suspension polymerization is used in the production of polyvinylchloride, polystyrene, polymethylmethacrylate, and others. For the production of rubbery, sticky, polymers (for example, the polyacrylates) pearl polymerization is less suited. During the polymerization of vinylchloride, which is carried out in the United States almost entirely by suspension polymerization and in Germany for the most part by suspension polymerization, one obtains either glassy, transparent pearls, or opaque, white, porous pearls, or even both types of pearls together (depending on polymerization conditions, type and concentration of the protective colloid, temperature, and monomer: water ratio). Because polyvinylchloride is insoluble in its own monomer, a precipitation polymerization is going on inside each pearl, so that temporarily there is a three phase system consisting of water, monomeric vinylchloride, and, inside the latter, as a solid phase, polyvinylchloride.

Emulsion Polymerization

If under light stirring, one disperses a water-insoluble monomer (styrene, acrylic esters, vinylchloride, vinylacetate) in an aqueous soap solution (aqueous phase to monomer phase = 2:1 − 1:1), one obtains an emulsion with a milky appearance. If one adds to this emulsion a water-soluble initiator (for example, potassium persulfate) this starts (in the absence of oxygen, or other inhibitors) a polymerization at 40°C-70°C, which can be often recognized through development of heat and a bluish or colored "fluorescence" of the latex at the side of the reaction vessel caused by the tendency of the spherical latex particles to form regular three-dimensional lattices. The visible light is diffracted by these lattices in the same manner as X-rays are diffracted by crystals (Wesslau and Luck). The latex character remains during the polymerization, and one obtains a polymer dispersion very similar to a natural rubber latex, such as is obtained from the Hevea tree. Such lattices prepared by emulsion polymerization can be stable for months and years, if one uses suitable emulsifying agents.

On the other hand, such a polymer dispersion can also be easily coagulated if one adds methanol or electrolytes (sodium chloride, sodium sulfate, aluminum chloride, formic acid, or acetic acid), or if one freezes it. This allows the preparation of the polymer in pure form after suitable washing. The purity of such an emulsion polymer is, of course, less than that of a pearl polymer and especially of a bulk polymer, because, particularly on a technical scale, it is never possible to remove the soap residue completely. In many cases one does not even attempt to remove the soap (one usually speaks about emulsifiers) and simply evaporates the water on the mill or by spray-drying, in which case the polymer is obtained as a finely divided powder. Of course this is only possible with hard and nonsticky polymers.

In all cases where the presence of emulsifier is not disturbing, emulsion polymerization can be used to advantage. In comparison with the other polymerization techniques it has the following advantages: the polymerization heat can be removed very easily; and the viscosity of the lattices even with high concentration (up to 60%) is low in comparison with corresponding solutions. One can also prepare rubbery and sticky polymers in this way.

One large-scale use of emulsion polymerization is the production of synthetic rubber and, on a smaller scale, of polyvinylchloride and polystyrene. The other large use is the production of plastics dispersions used as such, without first coagulating them, for the production of paints, pigment inks, coatings, impregnations, floor coverings, and adhesive pastes (polyvinylacetate, polyvinylpropionate, and polyacrylic ester dispersions).

The emulsifiers are used at a concentration of 0.2-1% (based on the monomer).

The following compounds are suitable:

1. Anionic soaps: alkali salts of palmitic acid, stearic acid, oleic acid, sodium salts of sulfonic acids with 10-16 carbon atoms [such as are obtained by sulfo-chlorination of Fischer-Tropsch paraffins (Mersolate)], and alkali salts of half-esters of phthalic acid with C_8-C_{18} alcohols.

2. Cationic soaps: cetylamine hydrochloride and salts of other fatty acid amines with concentrated acids. The cationic soaps are only seldom used as emulsifiers in emulsion polymerization.

3. Nonionic emulsifiers, which are obtained by the reaction of long-chain alcohols or alkylated phenols with ethylene oxide (ethoxylation). In these compounds the polyethoxy chain (4-100 ethylene oxide residues) represents the hydrophilic component of the emulsifier. The nonionic emulsifiers are characterized by their resistance against the addition of electrolytes and partly also their resistance against cold temperatures. Their emulsifying action is, however, lower than that of the soaps, and consequently one has begun to use emulsifiers which contain sulfonic acid groups as well as polyethylene oxide residues.

In addition to the emulsifiers, one usually adds other compounds, such as sodium pyrophosphate and sodium polyphosphates, which bring about a thickening of the dispersion. In order to obtain a satisfactory polymerization one usually has to maintain a certain pH. Styrene polymerizes more readily in a slightly alkaline medium, whereas vinylacetate polymerizes better in a slightly acidic one.

In addition to the normal emulsion polymerization with soap-like emulsifiers, one can utilize emulsion polymerization with protective colloids for the preparation of wood glues and binders for paints. In such cases one uses higher concentrations (4-5%) of water-soluble macromolecular compounds such as gelatins, polyvinyl alcohol, or polyvinylpyrrolidone, and one obtains paste-like polymer dispersions with spherical particles whose sizes are approximately 1μ (just barely seen in the optical microscope). The particle size, with soap dispersions, is only of the order of 0.05-0.15μ, and this cannot be seen in the optical microscope. In emulsion polymerizations with water-soluble protective colloids as emulsifiers, we have to assume the formation of graft copolymers between the macromolecules of the emulsifier and the growing polymer chains. This was demonstrated in some cases.

The initiators commonly used in emulsion polymerization are such water-soluble compounds as H_2O_2 and potassium persulfate. With the production of synthetic rubber, one generally uses redox systems consisting of a water-insoluble hydroperoxide and a water-soluble reducing agent, which produce a rapid emulsion polymerization at $0°C$ and even lower.

Kinetics of Emulsion Polymerization
Mechanism of Emulsion Polymerization[20]

With regard to mechanism, one must distinguish between two types of emulsion polymerization: emulsion polymerization with water-insoluble initiators (azobisisobutyronitrile, benzoylperoxide, and cumenehydroperoxide) and true emulsion polymerization with water-soluble initiators or initiator systems (potassium-persulfate, redox systems with water-soluble components, such as cumenehydroperoxide + Fe^{++}, ascorbic acid, or polyamines).[21]

In emulsion polymerization with water-insoluble initiators, the polymerization occurs mostly in the monomer droplets which in the course of the polymerization pass through a highly viscous stage and then, depending on the type of monomer, become hard or rubbery polymer beads of about the same size as the monomer droplets (0.5-10μ). Just as is the case with suspension polymerization, this type of emulsion polymerization has the same kinetics as bulk polymerization. In addition to these pearls of the size of the monomer droplets, emulsion polymerization with monomer-soluble initiators yields more or less large amounts (usually about 50% of the polymer) of a stable polymer latex whose particles are smaller than 0.5μ. This material is formed because a part of the initiator or of the initiator radicals is soluble in the aqueous soap phase and initiates polymerization there.

Using water-soluble initiators leads to a completely different polymerization mechanism (see Figure 155). While the monomer droplets have a diameter of the order of 0.5-10μ, the polymer particles are in this case smaller by an order of magnitude (0.05-0.1μ and in exceptional cases up to 0.5μ) and therefore cannot be seen under the light microscope. From this one has to conclude that the polymer particles in the aqueous phase, starting from the smallest primary particle, are newly formed. And it is especially important to recognize that the number of primary seeds per cubic centimeter is much greater than the number of monomer droplets per cubic centimeter. The fact that the number of particles per cubic

---

[20]The mechanism of emulsion polymerization was correctly interpreted for the first time by H. Fikentscher in a qualitative manner and by Smith and Ewart quantitatively.

[21]Type: $H_2N-CH_2-CH_2[NH-CH_2-CH_2]_nNH_2$.

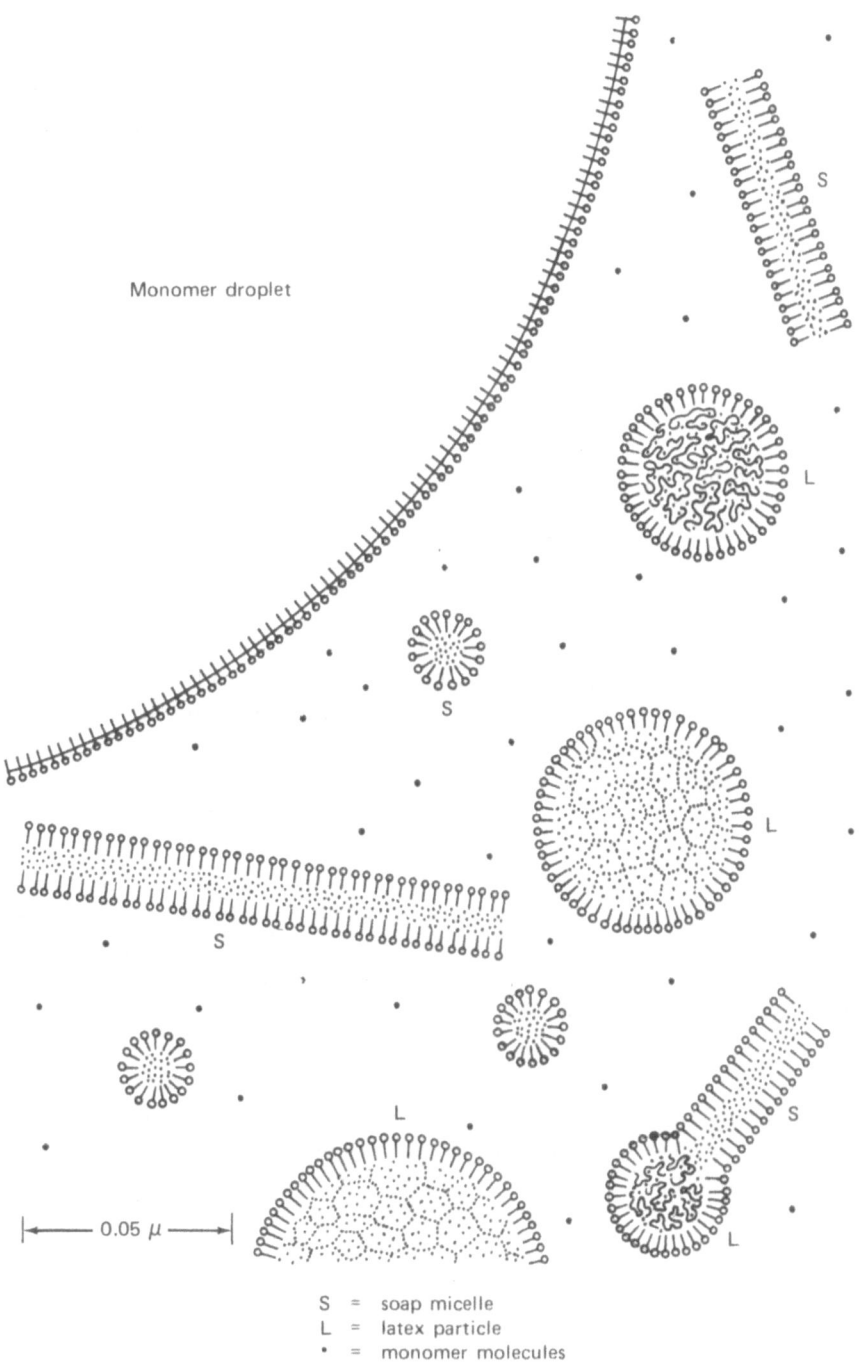

Monomer droplet

├── 0.05 μ ──┤

S = soap micelle
L = latex particle
• = monomer molecules

FIG. 155 – Schematic representation of emulsion polymerization

centimeter is the greater (and the diameter of the polymer spheres, of course, the smaller), the higher the soap concentration, leads to the conclusion that the formation of the primary seeds occurs within the soap micelles. This also explains the observation that the surface tension of the emulsion, which is much lower than that of water because of the presence of free soap micelles, increases very rapidly after the start of polymerization.

The soap-forming salts (sodium palmitate, sodium lauryl sulfonate, etc.) are not molecularly dispersed in aqueous solution, but they form aggregates called micelles. If one imagines that the individual molecules are parallel or radially ordered with respect to each other so that the hydrophilic groups (COO-or SO_3) form the outer surface, then one obtains the structures shown in Figure 155.

Between the lamellae of the micelles (where the paraffin chains come together) organic solvents or monomers can be incorporated. Thus one can explain that the solubility of styrene in water (at 20°C soluble to 0.02%) can be increased by a factor of 10 through the addition of 2% soap. Even though one still has not come to a uniform picture of the details of the micelle structure,[22] it is clear that the size of the micelles increases in the presence of water-insoluble monomers or solvents. This can only be explained by the assumption that a certain amount of monomer is taken up by the micelles.

Because of the high monomer concentration in the soap micelles, the initiation of a chain through absorption by the micelle of a radical formed in the aqueous phase becomes very likely. One has to leave open the possibility that also outside the soap micelles, somewhere in the aqueous phase, a chain can be initiated. However, these polymer particles formed outside the soap micelles are also immediately covered with a soap skin, which consists of soap molecules from the micelles. Wherever the polymer particles are formed, they can exist as single particles only if there is sufficient soap available. Therefore, regardless of whether the polymer particles are in the micelles or outside, it is the soap concentration which determines their number.

The monomer necessary for propagation of the growing chain is found inside the soap micelle. After this monomer concentration in the micelle decreases because of the polymerization, monomer is transferred from the monomer droplet in the aqueous phase into the micelles and therefore replenishes the monomer concentration in the micelle. As long as there are monomer droplets left in the aqueous phase, the monomer concentration has a constant saturation value. The newly formed polymer molecules, or those in the process of being formed, suck up monomer just like small sponges. The small particles have the consistency of a gel which has the desire, by absorption of monomer, to swell further. However, the gel never becomes highly diluted, because the monomer which is taken up is continually used up through conversion to polymer.

---

[22]The structure of the micelles depends on the soap concentration. At high soap concentrations the micelles are rod-like and have a length of 1000-3000 Å. At low soap concentrations (1-2%) the micelles are smaller and have a spherical shape (50-100 Å diameter).

The soap micelles are rapidly enlarged by the growing polymer particles, and eventually the micelle bursts and the individual soap molecules surround the polymer particle only as a protecting layer. Because of the ionization of the soap molecules, this electrically charged soap skin prevents a coagulation of the polymer particles with each other or with the monomer droplets, which themselves are also covered with such a soap skin.

After the soap micelles have been thus converted to polymer particles, further polymerization takes place only in these particles. The monomer diffuses from the monomer droplets into the aqueous phase at the same rate as it is removed from the aqueous phase by the growing polymer particles.

Inside the soap micelle (i.e., inside a polymer particle surrounded by a soap skin) a growing chain can grow until a second initiator radical (or a growing chain existing in isolation in the aqueous phase) is absorbed by the micelle. The growing chain cannot react with a growing chain in *another* micelle, or polymer particle, because the particles are separated from each other by the soap skin surrounding both of them. Because of the small size of the particles (seeds) and the strong tendency of radicals to react with each other, the existence of two or more radicals within one soap micelle or polymer particle is highly improbable. The new radical entering the micelle, therefore, (in almost all cases) terminates the growing chain radical before it can initiate a new chain of its own. A new chain will start only when a further initiator radical enters the polymer particle.

This special type of chain growth in individual polymer particles (which are isolated from each other by a soap skin) and the resulting cycle of growth followed by stagnation determine a characteristic kinetics for emulsion polymerization which is different from that of bulk polymerization.

Since the probability of radical absorption is independent of the state in which the polymer particles finds itself (i.e., growth or stagnation), both periods are on the average equally long. Statistically, therefore, at a given time always half the particles are occupied by a radical while the other half are without a radical. If N is the number of particles present in 1 cm^3 of emulsion, then the number of radicals present in 1 cm^3 is equal to N/2. As quantitative considerations have shown, the average length of a growth or stagnation period is somewhere between 1 and 100 seconds, depending upon the initiator concentration and the number of polymer particles per cm^3.

Rate of Polymerization

As has been derived earlier (p. 86), the number n_m of monomer molecules which add to a single chain radical per second is equal to $k_p \cdot [M]$, and therefore $N/2 \cdot k_p \cdot [M]$ monomer molecules are added to N/2 radicals per second. N/2 is the number of radicals present in 1 cm^3 of emulsion and this amount remains constant during the polymerization. One can define, therefore, the rate of polymerization in an emulsion polymerization as

$$r_P = (N/2) \cdot k_p \cdot [M] \text{ molecules} \cdot cm^{-3} \cdot sec^{-1} \tag{40}$$

This is equal to the number of monomer molecules in 1 cm³ of emulsion which are used up in the polymerization in 1 second. If one divides this by Avogadro's Number, one obtains the number of moles, instead of the number of molecules, which are added per 1 cm³ of emulsion per second.

$$r_P = \tfrac{1}{2}(N/N_A) \cdot k_p \cdot [M] \text{ moles} \cdot cm^{-3} \cdot sec^{-1} \tag{40a}$$

Of course, one could again write [R $\sim\sim\sim$ M·] instead of $N/2N_A$. However, by using N, the number of radicals, instead of the molar concentration, we make clear that the number of chain-radicals in an emulsion polymerization is an experimentally measurable quantity (for example, by counting of particles in the electron microscope, (Compare Figure 497) or from the sedimentation rate in the ultracentrifuge). Furthermore, this expression shows that the concentration of chain radicals in emulsion polymerization is strongly related to the number N of polymer particles in the dispersion. This means that with emulsion polymerization—unlike bulk and solution polymerization—the chain radical concentration is independent of the rate constants of the termination reaction. The number N of polymer particles is determined by two quantities: by the emulsifier concentration and by the rate of initiator decomposition.

That the number of polymer particles formed is larger (and, therefore, the diameter of the particles smaller) the larger the emulsifier concentration can be seen readily if one considers that the total surface stabilized by the emulsifier is the larger, the smaller and more numerous the particles. If much emulsifier is available, the system will have a large surface area (i.e., many small polymer particles). With small emulsifier concentration, the total surface for which emulsifier is available is smaller, and therefore from the same amount of monomer one obtains a dispersion with fewer polymer particles but with larger diameters.

The influence of the initiator radical concentration on the number of particles is based on the fact that an equilibrium exists between the growing polymer particles and the unoccupied soap micelles. This means that only a certain part of the emulsifier is used to cover the surface of the polymer particles, whereas the rest is present in the form of unoccupied micelles. Both the polymer particles and the micelles have the same capability of absorbing initiator radicals and therefore compete with each other for these radicals. The more radicals are formed per second, the sooner the number of polymer particles will increase and the number of free micelles will decrease. This leads to an increase in the number of polymer particles because the probability of initiator radical absorption by an unoccupied micelle will increase with increasing concentration of initiator radicals.

As quantitative considerations have shown, the number of particles increases according to the relation $N \propto n_R^{2/5}$, where n_R is the number of newly formed radicals in 1 cm³ of dispersion per second. N is also proportional to $O_S^{3/5}$, where O_S is the surface which can be covered by the soap present in 1 cm³ of emulsion.

The number of polymer particles per cm³, N, under the usual technical conditions (soap concentration about 1%, initiator concentration 0.1%, both based on monomer) can vary between 10^{14} and 10^{15} cm^{-3}. Both parameters, N and [M], hardly change during the course of the emulsion polymerization so that the rate of polymerization r_P remains constant up to high conversions (60-80%) as has been demonstrated. Only at conversions higher than 80% (when no more monomer droplets are present) does the monomer concentration [M] in the polymer particles decrease and r_P become slower.

In bulk and solution polymerization the radical concentration increases at the beginning of the polymerization until the rate of chain termination becomes equal to the rate of chain initiation (steady state). At that stage just as many radicals disappear in unit time through recombination as are formed as a result of initiation, and therefore the upper limit of the radical concentration has been reached. Accordingly, the rate of polymerization (at constant temperature) cannot increase beyond a certain limit. On the other hand, in emulsion polymerization there is no automatic self-control of the chain radical concentration by recombination, because the polymer particles are isolated from each other by the soap layer. Thus the chain radical concentration can be made considerably higher than is possible with other polymerization techniques under similar conditions.

The steady state is characterized by $r_t = r_i$. If we assume that the rate of initiator decomposition in bulk (B) and in emulsion (E) polymerizations is the same (by suitable choice of conditions), then we can write:

$$r_{t(B)} = k_{t(B)}\ [\text{R}\leadsto\text{M}\cdot]_B^2 = r_{t(E)} = k_{t(E)}\ [\text{R}\cdot]_E \cdot [\text{R}\leadsto\text{M}\cdot]_E = r_i \tag{40b}$$

If we assume further that the rate constants of both of the chain terminations are in the same order of maguitude, $k_{t(E)} \approx k_{t(B)}$, we obtain:

$$[\text{R}\leadsto\text{M}\cdot]_E \cdot [\text{R}\cdot]_E = [\text{R}\leadsto\text{M}\cdot]_B \cdot [\text{R}\leadsto\text{M}\cdot]_B$$

$$[\text{R}\leadsto\text{M}\cdot]_E = \frac{[\text{R}\leadsto\text{M}\cdot]_B}{[\text{R}\cdot]_E} \qquad [\text{R}\leadsto\text{M}\cdot]_B = x\ [\text{R}\leadsto\text{M}\cdot]_B \tag{40c}$$

$$x = [\text{R}\leadsto\text{M}\cdot]_B/\ [\text{R}\cdot]_E \text{ and } [\text{R}\cdot]_E = \frac{1}{x}\ [\text{R}\leadsto\text{M}\cdot]_B$$

According to Equation (40c), [R$\leadsto$M$\dot{\text{E}}$] means that $[\text{R}\cdot]_E$ is always smaller than $[\text{R}\leadsto\text{M}\cdot]_B$ by the same factor x, by which $[\text{R}\leadsto\text{M}\cdot]_E$ is larger than $[\text{R}\leadsto\text{M}\cdot]_B$. Since [R$\cdot$] with all polymerizations is considerably smaller than $[\text{R}\leadsto\text{M}\cdot]$ (1/x = 100-10,000, depending on the molecular weight), $[\text{R}\leadsto\text{M}\cdot]_E$ is much larger than $[\text{R}\leadsto\text{M}\cdot]_B$. It is not possible to give precise values, because $[\text{R}\cdot]_E$ contains the total amount of radicals which are possible reaction partners of $[\text{R}\leadsto\text{M}\cdot]_E$, that is, all radicals able to enter a polymer particle. One can assume that in this group are not only the initiator radical R$\cdot$ but also the oligomer-radicals R–M$\cdot$, R–M–M$\cdot$, R–M–M–M$\cdot$, etc. We do not exactly know how long the chain may be until it becomes unable to penetrate the soap layer of the polymer particles, but we can suppose that the limit lies between 10 and 100 structural units.

We have to remember that $[R\text{-}\!\sim\!\!\sim\!\text{-}M\cdot]_E$ is given by the number of polymer particles per cm^3, N. Also, the product $k_t \cdot [R\cdot]_E \cdot [R\text{-}\!\sim\!\!\sim\!\text{-}M\cdot]_E$ is fixed by r_i:

$$r_i = r_t = n_R/2 = k_t \, [R\cdot]_E \cdot [R\text{-}\!\sim\!\!\sim\!\text{-}M\cdot]_E \qquad (40d)$$
$$[R\text{-}\!\sim\!\!\sim\!\text{-}M\cdot]_E = N/2\,N_A,$$

where n_R is the number of radicals $R\cdot$ formed per second in 1 cm^3 of emulsion, of which the one half is used to initiate and the other half to terminate polymerization.

N depends on the concentration of emulsifier and on n_R, the number of radicals formed per second in 1 cm^3. Therefore, the soap concentration and the initiator concentration determine the magnitude of $[R\text{-}\!\sim\!\!\sim\!\text{-}M\cdot]_E$ and $[R\cdot]_E$. According to Equation (40d), $[R\cdot]_E$ is small with high concentration of emulsifier and vice versa. This seems reasonable because the larger the number of polymer particles, the faster the newly formed radicals are used up $(R\cdot + M \rightarrow R - M\cdot$ and $R\cdot + R\text{-}\!\sim\!\!\sim\!\text{-}M\cdot \rightarrow$ $R\text{-}\!\sim\!\!\sim\!\text{-}M - R)$ and the smaller becomes their concentration $[R\cdot]_E$, $[R-M\cdot]_E$, $[R-M-M\cdot]_E$, etc. Thus also the magnitude of the factor $x = [R\text{-}\!\sim\!\!\sim\!\text{-}M\cdot]_B / [R\cdot]_E$ depends on the concentration of emulsifier.

Because of $r_p = k_p \, [R\text{-}\!\sim\!\!\sim\!\text{-}M\cdot] \, [M]$, the polymerization rate of emulsion polymerizations should be larger by the same factor x than the polymerization rate of bulk polymerizations of the same monomer under comparable conditions. Indeed, this is a well-known, experimentally observed phenomenon. The bulk or pearl polymerization of styrene, for instance, at 80°C requires 30-40 hours to reach a conversion of 97 to 98%, while with emulsion polymerization of styrene at the same initiator concentration and the same temperature this conversion is reached in 3-4 hours. In this example, the factor is $x \approx 10$, if we can assume that k_p and $[M]$ have the same order of magnitude in both cases.

It is characteristic of emulsion polymerizations that—exactly contrary to other types of radical polymerization—the higher polymerization rate is not coupled with a lowering of the degree of polymerization (molecular weight). The same mechanism that affects the higher polymerization rate causes the molecular weight of polymers prepared by emulsion polymerization to be higher than that one of polymers prepared by other techniques.

Because of the influence of the emulsifier concentration on the number of particles and thus on $[R\text{-}\!\sim\!\!\sim\!\text{-}M\cdot]_E$, the polymerization rate and the degree of polymerization can be arbitrarily selected to fall within a certain range.

Degree of Polymerization

N is the number of polymer particles in 1 cm^3 of emulsion and n_R is the number of initiator radicals which are formed per second in 1 cm^3 of emulsion and are absorbed by the polymer particles. If N polymer particles on the average absorb n_R initiator radicals per second, then one polymer particle absorbs on the average n_R/N initiator radicals per second. However, if n_R/N initiator radicals are absorbed by a polymer particle per second, then, on the average, one initiator radical enters a polymer particle every $1/(n_R/N) = N/n_R$ seconds.

n_R/N is the frequency with which a polymer particle catches radicals and $1/(n_R/N) = N/n_R$ is the length of the period which lies between two absorption acts. This is the average length of the undisturbed chain growth period of the polymer chain in the polymer particle and also the average time of stagnation (when there is no growth). This, of course, corresponds to the average lifetime of a growing chain-radical. The number of monomer molecules which in this length of time N/n_R add to a chain radical, is by definition the kinetic chain length L_{kin}. In this case, because recombination of growing chain radicals can be neglected in emulsion polymerization, it is identical with the average degree of polymerization of the macromolecules formed (assuming that no chain transfer occurs).

Since in one second, $k_p \cdot [M]$ monomer molecules add to a chain radical [(derivation see p. 86 Equation (23)], in N/n_R seconds, i.e., during the average lifetime of a radical, $(N/n_R)k_p \cdot [M]$ molecules will add and one can write for emulsion polymers that:

$$L_{kin} = \bar{P} = (N/n_R) k_p \cdot [M] , \tag{41}$$

where $\bar{P}$ (average degree of polymerization) is the number of structural units in the macromolecule; k_p, the reaction rate constant of the growth reaction (chain propagation); $[M]$, the monomer concentration in the polymer particles during an emulsion polymerization in moles$\cdot l^{-1}$; N, the number of polymer particles in 1 cm^3 of emulsion; and n_R, the number of initiator radicals which are formed per second in 1 cm^3 of emulsion.

The degree of polymerization $\bar{P}$ is therefore dependent only on the number of polymer particles N and the number n_R of the radicals formed per second in 1 cm^3 (i.e., the initiator decomposition rate), (k_p and $[M]$ are constant for a given system.)

The number of particles, N, can be varied arbitrarily in a certain range by the emulsifier concentration. Therefore, according to Equation (41), also the average degree of polymerization can be influenced by the soap concentration: increasing the soap concentration results in a higher molecular weight. It is a noticeable fact that emulsion polymers always have considerably higher molecular weights than the corresponding bulk polymers prepared under comparable conditions.

Equation (42b) should not lead to the conclusion that $L_{kin} = r_P/r_t$ is larger with emulsion polymerization because the termination rate, $r_{t(E)} = k_t [R\cdot] [R\text{\textasciitilde\textasciitilde}M\cdot]_E$ is smaller than the termination rate of a bulk polymerization under comparable conditions ($r_{t(B)} = k_t [R\text{\textasciitilde\textasciitilde}M\cdot]_B \cdot [R\text{\textasciitilde\textasciitilde}M\cdot]_B$). It is true that $[R\cdot]$ is much smaller than $[R\text{\textasciitilde\textasciitilde}M\cdot]$, however $[R\text{\textasciitilde\textasciitilde}M\cdot]_E$ is correspondingly larger than $[R\text{\textasciitilde\textasciitilde}M\cdot]_B$. The number of disappearing radicals at the steady state is always, regardless of whether in bulk or emulsion polymerization, equal to the number of the newly formed radicals. Since the initiation rate (initiator decomposition) is approximately equal with emulsion and with bulk polymerization, we must have in the steady state also $r_{t(E)} \approx r_{t(B)}$.

The real reason for the higher degree of polymerization with emulsion polymers is the higher chain radical concentration $[R\text{\textasciitilde}M\cdot]_E$ and, therefore, the higher polymerization rate ($L_{kin} = v_P/v_t$). The fact that $[R\text{\textasciitilde}M\cdot]_E$ can be canceled out in Equation (42b) should not mislead us. It simply means that $[R\cdot]$ is always smaller than $[R\text{\textasciitilde}M\cdot]_B$ by the same factor, by which $[R\text{\textasciitilde}M\cdot]_E$ is larger than $[R\text{\textasciitilde}M\cdot]_B$.

To summarize: *one* peculiarity of emulsion polymerization is that the number N of polymer particles per cm^3, and therefore the chain radical concentration, can be increased without increasing the temperature or the initiator concentration. *The other* peculiarity lies in the fact that the chain radicals grow isolated from each other in separate polymer particles and therefore do not recombine or disproportionate. The consequences of this are the following:

1. Higher polymerization rate (with redox initiators one can carry out an emulsion polymerization even at temperatures below 0°C with an industrially useful rate, for example, for the production of synthetic rubber).

2. Higher molecular weight of emulsion polymers, and

3. The possibility of increasing the rate of polymerization and at the same time the degree of polymerization, by increasing the emulsifier concentration (in analogy to the Norrish-Trommsdorf effect, see p. 89).

With styrene, for example, bulk polymerization at 80°C leads to a polystyrene with a molecular weight of approximately 300,000. With emulsion polymerization, at the same temperature, one obtains a polystyrene with approximately five times that molecular weight.

This is understood immediately on considering the following relationships:

Bulk polymerization:

$$L_{kin} = \frac{r_P}{r_t} = \frac{k_p [M] \cdot [R\text{\textasciitilde}M\cdot]_B}{k_t [R\text{\textasciitilde}M\cdot]_B^2} = \frac{k_p [M]}{k_t [R\text{\textasciitilde}M\cdot]_B} \tag{42a}$$

$$[B = bulk]$$

Emulsion polymerization:

$$L_{kin} = \frac{r_P}{r_t} = \frac{k_p [M] \cdot [R\text{\textasciitilde}M\cdot]_E}{k_t [R\cdot] [R\text{\textasciitilde}M\cdot]_E} = \frac{k_p \cdot [M]}{k_t \cdot [R\cdot]} \tag{42b}$$

$$[E = emulsion]$$

Since with emulsion polymerization chain termination does not occur by reaction of two growing chain radicals $R\text{\textasciitilde}M\cdot$ ($r_t = k_t \cdot [R\text{\textasciitilde}M\cdot]^2$), but through reaction of a chain radical $R\text{\textasciitilde}M\cdot$ with an entering initiator radical $R\cdot$, the rate of termination $r_t = k_t [R\cdot] [R\text{\textasciitilde}M\cdot]$; and in place of the chain radical concentration [R⌇⌇⌇⌇M·] in Equation (42a), one has to substitute the initiator radical concentration [R·]. Since [R·] is always smaller than [R⌇⌇⌇⌇M·], L_{kin}, and therefore the degree of polymerization $\bar{P}$, with emulsion polymerization (all other conditions being equal) must always be larger than with bulk polymerization.

The mechanism of precipitation polymerization is, in principle, very similar to that of emulsion polymerization. It occurs whenever the polymer is insoluble in the monomer, or in the solvent used during the polymerization. Soon after the start of the polymerization, the polymer precipitates out in the form of small particles (compare p. 150). Even if complete precipitation of the polymer particle does not take place, a certain isolation of the chain radicals occurs, brought about by a more or less strongly restricted mobility of the polymer molecules, as is the case in bulk polymerization after the conversion reaches a certain value and the polymerization medium becomes gel-like and highly viscous. The resulting reduction of chain termination due to restricted combination of chain radicals has the result that at this stage of the polymerization (i.e., at high conversion) the polymerization rate increases very rapidly and chains with higher degree of polymerization are formed (self-acceleration; compare p. 89). The same phenomenon is observed in solution polymerization in poor solvents (gel effect) (compare pp. 343, 454, 491, and 517).

212 Ionic Polymerization

Ionic polymerization, similar to radical polymerization, also has the mechanism of a chain reaction. The kinetics of ionic polymerization are, however, considerably different from that of radical polymerization:

1. The initiation reaction of ionic polymerization needs only a small activation energy. Therefore, the rate of polymerization depends only slightly on the temperature. Ionic polymerizations occur in many cases with explosive violence even at temperatures below $50°C$ (for example, the anionic polymerization of styrene at $-70°C$ in tetrahydrofuran, or the cationic polymerization of isobutylene at $-100°C$ in liquid ethylene).

2. With ionic polymerization there is no compulsory chain termination through recombination, because the growing chains can not react with each other. Chain termination takes place only through impurities, or through the addition of certain compounds such as water, alcohols, acids, amines, or oxygen, and in general through compounds which can react with the polymerizing ions under the formation of neutral compounds or inactive ionic species. If the initiators are only partly dissociated, the initiation reaction is an equilibrium reaction, where reaction in one direction gives rise to chain initiation and in the other direction to chain termination.

In general ionic polymerization can be initiated through acidic or basic compounds. For cationic polymerization, complexes of BF_3, $AlCl_3$ $TiCl_4$, and $SnCl_4$ with water, or alcohols, or tertiary oxonium salts have shown themselves to be particularly active. The positive ions are the ones that cause chain initiation. For example:

$$BF_3 + HO - R \rightleftharpoons \left[\begin{matrix} F \\ | \\ F - B - F \\ | \\ OR \end{matrix} \right]^{(-)} + H^{(+)} \rightleftharpoons H[BF_3OR]$$

$$[(C_2H_5)_3O]BF_4 \rightleftharpoons (C_2H_5)_3O^{(+)} + BF_4^{(-)}$$

triethyloxonium- borofluoride

However, also with HCl, H_2SO_4, and $KHSO_4$, one can initiate cationic polymerization. Initiators for anionic polymerization are alkali metals and their organic compounds, such as phenyllithium, butyllithium, phenyl sodium, and triphenylmethyl potassium, which are more or less strongly dissociated in different solvents. To this group belong also the so called Alfin catalysts, which are a mixture of sodium isopropylate, allyl sodium, and sodium chloride.

With BF_3 (and isobutylene as the monomer), it was demonstrated that the polymerization is possible only in the presence of traces of water or alcohol. If one eliminates water quantitatively, BF_3 alone does not give rise to polymerization. Water or alcohols are necessary in order to allow the formation of the BF_3-complex and the initiator cation according to the above reactions. However, one should not describe the water or the alcohol as a "cocatalyst."

Table 165 lists a series of monomers which can be polymerized ionically.

Just as by radical polymerization, one can also prepare copolymers by ionic polymerization: for example, anionic copolymers of styrene and butadiene, or cationic copolymers of isobutylene and styrene, or isobutylene and vinyl ethers. As has been described in detail with radical polymerization, one can characterize each monomer pair by two parameters r_1 and r_2 (compare p. 97). The actual values of these parameters are, however, different from those used for radical copolymerization.

2121   Ionic Polymerization With Quantitative Dissociation of the Initiator (Stoichiometric Polymerization)

The Mechanism of Stoichiometric Polymerization

The simplest case of an ionic polymerization occurs if the initiator is already quantitatively transformed into the active, dissociated[23] form before the start of the polymerization, and if even during the course of the polymerization, it shows no tendency towards the formation of the neutral form. This is the case, for example, if one uses as initiators alkali organic compounds (for example, phenyllithium, butyllithium, or sodium naphthalene)[24] in solvents which have unshared

---

[23]The term "dissociation" does not denote any special kind of charge separation. Probably we have to regard this dissociation as a separation of ion pairs by solvent molecules, but not a separation in the sense of forming single ions which can move independently from the gegenion.

[24]Sodium naphthalene can be very conveniently prepared by reacting pieces of sodium for several hours with naphthalene dissolved in tetrahydrofuran. The sodium dissolves and the solution assumes a deep green color because of the formation of the colored radical.

TABLE 165. Examples of monomers which polymerize ionically.

anionic	anionic and cationic	cationic
$CH_2=\overset{\overset{\displaystyle CH_3}{\vert}}{C}-CH=CH_2$ Isoprene*	$CH_2=CH$  (phenyl) Styrene*	$CH_2=\overset{\overset{\displaystyle CH_3}{\vert}}{\underset{\underset{\displaystyle CH_3}{\vert}}{C}}$ Isobutylene
$CH_2=CH-CH=CH_2$ Butadiene*	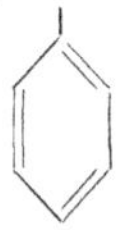 $CH_2=\overset{\overset{\displaystyle CH_3}{\vert}}{C}$ (phenyl) α-Methylstyrene	$CH_2=CH-\overset{\overset{\displaystyle CH_3}{\vert}}{CH}-CH_3$ 3-Methylbutene-1
$CH_2=\overset{\overset{\displaystyle COOR}{\vert}}{\underset{\underset{\displaystyle COOR}{\vert}}{C}}$ Methylenemalonate*	$\left[\begin{array}{c} CH_2=O \\ \text{Formaldehyde} \end{array}\right]$	$CH_2=CH-CH_2-\overset{\overset{\displaystyle CH_3}{\vert}}{CH}-CH_3$ 4-Methylpentene-1
$CH_2=\overset{\overset{\displaystyle CN}{\vert}}{\underset{\underset{\displaystyle COOR}{\vert}}{C}}$ α-Cyanoacrylate*	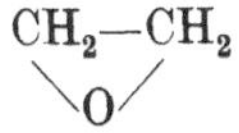 $\left[\begin{array}{c} CH_2-CH_2 \\ \diagdown O \diagup \\ \text{Ethyleneoxide} \\[4pt] \text{and}\ \ CH_2-CH-R \\ \diagdown O \diagup \\ \text{Epoxy-} \\ \text{compounds} \end{array}\right]$	$CH_2=CH$ $\vert$ O $\vert$ R Vinylethers
$CH_3-CH=CH-CH=\overset{\overset{\displaystyle CN}{\vert}}{\underset{\underset{\displaystyle COOR}{\vert}}{C}}$ α-Cyanosorbate*	$\begin{array}{c} CH_2-CH_2 \\ \diagdown S \diagup \\ \text{Ethylenesulfide} \end{array}$	$\left[\begin{array}{c} Cl-CH_2 \quad CH_2-Cl \\ \diagdown C \diagup \\ CH_2 \quad CH_2 \\ \diagdown O \diagup \\ \text{Oxacyclobutane-} \\ \text{derivatives} \end{array}\right]$
$CH_2=CH$ $\vert$ CN Acrylonitrile*	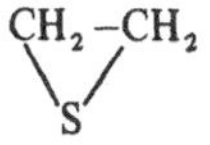	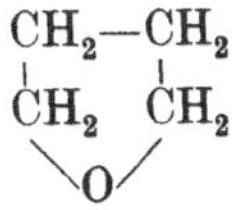 $\left[\begin{array}{c} CH_2-CH_2 \\ \vert \qquad \vert \\ CH_2 \quad CH_2 \\ \diagdown O \diagup \\ \text{Tetrahydrofuran} \end{array}\right]$
$CH_2=\overset{\overset{\displaystyle CH_3}{\vert}}{\underset{\underset{\displaystyle COOR}{\vert}}{C}}$ Methacrylic acid esters*		

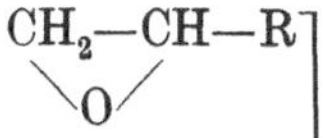

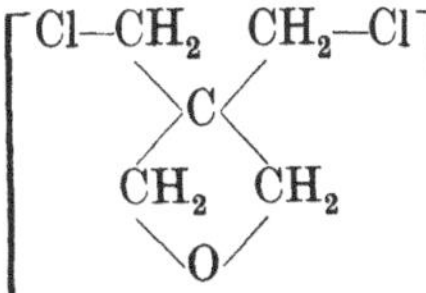

*indicates momoners which also polymerize by radical initiation.

The cyclic ethers in brackets [ ] polymerize by the same mechanism as the olefins under which they are listed.

electron pairs (Lewis bases). In this case the alkali forms stable positively charged complex ions with the Lewis base, so that the organic residue is negatively charged (carbanion) and an ionic polymerization can be initiated by this carbanion:

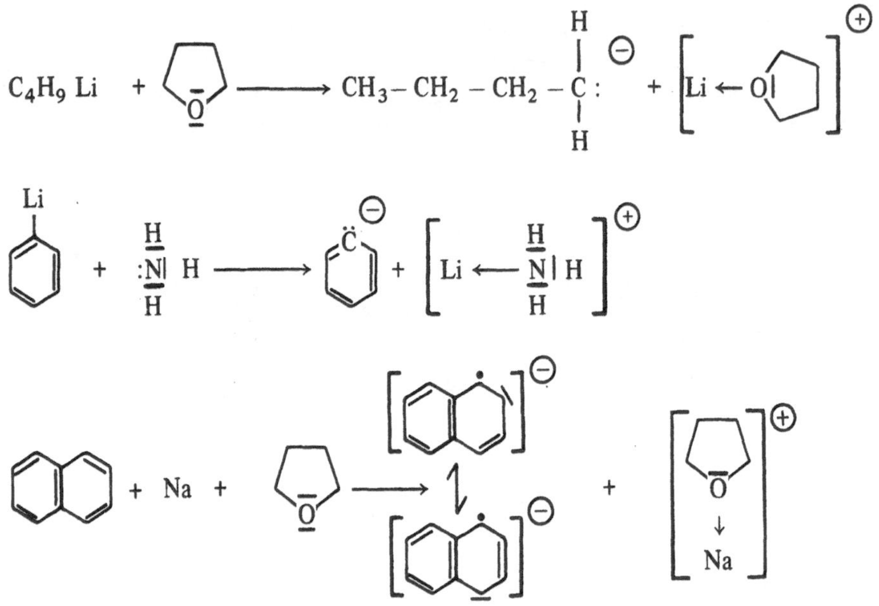

The complex counterion (here the cation) is usually left out for simplicity in writing ionic polymerization reactions.

With styrene the polymerization occurs according to the following scheme:

a) with phenyllithium as initiator:

Initiation:

Propagation:

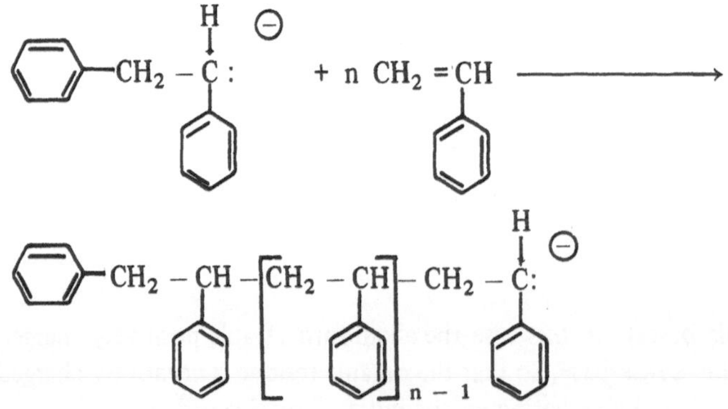

b) with sodium naphthalene as initiator (Szwarc):

Initiation:

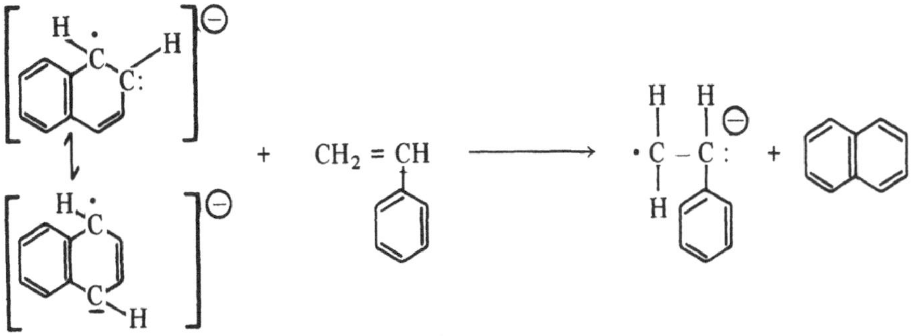

Chain Growth (propagation) (I):
Anionic:

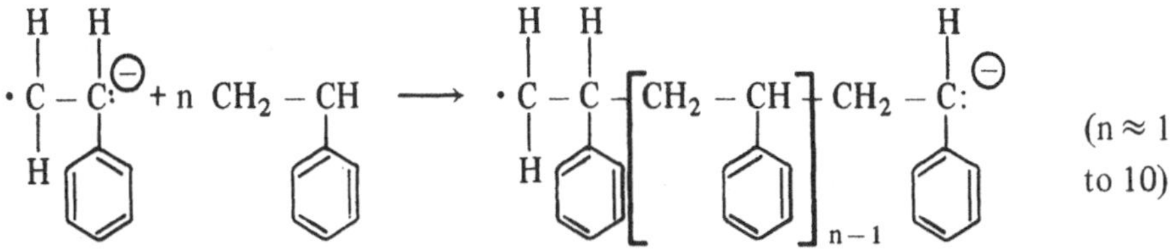

$(n \approx 1$ to 10$)$

Radical:

$(n \approx 1$ to 10$)$

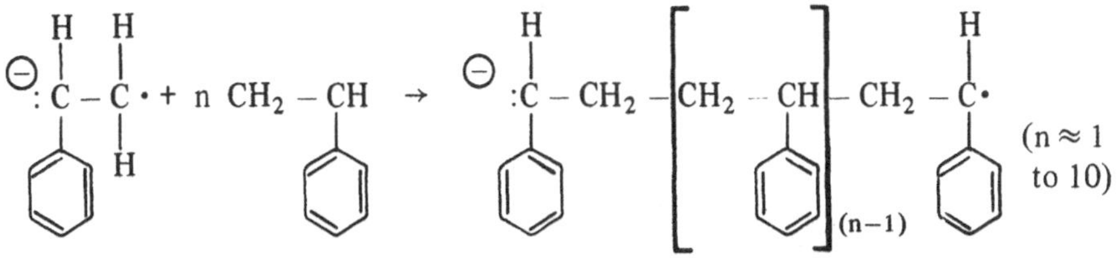

Radical Combination:

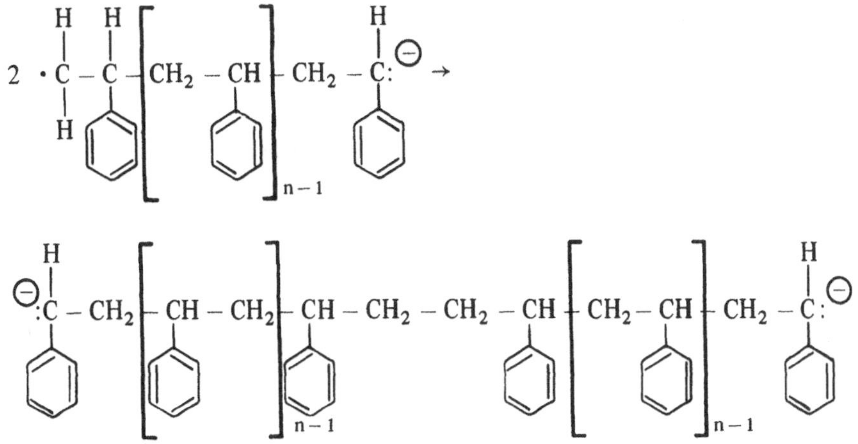

Chain Growth (propagation) (II):

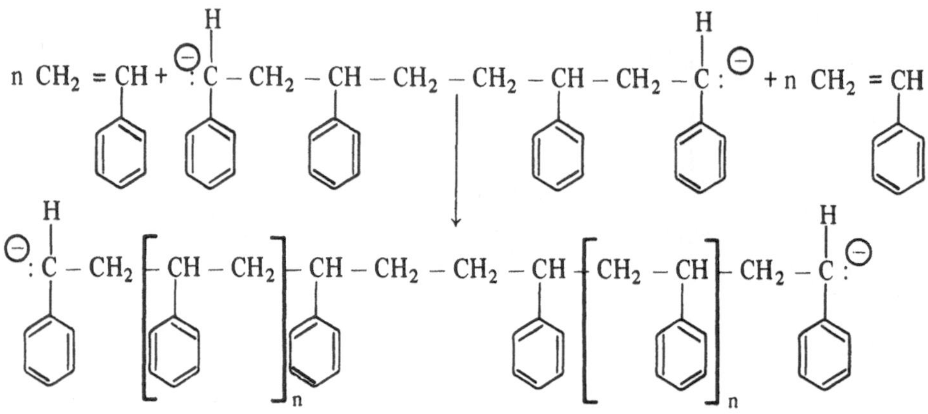

$$(n = 10^2 \text{ to } 10^5)$$

[The reaction of naphthalene with sodium is an addition reaction in the formal sense, unlike the preparation of phenyllithium, which is formed from chloro-benzene or bromobenzene by substitution of the halogen. The polymerization of styrene with sodium naphthalene was first described by Scott (1939), but the mechanism was only elucidated about 20 years later by Szwarc].

As with radical polymerization, one can also raise the formal question here whether the dissociation of the initiator or the addition of the first monomer molecule constitutes the initiation of the chain reaction.

Polymerization with sodium naphthalene as the initiator occurs in a somewhat different manner from that with phenyllithium. At the initiation step the sodium naphthalene transfers its excess electrons to the styrene molecule, naphthalene is reformed, and a styrene radical-ion is created, which then starts the chain. Chain growth occurs in two steps because of the initial dimerization of the two radicals: first, from one side only, and after the radical combination, from both sides. At what degree of polymerization the combination of the two radicals occurs cannot be said with certainty. However, in comparison with radical polymerization, the present system has a very high radical concentration, and furthermore, the high reactivity of the radical also contributes to combination of relatively short chains [see Chain Growth (I)]. To a certain extent the chain length in reaction (I) may be influenced by the conditions of polymerization (concentration of sodium naphtha-lene and rate of addition of the reactants). A combination of two monomer radicals should be possible (compare the formation of tetraphenylbutane from asymmetric diphenylethylene and sodium, discussed in the next section). However, with compounds that can polymerize, (because of their great rate of addition with ionic chain growth, and perhaps also because of the nearness of the two negative charges), a combination of oligomeric and polymeric radicals will be preferred. Since the combination does not occur at the same time for all chains, the number of additions steps in (I) is quite arbitrary. Therefore one obtains a mixture of chains (I) with different lengths. For the total length of the chains, the determining factor is the number of additions in (I) and (II) and since (I)≪≪(II) the homogeneity of the final degree of polymerization is not influenced by (I).

If, and to what extent, next to the anionic addition of monomeric styrene, there is at the same time a radical addition, is a question of the temperature. For temperatures of $-50°C$ or $-70°C$, where anionic chain growth still occurs very rapidly, the amount of radical chain growth is very small, whereas at higher temperatures (0 to $100°C$) radical and ionic polymerizations can occur side by side and therefore, theoretically, the homogeneity of the polymer is disturbed. The practical significance of a radical chain growth which might occur at the same time in the sodium naphthalene-initiated styrene polymerization, is always small, for two reasons: first, the radicals, because of their relatively high concentration, combine at the beginning of the reaction. Second, by regulation of the monomer addition, one can determine the rate of the anionic polymerization; one therefore has the possibility of giving the combination reaction sufficient time to occur before any addition reactions can take place. If the monomer is added slowly enough, one can be sure that the radical combination reaction has already occurred in the early oligomer state.

The confirmation for the formulated reaction scheme with the radical anion as the initial reactive species is difficult to demonstrate with styrene because of immediately continuing polymerization. However, through the work of Schlenk we know that from asymmetric diphenylethylene through the addition of sodium and the subsequent reaction with water or CO_2, one obtains tetraphenylbutane or tetraphenylbutanedicarboxylic acid, respectively:

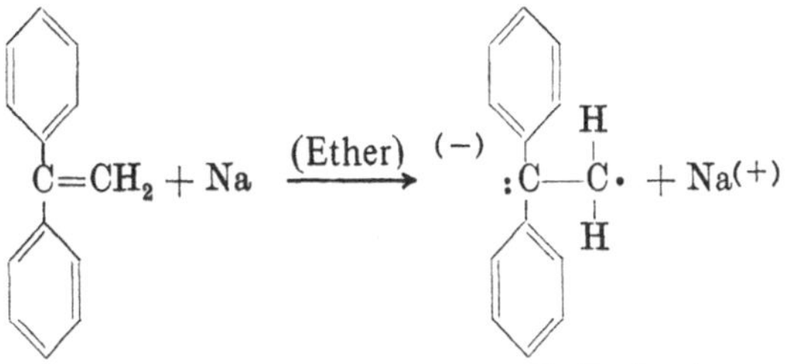

Radical anion

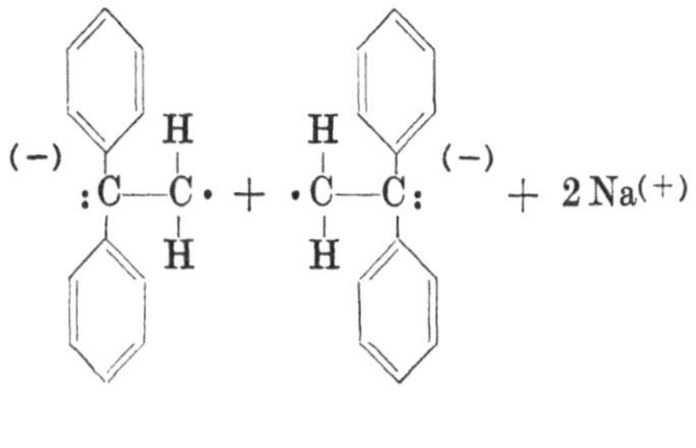

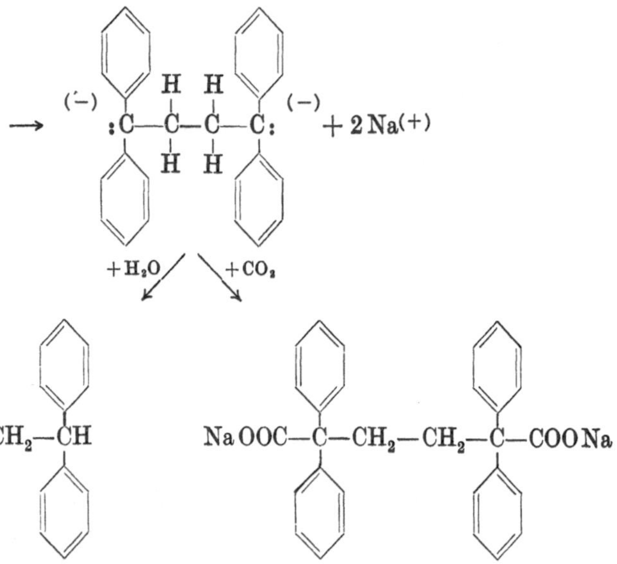

One can take this reaction as a model for the first steps of the anionic polymerization of styrene. The radical anion, or dianion, resulting from the reaction of diphenylethylene with sodium, instead of naphthalene sodium, can also be used as an initiator.

α-methylstyrene can react just as readily as asymmetric diphenylethylene (=α-phenylstyrene) with sodium in the presence of such ethers as tetrahydrofuran. Because of the greater polymerization tendency of α-methylstyrene (greater than that of α-phenylstyrene, but less than that of styrene), one obtains (at 70°C) the dimeric radical anion and from that, through radical combination, the tetrameric dianion (Szwarc):

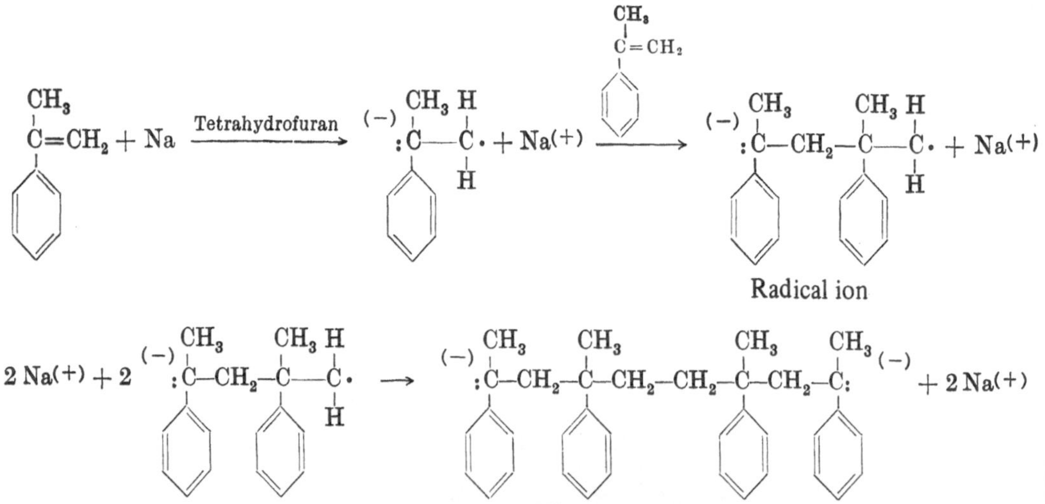

α- Methyl-styryl-sodium

Actually the polymerization occurs in the following way: Under the absolute exclusion of water and oxygen, one lets styrene drop slowly into a solution of the initiator (phenyllithium, sodium naphthalene, or α-methylstyrene sodium) in tetrahydrofuran. The styrene is immediately added according to the above reaction under the development of heat. The color of the solution, if one uses sodium naphthalene as initiator, changes immediately from green to dark red. Since there is no chain termination, the polymerization continues as long as one adds styrene. One can even interrupt the addition of styrene for several hours. If after that more styrene is added, the polymerization continues with unchanged rate[25] until water, acids (also CO_2), or oxygen is added to terminate the reaction. The chain termination may be recognized from the disappearance of the dark red color caused by the styrene anion:

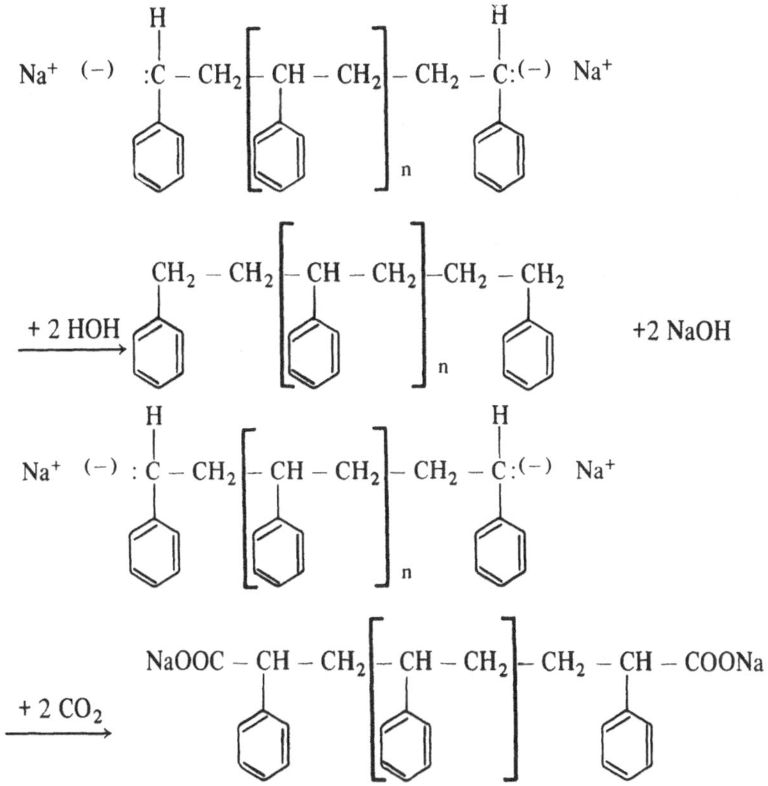

$$Na^+ \ ^{(-)} \ :C - CH_2 \left[CH - CH_2 \right]_n CH_2 - C:^{(-)} \ Na^+$$

$$\xrightarrow{+ \ 2 \ HOH} \quad CH_2 - CH_2 \left[CH - CH_2 \right]_n CH_2 - CH_2 \qquad +2 \ NaOH$$

$$Na^+ \ ^{(-)} \ : C - CH_2 \left[CH - CH_2 \right]_n CH_2 - C:^{(-)} \ Na^+$$

$$\xrightarrow{+ \ 2 \ CO_2} \quad NaOOC - CH - CH_2 \left[CH - CH_2 \right]_n CH_2 - CH - COONa$$

---

[25]Because of this behavior, this type of polymerization is called a "living" polymerization, and the resulting polymers are called "living polymers." If one remembers that the growth of the macromolecules with an ionic polymerization in tetrahydrofuran can only be terminated by external interference, and therefore at one and the same time all individual chains are terminated, one finds that this is much less an analogy with "living" and "dying" than is the case with radical polymerization, or with the normal ionic polymerization which has a well-defined dissociation equilibrium (see below). It therefore seems more reasonable to call this type of polymerization a "stoichiometric" polymerization.

In addition to styrene, different styrene derivatives, such as α-methylstyrene and ring-methylated styrenes, as well as butadiene, isoprene, and methylmethacrylate, may be polymerized according to this stoichiometric mechanism. With methylmethacrylate, and even more with acrylonitrile and esters of acrylic acid, certain side reactions occur which gradually deactivate the initiator. In place of tetrahydrofuran one can also use dimethylether, or glycol dimethylether (but not diethylether) as a polymerization medium.

The stoichiometric mechanism is not restricted to anionic polymerization. In suitable strongly polar solvents such as liquid SO_2 or CO_2 one can also carry out cationic polymerization stoichiometrically. In order to avoid chain transfer, the polymerization has to occur at low temperatures. The cationic living polymerization is still relatively little investigated.

A characteristic difference exists between the chain structure of radical copolymers and "living" copolymers, if the copolymerization is carried out to high conversion and the r_1-r_2 parameters are different from 1. With radical copolymerization in such cases one always obtains a mixture of polymer chains with a composition which changes as the conversion increases. The composition of the individual chains however is always statistically uniform. With stoichiometric (living) polymerization on the other hand, all chains have the same composition (each one is like any other); however, the composition of the chains changes systematically from the beginning of the chain to its end. This is a result of the fact that, in the ideal case, all chains grow simultaneously (see Figure 38). Of course, it is not necessary to leave the formation of such polymers (where the chain molecules have a different composition at the beginning and at the end of the polymer chains) simply to the degree of deviation of r_1 and r_2 from 1. Thus by controlling and changing the monomer composition during the reaction, one can arrange which monomers are preponderant at the beginning of the chain and which ones are preponderant at the end of the chain.

In the extreme case, after a certain stage is reached in stoichiometric polymerization, one can completely replace the monomer being added by a different monomer, and in this elegant manner prepare block-copolymers (segment polymers), where the length of each segment can be determined at will. For example, first one polymerizes styrene, and at a later stage adds butadiene. Or one begins the polymerization with α-methylstyrene, and, after a certain amount is added, continues the polymerization with styrene. In each case the length of the segments can be determined by the amount of monomer added. This procedure is used for the industrial synthesis of styrene-butadiene block-copolymers. Depending on whether the polymerization is initiated with sodium naphthalene or with phenyllithium, one obtains symmetrical or unsymmetrical distribution of the segments:

INITIATION WITH Na-Naphthalene

POLYSTYRENE   POLY - α - METHYLSTYRENE   POLYSTYRENE

INITIATION WITH
PHENYLLITHIUM

$\downarrow$ POLYSTYRENE   POLYBUTADIENE

Kinetics of Stoichiometric Polymerization

The kinetics of stoichiometric polymerization are particularly simple (if one regards only the ideal case with one type of chain initiation). An initiation reaction which continues during the chain growth stage and therefore keeps on starting new chains does not exist. Chain growth starts all at once, because the number of active centers from which chain growth starts has already attained its maximal level *before* the start of polymerization and does not change during the polymerization (complete exclusion of chain-terminating substances is assumed).

The propagation reaction is a reaction of second order, just as with radical polymerization, and the rate of polymerization, r_P, is accordingly given by:

$$r_P = k_p \cdot [A^{(-)}] \cdot [M] \tag{43}$$

where $[A^{(-)}]$ is the anion concentration, and $[M]$ the monomer concentration.

In order to avoid an explosion-like course of the reaction, one keeps the monomer concentration low by adding the styrene continuously. Since the added styrene is instantaneously used up (as with a titration), the rate of the polymerization is in practice determined by the rate of addition of the monomer, which in turn is limited by the rate at which the heat can be removed.

Since none of the anions is preferred over another, under ideal mixing conditions the added styrene is distributed uniformly over the ions present. This means that all chains become essentially equally long. This means further that the molecular weight of a polystyrene prepared in this manner is determined by the amount of styrene added to a given predetermined amount of initiator. For example, if in a solution in tetrahydrofuran there are 10 initiator anions and one adds 10,000 styrene molecules, then 1,000 styrene molecules can add to each anion, and therefore the degree of polymerization equals 1,000 and the molecular weight 104,000 ($M_{pol} = P \cdot M_{monomer}$). If one adds twice this amount of styrene, then each anion adds twice as many styrene molecules and M is therefore twice as great. On the other hand, if one adds the same 1,000 styrene molecules to twice the number of initiator anions, then each anion receives $1,000/20 = 500$ styrene molecules and the molecular weight is therefore only 52,000. In general, one can write

$$\overline{P} = n_{monomer}/n_{initiator} \tag{44}$$

$$\overline{M}_{pol} = M_{mon} \cdot n_{mon}/n_{init} \tag{45}$$

If one replaces the number of moles, n, by the mass of the materials in grams $n=m/M$ one obtains:

$$\bar{P} = (m_{mon}/m_{init}) \cdot (M_{init}/M_{mon}) \tag{46}$$

$$\bar{M}_{pol} = m_{mon} \cdot M_{init}/m_{init} \tag{47}$$

$\bar{P}$= degree of polymerization (number of structural units per macromolecule); M= molecular weight; n= number of moles; and m= weight in grams.

In case b (see p. 167), where one uses sodium metal or sodium naphthalene as the initiator, two radical ions always combine to a double ion, and the molecular weight is therefore doubled. In this case all equations should be multiplied by a factor 2, thus:

$$\bar{P} = 2n_{mon}/n_{init} \quad \text{etc.} \tag{47a}$$

Through a simple stoichiometric calculation one can therefore determine the molecular weight of a polymer. Vice versa, for a desired molecular weight, one can calculate ahead of time the necessary amount of initiator. Since these simple stoichiometric relationships between initiator and molecular weight (for ionic polymerizations with quantitative *dissociation* of initiator) characterize the polymerization kinetics, one can describe this sort of polymerization as "stoichiometric polymerization."

Thus, if one wants to prepare a polystyrene with a molecular weight of 300,000 using sodium naphthalene as the initiator, Equation (47) shows that for 1 kg of styrene, one needs 0.153 g of sodium:

$$m_{init} = 2 \cdot m_{mon} \cdot M_{init}/M_{pol} = 2 \cdot 1{,}000 \cdot 23/300{,}000 = 0.153 \text{ g Na}$$

The sodium content of the sodium naphthalene solution in tetrahydrofuran can be obtained by titration with acid.

Different experiments have shown that the stoichiometrically calculated molecular weight corresponds very well with the experimentally determined one (by methods such as osmotic pressure and ultracentrifuge), which demonstrates the correctness of the formulated polymerization mechanisms. It was also experimentally confirmed that, as required by the simple kinetics of the stoichiometric polymerization, all polymer molecules are equally long (for all practical purposes).

The ratio of $\bar{M}_w/\bar{M}_n$, which is a measure of the molecular homogeneity of the polymer samples (compare p. 411), for anionically polymerized polystyrene in tetrahydrofuran, has values of 1.05-1.2, depending on the accuracy in preparing the polymerization. With other polymers prepared by the usual radical methods, one finds that $\bar{M}_w/\bar{M}_n$ = 2-5, in extreme cases up to 10. Polymers prepared by stoichiometric polymerization therefore have better homogeneity than a carefully prepared fraction and are practically molecularly homogeneous, even though not in the exact theoretical sense.

As one can see from the small amounts of initiators used (0.015%), the polymerization has to be carried out under careful exclusion of water and oxygen. Even 0.012% of water is sufficient to completely prevent polymerization in the above example, and even much smaller amounts are sufficient in order to greatly affect the molecular homogeneity of the polymer.

Chain transfer reactions (compare p. 63) are also possible with ionic polymerizations, and sometimes become quite important. However, the fact that with anionic polymerization of styrene and α-methylstyrene in tetrahydrofuran one obtains polymers with a nearly uniform chain length shows that at least with some anionic polymerizations chain transfer is negligible. One can therefore regard polymers prepared by stoichiometric polymerization as unbranched (linear). The particular properties of the stoichiometric polymers, their molecular homogeneity, and the absence of branching make them particularly suitable for specific investigations, for example, the determination of the relationship between physical properties and the molecular weight of polymers.

Anionic polymerization has been described here in a highly simplified manner. Actually there is an equilibrium between several states of the initiator and the growing chain, all with different degrees of activity (G. V. Schulz, M. Szwarc). This is why polymers obtained by stoichiometric polymerization are not completely monodisperse. Other reasons for a certain polydispersity of stoichiometric polymers are (1) the great difficulty in obtaining completely uniform mixing of the initiator solution and the monomer, and (2) the impossibility of quantitatively removing, before and during the polymerization, all impurities capable of destroying the initiator or inactivating the growing chain.

The Polymerization Equilibrium With α-Methylstyrene

Of particular interest is the anionic polymerization of α-methylstyrene, because it leads to a temperature-dependent, reversible equilibrium between monomeric and polymeric α-methylstyrene. If one adds α-methylstyrene to a solution of sodium naphthalene in tetrahydrofuran, the deep green color immediately changes to dark red, without the evolution of a significant amount of heat or an increase in viscosity. If one cools the solution to from −40°C to −70°C, polymer is formed and the solution becomes highly viscous. On warming, the poly-α-methylstyrene depolymerizes, and if one then cools the solution, it polymerizes again. Obviously the enthalpy, ΔH, between the monomeric and polymeric state with the system α-methylstyrene/poly-α-methylstyrene is so small that at normal temperatures the entropy term $T \cdot \Delta S$ has the same order of magnitude as ΔH. If one lowers the temperature, $T \cdot \Delta S$ becomes smaller, and the equilibrium condition $\Delta F = \Delta H - T \cdot \Delta S = 0$ is disturbed. As a result of this, polymer is formed until for the new concentration ratio of polymer to monomer ΔF is again equal to 0. The temperature for which at the starting point (100% monomer) $T \cdot \Delta S = \Delta H$, so that $\Delta F = \Delta H - T \cdot \Delta S = 0$, is called the ceiling temperature. At this temperature, the concentration of polymer in the equilibrium "monomer ⇌ polymer" is equal to 0. With α-methylstyrene, the ceiling temperature is between +50°C and +70°C, whereas for most other monomers it is above 300°C.

In spite of its thermodynamic instability, one can prepare poly-α-methylstyrene of high degree of polymerization, if one adds water or methanol at temperatures below −70°C to the polymer solution and terminates the chains. Through the neutralization of the ionic chain ends, the depolymerization reaction is prevented by a potential barrier and the polymer can therefore exist even at normal temperatures, that is, the depolymerization reaction occurs much more slowly. Because of its low ceiling temperature and the resulting characteristic thermodynamic instability, poly-α-methylstyrene will degrade rapidly only at temperatures over 200°C. For this reason pure poly-α-methylstyrene is unsuitable for injection molding.

Because it polymerizes only at low temperatures, one can prepare very homogeneous, almost monodisperse, polymers from α-methylstyrene by mixing the initiator (α-methylstyrene sodium or sodium naphthalene) at +50°C to 60°C with the monomeric α-methylstyrene, then cooling the reaction mixture slowly and continuously to −70°C, and terminating the polymerization by methanol. Poly-α-methylstyrene prepared in this manner behaves in the ultracentrifuge as a monodisperse system. Its remarkable homogeneity is obviously caused by the fact that the inhomogeneous distribution of monomer added to the reaction mixture, which is caused by faulty stirring during the polymerization, is not a factor in this case.

2122   Ionic Polymerization With Initiators Which Dissociate Only Partly

The initiators used in ionic polymerization are not always completely dissociated.[26] Thus they can be in equilibrium with the neutral form in such a way that only a small portion of the initiator dissociates. This is the case, for example, with phenyllithium or butyllithium in benzene:

$$C_4H_9 - Li \rightleftharpoons C_4H_9^{(-)} + Li^{(+)}$$

Chain growth occurs in the same way as with stoichiometric polymerization, with the difference that here the growing chain anion also takes part in the dissociation equilibrium:

$$C_4H_9 - M - M \text{\~\~\~\~\~} M:^{(-)} \ + \ Li^{(+)} \rightleftharpoons C_4H_9 - M - M \text{\~\~\~\~\~} M - Li$$

It can therefore happen that a chain stops growing spontaneously, and that another one, which was previously inactive (because it was not dissociated), now dissociates and continues to grow. This chain termination is, however, only temporary, because the neutral chain end can at any time dissociate again and, if there are still monomers present, continue to grow. The growth of the chain is therefore only interrupted for a more or less short period. Since the dissociation is only a matter of chance, each chain (if one considers a sufficiently long period) spends the same amount of time in the dissociated and in the neutral state. Thus one can count on the formation of equally long macromolecules, provided that the equilibrium concentration of the dissociated form is not too low and that the change from the neutral to the dissociated stage, or vice versa, does not occur too slowly, i.e., that all initiator molecules have the same starting conditions. If, however, the time the molecules spend in the two states is long and the concentration of the ions in the equilibrium is low (for example, such that half of the monomer has already

---

[26]In the sense of forming ion pairs.

polymerized before the last initiator molecule dissociates for the first time), then the initiator molecules that have by accident dissociated first have an advantage in the addition of monomer molecules over those which happen to have dissociated last. Under such conditions the butyllithium molecules dissociated last can not catch up, and one obtains a mixture of differently sized macromolecules, as is the case with radical polymerization.

The dissociation does not always occur spontaneously; it can also be induced by the monomer:

$$C_4H_9 - Li + CH_2 = CH \rightarrow C_4H_9 - CH_2 - \overset{\cdot\cdot}{CH}^{(-)} + Li^{(+)}$$

butyllithium   styrene

How far an ionic polymerization differs from the stoichiometric mechanism usually depends on the initiator and the solvent used since these determine the dissociation equilibrium. In polar solvents the dissociation of the initiator is favored. This results in the chain interruption, i.e., termination, occurring less often and the chain molecules growing to a longer chain. The rate of polymerization is higher and the molecular weight is greater provided no chain transfer occurs. Polar solvents for anionic polymerization are, for example, liquid ammonia, amines, and ethers. Polar solvents for cationic polymerization are such chlorinated hydrocarbons as methyl chloride, methylenechloride, chloroform, or carbon tetrachloride, CS_2, liquid SO_2, and liquid CO_2.

Many solvents have a strong chain transfer effect in ionic polymerization, which is noticeable in the lowering of the molecular weight and the strong dependence of the molecular weight on the temperature in such solvents. This effect is explained by the fact that the transfer reaction has a considerably greater activation energy than the ionic initiation and growth reactions, so that on decreasing the temperature the rate of the transfer reaction decreases rapidly whereas the rate of polymerization is only slightly effected. Since for most technical products a relatively high molecular weight is required, one usually carries out ionic polymerizations at temperatures between $-50°C$ and $-100°C$, as, for example, the cationic polymerization of bis-chloromethyl oxacylobutane (in liquid SO_2), vinylisobutyl ether (in liquid butane) or isobutylene (in liquid ethylene).

The solvent has an influence not only on the dissociation equilibrium and the size of the molecules, but also, one can affect the kind of chain propagation and, therefore, the structure of the polymer. For example, the anionic polymerization of isoprene with butyl or phenyllithium in tetrahydrofuran yields 60-70% 1,2- and 3,4-polyisoprene, and 30-40% trans-1,4-polyisoprene (no cis-1,4-polyisoprene). However the polymerization of isoprene with the same initiator in cyclohexane gives 94% cis-1,4-polyisoprene and 6% 3,4-polyisoprene (according to Tobolsky and Rogers):

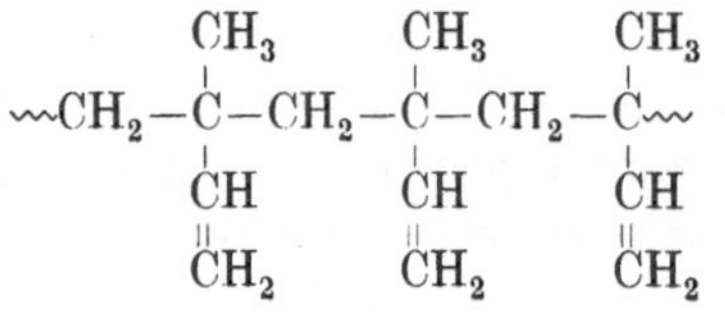

$$\underset{\substack{\displaystyle \\ \mathrm{CH_2}}}{\overset{\substack{\mathrm{CH_3} \\ \displaystyle}}{\mathrm{\sim\!\!CH_2\!-\!C\!-\!CH_2\!-\!C\!-\!CH_2\!-\!C\!\sim}}}$$

1,2-Polyisoprene

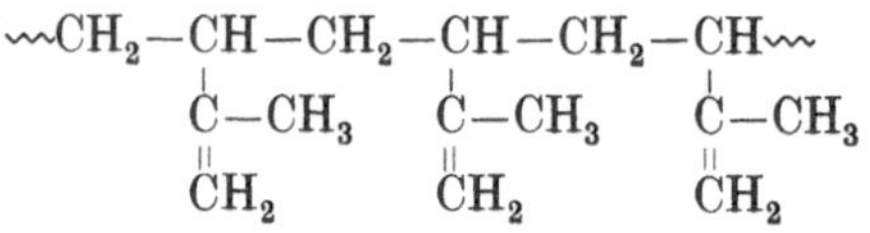

3,4-Polyisoprene

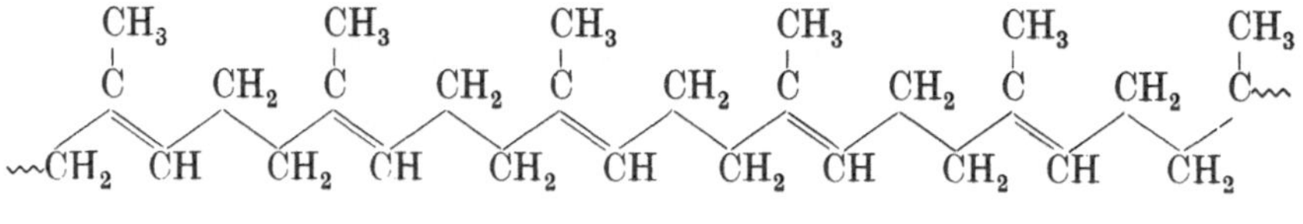

trans-1,4-Polyisoprene

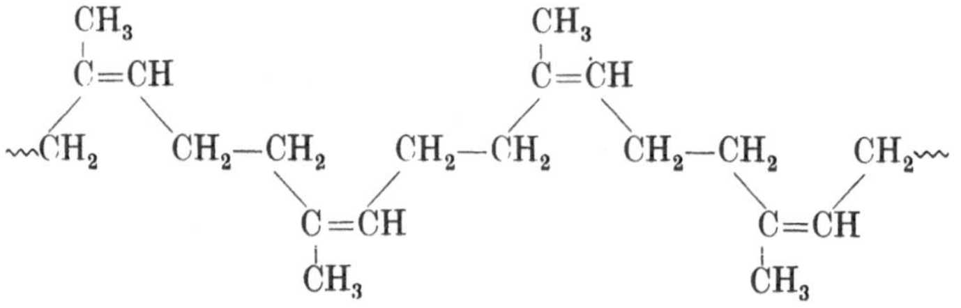

cis-1,4-Polyisoprene

As an example for a commercially important cationic polymerization, one might mention the polymerization of isobutylene, which is carried out at $-100°C$ with BF_3-complexes (e.g., $[BF_3OC_4H_9]^-H^+$) as initiators in liquid ethylene:

Chain-start (Initiation):

$$BF_3 + HOH \rightleftharpoons H[BF_3OH] \rightleftharpoons [BF_3OH]^{\ominus}\ldots H^{\oplus}$$

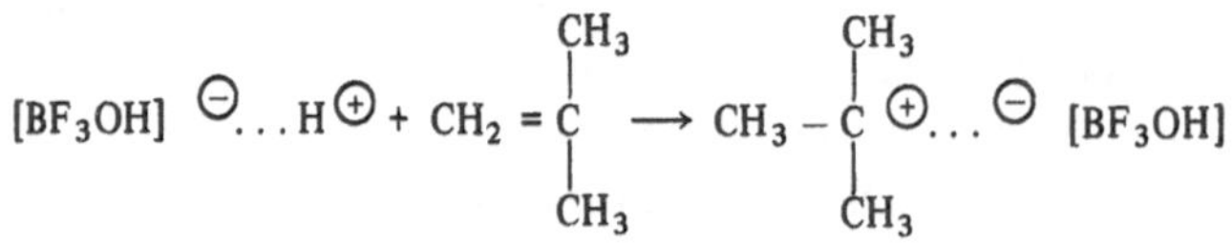

$$[BF_3OH]^{\ominus}\ldots H^{\oplus} + CH_2 = \underset{\mathrm{CH_3}}{\overset{\mathrm{CH_3}}{C}} \longrightarrow CH_3 - \underset{\mathrm{CH_3}}{\overset{\mathrm{CH_3}}{C}}{}^{\oplus}\ldots{}^{\ominus}[BF_3OH]$$

Chain-propagation:

$$CH_3 - \underset{\mathrm{CH_3}}{\overset{\mathrm{CH_3}}{C}}{}^{\oplus}\ldots{}^{\ominus}[BF_3OH] + CH_2 = \underset{\mathrm{CH_3}}{\overset{\mathrm{CH_3}}{C}} \longrightarrow CH_3 - \underset{\mathrm{CH_3}}{\overset{\mathrm{CH_3}}{C}} - CH_2 - \underset{\mathrm{CH_3}}{\overset{\mathrm{CH_3}}{C}}{}^{\oplus}\ldots{}^{\ominus}[BF_3OH] \overset{\text{etc}}{\longrightarrow}$$

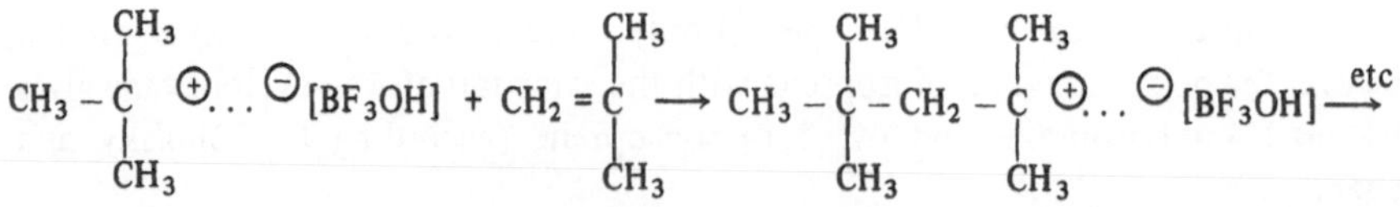

Chain-transfer:

$$\sim CH_2 - \underset{\underset{CH_3}{|}}{\overset{\overset{CH_3}{|}}{C}} \oplus \cdot\cdot \ominus [BF_3OH] + CH_2 = \underset{\underset{CH_3}{|}}{\overset{\overset{CH_3}{|}}{C}} \longrightarrow \sim CH_2 - \underset{\underset{CH_3}{|}}{\overset{\overset{CH_2}{||}}{C}} + CH_3 - \underset{\underset{CH_3}{|}}{\overset{\overset{CH_3}{|}}{C}} \oplus \cdot\cdot \ominus [BF_3OH] \xrightarrow{\text{Chain prop.}}$$

$$\sim CH_2 - \underset{\underset{CH_3}{|}}{\overset{\overset{CH_3}{|}}{C}} \oplus \cdot\cdot \ominus [BF_3OH] + CH_2 = \underset{\underset{CH_3}{|}}{\overset{\overset{CH_3}{|}}{C}} \longrightarrow \sim CH = \underset{\underset{CH_3}{|}}{\overset{\overset{CH_3}{|}}{C}} + CH_3 - \underset{\underset{CH_3}{|}}{\overset{\overset{CH_3}{|}}{C}} \oplus \cdot\cdot \ominus [BF_3OH] \xrightarrow{\text{Chain prop.}}$$

$$\sim CH_2 - \underset{\underset{CH_3}{|}}{\overset{\overset{CH_3}{|}}{C}} \oplus \cdot\cdot \ominus [BF_3OH] \longrightarrow \sim CH_2 - \underset{\underset{CH_3}{|}}{\overset{\overset{CH_2}{||}}{C}} + H[BF_3OH] \rightleftharpoons H \oplus \cdot\cdot \ominus [BF_3OH] \xrightarrow{\text{Chain prop.}}$$

$$\sim CH_2 - \underset{\underset{CH_3}{|}}{\overset{\overset{CH_3}{|}}{C}} \oplus \cdot\cdot \ominus [BF_3OH] + HOH \rightarrow \sim CH_2 - \underset{\underset{CH_3}{|}}{\overset{\overset{CH_3}{|}}{C}} - OH + H \oplus \cdot\cdot \ominus [BF_3OH] \xrightarrow{\text{Chain prop.}}$$

Chain-termination:

$$\sim CH_2 - \underset{\underset{CH_3}{|}}{\overset{\overset{CH_3}{|}}{C}} \oplus \cdot\cdot \ominus [BF_3OH] + NH_3 \rightarrow \sim CH_2 - \underset{\underset{CH_3}{|}}{\overset{\overset{CH_3}{|}}{C}} - OH + H[BF_3NH_2]$$

2123 Polymerization With Hydride Shift

Polymerization does not occur through 1,2-addition with all cationic polymerizations. For example, 3-methylbutene-1, which normally polymerizes at $-78°C$ to a 1,2-polymer, polymerizes at $-130°C$ through 1,3-addition. One must assume that as a primary step the "normal" cation forms through 1,2-addition, and that then at $-130°C$, because of the slower 1,2-addition, it finds time to form the more stable tertiary cation through a hydride shift (J. P. Kennedy):

At $-78°C$:

$$\sim CH_2 - \underset{\underset{\underset{CH_3 \quad CH_3}{\diagup \diagdown}}{CH}}{\overset{\overset{H}{|}}{C}} {}^{(+)} + CH_2 = \underset{\underset{\underset{CH_3 \quad CH_3}{\diagup \diagdown}}{CH}}{CH} \rightarrow \sim CH_2 - \underset{\underset{\underset{CH_3 \quad CH_3}{\diagup \diagdown}}{CH}}{CH} - CH_2 - \underset{\underset{\underset{CH_3 \quad CH_3}{\diagup \diagdown}}{CH}}{CH} {}^{(+)}$$

Poly-1,2-Methylbutene

At − 130°C:

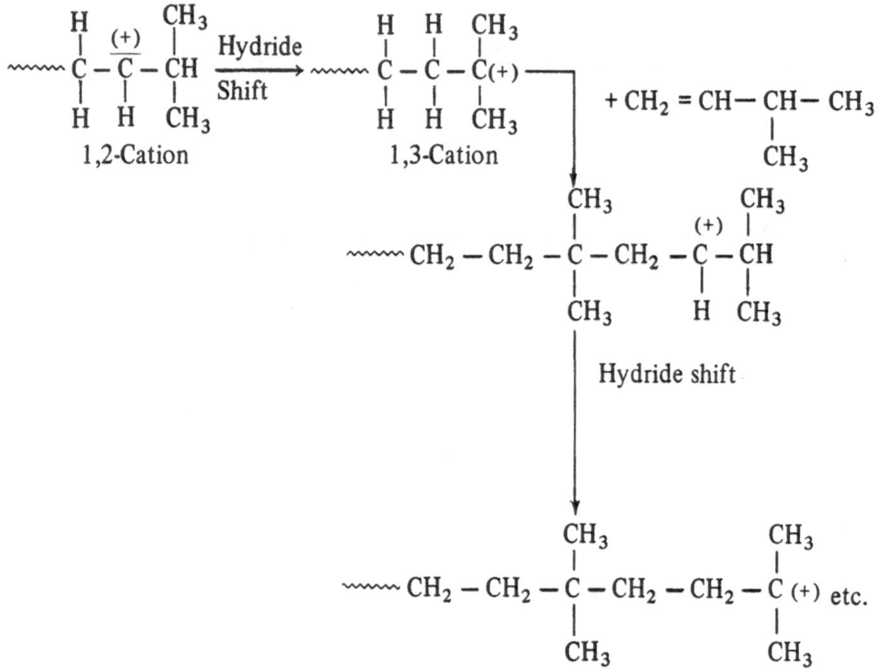

Poly-1,3-Methylbutene

Also compare the formation of poly-1,4-methylpentene:

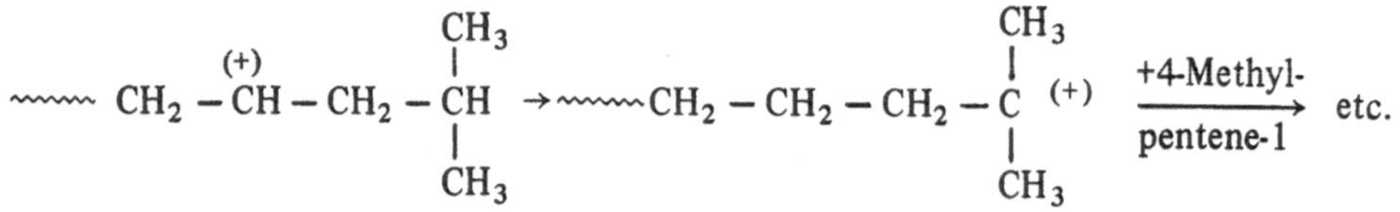

In a similar manner (see reaction above), one obtains up to 50% 1,4-polymer from 4-methylpentene-1 at −100°C and up to 30% of a 1,5-polymer from 5-methylhexene-1. This probably occurs in such a way that there is statistical distribution of 1,2-and 1,4-units, and 1,2- and 1,5-units, respectively, in the chain. A hydride shift can thus occur over several carbon atoms.

A similar hydride shift mechanism is also encountered in anionic polymerization of acrylamide (compare p. 247).

Cationic Polymerization With Dehydrogenation

Benzene can be polymerized with aluminum chloride, i.e., $H[AlCl_3OH]$, to polyphenylenes in the presence of Cu^{II}–Chloride. The reaction probably occurs in such a way that each addition step is followed by a dehydrogenation step:

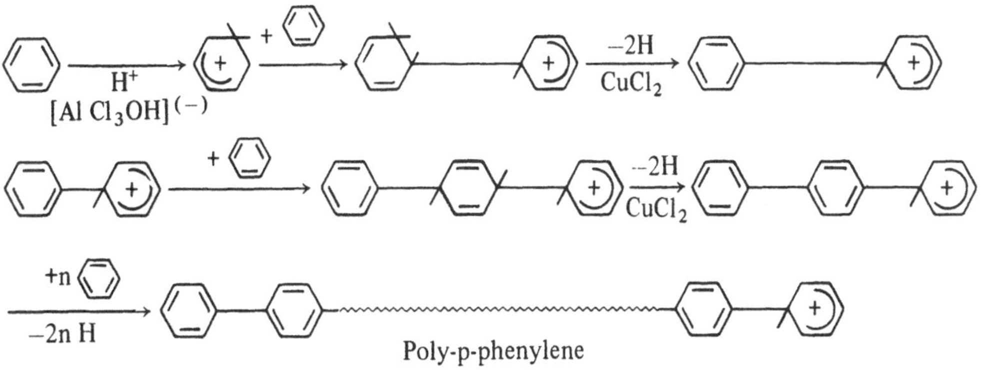

$$\text{(benzene)} \xrightarrow[\text{[Al Cl}_3\text{OH]}^{(-)}]{H^+} \quad \xrightarrow{+ \text{(benzene)}} \quad \xrightarrow{-2H}{CuCl_2}$$

$$\xrightarrow{+ \text{(benzene)}} \quad \xrightarrow{-2H}{CuCl_2}$$

$$\xrightarrow[-2n\,H]{+n\,\text{(benzene)}} \qquad \text{Poly-p-phenylene}$$

The resulting polyphenylene is a solid brown substance that is insoluble in all solvents.

By using chlorine as the dehydrogenation agent and introducing sulfonicacid-groups, it was possible to obtain soluble derivatives (molecular weights up to 50,000.)

Types of Ionic Initiators

In addition to BF_3 (that is, $H[BF_3OH]$), there are other Lewis acids which are good initiators for cationic polymerization. Depending on the type of initiator used, the position of the initiator dissociation equilibrium may be quite different. Furthermore, the dielectric constant of the solvent used will have a decided influence on the degree of dissociation of the intiator. It is quite possible that with a cationic polymerization under favorable conditions, the dissociation of the initiator may be quantitative, so that it proceeds by a stoichiometric mechanism (cationic "living" polymerization). But it is also possible that under certain circumstances only little or no dissociation of the initiator occurs. One could therefore ask whether the above polymerization of isobutylene with BF_3, which is always used as a classic example of cationic polymerization, occurs by an ionic mechanism under the usual technical conditions (hydrocarbon solvent), or whether it occurs via an undissociated complex.

In many cases one can use the influence of the counter-ion (that is the complex partner) on the polymerization and the properties of the resulting polymer as indications of the mechanism. Thus for a pure ionic polymerization, the type of counter-ion should have no or only a very slight influence, since it is not taking part directly in the reaction. The smaller the dissociation, the larger will be the influence of the counter-ions. As far as the properties of the polymers are concerned, the transition from dissociated ion to a donor-acceptor complex, may make itself felt in the appearance of stereoregular structures which can markedly change the properties of the polymers. This will be discussed further in a later section (compare 2131, p. 185).

With the aid of an earlier scheme proposed by Bawn and Ledwith, M. Roha has tried to consider the entire range of ionic and complex polymerizations from a uniform point of view, using the isotactic and syndiotactic components of the polymer produced as criteria for the more or less pronounced ionic character of the polymerization. Thus one obtains a whole range of ionic polymerizations and ionic initiators, at one end of which one has a clearly cationic initiator, and at the other end an initiator which acts in a typically anionic manner. In the middle there are the undissociated complex initiators (compare Figure 183). It should be pointed out that there are no sharp boundaries between cationic and complex-polymerization on the one hand, and anionic and complex-polymerization on the other hand. Rather, an ionic polymerization takes on more of the character of a complex-polymerization, the more the dissociation is repressed, or in other words, the more the independence or mobility of the separated ions disappears. If one approaches the situation from the cationic side of the range of initiators, where the cation starts the chain and the anion plays the role of an indifferent counter-ion, one then comes to the covalent donor-acceptor complexes and finally to the initiators which act as anions and therefore have the opposite role from the cations, namely, the anion starts the chain and the cation plays the role of the counter-ion with little or nothing to do in the reaction.

The sequence of initiators shown in Figure 183 is not based on quantitative experiments, but simply gives a rough estimate based on the properties of the polymers obtained. One should not consider the position of the initiators in Figure 183 as unchangeable. Rather, by different combinations of initiators and solvents, the position of the initiator may be moved or exchanged. The strong influence of the solvent can be seen particularly clearly with the polymerization of butadiene initiated by butyllithium. In hydrocarbon solvents the polymerization definitely follows a complex-mechanism, whereas in polar solvents such as tetrahydrofuran (THF), the same initiator will bring about an anionic polymerization. This can be recognized by the fact that in hydrocarbon solvents stereoregular (cis-1, 4 or trans-1, 4) structures result, whereas in THF polymers with mixed cis-, trans-1, 4, and 1, 2-structures are formed.

Comparison of ionic polymerization with radical polymerization reveals that the radical polymerization is easier to define and has a less complex mechanism. Ionic polymerization is a stoichiometric polymerization only in the ideal limiting case. Whereas stoichiometric polymerization is a homogeneous and easily described phenomenon, the normal ionic polymerization is very complex and has not been nearly as well investigated as radical polymerizations.

It is also interesting to compare the behavior of different monomers toward ionic (and complex) initiators. Monomers such as styrene, butadiene, and ethylene oxide, are not at all specific with regard to the initiator used; others, such as ethylene or propylene are polymerized only by complex initiators, not by ions. This is probably determined by the polarizability of the monomer molecule, which with styrene, (because of the presence of the aromatic ring) is easily possible in any direction because the phenyl residue is capable of accepting electrons as well as giving them up (acceptor and donor).

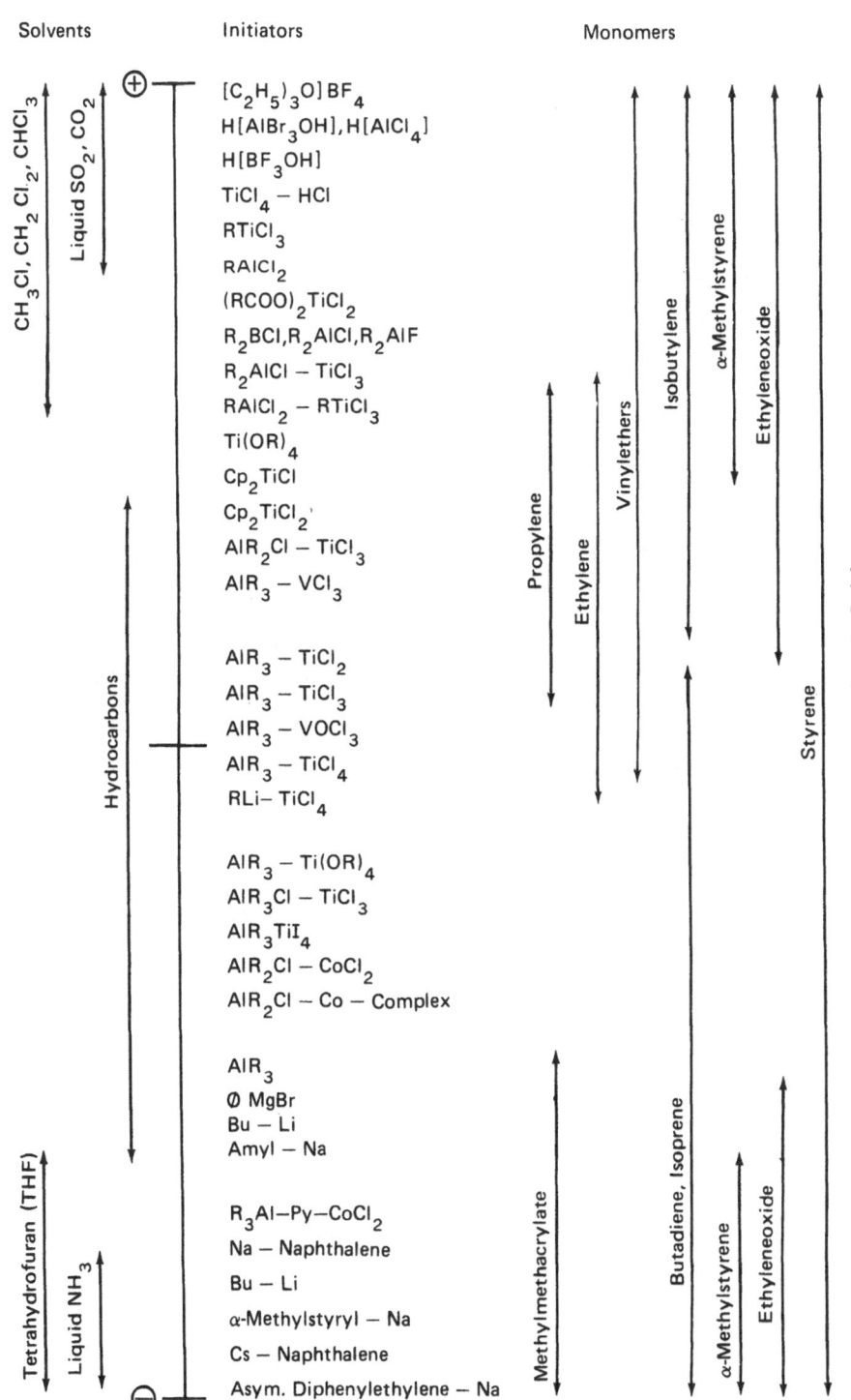

FIG. 183 — The Range of Ionic Initiators

2124. Techniques of Ionic Polymerization

Ionic polymerizations are mainly carried out in solution. In many cases, in order to prevent chain transfer and to obtain a high molecular weight, one has to polymerize at low temperatures. In the polymerization of isobutylene in liquid ethylene it has been possible to carry out such a low-temperature polymerization on a continuous basis. One lets the liquefied isobutylene, as a mixture with liquid ethylene and with the initiator solution (BF_3 + alcohol in liquid ethylene) flow onto a moving belt in a closed chamber. Through the heat of polymerization the ethylene evaporates and the polyisobutylene can be removed continuously as a rubbery sheet from the belt.

In general, ionic polymerization is more difficult to carry out on a technical scale than radical polymerization. Furthermore, because radical polymerization can be carried out in aqueous systems, it has greater flexibility. One therefore uses ionic polymerization only if the monomer does not polymerize by radical mechanism (for example, with the vinyl ethers or isobutylene) or if the polymers prepared by ionic polymerization show certain advantages. Thus synthetic rubber was initially prepared by anionic polymerization with sodium, or potassium, as initiator. Today it is almost exclusively produced by radical copolymerization of butadiene and styrene with emulsion systems. It was only relatively recently that polymerization of butadiene and isoprene with lithium or Ziegler catalysts in hydrocarbon solvents has been carried out on a technical scale. These catalyst systems yield a polymer which has essentially an all-*cis*-1, 4 structure and which resembles natural rubber in most of its properties.

In general, ionic polymerizations are carried out under the exclusion of water. There are, however, a few monomers, such as the esters of α-cyanosorbic acid, nitroethylene, and the methylenemalonates, which can polymerize to macromolecular substances even with aqueous alkali:

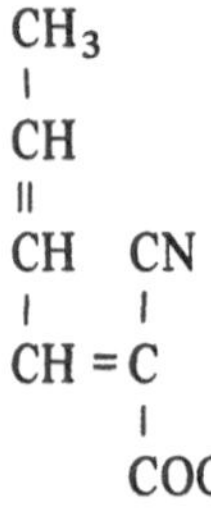

$$CH_3$$
$$|$$
$$CH$$
$$||$$
$$CH \quad CN$$
$$| \qquad |$$
$$CH = C$$
$$|$$
$$COOR$$

α - cyanosorbic
acid esters

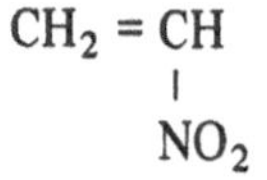

$$CH_2 = CH$$
$$|$$
$$NO_2$$

nitroethylene

$$\overset{O}{\underset{||}{}} \qquad \overset{O}{\underset{||}{}}$$
$$R - O - C - C - C - O - R$$
$$||$$
$$CH_2$$

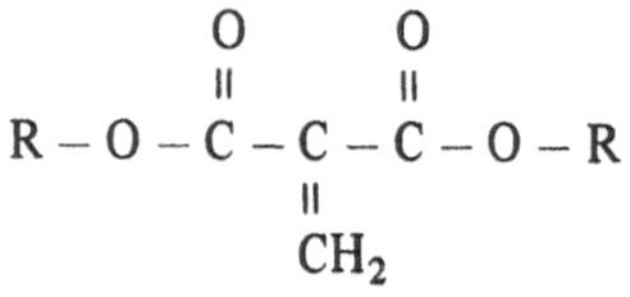

methylene malonates

213   Polymerization with Complex Catalyst Systems

Complex initiators have in recent years attained great technical importance, especially for the polymerization of ethylene and propylene. They consist of a combination of main and subgroup elements in the periodic table and are often called Ziegler catalysts. The most important (from a technical point of view) is the combination of titaniumtetrachloride, or trichloride, with aluminum alkyls. In mixing solutions of the components of the complex, the active initiator usually precipitates in the form of a finely divided suspension. However, it seems that the solid surface may not be required for initiator action, because a number of soluble complex compounds have been described which are also able to polymerize ethylene and propylene (even if the molecular weight is not so high), for example, complexes of cyclopentadiene and titanium in combination with aluminum alkyls:

$$(C_5H_5)_2 TiCl [(C_2H_5)_n AlCl_{3-n}]$$

These are formed from bis(cyclopentadienyl)-titaniumdichloride and ethyl-aluminum-dichloride, or di-ethyl-aluminum-chloride.

In addition to ethylene and propylene, a number of other olefins and other unsaturated compounds can be polymerized with Ziegler catalysts: for example, styrene, vinyl ethers, methylmethacrylate, butadiene, and isoprene. While ethylene and the last named monomers can also be polymerized either by radicals or ionically, and some of them both radically and ionically, propylene and butene can only be polymerized by Ziegler catalysts.

Polymers prepared by complex initiators are often characterized by their high melting points in comparison to the corresponding polymers prepared by radical or ionic polymerization, for these polymers have a high degree of crystallinity, as can be shown by X-ray diagrams. The tendency toward crystallization in turn follows from the highly stereo-regular arrangement of the structural units in the chain.

2131   Stereochemistry of Vinyl Polymers

With vinyl polymers every second carbon atom in the chain is asymmetric (pseudoasymmetric). There are therefore similar possibilities for stereoisomers (mirror image isomers), as with the sugars. Of course, with polymer chains consisting usually of more than 1,000 carbon atoms, one cannot think of isolating any one of the fantastically large number of possible isomers. However, in the polymerization of different monomers with Ziegler catalysts, Natta has found that certain particularly simple, limiting structures—easily recognizable by their high crystallizability—are sometimes preferred in the polymerization. Thus by polymerization with aluminumtriethyl and titaniumtrichloride, Natta has been able to prepare a crystalline polystyrene with a melting point of 230°C, in which the neighboring phenyl residues in the chain all have the same steric position. This type of polymer, where with atomic models the substituents can be turned in such a way that they (i.e., the phenyl residues with polystyrene) all lie over, or beside, each other with respect to a regular zig-zag chain, is called an *isotactic* polymer (compare Figures 186 and 187). On the other hand, if one exchanges in such a model chain the position of the hydrogen and the substituent R in every second repeat unit, one obtains a so-called *syndiotactic* structure. With the aid of the formulas in Figure 186 or with atomic models one can easily see how these different forms compare with each other. In addition to the syndiotactic structure, one can think of other

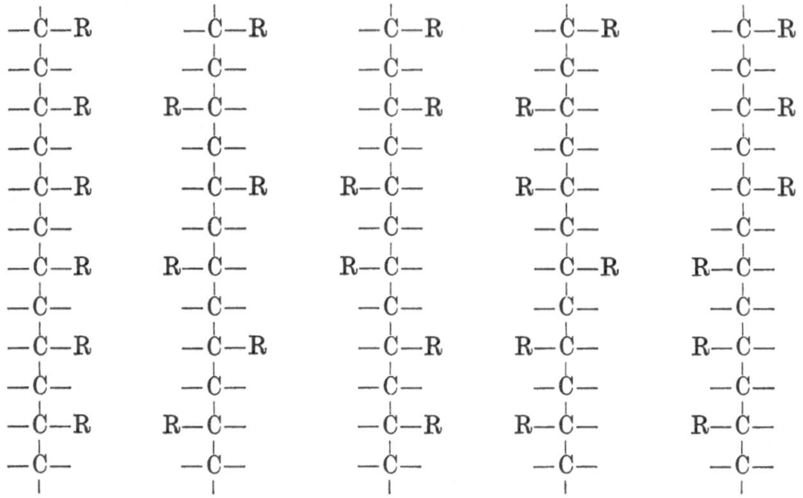

| Isotactic structure | Syndiotactic structure | Steric structures which exist probably only in random sequences along the polymer chains |

FIG. 186 – Some different possibilities of arranging of substituents R along a vinyl chain.

structures with a regularly alternating steric sequence of the substituents. However, so far no confirmation has been obtained for their existence. Rather, one must assume that in the polymer chains under normal conditions (i.e., when there is no isotactic or syndiotactic structure present) the steric sequence of the substituents along the chain changes in a random manner. Such polymers where there is no recognizable regularity in the steric structure are called *atactic* polymers. Since an atactic structure is always more probable than an isotactic or a syndiotactic structure, atactic polymer chains are preferred in polymerization unless there are special circumstances which favor stereospecific addition to form iso—or syndio-tactic chains.

In reality such a fully stretched-out chain cannot exist because of the steric hindrance caused by the substituents. Even the methyl group is too large to permit the formation of fully stretched-out chains in the sense of Figure 187a (hindrance of rotation of the chain C-atoms). Instead, especially in the crystalline state, the substituents arrange themselves around the C-C chain along a spiral, i.e., the polymer crystallizes in the form of a helix (compare Figures 188b, 478/1, 595, and 597).

In general, one observes that low temperature favors the formation of a syndiotactic structure. For example, during ultraviolet polymerization of vinyl-chloride at −20°C to −40°C, one obtains polymers which, because of their high degree of syndiotactic structure, have a softening point approximately 20°C above that of the atactic polyvinylchloride. Perhaps the activation energy of the two possible addition reactions:

a) $\sim\sim\sim CH_2 - \underset{\underset{R}{|}}{\overset{\overset{H}{|}}{C}}\cdot \ + \ CH_2 = \underset{\underset{R}{|}}{CH} \ \rightarrow \ \sim\sim\sim CH_2 - \underset{\underset{R}{|}}{\overset{\overset{H}{|}}{C}} - CH_2 - \underset{\underset{H}{|}}{\overset{\overset{R}{|}}{C}}\cdot$

and

b) $\sim\sim\sim CH_2 - \underset{\underset{R}{|}}{\overset{\overset{H}{|}}{C}}\circ + CH_2 = \underset{\underset{R}{|}}{CH} \rightarrow \sim\sim\sim CH_2 - \underset{\underset{R}{|}}{\overset{\overset{H}{|}}{C}} - CH_2 - \underset{\underset{R}{|}}{\overset{\overset{H}{|}}{C}}\circ$

is different. Since reaction a) occurs more rapidly than reaction b) at lower temperature ($-20°C$ to $-80°C$), whereas at higher temperatures ($+50°C$ to $+100°C$) both reactions have the same rate, b) must have a stronger temperature dependence of the reaction rate and therefore reaction b) must have a greater activation energy.

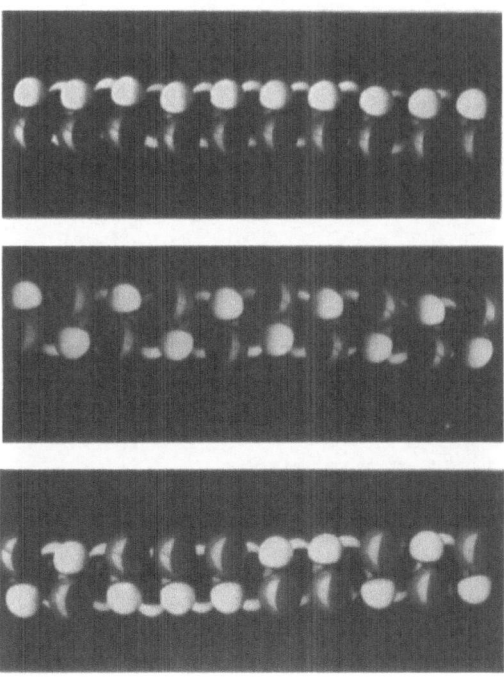

black = carbon   white = hydrogen
gray = any substituent R

FIG. 187a-c — Molecular models of a) an iso-tactic, b) a syndiotactic, and c) an atactic polymer with a regular fully extended zig-zag chain.

On the other hand, the formation of stereoregular structures may also be connected to a special mechanism of the growth reaction, such as is the case in the formation of isotactic polymers using complex initiators.

An isotactic chain structure means that with each structural unit in the chain there is always the same sterical position of the substituents. However, this still says nothing about the *conformation* of the chain, because the steric regularity refers only to the structure and position of the hydrogen and the substituent R relative to the neighboring structural unit. The absolute position of the chain in space has not been fixed in any way, because of the free rotation around the carbon-carbon bonds in the chain. Therefore, an isotactic chain does not always have to have the form presented in Figure 187a. Instead, by turning the chain units with respect to each other, one can obtain a number of completely irregular chain conformations

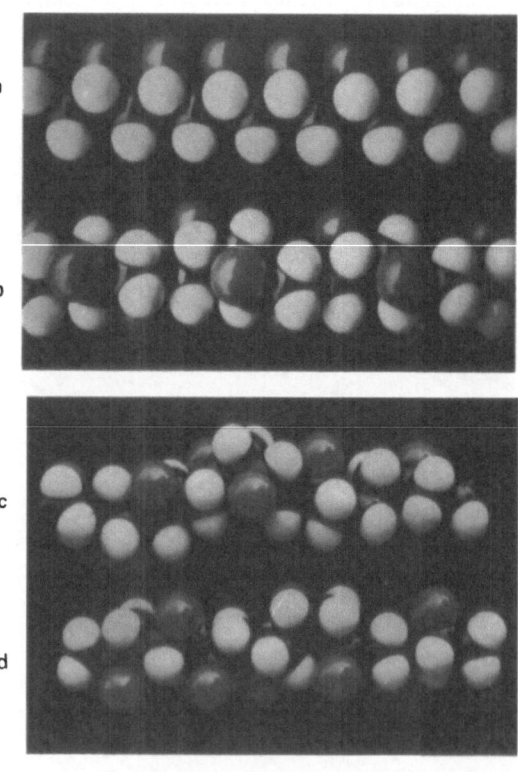

a) Fully extended chain (maximal length)
b) Regular spiral form (3-helix)
c-d) irregular chain conformations

FIG. 188a-d — Statistical (random) and regular chain conformations of one and the same isotactic polymer model. (gray = any substituent R, black = carbon, white = hydrogen)

(see Figure 188). In this respect there is absolutely no difference between these chains and an atactic chain. In solution, the isotactic polymer, just as the atactic one, forms random statistical coils which have the same properties (as can be shown by viscosity measurements) as the coils of an atactic polymer. The difference between isotactic and atactic polymers is only that the isotactic polymers can form completely regular structures which cannot be formed in any way from the atactic chains. Figure 188 shows a number of possible chain conformations which have been formed from the same isotactic chain model by turning the chain elements with respect to each other.

It should be realized that the completely extended chain is only an ideal structure, which cannot exist with real substituents for steric reasons. This is why fully extended zig-zag chains are only present in crystals of polyethylene, or in polyamide crystals of the nylon type, or with other polymers whose chains have no side-groups. In crystals of polypropylene, polybutene, and other isotactic vinyl polymers, however, the chains are in the form of a helix.

Because of the regular arrangement of the substituents along the polymer chain, the chains are able to organize themselves next to each other in a regular manner, i.e., they can form crystals. In this way the secondary valence forces between the atoms and groups of different chains can reach their maximum effect, and, therefore, the isotactic polymers have considerably higher melting points. This is shown for a few examples in Table 189.

Sometimes in the literature one finds different solubilities for isotactic and atactic species of the same polymer. This, however, cannot be really proved, because the differences in solubility could also be caused by different molecular weights.

The contrast just discussed between atactic and isotactic structures occurs only with polymers of unsymmetrically substituted olefins, i.e., with vinyl polymers, but

TABLE 189. Comparison of the properties of isotactic and atactic polymers (according to NATTA).

Polymer	Melting point or softening range, °C	Density
Polypropylene:		
isotactic, crystalline	160	0.92
atactic, amorphous	75	0.85
Poly-l-butene:		
isotactic, crystalline	128	0.91
atactic, amorphous	65	0.87
Polystyrene:		
isotactic, crystalline	230	1.08
atactic, amorphous	100	1.06

not with polyethylene or polyvinylidene compounds. With these polymers the chains are always regular, whereas with vinyl compounds only the syndiotactic and isotactic forms have a regular chain structure. Thus such polymers as polyethylene, vinylidenechloride, polytetrafluoroethylene, or polyisobutylene always show a more or less strong tendency for crystallization, unless there are irregularities of a constitutional type, such as branching, short side-chains etc., which prevent or reduce crystallization. This is, for example, the case with high pressure polyethylene, which has a branched structure and, therefore, a lower degree of crystallization, than the less branched, low pressure polyethylene.

Stereoregular structures do not always owe their origin to the presence of asymmetric carbon atoms (as in the case of polyvinyl compounds), but they can also result from the presence of double bonds in the polymer chain, as for example, with 1, 4-polybutadiene and 1, 4-polyisoprene. In these cases the CH_2-groups adjacent to the double bond in the chain can either have a *cis* or a *trans* position in relation to the double bond:

$$\text{~~~CH}_2 \qquad \text{CH}_2\text{—CH}_2 \qquad \text{CH}_2\text{—CH}_2 \qquad \text{CH}_2\text{—CH}_2 \qquad \text{CH}_2\text{—CH}_2\text{~~~}$$
$$\text{CH=CH} \qquad \text{CH=CH} \qquad \text{CH=CH} \qquad \text{CH=CH}$$

cis-1,4-Polybutadiene

$$\text{~~~CH}_2 \qquad \text{CH} \qquad \text{CH}_2 \qquad \text{CH} \qquad \text{CH}_2 \qquad \text{CH}$$
$$\text{CH} \qquad \text{CH}_2 \qquad \text{CH} \qquad \text{CH}_2 \qquad \text{CH} \qquad \text{CH}_2 \text{~~~}$$

trans-1,4-Polybutadiene

Practically pure *cis*-1, 4-polybutadiene is obtained using an initiator-combination such as $AlR_3/TiBr_4$, AlR_3/TiI_4, or $AlR_2Cl/CoCl_2$, in hydrocarbons as solvents. It is of particular technical interest that such an all-*cis*-1, 4-polybutadiene or polyisoprene can be prepared, because these materials have properties very similar to natural rubber. *Trans*-1, 4-polybutadiene can be obtained, for example, by polymerization of butadiene with $Al(C_2H_5)_3/VCl_3$ as the initiator. By radical polymerization of butadiene one usually obtains chains with a mixed structure. However, there are exceptions: for example, in emulsion polymerization of butadiene with $RhCl_3 \cdot 3\ H_2O$ as the initiator, one obtains nearly pure *trans*-1, 4-polybutadiene (compare p. 51).

2132   The Mechanism of Polymerization With Complex Initiators

One first assumed that polymerization with complex initiators, such as the Ziegler catalysts (aluminumalkyls plus titaniumhalides), works by a simple ionic mechanism. Since single aluminumalkyls normally cause an anionic, and titaniumhalides a cationic chain reaction, the two components of the initiator should neutralize each other and only the excess of one over the other could be active. If this were true, however, either one of the components alone should be able to initiate the polymerization of ethylene or propylene, but this is definitely not the case (except for the formation of low molecular weight polyethylene and polypropylene oils and greases when cationic initiators, such as $AlCl_3$, BF_3, or $TiCl_4$, are used). An anionic or cationic mechanism can therefore not explain the polymerization with Ziegler catalysts.

According to a theory of Patat and Sinn the really active catalysts are complexes which have an electron-deficient bond ($Ti\cdots C\cdots Al$):

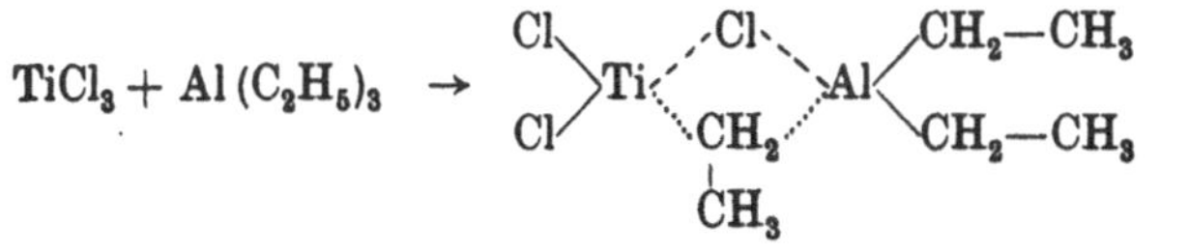

$$TiCl_3 + Al(C_2H_5)_3 \rightarrow$$

(not isolated) active

$$(C_5H_5)_2TiCl_2 + Al(C_2H_5)_3 \rightarrow$$
Titanium-bis-cyclo-
pentadienyl-dichloride

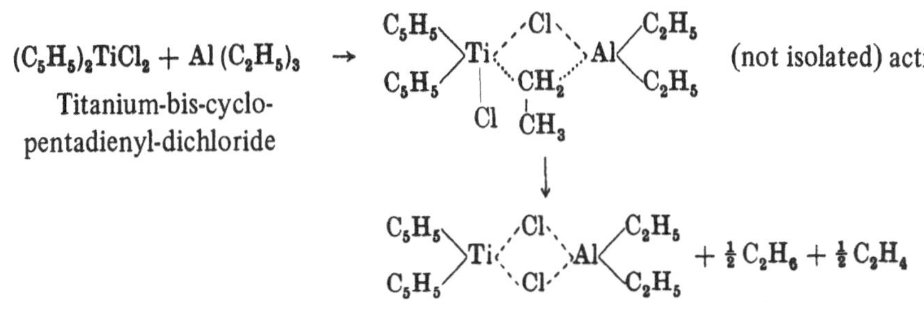

(not isolated) active

(can be isolated) inactive

The mechanism of the polymerication is presented by Patat and Sinn in the following way: a $Ti\cdots C$-bond opens as an olefin molecule enters the complex through reaction of its π-electron system with the $3d$ electron of the titanium and subsequent formation of a new $Ti\cdots C$-bond and of two unsaturated residual valences on two carbon atoms:

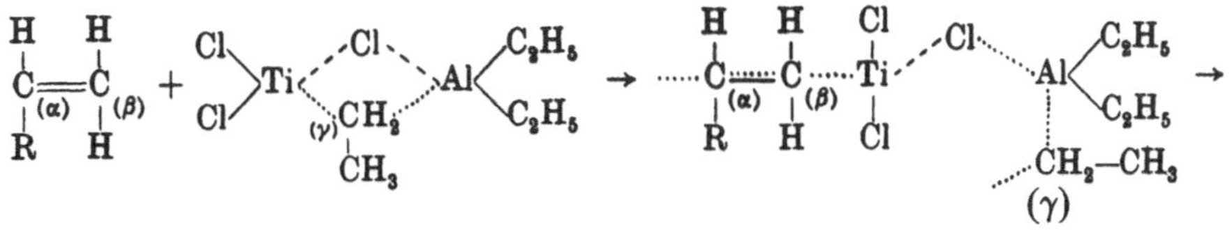

To explain this stereospecific activity of the catalysts, it is important that free rotation of the partially opened carbon-carbon double bond of the olefin remains suspended. In the same way there is no free rotation around the $Ti\cdots C$-bond (planar 2p-3d-overlap) and therefore in the further reactions the substituent R maintains its steric position.

In a further reaction step the free residual valences become saturated by ring formation (2p-2p-overlap):

$$\rightarrow \quad
\begin{array}{c}
\text{Cl}\diagdown \qquad \diagup\text{Cl}\diagdown \qquad \diagup\text{C}_2\text{H}_5 \\
\qquad \text{Ti}\diagup \qquad \diagdown \text{Al}\diagdown \\
\text{Cl}\diagup \qquad \qquad \text{C}_2\text{H}_5 \\[4pt]
\overset{\cdot}{\text{CH}_2}\ \ \text{H}\quad \overset{\cdot}{\text{CH}_2}-\text{CH}_3 \\
(\beta)\diagdown\qquad\ \ (\gamma) \\
\qquad\ \underset{\underset{\text{R}}{|}}{\text{C}}{\scriptstyle(\alpha)}
\end{array}
\quad\rightarrow\quad
\begin{array}{c}
\text{Cl}\diagdown \qquad \diagup\text{Cl}\diagdown \qquad \diagup\text{C}_2\text{H}_5 \\
\qquad \text{Ti}\diagup \qquad \diagdown \text{Al}\diagdown \\
\text{Cl}\diagup \qquad \qquad \text{C}_2\text{H}_5 \\[4pt]
\overset{\cdot}{\text{CH}_2}\ \ \text{H}\quad \text{CH}_2 \\
(\beta)\diagdown\qquad\ \ (\gamma)\diagdown \\
\qquad\ \underset{\underset{\text{R}}{|}}{\text{C}}{\scriptstyle(\alpha)}\qquad\qquad \text{CH}_3
\end{array}$$

$$\rightarrow\quad
\begin{array}{c}
\text{Cl}\diagdown \qquad \diagup\text{Cl}\diagdown \qquad \diagup\text{C}_2\text{H}_5 \\
\qquad \text{Ti}\diagup \qquad \diagdown\text{Al}\diagdown \\
\text{Cl}\diagup \qquad \qquad \text{C}_2\text{H}_5 \\[4pt]
\qquad\qquad \text{CH}_2\diagdown \\
\qquad\qquad\qquad \text{CH}-\text{CH}_2-\text{CH}_3 \\
\qquad\qquad\qquad\ \ \underset{}{|} \\
\qquad\qquad\qquad\ \ \text{R}
\end{array}$$

The 2p-2p-overlap (Cα, Cγ) is converted by an exothermic reaction with hybridization into a regular σ-hybrid bond. At the same time the Al$\cdots$C-bond is opened, and on the aluminum and on the Cβ residual valences are formed. Thus the original complex structure is reformed and another olefin molecule can add. The driving force of the reaction cycle, as with all polymerizations, is the conversion of the olefin double bond to a C-C single bond, which can occur with very low activation energy over the detour of the Ti$\cdots$C$\cdots$Al-complex. As a possible chain termination reaction, one can suggest the formation of a metal hydride and a double bond at the chain end, or homolytic decomposition of the Ti$\cdots$C-bond.

In order to explain the stereospecificity of the addition step, one assumes that the catalytically active complex with solid catalyst systems (only those have so far been able to form isotactic polymers) has a certain asymmetry as a result of the incorporation of the complex into a crystalline surface. This brings about that the two possible steric structures of the olefin addition complex are energetically somewhat different in that the addition according to step a) or step b) is favored one over the other.

$$
\begin{array}{c}
\text{H}\ \ \text{H} \\
\underset{\underset{\text{R}}{|}}{\overset{\overset{}{|}}{\text{C}}}{\scriptstyle(\alpha)}=\underset{\underset{\text{H}}{|}}{\text{C}}{\scriptstyle(\beta)} \ + \\
\end{array}
\begin{array}{c}
\text{Cl}\diagdown\quad\diagup\text{Cl}\diagdown\quad\diagup\text{C}_2\text{H}_5 \\
\ \text{Ti}\diagup\quad\diagdown\text{Al}\diagdown \\
\text{Cl}\diagup\ \ (\gamma)\text{CH}_2\quad\ \ \text{C}_2\text{H}_5 \\
\qquad\qquad |\\
\qquad\qquad\text{CH}_3
\end{array}
$$

$$
\overset{\nearrow}{}
\begin{array}{c}
\text{H}\ \ \text{H}\ \ \text{Cl} \\
\cdots\text{C}\text{------}\text{C}\text{------}\text{Ti}\diagup^{\text{Cl}\diagdown}\diagdown_{\text{Al}}\diagup^{\text{C}_2\text{H}_5} \\
|{\scriptstyle(\alpha)}\ |{\scriptstyle(\beta)}\ \ |\qquad\qquad\diagdown\text{C}_2\text{H}_5 \\
\text{R}\ \ \ \text{H}\ \ \text{Cl}\ \ \text{CH}_2 \\
\qquad\qquad\qquad\quad |{\scriptstyle(\gamma)} \\
\qquad\qquad\qquad\quad \text{CH}_3
\end{array}\ \ (a)
$$

$$
\overset{\searrow}{}
\begin{array}{c}
\text{R}\ \ \text{H}\ \ \text{Cl} \\
\cdots\text{C}\text{------}\text{C}\text{------}\text{Ti}\diagup^{\text{Cl}\diagdown}\diagdown_{\text{Al}}\diagup^{\text{C}_2\text{H}_5} \\
|{\scriptstyle(\alpha)}\ |{\scriptstyle(\beta)}\ \ |\qquad\qquad\diagdown\text{C}_2\text{H}_5 \\
\text{H}\ \ \ \text{H}\ \ \text{Cl}\ \ \text{CH}_2 \\
\qquad\qquad\qquad\quad |{\scriptstyle(\gamma)} \\
\qquad\qquad\qquad\quad \text{CH}_3
\end{array}\ \ (b)
$$

Since due to the lack of freedom of rotation exchange in the positions of H and R are no longer possible in the complex, the steric position of R must be the same with every structural unit in the growing chain, i.e., the corresponding polymer is isotactic. Since the publication of Patat and Sinn in 1958, many papers have dealt with the mechanism of Ziegler-Natta polymerizations, but it is still not possible to define the reaction steps in all details and to confirm a certain mechanism unequivocally. However, the basic concept, that the monomer is incorporated into a Ti-Al complex with an electron-deficient bond and that the polymer chain then grows out of this complex as the alkyl-ligand, has been confirmed.

Polymerization of Ethylene with Soluble Ti-Al Complexes

The complex initiators which are used technically for the polymerization of ethylene, propylene, and butadiene are insoluble in inert solvents. By using dicyclopentadienyltitaniumdichloride as the Ti component, one obtains with $AlCl_3$ or $AlEtCl_2$ soluble initiators which polymerize ethylene even at temperatures as low as 0°C. However, the molecular weight of the polymers obtained with such catalyst systems is lower than that of polyethylene produced by industrial processes.

In experiments with the polymerization of ethylene using such soluble Ti-Al complexes, G. Henrici-Olivée and S. Olivée, through careful kinetic measurements of reduction and polymerization at normal pressure and 0°C temperature, formed a rather concrete picture of the reaction. This comes quite close to the theoretical concepts developed by P. Cossee and will be described in the following section.

If one reacts 1 mole of dicyclopentadienyltitaniumdichloride, $[CpTiCl_2(I)]$, with approximately 2 moles of ethylaluminumdichloride, $[EtAlCl_2(II)]$, in toluene, a system of complexes is formed (accompanied by a color change) and the equilibria involved can be described by the scheme shown on the following page.

The reduced complex (IX) was confirmed by means of ESR spectroscopy. Complex (IX) can also be prepared in a different way in crystalline form and gives the same ESR spectrum. Since the reduction of the titanium occurs by the elimination of an exocyclic ethyl group on the titanium (confirmed by Sinn and Patat), along with the formation of ethylene and ethane, it must be true that compound (IX) is formed via compounds (III), (V) and (VII). Compound (V) can then react, instead of with $AlCl_3$, with the excess of ethylaluminumdichloride (II), resulting in the formation of Complex (VI). The latter is then reduced to compound (VIII) (ESR spectrum) and finally is transformed into compound (IX).

Of all the complexes in this equilibrium only compound (VI) is active as an initiator for the polymerization of ethylene. Compounds (VI) and (VII) can be prepared from the pure crystalline form of compound (V) by reaction with $AlCl_3$ and $AlEtCl_2$, respectively, and are obtained also in a homogeneous crystalline form. Only compound (VI) initiates ethylene polymerization, but not compound (VII), or the reduced complexes (VIII) and (IX). If one reacts Cp_2TiCl_2 and $AlEt_2Cl$ in an analogous equilibrium reaction, one obtains the complex $Cp_2Ti\,EtCl_2AlEt_2$ (X) with two ethyl-groups on the aluminum atom, which is even more active for ethylene polymerization than compound (VI). One therefore has the sequence in increasing activity of the initiator shown on p. 194.

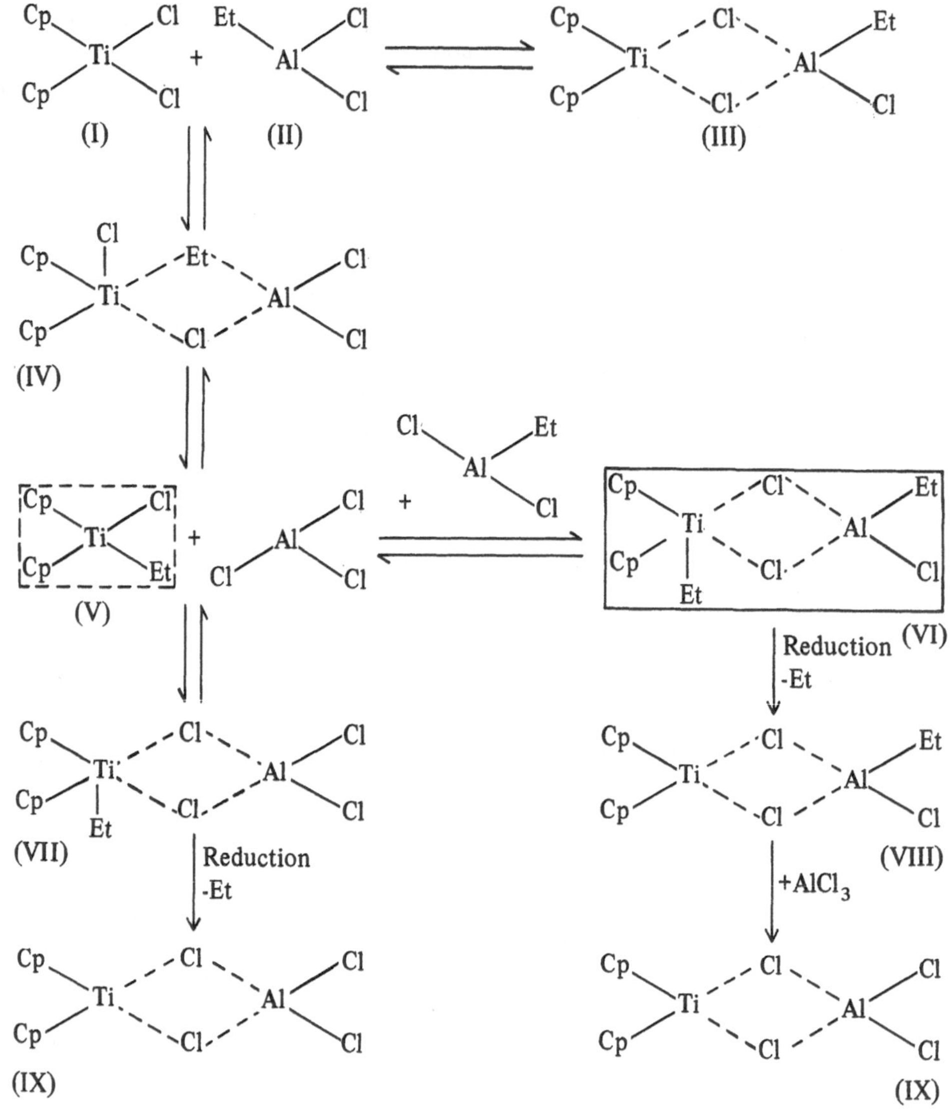

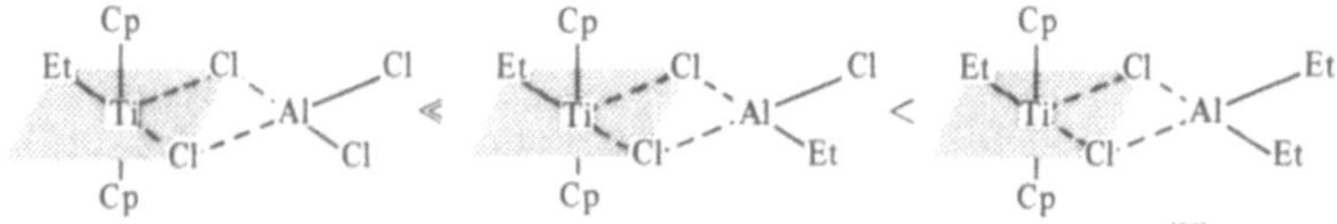

(I)   (II)   (III)

(IV)

(V)   (VI)

Reduction (VI)
-Et

(VII)   (VIII)

Reduction
-Et

+AlCl$_3$

(IX)   (IX)

(X)

With the transition from Cp$_2$TiCl$_2$ (Tetrahedral structure in the crystalline state) to the Ti-Al mixed complexes, which makes it necessary that one of the ligands of the aluminum atom enters into the coordination sphere of the titanium atom, probably the transition to an octahedral configuration is involved:

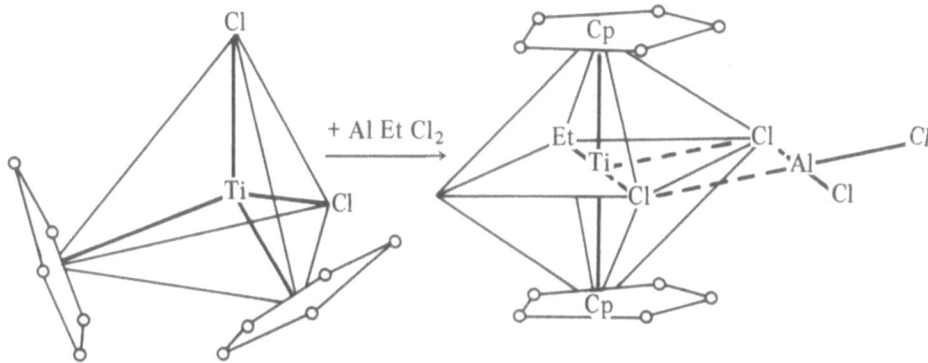

Through this transition the Et and Cl ligands of titanium and aluminum take up a planar arrangement and the Ti-Et bond is further weakened, the more Cl ligands on the aluminum are replaced by electron-donating ethyl groups. As the ESR spectra of the reduced complexes show, the ethyl substituents lower the spin density at the nucleus of the aluminum atom, which can be interpreted as resulting from the electron shift in the direction Et → Al → Cl → Ti, and which is responsible for the weakening of the Ti-Et bond.

The tendency towards the splitting-off of the ethyl radicals from the titanium, which increases in the sequence (VII) → (VI) → (X), can be observed from the strongly increasing rate of the reduction $Ti^{IV} \to Ti^{III}$ (Figure 196). This reduction is clearly a second order reaction and can therefore be formulated as follows:

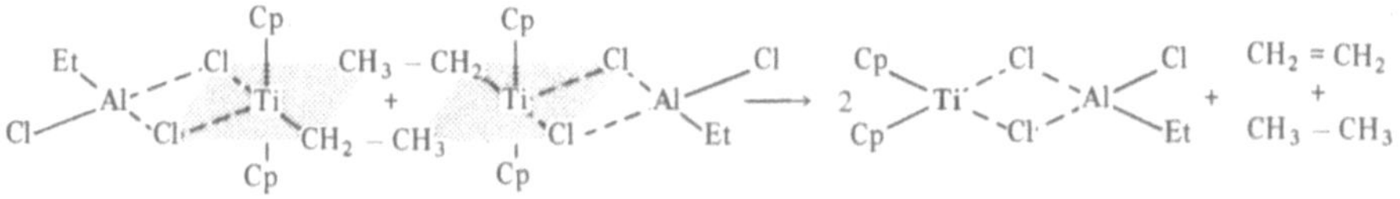

This reaction corresponds completely to a disproportionation reaction of two radicals, and even though there are no free radicals present, the loosely bound ethyl groups on the titanium can react similar to ethyl radicals.

If one considers the scheme of the reduction reaction and sees that the rate of the reduction and of the polymerization are influenced the same way by the ethyl group on the aluminum (see Figure 196), then this is a clear indication that the splitting-off of the ethyl group on the titanium also plays a deciding role in the polymerization reaction. It is easy to imagine that the unoccupied sixth coordination position on the titanium, which during the reduction reaction is temporarily occupied by the ethyl group of a second complex molecule, is occupied by an ethylene molecule during the polymerization reaction.

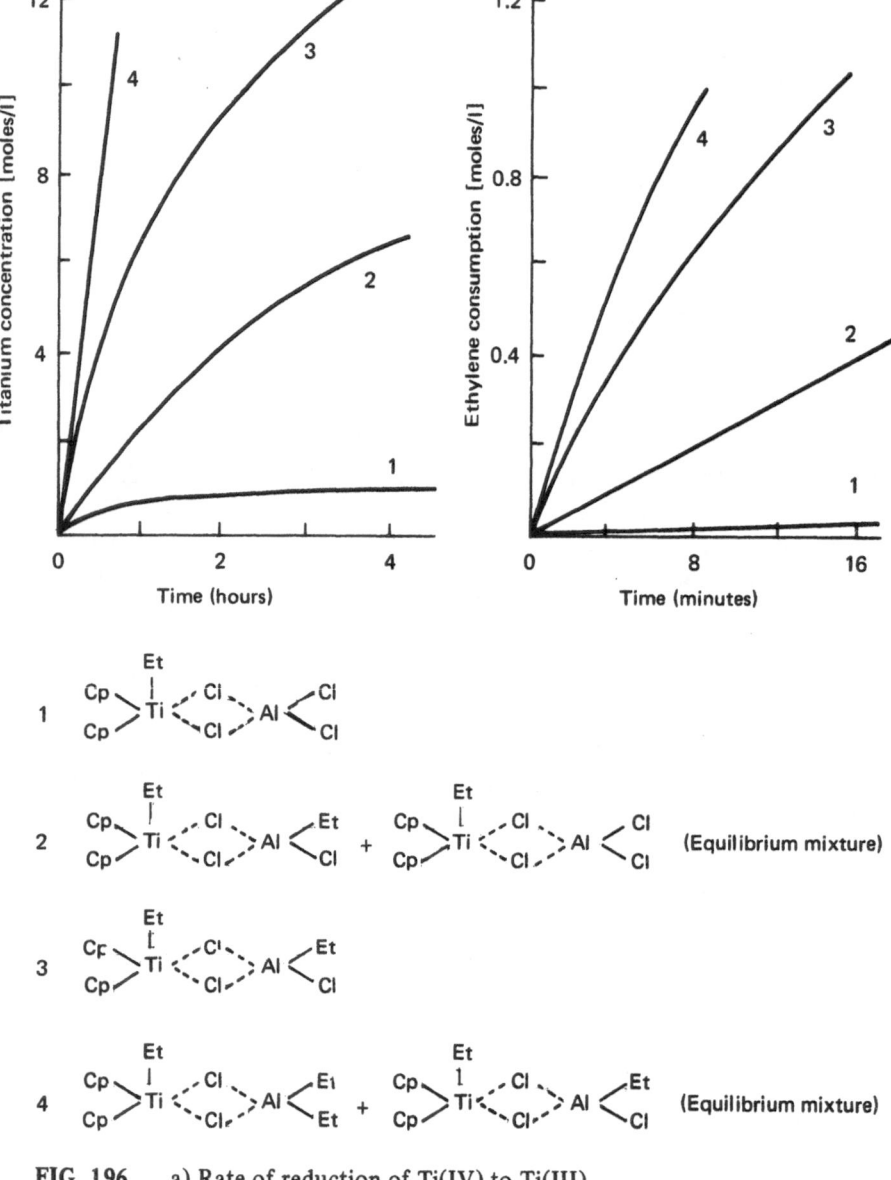

FIG. 196 — a) Rate of reduction of Ti(IV) to Ti(III)

        b) Rate of polymerization of ethylene for different Ti-Al-complexes
(Data of G. HENRICI − OLIVÉE and S. OLIVÉE)

Ethylene, because of its electron configuration, is particularly suitable for complex formation with a transition metal (compare Figure 197b). The Ti-Et bond, as has been shown by a molecular orbital representation of the complex, (see Figure 200), is weakened even further through the coordination with ethylene.

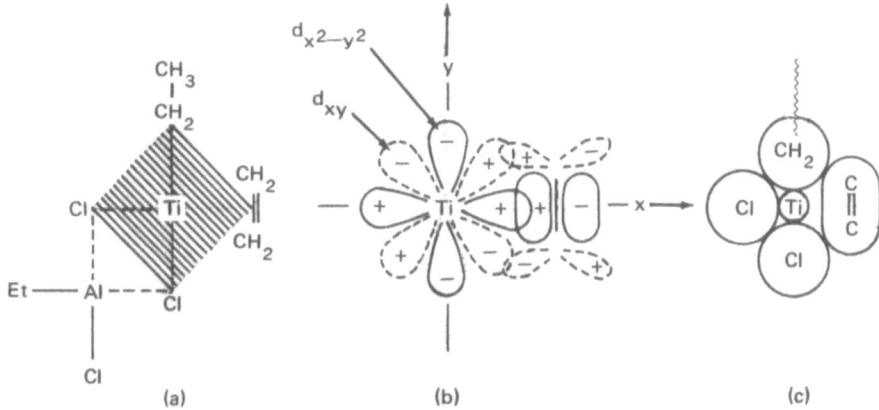

(a) structural formula (Ti–Cp vertical to the plane of the paper)

(b) schematic representation of the orbitals in the x − y − plane of the complex*

(c) x − y − plane of the complex showing the actual size relationships of the metal − ligands (after P. COSSEE)

*The bonding π- orbital of the ethylene molecule is σ -symmetric with respect to the x-axis of the complex and therefore corresponds to the (empty) $d_{x^2-y^2} - \sigma$ – orbital of the titanium. Similarly, the anti-bonding π^*-orbital of the olefin corresponds to an orbital of the titanium – the (also empty) $d_{xy} - \pi$ – orbital (π-symmetry with respect to the x – axis). A certain part of the electron density is transferred to the metal by the bonding π – orbital of the ethylene, which causes a weakening of the ethylene double-bond (compare also Fig. 200)

FIG. 197 − x-y plane of the active Ti-Al-complex (VI) with ethylene as the sixth Ti− ligand

Ethylene does not have the opportunity to react with the ethyl group under formation of ethane or ethylene, and therefore it is satisfied to form with the ethyl group a butyl group, which now, in place of the ethyl group, occupies the ligand on the titanium atom. The sixth coordination position is again free for a new ethylene, which then, in the same way, reacts with the butyl group to form a hexyl group. In this way, through the back and forth movement (*cis*-migration) of the ethyl group i.e., the growing chain, between the two neighboring coordination positions on the titanium atom, the polyethylene chain grows longer and longer until, through reaction with a second complex molecule, the chain is split-off under formation of an olefinic and a paraffinic chain end, and the titanium goes over into the trivalent state:

Chain growth:

Chain termination:

$$
\begin{array}{ccc}
\text{Cl} & \text{Cl} & \text{Cp} \\
\text{Al} & \text{Ti} & \text{CH}_2 \\
\text{Et} & \text{Cl} & \text{CH}_2
\end{array}
\quad + \quad
\begin{array}{c}
\text{Cp} \\
\text{Ti} \\
\text{Cp}
\end{array}
\quad \longrightarrow \quad
\begin{array}{c}
\text{CH}_2=\text{CH} \\
+ \\
\text{CH}_3-\text{CH}_2
\end{array}
\quad + 2 \quad
\begin{array}{c}
\text{Cp} \\
\text{Ti} \\
\text{Cp}
\end{array}
$$

In this way one can picture the chain growth step in the polymerization of ethylene with soluble Ti-Al complexes as a quasi-radical polymerization within the coordination sphere of the titanium. Thus the chain end, which is bound by coordination to the titanium atom, wanders as a radical to the newly coordinated ethylene (in *cis*-position), whereby the coordination position it has just left becomes available to take up a new ethylene. The chain radical is, of course, not a free radical, but remains as a ligand bound to the titanium during the entire chain growth. The *cis*-migration of the $\cdot$CH$_2$ (i.e., the growing chain) to the coordinated monomer is simplified by the steric arrangement (compare Figure 197), and by the resulting overlap of the orbitals of the CH$_2$ group with those of the olefin. Furthermore, through the acceptance of the ethylene into the empty position on the complex, not only the Ti-Et bond is weakened, but also the olefin double bond (by partial electron transfer from the bonding π-orbital to the titanium atom). It is hard to say how much of this study of soluble Ti^{IV}-Al complexes can be applied to the mechanism of ethylene polymerization by heterogeneous Ti-Al systems. In industry, the Ti^{III}-Al combinations [for example, TiCl$_3$-Al(C$_2$H$_5$)$_3$] are very active initiators for the polymerization of olefins such as propylene. As one can see from models, the fracture surface of a TiCl$_3$ crystal (hexagonal lattice) has a structure which allows the formation of Al complexes with octahedral arrangement of the ligands and an unoccupied coordination position, so that a samilar polymerization reaction, as described above, may also be possible on the surface of the TiCl$_3$ crystals.

Just as with other polymerizations, a part of the initiator is also incorporated into the polymer chain, i.e., an ethyl group of the aluminumtriethyl. If aluminumtriphenyl is used instead of aluminumtriethyl, one finds, as is expected, a phenyl residue at the chain end. Hydrogen acts as a modifier and reduces the molecular weight of the polymer.

In contrast to other types of polymerization, where the growing chain end has the power to add a monomer only because of the excess or the lack of an electron, in the Ziegler-type polymerization it is the metal complex which brings about an addition at the growing chain end. The chain grows in such a way that one monomer molecule after the other introduces itself between the chain and the metal complex. The metal complex therefore resembles in its function some automatic mechanism (for example, the moving part of a zipper), which combines the added units into a chain. In contrast to radical and ionic polymerization, where the initiator transfers its functions at the moment of chain initiation to the growing chain, the metal complex enters into the reaction at every single addition of the monomer. Thus one can talk with more justification about polymerization catalysts with these complex initiators than with ionic or radical initiators. The mechanism of this type of chain growth is very similar to that involved in the formation of natural macromolecules, brought about by enzymes.

Phillips Catalysts

Certain heavy metal oxides such as $Cr^{VI}O_3$ (Phillips Petroleum Co.) and $Mo^{VI}O_3$ (Standard Oil of Indiana) are excellent catalysts, which in their reaction mechanism and in their technical importance are similar to the Ziegler catalysts. This is especially true for the polymerization of ethylene at low pressure. The oxides are deposited on carriers such as Al_2O_3, SiO_2, or SiO_2/Al_2O_3, and they are then activated at temperatures between 400°C and 500°C, sometimes with the addition of alkaline or alkaline-earth metals. This activation brings about a partial reduction to lower valence state oxides. These catalysts are used in the form of small particles in suspension in hydrocarbon solvents such as cyclohexane. Their activity has been increased greatly during the last years, and because of the low concentration now required, they do not have to be removed from the final polymerization product.

Process Technology of Metal Complex Polymerizations

Since their discovery between 1950 and 1960, polymerizations initiated by metal complexes [Ziegler catalyst (Ti/Al), Phillips catalysts (Cr_2O_3), Standard Oil catalyst (MoO_3)] are of increasing technical interest for the polymerization of ethylene, propylene, butene-1, and butadiene.

The polymerizations can be carried out by three different processes: solution polymerization, suspension polymerization, and gas-phase polymerization. The solution polymerization occurs in solvents in which the polymer remains dissolved, e.g., in cyclohexane. In a continuously running process, solvent, monomer, and the catalyst suspension are fed into an agitation vessel at the same rate as the polymer solution is leaving the vessel. By stripping with steam, the polymer is precipitated, filtered, and dried. Solvent and nonreacted monomer are recycled.

The suspension process is quite similar: one takes mixtures of hydrocarbons as solvents, in which the polymer is formed as an insoluble filterable suspension. In recent years another suspension process has reached technical maturity, in which the monomer, without admixture of solvents, is polymerized directly out of the gas phase (Wisseroth, BASF). The polymerization temperature lies below the softening point, and the polymer exists in the reactor as a fine powder, which has the properties of a fluid with regard to its rheological behavior. The main problem in this process is the removal of the reaction heat, because the fine polymer-in-gas-dispersion, unlike a true liquid, is an extremely bad conductor of heat. This problem has been solved in an elegant way: the recycled monomer is compressed and cooled outside the reaction vessel and then, during pressure drop, fed again into

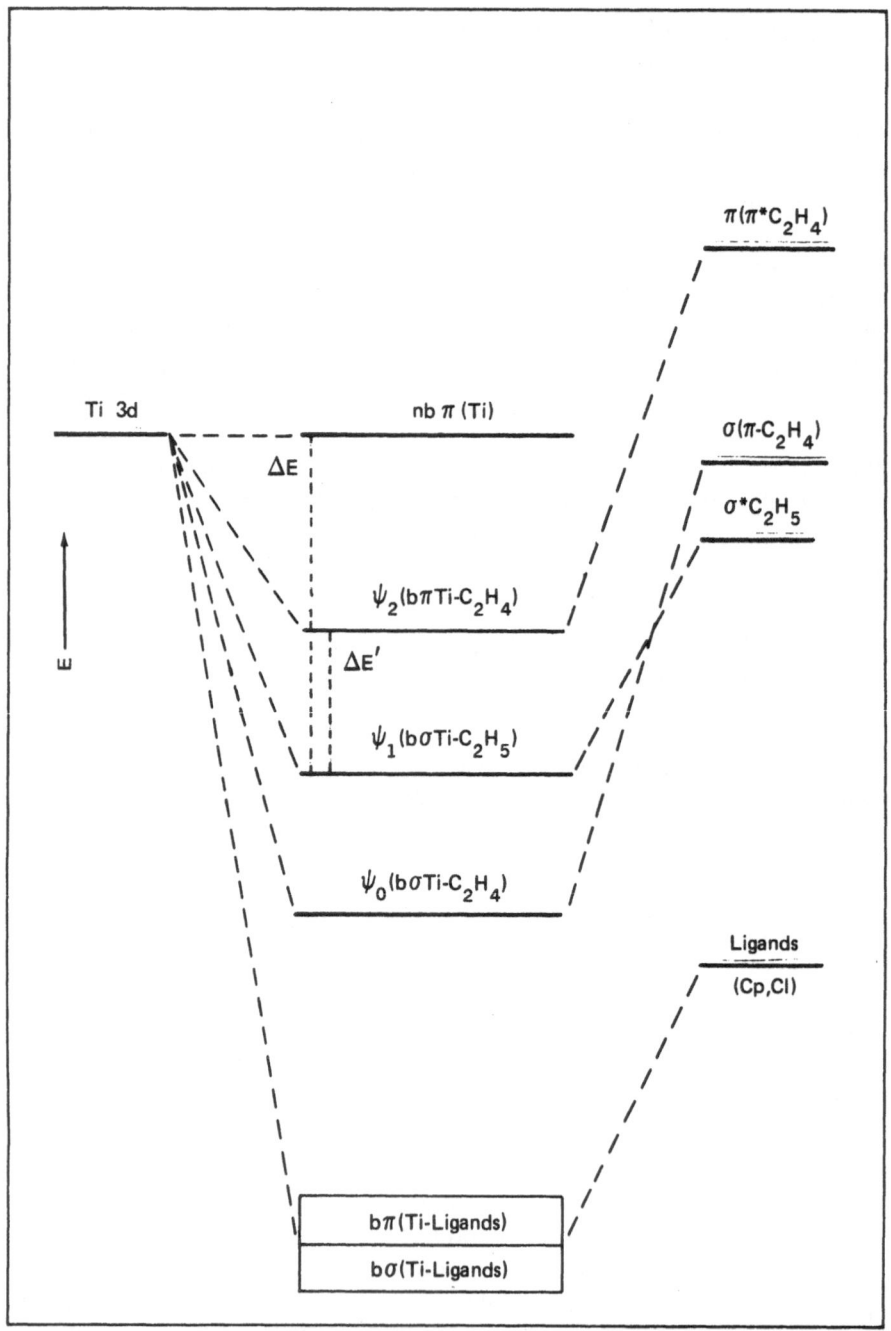

FIG. 200 — Simplified qualitative MO– diagram showing the influence of the ethylene and coordination on the strength of the Ti–Et – bond (after G. HENRICI OLIVÉ and S. OLIVÉ)

the reactor; because of the Joule-Thomson effect, it causes so much cooling that the reaction heat is removed.

214   Other Methods for the Synthesis of Macromolecules with Carbon-Carbon Chains

Although polymerization is one of the most important technical reactions for the formation of macromolecules with carbon-carbon chains, it is not the only possibility. In principle, all reactions of bifunctional compounds which lead to the formation of carbon-carbon bonds may be used. For example, the Wurtz-Fittig reaction, the Friedel-Crafts reaction, and several others which will be described below.

Wurtz-Fittig Reaction

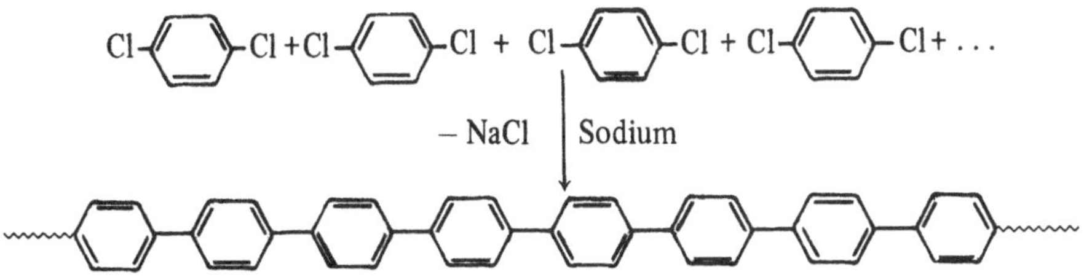

This reaction does not proceed in a uniform manner and does not constitute a useful method for the preparation of polyphenylene. However, polyphenylene can

---

(Legend for Fig. 200 continued)

Explanations:

bσ and bΠ :   Stable, occupied $\sigma-$ and $\Pi-$ bonding orbitals, which link the Ti with the ligands Cp and Cl.

ψ_1 :   Bonding σ − MO, which is responsible for the Ti−Et bond. ψ_1 is the energy-richest occupied orbital of the penta-coordinated complex (before the incorporation of the ethylene molecule).

In addition, the titanium has a non-bonding σ − orbital (empty), and non-bonding Π − orbitals.

When an ethylene molecule approaches the empty orbital, two new M−orbitals are formed:

ψ_0 :   A Bonding σ − MO, which results from overlapping of the bonding Π − orbital of the ethylene with the formerly non-bonding Ti − σ − orbital (energy still under investigation)

ψ_2 :   Unoccupied Π − MO, arising from a non-bonding Ti − Π − orbital (before incorporation of the ethylene) and the anti-bonding Π^* − orbital of the ethylene.

ΔE :   Energy, which (before the coordination of the ethylene) would be required to raise an electron by thermal excitation from ψ_1 to the next-higher unoccupied orbital.

$\Delta E'$:   Energy, which (after entry of the ethylene into the empty position in the complex) is still required in order to make the ethyl-group available as a radical.

be prepared through cationic polymerization of benzene under simultaneous dehydrogenation (compare p. 181). Similarly, the following reactions do not represent useful preparative reactions for the synthesis of polyethylene, respectively poly-p-xylylene:

$$Br\,(CH_2)_{10}\,Br + Br\,(CH_2)_{10}\,Br + Br\,(CH_2)_{10}\,Br + \ldots\ldots$$

$$\text{– NaCl} \quad\Big|\quad \text{Sodium}$$

$$\sim\sim\sim CH_2 - CH_2 - CH_2 - CH_2 - CH_2 - CH_2 - CH_2 - CH_2 - CH_2 - CH - CH_2 - CH_2 - CH_2 - CH_2 - CH_2\sim\sim\sim$$

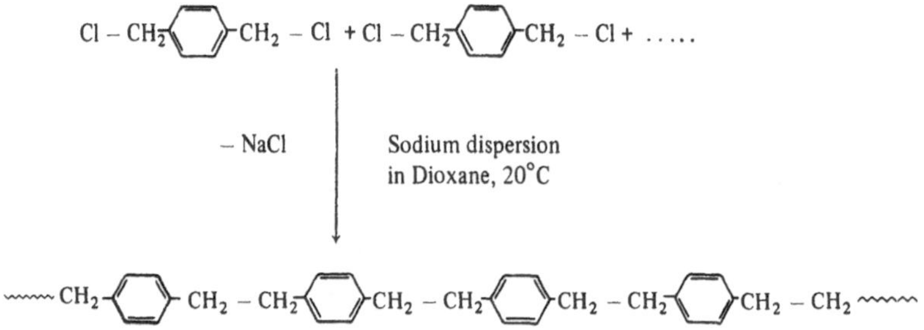

$$Cl - CH_2\!\!\left\langle\bigcirc\right\rangle\!\!CH_2 - Cl + Cl - CH_2\!\!\left\langle\bigcirc\right\rangle\!\!CH_2 - Cl + \ldots\ldots$$

$$\text{– NaCl} \quad\Big|\quad \begin{array}{l}\text{Sodium dispersion}\\ \text{in Dioxane, 20°C}\end{array}$$

$$\sim\sim CH_2\!\!\left\langle\bigcirc\right\rangle\!\!CH_2 - CH_2\!\!\left\langle\bigcirc\right\rangle\!\!CH_2 - CH_2\!\!\left\langle\bigcirc\right\rangle\!\!CH_2 - CH_2\!\!\left\langle\bigcirc\right\rangle\!\!CH_2 - CH_2\!\sim\sim$$

For the synthesis of poly-p-xylylene compare page 203.

Friedel-Crafts Reaction

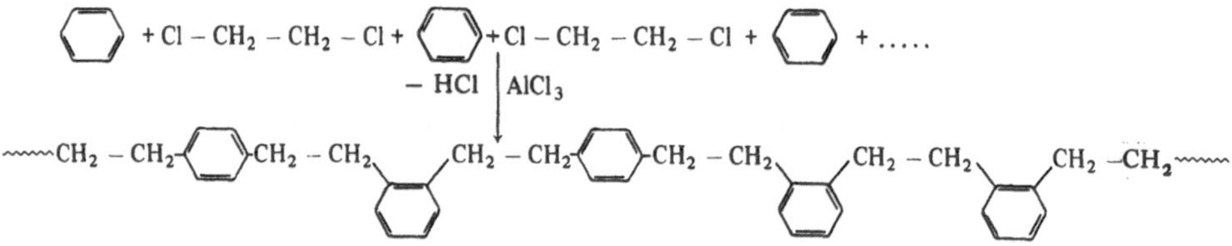

$$\bigcirc + Cl - CH_2 - CH_2 - Cl + \bigcirc + Cl - CH_2 - CH_2 - Cl + \bigcirc + \ldots..$$

$$\text{– HCl} \quad\Big|\quad AlCl_3$$

$$\sim\sim CH_2 - CH_2\!\!\left\langle\bigcirc\right\rangle\!\!CH_2 - CH_2 \diagdown \quad CH_2 - CH_2\!\!\left\langle\bigcirc\right\rangle\!\!CH_2 - CH_2 \diagdown \quad CH_2 - CH_2 \diagdown \quad CH_2 - CH_2 \sim\sim$$

or:

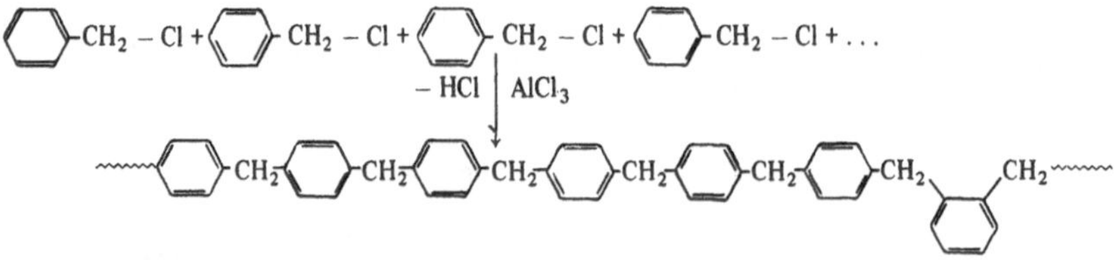

$$\bigcirc\!\!-CH_2 - Cl + \left\langle\bigcirc\right\rangle\!\!-CH_2 - Cl + \left\langle\bigcirc\right\rangle\!\!-CH_2 - Cl + \bigcirc\!\!-CH_2 - Cl + \ldots$$

$$\text{– HCl} \quad\Big|\quad AlCl_3$$

$$\sim\sim\!\!\left\langle\bigcirc\right\rangle\!\!-CH_2\!\!\left\langle\bigcirc\right\rangle\!\!-CH_2\!\!\left\langle\bigcirc\right\rangle\!\!-CH_2\!\!\left\langle\bigcirc\right\rangle\!\!-CH_2\!\!\left\langle\bigcirc\right\rangle\!\!-CH_2\!\!\left\langle\bigcirc\right\rangle\!\!-CH_2 \diagdown \quad CH_2\sim\sim$$

If these reactions have not found any commercial interest for the preparation of polymers, it is mainly for two reasons: (1) it is not easy with Friedel-Crafts or Wurtz-Fittig reactions to prepare polymers with a *high* molecular weight; and (2) these reactions usually lead to cross-linked products, because they cannot be limited to only two positions on the monomer molecule.

Diazoalkane Decomposition

A reaction which leads to polymers with high molecular weight is the decomposition of diazoalkanes. For example, diazomethane is catalyzed by boron compounds (BF_3, BH_3, $B(OR)_3$, BR_3) and gives rise to a highly crystalline, unbranched, polymethylene. This polymer is essentially identical in its properties with low pressure polyethylene, as is obtained by polymerization with metal complex catalysts (G. D. Buckley):

$$CH_2=N\equiv N \; + \; CH_2=N\equiv N \; + \; CH_2=N\equiv N \; + \; \cdots$$

$$\xrightarrow{BF_3} \quad \sim\sim CH_2-CH_2-CH_2-CH_2-CH_2\sim\sim + N_2 + N_2 + \cdots$$

If diazoethane and other diazoalkanes are used as comonomers, rubber-like copolymers can be prepared in the same way. In the polymerization of a mixture of equimolar amounts of diazomethane and diazophenylmethane, one obtains a copolymer (using boron compounds as initiators), which compares in its properties with isotactic polystyrene (W. W. Korschak):

$$CH_2=N\equiv N \; + \; CH=N\equiv N \; + \; \cdots \xrightarrow{B(OR)_3} \sim\sim CH_2-CH-CH_2-CH\sim\sim + N_2$$

The decomposition and polymerization of diazomethane with the aid of catalytic amounts of boron compounds are not, as one might assume, an ionic chain polymerization, but rather a radical polymerization. The mechanism, however, is still not quite clear. Diazomethane also decomposes without boron fluoride under formation of polymethylene. This occurs especially on porous surfaces (boiling chips) and in the presence of light.

Poly-p-Xylylene

A further decomposition reaction which leads to polymers with high molecular weight is the pyrolytic decomposition of p-xylene at 950°C in the presence of water vapor. As Gorham has found, one can carry out this reaction in such a manner that one first forms a crystalline cyclic dimer, which at 550°C in vacuum rearranges to poly-p-xylylene, in the form of films precipitated on cooled polished surfaces. These films polymerized in the vapor phase are characterized by their remarkable uniformity. Poly-p-xylylene is the most expensive industrially produced polymer:

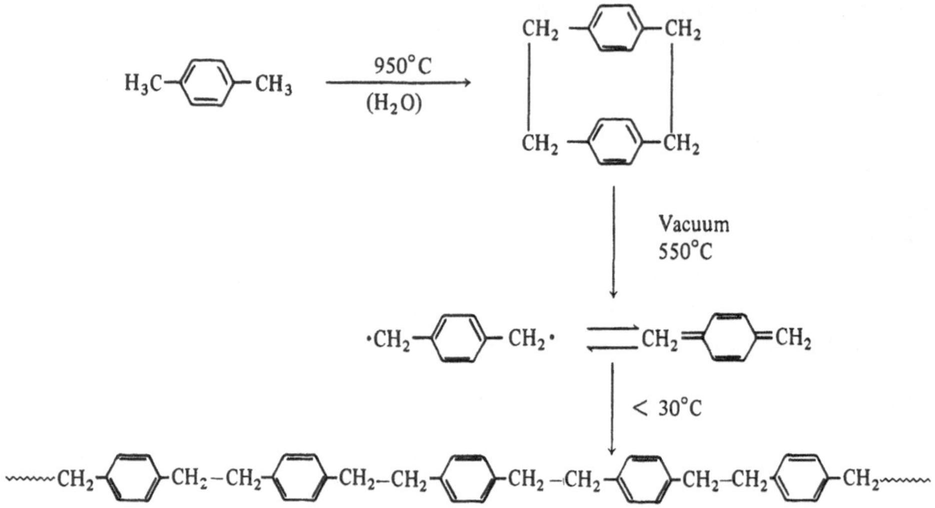

Poly-p-xylene

The quinoid monomeric reaction intermediate cannot be isolated as a solid, however, if one bubbles the hot reaction gases into a cold organic solvent, (e.g., $-70°C$), one can isolate it as a solution. It reacts with oxygen to a macromolecular peroxide, which on warming decomposes to terephthalaldehyde with the formation of hydrogen.

Some additional methods for the preparation of poly-p-xylylene have been described:

According to Young, for example, one pours a 50% aqueous solution of trimethyl-p-methylbenzyl-ammoniumchloride into hot concentrated sodium hydroxide under stirring. A Hoffmann degradation occurs, and the polymer precipitates (but no information about the molecular weight is given):

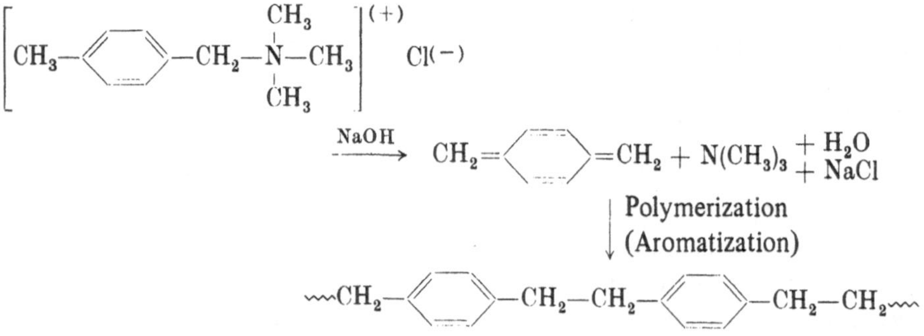

Phenol-Formaldehyde Resins

Among the polymer syntheses carried out on a large scale are the condensation reactions of phenol and formaldehyde, which lead also to the formation of carbon-carbon chains. For example:

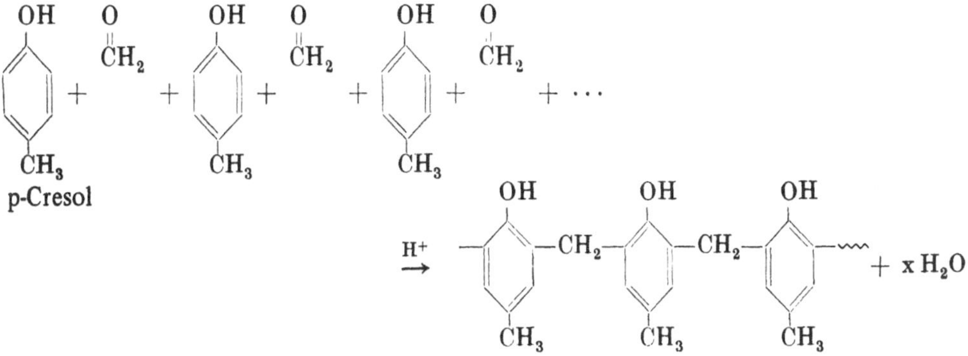

This reaction involves methylol compounds as intermediates, e.g.

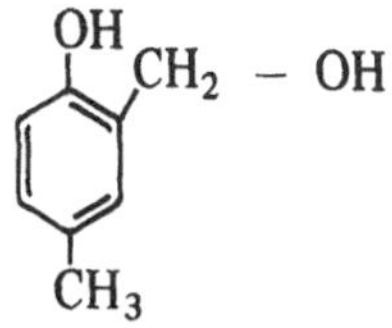

which cannot be isolated under acid conditions. In the technical use of this polycondensation one employs phenol or mixtures of phenols and cresols, and one obtains branched, and, after curing at high temperatures, completely insoluble, cross-linked resins: the phenol-formaldehyde resins, Resits and Bakelites (soluble preliminary condensates are called Novolaks):

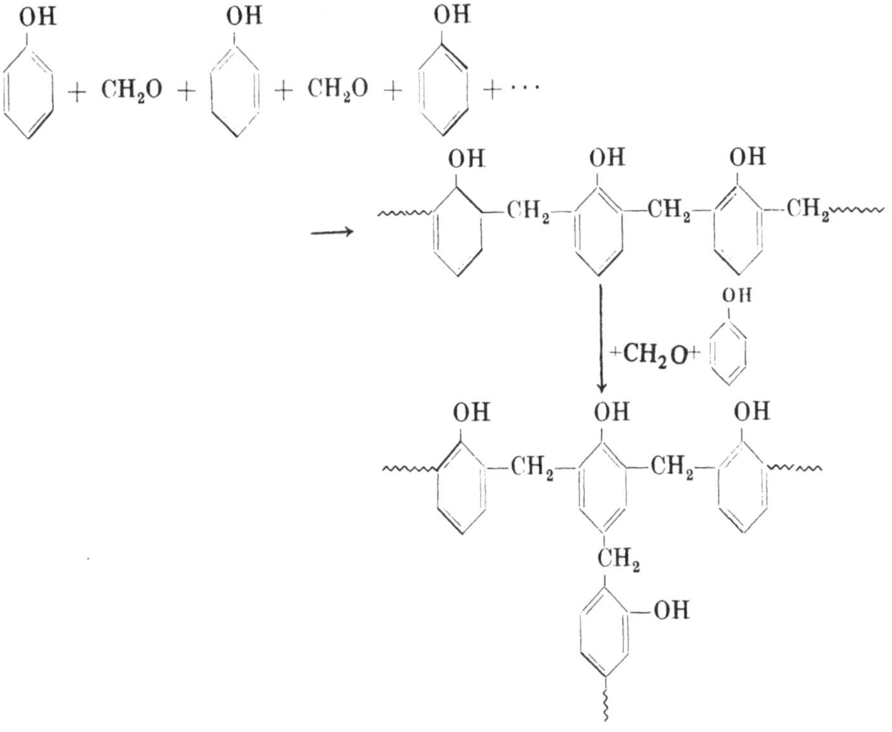

This condensation can also be carried out in an alkaline medium. One then obtains products, which contain ether bridges in addition to the C-C-cross-links.

22   SYNTHESIS OF MACROMOLECULES WITH HETEROATOMS IN THE CHAIN

The possibilities for combining molecules through heteroatoms (O, N, S) are very numerous. Actually such reactions constitute a large part of organic chemistry. Different from the polymerization reactions, which, in a single run so to speak, prepare macromolecules with usually very high molecular weight (10^5 to 10^6), the

TABLE 206. Reactions used for the preparation of polymers with heteroatoms in the chain.

	Reaction	Type of Chain	Name of Product
1.	Dicarboxylic acids + glycols	→ polyesters	Dacron, Mylar, poly-
	Dicarboxylic acid esters + glycols	→ polyesters	ester-casting resins
	Dicarboxylic acid anhydrides + glycols	→ polyesters	
2.	Phthalic anhydride + glycerol	→ polyesters	alkyd resins (raw materials for coatings)
3.	Dicarboxylic acids + diamines	→ polyamides	6,6-Nylon
	ω-aminoacids + diamines	→ polyamides	Rilsan
4.	Urea + formaldehyde	→ polyureas	aminoplastics (foam for insulation)
5.	Melamine + formaldehyde	→ polyamines	melamine resins (Melmac, etc.)
6.	Dichlorosilane + water	→ polysiloxanes	Silicones (oils, elastomers, resins)
7.	Bisphenols + phosgene	→ polycarbonates	Lexan
8.	Bischlorocarbonates + diamines	→ polyurethanes	elastomers, foams
9.	Diisocyanates + glycols	→ polyurethanes	Vulcollan (elastomers)
10.	Diisocyanates + dicarboxylic acids	→ polyamides	Lycra (elastic fiber)
11.	Diisocyanates + water	→ polyureas	Foams (Moltopren)
12.	Bisphenols + bisepoxides	→ polyethers	epoxy-resins (Epon, Araldite)
13.	Lactams	→ polyamides	6-Nylon (Caprolan)
14.	Cyclic ethers	→ polyethers	Penton, polyethyleneoxide
15.	Formaldehyde	→ polyacetals	polyformaldehyde, polyoxymethylene (Delrin, Celcon)

reactions which we will discuss now lead to macromolecular compounds only in a stepwise manner over intermediate products such as dimers and trimers, which can readily be isolated. Actually high molecular weight compounds (over 20,000), can only be obtained if one employs special preparation methods. Of the different types of reactions available for this purpose, not all are equally useful for the synthesis of macromolecules. They are more suitable, the more homogeneously they occur. The more side products and the greater the fraction of these side products, the less suitable is the reaction for the synthesis of macromolecular compounds. The reason is that in the synthesis of low molecular weight compounds, one can usually remove the side products in a relatively simple manner: for example, by distillation and recrystallization. This is usually not possible with macromolecular syntheses, because with macromolecular reactions, such side reactions usually lead to "defective" macromolecules (branched or cross-linked), or to the formation of short chain molecules and discolored low molecular weight impurities, whose separation causes expense and effort and, in many cases, is not even possible. Even though there are technical products, where discoloration can be accepted (for example, with the phenolformaldehyde resins and cumarone resins), usually the homogeneous syntheses lead to much more valuable products (synthetic fibers and films), which can be cross-linked, dyed, or otherwise chemically reacted in a controlled fashion at a later stage.

A number of reactions can be used for the preparation of important industrial macromolecular products with heteroatoms in the chain (see Table 206).

The number of technically useful reactions is constantly being enlarged. New reactions are found, and old reactions are improved by careful study of the reaction conditions and purification of the monomeric starting materials, so that they lead to valuable new polymeric products. On the other hand, certain reactions lose their importance with time, because the raw material basis is changed, or because they can not compete economically with other similar reactions.

Usually the reactions used for the preparation of polymers with heteroatoms in the chain can be divided into three groups: polycondensation (1-8), polyaddition (9-12), and ring opening reactions (13-15). The common characteristic of all these polymer syntheses which involve reactions of functional groups is the correlation between the molecular weight of the resulting polymer and the degree of conversion of the synthesis reaction (see Table 206).

221 Polycondensations and Polyadditions

2211 Polyesters and Polyamides

Linear Polyesters and Polyamides

Condensation reactions (ester formation from carboxylic acids and alcohols, amide formation from carboxylic acids and amines) were first utilized by Carothers for the formation of such macromolecular compounds as the polyesters and polyamides. They are not different in their mechanisms from the condensation and addition reactions of low molecular weight compounds. Thus, when an acid and an alcohol react with each other, they form an ester molecule accompanied by the splitting out of water. If a dicarboxylic acid reacts with a dialcohol or diol, one obtains an ester molecule that still has a free carboxyl group and an OH group:

$$HOOC—R—COOH + HO—R'—OH \rightarrow HOOC—R—COO—R'—OH + H_2O$$

This ester molecule can react further, either with another glycol molecule or with another dicarboxylic acid molecule:

$$HOOC—R—COO—R'—OH + HOOC—R—COOH$$

$$\rightarrow HOOC—R—COO—R'—OOC—R—COOH + H_2O$$

$$HOOC—R—COO—R'—OH + HO—R'—OH$$

$$\rightarrow HO—R'—OOC—R—COO—R'—OH + H_2O$$

This leads to the formation of ester molecules which have either carboxyl end groups or OH groups at both ends; these molecules can therefore again react with glycols or carboxylic acids, which in turn leads again to ester molecules with an OH group at one end, and a carboxyl group at the other. Every time an additional glycol or dicarboxylic acid molecule condenses with the previous one under the formation of an ester molecule, the resulting molecule remains able to react, and finally a long chain polyester molecule is formed where the groups R and R' alternate regularly with each other. Naturally, such polyester molecules (*oligomers*) are also capable of reacting with each other:

$$2\,HOOC—R—COO—R'—OOC—R—COO—R'—OH$$

$$\rightarrow HOOC—R—COO—R'—OOC—R—COO—R'—OOC—R—COO—R'—OOC—R—COO—R'—OH + H_2O$$

or:

$$HOOC—R—COO—R'—OOC—R—COOH + HO—R'—OOC—R—COO—R'—OH$$

$$\rightarrow HOOC—R—COO—R'—OOC—R—COO—R'—OOC—R—COO—R'—OH + H_2O$$

If one replaces the glycol with a diamine, one obtains in an analogous manner a polyamide.

In place of the free dicarboxylic acid, one can also use dicarboxylic acid esters for the formation of polyesters. In that case the chain is formed by ester interchange reactions:

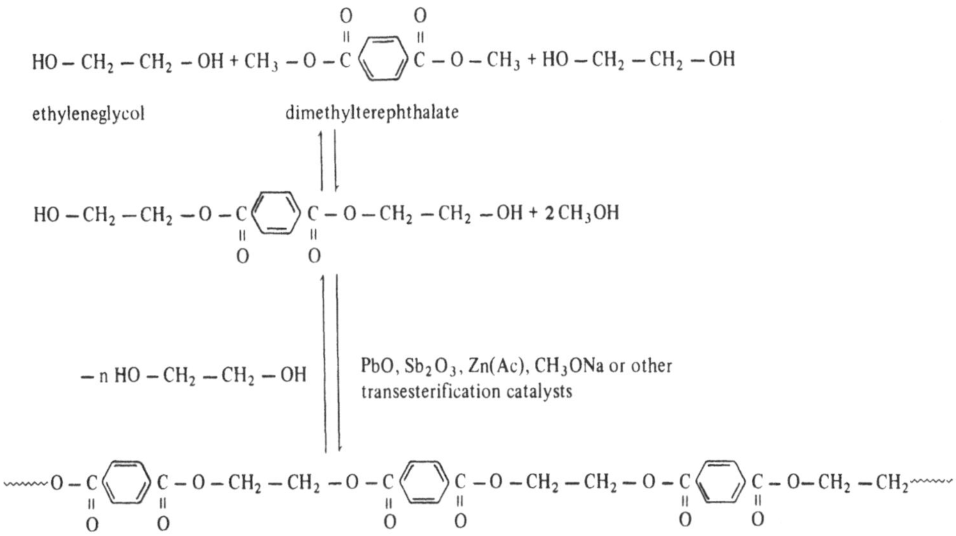

$$HO - CH_2 - CH_2 - OH + CH_3 - O - C \overset{O}{\underset{}{\|}} \bigcirc \overset{O}{\underset{}{\|}} C - O - CH_3 + HO - CH_2 - CH_2 - OH$$

ethyleneglycol   dimethylterephthalate

$$HO - CH_2 - CH_2 - O - C \bigcirc C - O - CH_2 - CH_2 - OH + 2\,CH_3OH$$
$$\overset{}{\underset{O}{\|}} \qquad \overset{}{\underset{O}{\|}}$$

$- n\,HO - CH_2 - CH_2 - OH$ $\quad$ PbO, Sb$_2$O$_3$, Zn(Ac), CH$_3$ONa or other transesterification catalysts

$$\sim\sim\sim O - C \bigcirc C - O - CH_2 - CH_2 - O - C \bigcirc C - O - CH_2 - CH_2 - O - C \bigcirc C - O - CH_2 - CH_2 \sim\sim$$

Polyethyleneterephthalate

(Trade names: Dacron, Mylar, Terylene, Trevira, Diolen, etc.)

Dicarboxylic acid esters usually have the advantage that they are more soluble in the reaction mixture. Furthermore, the problem of the accurate equivalence of the corresponding functional groups (see p. 214) is solved in an elegant way by removing the glycol which is split off out of the equilibrium system.

In addition to the actual chain growth reactions, one always finds at the same time ester interchange, or transamidation reactions, of the polyester- or polyamide-chains with the end groups of other polyester- or polyamide-chains, or with dicarboxylic acids and glycols, or diamines. Furthermore, hydrolysis of polyester chains or polyamide chains by the water formed during the reaction, can occur.

In order to form linear polymers by means of polycondensation, one requires bifunctional compounds which react with each other to form covalent bonds. The necessary functional groups, ($-COOH$, $-OH$, and $-NH_2$, etc.) can be found on two different molecules, such as in the previous example of polyester formation by reaction of a dicarboxylic acid with a glycol. However, they can also be present in one and the same molecule, as in ω-aminocarboxylic acids or in hydroxy acids. The mechanism of the polymerization in both cases is essentially the same, only one has to realize that in the latter case, it is not possible to form polymer chains with two identical end groups. This may be important if one wants to react such polycondensates further. In addition, the structure of the two polycondensates is different, which has an influence on such properties as softening point and crystallizability. In one case the ester or amide groups in the chain have a different structure from group to group, whereas in the polycondensates made from a single starting material, each group has the same structure. There is still another difference between the two types of polymers: in type (II) the unit R is the same in the whole chain, whereas in type (I) usually two different units, R and R', alternate.

Polyamide from a dicarboxylic acid and a diamine:

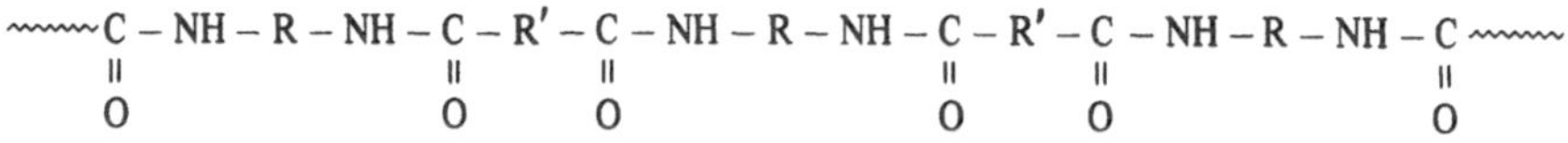

~~~~C – NH – R – NH – C – R' – C – NH – R – NH – C – R' – C – NH – R – NH – C ~~~~ (I)
    ‖   ‖   ‖   ‖   ‖   ‖
    O   O   O   O   O   O

Polyamide from an ω-amino acid:

~~~~C – NH – R – C – NH – R – C – NH – R – C – NH – R – C – NH – R – C – NH ~~~~ (II)
  ‖   ‖   ‖   ‖   ‖   ‖
  O   O   O   O   O   O

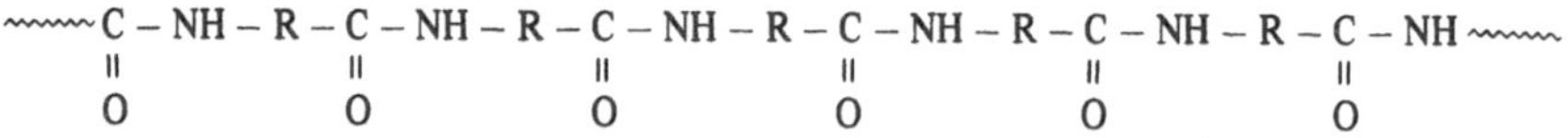

The polymers formed by polycondensation, just as those formed by chain polymerization, are mixtures of polymer molecules of different size. The degree of inhomogeneity is larger (i.e., the molecular weight distribution broader), the higher the molecular weight. Molecules of average size are present in the greatest quantity, while the amounts of the larger and the smaller molecules decrease the larger, or the smaller, they are. The distribution curve can be calculated on a theoretical basis (Schulz-Flory distribution) and corresponds generally rather well with the experimentally determined distribution curves (compare pp. 393-403).

By variation of the residues R and R', one can prepare polycondensates with very different properties. One can increase these possibilities for variation even further, if, instead of a single dicarboxylic acid or a single glycol or amine, one uses different ones, so that one obtains copolymers. Actually all polycondensates from dicarboxylic acids, (M_1), and glycols or diamines, (M_2), are copolymers where the two monomers M_1 and M_2 alternate in a perfectly regular manner in the chain (compare p. 23).

Branching, Cross-linking and Ring Formation

If one includes tri- and tetrafunctional compounds, such as glycerin, or pentaerythritol, one can incorporate branches and cross-links into the molecule:

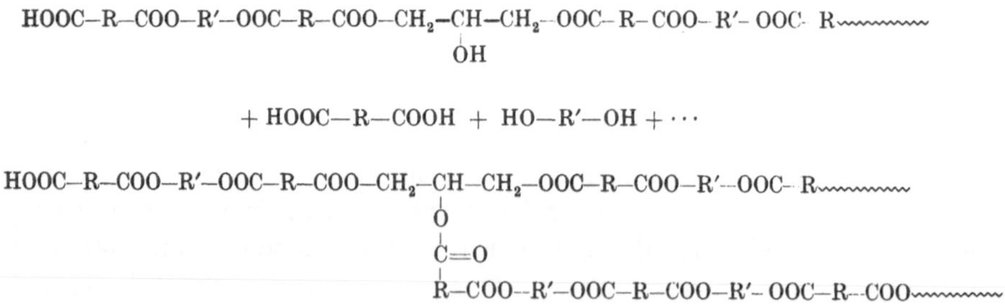

HOOC–R–COO–R'–OOC–R–COO–CH₂–CH–CH₂–OOC–R–COO– R'– OOC· R~~~~
                                           OH

+ HOOC–R–COOH + HO–R'–OH + · · ·

HOOC–R–COO–R'–OOC–R–COO–CH₂–CH–CH₂–OOC–R–COO–R'–OOC– R~~~~
                                           O
                                           C=O
                                R–COO–R'–OOC–R–COO–R'– OOC–R–COO~~~~

Through reaction of branched molecules one obtains cross-links in the following manner:

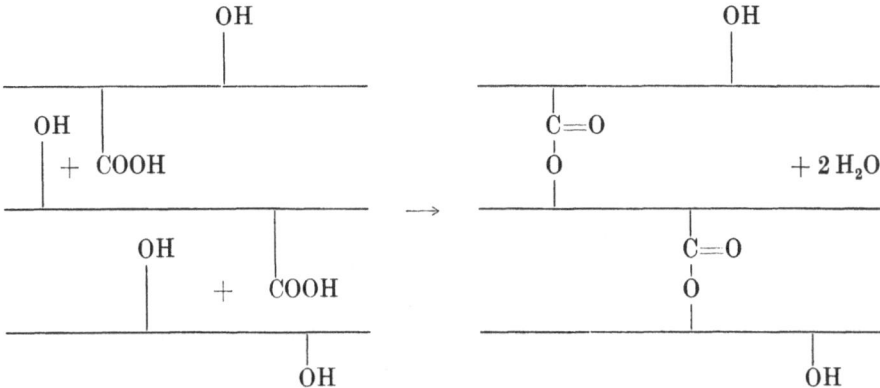

Technical products based on this principle of formation are the alkyd resins, which can be formed through condensation of phthalic anhydride and glycerol, mostly in the presence of unsaturated fatty acids.

In addition to the linear polymers, the polycondensation of diamines, or glycols, with dicarboxylic acids, or of ω-amino (or hydroxy) acids, always results in small amounts of cyclic oligomers,[27] which are formed through the reaction of two different end groups on one and the same polymer chain. The formation of such a cyclic compound means that, for the moment at least, the chain stops growing because with the formation of the ring the reactive end groups disappear. Through hydrolysis or transesterification (or transamidation) reactions, the ring may, however, be opened again. In each stage of a polycondensation there is an equilibrium between rings and linear polymers which depends on the type of monomer and the conditions of the polycondensation. With the polycondensation of ω-aminocaproic acid, for example, one obtains a polycondensate which at 200°C, in addition to the polymer, still contains approximately 6% of the cyclic monomer (caprolactam). Some macrocyclic oligomers have also been isolated and identified (Rothe, H. Zahn).

Influence of Water on the Size of the Molecules

In view of the fact that condensation reactions, such as esterification and amidation, are equilibrium reactions where one mole of water is formed per mole of ester or amide, the conversion, and therefore also the length of the polymer chains, depends on the water content of the reacting systems. Thus if one reacts for example terephthalic acid with ethylene glycol, then theoretically, two moles of water can be split off per mole of acid and glycol. This amount of water, which corresponds to a maximum conversion, is however hardly ever reached in practice, because a more or less small part of the water remains in the form of carboxyl and

[27]Polymers of $P \leqslant 20$ are called oligomers.

OH end groups as part of the polymer molecules. The molecular weight of the ethylene glycol is 62, that of the terephthalic acid, 166. If one mole of glycol and one mole of terephthalic acid react under the formation of one mole of water, then one obtains 228 − 18 = 210 g of ester with a degree of polymerization of 2 and a molecular weight of 210. Of the theoretically possible 36 g of water, only 18 g of water are split off and the conversion is therefore 50%. If one would like to obtain a molecule with a degree of polymerization of 4, then (as one can easily see from the following Table 213) one has to split off 3 moles of water of the theoretically possible 4 moles, which corresponds to a conversion of 75%. The molecular weight is then 2 x 228 − 3 x 18 = 402. If one wants to obtain products with higher molecular weight, one has to remove the water from the equilibrium to increase the conversion. Table 213 and Figure 212 show how the molecular weight and the corresponding degree of polymerization increase with the effective conversion.

The range of molecular weights of polyesters and polyamides of technical interest lies between 10,000 and 30,000. In order to reach such a molecular weight, the reaction has to be carried to a conversion of >99%. This is only possible if one carefully removes the water formed, and, of course, any other kind of moisture from the reaction medium and also from the room where the reaction is carried out.

This, however, does not mean that the *total* amount of water split off up to a certain conversion has been separated from the reaction medium. A more or less small amount of the water split out during the reaction belongs to the equilibrium, even with high molecular weights. How high this water content is, depends on the equilibrium constants [compare p. 220].

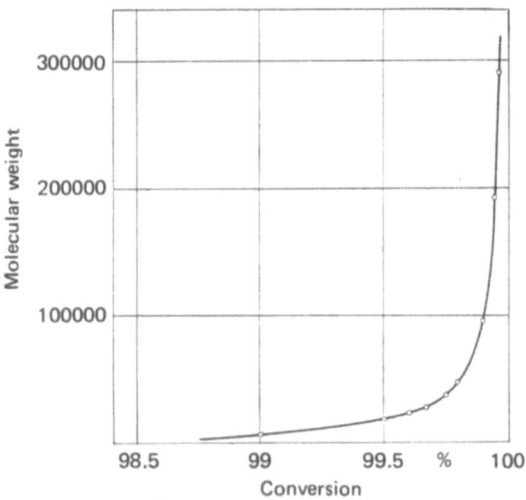

FIG. 212 − Molecular weight of poly-ethyleneglycolterephthalate as a function of conversion.

TABLE 213. Molecular weight of polyethyleneglycolterephthalate
as a function of the conversion.

Moles Tereph-thalic acid	Moles Glycol	Composition of the ester: − Ethyleneglycol; 〰 Tereph-thalic acid; ○ Esterlinkage	Moles Water Split-out	$\bar{P}$	$\bar{M_P}^*$	Conversion in %
1	1	—○〰	1	2	210	50
2	2	—○〰—○—○〰	3	4	402	75
3	3	—○〰○—○〰○—○〰	5	6	594	83.3
4	4	—○〰○—○〰○—○〰○—○〰	7	8	786	87.5
5	5	—○〰○—○〰○—○〰○—○〰○—○〰	9	10	978	90
10	10		19	20	1938	95
50	50	usw.	99	100	9618	99
100	100		199	200	19218	99.5
150	150		299	300	28812	99.7
1000	1000		1999	2000	192018	99.95
1500	1500		2999	3000	288018	99.97

$$* \; M_P = P/2 \, (M_{M\,(1)} + M_{M\,(2)}) - 18 \, (P - 1).$$

With high molecular weights the end-groups may be neglected and then:

$$M_P = P/2 \, (M_{M\,(1)} + M_{M\,(2)} - 36).$$

Influence of an Excess of One of the Reaction Components

In the polycondensation of dicarboxylic acids and diamines or glycols, the excess of one of the components over and above the stoichiometric amount causes a decrease in the molecular weight. For example, if one reacts one mole of dicarboxylic acid with two moles of diamine or glycol (i.e., 100% excess of glycol), then one obtains as the main reaction product an ester with a degree of polymerization P = 3:

$$1 \; HOOC-R-COOH + 2 \; HO-R'-OH \rightarrow HO-R'-OOC-R-COO-R'-OH$$

Correspondingly, from two moles of dicarboxylic acid and three moles of glycol (50% excess), one may expect an ester with an average degree of polymerization P = 5:

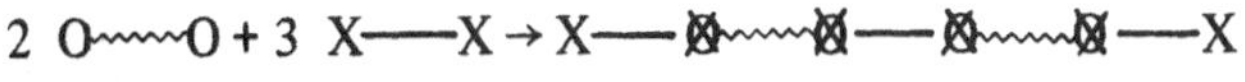

$$2 \; O\text{〰}O + 3 \; X\text{——}X \rightarrow X\text{——}\!\boxtimes\!\text{〰}\!\boxtimes\text{——}\!\boxtimes\!\text{〰}\!\boxtimes\text{——}X$$

In the following table, (Table 214), this series is expanded and the maximum (with 100% conversion) attainable degree of polymerization (i.e., molecular weight) is shown for the system terephthalic acid—ethylene glycol as a function of the excess of glycol. Obviously the degrees of polymerization and molecular weight shown in this table are also averages.

It can be seen that with a larger excess of glycol (the same holds for dicarboxylic acid), it becomes completely impossible to prepare polyesters with high molecular weight. Only with less than 1% excess is it possible to reach a technically useful range of molecular weights.

However, it should be understood that the values of $\overline{M}_{max}$ cannot be reached because the conversion is never 100%. This is why the $\overline{M}$ values for 99% and 99.5% conversion are listed. These are compared in the last row of the table with the corresponding values of exactly equivalent amounts (from Table 213).

A comparison of the values for $\overline{M}$ at 99% and 99.5% conversion without excess of glycol, with the values for $\overline{M}_{max}$ at 2% and 1% glycol excess, shows that a glycol or dicarboxylic acid excess has the same effect on the molecular weight of the polyester as the presence of half of that amount of water.

In general, the influence of an excess of glycol or dicarboxylic acid is the stronger, the higher the conversion; with 99% conversion the molecular weight $\overline{M}$ is reduced to half its value (9618 → 4831) by a glycol excess of 2%. With 99.5% conversion, already 1% excess is sufficient in order to produce the same effect (19218 → 9631), and even 0.1% excess is sufficient in order to reduce the molecular weight from 19,000 to 17,000.

This shows how important it is with polycondensations to achieve an exact equivalence of the functional groups. With hydroxy acids and amino carboxylic acids, this equivalence is given automatically. In the production of 6, 6-nylon (from adipic acid and hexamethylene diamine), one makes use of the fact that adipic acid forms an easily crystallizable salt with hexamethylenediamine in the mole ratio 1:1,

TABLE 214. Dependence of the molecular weight of polyethyleneglycol-terephthalate on the excess of glycol.

Moles of dicar-boxylic acid	Moles glycol	Excess in %	Composition of the polyester: ∿ terephthalicacid residue; — glycol residue; O ester linkage	$\overline{P}_{max}$	$\overline{M}_{max}^{*}$ at 100% Conversion	$\overline{M}$ at 99% Conversion **	$\overline{M}$ at 99.5% Conversion
1	2	100	—O∿O—	3	254		
2	3	50	—O∿O—O∿O—	5	446		
3	4	$33\frac{1}{3}$	—O∿O—O∿O · · O∿O · ·	7	638		
4	5	25	—O∿O—O∿O—O∿O—O∿O—	9	830		
5	6	20		11	1022	930	
50	51	2		101	9662	4831	6441
100	101	1		201	19262	6420	9631
1000	1001	0.1		2001	192062	9146	17460
1	1	0.0			As large as desired	9618	19218

$$* \; M_{P_{max}} = \left[\frac{P-1}{2} (M_{M(1)} + M_{M(2)}) \right] - [(P-1) \cdot 18] + M_{M(excess)}$$

**99% conversion means that of 100 possible ester linkages only 99 have been formed (i.e. 1 remains open). For a maximum degree of polymerization $\overline{P}_{max} = 101$ (i.e. 100 ester linkages) the molecular weight therefore drops to one-half. For $\overline{P}_{max} = 201$, two ester bonds have to be opened again to reach 99% conversion, i.e. $\overline{M}$ drops to 1/3 of $\overline{M}_{max}$, and so forth. At 99.5% conversion, of 200 possible ester linkages, only 199 are closed and 1 remains open, i.e. for $\overline{P} = 200$, $\overline{M} = (1/2) \overline{M}_{max}$.

which can be brought to a high degree of purity (AH salt) through recrystallization. In other cases, one has to achieve a molar ratio of exactly 1:1 by careful weighing. Obviously it is not sufficient to obtain this exact equivalence in the beginning of the reaction; it must also be maintained during the entire course of the polycondensation. This is particularly difficult if one of the components is relatively volatile. (For an elegant technical solution of this problem with the synthesis of polyethyleneglycolterephthalate (Dacron) see p. 209).

Monocarboxylic acids, alcohols, amines, and all other substances which can react with one of the components in a polycondensation, have the same effect as an excess of dicarboxylic acid, glycols, or diamines. For this reason, the purity of the monomeric compounds is an absolute prerequisite if one wants to prepare polymers with high molecular weight by means of condensation. (See, however, section on interfacial polycondensation).

Influence of the Equilibrium Constant on the Molecular Weight

A polycondensation reaction, when considered in detail, consists of a large number of equilibria:

$$X-O + X-O \rightleftharpoons X-\boxtimes-O + H_2O \tag{a}$$

$$X-\boxtimes-O + X-O \rightleftharpoons X-\boxtimes-\boxtimes-O + H_2O \tag{b}$$

$$X-\boxtimes-O + X-\boxtimes-O \rightleftharpoons X-\boxtimes-\boxtimes-\boxtimes-O + H_2O \tag{c}$$

$$X-\boxtimes-\boxtimes-\boxtimes-O + X-O \rightleftharpoons X-\boxtimes-\boxtimes-\boxtimes-\boxtimes-O + H_2O \tag{d}$$

$$X-\boxtimes-\boxtimes-\boxtimes-O + X-\boxtimes-O \rightleftharpoons X-\boxtimes-\boxtimes-\boxtimes-\boxtimes-\boxtimes-O + H_2O \tag{e}$$

From a formal point of view, these reactions are, of course, all different, because the reaction partners have a different size. However it has been shown that—within a certain range of molecular weights[28]—the difference in the reaction rate between large and small molecules is almost insignificant, so that one can disregard the difference in size between the reaction partners and can describe the polycondensation reaction simply in terms of the functional groups involved, regardless of the size of the chain residue to which they are attached. A polyamide synthesis can then be formulated in the following way:

$$-COOH + NH_2 \rightleftharpoons -\underset{\underset{O}{\|}}{C}-NH- + H_2O$$

$$\text{or in general: } 0 + X \rightleftharpoons \boxtimes + H_2O$$

It is unimportant whether the synthesis starts with an ω-amino acid, or with a diamine + dicarbocylic acid. For the equilibrium constants we can write:

---

[28]Since at higher molecular weights the polymer coils are impenetrable for each other, the reaction rate becomes strongly dependent on the molecular weight (see p. 551).

$$K_P = \frac{[-CO-NH-] \cdot [H_2O]}{[-COOH] \cdot [-NH_2]} \tag{f}$$

or in general: $$K_P = \frac{[\otimes] \cdot [H_2O]}{[O] \cdot [\times]} \tag{f'}$$

It is clear that these equations do not differ in any way from those for a low molecular amide synthesis from mono-functional reaction components. The speciality of a polycondensation is not based on the equilibrium constants, (since in both cases the same functional groups are involved), but rather on the dependence of the degree of polymerization (and therefore also the properties of the reaction product!) on the conversion and therefore on the equilibrium constant.

If one considers reactions (a) to (e), one sees that the conversion and the chain length (degree of polymerization) increases regularly. With each newly split-off water molecule one of the polymer molecules becomes larger by one or more structural units. And each polymer molecule, therefore, through its chain length, represents a certain fixed concentration ratio of the functional groups involved in the equilibrium (including H_2O), and obviously also a certain state of the equilibrium. Thus one can imagine, that with a particular polycondensation, on heating in a closed reaction vessel (so that the water of reaction cannot evaporate) and at a certain temperature, the reaction, on the basis of the equilibrium constant of the system, has progressed to the point where in the reaction vessel one only finds the dimer in equilibrium with water. In reality, of course, one has a polydisperse mixture of the monomer and oligmers of different degrees of polymerization. It simply means that the *average* degree of polymerization will be 2. This equilibrium is completely described by the symbol O-⊠-X, which signifies that there are two functional groups here, O and X, (for example, $-COOH$ and $-NH_2$, or $-COOH$ and $-OH$), in equilibrium with a group ⊠ (for example, an ester or an amide group), and one molecule of water. In the same fashion the symbol O-⊠-⊠-⊠-⊠-⊠-⊠-⊠-⊠-⊠-⊠-X means that in the closed reaction vessel we have in equilibrium 1 mole each of the groups O and X as well as 10 moles of the group ⊠ and 10 moles of H_2O. The equilibrium degree of polymerization in this case is then $\bar{P}_{eq} = 11$.

For a given concentration of the functional groups, one can immediately determine the corresponding equilibrium constant for each equilibrium degree of polymerization. By concentration one understands the amount (or fraction) of one of the components in a system consisting of several components. If one multiplies this by 100, one obtains the percentage which this component represents in the total mixture (= concentration in percent). One can express the fraction as the weight-, volume-, or number-fraction. For example, the concentration of component A in a two component system, consisting of components A and B is given by:

$$\frac{\text{concentration}}{\text{of component A}} = \frac{\text{Number of molecules of Component A}}{\text{Total number of molecules A plus B}} \equiv$$

If one selects the number-fraction in the form of the molar ratio (as is often done with chemical problems), then it is easy to define the concentrations for the free groups O and X and for the reacted groups ⊗. These concentrations can then be related to the conversion and to the degree of polymerization.

By definition the following relationships will hold: For the concentrations [O] and [X] of the free (i.e., unreacted) end groups O and X, respectively, in a polymer system consisting of n_K polymer chains (compare Figure 218):

$$[O] = \frac{\text{Number (or number of moles) of the free O-groups (in a system of } n_K \text{ polymer chains)}}{\text{Number of free and reacted O-groups}} =$$

$$\frac{N_F}{N_F + N_R} = \frac{N_F}{N_T} \qquad (\text{F = free; R = reacted; T = total}) \tag{g}$$

$$[X] = \frac{\text{Number of free X-groups (in a system of } n_K \text{ polymer chains)}}{\text{Number of free and reacted X-groups}} =$$

$$\frac{N_F}{N_F + N_R} = \frac{N_F}{N_T} \tag{h}$$

Correspondingly, we can write [⊗] for the concentration of the ⊗-groups formed by reaction of O and X:

$$[⊗] = \frac{\text{Number of } ⊗ \text{-groups in a system of } n_K \text{ polymer chains}}{\text{Number of } ⊗ \text{-groups at complete conversion}} =$$

$$= \frac{\text{Number of reacted O-groups}}{\text{Number of reacted and free O-groups}} = \frac{N_R}{N_F + N_R} = \frac{N_R}{N_T} \tag{i}$$

In a closed system the concentration [⊗] is always equal to the H_2O concentration:

$$[H_2O] = \frac{\text{Number of split-off } H_2O \text{ molecules}}{\text{Number of } H_2O \text{ molecules at complete conversion}} = \frac{N_R}{N_T} \tag{j}$$

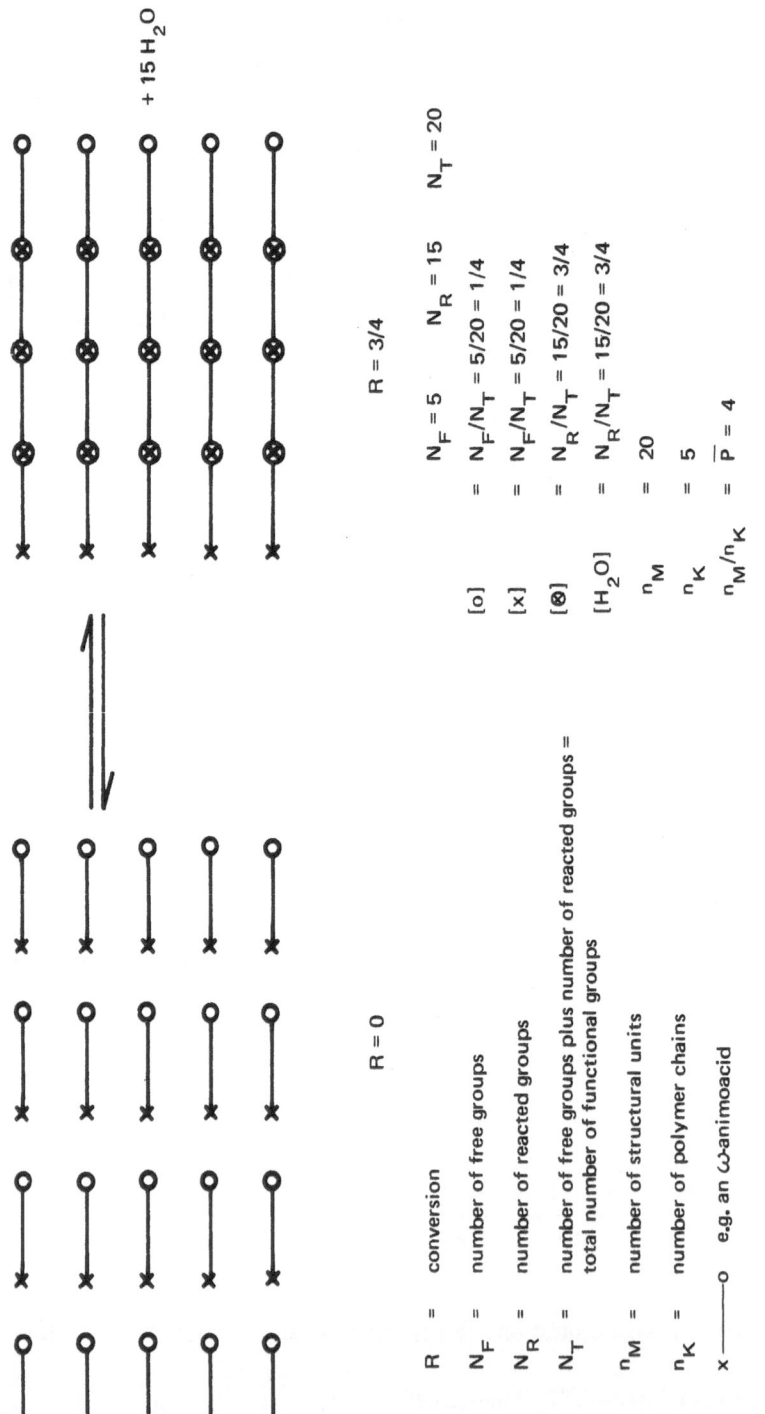

FIG. 218 — Model for a polycondensation equilibrium.

Figure 218 shows a system of five polymer chains with the degree of polymerization $P = 4$ as a model for a polycondensation equilibrium. With such a model it is easy to see the relationship between the concentration of functional groups, the conversion, and the molecular weight (degree of polymerization).

As the conversion R (relative yield, extent of a reaction), one usually designates the fraction of the reaction product obtained in relation to the theoretically possible amount or number of moles at complete conversion:

$$R = \frac{\text{Number of reacted groups O (or X groups)}}{\text{Maximum number of groups O (or X groups) which can react}} = \frac{N_R}{N_T} \quad (k)$$

Correspondingly, the fraction F of free, unreacted groups is given by:

$$F = \frac{\text{Number of free (unreacted) groups O (or X)}}{\text{Maximum number of groups O (or X) which can react}} = \frac{N_F}{N_T} \quad (l)$$

Since the total system consists only of the components R and F, the sum of both must be equal to 1:

$$F + R = 1, \text{ and } F = N_F/N_T = 1 - R \quad (m)$$

As can be seen by comparing Equations (l) and (m), with (g) and (h), the concentrations [O] and [X] of the free functional groups O and X is given by:

$$[O] = [X] = N_F/N_T = F = 1 - R \quad (n)$$

and correspondingly, the concentration [⊠] of the groups ⊠ formed in the polycondensation and present in the chain [compare Equations (k) and (l)] is:

$$[⊠] = \frac{N_R}{N_T} = R \quad (n^1)$$

One can write, therefore, the equilibrium constant according to Equation (f) also in the following way:

$$K_P = \frac{[⊠] \cdot [H_2O]}{[O] \cdot [\times]} = \frac{R \cdot [H_2O]}{(1 - R)^2} \quad (o)$$

Since the number of water molecules formed is always equal to the number of ester or amide groups formed (symbol: ⊠), one can write for the undisturbed equilibrium water concentration in closed systems:

$$[H_2O] = N_R/N_T = R \quad (p)$$

and correspondingly for K_p in closed systems

$$K_{P(equ.)} = \frac{[H_2O]^2_{equ.}}{(1 - R_{equ})^2} = \frac{R^2_{equ}}{(1 - R_{equ})^2} \qquad (q)$$

The degree of polymerization is defined as the number of structural units per polymer chain. Therefore, if there are n_K polymer chains in a given system and these consist of a total of n_M monomer units (structural units), then there will be in one chain on the average n_M/n_K structural units. This is, by definition, the degree of polymerization $\bar{P}_n$:

$$\bar{P}_n = n_M/n_K \qquad (r)$$

It is readily understood [compare Figure 218] that n_M (the number of structural units) is identical with the total number of (free and already reacted) functional groups O (or X); therefore $n_M = N_F + N_R = N_T$. And n_K, the number of polymer chains, is identical with the number of still unreacted end groups O (or X), therefore $n_K = N_F$, so that one can write for the degree of polymerization:

$$\bar{P}_n = n_M/n_K = N_T/N_F. \qquad (r')$$

As can be seen by comparison of Equation (r') with Equation (m), N_T/N_F = the reciprocal value of 1-R, and therefore the degree of polymerization is:

$$\bar{P}_n = N_T/N_F = 1/(1\text{-}R). \qquad (s)$$

According to Equation (p), one can write for the relation between the equilibrium constant K_p and the conversion R:

$$\left(\frac{1}{1-R}\right)^2 = \frac{K_P}{R \cdot [H_2O]} \qquad (t)$$

or

$$\frac{1}{1-R} = \sqrt{\frac{K_P}{R \cdot [H_2O]}} \qquad (t^1)$$

However, according to Equation (s) this is identical with the degree of polymerization $\bar{P}_n$; therefore:

$$\bar{P}_n = \sqrt{\frac{K_P}{R \cdot [H_2O]}} \cdot {}^{29} \qquad (u)$$

---

[29]We have to consider that R and $[H_2O]$ are not independent of each other. Therefore Equation (u) does not allow the conclusion: "P_n decreases with increasing R." With increasing R, the product $R \cdot [H_2O]$ decreases.

In closed systems $R_{eq} = [H_2O]_{eq}$, and one can write for the equilibrium degree of polymerization $\bar{P}_{eq}$:

$$\bar{P}_{eq} = \frac{1}{R_{eq}} \sqrt{K_P} = \frac{1}{[H_2O]_{eq}} \sqrt{K_P} . \tag{v}$$

For polymers with high molecular weight ($\bar{M} > 10{,}000$) $R \approx 1$, (as is shown by Figure 212), and one can therefore write:

$$\bar{P}_n = \sqrt{K/[H_2O]} . \tag{w}$$

In Equation (w), $[H_2O]$ is a water concentration which, by selection of suitable reaction conditions, should be as small as possible if one wants to obtain polymers with high molecular weights. The smaller the equilibrium constant K, the smaller the water concentration has to be in order to obtain high molecular weight polymers [see Equation (w)].

To interpret Equation (w) correctly, one has to distinguish between the equilibrium degree of polymerization in a closed system, the degree of polymerization at any particular stage in an open polycondensation reaction, and finally the maximum possible degree of polymerization, $\bar{P}_{max}$. The equilibrium degree of polymerization in a closed system, $\bar{P}_{eq}$, is determined by K and R_{eq} and is therefore not subject to external influences. Both parameters are related by Equation (v) at equilibrium, and are therefore determined only by the temperature. If one removes water from the system, the conversion, i.e., the ⨂ concentration and, therefore, the formation of long polymer chains increases, and a new equilibrium is established. K is a constant that is dependent only on the particular system and the temperature; $K/[H_2O]$, however, is a parameter that can be changed at will through experimental manipulation, so that $\bar{P}_{max}$ is limited only by the quality of the reaction equipment.

It would therefore be incorrect to say that the maximum degree of polymerization that one can obtain in a polycondensation is dependent on the equilibrium constant, because the negative influence of the low equilibrium constant on the degree of polymerization can be compensated by a corresponding lowering of the water concentration in the reaction medium. This is assuming, of course, the availability of a good polycondensation unit. The lower the H_2O concentration has to be in order to reach a certain degree of polymerization, the higher is the required quality of the equipment used in the preparation of a polymer of the polyester or polyamide type. The required equipment is a function of the equilibrium constant: thus the reaction of diamines and dicarboxylic acids, because of their relatively high equilibrium constant $K \approx 400$, leads relatively easily to high molecular weights, whereas with polyesters the equilibrium constant is equal to only about 10, and it is much more difficult to obtain polymers with high molecular weights. The chemical industry has made such progress in the design of polycondensation equipment that the degree of polymerization is usually predetermined with both polyamides and polyesters by the addition of monofunctional compounds, in order not to obtain polyamides and polyesters having too high a molecular weight (which, because of their high melt-viscosity, would be difficult to process).

2212 Other Polycondensations

Paraformaldehyde

On standing, aqueous formaldehyde solutions form white precipitates of polymeric formaldehyde. This occurs even more rapidly if these solutions are evaporated.

$$CH_2=O + H_2O \rightleftharpoons HO-CH_2-OH$$

$$2\,HO-CH_2-OH \rightarrow HO-CH_2-O-CH_2-OH + H_2O$$

$$\xrightarrow{+ HO-CH_2-OH} HO-CH_2-O-CH_2-O-CH_2-OH + H_2O$$

$$\xrightarrow[\text{etc.}]{+ HO-CH_2-OH} \text{Paraformaldehyde}$$

In addition to these reactions, one also has to expect an addition reaction of formaldehyde according to the following reaction:

$$HO-CH_2-OH + CH_2=O \rightarrow HO-CH_2-O-CH_2-OH$$

$$\xrightarrow{+ CH_2=O} HO-CH_2-O-CH_2-O-CH_2-OH$$

Paraformaldehyde has only a relatively low degree of polymerization and can be depolymerized again to monomeric formaldehyde through simple heating. Polyformaldehyde with high molecular weight, (which has become an important plastic), can be obtained through anionic polymerization of pure formaldehyde in water-free solvents at low temperatures and by cationic polymerization of trioxane (compare pp. 253-255).

Urea-Formaldehyde and Melamine-Formaldehyde Resins

Formaldehyde reacts with urea under the formation of methylolureas, which can react further with additional urea, under splitting out of water, to form chains and finally cross-linked products:

$$H_2N-C-NH_2 + CH_2O$$
$$\underset{O}{\overset{\|}{}}$$

$$\rightarrow H_2N-\underset{\underset{O}{\|}}{C}-NH-CH_2-OH \xrightarrow{+ CH_2O} HO-CH_2-NH-\underset{\underset{O}{\|}}{C}-NH-CH_2-OH$$
$$\qquad\qquad\text{Methylolurea} \qquad\qquad\qquad\qquad\qquad \text{Dimethylolurea}$$

$$HO-CH_2-NH-\underset{\underset{O}{\|}}{C}-NH-CH_2-OH + H_2N-\underset{\underset{O}{\|}}{C}-NH_2$$

$$\xrightarrow{- H_2O} HO-CH_2-NH-\underset{\underset{O}{\|}}{C}-NH-CH_2-NH-\underset{\underset{O}{\|}}{C}-NH_2$$

These reactions are used on a large technical scale for the preparation of plastics, without isolation of the intermediate methylol compounds, however. In summary, the formation of these resins can be formulated according to the following scheme:

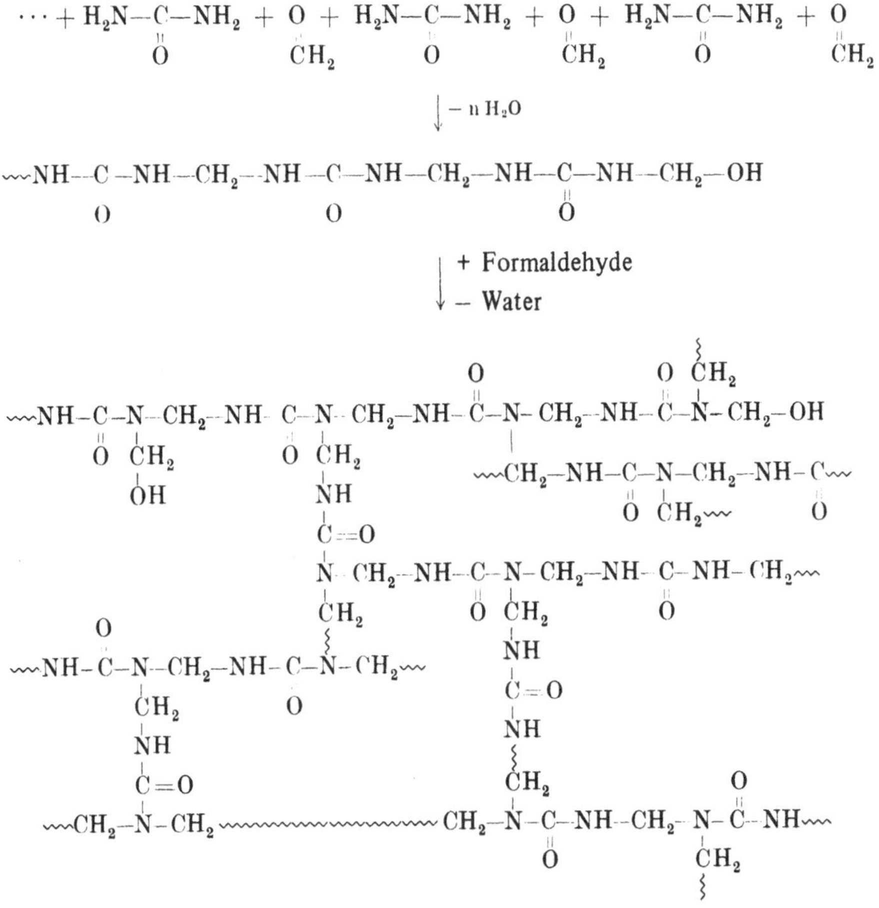

$$\cdots + H_2N-\underset{\underset{O}{\|}}{C}-NH_2 + \underset{\underset{O}{\|}}{\overset{CH_2}{}} + H_2N-\underset{\underset{O}{\|}}{C}-NH_2 + \underset{\underset{O}{\|}}{\overset{CH_2}{}} + H_2N-\underset{\underset{O}{\|}}{C}-NH_2 + \underset{\underset{O}{\|}}{\overset{CH_2}{}}$$

$$\downarrow -n\,H_2O$$

$$\sim\!\!\sim NH-\underset{O}{C}-NH-CH_2-NH-\underset{O}{C}-NH-CH_2-NH-\underset{\underset{O}{\|}}{C}-NH-CH_2-OH$$

$$\downarrow \quad \begin{array}{l}+ \text{ Formaldehyde}\\ - \text{ Water}\end{array}$$

[Complex branched polymer network structure with urea-formaldehyde linkages]

The course of the condensation reaction is strongly dependent on the pH value. The formulated reactions apply to a neutral, or weakly alkaline, medium. In an acidic medium one can not isolate the methylol compounds. These split out water intramolecularly, so that unsaturated compounds are formed.

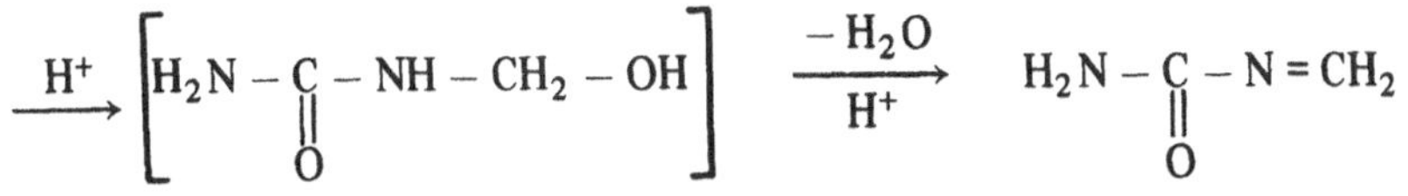

$$\xrightarrow{H^+} \left[H_2N-\underset{\underset{O}{\|}}{C}-NH-CH_2-OH \right] \xrightarrow[H^+]{-H_2O} H_2N-\underset{\underset{O}{\|}}{C}-N=CH_2$$

These unsaturated compounds may then react further by a chain polymerization. One also has to take into account that methylol intermediates may react with each other under water formation, so that ether bridges are formed:

$$\text{\raisebox{0pt}{$\sim$}HN} - \underset{\underset{O}{\|}}{C} - NH - CH_2 - OH + HO - CH_2 - NH - \underset{\underset{O}{\|}}{C} - NH\text{\raisebox{0pt}{$\sim$}}$$

$$\downarrow -H_2O$$

$$\text{\raisebox{0pt}{$\sim$}HN} - \underset{\underset{O}{\|}}{C} - NH - CH_2 - O - CH_2 - NH - \underset{\underset{O}{\|}}{C} - NH\text{\raisebox{0pt}{$\sim$}}$$

Urea-formaldehyde resins (aminoplasts) and phenol-formaldehyde resins (phenoplasts) belong to the oldest commercial plastics. The polycondensation is carried out in two steps: in the first step, the condensation is only carried out to the point where the reaction products are still soluble and fusible. Only in the second step, which can be described as hardening, does one obtain complete cross-linking. After this hardening, no further plastic deformation is possible; therefore in the production of plastic materials, this hardening step has to be carried out in the mold.

Instead of urea, for the production of especially high quality materials, one uses melamine (triaminotriazine=cyanuric acid amide) which can react with formaldehyde, while splitting off water, to form cross-linked polyamines:

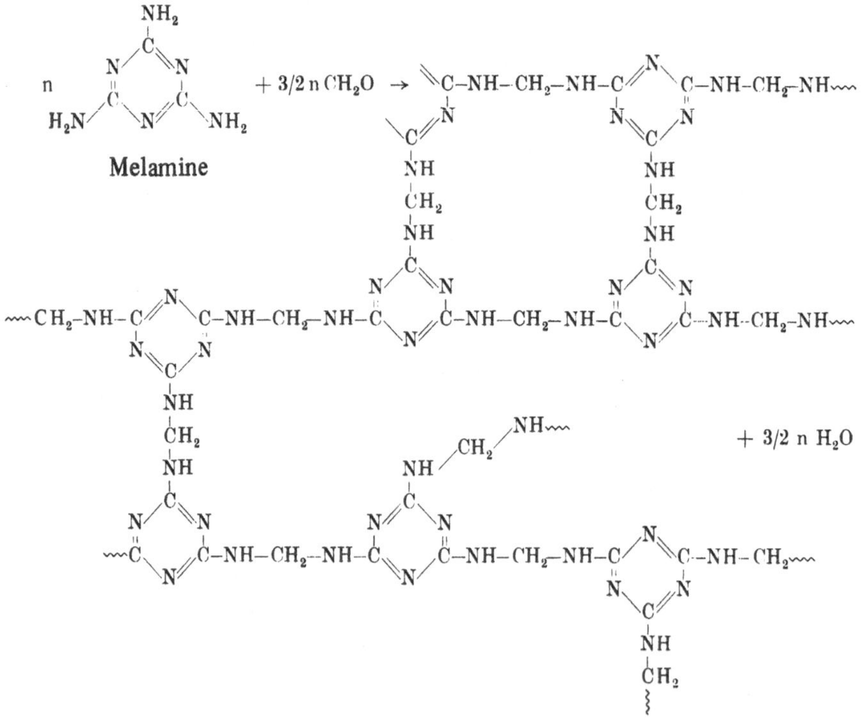

Polycarbonates

The reaction of p,p'-dioxy-2,2-diphenyl-propane (Bisphenol-A, which one obtains from phenol and acetone) with phosgene or carbonic acid esters, has become important for the production of plastics characterized by especially high mechanical strength as well as their transparence and their high softening point:

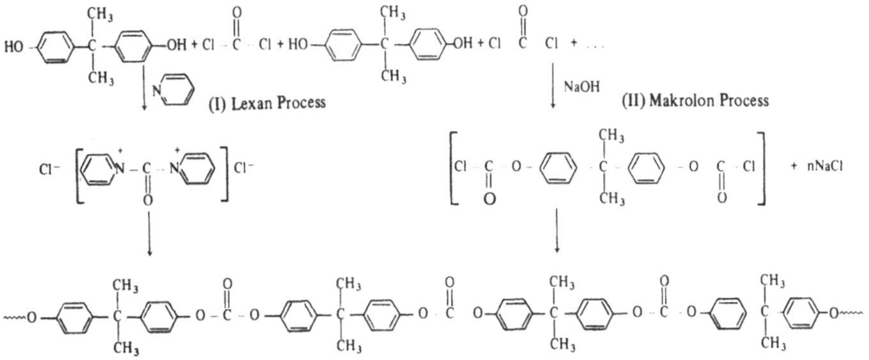

Polycarbonate (Lexan, Makrolon)

This reaction is carried out on a commercial scale by solution polycondensation (I), with tertiary amines (pyridine) as catalysts and HCl acceptors, and also by a special kind of interfacial polycondensation (II) (see p. 235) with the sodium-salt of bisphenol-A and excess NaOH in the water phase and the bischloro-carbonic acid ester (formed as an intermediate) in the dichloromethane phase, in which the resulting polycarbonate is also dissolved (H. Schnell).

Polybenzimidazoles

Through the reaction of aromatic *o*-diamines with carboxylic acids, one obtains imidazoles:

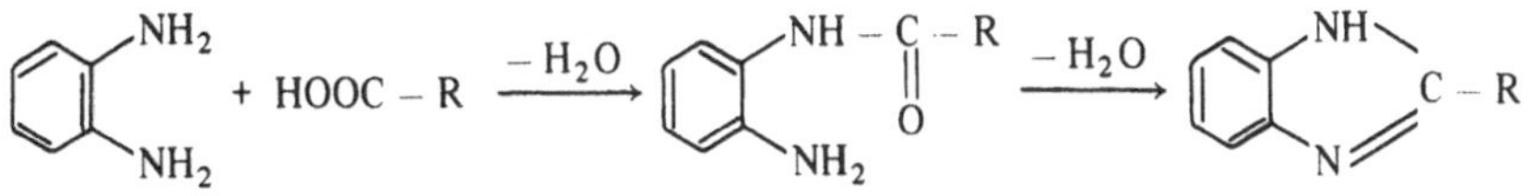

If one carries out the same reaction with aromatic tetramines and dicarboxylic acids, one obtains polybenzimidazoles (Marvel and Vogel):

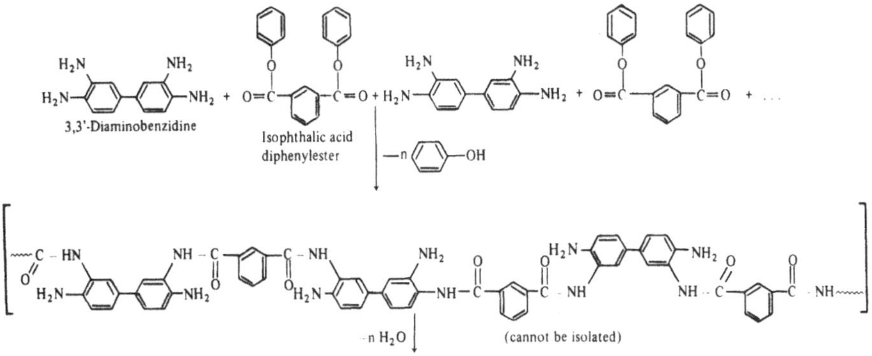

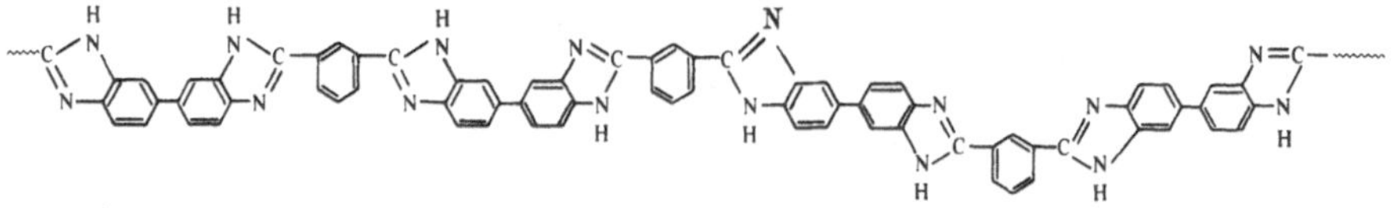

Polybenzimidazoles

If instead of diaminobenzidine, one uses 3,3-dimercaptobenzidine one obtains the corresponding polybenzothiazoles (P. M. Hergenrother et al.)

For the preparative reaction, one heats the tetramines with the phenyl esters of the dicarboxylic acid under strict exclusion of oxygen and under high vacuum for about half an hour at 260°C. Then the resulting mass is pulverized and heated under vacuum for several hours at 400°C. Because of the high melting temperature of polybenzimidazoles, they are never present as they melt in any of the stages of the second reaction step. The powder sinters but does not melt. One can carry out the entire reaction even better in a ball-mill. Polybenzimidazoles are soluble in dimethysulfoxide.

Polyimides

The reaction of tetracarboxylic acid anhydrides and diamines is much easier to carry out and, therefore, much more interesting for industry. In the first reaction step, which is exothermic at normal temperature, one obtains polyamide-carboxylic acids. When these are heated at 150°C to 250°C in a drying oven, polyimides are formed, accompanied by splitting-off of water and simultaneous evaporation of the solvent (Du Pont). The polyimides formed in this way can be processed to strong and tough films with high temperature resistance.

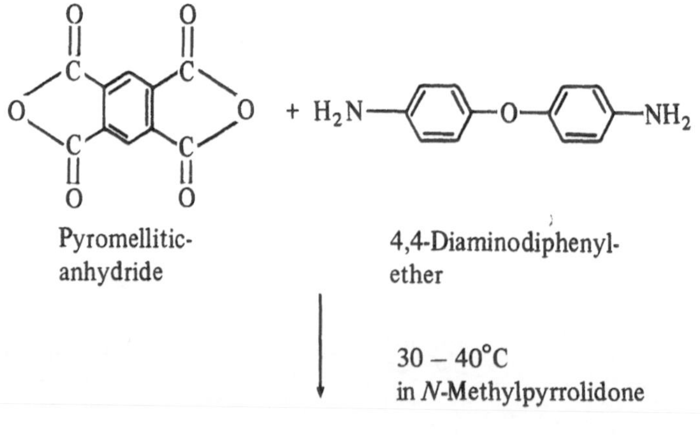

Pyromellitic-
anhydride

4,4-Diaminodiphenyl-
ether

30 – 40°C
in N-Methylpyrrolidone

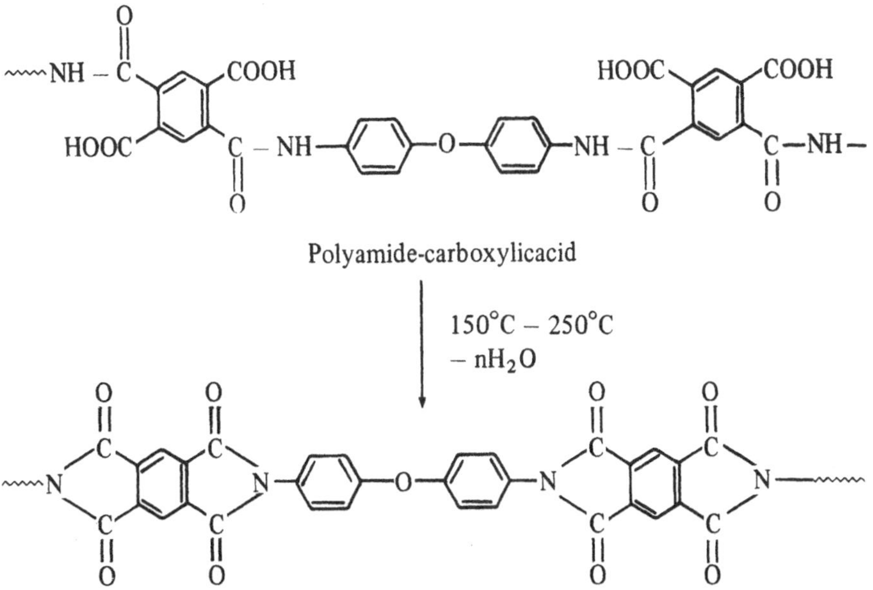

Polyamide-carboxylicacid

$$150°C - 250°C$$
$$- nH_2O$$

Polypyromellitimide

The polyamide-carboxylic acids formed as intermediates can be isolated, but they are strongly sensitive to hydrolysis, and therefore their viscosity in solution decreases on standing. The extent of the viscosity decrease depends in a remarkable way on the solvent used. Addition of toluene, benzene, or xylene, diminishes the sensitivity toward hydrolysis. Polypyromellitimide is commercially available under the trade names Kapton, formerly H-Film (Du Pont). It has a continuous temperature stability of approximately 250°C.

The synthesis of polyimides is also rather interesting because in this case on heating of a polymer, namely the polyamide-carboxylic acid, there occurs a chemical change in the polymer chain at high temperature which brings about a considerable improvement of the properties of the polymer. With most polymers no such improvement occurs on heating. In fact, in most cases the polymer degrades on heating and therefore one has to add stabilizers which prevent such reactions, for example: the oxidation of polyolefins, or HCl formation from polyvinylchloride.

Polyimidazopyrrolones (Pyrrones)

Dicarboxylic acids + tetramines ⟶ polyimidazoles
Tetracarboxylic acids + diamines ⟶ polyimides
Tetracarboxylic acids + tetramines ⟶ polyimidazopyrrolones

The above reactions show how by a different combination of di- and tetra-functional carboxylic acids and amines three types of polymers can be prepared.

The pyrrones are formed by the following reactions:

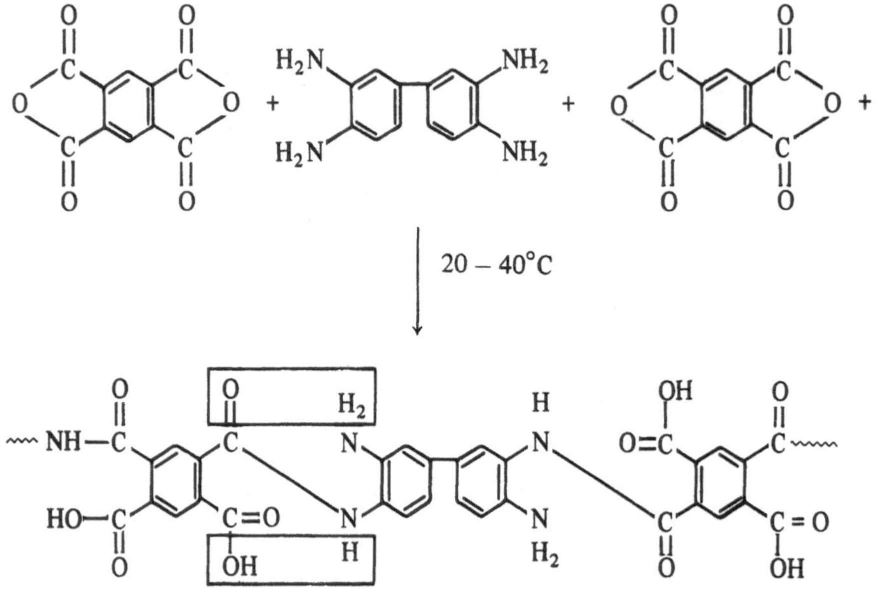

$$20 - 40°C$$

Polyaminoamide-carboxylic acid (A – A – A-Polymer)

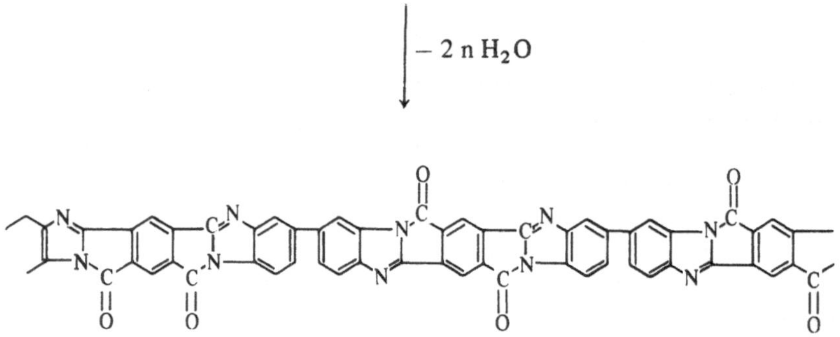

$$- 2 \, n \, H_2O$$

Polyimidazopyrrolone (Pyrrone)

The synthesis reaction occurs in two steps, just as with the polyimides. The solution of the polyaminoamide carboxylic acid, (this product can be isolated), which is formed at room temperature on the addition of the compounds in DMF solution, is poured out on a glass plate and slowly heated to 300°C. Under splitting-out of two water molecules per unit, a deep red polymer is formed as a film which always contains considerable amounts of the solvent. Pyrrones, because of their high radiation resistance, are particularly of interest for space technology. If one uses tetraminobenzene in the place of diaminobenzidine, one obtains a ladder polymer:

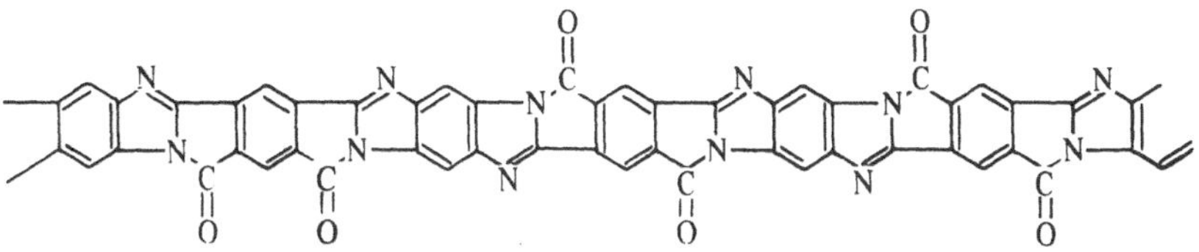

Polyamide sulfonic acid

Recently a new monomer has been prepared from toluene and H_2SO_4 by Sass and Vollmert:

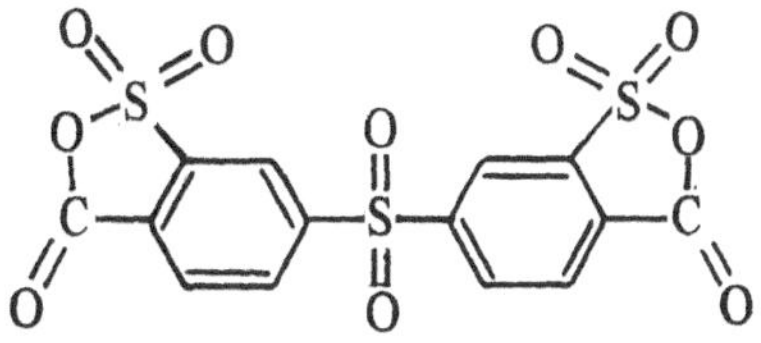

DSCA
(Di-sulfocarboxyanhydride)

By reaction of this sulfocarboxylic anhydride with aromatic diamines such as benzidine, 4,4' diaminodiphenylether, or 4,4' diaminodiphenylsulfone, a poly-amidesulfonic acid of high molecular weight (> 30,000) could be prepared. These polymers are strong polyelectrolytes with regular chain structures:

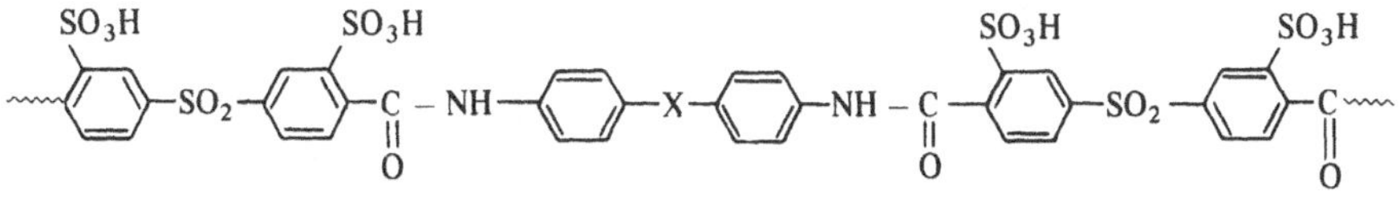

$$X = -, O, SO_2$$

Polyphenylene Ethers by Oxidative Coupling

Unsubstituted polyphenylene ethers are insoluble and infusible substances, and in spite of their high temperature stability have, therefore, not found any application. The dimethyl derivative (known as PPO), however, is thermo-plastic and soluble in many organic solvents. It can be prepared in an elegant synthesis from 2,6-dimethyl-phenol in a simple matter, (A. Hay and collaborators):

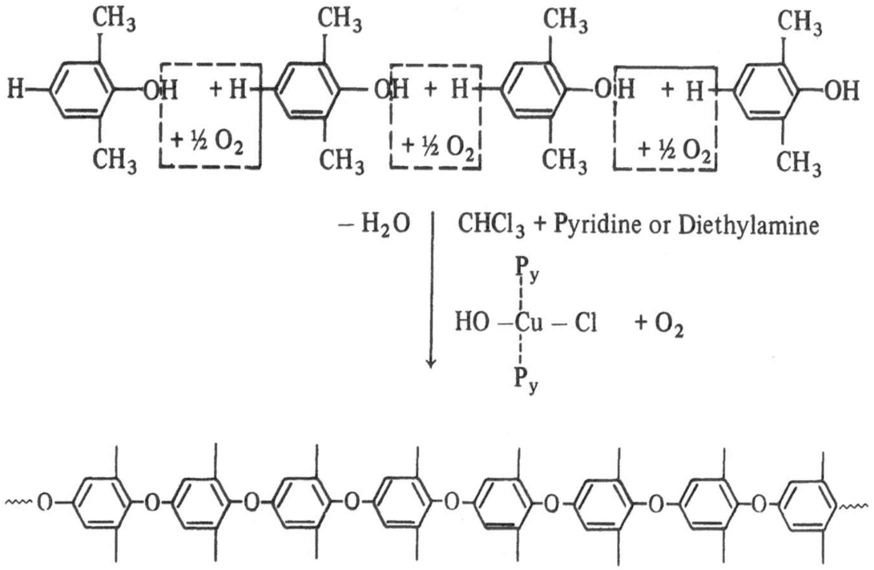

As oxygen is passed into a solution of the phenol, dehydrogenation occurs, accompanied by the splitting-out of water. The reaction is slightly exothermal, and therefore can be carried out without additional heating. In order to avoid cross-linking, one has to remove the heat of reaction. When oxygen uptake is terminated, the viscous solution of the polymer is stirred into methanol and the PPO precipitates as a white, fibrous substance. If one adds butanol or any other nonsolvent to the reaction mixture, the polymerization reaction behaves like a precipitation polymerization.

The determining factor for the course of the synthesis is the presence of a copper-II complex, which catalyzes the reaction, and also a certain amount of amine. In this case the designation "catalyst" is correct, because there is no initiation step, such as with chain reactions. The reaction is clearly a step reaction, as the dependence of the degree of polymerization on the conversion shows. Probably the copper-II-amine complex first forms a complex salt with phenol, which then reacts with the monomer while copper-I is split out. Through the oxygen present, the Cu^I is transformed again into Cu^{II}. Longer chains can also react with each other in the same way. The current interpretations of the mechanism of this reaction assume that a phenoxy radical appears as an intermediate. This assumption is unsatisfactory, however, because a phenoxy radical, once it is formed, immediately goes over into the tetramethyldiphenoquinone by way of the mesomeric limiting form of the phenyl radical, and therefore there will always be small amounts of undesirable side products. When the Cu^{II} concentration is too low, or the amine concentration is too low (pyridine or other tertiary and secondary amines are used), one often obtains the quinone as the main product of the synthesis:

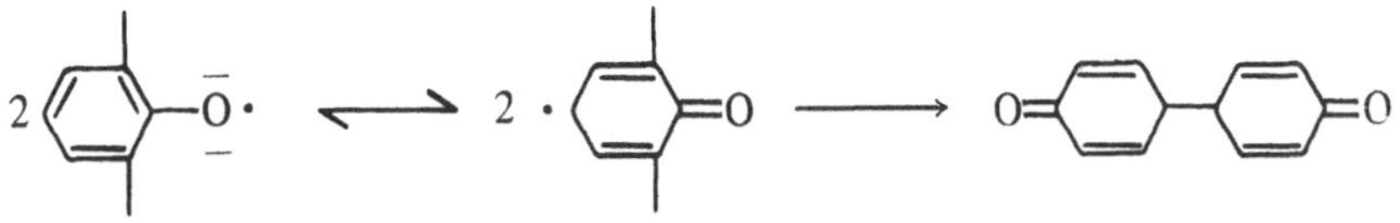

It is probably, therefore, the main function of the Cu^{II}- complex in this reaction to prevent the formation of the phenoxy radicals and the resulting undesirable carbon-carbon coupling.

Polyphenyleneoxide (PPO, trade name Noryl, General Electric Co.) is a thermoplastic material with a softening temperature of approximately 220°C. It forms hard and tough films which do not show any crystallinity in the x-ray diagram. Possibly the PPO molecules are slightly cross-linked. It is remarkable that the 2,6-dimethylphenol permits, preferably, the synthesis of polymers with high molecular weights. Phenol itself and cresol lead to cross-linked polymers, if one does not use complex-forming amines which are sterically hindered, such as pyridine derivatives with methyl or ethyl groups. Chlorophenols, as well as phenol derivatives with larger substituents (ethyl, propyl, and butyl), lead to polymers with low molecular weight.

Polymers with aromatic and heterocyclic structural units, especially polyimides, have become of increasing industrial importance because of their stability at higher temperatures. This stability refers not only to their resistance to a change in the state of aggregation (aromatic polyimides have softening points of about 800°C), but also to their resistance to chemical alterations at higher temperatures (e.g., chain degradation by oxygen, irradiation, or other chemical reactions).

Polyethylenesulfide (Thiokol)

Through the reaction of ethylenechloride with sodium tetrasulfide, one can form rubber-like polymers. The synthesis probably occurs according to the following reactions:

$$\ldots + Cl-CH_2-CH_2-Cl + Na-S-S-S-S-Na + Cl-CH_2-CH_2-Cl +$$

$$\downarrow - n\ NaCl$$

$$\sim\sim\sim CH_2-CH_2-S-S-S-S-CH_2-CH_2-S-S-S-S-CH_2-CH_2-S-S-S-S\sim$$

It is not certain that the number of sulfur atoms per segment is always exactly four, and different structures for these polymers may be postulated.

Thiokol is a rubber which resists almost all organic solvents. It is therefore valuable as a caulking material. Disadvantages are its poor aging properties, its cold flow (compare p. 564), and its unpleasant odor.

Silicones

These polymers, which can be produced in the form of oils, rubbers, and hard lacquers, are obtained through the reaction of dichlorosilane with water, and the condensation of the dialkyl silicic acids, (which are formed in this condensation and which cannot be isolated in their monomeric state) among themselves, or with alkylchlorosilanes. In this process one obtains macromolecules which do not contain any carbon in the chain.

$$
\underset{\text{Dimethyldichlorosilane}}{\text{Cl}-\underset{\overset{|}{\text{CH}_3}}{\overset{\overset{\text{CH}_3}{|}}{\text{Si}}}-\text{Cl} + \text{H}_2\text{O}} \;\rightarrow\; \text{Cl}-\underset{\overset{|}{\text{CH}_3} + \text{HCl}}{\overset{\overset{\text{CH}_3}{|}}{\text{Si}}}-\text{OH} \;\xrightarrow{+\;\text{Cl}-\underset{\overset{|}{\text{CH}_3}}{\overset{\overset{\text{CH}_3}{|}}{\text{Si}}}-\text{Cl}}\; \text{Cl}-\underset{\overset{|}{\text{CH}_3}}{\overset{\overset{\text{CH}_3}{|}}{\text{Si}}}-\text{O}-\underset{\overset{|}{\text{CH}_3}}{\overset{\overset{\text{CH}_3}{|}}{\text{Si}}}-\text{Cl} + \text{HCl}
$$

$$
\text{Cl}-\underset{\overset{|}{\text{CH}_3}}{\overset{\overset{\text{CH}_3}{|}}{\text{Si}}}-\text{O}-\underset{\overset{|}{\text{CH}_3}}{\overset{\overset{\text{CH}_3}{|}}{\text{Si}}}-\text{Cl} + 2\,\text{H}_2\text{O} \;\rightarrow\; \left[\text{HO}-\underset{\overset{|}{\text{CH}_3}}{\overset{\overset{\text{CH}_3}{|}}{\text{Si}}}-\text{O}-\underset{\overset{|}{\text{CH}_3}}{\overset{\overset{\text{CH}_3}{|}}{\text{Si}}}-\text{OH}\right] + 2\,\text{HCl}
$$

$$
2\left[\text{HO}-\underset{\overset{|}{\text{CH}_3}}{\overset{\overset{\text{CH}_3}{|}}{\text{Si}}}-\text{O}-\underset{\overset{|}{\text{CH}_3}}{\overset{\overset{\text{CH}_3}{|}}{\text{Si}}}-\text{OH}\right] \;\rightarrow\; \text{HO}-\text{Si}-\text{O}-\text{Si}-\text{O}-\text{Si}-\text{O}-\text{Si}-\text{OH} + \text{H}_2\text{O}
$$

(with CH₃ groups on each Si)

$$
\Big\downarrow\; +\;\text{Cl}-\underset{\overset{|}{\text{CH}_3}}{\overset{\overset{\text{CH}_3}{|}}{\text{Si}}}-\text{Cl}
$$

$$
\text{HO}-\underset{\overset{|}{\text{CH}_3}}{\overset{\overset{\text{CH}_3}{|}}{\text{Si}}}-\text{O}-\underset{\overset{|}{\text{CH}_3}}{\overset{\overset{\text{CH}_3}{|}}{\text{Si}}}-\text{O}-\underset{\overset{|}{\text{CH}_3}}{\overset{\overset{\text{CH}_3}{|}}{\text{Si}}}-\text{O}-\underset{\overset{|}{\text{CH}_3}}{\overset{\overset{\text{CH}_3}{|}}{\text{Si}}}-\text{O}-\underset{\overset{|}{\text{CH}_3}}{\overset{\overset{\text{CH}_3}{|}}{\text{Si}}}-\text{Cl} + \text{HCl}
$$

etc.   Polysiloxane

If one adds trichlorosilanes to the reaction mixture, one obtains cross-linked polymers, and if one adds monochlorosilanes, one obtains unreactive chain ends:

$$
\underset{\text{Monochlorosilane}}{\text{H}_3\text{C}-\underset{\overset{|}{\text{CH}_3}}{\overset{\overset{\text{CH}_3}{|}}{\text{Si}}}-\text{Cl}} + \text{HO}-\underset{\overset{|}{\text{CH}_3}}{\overset{\overset{\text{CH}_3}{|}}{\text{Si}}}-\text{OH} + \text{Cl}-\underset{\overset{|}{\text{Cl}}}{\overset{\overset{\text{CH}_3}{|}}{\text{Si}}}-\text{Cl} + \text{HO}-\underset{\overset{|}{\text{CH}_3}}{\overset{\overset{\text{CH}_3}{|}}{\text{Si}}}-\text{OH} + \cdots
$$

$$
+ \text{H}_3\text{C}-\underset{\overset{|}{\text{OH}}}{\overset{\overset{\text{OH}}{|}}{\text{Si}}}-\text{CH}_3
$$

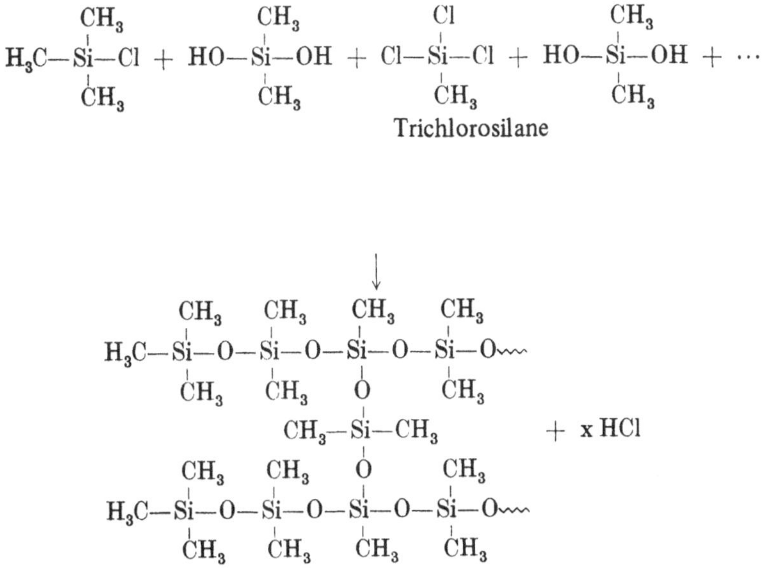

$$
\begin{array}{ccccc}
& CH_3 & CH_3 & Cl & CH_3 \\
& | & | & | & | \\
H_3C-\underset{\underset{\displaystyle CH_3}{|}}{Si}-Cl & + & HO-\underset{\underset{\displaystyle CH_3}{|}}{Si}-OH & + & Cl-\underset{\underset{\displaystyle CH_3}{|}}{Si}-Cl & + & HO-\underset{\underset{\displaystyle CH_3}{|}}{Si}-OH & + & \cdots
\end{array}
$$

Trichlorosilane

$\downarrow$

$$
\begin{array}{cccc}
CH_3 & CH_3 & CH_3 & CH_3 \\
| & | & | & | \\
H_3C-Si-O-Si-O-Si-O-Si-O\sim \\
| & | & | & | \\
CH_3 & CH_3 & O & CH_3
\end{array}
$$

$$CH_3-Si-CH_3 \qquad + \ x\,HCl$$

$$
\begin{array}{cccc}
CH_3 & CH_3 & O & CH_3 \\
| & | & | & | \\
H_3C-Si-O-Si-O-Si-O-Si-O\sim \\
| & | & | & | \\
CH_3 & CH_3 & CH_3 & CH_3
\end{array}
$$

According to their structure and properties, the silicones lie between the organic and the inorganic polymers (glass, asbestos, and waterglass). Just as with the glasses, a part of the silicon atoms in the chain may be replaced by the atoms of other elements, such as aluminum, boron, or lead.

Polymeric silicic acid is obtained through polycondensation of silicic acid, for example, if one carefully hydrolyzes silicontetrachloride and removes the HCl which is formed by dialysis:

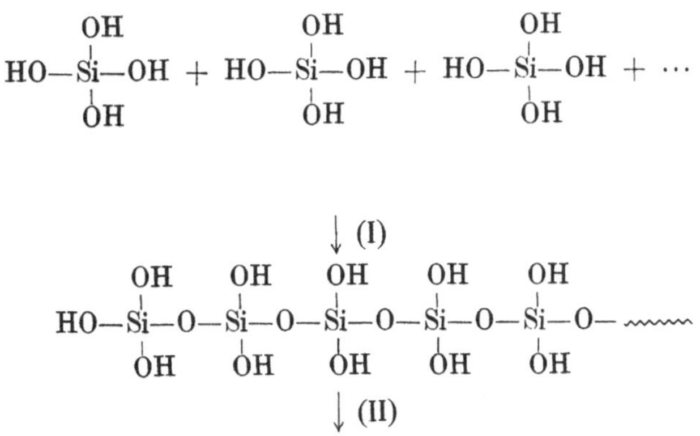

$$
\begin{array}{ccc}
OH & OH & OH \\
| & | & | \\
HO-\underset{\underset{\displaystyle OH}{|}}{Si}-OH & + HO-\underset{\underset{\displaystyle OH}{|}}{Si}-OH & + HO-\underset{\underset{\displaystyle OH}{|}}{Si}-OH & + \cdots
\end{array}
$$

$$\downarrow \text{(I)}$$

$$
\begin{array}{ccccc}
OH & OH & OH & OH & OH \\
| & | & | & | & | \\
HO-Si-O-Si-O-Si-O-Si-O-Si-O-\sim\sim\sim \\
| & | & | & | & | \\
OH & OH & OH & OH & OH
\end{array}
$$

$$\downarrow \text{(II)}$$

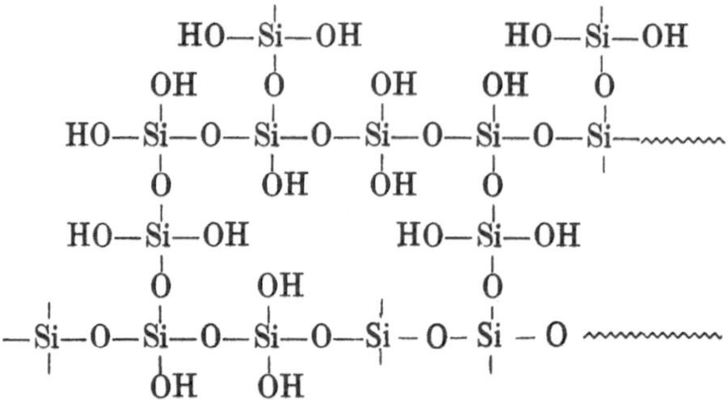

$$
\begin{array}{cccc}
& \text{HO}-\overset{|}{\text{Si}}-\text{OH} & & \text{HO}-\overset{|}{\text{Si}}-\text{OH} \\
& \text{OH} \quad \text{O} & \text{OH} \quad \text{OH} \quad \text{O} \\
\text{HO}-\overset{|}{\underset{|}{\text{Si}}}-\text{O}-\overset{|}{\underset{|}{\text{Si}}}-\text{O}-\overset{|}{\underset{|}{\text{Si}}}-\text{O}-\overset{|}{\underset{|}{\text{Si}}}-\text{O}-\overset{|}{\text{Si}}-\!\!\!\sim\!\!\sim \\
& \text{O} \quad \text{OH} \quad \text{OH} \quad \text{O} \\
& \text{HO}-\overset{|}{\text{Si}}-\text{OH} & & \text{HO}-\overset{|}{\text{Si}}-\text{OH} \\
& \text{O} \quad \text{OH} & & \text{O} \\
-\overset{|}{\underset{|}{\text{Si}}}-\text{O}-\overset{|}{\underset{|}{\text{Si}}}-\text{O}-\overset{|}{\underset{|}{\text{Si}}}-\text{O}-\overset{|}{\text{Si}}-\text{O}-\overset{|}{\text{Si}}-\text{O}\sim\!\!\sim\!\!\sim \\
& \text{OH} \quad \text{OH}
\end{array}
$$

Through further condensation (step II) one obtains branched and cross-linked silicic acids which precipitate in the form of gels.

The formation of inorganic glasses can be described as the transcondensation of SiO_2 (quartz-sand) with the aid of Na_2CO_3 and CaO:

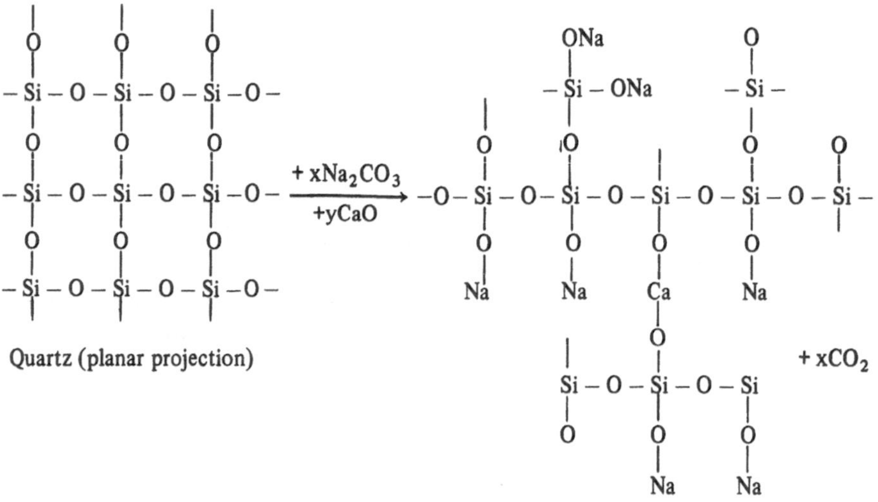

Quartz (planar projection) $\xrightarrow[+yCaO]{+\,xNa_2CO_3}$ Glass $+\,xCO_2$

2213   The Technique of Polycondensation

Because of the different ways these compounds interact, there is no universal method for carrying out polycondensation reactions.

Melt Polycondensation

For the preparation of polyesters and polyamides (e.g., nylon 66) one usually heats the components in a dry inert gas stream (usually nitrogen) for a longer time (24-48 hours) at higher temperatures in such a way that the temperature is slowly increased up to 250°C during the course of the reaction, while the water of the reaction distills off (melt polycondensation). Often, the last part of the reaction is carried out under vacuum, which lowers the water concentration still more and therefore brings about an increase in the molecular weight. This technique can be carried out continuously (e.g., in the production of nylon 6).

Solution Polycondensation

Relatively high molecular weights can be reached, if one heats the components in an inert solvent such as toluene or xylene under addition of an esterification catalyst (paratoluenesulfonic acid) and removes the water of reaction continually with a boiling solvent azeotrope. The carefully dried solvent can be recycled through the reaction chamber, as with usual esterifications. The more the recycling solvent is dried, the higher the molecular weight of the polymer condensate. In this way it is possible to prepare polyesters with molecular weights of about 30,000 (H. Batzer).

With another type of solution polycondensation of highly reactive bifunctional monomers, we find that dicarboxylicacid dichlorides, for example, react even at low temperatures with diamines in the presence of weakly basic solvents, such as dimethylformamide or dimethylacetamide, sometimes with the addition of LiOH. The LiCl formed improves the solubility of the resulting polyamides, so that certain aromatic polyamides [e.g., from isophthalic acid and phenylenediamine (Nomex)] can be produced in this way.

Interfacial Polycondensation

Polyesters and polyamides are usually prepared by the reaction of glycols or diamines with dicarboxylic acids. If instead of the dicarboxylic acid, one uses the much more reactive acid chloride, the polycondensation can be carried out in a particularly elegant manner[30] in a two-phase system (Morgan, Wittbecker, et al.). Thus one has a layer of a dilute aqueous solution of diamine on top of a dilute solution of the dicarboxylicacid dichloride in a solvent such as carbontetrachloride, which is not miscible with water. At the interface, a very thin film of polyamide is formed instantaneously, and this film can be removed with a glass rod wound up on the rod, or on some other roller (Figure 236). If the concentration of diamine and acid chloride has been properly selected, the thread of polymer formed in this way does not tear, and one can therefore wrap it up on the roller continuously since the film at the interface is always reformed the instant at which it is pulled away. Finally, when the solution is very dilute the thread breaks. The HCl formed during the reaction diffuses into the aqueous phase and is bound by the diamine in the form of a salt. The molecular weights of the polyamides formed in this way can be relatively high.

---

[30]The first interfacial polycondensation is described by Einhorn (1898), who reacted a solution of phosgene in toluene with an aqueous solution of hydroquinone-sodium.

This technique for carrying out an interfacial polycondensation is suitable for laboratory demonstrations (sebacoyl chloride in the CCl_4-phase and hexamethylene

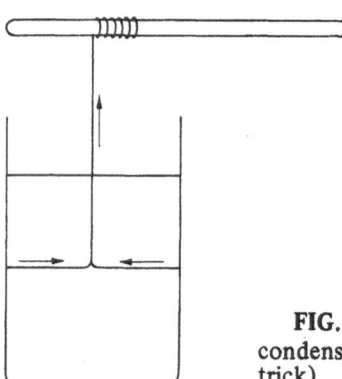

FIG. 236 — Interfacial poly-condensation. (The Nylon Rope-trick)

diamine in the water-phase is suggested), whereas for preparative purposes the two phases are usually mixed in a highly effective stirring apparatus. For example, the synthesis of polyamides with high melting points can often be carried out advantageously by interfacial polycondensation. With the exception of the poly-carbonate synthesis, however, no industrial use has been made of this technique. The melt and the solution polycondensation are generally preferred.

2214 Polyaddition

According to the same principle which pertains to reactions of bifunctional compounds, one can prepare chain-like polymer molecules by the addition of active hydrogen-containing bifunctional monomers to unsaturated or cyclic compounds. In this case the splitting-off of the water has already occurred in the formation of the monomer.

Of the many compounds theoretically capable of polyaddition, the diisocyanates and the bis-epoxides have so far found the greatest technical application.

Isocyanates

The diisocyanates readily add to all compounds with acidic hydrogen. With water, alcohols, amines, and carboxylic acids, the reaction occurs in the following way:

$$2\,R-N=C=O+HOH \rightarrow 2\,[R-NH-\underset{\underset{O}{\|}}{C}-OH] \;\rightarrow\; R-NH-\underset{\underset{O}{\|}}{C}-NH-R + CO_2$$

urea

$$R-N=C=O + HO-R' \;\rightarrow\; R-NH-\underset{\underset{O}{\|}}{C}-O-R'$$

urethane

$$R-N=C=O + H_2N-R' \rightarrow R-NH-\underset{\underset{O}{\|}}{C}-NH-R'$$
urea

$$R-N=C=O + HOOC-R' \rightarrow R-NH-\underset{\underset{O}{\|}}{C}-R' + CO_2$$
amide

In a completely analogous fashion, one obtains polyurethanes, polyureas, and polyamides in the reaction of diisocyanates with glycols, diamines, and dicarboxylic acids. For example:

$$O=C=N-R-N=C=O + HO-R'-OH +$$

$$+ O=C=N-R-N=C=O + HO-R'-OH + \cdots$$

$$\downarrow$$

$$O=C=N-R-NH-\underset{\underset{O}{\|}}{C}-O-R'-O-\underset{\underset{O}{\|}}{C}-NH-R-NH-\underset{\underset{O}{\|}}{C}-O-R'-\backsim OH$$

Polyurethane

Usually the polyureas and also, to a lesser degree, the polyamides and polyurethanes prepared by addition of diisocyanates are cross-linked because the CONH groups in the polymer chain can also react with isocyanate groups:

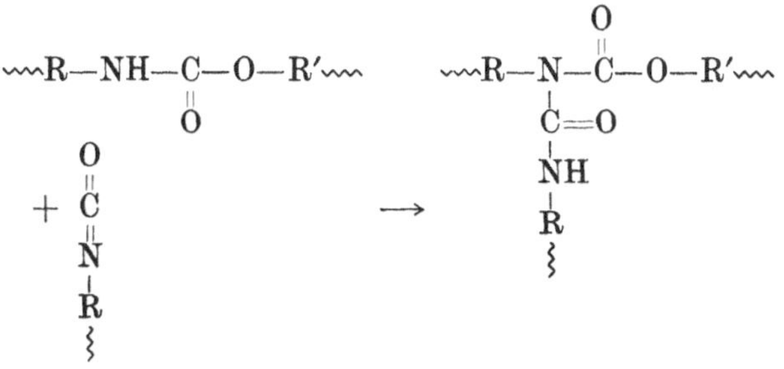

$$\backsim R-NH-\underset{\underset{O}{\|}}{C}-O-R'\backsim \quad\quad \backsim R-N-\overset{\overset{O}{\|}}{C}-O-R'\backsim$$

$$+ \underset{\underset{R}{\underset{|}{N}}}{\overset{\overset{O}{\|}}{C}} \quad\quad \longrightarrow \quad\quad \underset{\underset{R}{\underset{|}{NH}}}{\underset{\|}{C=O}}$$

In the reaction of isocyanates with carboxyl groups or water, CO_2 is formed. In the first case, one obtains amide groups and in the second case, urea groups:

1. $R-N=C=O + HOOC-R' \rightarrow R-NH-\underset{\underset{O}{\|}}{C}-R' + CO_2$

2. $R-N=C=O + HOH \rightarrow R-NH-\underset{\underset{O}{\|}}{C}-OH$

$$\xrightarrow{\;\;O=C=N-R\;\;} R-NH-\underset{\underset{O}{\|}}{C}-NH-R + CO_2$$

TABLE 238. Monomers used in polycondensations.

Monomer	Formula	Melting Point °C	Boiling Point °C/atm	Application
Adipic acid	$HOOC-(CH_2)_4-COOH$	152°	$205_{(10)}$	Polyamide fiber (Nylon 6,6) Polyamide plastics
Sebacic acid	$HOOC-(CH_2)_8-COOH$	134°	$295_{(100)}$	Polyesters, Polyamides (Nylon 6,10)
Maleic-anhydride	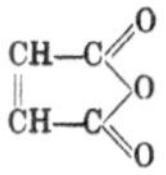	53°	202°	Polyesters (Polyester casting resins)
Phthalic-anhydride	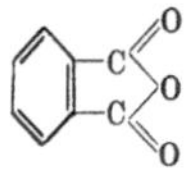	131°	285°	Polyesters coating raw materials (Alkyd resins)
Pyromellitic acid dianhydride (PMDA)	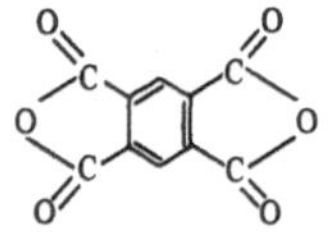	286°		Polyimides (Kapton)
Terephthalic-acid	 $HOOC-\langle\bigcirc\rangle-COOH$			Polyester fibers (Dacron, Terylene)
Dimethyl terephthalate	 $H_3COOC-\langle\bigcirc\rangle-COOCH_3$	141,6°	$177_{(45)}$	Polyester fibers, films (Dacron, Mylar)
Hexamethylenediamine	$H_2N-(CH_2)_6-NH_2$	40,7° 184°	200°	Polyamide fibers (Nylon)
Ethylene glycol	$HO-CH_2-CH_2-OH$	−11°	197°	Polyester fibers (Terylene)
Glycerol	$HO-CH_2-CH-CH_2-OH$ with OH	20°	290°	Coating raw materials (Alkyd resins)
ω-Aminoundecanoic acid	$HOOC-(CH_2)_9-CH_2-NH_2$	185°		Polyamide fiber (Rilsan)
Dioxy-diphenyl-Propane (Diphenolacetone Bisphenol-A)	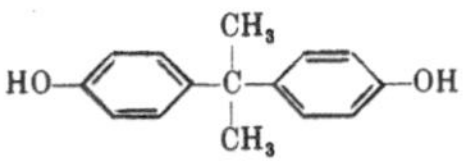	153–155°		Polycarbonates
Phosgene	$Cl-C-Cl$ with O	−118°	8°	Polycarbonates
Dimethyl-dichlorosilane	CH_3 $Cl-Si-Cl$ CH_3	−76°	70°	Silicones
Diethyldichlorosilane	C_2H_5 $Cl-Si-Cl$ C_2H_5		130°	Silicones
4,4′-Diamino-dicyclohexylmethane	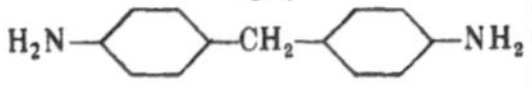 $H_2N-\langle\bigcirc\rangle-CH_2-\langle\bigcirc\rangle-NH_2$	35–40°	$150-155_{(5)}$	Polyamides
4,4′-Dichloro-diphenylsulfone	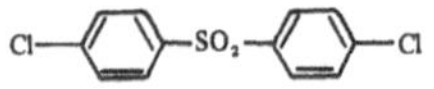 $Cl-\langle\bigcirc\rangle-SO_2-\langle\bigcirc\rangle-Cl$	149°		Polysulfone
4,4′-Diamino diphenylether	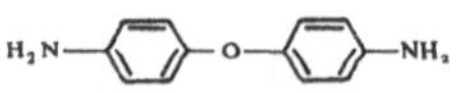 $H_2N-\langle\bigcirc\rangle-O-\langle\bigcirc\rangle-NH_2$	187°		Polyimides (Kapton)
2,6-Dimethyl-phenol	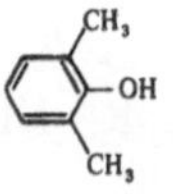	49°	212°	Polyphenyleneoxide (PPO, Noryl)

Correspondingly, in the reaction of diisocyanates with dicarboxylic acids, one obtains polyamides, and in the reaction of diisocyanates with water, one obtains polyureas. The carbon dioxide formed during the reaction expands the entire reaction mixture to a foam consisting of many small pores. This happens because the CO_2 bubbles cannot escape because of the rapidly increasing viscosity of the reaction medium. It is therefore possible to use this reaction for the preparation of elastic and hard foams (e.g., Moltoprene).

Isocyanate groups are particularly reactive and their reaction with glycols, water, or amines proceeds quickly and completely at room temperature and in relatively high dilution. These are the reasons behind the great technical importance of the diisocyanates as starting materials for hard or elastic foams (Moltoprene), specialty rubbers (Vulcollans), and elastic fibers (Lycra [Spandex]) and also high quality lacquers. One important advantage is that it is now possible to use macromolecules or at least higher molecular weight glycols (2,000 to 5,000) for the reaction instead of low molecular glycols. These are usually polyesters and polyethers with OH groups at both chain ends, which are prepared by polycondensation and by cationic polymerization, respectively, from ethylene oxide, propylene oxide, or tetra-hydrofuran. The following two examples will illustrate the reactions involved:

1) Synthesis of a polyurethane foam:

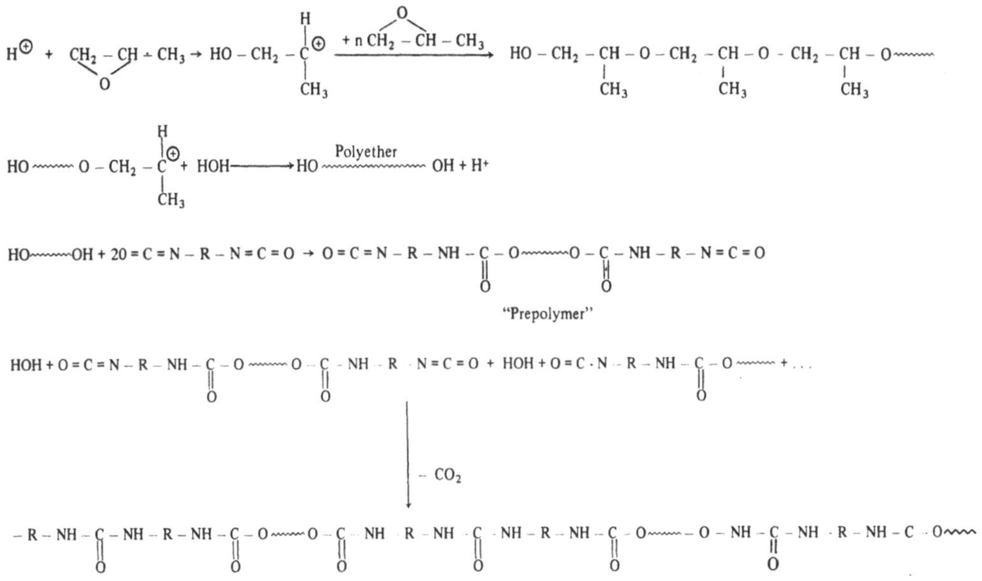

These reactions can be carried out separately by first reacting the polyether (or polyester) with a calculated excess of diisocyanate, and then—after removal of the excess of diisocyanate—reacting the polyether, which now has isocyanate groups at the chain ends, with water. However, it is simpler to carry out the entire reaction sequence in one single process ("one-shot"), by mixing the calculated amount of water with the polyether or polyester solution including the required diisocyanate. In order to accelerate the reaction, one adds tertiary amines and tin salts of the higher fatty acids. One also adds certain silicone oils to stabilize the foams and to achieve a foam with small pores. The most commonly used accelerators are diazobicyclooctane (DABCO) and tin octoate:

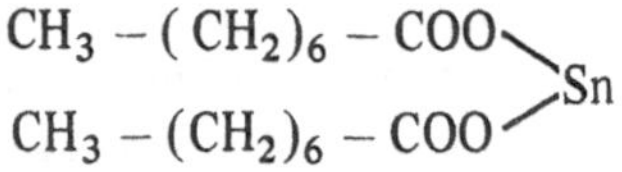

$$CH_3 - (CH_2)_6 - COO$$
$$CH_3 - (CH_2)_6 - COO$$

Sn

tin^{II} octoate

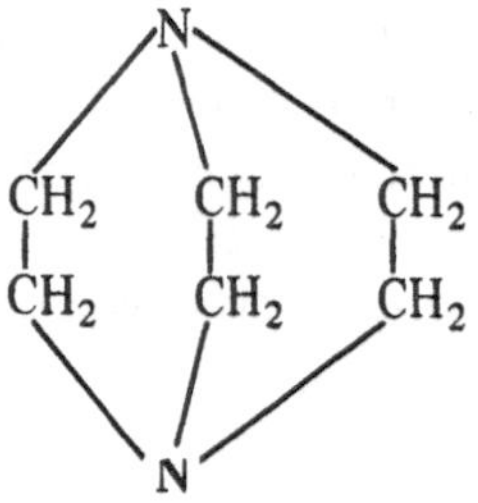

"DABCO"

The preceding reactions were based on linear polyesters in order to make the reactions simpler to follow. In practice, however, one uses branched polyethers or polyesters for the preparation of foams. Using lightly branched polyethers and difunctional isocyanates, one obtains rubbery, soft foams, whereas if one uses strongly branched polyethers or polyesters and tri- or poly-valent isocyanates ("polyisocyanates"), one obtains hard foams.

Lightly branched polyethers are prepared by the poly-addition of ethylene oxide to glycerol or trimethylolpropane.

"Polyisocyanates" ("poly" refers here to the number of isocyanate groups) are prepared by reacting diisocyanates (in excess) with trimethylolpropane:

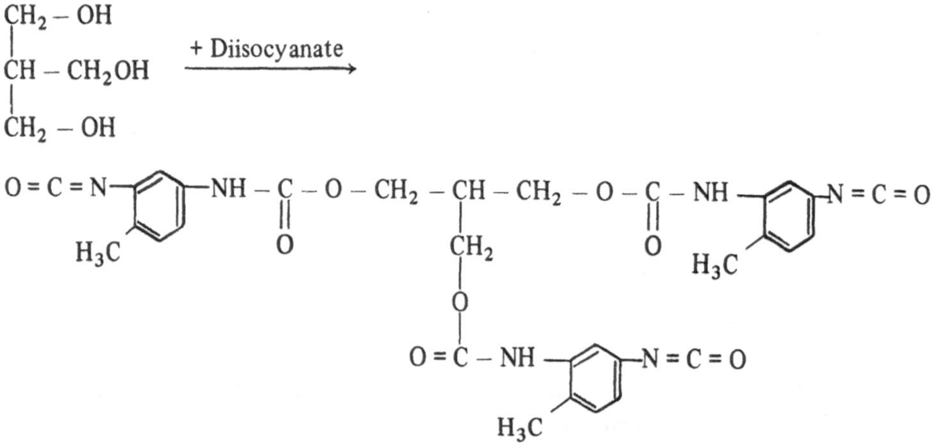

$$CH_2 - OH$$
$$CH - CH_2OH \quad \xrightarrow{+ \text{ Diisocyanate}}$$
$$CH_2 - OH$$

or by phosgenation of polyamines, which are prepared by reaction of aniline and formaldehyde in acid solution:

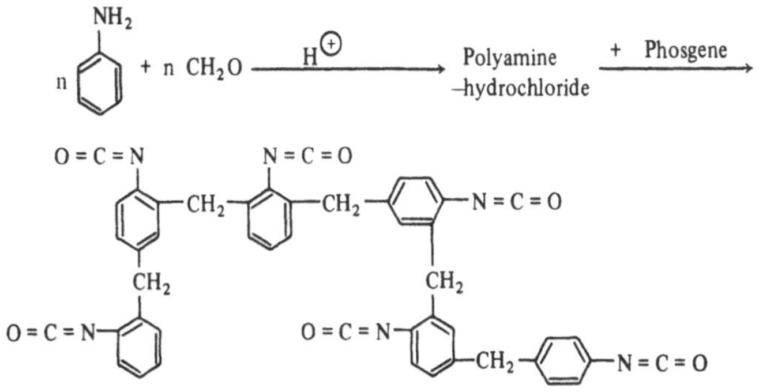

$$\text{n} \underset{\text{NH}_2}{\bigcirc} + \text{n CH}_2\text{O} \xrightarrow{\text{H}^{\oplus}} \begin{matrix}\text{Polyamine} \\ \text{-hydrochloride}\end{matrix} \xrightarrow{+ \text{ Phosgene}}$$

$$O = C = N \qquad N = C = O$$

$$\bigcirc -CH_2 - \bigcirc -CH_2 - \bigcirc -N = C = O$$

$$\qquad \qquad CH_2 \qquad \qquad CH_2$$

$$O = C = N - \bigcirc \qquad O = C = N - \bigcirc -CH_2 - \bigcirc -N = C = O$$

2. Synthesis of an elastic fiber (Lycra):

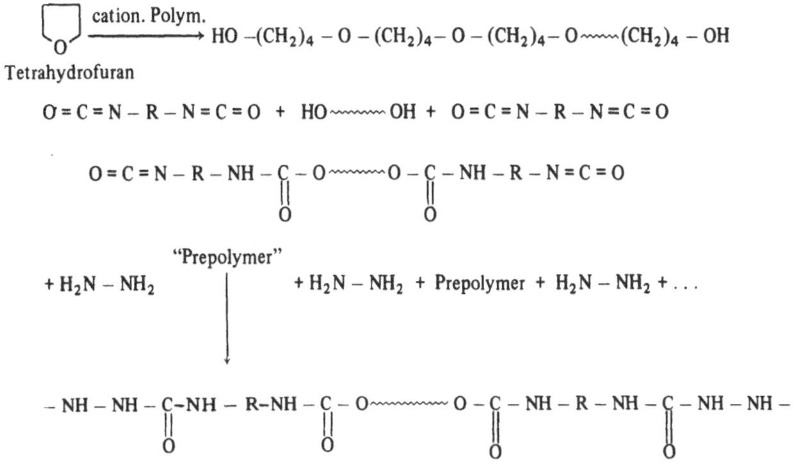

$$\underset{\text{Tetrahydrofuran}}{\bigcirc_O} \xrightarrow{\text{cation. Polym.}} HO -(CH_2)_4 - O - (CH_2)_4 - O - (CH_2)_4 - O \sim\sim\sim (CH_2)_4 - OH$$

$$O = C = N - R - N = C = O \ + \ HO \sim\sim\sim OH \ + \ O = C = N - R - N = C = O$$

$$O = C = N - R - NH - \underset{O}{\overset{\|}{C}} - O \sim\sim\sim O - \underset{O}{\overset{\|}{C}} - NH - R - N = C = O$$

$$\overset{\text{"Prepolymer"}}{\underset{+ H_2N - NH_2}{\Big|}} \qquad + H_2N - NH_2 \ + \ \text{Prepolymer} \ + \ H_2N - NH_2 \ + \ldots$$

$$- NH - NH - \underset{O}{\overset{\|}{C}} - NH - R - NH - \underset{O}{\overset{\|}{C}} - O \sim\sim\sim O - \underset{O}{\overset{\|}{C}} - NH - R - NH - \underset{O}{\overset{\|}{C}} - NH - NH -$$

For the preparation of elastic fibers from polyesters (or polyethers) and diiso-cyanate, one always first prepares the prepolymer and then reacts this in solution (dimethyl formamide) with hydrazine or aliphatic diamines. The resulting polymer has a periodic succession of polar and unpolar segments in the chain and is therefore often called a segment- or block-copolymer. It remains soluble and is spun by a dry spinning process. The polyester or polyether segments in the chain also remain flexible in the solid state (without solvent) whereas the polar groups of the polyurethane-polyurea segments act as cross-links of the macromolecules via hydrogen bonds.

Epoxides

Epoxides, just like isocyanates, are also capable of reacting with amines, acids, and alcohols (phenols). Epichlorohydrin and its addition products with bis-phenol-A have especially become technically important for the synthesis of polymers:

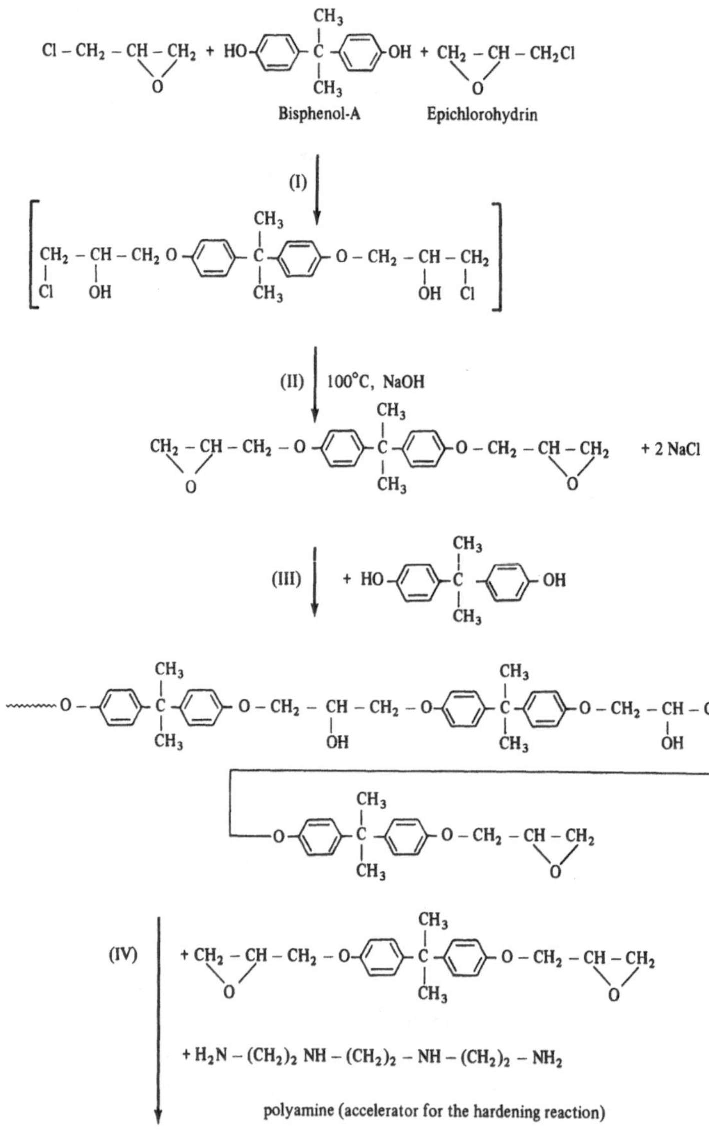

$$Cl-CH_2-CH-CH_2 + HO-\underset{CH_3}{\overset{CH_3}{C}}-OH + CH_2-CH-CH_2Cl$$

Bisphenol-A   Epichlorohydrin

(I)

$$\left[CH_2-CH-CH_2O-\underset{CH_3}{\overset{CH_3}{C}}-O-CH_2-CH-CH_2 \right]$$
$$\;\; Cl \quad OH \qquad\qquad\qquad\qquad OH \quad Cl$$

(II)   100°C, NaOH

$$CH_2-CH-CH_2-O-\underset{CH_3}{\overset{CH_3}{C}}-O-CH_2-CH-CH_2 \quad + 2\,NaCl$$

(III) $+ HO-\underset{CH_3}{\overset{CH_3}{C}}-OH$

$$\sim\!\!\sim\!\!\sim O-\underset{CH_3}{\overset{CH_3}{C}}-O-CH_2-CH-CH_2-O-\underset{CH_3}{\overset{CH_3}{C}}-O-CH_2-CH-CH_2$$
$$OH \qquad\qquad\qquad\qquad\qquad\qquad OH$$

$$-O-\underset{CH_3}{\overset{CH_3}{C}}-O-CH_2-CH-CH_2$$

(IV) $+ CH_2-CH-CH_2-O-\underset{CH_3}{\overset{CH_3}{C}}-O-CH_2-CH-CH_2$

$+ H_2N-(CH_2)_2\,NH-(CH_2)_2-NH-(CH_2)_2-NH_2$

polyamine (accelerator for the hardening reaction)

(*continued*)

FIG. 242 — Synthesis reactions of poly-epoxide-resins.

FIG. 242 – (concluded).

According to this (simplified) reaction scheme, one obtains epoxy resins (ethoxy-lene resins) which can be used for the preparation of glass fiber reinforced plastics and as raw materials for lacquers and top-grade adhesives. The reactions (I) to (III) are carried out in the chemical plant, whereas the hardening-reaction (IV) is carried out at the place where the epoxy-resins are used (as adhesives, etc.) after mixing the two components, the prepolymer [formed by reaction (III)], and the hardening agent.

For the reaction of ethylene oxide with monofunctional amines and phenols, see p. 251.

222   Polymers from Heterocyclic Compounds and Formaldehyde

Of the heterocyclic compounds, the lactams and cyclic ethers are especially important for the preparation of polymers. According to its polymerization behavior, one can also consider formaldehyde among the cyclic ethers (i.e., as a cyclic ether which has already been opened or one which was unable to cyclize).

2221   Lactam Polymerization With Water

If one heats lactams of more than 6-ring atoms in the presence of traces of water, or aqueous carboxylic acids, they are transformed into polyamides. From ε-caprolactam, one obtains in this way 6-nylon (Perlon, Caprolan):

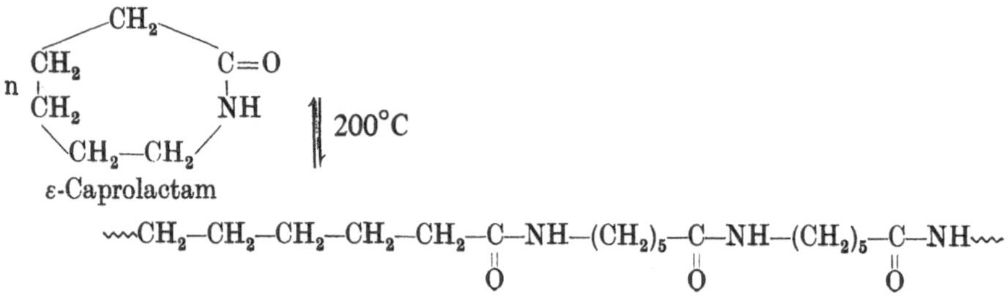

$$n \quad \begin{array}{c} {}_{\diagup}CH_2{}_{\diagdown} \\ CH_2 \qquad C=O \\ CH_2 \qquad NH \\ {}_{\diagdown}CH_2-CH_2{}_{\diagup} \end{array} \quad \Big\| \; 200°C$$

ε-Caprolactam

$$\text{ᴠᴠᴠ}CH_2-CH_2-CH_2-CH_2-CH_2-\underset{O}{\overset{\|}{C}}-NH-(CH_2)_5-\underset{O}{\overset{\|}{C}}-NH-(CH_2)_5-\underset{O}{\overset{\|}{C}}-NH\text{ᴠᴠᴠ}$$

This reaction is an equilibrium reaction (ring-chain equilibrium), which always leads to a mixture of polymer and cyclic monomer.

The amount of monomer depends on the temperature and is approximately 6% at a temperature slightly above the melting point of 6-nylon (about 200°C).

Also if one heats monomer-free polycaprolactam, which has been obtained by washing of the polymer with water, above its melting point, one obtains again a polymer-monomer mixture with 6% of the cyclic monomer. The same equilibrium mixture is obtained through polycondensation of ω-aminocaproic acid.

Caprolactam polymerization can be summarized in the following manner. First, through hydrolysis of the lactam, one obtains ω-aminocaproic acid, which then forms the polyamide through polycondensation. The water formed in this way is capable of immediately hydrolyzing fresh lactam to aminocaproic acid, which then again becomes available for the formation of the polyamide.

As extensive investigations of the kinetics of lactam polymerization (Wiloth, Kruissink, and Hermans) have shown, the lactam adds to the growing chain through transamidation. The water produced gives rise only to small amounts of aminocaproic acid, which function as initiator of the addition reaction:

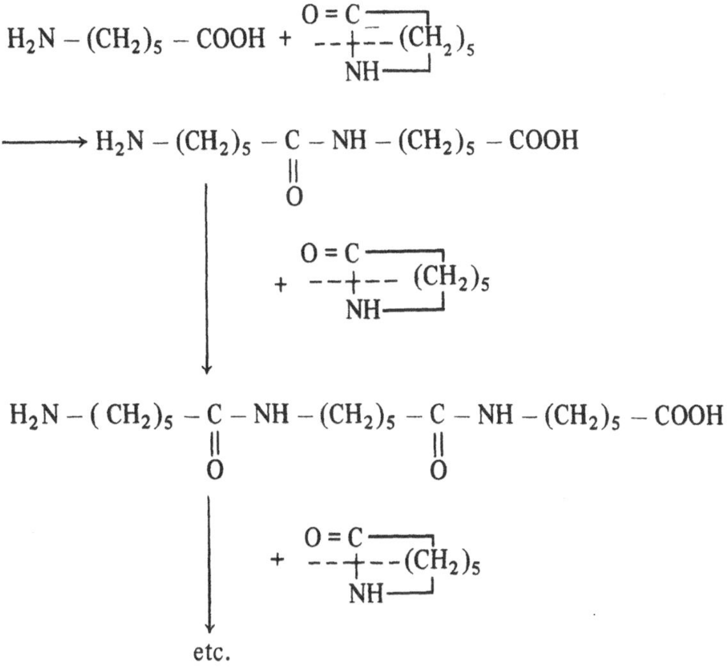

$$H_2N - (CH_2)_5 - COOH \; + \; \begin{matrix} O=C \\ \\ NH \end{matrix}\!\!-\!\!(CH_2)_5$$

$$\longrightarrow H_2N - (CH_2)_5 - \underset{\underset{O}{\|}}{C} - NH - (CH_2)_5 - COOH$$

$$+ \; \begin{matrix} O=C \\ \\ NH \end{matrix}\!\!-\!\!(CH_2)_5$$

$$H_2N - (CH_2)_5 - \underset{\underset{O}{\|}}{C} - NH - (CH_2)_5 - \underset{\underset{O}{\|}}{C} - NH - (CH_2)_5 - COOH$$

$$+ \; \begin{matrix} O=C \\ \\ NH \end{matrix}\!\!-\!\!(CH_2)_5$$

etc.

Because caprolactam polymerization cannot be started with water-free carboxylic acids (for example, benzoic acid) or with water-free amines, but starts easily with pure aminocaproic acid, one assumes that the inner salt of the aminocaproic acid, $H_3N^{(+)}-(CH_2)_5-COO^{(-)}$, must be the real carrier of the reaction.

The kinetics, final equilibrium, and dependence of the degree of polymerization on the conversion (according to Figure 212) are clear indications that the lactam polymerization initiated by water or dilute acetic acid must be a step reaction and not a chain reaction. On the other hand, the anionic lactam polymerization described in the following section is a typical chain reaction. One can also prepare copolyamides from adipic acid, hexamethylene diamine (AH-salt), and caprolactam, which have a better solubility than the pure polyamides. Also here caprolactam behaves exactly in the same manner as ω-aminocaproic acid.

2222 The Anionic Polymerization of Caprolactam

Lactams can polymerize much more rapidly using water-free alkali, metallic sodium, sodium methylate, water-free Na_2CO_3, and a number of other alkaline systems as initiators than with water as the only initiator. The addition of acetic anhydride or acetyl caprolactam increases the rate of polymerization even further. Even the five-membered pyrrolidone, which does not polymerize with water, can be transformed into a polyamide with alkali and acetic anhydride. The reaction most probably occurs according to the mechanism of an anionic polymerization, assuming that sodium methylate acts as the initiator (W. Griehl and S. Schaaf):

Initiation (CH_3OH is distilled off):

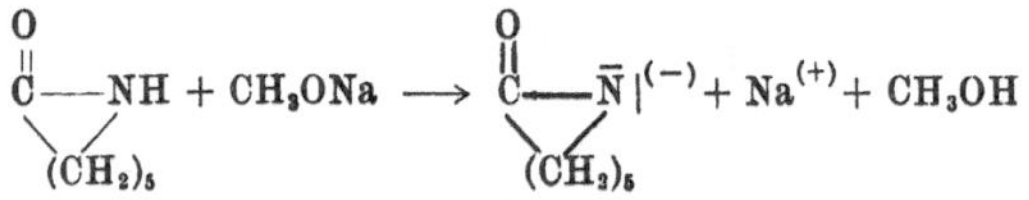

$$\underset{(CH_2)_5}{\overset{O}{\overset{\|}{C}}}{-}NH + CH_3ONa \longrightarrow \underset{(CH_2)_5}{\overset{O}{\overset{\|}{C}}}{-}\bar{N}|^{(-)} + Na^{(+)} + CH_3OH$$

Caprolactamanion

$$\underset{(CH_2)_5}{\overset{O}{\overset{\|}{C}}}{-}\bar{N}|^{(-)} + \underset{(CH_2)_5}{\overset{O}{\overset{\|}{C}}}{:}{-}NH \longrightarrow \left[\underset{(CH_2)_5}{\overset{O}{\overset{\|}{C}}}{-}N{-}\overset{O}{\overset{\|}{C}}{-}(CH_2)_5{-}\overset{H}{\underset{}{\bar{N}}}|^{(-)} \right]$$

$$\downarrow + \text{Caprolactam}$$

$$\underset{(CH_2)_5}{\overset{O}{\overset{\|}{C}}}{-}N{-}\overset{O}{\overset{\|}{C}}{-}(CH_2)_5NH_2 + \underset{(CH_2)_5}{\overset{O}{\overset{\|}{C}}}{-}\bar{N}|^{(-)}$$

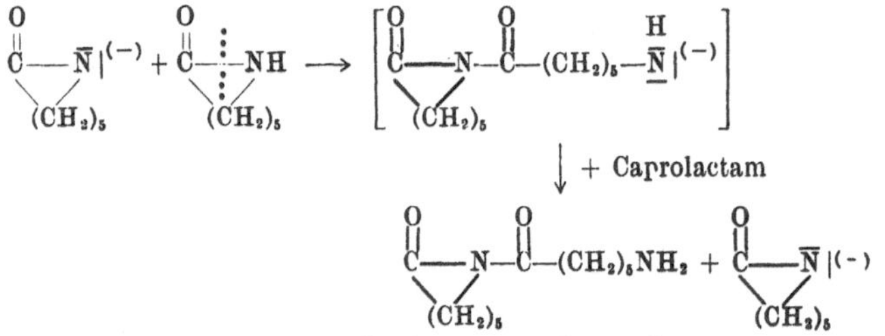

ω-Amino-caproylcaprolactam

Propagation:

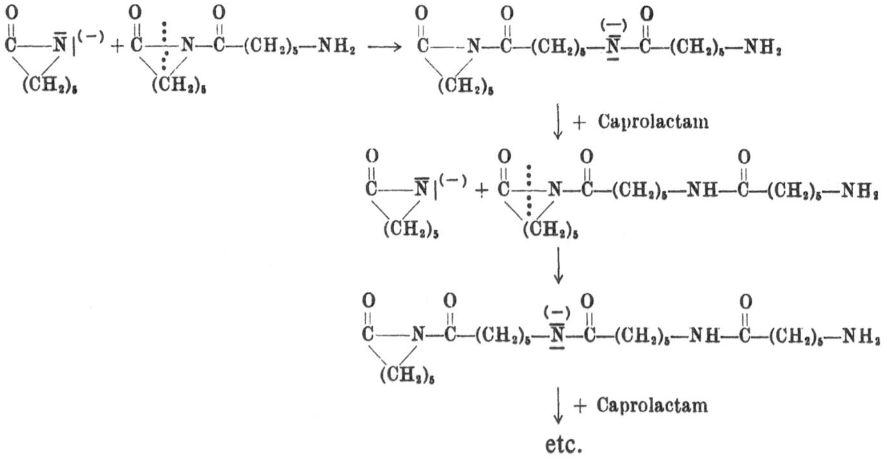

$$\underset{(CH_2)_5}{\overset{O}{\overset{\|}{C}}}{-}\bar{N}|^{(-)} + \underset{(CH_2)_5}{\overset{O}{\overset{\|}{C}}}{:}{-}N{-}\overset{O}{\overset{\|}{C}}{-}(CH_2)_5{-}NH_2 \longrightarrow \underset{(CH_2)_5}{\overset{O}{\overset{\|}{C}}}{-}N{-}\overset{O}{\overset{\|}{C}}{-}(CH_2)_5{-}\overset{(-)}{\bar{N}}{-}\overset{O}{\overset{\|}{C}}{-}(CH_2)_5{-}NH_2$$

$$\downarrow + \text{Caprolactam}$$

$$\underset{(CH_2)_5}{\overset{O}{\overset{\|}{C}}}{-}\bar{N}|^{(-)} + \underset{(CH_2)_5}{\overset{O}{\overset{\|}{C}}}{:}{-}N{-}\overset{O}{\overset{\|}{C}}{-}(CH_2)_5{-}NH{-}\overset{O}{\overset{\|}{C}}{-}(CH_2)_5{-}NH_2$$

$$\downarrow$$

$$\underset{(CH_2)_5}{\overset{O}{\overset{\|}{C}}}{-}N{-}\overset{O}{\overset{\|}{C}}{-}(CH_2)_5{-}\overset{(-)}{\bar{N}}{-}\overset{O}{\overset{\|}{C}}{-}(CH_2)_5{-}NH{-}\overset{O}{\overset{\|}{C}}{-}(CH_2)_5{-}NH_2$$

$$\downarrow + \text{Caprolactam}$$

etc.

Termination:

$$\underset{(CH_2)_5}{\overset{O}{\overset{\|}{C}}}{-}\bar{N}|^{(-)} + H_2O \longrightarrow \underset{(CH_2)_5}{\overset{O}{\overset{\|}{C}}}{-}NH + OH^{(-)}$$

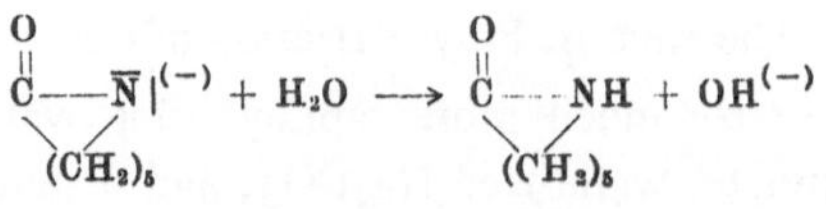

According to this mechanism (which has not yet been experimentally demonstrated in all its details) the initiation of the reaction occurs with the formation of aminocaproylcaprolactam. The chain grows through the opening up of the caproyl-caprolactam ring by addition of the lactam anion, and thus there is always a new caproylcaprolactam group formed at the chain end. The caproylcaprolactam can be replaced by other acyllactams (for example, acetylcaprolactam), which can add a lactam anion in a completely analogous manner to that of caproylcaprolactam. Since the formation of acetyllactam by reaction of lactam with acetic anhydride, acetylchloride, ketene, or isocyanates, occurs much faster than the formation of aminocaproylcaprolactam from the lactamion, one can understand the activation function of the above mentioned compounds during lactam polymerization.

The polymerization of acrylamide, with sodium, sodamide, or sodium methylate, to a polyamide may occur (according to Breslow) by a mechanism similar to the alkaline lactam polymerization via amide anions (which probably are formed by a $H^{(+)}$ shift [compare p. 180]):

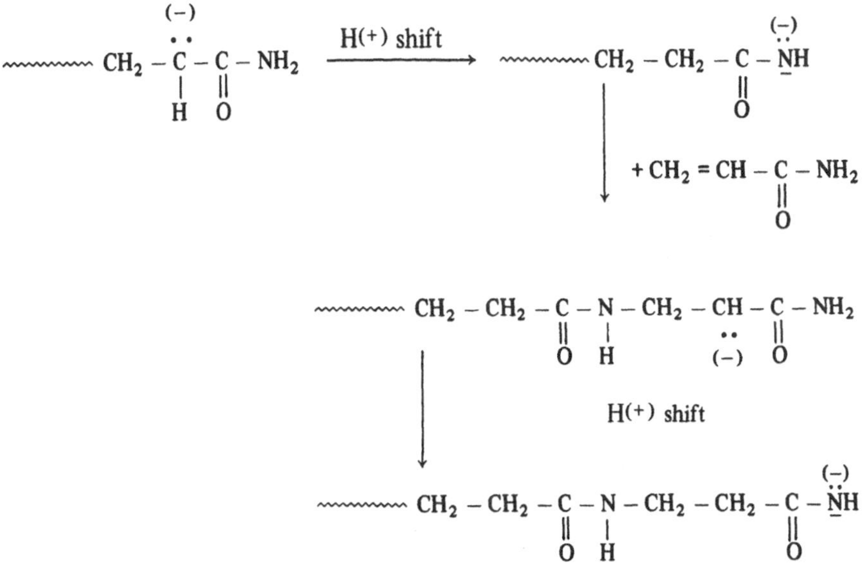

With radical initiators (azobisisobutyronitrile or peroxides), acrylamide can polymerize similar to other vinyl compounds to polymers with C–C chains:

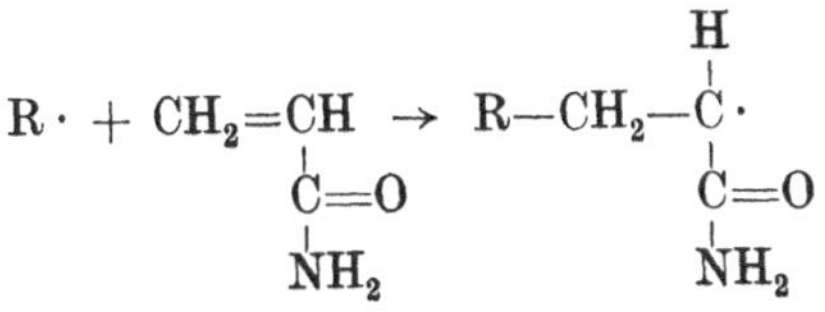

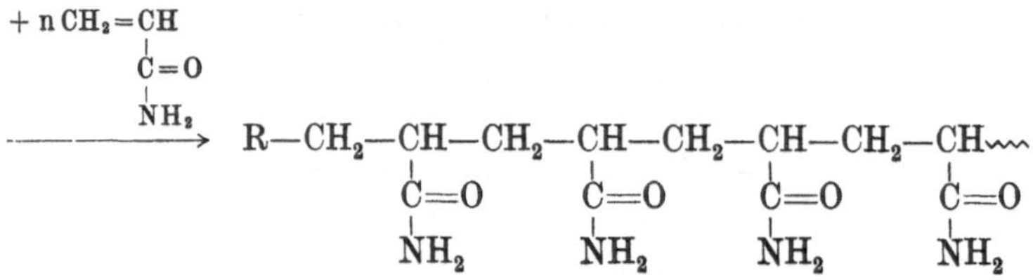

$$+ \, n \, CH_2 = CH$$
$$\qquad\qquad \underset{\displaystyle NH_2}{\overset{\displaystyle C=O}{|}}$$

$$\xrightarrow{\hspace{2cm}} \quad R-CH_2-CH-CH_2-CH-CH_2-CH-CH_2-CH\rightsquigarrow$$
$$\qquad\qquad\qquad \underset{NH_2}{\overset{C=O}{|}} \quad \underset{NH_2}{\overset{C=O}{|}} \quad \underset{NH_2}{\overset{C=O}{|}} \quad \underset{NH_2}{\overset{C=O}{|}}$$

In the presence of sodium, however, acrylamide reacts similar to the difficult-to-obtain 4-ring lactam:

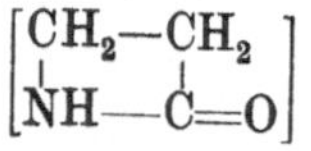

$$\begin{bmatrix} CH_2 - CH_2 \\ | \qquad | \\ NH - C = O \end{bmatrix}$$

The 5-ring lactam, which is produced on a technical scale (i.e., pyrrolidone), can be polymerized to polyamides in the presence of acetic anhydride or isocyanates, with alkali metals or alkali metal compounds ($NaOCH_3$). Like polycaprolactam, one can also use polybutyrolactam (=polypyrrolidone) for the production of synthetic fibers (nylon-4). Nylon-4 distinguishes itself from nylon-6 and nylon-6,6 by its greater takeup of water and its stronger tendency to depolymerization. Also higher lactams, such as:

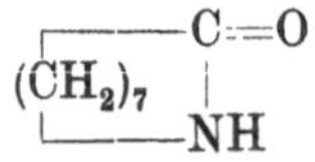

$$(CH_2)_7 \underset{\displaystyle \overline{NH}}{\overset{\displaystyle \overline{}C=O}{|}} \qquad \text{from Cyclooctatetraene}$$

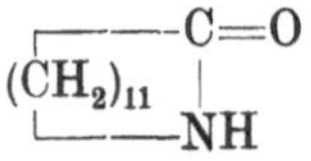

$$(CH_2)_{11} \underset{\displaystyle \overline{NH}}{\overset{\displaystyle \overline{}C=O}{|}} \qquad \text{from Dodecatriene}$$

can be polymerized. The polyamides formed are softer, have lower melting points, and take up less water than nylon-6 or -6,6.

2223   The Polymerization of N-carboxyl-amino acid anhydrides

Carboxyanhydrides can be polymerized by cationic or anionic initiators (for example, oxonium borofluoride ($[(C_2H_5)_3O] BF_4$) or amines), to polyamides with high molecular weight accompanied by the formation of CO_2:

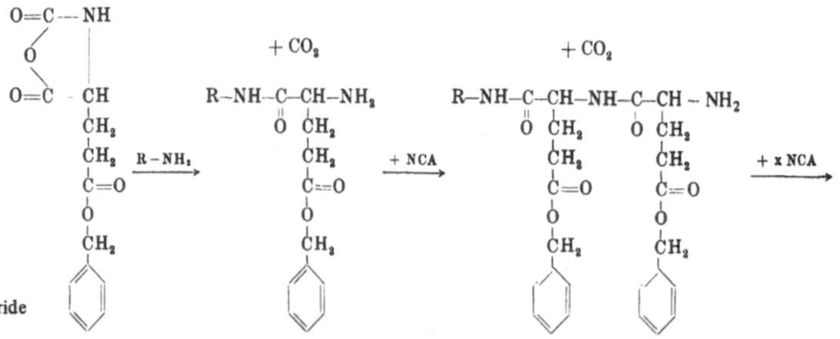

N-carboxyanhydride
of glutamic acid
benzylester (NCA)

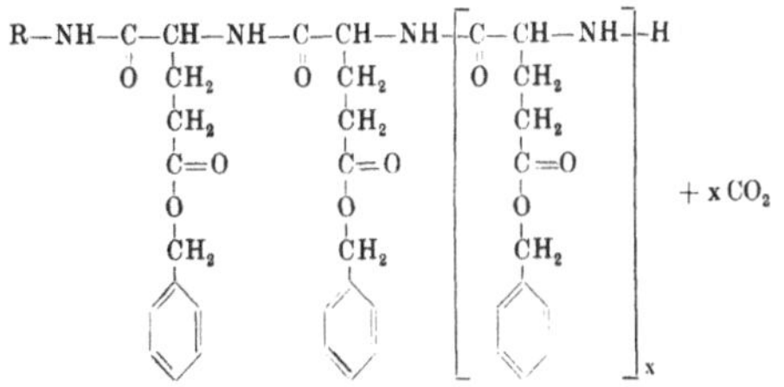

$$R-NH-C-CH-NH-C-CH-NH-\left[C-CH-NH\right]-H$$

(structure with O, CH_2, CH_2, $C=O$, O, CH_2, and phenyl groups repeated)

$$+ x\,CO_2$$

Since these N-carboxyanhydrides can be prepared from many different amino acids, this type of polymerization is used often for the preparation of protein model substances.

2224 The Polymerization of Monoisocyanates

Another process for preparing polymers with amide groups in the chain is the polymerization of monoisocyanates at $-20°C$ to $-50°C$, initiated by sodium naphthalene or sodium cyanide in dimethylformamide. We have to assume that this polymerization occurs by an anionic mechanism (Shashoua). It can be used for the preparation of fibers:

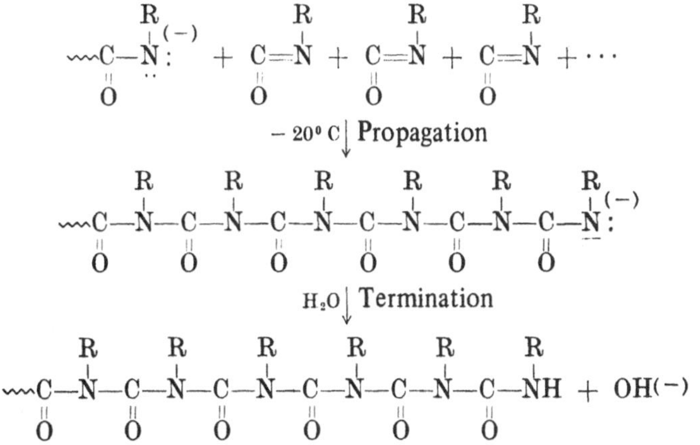

$$\sim\!\!\!\sim C-\overset{R}{\underset{O}{N}}\!\!:^{(-)} + \overset{R}{\underset{O}{C}}\!\!=\!\!N + \overset{R}{\underset{O}{C}}\!\!=\!\!N + \overset{R}{\underset{O}{C}}\!\!=\!\!N + \cdots$$

$$-20°\,C \downarrow \text{ Propagation}$$

$$\sim\!\!\!\sim C-\overset{R}{N}-C-\overset{R}{N}-C-\overset{R}{N}-C-\overset{R}{N}-C-\overset{R}{N}-C-\overset{R}{N}\!\!:^{(-)}$$

$$H_2O \downarrow \text{ Termination}$$

$$\sim\!\!\!\sim C-\overset{R}{N}-C-\overset{R}{N}-C-\overset{R}{N}-C-\overset{R}{N}-C-\overset{R}{N}-C-\overset{R}{N}H + OH^{(-)}$$

At higher temperatures one obtains not the linear polymer, but rather a cyclic trimer.

2225 The Polymerization of Cyclic Ethers

Cyclic ethers (such as ethylene oxide and tetrahydrofuran) and oxacyclobutane derivatives (such as, 3,3-bis-chloromethyl-oxacyclobutane) can be polymerized by initiators commonly used in cationic polymerization: various BF_3 derivatives, $H[BF_4]$, $H[BF_3OH]$, $[(C_2H_5)_3O] BF_4$, and complexes of $AlCl_3$, $FeCl_3$, $SnCl_4$, and $TiCl_4$.

Initiation:

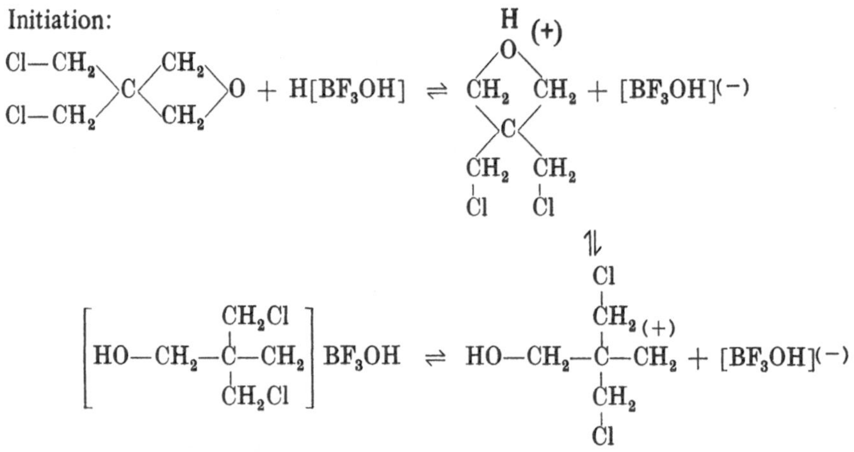

$$
\begin{array}{c}
Cl{-}CH_2 \\
\quad\quad\quad\diagdown C \diagup \\
Cl{-}CH_2
\end{array}
\begin{array}{c}
CH_2 \\
\diagdown O \\
CH_2
\end{array}
+ H[BF_3OH] \;\rightleftharpoons\;
\underset{\underset{\displaystyle Cl \quad\; Cl}{CH_2 \;\; CH_2}}{\overset{\displaystyle \overset{H}{O}\,(+)}{CH_2 \;\; CH_2}}\;
\;C\;
+ [BF_3OH]^{(-)}
$$

$$
\left[HO{-}CH_2{-}\underset{\underset{\displaystyle CH_2Cl}{\displaystyle CH_2Cl}}{\overset{\displaystyle CH_2Cl}{C}}{-}CH_2 \right] BF_3OH \;\rightleftharpoons\; HO{-}CH_2{-}\underset{\underset{\displaystyle Cl}{\displaystyle CH_2}}{\overset{\displaystyle \overset{Cl}{CH_2\,(+)}}{C}}{-}CH_2 + [BF_3OH]^{(-)}
$$

Propagation:

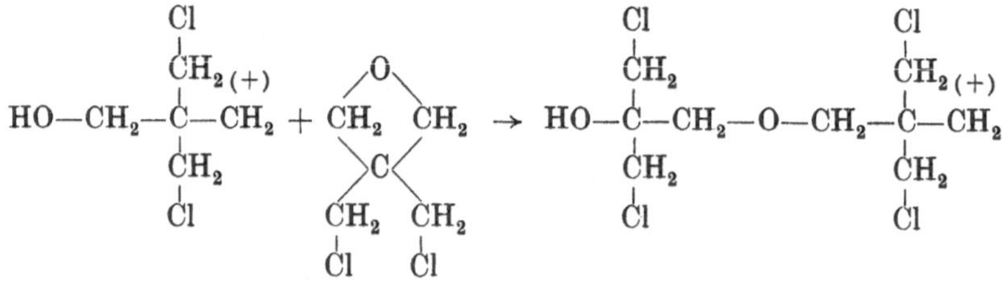

$$
HO{-}CH_2{-}\underset{\underset{\displaystyle Cl}{\displaystyle CH_2}}{\overset{\displaystyle \overset{Cl}{CH_2\,(+)}}{C}}{-}CH_2 +
\begin{array}{c}
O \\
\diagup \;\; \diagdown \\
CH_2 \;\; CH_2 \\
\diagdown C \diagup \\
CH_2 \;\; CH_2 \\
| \quad\;\; | \\
Cl \quad\; Cl
\end{array}
\rightarrow
HO{-}\underset{\underset{\displaystyle Cl}{\displaystyle CH_2}}{\overset{\displaystyle \overset{Cl}{C}}{C}}{-}CH_2{-}O{-}CH_2{-}\underset{\underset{\displaystyle Cl}{\displaystyle CH_2}}{\overset{\displaystyle \overset{Cl}{CH_2\,(+)}}{C}}{-}CH_2
$$

Termination:

$$
\sim\!\!\sim\!\!O{-}CH_2{-}\underset{\underset{\displaystyle Cl}{\displaystyle CH_2}}{\overset{\displaystyle \overset{Cl}{CH_2\,(+)}}{C}}{-}CH_2 + OH^{(-)} \rightarrow \sim\!\!\sim\!\!O{-}CH_2{-}\underset{\underset{\displaystyle Cl}{\displaystyle CH_2}}{\overset{\displaystyle \overset{Cl}{CH_2}}{C}}{-}CH_2{-}OH
$$

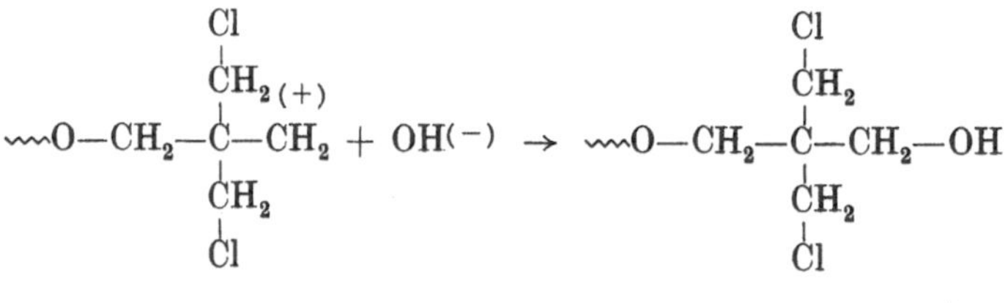

The above reaction is used for the formation of plastics with excellent properties (Penton).

The cationic polymerization of other cyclic ethers can also be formulated according to the same scheme. Only the 6-membered ring ether, oxacyclohexane, does not polymerize.

It is not simple to prepare polymers with high molecular weight (over 100,000) from cyclic ethers by cationic polymerization. As with most polymerizations, the starting reaction is an equilibrium reaction (dissociation-equilibrium), and the chain length depends on the equilibrium concentrations, because the reverse reaction is also a termination reaction (combination of the ions to neutral complexes). The equilibrium can be made more favorable toward the ions by the use of polar solvents where one obtains polymers with higher molecular weights. Polar solvents suitable for cationic polymerization are chlorinated hydrocarbons, such as methylchloride, methylene chloride, chloroform, carbon disulfide, and liquid CO_2 and SO_2. To prevent chain transfer, one has to polymerize at low temperatures ($-20°C$ to $-100°C$).

In addition to the polymerization by acid catalysts (Friedel-Crafts catalysts), ethylene oxide can also be polymerized by alcohols [phenols, amines, and mercaptans (poly-addition)].

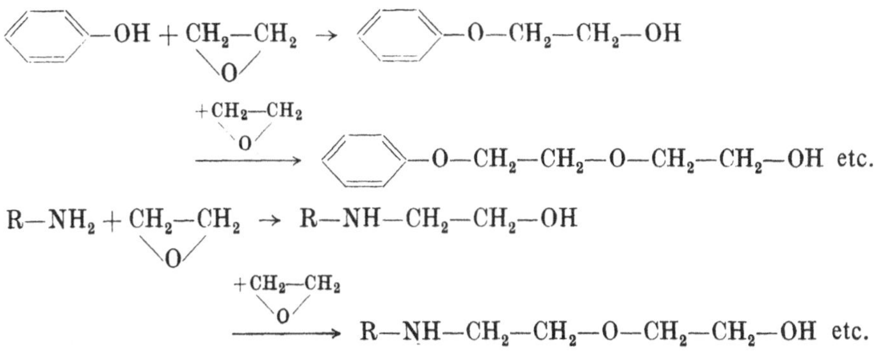

$$\langle\!\!=\!\!\rangle\!-OH + CH_2\!-\!CH_2 \xrightarrow{\ \ } \langle\!\!=\!\!\rangle\!-O\!-\!CH_2\!-\!CH_2\!-\!OH$$
$$\underset{O}{\diagdown\!\diagup}$$

$$\xrightarrow[O]{+CH_2-CH_2} \langle\!\!=\!\!\rangle\!-O\!-\!CH_2\!-\!CH_2\!-\!O\!-\!CH_2\!-\!CH_2\!-\!OH \ \ etc.$$

$$R\!-\!NH_2 + CH_2\!-\!CH_2 \xrightarrow{\ \ } R\!-\!NH\!-\!CH_2\!-\!CH_2\!-\!OH$$
$$\underset{O}{\diagdown\!\diagup}$$

$$\xrightarrow[O]{+CH_2-CH_2} R\!-\!NH\!-\!CH_2\!-\!CH_2\!-\!O\!-\!CH_2\!-\!CH_2\!-\!OH \ \ etc.$$

This reaction is called an ethoxylation. Each reaction step consists of the addition of an alcohol (the first step consists of the addition of a phenol or an amine) to ethyleneoxide. In every reaction a new alcohol (polyether with OH end groups) is formed, which can again add fresh ethyleneoxide. Thus this is a special case of polyaddition, similar to lactam polymerization with water, or amino acids, or the polymerization of N-carboxyanhydrides with amines.

Neither through cationic polymerization nor through ethoxylation of phenols or amines can one obtain polyethyleneoxides with molecular weights over 50,000. Polyethyleneoxide with molecular weights of 10^6 and greater, can be prepared, however, through anionic polymerization of ethyleneoxide using metal alcoholates, aluminum alkyls, and especially $SrCO_3$ or $CaCO_3$, as initiators. The carbonates must be prepared by precipitation of purest hydroxide solutions with CO_2 (S.N. Hill et al.). The initiation reaction of this relatively recent polymerization is still not completely clear.

2226 Polyformaldehyde

Polyformaldehyde can be prepared from pure formaldehyde both by polyaddition (polycondensation) and by anionic or cationic polymerization. The more pure the formaldehyde, the higher the molecular weights which can be obtained.

Anionic Polymerization of Formaldehyde

Anionic polymerization of formaldehyde can be initiated by amines (for example, butylamines or phosphines) and occurs rapidly and with strong heat formation even at $-70°C$ (in toluene).

Initiation:

$$R-\overset{\displaystyle R}{\underset{\displaystyle R}{\overset{|}{\underset{|}{P}}}}: \; + \; CH_2O \; \rightarrow \; \overset{\oplus}{R_3P}-CH_2-O:\overset{\ominus}{}$$

Chain growth:

$$\overset{\oplus}{R_3P}-CH_2-O:\overset{\ominus}{} \; + nCH_2O \; \rightarrow \; \overset{\oplus}{R_3P}-CH_2-O-CH_2-O\sim\sim\sim CH_2-O:\overset{\ominus}{}$$

Chain termination: $\downarrow$ + HOH

$$\overset{\oplus}{R_3}\overset{\displaystyle OH}{\overset{|}{P}}-CH_2-O-CH_2-O\sim\sim\sim CH_2-OH$$

With amines and traces of water or alcohol, the anionic polymerization of formaldehyde can be expressed as follows:

Initiation:

$$R-NH_2 + HOH \rightleftharpoons R-N\overset{\oplus}{H_3} \; OH \rightleftharpoons R-N\overset{\oplus}{H_3} \; + \; OH\overset{\ominus}{}$$

$$HO\overset{\ominus}{} \; + \; CH_2=O \; \rightarrow HO-CH_2-O:\overset{\ominus}{}$$

$$\underset{\displaystyle R-N\overset{\oplus}{H_3}}{} \qquad\qquad\qquad \underset{\displaystyle R-N\overset{\oplus}{H_3}}{}$$

Chain growth:

$$HO-CH_2-O:\overset{\ominus}{} + nCH_2O \; \rightarrow \; HO-CH_2-O-CH_2-O\sim\sim\sim CH_2O:\overset{\ominus}{}$$

$$\underset{\displaystyle R-N\overset{\oplus}{H_3}}{} \qquad\qquad\qquad\qquad\qquad \underset{\displaystyle R-N\overset{\oplus}{H_3}}{}$$

Chain termination: $\downarrow$ + HOH

$$HO-CH_2-O-CH_2-O\sim\sim\sim CH_2-OH + RNH_3OH$$

With triphenylphosphine as the initiator, one finds phosphorus in the polymer. One therefore assumes that the first reaction sequence is the correct one (zwitter ion mechanism). With amines, the polymerization probably runs according to the second reaction scheme.

If one uses carefully purified formaldehyde, this reaction yields molecular weights up to an order of magnitude of 10^6. One obtains the polyformaldehyde as a white powder, which always smells of formaldehyde, because formaldehyde is split off from the hemiacetal chain ends. This can be prevented, if one reacts the hemiacetal OH group by esterification or etherification:

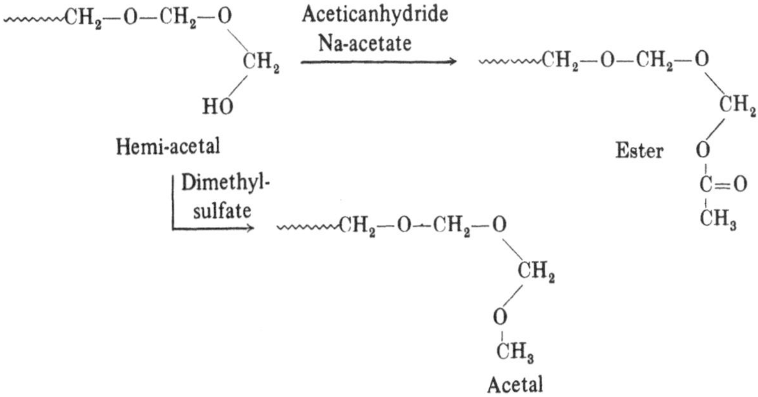

The polyformaldehyde, which is stabilized in this or some similar fashion, can be transformed to transparent and hard plastics with high mechanical strength (Delrin, Celcon).

Cationic Polymerization of Trioxane

With cationic polymerization, which has been known for a long time (first studied by H. Staudinger), one starts with cyclic trioxymethylene (trioxane), which can be purified easily by recrystallization. This reacts with BF_3, perchloric acid, or acetylperchloride, and produces polyformaldehyde. The mechanism is the same as that in the cationic polymerization of ordinary cyclic ethers and vinyl compounds, which also can form copolymers with trioxane. For example, ethylene oxide, propylene oxide, styrene oxide, β-propiolactone, γ-butyrolactone, dioxolane, ethylenecarbonate, and styrene (Kern and Jaacks).

Initiation:

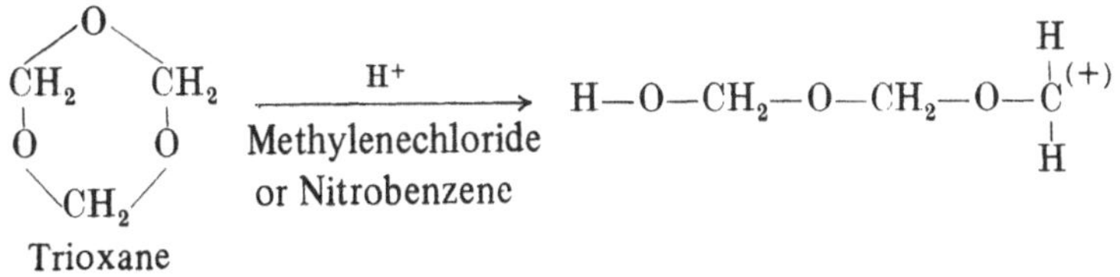

Propagation:

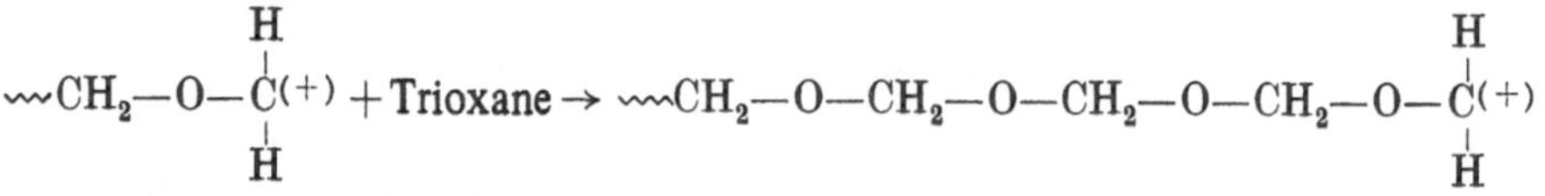

$$\sim CH_2-O-\overset{\underset{\displaystyle H}{|}}{\underset{\underset{\displaystyle H}{|}}{C}}(+) + \text{Trioxane} \rightarrow \sim CH_2-O-CH_2-O-CH_2-O-CH_2-O-\overset{\underset{\displaystyle H}{|}}{\underset{\underset{\displaystyle H}{|}}{C}}(+)$$

Chain transfer with ethers or acetals:

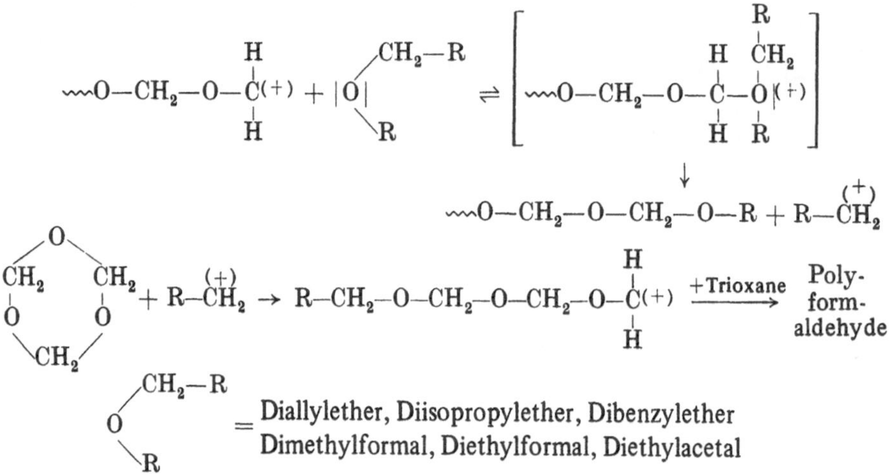

$$\sim O-CH_2-O-\overset{\underset{\displaystyle H}{|}}{\underset{\underset{\displaystyle H}{|}}{C}}(+) + |\overset{\displaystyle CH_2-R}{\underset{\displaystyle R}{O}}| \rightleftharpoons \left[\sim O-CH_2-O-\overset{\underset{\displaystyle H}{|}\overset{\displaystyle R}{|}}{\underset{\underset{\displaystyle H}{|}\underset{\displaystyle R}{|}}{C}}-O(+) \right]$$

$$\downarrow$$

$$\sim O-CH_2-O-CH_2-O-R + R-\overset{(+)}{CH_2}$$

$$\overset{\displaystyle O}{\underset{\displaystyle \underset{CH_2}{\overset{CH_2 \quad CH_2}{\underset{O \qquad O}{}}}}{}} + R-\overset{(+)}{CH_2} \rightarrow R-CH_2-O-CH_2-O-CH_2-O-\overset{\underset{\displaystyle H}{|}}{\underset{\underset{\displaystyle H}{|}}{C}}(+) \xrightarrow{+\text{Trioxane}} \begin{array}{l}\text{Poly-}\\\text{form-}\\\text{aldehyde}\end{array}$$

$$O \overset{CH_2-R}{\underset{R}{\diagdown}} = \begin{array}{l}\text{Diallylether, Diisopropylether, Dibenzylether}\\\text{Dimethylformal, Diethylformal, Diethylacetal}\end{array}$$

Termination:

$$\sim O-CH_2-O-\overset{\underset{\displaystyle H}{|}}{\underset{\underset{\displaystyle H}{|}}{C}}(+) \xrightarrow{\text{OH}(-)} \sim O-CH_2-O-CH_2-OH$$

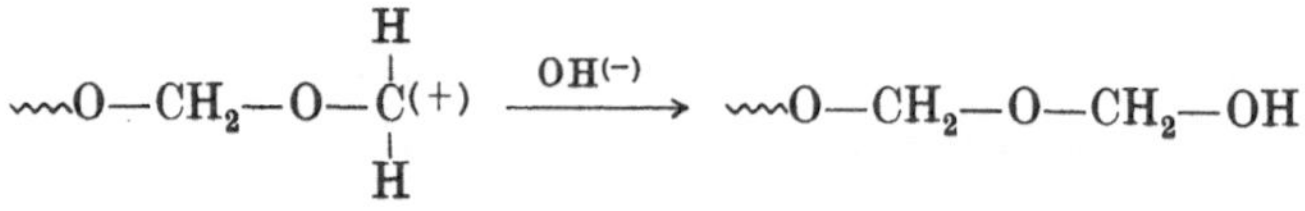

By copolymerization of trioxane with small amounts of cyclic ethers, such as ethylene oxide, propylene oxide, or dioxolane, it is possible to interrupt the regular sequence of $-CH_2-O-$ groups in smaller or larger intervals by $-CH_2-CH_2-O-$ groups. If such copolymers are heated, there is a thermal depolymerization from the hemiacetal chain ends until a structural unit with two CH_2 groups is reached. At that point there is no longer an acetal structure at the chain end, but a much more thermally stable ether-alcohol group:

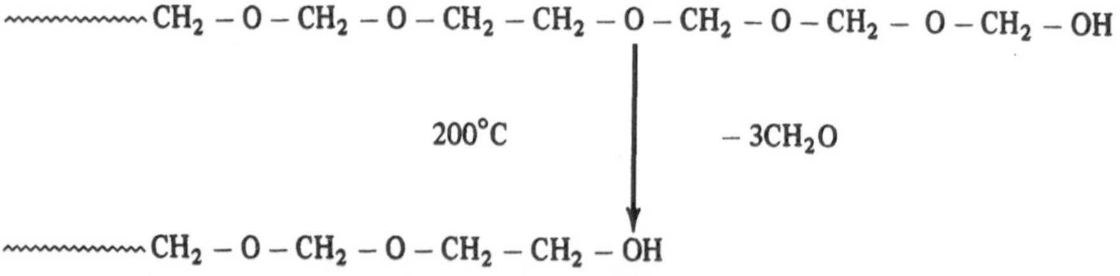

$$\sim\sim\sim CH_2 - O - CH_2 - O - CH_2 - CH_2 - O - CH_2 - O - CH_2 - O - CH_2 - OH$$

$$200°C \qquad\qquad \Big\downarrow \qquad -3CH_2O$$

$$\sim\sim\sim CH_2 - O - CH_2 - O - CH_2 - CH_2 - OH$$

This possibility of stabilizing the polymer against degradation from the chain ends is used to produce polyformaldehyde industrially. The copolymer is passed through an extruder with vacuum chambers, and stabilization and granulation occur in one step.

Both types of polymerization, the anionic (Delrin) and the cationic (Celcon, Hostaform), are used in industry. It is also known that by gamma-ray irradiation, polyformaldehyde of high molecular weights can be prepared from trioxane (Okamura).

It was recently found that also polymerization of formalhehyde in aqueous phase leads to polyformaldehyde with a high molecular weight, provided the reaction is carried out at high temperature and under pressure. In that case the polymerization occurs with simultaneous crystallization, and it is assumed that the monomer molecules during polymerization are built directly into the crystal lattice.

23 ENZYMATIC SYNTHESIS

In the living organism, the synthesis of macromolecules is carried out by enzymes. The reactions are equilibrium reactions which can be recognized because sometimes the same enzymes which permit the growth of the chain can also, under different conditions, catalyze their degradation. In many cases the enzymes can be isolated, which permits carrying out of the corresponding synthesis also *in vitro*. It has even been possible to synthesize polysaccharides and nucleic acids in the absence of enzymes (G. Schramm).

With respect to their chemical nature and chain structure, enzymes are proteins, i.e., linear or branched polypeptide copolymers consisting of 20 different structural units corresponding to the 20 different natural monomeric amino acids which are arranged in the chain in an exactly defined peptide sequence which is identical in all macromolecules of the same species (compare the structure of insulin, p. 282). This statement therefore includes the fact that enzyme proteins are monodisperse in the true sense, i.e., all protein macromolecules of a certain type have exactly the same degree of polymerization. The synthesis of proteins occurs through enzymes already present in the germ cells of organisms, with the aid of certain polymer molecules (DNA) containing the information for the sequence of the amino acids (see p. 274). The enzymatic activity becomes possible through a well-defined tertiary structure of the protein chain, i.e., a definite spatial arrangement of the α-helix, which is interrupted at different but identical distances for all of the chain molecules. The tertiary structure (see also p. 30) is determined by the primary structure (i.e., the amino acid sequence in the chain) and by the possibility of hydrogen-bond bridges or S—S bridges between different chain segments. The tertiary structure of myoglobin (the carrier of oxygen in the muscle), was elucidated by X-ray structure analysis (Perutz, Kendrew).

Enzymes are automatically effective tools or instruments, which, because of their special geometrical shape, are able to seize single molecules (more precisely, only a quite special type of molecule) and to bring and hold them in a position in which a certain special reaction is optimally favored. In other words, the activation energy of enzymatic reactions is rigorously decreased by the formation of enzyme-substrate complexes as intermediates, so that in the living cell all necessary reactions are possible at about 30°C. The second consequence of the enzyme mechanism, based on the special sterical arrangement of the protein chain, is the extremely high specificity of enzymatic reactions. The relation between enzyme and substrate has been compared with the relation between lock and key. Enzymes are able to pick their substrate out of a large number of similar molecules. The same structure, which renders possible this high specificity, allows also termination of the catalytic activity through use of certain blocking agents, which can prevent entrance of the substrate molecules. These blocking agents are molecules with a chemical constitution and geometrical shape similar to that of the substrate. Thus each single step of a sequence of reactions, which are necessary to synthesize a special compound, is catalyzed by a special enzyme. This is possible only, if a very large number of different enzymes is available. And in fact, there is no deficiency of different enzymes, because of the fantastically high number of possible different tertiary structures of protein chains.

In the following chapter some examples of enzymatic polymer syntheses are described which will show the amazing efficiency of enzymes. The detailed mechanism of an enzymatic polymer synthesis is unclear, and in some cases may be even considered mysterious.

231   Synthesis of Polysaccharides

Starch is degraded by phosphorylase (an enzyme present in yeast and potatoes, and also occurring in muscle) to glucose-1-phosphate (Cori-Ester). The same enzyme is capable of forming a polysaccharide from glucose-1-phosphate, which in its properties corresponds to amylose[31] (H. Husemann):

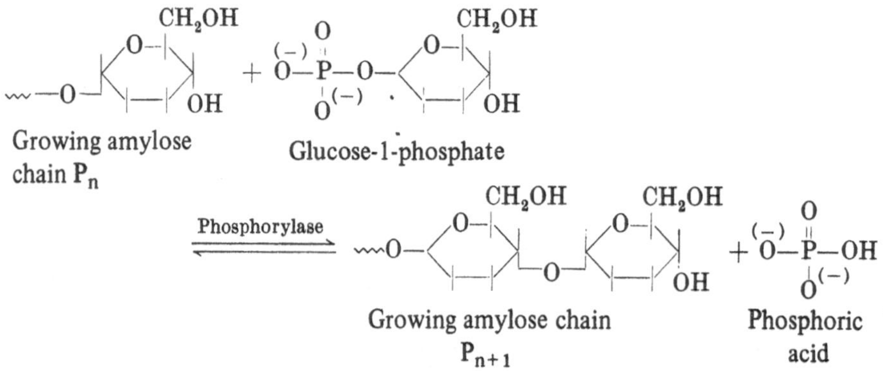

Growing amylose chain P_n

Glucose-1-phosphate

Phosphorylase

Growing amylose chain P_{n+1}

Phosphoric acid

---

[31]Starch is a mixture of amylose (a linear α-1,4-polyglucoside) and amylopectin (branched α-1,4 and α-1,6-polyglucosides). The chain structure of amylose is given in Table 12.

Phosphorylase forms only α-1,4-linkages, so that the resulting polymer is un-branched. There are other enzymes which produce 1,6-glucosidic linkages, so that a mixture of enzymes can also form branched polysaccharides, similar to glycogen.[32]

Polyglucosides (glucosans) can also be formed enzymatically from glucose. This is shown in the synthesis of pullulan with the aid of a yeast fungus: *Pullularia pullulans.* Pullulan is a water-soluble polysaccharide, where the chain molecules contain in regular alternation an α-1,6-linkage, following every three α-1,4-glucosidic linkages:

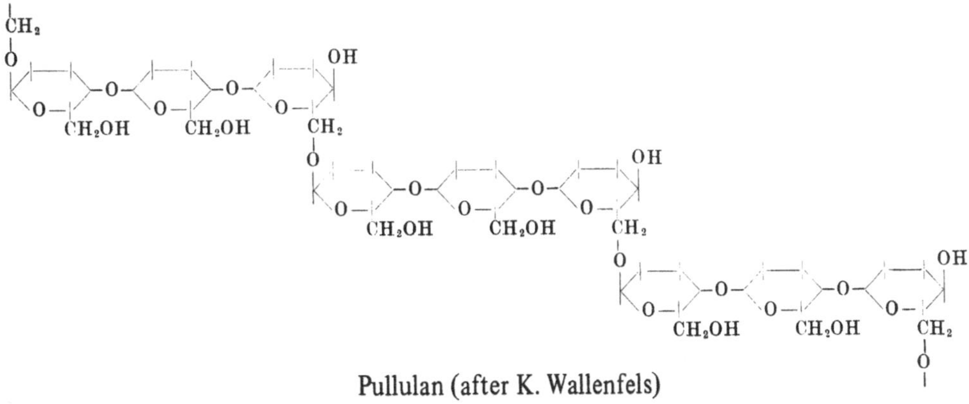

Pullulan (after K. Wallenfels)

The pullulan synthesis also occurs presumably over a phosphate as intermediary, because a direct combination of glucose to polyglucoside cannot lead to long chains in an aqueous system on thermodynamic grounds. This reaction is different from the synthesis of dextran which is possible with the aid of different enzymes, for example, those from *Leuconostoc mesenteroides.* The reaction is also carried out on a technical scale, because dextran solutions have shown themselves to be useful as blood plasma substitutes and, in a cross-linked state, as Sephadex gel for gel permeation chromatography (GPC, see p. 406). In this synthesis, the substrate is not glucose but saccharose, so that only the glucose-fructose bond has to be converted to a glucose-glucose bond. Probably the chain grows in such a manner that, with each step, first a saccharose residue enters between the enzyme and the chain, and then from this intermediary compound a molecule of fructose is split out (F. Patat).

enzyme-dextran chain (P_n) ⇌ [enzyme-saccharose-dextran chain (P_n)]

↓

enzyme-dextran-chain (P_{n+1}) + fructose

---

[32]Glycogen, the reserve polysaccharide of the liver, is more branched than amylopectin.

In a probably analogous fashion, cellulose might be synthesized from sugar by an enzyme which occurs in *acetobacter xylinum* (bacterial cellulose).

232   The Enzymatic Synthesis of Nucleic Acids

2321   Structure of DNA

Deoxyribonucleic acid (DNA) is the main component of the chromosomes. As the carrier of genetic information it plays a central role in all living matter. For this reason, DNA in the living organism is not subject to the transformation cycle where polymer chains are continuously built up and degraded side by side. (This can be followed by radioactive tagged monomers.) The amount of DNA in the different cells of the body—except the germ cells—is constant.

The constitution of DNA corresponds to that of a linear polyester, with phosphoric acid as the component corresponding to the dicarboxylic acid (the third function is not used in chain synthesis) and a substituted deoxyribose (pentose) as the component corresponding to the glycol. DNA is molecularly homogeneous, and its molecular weights are extremely high (of the order of 10^6 to 10^9). The single name DNA only applies to the chains from the point of view of their composition and the type of their structural units (one acid and four different "glycols"). However, the single name DNA is not justified with regard to the arrangement of the structural units in the chain because in nature there are as many different types of DNA as there are different types of organisms. Each organism has its own DNA whose structure and chain length are completely predetermined.

Figure 259 shows the structural principle of the DNA molecule (according to Chargaff). On hydrolysis with the enzyme deoxyribonuclease, the chain is split into its structural components, the four mononucleotides (this is shown by the dashed lines in the diagram). The nucleotides: adenosine monophosphate, guanosine-phosphate, thymidinephosphate and cytidinephosphate are further split into phosphoric acid and the corresponding nucleoside by the enzyme nucleotidease. The nucleosides are *N*-glycosidic compounds of ribose, with one of the four bases adenine (A), guanine (G), cytosine (C), and thymine (T), (compare Table 260).

In addition to DNA there are also other nucleic acids, such as ribonucleic acid (RNA), which is present in different forms in the cell (not only in the cell nucleus). RNA differs from DNA in so far as there is an OH group at the carbon atom C_2 of the ribose, and instead of the base thymine it contains the base uracil. The synthesis of RNA and its function in the enzymatic synthesis of proteins is discussed under that section (see p. 277).

X-ray structure analysis of DNA (Wilkins) has shown that the DNA molecule forms a plectonemic double helix which is also maintained in the dissolved state. (Plectonemic double helices consist of two intertwined helices. In contrast, a paranemic double helix is one which is formed by taking two single helices and moving one inside the other sideways.) The strands of the helix are formed by the polyphosphate chains, whereas the nucleoside residues fill the interior of the helix

(Figure 261). As a physical model, one can imagine a ladder made of wire which has been twisted into a helix. The most plausible explanation for this structure and its stability, from which one then immediately sees the possibilities for replication, has been given by J. D. Watson and F. H. C. Crick. The double helix of the DNA molecule (double-stranded helix) is therefore usually called the Watson-Crick helix. The explanation for the structure lies in the tendency for the

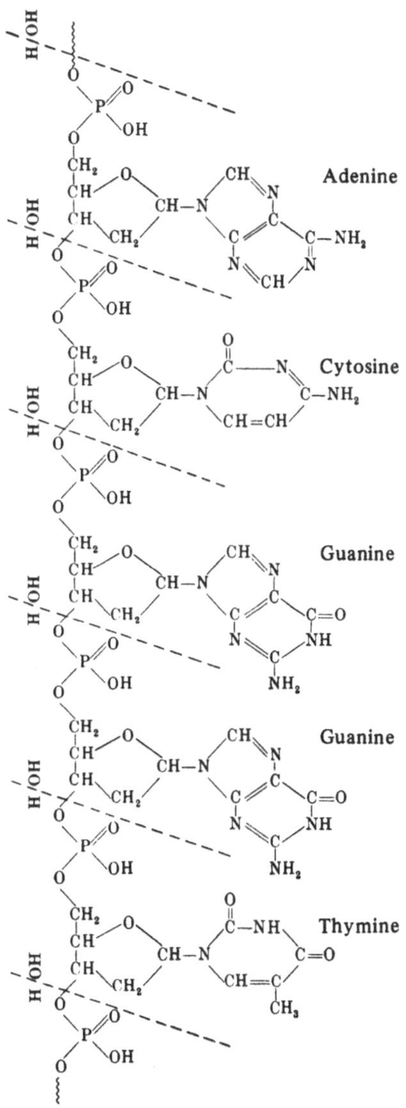

FIG. 259 – Chemical structure of DNA. (after CHARGAFF)

formation of particularly stable hydrogen bonds between the base pairs adenine-thymine and guanine-cytosine. Corresponding secondary valence bonds between adenine and cytosine, and guanine and thymine would not be as stable because of the different steric circumstances and the resulting distances between the groups forming the hydrogen bonds (Figure 262). Therefore, in the double helix, the base pairs which form the more stable hydrogen bonds face each other and form, so to say, the rungs of a twisted ladder.

2322 Synthesis of DNA

The fact that for each single strand of the DNA molecule there is always a complementary strand, and that the two then combine into a double spiral in which each base residue A is faced by a base residue T, and correspondingly each base residue G is faced by a base residue C, is brought about by the special way in which enzymatic synthesis proceeds: thus each base strand is copied by the newly growing one in such a way that along one strand the new chain grows by adding the next

TABLE 260. Chemical composition of DNA and RNA.
(Structural units and nomenclature).

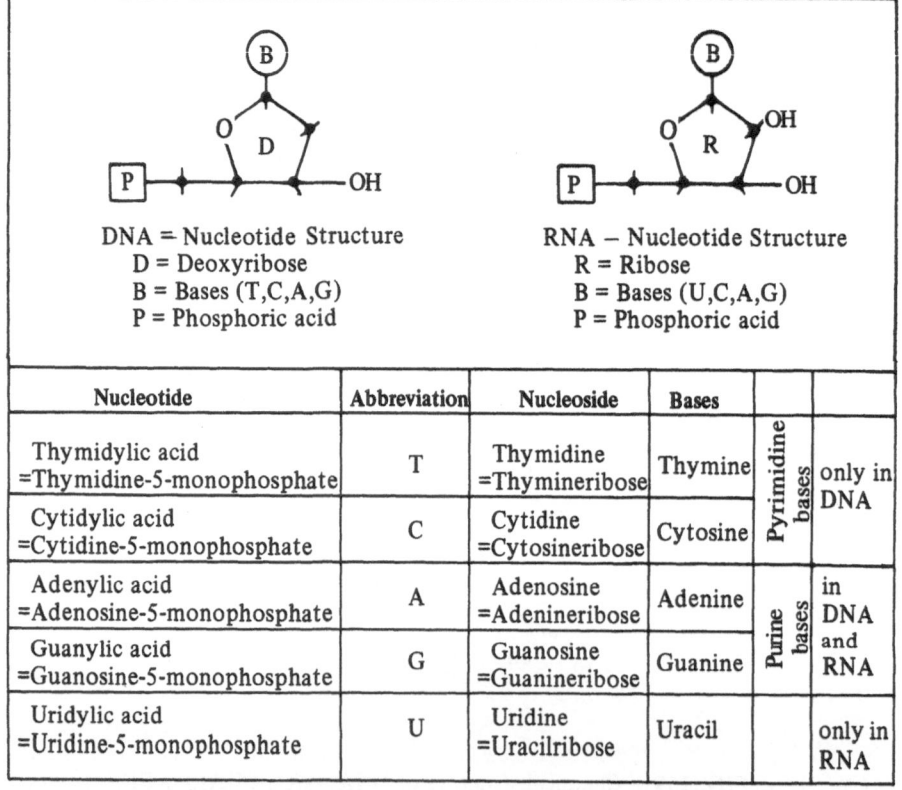

DNA = Nucleotide Structure
D = Deoxyribose
B = Bases (T,C,A,G)
P = Phosphoric acid

RNA – Nucleotide Structure
R = Ribose
B = Bases (U,C,A,G)
P = Phosphoric acid

Nucleotide	Abbreviation	Nucleoside	Bases		
Thymidylic acid =Thymidine-5-monophosphate	T	Thymidine =Thymineribose	Thymine	Pyrimidine bases	only in DNA
Cytidylic acid =Cytidine-5-monophosphate	C	Cytidine =Cytosineribose	Cytosine		
Adenylic acid =Adenosine-5-monophosphate	A	Adenosine =Adenineribose	Adenine	Purine bases	in DNA and RNA
Guanylic acid =Guanosine-5-monophosphate	G	Guanosine =Guanineribose	Guanine		
Uridylic acid =Uridine-5-monophosphate	U	Uridine =Uracilribose	Uracil		only in RNA

complementary nucleotide, which is already in the proper place for this addition step. (This is shown schematically in Figure 263.)

This type of replication (called semi-conservative replication) is possible only if the double helix separates into single strands at the locus of synthesis. There are several hypotheses to explain the mechanism of this untwisting of the double helix, but there are no experimentally confirmed results. Only the fact that such replication of DNA occurs during enzymatic synthesis has been confirmed by the well-known experiments of Kornberg. Thus, Kornberg was able to isolate an enzyme (DNA-polymerase) from *E. coli*. This enzyme is capable of synthesizing DNA from the four nucleotides, i.e., the deoxyribonucleosidetriphosphates, whereby both outer phosphate groups are split off as pyrophosphates (reaction equations in Figure 264).

The investigation of this synthesis confirmed the scheme of semi-conservative replication through the following important results:

(1) The synthesis occurs only if all four nucleosidetriphosphates are present at the same time. If one of the triphosphates is completely missing, the synthesis does not work.

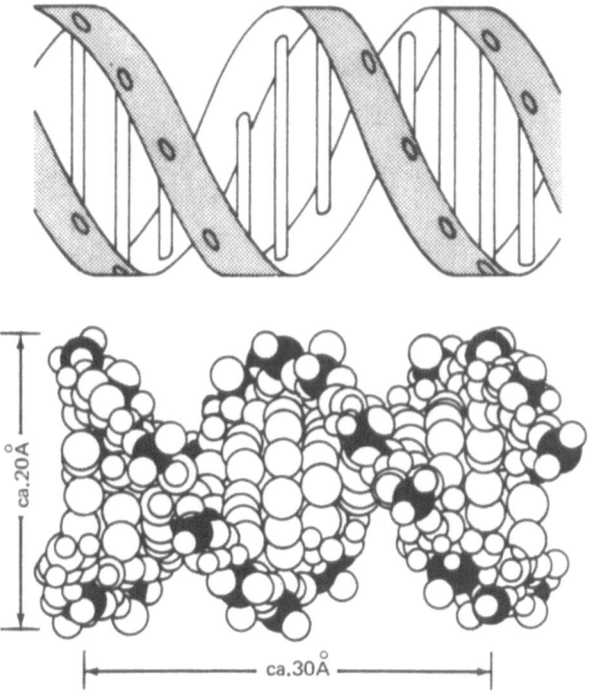

FIG. 261 – The WATSON-CRICK model of DNA. (Top: schematic representation; bottom: atomic model)

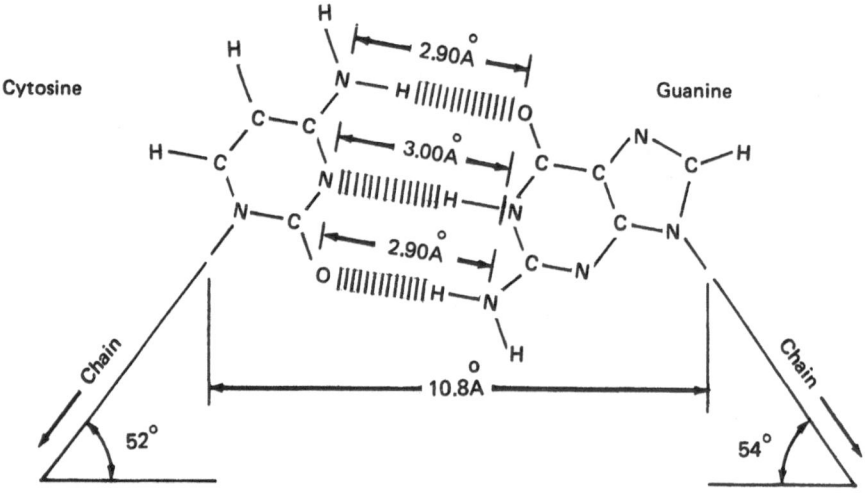

FIG. 262 — Hydrogen-bonds between complementary base-pairs in a DNA chain (after J.D. WATSON).

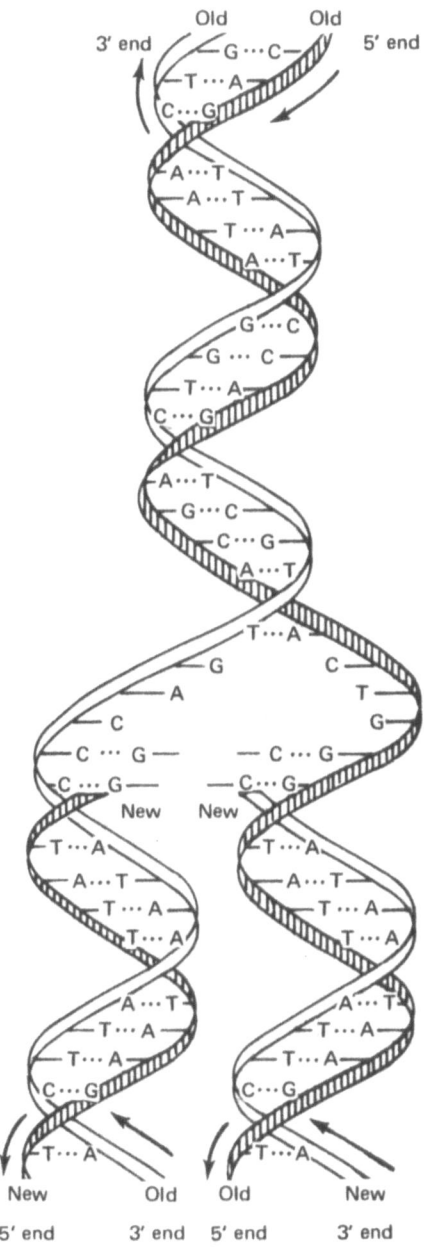

FIG. 263 – Schematic representation of DNA self – replication (after J.D. WATSON).

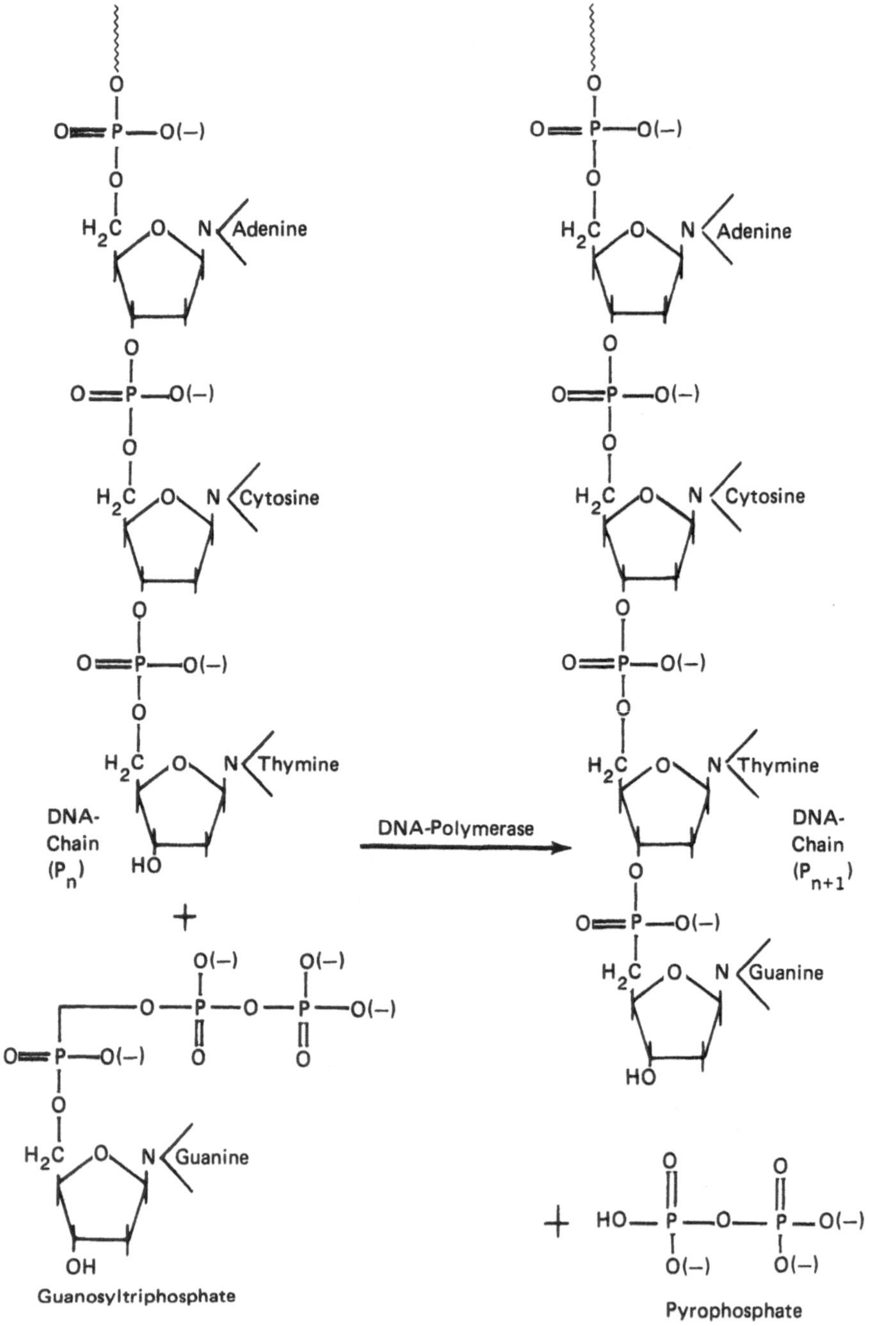

FIG. 264 — Addition of a mononucleotide to a DNA chain (KORNBERG-Synthesis)

(2) For a rapid synthesis one always needs the presence of DNA (matrix-DNA or primer-DNA), and it does not matter whether this DNA is from a plant, or of animal origin. Without the presence of DNA in the synthesis, one obtains after a long reaction time nucleic acids with alternating base sequences:

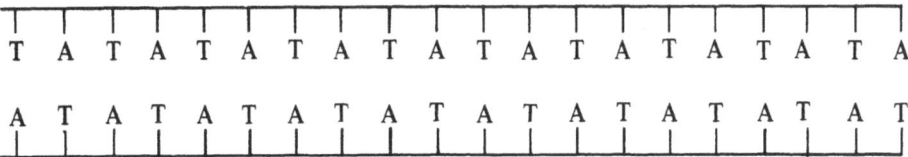

as well as nucleic acids consisting of homopolymeric strands, which are in the form of complementary double helices:

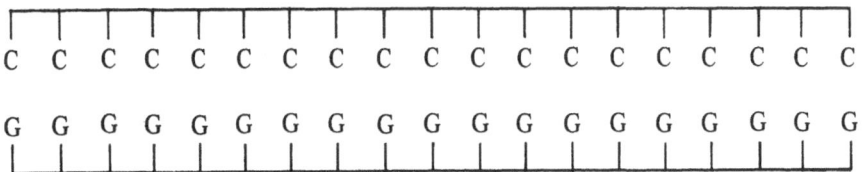

(3) The composition of the DNA formed in the Kornberg synthesis (independent of the concentration ratios of the four nucleosidetriphosphates) is always identical with the composition of the matrix-DNA (primer-DNA).

It should be noted that not only the composition of the synthetic DNA (the concentration ratio of the four bases), but also the sequence of the bases in the chain, is determined by the primer-DNA.

Analysis of Base Sequences by Means of the Kornberg Synthesis

In order to understand this analysis properly one has to realize that the DNA molecule has a sense of direction: AG is not identical with GA, nor is CT identical with TC. This is, as can be seen from Figure 266, a result of the chain structure of DNA. The phosphoric acid residue is always bound to the C-3 of one neighboring nucleoside and to the C-5 of the other neighboring nucleoside. This means that the sequence of the nucleosides must always be written in one direction only. Usually one designates the nucleosides by the first letter of the base and writes the sequence in the form shown in Figure 266. Thus, if a dinucleotide is designated by the letters ApC, this means that adenosine is esterified with phosphoric acid through its C-3 hydroxyl and cytidine through its C-5 hydroxyl.

From Figure 266 one can determine the frequency of the neighboring bases in a double-stranded DNA. Thus the sequence ApC must occur as frequently as the sequence GpT (and not as TpG!), or TpG will occur as frequently as CpA, or CpT as frequently as ApG, etc., ("complementary sequences"). The frequency of such neighboring base arrangements can be determined by using P^{32}-labeled nucleotides in the Kornberg synthesis. Actually four synthesis experiments have to be carried

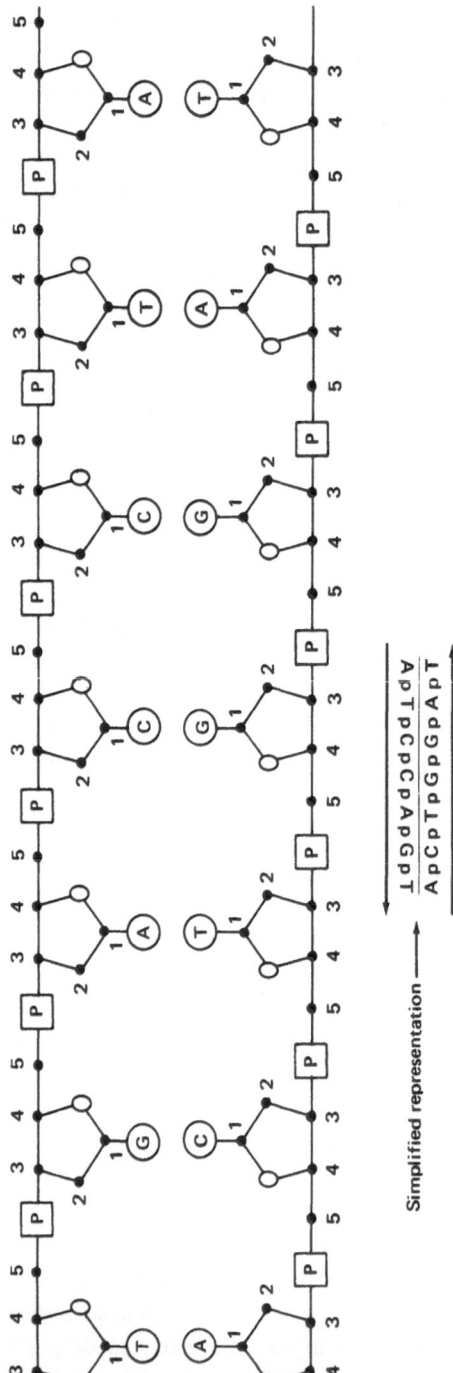

FIG. 266 — Schematic representation of the DNA molecule showing complementarity between two stands of opposite direction.

out and analyzed, and in each case only one type of nucleotide is labeled. In the nucleotidetriphosphates the radioactive phosphorus atom is attached to the OH of the C-5 of the ribose. If one assumes, for example, that in the scheme shown in Figure 266 the double helix is the primer for the Kornberg synthesis of the P^{32}-labeled thymidinetriphosphate, then one would obtain the DNA double chains shown in Figure 267. There are nucleases which can split the DNA chains between the C-5 and the phosphorus atom. In this way, as Figure 267 shows, the phosphorus can change from the nucleotide through which it was introduced into the chain during the synthesis to its chain neighbor counter to the direction of the arrow. After electrophoretic separation of the nucleotides (obtained through splitting by the nuclease) and measurement of the radioactivity, one knows the chain neighbors of the thymidine base and the relative frequency of such neighboring group arrangements (in the example in Figure 267: GpT, CpT, and ApT). If one repeats the same synthesis also with labeled guanosine-, cytidine-, and adenosine-triphosphate, one obtains in the same manner information about the frequency of the neighbor combinations of these nucleotides. The results of such a series of experiments are usually presented in the form of a table, so that the complementary sequences are shown as mirror images on each side of the diagonal (Table 268).

The mirror symmetry in this table shows that the synthetic DNA made up of complementary chains (strands) follows the structure shown in Figure 266. The fact that in the synthesis of DNA with double-stranded matrix-DNA, from the beginning of the synthesis until high conversions, the frequency of neighboring group arrangements remains unchanged, together with the proof that also this newly synthesized DNA is again acting as a primer-DNA, proves that primer-DNA and the newly synthesized DNA have the same complementary base sequence. The fact that the newly formed DNA also can act as a matrix-DNA has been shown by

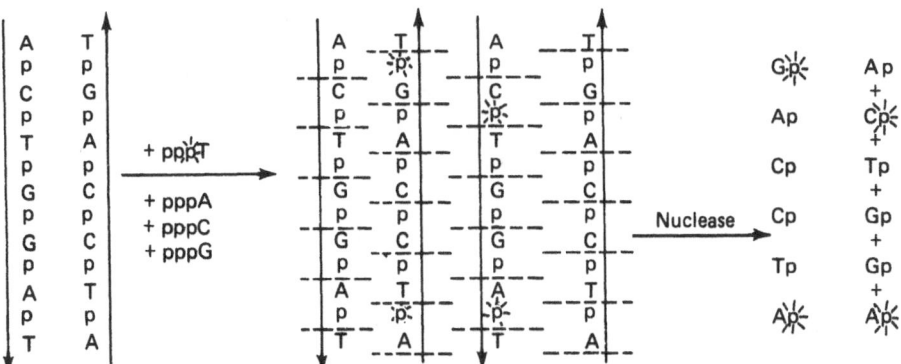

FIG. 267 – Enzymatic synthesis of DNA with a P^{32} –tagged nucleotide showing transfer of the P^{32} to its neighbor in the chain or enzymatic hydrolysis.

TABLE 268. Results of a base-pair
sequence analysis. Relative frequency of
XpY in a synthetic DNA
(after C. BRESCH)

X \ Y	T	A	C	G
A	12	24	63	65
T	26	31	45	60
G	63	45	139	90
C	61	64	90	122

experiments with single-stranded matrix-DNA. Such single-stranded DNA can be found in a few colonies of phages.

Analysis of the frequency of neighboring group arrangements with different conversions in a Kornberg synthesis with single-stranded X-phage-DNA as primer shows that at the beginning of the synthesis there is no mirror symmetry, but that after high conversions (up to 600% of the original DNA-primer) one obtains mirror symmetry. At the beginning of the synthesis, the single-stranded DNA is simply complemented to a double helix. However, later both strands are copied and the frequency table becomes mirror-symmetric.

All results of the neighboring group analysis can be explained through the principle of semi-conservative replication.

The Experiment of Meselson and Stahl

The results of Meselson and Stahl are one of the most convincing arguments for the validity of the semi-conservative replication principle. If one centrifuges DNA in a density-gradient ultracentrifuge (the method is described on p. 363) in a CsCl solution, one obtains a sharp concentration band at (I), which corresponds to the monodisperse character of the DNA macromolecules. If one repeats the same experiment with DNA which contains N^{15} instead of N^{14} (by culture of *E. coli* bacteria on $N^{15}H_4$-salts, contained in the nutrient substrate), then the band lies at (II), i.e., further away from the centrifuge axis. Then, if one takes the N^{15}-bacteria (both DNA strands contain N^{15}) back to normal N^{14}-containing nutrients, then, according to the mechanism of semi-conservative replication, after one cell division according to the scheme in Figure 269, one would expect DNA to be formed which has one strand containing N^{14} and another strand containing N^{15}. On centrifugation one would then expect only one band at a position between (I) and (II). After two cell divisions, according to Figure 269, one would expect equal amounts of light and half-heavy DNA. Exactly this result was found in the experiments of Meselson and Stahl.

Finally, Cairns, by means of autoradiography of DNA which contained along the entire chain thymine labeled with tritium (thymine-requiring mutants of *E. coli* were fed with tritiated thymine), provided a direct picture of the Y-scheme of semiconservative replication. From this last experiment it could be seen that the enzymatic synthesis of DNA occurs simultaneously on both strands. However, this means that the enzyme which is active in living cells is capable of adding the nucleotides both at the C_3-OH and at the C_5-OH of the ribose.

By means of the enzymatic DNA synthesis, which using the Kornberg enzyme can also occur in cell-free systems, one now has a clear picture of the passing-on of genetic information and characteristics from one generation to another. This is brought about by the automatic, and completely predetermined, replication of a macromolecule caused by the selective hydrogen bonding which can occur only between certain complementary base pairs.

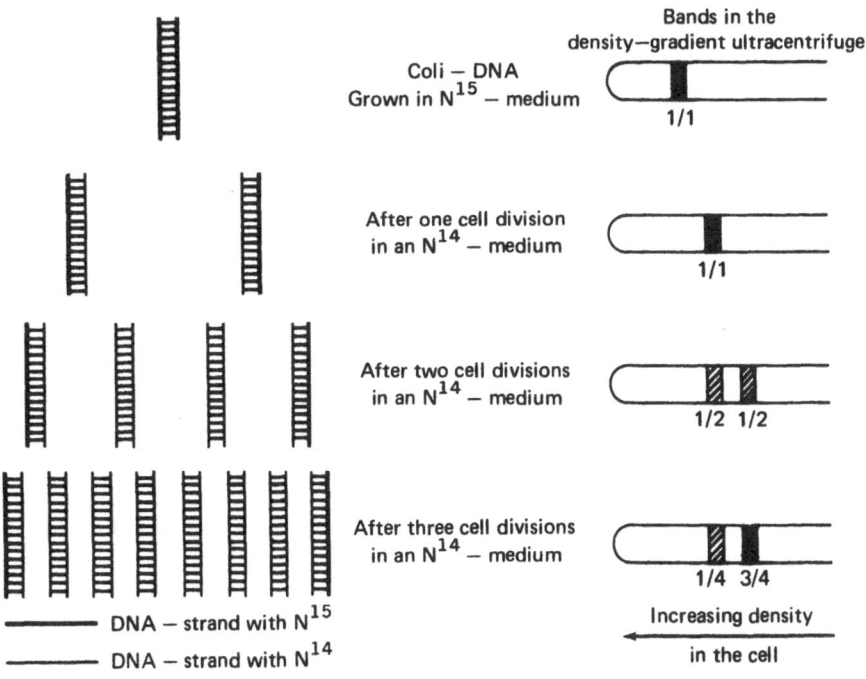

FIG. 269 — Schematic description of the MESELSON-STAHL experiment demonstrating self-replication.

2323   DNA as the Carrier of Genetic Information

Biological Significance of the Enzymatic DNA Replication

It is a well known fact that living beings pass on certain traits characteristic to their species and also many individual characteristics to successive generations by a genetic process. Organisms such as animals and plants, seen from the point of view of an architect, are extremely complicated forms. It does not seem that it will be possible at any time to construct a living being in exactly the same way as one copies a machine and expect the product to be capable of living. In nature, of course, this process occurs completely automatically during reproduction: from one body cell (with plant reproduction), or from one zygote (with sexual reproduction by union of an egg-cell and a sperm-cell), a new living being is formed through cell division and differentiation, and this new being has the same characteristics as the one from which it is formed. The cells must therefore contain a program which guides this automatic reproduction process. This program is called a genome, and it is the sum of the hereditary factors or genes which on sexual reproduction can be combined in a new way and which guide the formation of certain characteristics which are typical for that individual. The systematic examination of hereditary characteristics through cross-breeding experiments has led to the conclusion that the hereditary factors or genes (which often are only recognized as such through the appearance of mutants), are combined into coupling-groups where the genes are arranged linearly. During the miotic separation, i.e., during the segregation of the parental hereditary material, the genes are usually distributed (to the four haploid gametes) in the form of closed coupling-groups. While this is the preferred process, it does not occur in every case. Thus the coupling-group can be divided and one can have a crossover between the four homologous information copies (two from each parent) (see Figure 271). The probability that two genes are separated through crossover and combined with the corresponding genes of the partner is the greater, the more the genes are separated from each other in the coupling-group, so that the frequency of recombination becomes a measure of the distance of the genes from each other and their relative position can be determined.

The linearity of the gene arrangement in the coupling-group is determined from the additivity of the gene distances obtained by determination of the recombination frequency (they are given in recombination units = recombination frequency in percent). With microorganisms, especially phages, hundreds of independent mutants can be obtained, and therefore many crossover experiments can be carried out. It was therefore possible within one gene to localize up to three hundred different mutants on a linear "chart." One of the best known examples for such a fine-structure analysis of a gene, is the chart of the rII-region of the coli-phage T4. This was obtained by crossing of many independently derived rII-mutants and the determination of the random recombinations (see Figure 272).

Linear gene charts, as they are obtained from crossover experiments, are of course only abstract aids for the interpretation and visualization of the hereditary laws. They at least permit the conclusion that the gene, as the carrier of the hereditary information (information in the sense of instruction), in passing on this information through coupling, behaves as if it were a material structure consisting of long chains, which in their turn consist of groups or segments. These structures can then exchange smaller or larger segments with the homologous coupling-groups contributed by the partner.

Cytological research, especially the microscopic examination of processes in the cell nucleus before and during cell division, has confirmed this picture completely, and even as early as 1885, Weismann pointed out the parallel between the coupling-group and the chromosomes. The material representatives of the coupling-groups are the chromosomes of the cell nucleus, or conversely, the gene coupling-groups are models for the chromosomes, and the coupling-groups in Figure 273 represent the state of a homologous chromosome pair in the diploid state, or the metaphase (I), of the miotic division before the separation into four gametes.

The crossover combinations of characteristics are seen under the microscope as an overlap of chromatide threads (chiasmata) which are clearly visible at the beginning of the metaphase (I) of the miotic cell division.

Since every one of the gametes formed during miosis contains the entire genetic information, the genome must have been copied *before* the cell division. Since O. T.

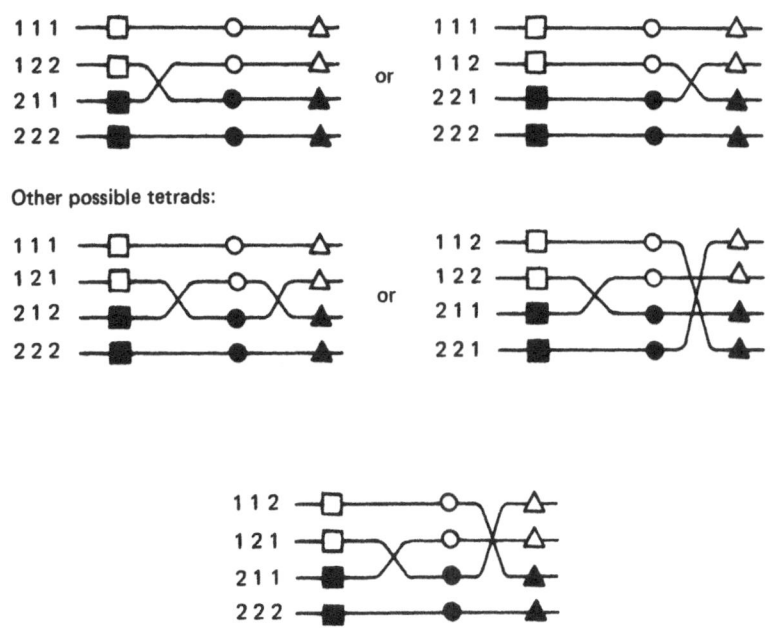

FIG. 271 — Some cross-over possibilities [involving 3 factors (genes)] (after C. BRESCH)

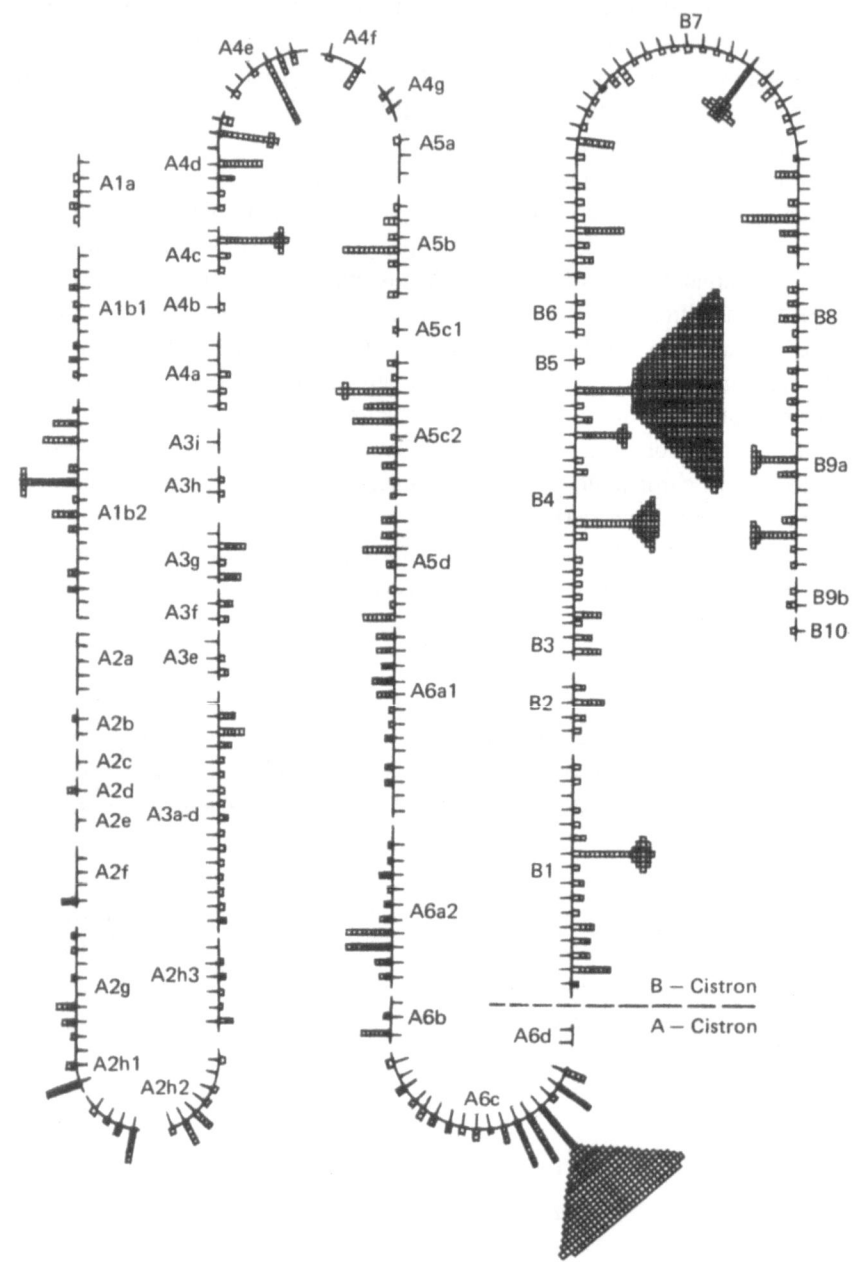

FIG. 272 — Chart of the rII-region of the coli-phage T4 (S. BENZER) (Each independently formed spontaneous mutants is represented by a small square. Mutants at the furthest edges of the chart yield approx. 6% recombinants, while the closest neighbors yield from 0.01 to 0.02% wild recombinants.

Avery (1944) has identified DNA as the transforming principle in the transformation of pneumococcus, one knows that the genetic information is deposited in the DNA molecules. Therefore one must look to the DNA synthesis as the mechanism where the genome is copied before the division into the sister cells. The gene fine-structure chart [obtained through crossover analysis (Figure 272)] with its linear arrangement of the mutation loci can be transferred to a DNA chain and one therefore obtains an abstract drawing (chart) in the form of a linear macromolecule with one-dimensional marks for certain characteristics (mutation loci).

Thus it is possible to identify genes directly with segments of the DNA molecules. How large the DNA segments are, which correspond to a particular gene,

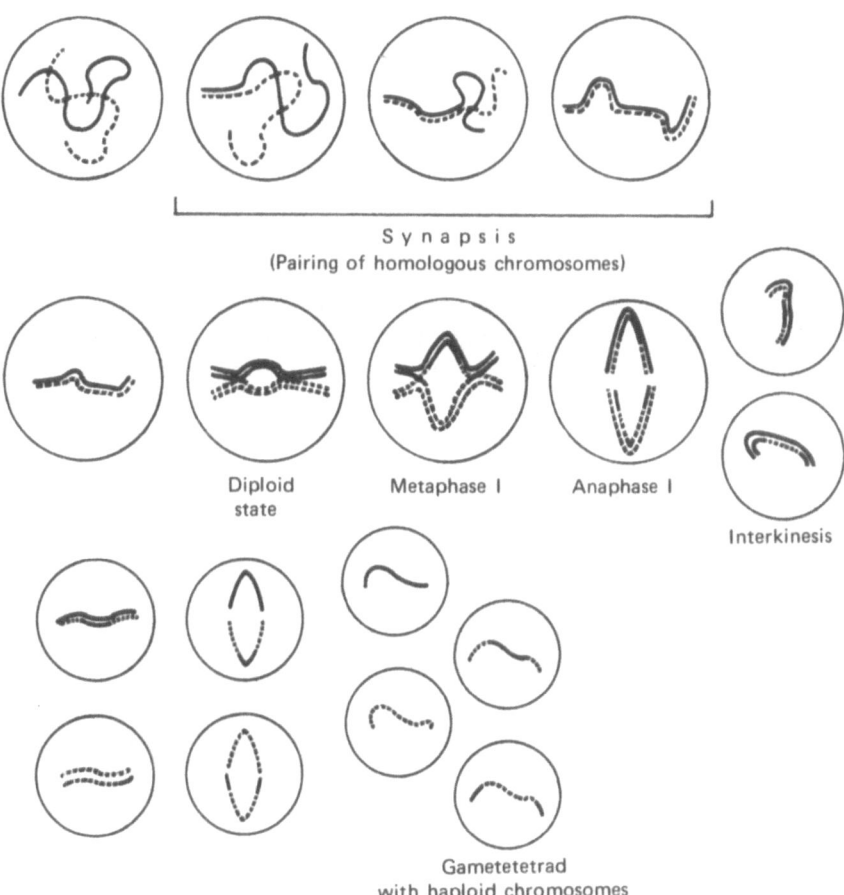

FIG. 273 — Illustration of meiotic cell-division leading to the formation of sex-cells (gametes).

depends on the definition of the gene. Contrary to the original definition of the gene as the unit for recombination and for mutation, which would lead to setting the gene equal to one nucleotide structural unit of the DNA chain, one now understands for the term gene the segment of the DNA chain which is necessary for the synthesis of an enzyme (see also "Enzymatic Protein Synthesis").

Also from the standpoint of cell division, DNA fills all the requirements for a carrier substance of the genome (sum of the genetic information). Each microscopically observable splitting of chromosomes during cell division is preceded by a semi-conservative replication of the DNA. Therefore enzymatic DNA synthesis may be considered as a propagation process on a molecular plane, a sort of elementary process of propagation, in which the instructions contained in the structure of the DNA molecule are copied by a mechanism in which both complementary spiral strands act as a matrix (negative form) for the formation of a new strand which is identical with the old complementary strand.

The Genetic Code and Its Interpretation

The basic principle of a retrievable data bank is the aperiodic sequence of certain signals (acoustic, optical, electric, magnetic, or any other form of signal which can be registered).

The only factor which is variable in the structure of the DNA chain is the sequence of the four nucleotide residues in the chain (the base sequence), and this must therefore contain the necessary instructions for the development of the organisms. The base sequence is a form of handwriting with four different characters. How can it be read? What is the form of the instructions it contains? How are these instructions carried out? The answer to these questions is one of the most imposing achievements and results of biochemical research in recent years. These questions cannot all be answered separately, because one only understands the method of reading after one understands the type of instructions and how these are carried out.

All reactions going on in the cell are controlled by the catalytic activity of the enzymes. Enzymes are proteins with quite specific steric structures. These structures permit the enzymes to react with certain molecules in the same way that an engineer or master mechanic works with the parts of his machine. The steric structure is the result of helix formation (secondary structure) and specific folding of the helices (tertiary structure), which in turn are the result of the amino acid sequences in the protein chain. The type and method of action of an enzyme are therefore determined only by the amino acid sequence in the chain, and therefore one can assume that the instructions of the DNA code will refer to the synthesis of enzymes, i.e., protein chains with certain amino acid sequences. The observation that there are defective mutants (of bacteria or yeast cells, for example), where certain enzymes are missing or are defective, i.e., that by a change in one of the genes or even in a part of a gene, the synthesis of the enzyme is disturbed, brought Beadle, Tatum, and Horowitz in 1945 to the theory known as the "one gene — one enzyme" theory, and has provided an important guideline for the subsequent work of others.

Recently, in a few cases by construction of a gene chart and designation of the mutation loci and also by análysis of the amino acid sequences of the enzyme, it was possible to show that a defective sequence in the protein chain of the enzyme corresponds to a mutation locus on the gene chart (colinearity between gene and protein chains [Yanofsky]).

If enzymatic protein synthesis is governed by the DNA molecule, then the four-character alphabet of DNA must make it possible that the four base symbols allow us to spell the names of 20 different amino acids. Said in other words, there must be a code which makes it possible to translate the DNA alphabet with its four different symbols, into a protein alphabet with 20 different symbols, just as there is a code which translates the Morse code with its two symbols (dash and dot), into our normal alphabet with its 26 letters:

A = · —	B = — · · ·	C = — · — ·
D = — · ·	E = ·	F = · · — ·
G = — — ·	H = · · · ·	I = · ·
J = · — — —	K = — · —	L = · — · ·
M = — —	N = — ·	O = — — —
P = · — — ·	Q = — — · —	R = · — ·
S = · · ·	T = —	U = · · —
V = · · · —	W = · — —	X = — · · —
Y = — · — —	Z = — — · ·	

As can be seen from the example of the Morse alphabet, this can be done by combining a number of symbols differing from one letter to the next to a new symbol, a codon. In addition, in a linear arrangement of the symbols, one introduces a further symbol, i.e., a larger space between the symbols so that the writing becomes clear. Thus · · · — — — · · · can have the following meanings: SOS, IAMS, SMDE, EIMB, VGI, EUZE, IATNI, etc. However, if space is put between the three combinations of dots and dashes, then only one interpretation becomes possible, namely SOS: · · ·   — — —   · · ·

Only if the number of symbols per codon is constant is it possible even without punctuation or spacings to read the handwriting right from the beginning. Since we must be able to represent 20 different amino acids as a combination of the nucleotides and a combination of 2 of the 4 different nucleotides only gives us 16 different combinations (which is not enough for 20 different amino acids), a system of equally long codons must contain at least three bases, (i.e., nucleotides) per codon.

On the basis of experimental results during the last few years one can describe the genetic code as follows: the genetic code is a punctuation-less triplet code. Each code word or codon (nucleotide triplet) corresponds only to a single amino acid, which, if that codon is present at the place where protein synthesis takes place, is allowed to add to the growing protein chain. Conversely, however, for nearly every amino acid there are several codons. This means that the code is strongly degenerate. In the DNA chain the codons are arranged next to each other without overlap. The beginning and end of a gene (i.e., the beginning and end of a particular protein synthesis) are designated by special codons. According to all known investigations with all organisms from the virus to the mammal, the genetic code is the same (the genetic code is universal).

A complete code-lexicon, with a precise designation of the position of the nucleotides within the triplets, has been constructed in recent years (see Figure 276). This code-lexicon is based on the work of Nirenberg, Matthaei, Ochoa, Khorana, and others, which has first permitted an assignment of the triplets to the amino acids, and later also the determination of the nucleotide sequence within the triplet. This work involved (1) planned mutation experiments (deamination with HNO_2), especially with tobacco mosaic virus, and analytical examination of the resulting changes in the amino acid sequence in the protein, and (2) synthesis of polypeptides with the help of RNA preparations with simple nucleotide sequences

Nucleotide 1 in the Triplet	Nucleotide 2 in the Triplet				Nucleotide 3 in the Triplet
	U	C	A	G	
U	Phe	Ser	Tyr	Cys	U
	Phe	Ser	Tyr	Cys	C
	Leu	Ser	−End	Cys(?)	A
	Leu	Ser	−End	Try	G
C	Leu	Pro	His	Arg	U
	Leu	Pro	His	Arg	C
	Leu	Pro	GluNH	Arg	A
	Leu	Pro	GluNH	Arg	G
A	Ileu	Thr	AspNH	Ser	U
	Ileu	Thr	AspNH	Ser	C
	Ileu	Thr	Lys	Arg	A
	Meth(start**)	Thr	Lys	Arg	G
G	Val	Ala	Asp	Gly	U
	Val	Ala	Asp	Gly	C
	Val	Ala	Glu	Gly	A
	Val	Ala	Glu	Gly	G

*) UAA and UAG are often called "nonsense-codons". In E.-coli they mean: "end of the chain".

**) Formyl-methionine is split off oafter the protein synthesis enzymatically.

FIG. 276. Genetic code lexicon

(UUUUUUUUU, ATATATATATAT, etc.), and the determination of the resulting polypeptides.

The course followed by these workers is very similar to that followed in the subsequent discussion of enzymatic protein synthesis and therefore represents a confirmation of the ideas one has formed about this synthesis.

233 Enzymatic Protein Synthesis

In the discussion of the chemical constitution of DNA it was already pointed out that, in addition to DNA in the cell, there is also a second nucleic acid, namely RNA, which has the same chain structure as DNA and which is also capable of assuming the functions of DNA (as can be seen from the fact that RNA occurs alone in certain viruses and phages). RNA differs from DNA in terms of its chemical constitution only through its OH group at the C-2 atom of the ribose, which is missing in DNA, and by the appearance of the base uracil instead of thymine. However, there are three completely different types of RNA, which are characterized by their different functions during the protein synthesis in the cell. The largest fraction of the RNA in the cell is the *ribosome*-RNA (85%). Approximately 10% of the total RNA in the cell is *transfer*-RNA and the rest is *messenger*-RNA.

The transfer-RNA (t-RNA, often also called soluble RNA) has a comparatively low molecular weight (about 25,000), which corresponds to approximately 70 nucleotide units. As one can see from X-ray data, there are parts of the t-RNA chain which are in the form of a double helix. However, since the chain length remains constant when the helix untwists, one has to assume that the chain is bent like a hairpin and is then spiralized. At one chain end there is the base sequence CCA and at the other chain end (the C-5 end of the chain) (compare Figure 266) there is usually a G. In the chain itself, in addition to the four nucleotides A, U, C, G, there are also small amounts (only a few percent) of nucleotides of other bases (for example, 1-methylguanine).

The second type of RNA, because of its function in protein synthesis, is called messenger-RNA (m-RNA). It is present in small amounts in the cell (about 3%) and always remains closely related to the DNA in the cell. It has much smaller and widely scattering molecular weights than DNA, (of the order of 100,000 to 800,000), but the nucleotide composition and nucleotide sequence of the RNA are identical with that of the DNA in the cell in which it is formed, or more specifically, with one strand of this DNA. When *E. coli* cultures are infected with phages, the single-stranded DNA of the phage is first completed by complementary replication. The m-RNA of these cells does not form a hybrid (double helix) together with the single-stranded phage DNA; however it does interact with the double strands formed after infection in the *coli* cells, when these are "melted," i.e., when by heating of the aqueous solution they are denatured (the helix is untwisted). Ochoa found an enzyme which is able, even in a test tube, to form RNA provided all four nucleosidephosphates (A, C, U, and G) are present and there is DNA available as a matrix. Composition and nucleotide sequence of this synthetic RNA corresponds strictly to that of the matrix-DNA, i.e., in a cell-free system both DNA strands are copied. Nothing is known about the mechanism of the copying process. However, one can assume that it corresponds to the process of semi-conservative replication. Under the influence of the enzyme (RNA-polymerase), the double helix opens at a certain well-defined spot and the synthesis proceeds along the chain while the double helix untwists (see Figure 278).

From the observed generation time with TMV (tobacco mosaic virus) infections (about 10^5 nucleotide pairs per minute) one calculates that the number of rotations of the chain during the replication process may be 10,000 turns per minute!

The third RNA type forms the main component of the ribosomes (2/3 RNA and 1/3 protein, depending on the type of organism; sometimes 50% protein). The ribosomes are small, submicroscopic particles which are present in large numbers on the membrane system in the cytoplasm of the cell, the so-called endoplasmatic reticulum, but also in the cytoplasm itself. They consist of a larger ($S = 50$ sec^{-1})

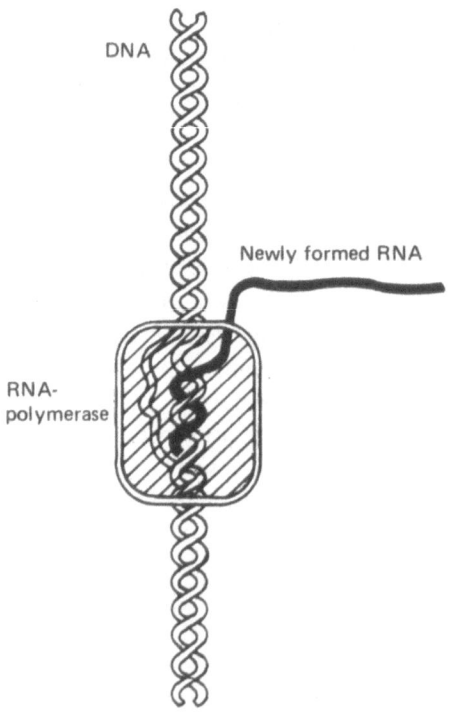

DNA

Newly formed RNA

RNA-
polymerase

FIG. 278 — Schematic representation of
the formation of a m—RNA at the DNA
double-helix (after G. SCHRAMM).

and a smaller ($S = 30 \text{ sec}^{-1}$) component [S = sedimentation constant in Svedberg units. This is a measure of the sedimentation velocity in the ultracentrifuge (see p. 362)]. The Mg^{++} concentration determines whether these parts are dissociated or aggregated. With phenol one can separate out the RNA. The RNA of the larger particles has a molecular weight which is twice as high as that of the smaller particles ($M = 1.1 \times 10^6$ vs. $M = 5.6 \times 10^5$). The inner structure of the ribosomes is still relatively unknown.

Protein synthesis requires coordination between the three RNA types and the DNA of the cell nucleus. If one homogenizes *E. coli* cells in an ultracentrifuge with such a high rate of rotation that only enzymes and t-RNA remain in solution, then on addition of C^{14}-labeled amino acids (and ATP [adenosinetriphosphate] as energy source) no further incorporation in proteins is found. If one then adds ribosomes, one first finds radioactivity on the ribosome and later in the proteins of the plasma. Protein synthesis therefore probably occurs on the ribosomes. When the course of the C^{14}-labeled amino acids was examined more closely, one found that the radioactivity (in centrifuged systems without ribosomes) is transferred to the t-RNA. As has been shown by Hoagland and Zamecnik, the first step is the formation of an anhydride from ATP and amino acids, which then with the help of enzymes (amino-acyl-RNA-synthetases) is bound to a t-RNA molecule at the C_3—OH chain end. For each t-RNA molecule, one always finds only one bound amino acid. As the following experiment demonstrates, each amino acid has its own t-RNA. If one adds ATP and an excess of an amino acid to the supernatant solution of the centrifuged *coli* cells, then only a small part of the t-RNA is bound to this amino acid. Only when another amino acid is added, another part of the t-RNA present is bound to amino acid, and so on.

The t-RNA seems to be the most intelligent partner of the whole system: it recognizes its amino acid, brings it to the locus of the synthesis, and sees that this amino acid is added to the growing protein chain as soon as the corresponding codon (nucleotide triplet) appears in or on one of the ribosomes. It was previously mentioned that when C^{14}-labeled amino acids are used, the radioactivity is transferred to the ribosomes. However, it was also shown (isotope marking with phage infection) that the m-RNA of the phage combines with the ribosomes in the cells. The ribosomes are therefore the meeting points for the t-RNA (which brings along its amino acid) on the one hand, and the m-RNA (which brings along the information for the amino acid sequence in the form of the triplet code copied from the DNA) on the other hand (see Figure 280).

t-RNA must, if this mechanism really functions, have at one place in the molecule the anticodon belonging (according to the code-lexicon) to an amino acid AX and at another place (or possibly even the same place), a region for recognition of the enzyme which catalyzes the coupling of the amino acid AX (amino-acyl-RNA-synthetase-AX). There must exist at least 20 different amino-acyl-RNA-synthetases, each one of which can couple a definite amino acid with the t-RNA which holds the complementary triplet of that amino acid. So far there has been no experimental proof for the type of structural characteristic which allows this recognition process. It is only certain that the enzymes which catalyze the coupling are involved in the choice of the correct amino acid. Once an amino acid is attached to t-RNA, then this amino acid by reduction can be transformed into a different amino acid and is incorporated into the growing protein chain at a point corresponding to the codon of the old amino acid. Thus the key RNA anticodon does not care which amino acid is at its CCA end, and it is left up to the enzyme (aminoacyl-RNA-synthetase) to see that it is the correct amino acid corresponding to the code-word. From this one realizes how important the tertiary structure of the proteins is, since it is the basis of this high degree of specificity. One also realizes how approximate the sketch is, which one presently has in order to explain protein synthesis.

After the preceding discussion, it is easy to see how the genetic code was decoded and how the code-lexicon was constructed.

Instead of m-RNA, a synthetic trinucleotide of exactly known structure was put into a system containing all amino acids. This system was cell-free and capable of enzymatic protein synthesis and consisted of a suspension or solution of ribosomes, enzymes, and adenosinetriphosphate (ATP). This did not result in a polypeptide, of

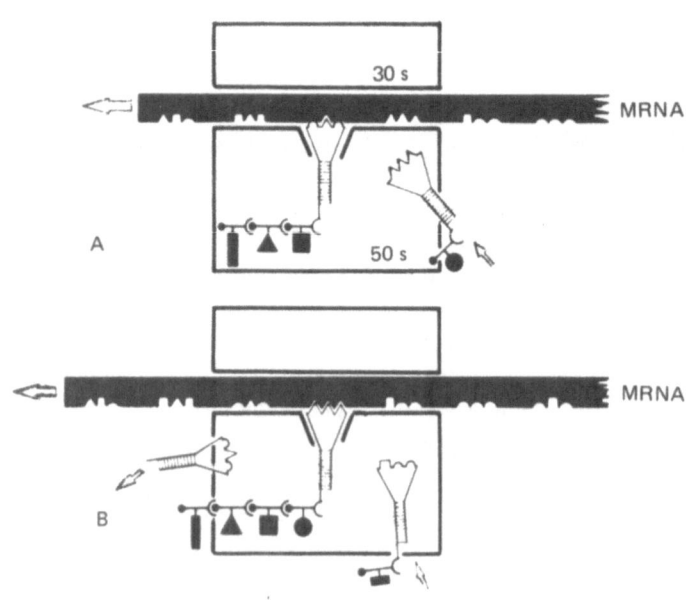

FIG. 280 — Illustration of an enzymatic protein synthesis A = before, and B = after
change of the transfer — RNA at the active site
on the ribosome after C. BRESCH.

course, but the synthetic trinucleotide was (exactly as m-RNA) absorbed by the ribosomes (see Figure 280) and combined with the t-RNA carrying its amino acid to a complex consisting of ribosome, trinucleotide, t-RNA, and amino acid which could be separated from the solution. In this way, by synthesis of all possible triplets in the form of trinucleotides of well-defined constitution, it was possible to isolate and identify the amino acid stimulated during the protein synthesis by the particular triplet used.

234 The Biosynthesis of Natural Rubber

Natural rubber, which is obtained on a large scale from the tree *Hevea brasiliensis,* is formed not by a polymerization of isoprene, as the synthetic *cis*-1,4-polyisoprene, but like the synthesis of polysaccharides and nucleic acids, by an enzymatic polycondensation under splitting off of pyrophosphate. The monomer is probably isopentenylpyrophosphate, which is formed by reaction of acetic acid over acetoacetic acid and mevalonic acid as intermediary products. Δ^3-isopentenylphosphate is polymerized to natural rubber (under splitting off of pyrophosphate) by an enzyme system, which is still present in active form in the fresh latex (F. Lynen):

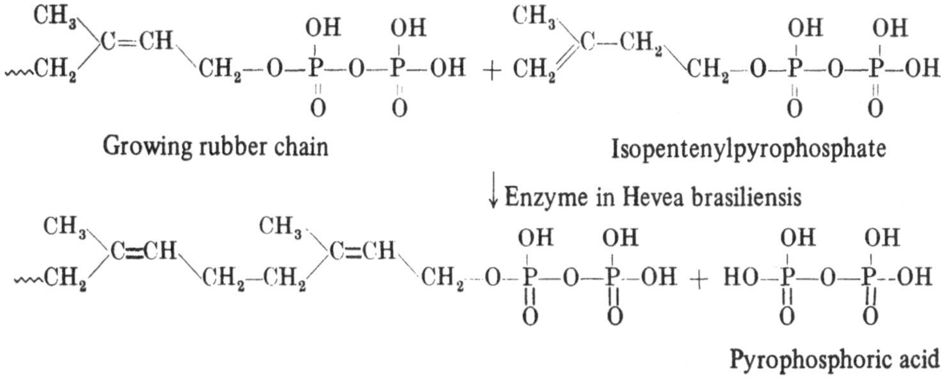

Growing rubber chain   Isopentenylpyrophosphate

↓ Enzyme in Hevea brasiliensis

Pyrophosphoric acid

24 THE STEPWISE SYNTHESIS OF PROTEINS

Whereas most industrial copolymers used as raw materials for plastics or coatings contain usually two or three, and only in a few cases as many as five or six, different monomers (because up to now there has been no necessity for systems with more monomer components), nature needs 20 different amino acids as the raw materials for the synthesis of proteins. These amino acids are the structural units of the protein chain, and they are arranged in a completely defined sequence which is identical with all the molecules of a particular protein. The first protein whose structure, including the amino acid sequence in the chain (actually there are A- and B-chains), was completely determined was insulin (F. Sanger). In the meantime a number of other proteins have been characterized with respect to their amino acid sequence and new ones are constantly being determined (sequence analysis, see p. 265).

All polymer syntheses which have been discussed up to now produce statistical or alternating copolymers, i.e., the different monomers are either randomly distributed in the chain (such as with copolymers prepared from monomer mixtures with finite r-values by radical or ionic polymerization) or the two monomers alternate regularly in the chain (such as with monomer pairs having $r_1 = 0$ and $r_2 = 0$, or nearly equal to zero). Such regularly alternating copolymers are also obtained with polyamides or polyesters from dicarboxylic acids and diamines, or glycols respectively.

Such syntheses are obviously unsuitable for the preparation of proteins with a well-defined structure. There is no other alternative than to couple one amino acid to the next in a stepwise manner in order to build up a certain sequence in the chain of a protein.

Since the work of Emil Fischer, the methods of peptide synthesis (see any organic chemistry textbook) have been constantly improved, so that in recent years it has been possible also to prepare longer peptide chains such as those present in proteins.

The synthesis of sheep insulin was published in 1963 and 1964 by two groups of workers: by H. Zahn (Aachen, Germany), and by P. G. Katsoyannis and G. H. Dixon (U.S.A.). In 1965, in China, beef insulin was synthesized and obtained in crystalline form by Niu, Wang Hsing, and others.

Insulin consists of two polypeptide chains, the A-chain and the B-chain, which are connected with each other over disulfide bridges. Sheep insulin has the following primary structure (according to Klostermann):

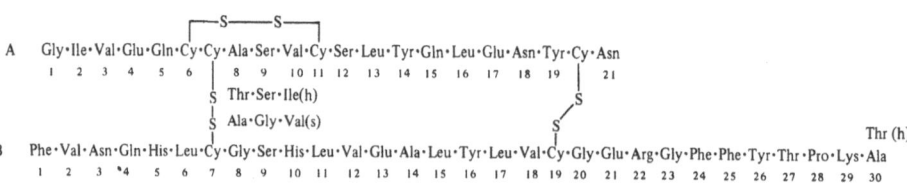

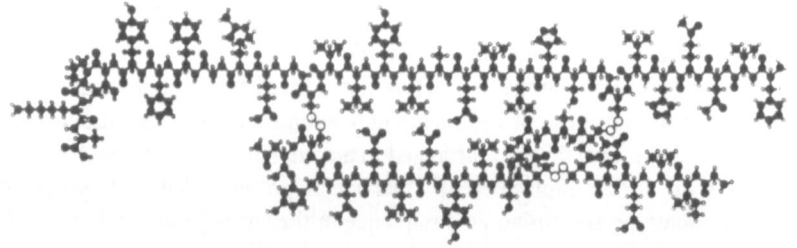

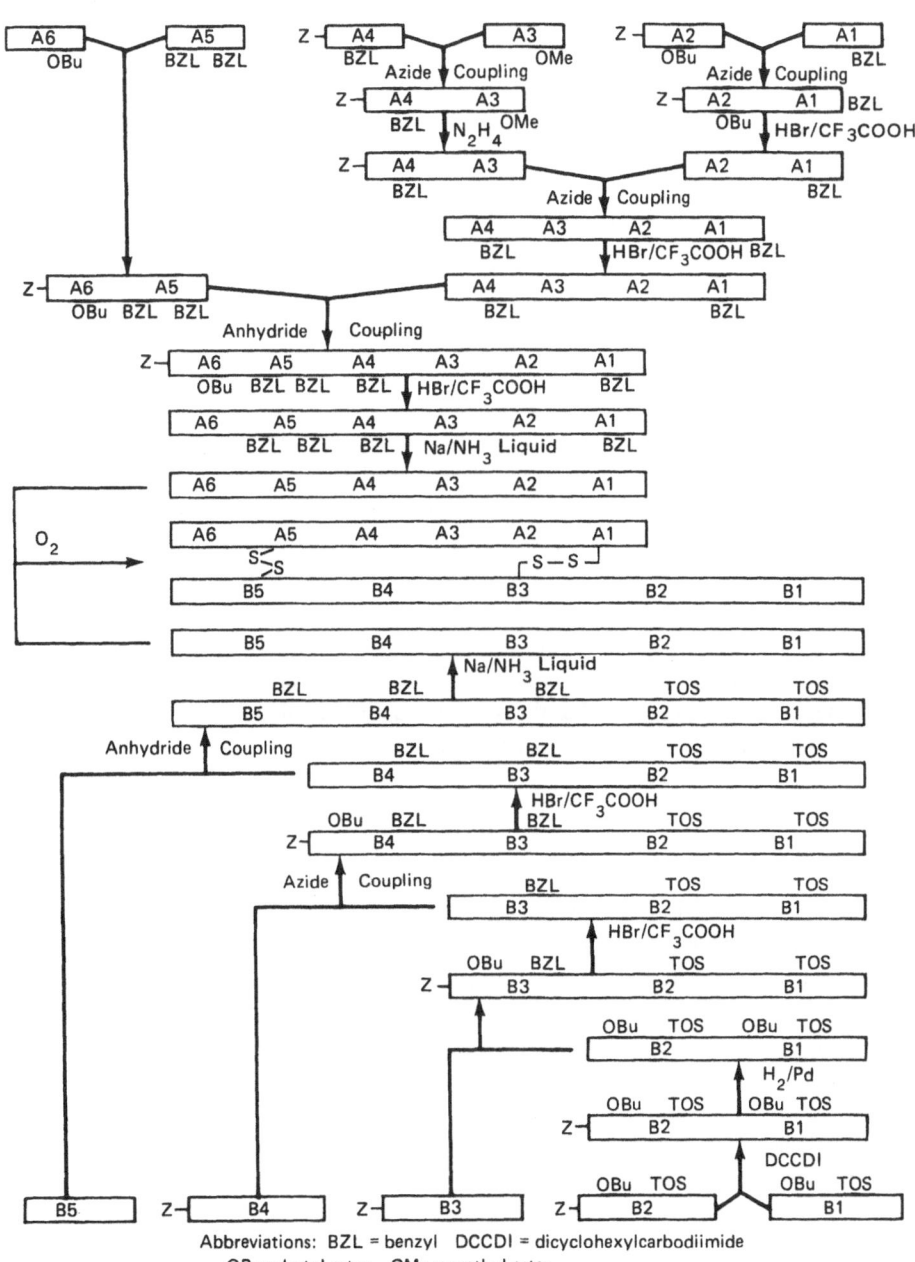

Abbreviations: BZL = benzyl   DCCDI = dicyclohexylcarbodiimide
OBu = butyl ester   OMe = methyl ester
TOS = p-toluenesulfonyl   Z = benzyloxycarbonyl
The chain-segments A1, A2, etc. and B1, B2, etc. were built-up
by step-wise chain-extension, an aminoacid at a time.

FIG. 283 — Scheme for the synthesis of insulin (H. ZAHN)

In the synthesis according to Zahn, the oligopeptides A1-A6 and B1-B5 were first built up by a stepwise increase of an amino acid residue at each step. These segments were then combined in a stepwise manner to the A-chain and the B-chain according to the plan of synthesis described in Figure 283. Finally, after removal of all the blocking groups (with sodium in liquid ammonia), the sulfur bridges were closed with oxygen. Altogether the synthesis consists of 224 separate steps, which shows that the synthesis plan described in Figure 283 is still only a rather rough approximation. The yields were 2.9% for the A-chain, 7% for the B-chain and between 0.25 and 0.55% for the complete insulin.

241 The Merrifield Procedure

An important advance in the stepwise synthesis of proteins was the procedure introduced by R. B. Merrifield in 1962. This consists of the stepwise heterogeneous graft-polymerization with subsequent splitting-off of the protein side chains after completed synthesis. The backbone polymer (one usually speaks of a "carrier") is usually a slightly cross-linked polystyrene (with divinylbenzene), which is easily prepared by pearl polymerization. (Diameter of the pearls 10 to 80 μ). The polystyrene is first chloromethylated and then nitrated. The CH_2 group allows one to tie the first amino acid onto the polystyrene by ester formation. Then the amino acids are hooked on step by step according to plan, using the standard methods of peptide chemistry. Figure 285 shows the reaction in detail up to the addition of the second amino acid, and Figure 286 shows the total scheme of a Merrifield synthesis.

As is seen in the diagram, this is a reaction which resembles the so-called "living" polymerization rather closely: all chains grow at the same time and at the same rate by addition of single structural units, so that the degree of polymerization is directly proportional to the conversion (see p. 173).

The Merrifield synthesis leads to high yields of a homogeneous polypeptide or protein only if the following conditions are fulfilled: (1) complete conversion in each reaction step; (2) complete removal of the excess amino acid after each coupling step; and (3) the reaction conditions of the coupling reaction and the type of binding of the first amino acid to the polystyrene chain have to be correlated in such a way that during the complete course of the coupling reaction the polystyrene-amino acid(1)-bond is not attacked.

If condition 1 does not hold, one obtains as side products chains in which certain amino acids are missing. If condition 2 does not hold, then the protein will contain some chains in which a few amino acids occur twice in a row. In each case, one obtains products with faulty sequences, whereas a partial fulfillment of condition 3 only decreases the yield. It is clear that the three conditions have to be fulfilled if one wants to obtain long protein chains. In order to obtain a high conversion in each step, the amino acids are added in large excess. After separation and washing, the reaction is repeated once more with fresh amino acid. As an accelerator one uses dicyclohexylcarbodiimide (DCCDI).

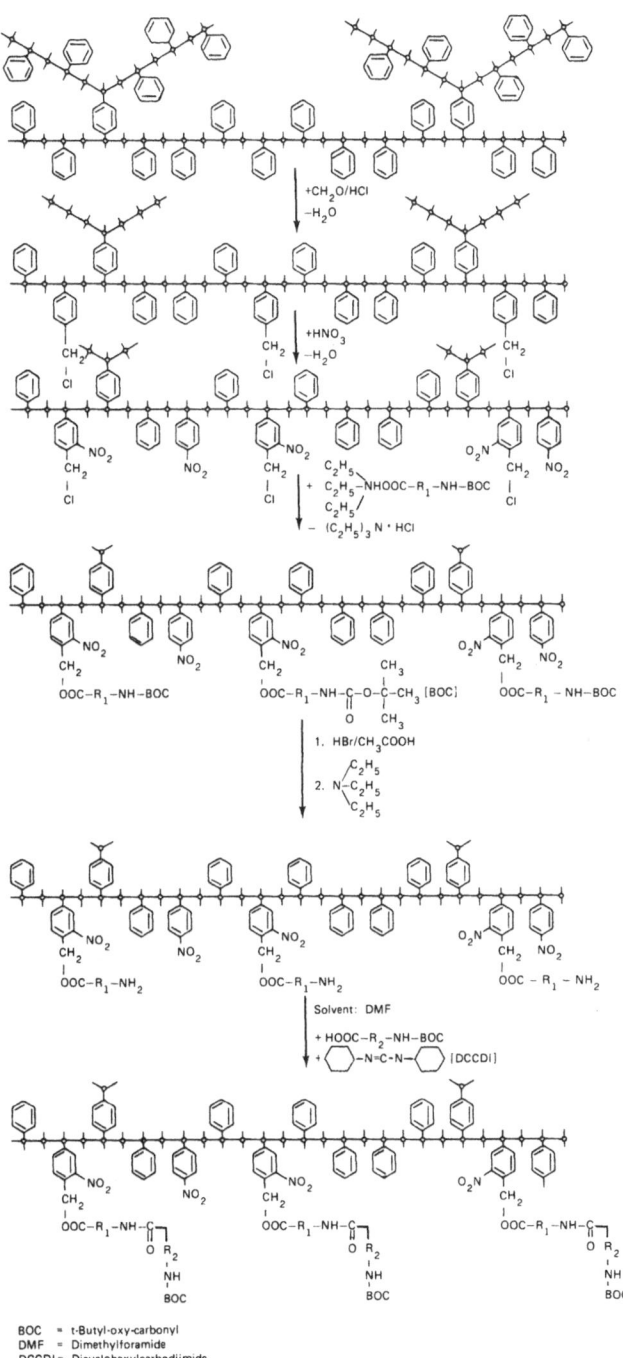

BOC   = t-Butyl-oxy-carbonyl
DMF   = Dimethylforamide
DCCDI = Dicyclohexylcarbodiimide

FIG. 285 – MERRIFIELD-Synthesis, Phase I.

The actual carrying-out of the syntheses is rather simple, because of the insolubility of the graft-copolymers: All operations are carried out in the same reaction vessel, one after the other (see Figure 287), with a suspension of polystyrene pearls which remains in the vessel during the entire synthesis. In this manner one avoids the otherwise necessary isolation and purification of the polypeptides after each step, and the entire synthetic process can be automatized. The operations only consist of the addition of fresh, and removal of used, solvents and reactants.

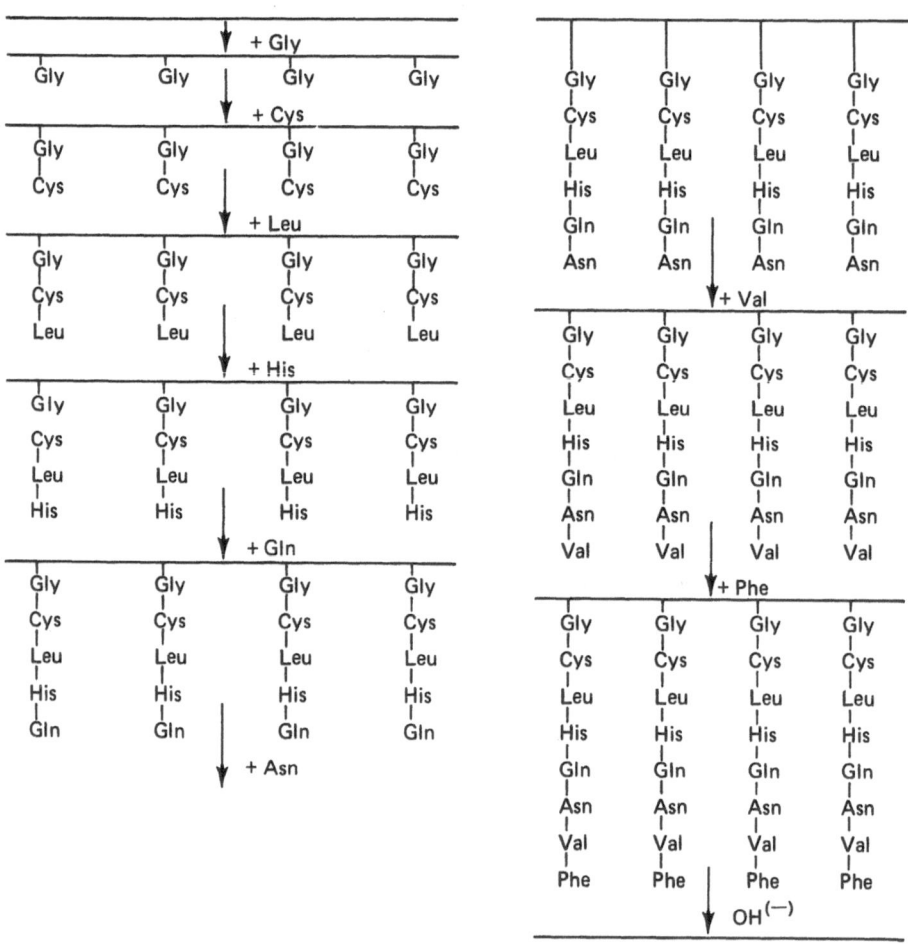

+n (HOOC)Gly—Cys—Leu—His—Gln—Asn—Val—Phe(NH$_2$)

FIG. 286 — MERRIFIELDS-Synthesis. (Total scheme).

The synthesis of the insulin chains (A- and B-chains), which according to the plan described in Figure 283 takes several months and only leads to yields of maximum 7%, can be carried out by means of the Merrifield procedure in a few weeks with 50-60% yield.

Until now, only proteins with relatively short chains (for example, insulin [P = 30 for the B chain], and glucagon [P = 29]) were synthesized. Difficulties increase considerably with longer chains, and therefore expectation for a synthetic preparation of proteins with high molecular weight are still rather slim. (The

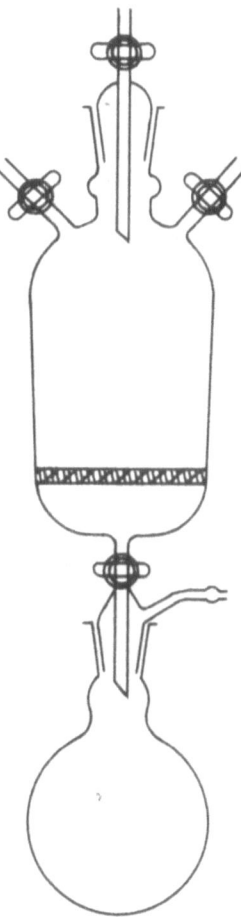

FIG. 287 — Reaction vessel used in the MERRIFIELD solid-phase peptide synthesis.

procedure used in nature by the cell for the production of proteins by means of messenger-RNA-program is still incomparably more elegant!) Perhaps more interesting than the copying of natural proteins, may be the synthesis of proteins with modified or completely new amino acid sequences, and especially the examination of such new proteins with respect to their activity in organisms.

25 GRAFT-COPOLYMERS AND BLOCK-COPOLYMERS

251   Synthesis of Graft-Copolymers

By "graft-copolymer," one understands a polymer with branched molecules in which the main chain is chemically different from the branches (compare pp. 28-30). The chemical nature, length, and concentration of the side chains may be different from case to case. If one looks at such a structure, one has to realize that the schematic representation has nothing to do with the real form of the macromolecules. The structural diagram is only a substitute for the structural formula and concerns the chemical constitution only, not the form of the macromolecule. The form of the molecule in dilute solution is not influenced by the presence of the branches, i.e., whether the chain is linear or branched, it is always present in the form of a random coil (statistical coil). Only the coil density is larger with branched molecules than with linear ones (compare pp. 516-517). As long as the side chains are formed from the same structural units as the main chain, the distinction between main chain and side chains is not very meaningful, because in a model of such a molecule it would be very difficult to distinguish the main chain from the side chains.

Since chain transfer (see pp. 63-69 and 90) to the polymer molecules which have been formed previously can not usually be avoided in the polymerization of unsaturated monomers, most polymer molecules are more or less strongly branched. As there are relatively few side chains per macromolecule, the properties

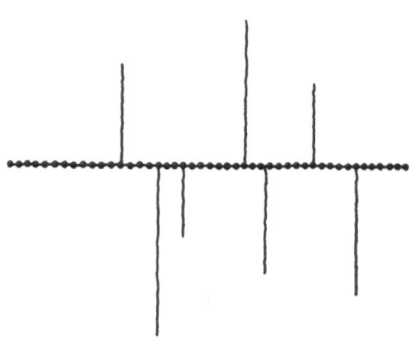

FIG. 288 — General structure of graft-copolymers.

of such polymers are not very different from the unbranched products. Only with polymers that are able to crystallize can the side chains affect the properties quite strongly, because the side chains disturb the regularity of the lattice and the crystalline content of such a polymer becomes smaller with increasing degree of branching. For example, polyethylene with strictly linear chains (produced by the decomposition of diazomethane, p. 203) represents one extreme; then there is polyethylene with very few side-chains per macromolecule (low pressure polyethylene, see p. 193,); and there is also polyethylene with strongly branched chains (high pressure polyethylene, which is prepared by radical polymerization at approximately 180°C and pressures of 1500 to 3000 atm). The branched polyethylene is softer and less stiff than the unbranched polyethylene, and its softening temperature is lower.

Whereas through moderate branching, where the side chains and the main chain have the same structural units, the physical properties of the polymer are usually modified only very slightly, with graft-copolymers one has a new type of polymer which usually has properties combining some of the properties of the two homopolymers. As experience has shown, most types of polymers are never miscible with each other in the sense of a homogeneous molecular mixture (incompatability of polymers, compare p. 559).

By the formation of graft-copolymers it is possible to attach immiscible polymers to each other. Because of the incompatibility of the components, however, one has to assume that even in a graft-copolymer molecule the two components are spatially separated from each other. Therefore, the structure of the graft-copolymer molecule can be different, depending on the ratio of the components, the density of the grafting points, the length of the side chains, and the method by which it was formed. (Figure 290 shows some graft-copolymer structures in a schematic fashion).

As a result, even graft-copolymers are by no means homogeneous, in the sense of a completely uniform mixing of macromolecules of both components, when they exist in a state of aggregation. The structure of such graft-copolymers permits molecular arrangements where the components form aggregates in which usually a large number of polymer coils of the same type are next to each other. Especially with graft-copolymers of type I or type II, this leads to aggregates with a large number of coils of the same type (F. M. Merret, G. Molau). The structural models shown in Figure 290 indicate, however, that with graft-copolymers the size of the aggregates is limited. Whereas with types I and II there are still aggregates possible, which consist of relatively many similar coils (see Figure 291), with type III, and with those with even more side chains, the components are distributed more or less homogeneously.

By formation of graft-copolymers (the same is true for block-copolymers), it has been possible to combine incompatible polymers in such a fashion that the components are either distributed homogeneously (for example, polyester casting resins) or they are at least firmly fixed to each other at the phase boundaries (for

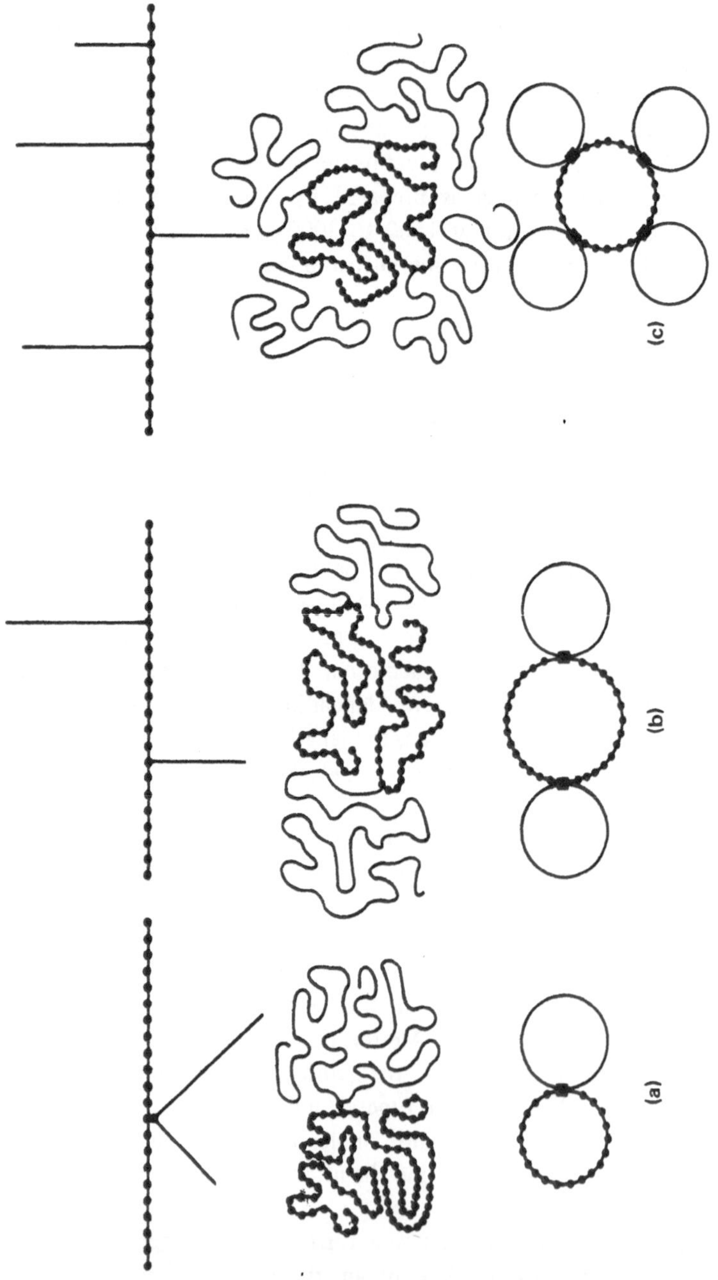

FIG. 290 — Coil-structure of graft-copolymer molecules.

example, with high-impact polystyrene, where the two components, polystyrene [about 90%] and rubber [about 10%], are combined through grafting bonds in such a way that a new material is formed, which has inherited from polystyrene its hardness and from rubber its toughness). In the following sections some procedures for the preparation of graft-copolymers are briefly discussed.

2511 Graft Copolymerization by Means of Radicals

Chain Transfer

If one dissolves a polymer (the primary polymer), in a monomer (the grafting monomer) (for example, a polyacrylate in styrene) and then adds an initiator and polymerizes, depending on the magnitude of the chain transfer constant of the polymer, a more or less strong chain transfer to the polymer occurs, and one obtains a mixture of graft-copolymer and linear polymer. In some cases such a mixture is sufficient for a particular technical application. (For the mechanism and kinetics of chain transfer, compare pp. 63 and 90). Since the chain transfer constants of the polymers are usually quite small (approximately 10^{-5}), the

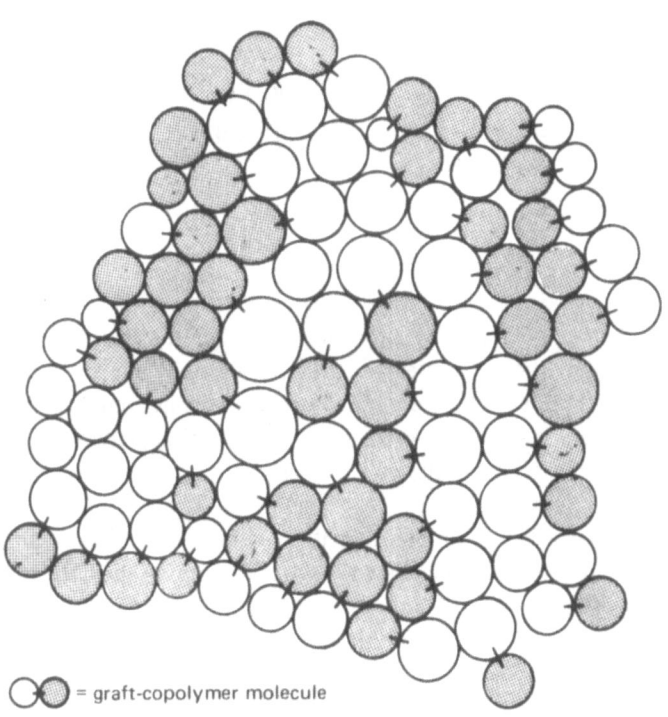

= graft-copolymer molecule

FIG. 291 — Aggregation with graft-copolymers.

amount of graft-copolymer which is formed by chain transfer is correspondingly small. As a result, the properties of the polymer mixtures which one obtains by polymerization of a monomer in the presence of another polymer differ only to a slight extent from the properties of the mixture of the two homopolymers.

The following reaction sequence describes the chain transfer which occurs when styrene is polymerized in the presence of a dissolved polyacrylate:

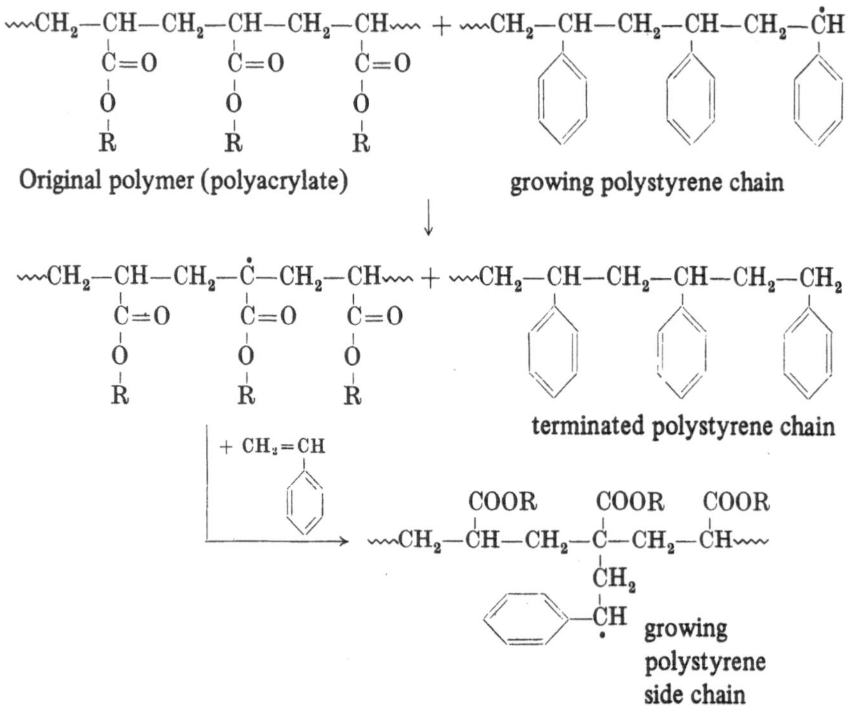

$$\text{~CH}_2\text{—CH—CH}_2\text{—CH—CH}_2\text{—CH~} + \text{~CH}_2\text{—CH—CH}_2\text{—CH—CH}_2\text{—ĊH}$$

with C=O, C=O, C=O groups (O—R) on the original polymer, and phenyl groups on the growing polystyrene chain.

Original polymer (polyacrylate)   growing polystyrene chain

$$\downarrow$$

$$\text{~CH}_2\text{—CH—CH}_2\text{—Ċ—CH}_2\text{—CH~} + \text{~CH}_2\text{—CH—CH}_2\text{—CH—CH}_2\text{—CH}_2$$

with C=O, C=O, C=O groups (O—R).

terminated polystyrene chain

$$+ \text{CH}_2\text{=CH}$$

$$\longrightarrow \text{~CH}_2\text{—CH—CH}_2\text{—Ċ—CH}_2\text{—CH~}$$

COOR   COOR   COOR

$$\overset{|}{\text{CH}_2}$$

—ĊH   growing
polystyrene
side chain

However, there are certain situations in which chain transfer to polymer reaches such an extent that the properties of the resulting polymers are influenced to a large degree. This is always the case if the polymer chains contain relatively labile hydrogen atoms which are easily abstracted by the attacking radicals. An example is the formation of cross-links in the bulk polymerization of acrylates which has been described on page 68.

This effect is even more pronounced with polybutadienes (natural and synthetic rubbers). In the polybutadiene chain the hydrogen atom of the CH_2 group is more labile because of the neighboring double bond and is therefore particularly easily attacked by a radical chain end. Although the emulsion copolymerization of butadiene and styrene (75:25) for the preparation of synthetic rubber is carried out at relatively low temperatures (−5°C to +40°C), cross-linking occurs at high conversions (high polymer concentrations). This leads to high gel content in the rubber, and one therefore has to cut short the polymerization already at conversions of 40-70%, in order to keep the gel content within desirable bounds.

The tendency of polybutadienes to undergo chain transfer reactions can be utilized advantageously for the preparation of impact-resistant polystyrene by dissolving a rubber in monomeric styrene (about 5 to 10% rubber) and polymerizing the solution. Chain transfer leads to graft-copolymers which give rise to a firm attachment between the polystyrene and rubber phases (for further details about the morphological structure of these graft-copolymers see p. 618).

a) Grafting Through Chain Train Transfer:

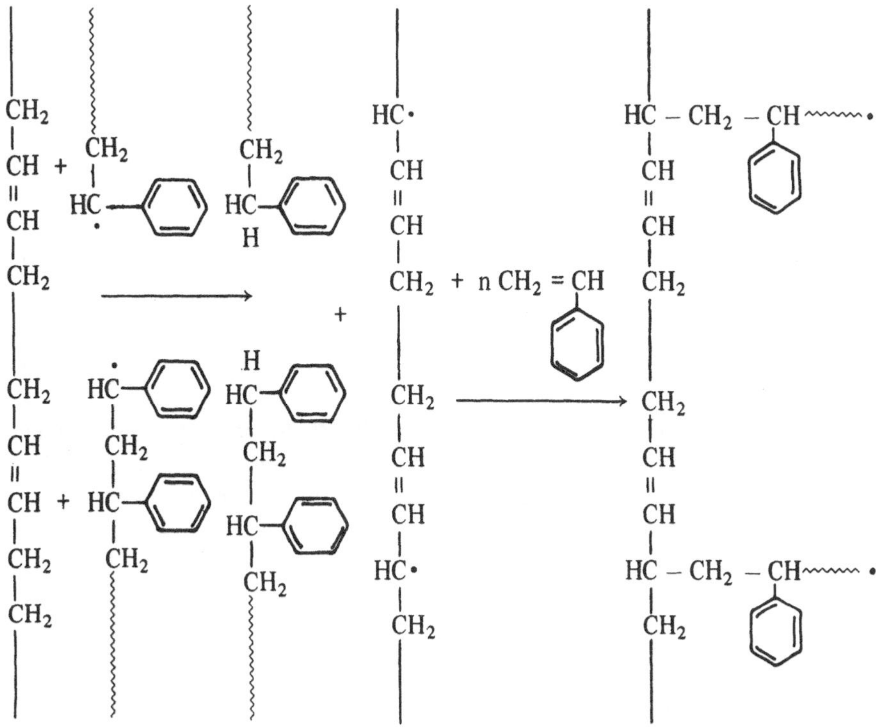

b) Crosslinking Through Radical Combination:

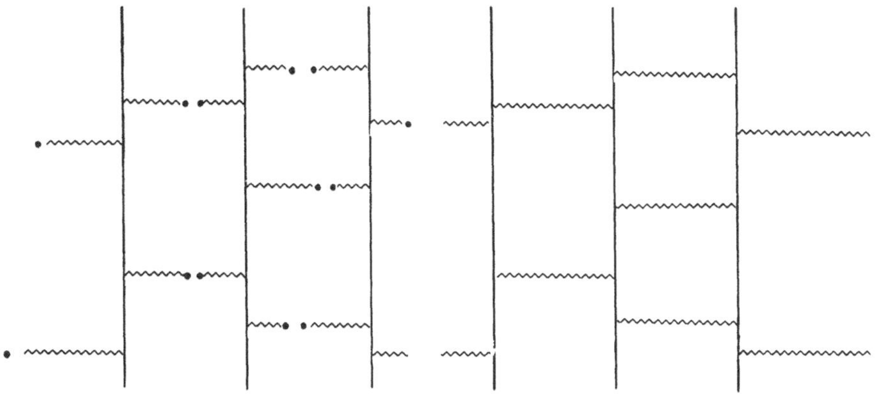

The tendency of chain transfer can "artificially" be increased by incorporation of groups with high transfer constants into the chain of the primary polymer, e.g., mercapto groups.

Grafting by Irradiation

X-rays, γ-rays, electrons, and radiation from such synthetic radioisotopes as Co^{60} are capable of producing radicals in organic molecules. If one irradiates the solution of a polymer in a monomer, one obtains the same type of reaction that occurs in chain transfer, i.e., on the polymer chain of the primary polymer radicals are formed which become the points of initiation for the side chains. At the same time the radiation initiates polymerization of the monomer, and one therefore obtains a mixture of graft-copolymer and pure homopolymer. One can also carry out the irradiation of the polymer in the absence of monomer, for example, if one irradiates films of polyethylene in air. In this case one obtains peroxide and hydroperoxide groups in the polymer, which decompose if one heats the film in the presence of a monomer. In this case where the film is dissolved or swollen in the monomer, one again obtains side chains. However, it is not always possible to clearly specify the mechanism by which they are formed. Also here one often obtains greater or smaller amounts of pure homopolymer in addition to the graft-copolymer.

If one incorporates bromine atoms in the polymer (for example, by copolymerization with parabromostyrene) even ultraviolet irradiation is sufficient to give rise to grafting, because the bromine atom is easily split off by the ultraviolet light and this leaves radicals on the chain which are capable of initiating polymerization of another monomer. In addition to bromine one can also use compounds which are sensitizers, such as azo compounds, benzoin, and naphthalene.

Incorporation of Peroxide- or Azo Groups

Without any type of radiation, one can form initiator radicals in a polymer chain if one creates peroxide- or azo-groups in the chain, or forms these later by means of oxygen. For example, polystyrene can be converted partially to poly-p-isopropylstyrene by reaction with isopropylchloride and $AlCl_3$. The resulting polymer can easily be converted with oxygen to a polymeric hydroperoxide similar to the formation of cumene hydroperoxide (Metz and Mesrobian):

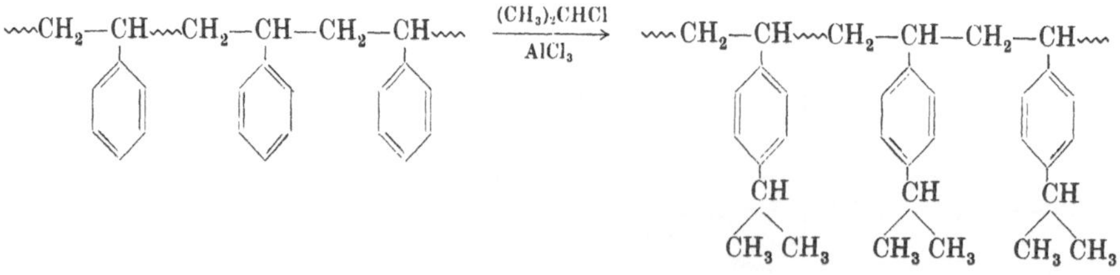

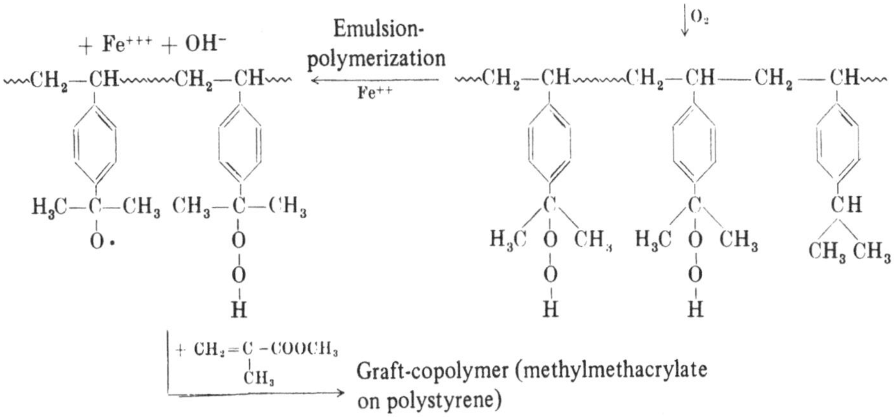

$$+ Fe^{+++} + OH^- \qquad \xleftarrow[Fe^{++}]{\text{Emulsion-polymerization}} \qquad \Bigg\downarrow O_2$$

~~CH$_2$—CH~~~CH$_2$—CH~~   ~~CH$_2$—CH~~~CH$_2$—CH——CH$_2$—CH~~

H$_3$C—C—CH$_3$ CH$_3$—C—CH$_3$ H$_3$C O CH$_3$ H$_3$C O CH$_3$ CH$_3$ CH$_3$

O·   O   O   O

          O   H   H

          H

$+ CH_2 = C - COOCH_3$
         |
         CH$_3$ Graft-copolymer (methylmethacrylate on polystyrene)

The resulting peroxide decomposes (in the presence of Fe++ as the redox component) into radicals which initiate the polymerization of the monomer, for example, methylmethacrylate, and produce pure graft copolymers.

It is even simpler to build azo-groups into a polymer which can then be used for the initiation of a graft copolymerization (Vollmert and Bolte):

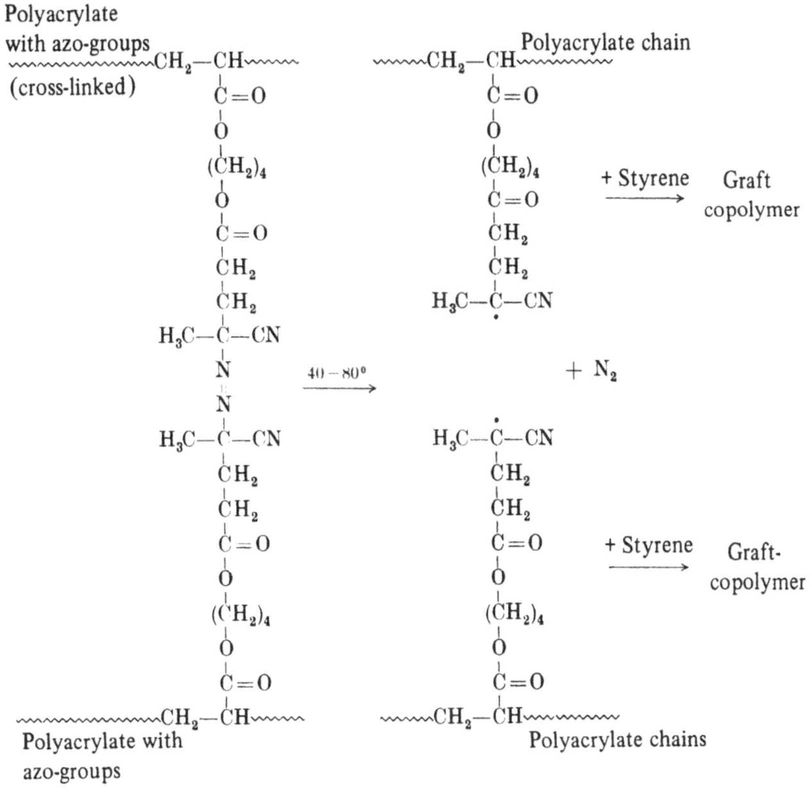

Polyacrylate with azo-groups (cross-linked)

~~~CH$_2$—CH~~~   Polyacrylate chain ~~~CH$_2$—CH~~~

C=O   C=O
O   O
(CH$_2$)$_4$ (CH$_2$)$_4$ + Styrene   Graft
O   C=O $\longrightarrow$ copolymer
C=O   CH$_2$
CH$_2$ CH$_2$
CH$_2$ H$_3$C—C—CN
H$_3$C—C—CN   ·
N $\xrightarrow{40-80°}$ + N$_2$
N
H$_3$C—C—CN   H$_3$C—C—CN
CH$_2$ CH$_2$
CH$_2$ CH$_2$
C=O   C=O   + Styrene   Graft-
O   O $\longrightarrow$ copolymer
(CH$_2$)$_4$ (CH$_2$)$_4$
O   O
C=O   C=O

~~~CH$_2$—CH~~~   ~~~CH$_2$—CH~~~
Polyacrylate with azo-groups   Polyacrylate chains

The azo-group containing primary polymer is then swollen in another monomer and heated under nitrogen to 60°C. It yields pure, transparent, graft-copolymers without presence of any linear homopolymer. The polymerization can also be carried out in bulk or in suspension.

Graft-Copolymers by means of Ethoxylation

If one reacts ethyleneoxide under pressure and at higher temperature with polymers containing NH_2 groups, then from these NH_2 groups one obtains side chains of polyether. Polymers with NH_2 groups on the chain can be formed by copolymerization with amino groups containing monomers, or reaction of ethylene-imine with polymers which contain carboxyl or OH groups on the chain. Similarly, one can also polymerize ethyleneoxide onto polymers with amide groups. For instance, one reacts ethyleneoxide with a finely divided polyamide and the formerly hard polyamide then becomes elastic through graft copolymerization:

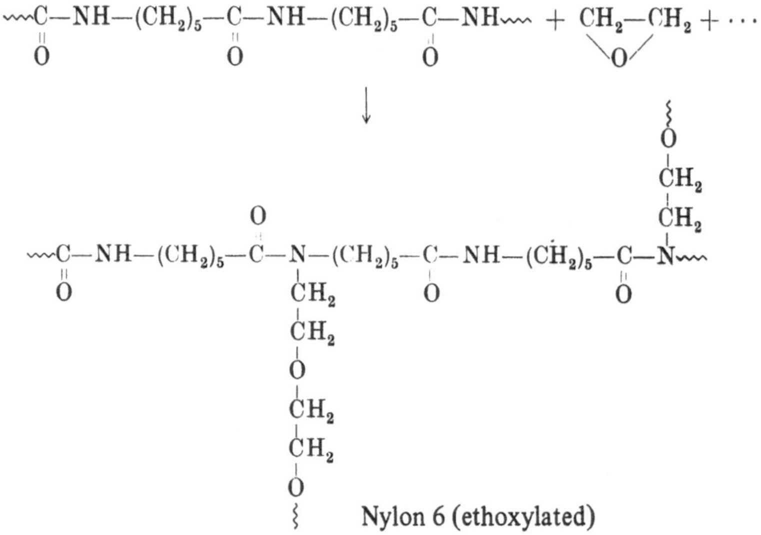

$$\sim\!\!C\!-\!NH\!-\!(CH_2)_5\!-\!C\!-\!NH\!-\!(CH_2)_5\!-\!C\!-\!NH\!\sim + \quad CH_2\!-\!CH_2 + \cdots$$

Nylon 6 (ethoxylated)

2512 Preparation of Graft-Copolymers with the Aid of Functional Groups

Graft-copolymers can be formed not only by growing side chains onto primary polymers, but also by attaching polymers, which have already attained their final length, onto other polymers with the aid of certain functional groups.

The most elegant method of preparing polymers with suitable functional groups is copolymerization with monomers which contain the desired functional groups. The following list contains a number of relatively easily obtainable monomers with functional groups:

Acryloyl chloride $\qquad$ $CH_2 = CH - COCl$

Acrylic acid $\qquad$ $CH_2 = CH - COOH$

Monoacrylate from glycols $\qquad$ $CH_2 = CH - COO - (CH_2)_n - OH$

Glycidylacrylate $\qquad$ $CH_2 = CH - COO - CH_2 - CH - CH_2$ with epoxide ring (O bridging CH and CH₂)

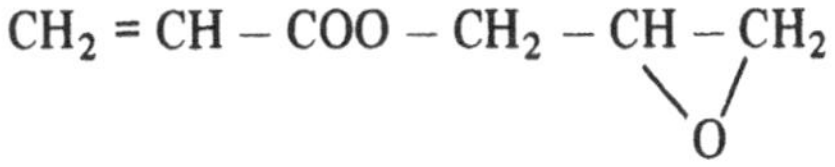

Isocyanato-acrylate $\qquad$ $CH_2 = CH - COO - CH_2 - CH_2 - N = C = O$

Dimethylamino-ethyl-acrylate $\qquad$ $CH_2 = CH - COO - CH_2 - CH_2 - N\begin{smallmatrix} CH_3 \\ CH_3 \end{smallmatrix}$

Maleicanhydride

$$CH\!=\!CH$$
$$O=\overset{|}{C}\qquad \overset{|}{C}=O$$
$$\diagdown O \diagup$$

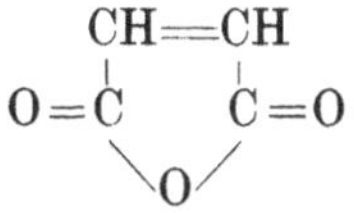

Acrolein $\qquad$ $CH_2\!=\!CH\!-\!CHO$

Ethylenecarbonate

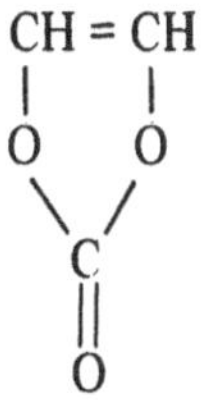

$$CH = CH$$
$$\overset{|}{O}\qquad \overset{|}{O}$$
$$\diagdown \underset{\underset{O}{\|}}{C} \diagup$$

One can also prepare polymers with functional groups by partial saponification of polyvinylesters (to give OH groups), or polyacrylic esters, or polyacrylonitrile (to give COOH groups). Polymers with functional groups at the chain ends are obtained if one uses initiators with functional groups: e.g., azo catalysts containing carboxyl groups. Because of chain termination by recombination, one usually obtains polymers with carboxyl groups at both chain ends:

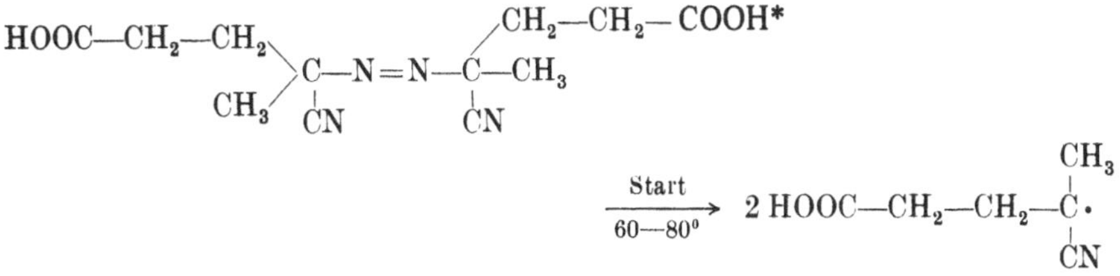

$$HOOC\!-\!CH_2\!-\!CH_2 \diagdown \underset{CH_3}{\overset{}{\underset{}{}}} \diagup \underset{CN}{\overset{|}{C}}\!-\!N\!=\!N\!-\!\underset{CN}{\overset{|}{C}}\!-\!CH_3 \diagdown \overset{CH_2\!-\!CH_2\!-\!COOH*}{}$$

$$\xrightarrow[60-80^0]{Start} 2\ HOOC\!-\!CH_2\!-\!CH_2\!-\!\overset{CH_3}{\underset{CN}{\overset{|}{\underset{|}{C}}}}\cdot$$

*Easily prepared by reaction of levulinic acid, $H_2N - NH_2$ and HCN

Polymers with one or two carboxyl end groups are obtained by anionic polymerization and subsequent chain termination with CO_2 and further reaction with mineral acids:

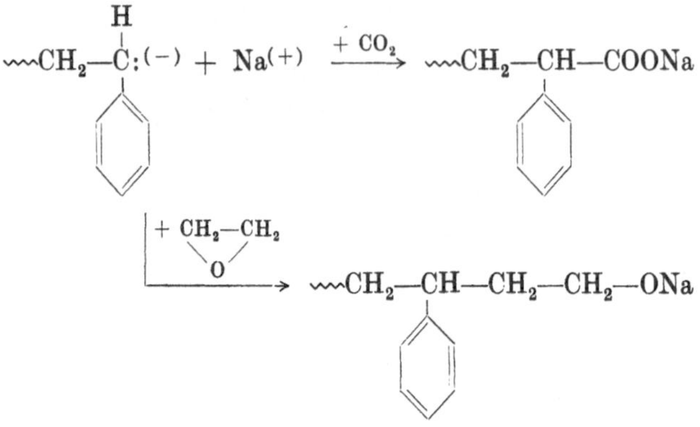

$$\text{\textasciitilde CH}_2\text{--}\overset{\overset{\displaystyle H}{|}}{\text{C}}\text{:}^{(-)} + \text{Na}^{(+)} \xrightarrow{+ \text{CO}_2} \text{\textasciitilde CH}_2\text{--CH--COONa}$$

$$\xrightarrow[\displaystyle \quad]{+ \; \underset{\displaystyle O}{\text{CH}_2\text{--CH}_2}} \text{\textasciitilde CH}_2\text{--CH--CH}_2\text{--CH}_2\text{--ONa}$$

If, instead of CO_2 ethyleneoxide is used to terminate the reaction, one obtains chains with OH groups at the ends.

For example, if a polymer A with carboxyl groups along the chain is reacted with a polymer B having OH groups, then one obtains graft-copolymers in which the chains of polymer A are combined with the chains of polymer B over ester groups:

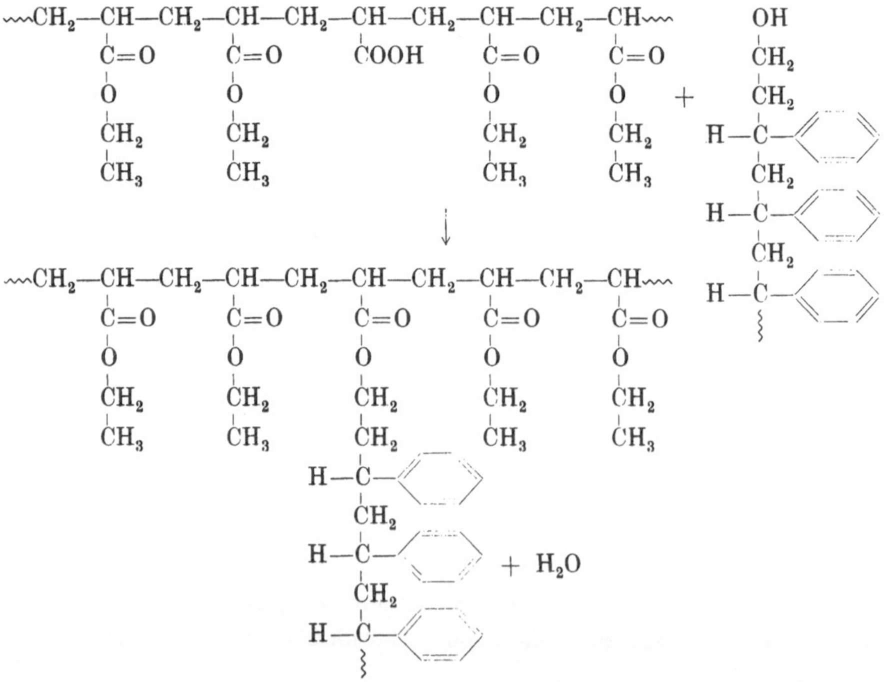

or, in general:

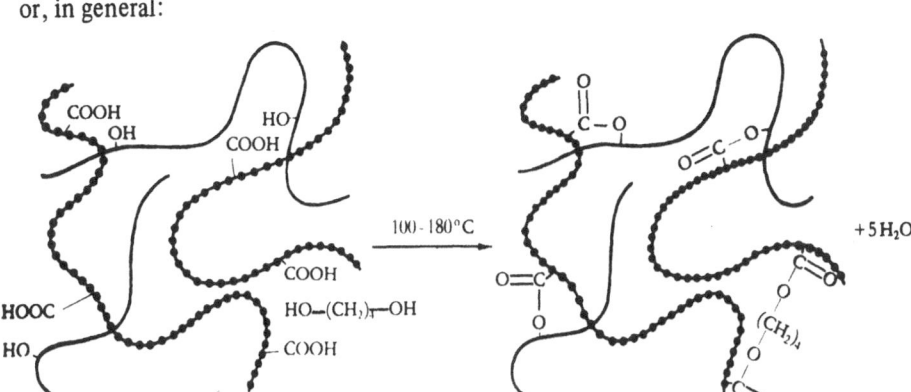

The structural scheme always differs somewhat from that shown in Figure 288, if the functional groups of the polymer which is to be attached, are not at the chain ends but somewhere in the chain.

Usually the graft-copolymers formed in this way are cross-linked. If one would like to obtain completely uncross-linked graft-copolymers with the aid of functional groups, one has to have side-chain copolymers which contain only *one* functional group per macromolecule. Such polymers can be obtained in the anionic polymerization of styrene (phenyl-Li as initiator), where CO_2 or ethyleneoxide is used for chain termination. Generally, however, one gets the functional groups through copolymerization, and one therefore obtains macromolecules with many unevenly distributed functional groups. Depending on the concentration of the functional groups along the polymer chains, one obtains more or less cross-linked graft-copolymers, which in certain cases can be degraded through milling to a point where they again become soluble and thermoplastic.[33] In practice this method of the formation of graft-copolymers has become important because it is relatively simple to carry out. It is used, for example, for preparation of impact-resistant polystyrene. This process is not suitable for the preparation of pure graft-copolymers (without the presence of any homopolymers).

2513 Graft-Copolymers Through Copolymerization

Cross-linked graft-copolymers are obtained in a particularly simple manner by copolymerization, where one dissolves unsaturated polymers in the monomer to be grafted on and allows this monomer to polymerize. The double bond of the polymer is then incorporated into the growing chain of the polymerizing monomer. Technically important examples for this type of graft polymerization are the unsaturated polyester casting resins.

---

[33] Thermoplastic = soft and malleable at elevated temperature.

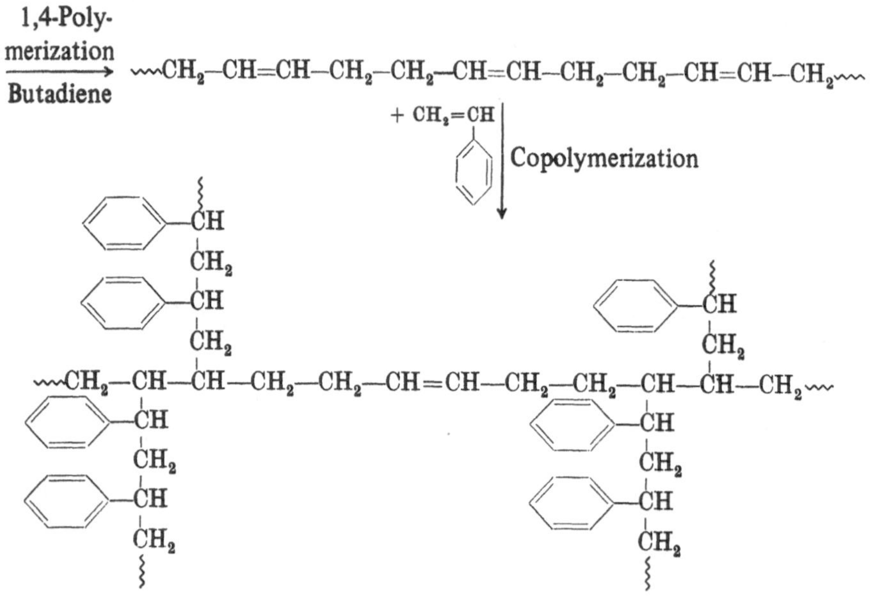

1,4-Poly-
merization
$\xrightarrow{\hspace{2cm}}$ ⌇CH$_2$—CH=CH—CH$_2$—CH$_2$—CH=CH—CH$_2$—CH$_2$—CH=CH—CH$_2$⌇
Butadiene

$+$ CH$_2$=CH

| Copolymerization

⌇CH$_2$—CH—CH—CH$_2$—CH$_2$—CH=CH—CH$_2$—CH$_2$—CH—CH—CH$_2$⌇

Or one dissolves an unsaturated polyester containing fumaric ester double bonds in the chain in styrene monomer and polymerizes the solution. Such unsaturated polyesters can easily be prepared from maleic anhydride and glycols, because at the high temperatures (up to 200°C) which are used for the preparation for this polyester, the *cis*-double bonds are converted to *trans*-double bonds. As one can see from the copolymerization parameters of styrene and fumaric ester ($r_1 = 0.3$, $r_2 = 0.07$, compare Table 107), this system has a tendency to produce copolymers of alternating structure, and one therefore obtains clear, transparent graft-copolymers, where the polyester chains are combined with polystyrene side chains if the azeotropic mixture is used. Instead of styrene one can also use other monomers such as methylmethacrylate, or acrylonitrile, or mixtures of several monomers.

All graft-copolymers which one produces by copolymerization are more or less cross-linked:

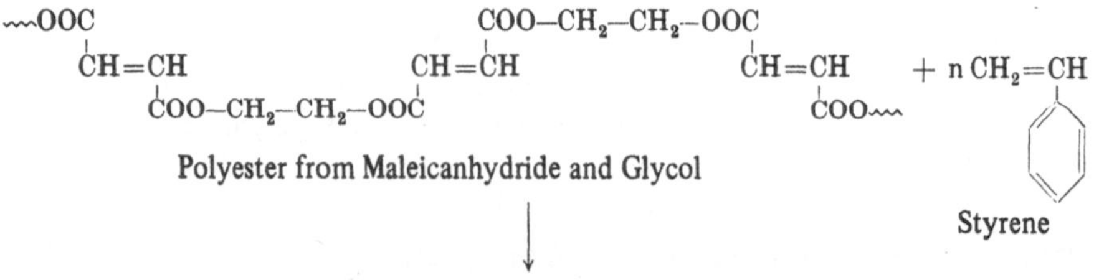

⌇OOC   COO—CH$_2$—CH$_2$—OOC

CH=CH   CH=CH   CH=CH $+$ n CH$_2$=CH

COO—CH$_2$—CH$_2$—OOC   COO⌇

Polyester from Maleicanhydride and Glycol

Styrene

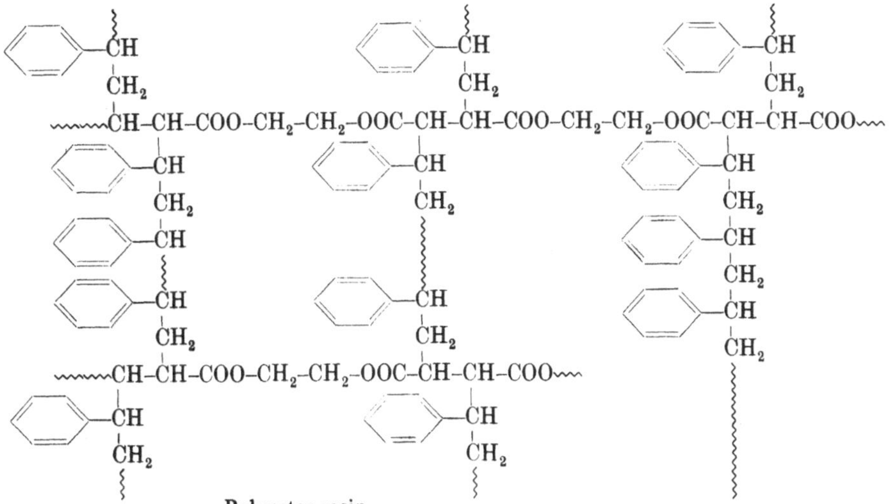

Polyester resin

This type of grafting is not restricted to unsaturated polyesters. One can also incorporate double bonds into a polymer by reaction of a suitable polymer with acrylic acid, or in general, with monomers which have a functional group suitable for grafting on of other chains:

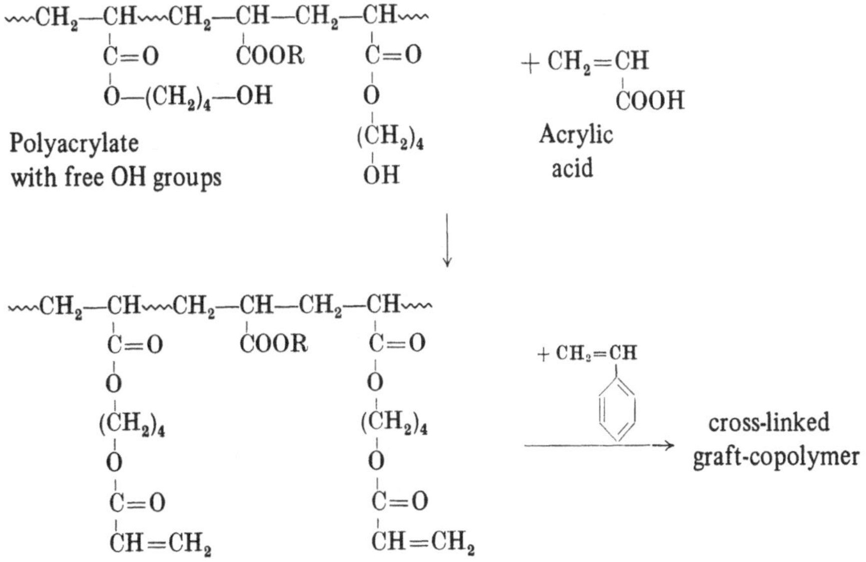

252 Preparation of Block-Copolymers

There are two possibilities for the formation of block-copolymers, which are similar to the formation of graft-copolymers:

(1) One can polymerize a second monomer onto an existing polymer chain, or, (2) one can combine preformed polymer chains with other polymer chains with the aid of functional groups.

The former procedure can be carried out most simply by means of stoichiometric polymerization where except for intentional chain termination, there is no termination. For example, one first polymerizes α-methylstyrene with the aid of phenyllithium or sodiumnaphthalene (compare p. 170). The chain length can be predetermined arbitrarily by means of the ratio of monomer to initiator. Then one adds styrene monomer and the chain continues to grow without recognizing the change in the monomer. Instead of styrene, methylmethacrylate, isoprene, or butadiene may be used. One can also add these monomers, one after the other, however a certain sequence must be followed so that the basicity of the chain end is never less than that of the new one which is formed by addition of the next monomer:

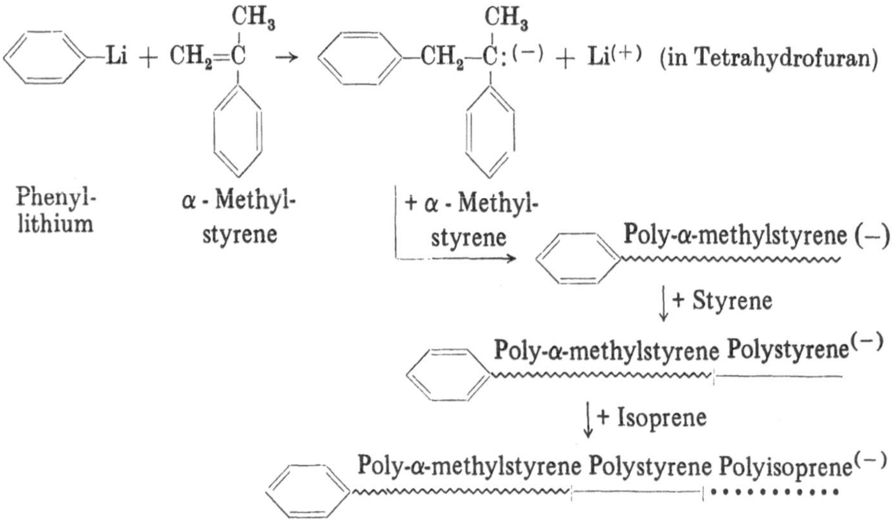

Phenyl-   α - Methyl-
lithium   styrene

If instead of phenyllithium, sodiumnapthalene is used as the initiator, then the chains grow in both directions and one obtains a block-copolymer in which the primary polymer is in the middle: Polyisoprene Polystyrene Poly-α-methylstyrene Polystyrene Polyisoprene. Other examples of procedure (1) are quite similar to those which have been discussed in the preparation of graft-copolymers. However, the initiator groups should not be part of the chain, but must be at the chain ends:

Polyester from Adipic Acid and Glycol

$$HO-\left[CH_2-CH_2-O\right]_{x-y}-CH_2-CH_2-\left[O-\underset{O}{\overset{\|}{C}}-(CH_2)_4-\underset{O}{\overset{\|}{C}}-O-(CH_2)_2\right]_n-O-\left[CH_2-CH_2-O\right]_y-H$$

Polyether   Polyester   Polyether

Instead of the polyesters, one can also graft ethyleneoxide onto vinylpolymers with OH end groups in order to produce block-copolymers with polyether segments at both ends and a polystyrene or polyacrylic ester segment in the middle. Principally, it is also possible to incorporate radical initiator residues at the chain ends, for example, azobisisobutyronitrile (compare p. 295) which can then be used to polymerize new monomer:

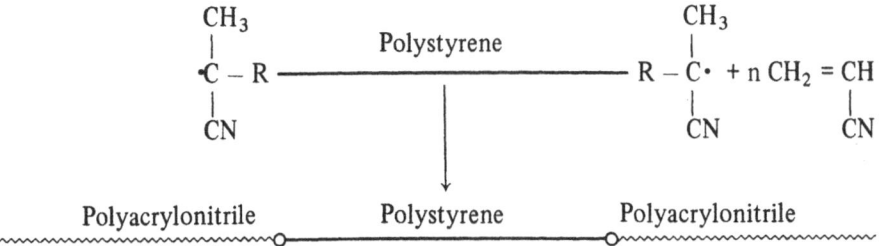

All of these reactions are really graft copolymerizations with the grafting restricted to the chain end.

With procedure (2), one starts with preformed polymer chains and combines these with the aid of functional groups. The reaction is a sort of polyaddition of polymers and one obtains chains with an undetermined and different (from chain to chain) number of segments:

Polystyrene ————————— OH   + HO ————————— Polystyrene

+   HOOC ~~~~~ Polyacrylic ester ~~~~~ COOH

Esterification
(Dicyclohexylcarbodiimide)

Polystyrene ————————— O — C ~~~~~ Polyacrylic ester ~~~~~ C — O ————————— Polystyrene
                            ‖   ‖
                            O   O

Polystyrene with hydroxyl end groups can easily be prepared through stoichiometric polymerization with sodiumnaphthalene and subsequent chain termination with ethyleneoxide (compare p. 298). Polyacrylates with carboxyl end groups can be prepared, for example, by polymerization with an azo-catalyst containing carboxyl groups.

$$2 \ CH_3—\overset{\overset{\displaystyle \|}{O}}{C}—CH_2—CH_2—COOH$$

Levulinic acid

$$\xrightarrow[\substack{+ \ 2 \ HCN \\ - \ 2 \ H_2O \\ - \ 2 \ H}]{+ \ H_2N—NH_2} \quad HOOC—(CH_2)_2—\underset{\underset{\displaystyle CN}{|}}{\overset{\overset{\displaystyle CH_3}{|}}{C}}—N=N—\underset{\underset{\displaystyle CN}{|}}{\overset{\overset{\displaystyle CH_3}{|}}{C}}—(CH_2)_2—COOH$$

$$\xrightarrow{60°} \quad 2 \ HOOC—(CH_2)_2—\underset{\underset{\displaystyle CN}{|}}{\overset{\overset{\displaystyle CH_3}{|}}{C}} \cdot \ + \ N_2$$

$$\downarrow \quad {+ \ CH_2=CH \atop \underset{\displaystyle COOR}{|}}$$

Polyacrylate with COOH-Endgroups

The two polymers which are to be combined do not have to have different end groups. It is also possible to combine polymers with the same end groups by addition of functional cross-linking agents, for example, polyesters with OH end groups and polyethers with OH end groups can be combined by reaction with a diisocyanate. In such a case, one cannot prevent like polymers being combined by the diisocyanate as well as the unlike ones. (For a commercial block copolymer of this type see p. 239).

In view of the large number of polymers with very different properties (soft, elastic, hard, brittle, water-soluble, benzene-soluble), the synthesis of graft- and block-copolymers leads to an infinite number of possible combinations, especially if one remembers that the branches and blocks do not always have to consist of homopolymers, but can also consist of normal statistical copolymers. With regard to the secondary structure of block-copolymers, we have to assume the same separation of the different segments of the chain as with graft-copolymers (see p. 289):

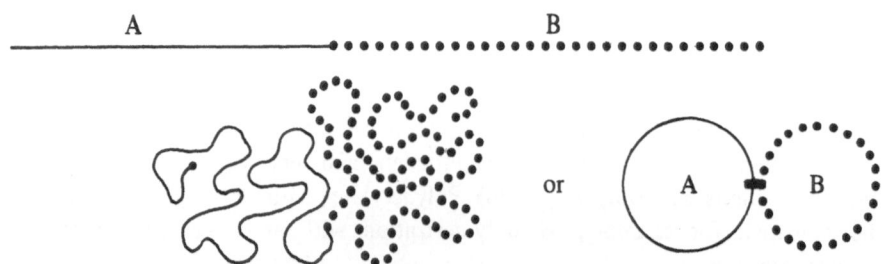

26 PURIFICATION OF POLYMERS

The usual purification operations used with low molecular weight compounds (for example, distillation and recrystallization) are usually not applicable to macromolecular compounds.

Macromolecular compounds are not volatile and cannot simply be recrystallized from saturated solutions. One therefore has to be satisfied with extracting them in suitable solvents and thus removing the impurities.

The purification effect of the extraction is in most cases rather slight, because in many instances the impurity is held to the polymer by strong secondary valence forces. Consequently, one usually first dissolves the macromolecular compound and then precipitates it by addition of a nonsolvent. The precipitate comes out as a more or less strongly swollen gel, and the impurities remain at least partly in solution (compare with section on fractional precipitation, p. 389). The gel can be removed by sedimentation if necessary, in the centrifuge. Usually it is necessary to repeat this precipitation 10 to 20 times. The solvents and precipitating agents have to be carefully selected from case to case. Many vinyl polymers are easily dissolved in benzene and toluene and precipitate under addition of methanol. Methanol is a sort of general precipitating agent in which most polymers, both the benzene-soluble ones and the water-soluble polymers, are insoluble. Since methanol is miscible in all proportions both with benzene and with most other organic solvents and also with water, it can be used as a precipitant in both cases, i.e., with organic and with aqueous polymer solutions.

One has to expect that the polymer is somehow affected by the extraction, regardless of whether this is done by heating or by precipitation. Thus if one extracts with acids or bases, hydrolytic degradation usually causes a lowering in the degree of polymerization (for example, hydrolytic degradation of polysaccharides). In the process of dissolving and reprecipitation, one usually finds that low molecular weight components remain in solution (fractionation), and therefore the precipitated part has a higher average molecular weight and a more uniform molecular weight distribution. This may or may not be desirable, but in any case one should be aware of it.

With synthetic polymers, purification may be avoided sometimes by choosing a method of synthesis where the polymer is immediately obtained in pure form. Usually an extreme purification of the monomer is simpler and more effective than a later purification of the polymer. Initiators and other additives, such as emulsifying agents and protective colloids, are also selected wherever possible in such a way that they can remain in the polymer without causing problems.

Dispersions, as obtained by emulsion polymerization, can be coagulated or precipitated by the addition of electrolytes. Formic acid or sodium chloride is used most often. The purification effectiveness of such a precipitation is rather incomplete, because the emulsifying agents are not completely removed; their content is only lowered. However, even this is often sufficient to make precipitation of the polymer dispersion preferable to a direct drying on the mill or by spraying. Polymer dispersions can also often be precipitated and purified by pouring them into methanol, by adding methanol, or by cooling the dispersion in some kind of a cooling mixture (−20°C to −40°C).

Of special importance is the purification of natural macromolecular compounds. These usually contain low molecular weight materials or other macromolecular compounds, and in some cases there is even a chemical combination between the different compounds: for example, cellulose with lignin (in wood), or pectins (in flax). In such cases it is justified to consider this combination as a new macromolecular substance. Since it is usually not possible to isolate the macromolecules without chemical reaction and therefore usually one does not obtain them in an unchanged form, one does not give these compounds a proper name but calls them *proto*pectin or *proto*cellulose (etc.). The isolation of the pure compounds (pectin or cellulose) can be brought about by acid, or alkaline, or enzymatic hydrolysis (wood pulp by the sulfite process or by alkali treatment).

Low molecular weight impurities (for example, residues of solvents or monomers) can usually be removed by heating in vacuum. Since one cannot mix the molten polymer in some kind of a stirred vessel, (because of its high viscosity) one has to use special techniques and apparatuses, depending on the polymer. Macromolecular substances with a high softening point (over 100°C) can often be dried in the form of a powder or in granules. This is done by spreading them out in a vacuum oven and heating them, usually with an infrared lamp. If the material which is to be removed is volatile only at a temperature which is higher than the softening point of the polymer, one can use certain extruders in which it is possible to apply vacuum at certain places along the machine and thereby remove the solvent. In such vacuum extruders (welding machines) one can treat thermoplastic materials, such as polystyrene or polymethylmethacrylate (at 180°C to 250°C). As a result of constant renewing of surface by the milling process in the vacuum-zones of the extruder, a very rapid and effective removal of the volatile compounds is obtained. Great care must be taken in the isolation of proteins from their aqueous solutions because they usually degenerate even at rather low temperatures. Such solutions can often be concentrated by freeze-drying. This is done by bringing the surface of the solution in close contact with a cold condenser and applying high vacuum to the entire apparatus so that the water from the solution forms as ice on the condenser. The solution to be concentrated can be in a frozen state during this process.

Low molecular weight impurities which are not volatile, such as inorganic salts in natural products, can be removed by dialysis, or electrodialysis, of the polymer solution.

The separation of different macromolecular compounds which are present in the mixture is particularly difficult. This is especially so if the solubility is the same, so that extraction or precipitation does not bring about a separation. This problem is often encountered in protein chemistry. Since the protein molecules are usually present in the solution in the form of ions, one can use the difference in their rate of diffusion in an electric field, (which depends on the number of acid and basic amino acids in the macromolecule), for the separation of different proteins. This process is called electrophoresis. For an electrophoretic separation, one needs a special apparatus which permits optical control and registration of the separation process, and which in many cases also permits a preparative separation of the components. Figure 629-1 shows the electrophoresis diagram of blood plasma.*

Another process, gel permeation chromatography (GPC), has been used with great success for the purification of polymers, especially for the separation of different polypeptides. Because of its general importance for the fractionation of polymers, this process will be discussed later, together with the other methods of fractionation (see p. 406).

27 CHEMICAL TRANSFORMATION OF POLYMERS

271 Degradation Reactions (Chain Scission)

The term "degradation" indicates those reactions in which the polymer chain is split into smaller parts. There are two basic kinds of degradation: (1) degradation starting from the chain ends (i.e., the opposite of a polymerization reaction), where one monomer residue after the other is split off from the chain ends; and (2) statistical degradation, where a bond is split somewhere at random in the chain, so that fragments result which, on an average of one split per molecule, are half as large as the starting molecules. Whether case (1) or case (2) applies can in general be determined relatively easily if one follows the degradation by molecular weight determinations (e.g., by viscosity measurements) and characterization of the split-off monomer at different stages of degradation. In the case of degradation from the chain end, one rapidly obtains a larger amount of monomer, and the molecular weight decreases only rather slowly. With statistical degradation on the other hand, one finds rapid decrease of the molecular weight at the beginning, without obtaining significant amounts of monomer.

2711 Degradation from the Chain Ends (Unzipping)

Degradation from the chain ends is restricted to a few cases, of which the following are the most important:

(1) If one heats a solution of non-terminated poly-α-methylstyrene prepared anionically at −70°C in tetrahydrofuran (or some other suitable solvent), from −70°C to +60°C, one obtains degradation from the chain ends. Polymerization and degradation represent in this case a reversible equilibrium reaction in the temperature region from −70°C to +60°C. At −70°C there is practically no monomer present, and at 60°C there is practically no polymer. By simple cooling or heating (of course, under strict exclusion of air and moisture in order to reduce chain termination), one can bring this mixture to polymerize or depolymerize as often as desired. If at temperatures between −40°C and −70°C the chains are terminated

---

*This figure appears on Page 629.

through introduction of water, and if the solvent is removed, then degradation only takes place at higher temperatures, i.e., in the region between 150°C to 300°C and there with increasing rate. Degradation from the chain ends in that case is related to a statistical degradation.

(2) Dry distillation. If one heats polymethylmethacrylate to approximately 300°C, the monomeric methylmethacrylate distills and can be obtained almost quantitatively from the polymer under vacuum. With a lower yield, one can obtain the corresponding monomers of poly-α-methylstyrene, polystyrene, polybutadiene, polyisoprene (natural rubber), and other polymers. This may also be of interest for qualitative analysis of these polymers.

(3) On heating of polyformaldehyde to temperatures over 100°C, one obtains monomeric formaldehyde. This degradation can be prevented by blocking the terminal OH groups, i.e., by acetylation or methylation.

(4) The enzymatic degradation of natural products, for example, the degradation of starch (amylose) with β-amylase or phosphorylase. β-amylase degrades amylose to maltose, beginning with the chain ends. On the other hand α-amylase degrades the amylose chain statistically. Thus there are enzymes which degrade from the chain end but also others which degrade statistically (i.e., by random scission).

(5) If one heats polycaprolactam, which has been purified of monomer by extraction with water, over its melting point, then degradation takes place until the caprolactam concentration is approximately 6%.

2712 Statistical Chain Degradation (Random Scission)

Statistical chain degradation, which is generally possible with all polymers, may be accomplished by thermal and mechanical techniques, by ultrasound, and by the action of such chemicals as oxygen, acids, and alkalis.

Thermal Degradation

Unless thermal degradation occurs by regeneration of the monomer (as is the case with polymethylmethacrylate, poly-α-methylstyrene, and partially also with polystyrene and natural rubber), it usually is accompanied by charring and by formation of such small molecules as acetic acid, formic acid, acetone, methanol, methane, ethylene, hydrogen, and CO_2. The nature of the resulting products depends on whether the degradation is carried out in an inert gas atmosphere (nitrogen or CO_2), or in the presence of oxygen. In the latter case, in addition to the purely thermal degradation (cracking), one also obtains oxidative degradation.

Oxidative Degradation

Oxidative degradation plays an important role in the so-called aging of polymers. By "aging" one means slow (in terms of months and years) degradation, influenced by light, air, CO_2, and water, which results in a more or less rapid decrease in the mechanical properties of the polymer and finally in its complete breakdown. With most polymers, aging results in yellowing and increasing brittleness.

Unsaturated polyolefins are especially susceptible to the attack of oxygen (autoxidation). The polymers of butadiene and isoprene, for example, can be attacked relatively easily by oxidative degradation, and they finally decompose completely. Oxidative chain scission may occur through a cyclic peroxide as an intermediate step according to the following scheme:[34]

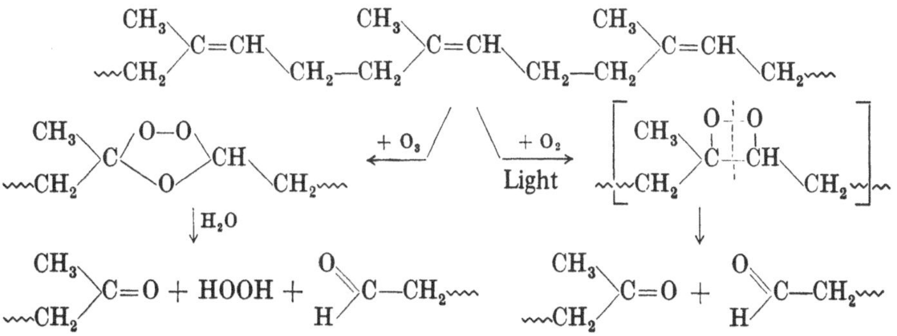

The chains of diene polymers are attacked by ozone even faster than by oxygen. Ozonides are formed as intermediate products, and these then decompose through hydrolysis. Complete ozone degradation of natural rubber yields levulinic aldehyde.

Not only unsaturated polymers, such as natural rubber, but also saturated polymers, such as polyethylene, polypropylene, polystyrene, polyamides, and cellulose, are affected in an undesirable manner by oxidation. In most cases hydroperoxide groups are formed in the first reaction step (chain reaction):

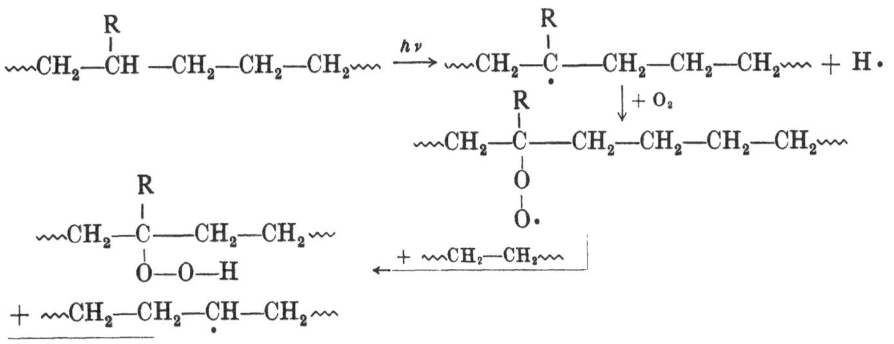

Analogous to the reactions used in the industrial production of phenol (and acetone) from cumene hydroperoxide, the hydroperoxides probably react further through chain scission:

[34]This is analogous to the oxidative scission of 1,1-diphenylethylene into benzophenone and formaldehyde. Similarly α-methylstyrene decomposes in the presence of air into acetophenone and formaldehyde.

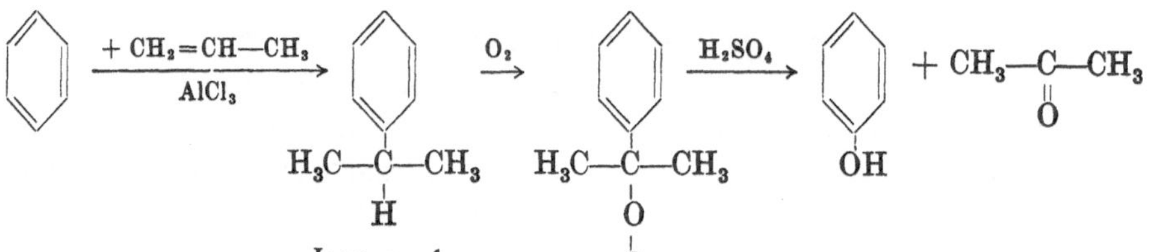

$$\text{benzene} \xrightarrow[\text{AlCl}_3]{+ CH_2=CH-CH_3} \text{Isopropyl-benzene} \xrightarrow{O_2} \text{Cumenehydroperoxyde} \xrightarrow{H_2SO_4} \text{phenol (OH)} + CH_3-\underset{O}{\underset{\|}{C}}-CH_3$$

Isopropyl-
benzene

Cumenehydroperoxyde

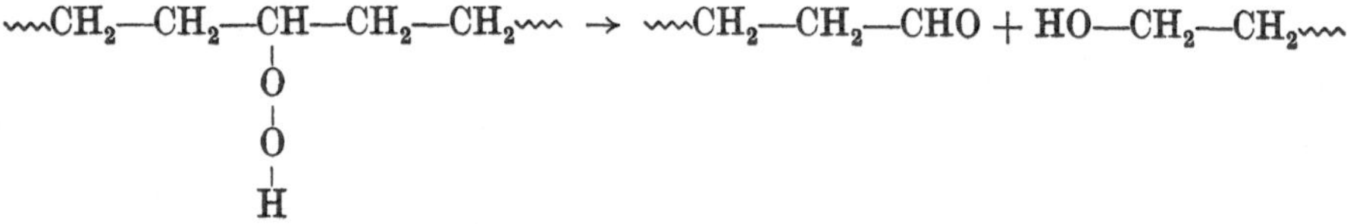

$$\text{wwCH}_2-CH_2-\underset{\underset{H}{\overset{\displaystyle O}{|}}{\overset{\displaystyle |}{\underset{}{\overset{O}{|}}}}{CH}-CH_2-CH_2\text{www} \rightarrow \text{wwCH}_2-CH_2-CHO + HO-CH_2-CH_2\text{www}$$

or with high pressure polyethylene with C_4 — side chains:

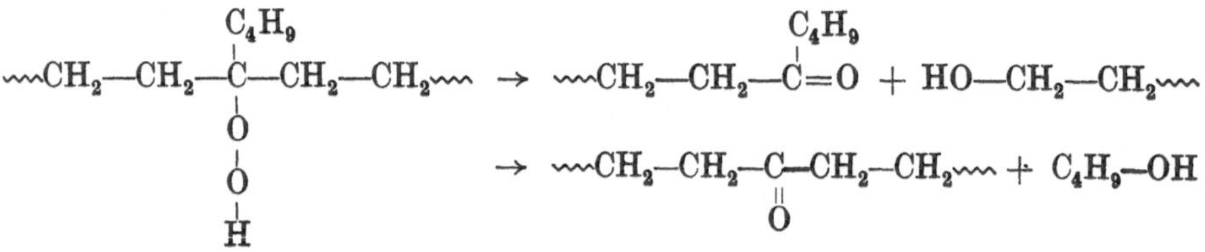

$$\text{wwCH}_2-CH_2-\underset{\underset{H}{\overset{\displaystyle O}{|}}{\overset{\displaystyle |}{\overset{O}{|}}}}{\overset{C_4H_9}{\underset{}{C}}}-CH_2-CH_2\text{www} \rightarrow \text{wwCH}_2-CH_2-\overset{C_4H_9}{\underset{}{C}}=O + HO-CH_2-CH_2\text{www}$$

$$\rightarrow \text{wwCH}_2-CH_2-\underset{O}{\underset{\|}{C}}-CH_2-CH_2\text{www} + C_4H_9-OH$$

One therefore adds stabilizers to these technical products which prevent the reaction by interfering with the mechanism. Such stabilizers are compounds which react readily with radicals or which absorb ultraviolet light (antioxydants or ultraviolet absorbers). Stabilizers which have been used are phenol or cresol derivatives, especially those which do not cause any darkening of the polymer, for example:

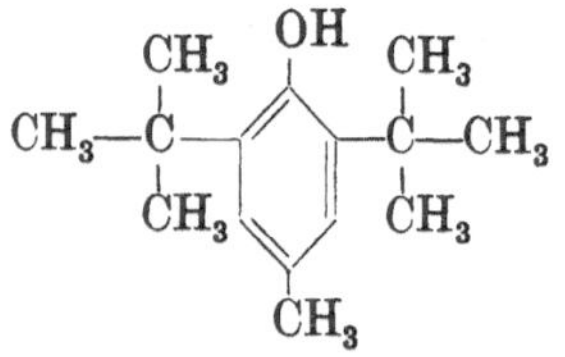

2,6-Di-t-butyl-p-cresol

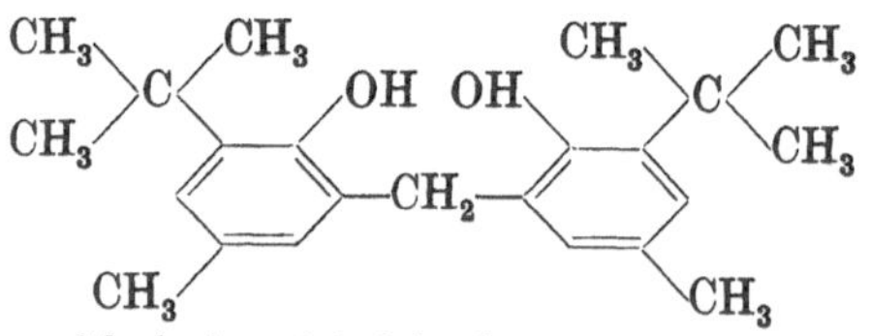

Methylene-bis-2-hydroxy-3-t-butyl-
5-methyl-benzene (Antioxydant)

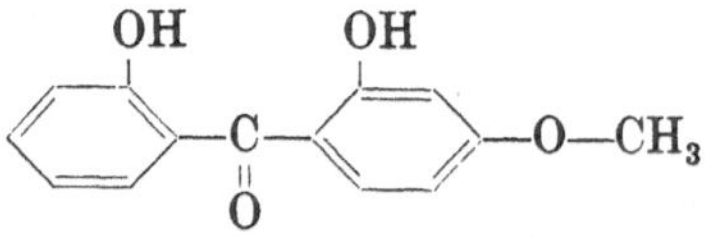

2,2'-Dihydroxy-4-methoxy-benzophenone
(UV-Absorber)

$P(OC_6H_5)_3$

Triphenyl-phosphite
(Antioxydant)

In addition to the phenolic compounds, there are a number of other compounds, especially amines, which have been proposed as antioxydants; however, most of them are applicable only in specific cases.

With isoprene and butadiene polymers, the sulfur which has been added for vulcanization purposes also has a stabilizing function. The sulfur is milled into the rubber at higher temperature and adds to the double bonds in the rubber chains. The carbon black which is added to rubber also carries out a stabilizing function.

Mechanical Degradation

Polymer chains can also be broken mechanically—for instance, by exposing a polymer solution to a strong shear gradient, by passing the solution through narrow orifices under high pressure, or by agitating it rapidly in a cylinder. When rubbery substances, or thermoplastic melts, are processed on a mill[35] or in an extruder, they are exposed to strong shearing forces and are mechanically degraded. The reduction in molecular weight brought about by this mechanical degradation is greater, the greater the molecular weight of the polymer before degradation. This is why mechanical degradation always leads to a certain leveling-off of the degree of polymerization after which no significant degradation occurs. In solution, the degree of mechanical degradation and the rate of degradation are dependent on the solvent: the better the solvent (the smaller the coil density and the greater the coil volume), the more rapid the degradation and vice versa.

Cross-linked polymers can be made thermoplastic and soluble by mechanical degradation, such as in the mastication of rubber. In this case, mastication is carried out intentionally to degrade the natural rubber, i.e., to make it more easily deformable and processable. In most other cases degradation is undesirable because it causes a deterioration of properties.

Degradation can also be obtained by intensive grinding of hard polymers in colloidal mills (ball mills) as was shown for the first time by K. Hess in the grinding of cellulose (more recently also by Battista).

It is hard to distinguish the molecular mechanism of mechanical degradation from that of thermal degradation. Especially with the mechanical degradation of solids or rubbery materials, it is possible to assume that chain scission occurs as a result of local overheating from the kinetic energy which has been imposed on the molecule through a sudden application of force. The clearest case of mechanical scission of macromolecular coils is demonstrated in the shearing of polymer solutions in a shear gradient, for example, when a solution is extruded through a narrow orifice.

---

[35]The shearing forces are the larger, the tighter the "nip" between the rolls. In order to achieve good mixing, one usually lets the rolls rotate at different speeds (in order to create friction). This leads to a considerable increase in the mechanical breakdown of the material being milled.

Degradation through Ultrasound

Chain degradation of dissolved polymers through ultrasound can be considered a special case of mechanical degradation. Since degradation is only found when irradiation of polymer solutions with ultrasound is carried out in the presence of dissolved gases (argon or H_2), one assumes that small pulsating gas bubbles (in resonance with the ultrasonic waves i.e., cavitation), cause the degradation. Through the rapid pulsation of these gas bubbles, one obtains strong local shear gradients, in which the coils are mechanically torn apart. At the places of chain scission, radicals are formed which react with oxygen or other radical absorbers (for example, diphenylpicrylhydrazil or quinone); but they are also able, in the presence of monomers, to initiate polymerization which gives rise to block-copolymers.

As with other cases of mechanical chain degradation, with degradation caused by ultrasound, the rate of degradation is the larger, the higher the molecular weight of the polymer. This was quantitatively expressed by G. Schmid:

$$dx/dt = k\,(P - P_{lim}),$$

according to which the number of bonds split in unit time (dx/dt) is proportional to the quantity $(P - P_{lim})$ by which the degree of polymerization, P, of the polymer being irradiated is larger than the limiting degree of polymerization, P_{lim}. The limiting degree of polymerization, P_{lim}, can never be decreased, even by a long period of irradiation, and is independent of the magnitude of the degree of polymerization of the starting material. Figure 629-2 shows the typical decrease in the viscosity of two polymers of different starting degree of polymerization being irradiated with ultrasound.[*]

Because of preferred degradation of molecules with high molecular weight, one finds that with ultrasound irradiation, just as with any other mechanical degradation, the polymers become more homogeneous.

2713   Chain Degradation through Hydrolysis

Polyethylene, polypropylene, polyisobutylene, and the chains of the vinyl polymers are not readily attacked by chemical reagents, except by oxygen or ozone under the influence of light. This reaction occurs more rapidly with unsaturated polymers (polyisoprene and polybutadiene) than with saturated polymers (compare p. 309). Well-defined, homogeneous degradation products are seldom obtained with oxidative degradation (as, for example, in the ozonization of natural rubber).

Hydrolytic Degradation through Acids and Bases

Chemical degradation of polymers whose chains are held together by functional groups (such as polyesters, polyamides, polyacetals, polysaccharides, and polyformaldehyde) can be more readily understood. Such polymers can be hydrolytically degraded with strong acids or bases, and, in the case of natural products, also by enzymes. Monomers are formed in all these processes. It is possible in this

---

[*]This figure appears on Page 629.

manner to produce glucose on a technical scale from starch and from cellulose (wood).

Hydrolysis of the polymer chains can be carried out very readily (as, for example, with starch) only if the polymers are soluble, or at least swellable, in water. This applies to most polysaccharides, but not to synthetic polymers with heteroatoms in the chain. Although chain bonds can be readily split through hydrolysis with acids or bases, the polymers are often rather resistant, because the attack can only occur at their surface, especially if the polymers are crystalline. If the polymers are finely divided (for example, through reprecipitation), or if one uses alcoholic solutions of acids or bases, the degradation can be speeded up.

Degradation of polysaccharides by acids can be easily understood because the linkages between the structural units in the chain are acetal bonds, and such acetals are readily hydrolyzed by acids. More difficult to understand is the fact that polysaccharides are degraded by alkalis. In this case the hydrolysis is preceded by an oxidation process in which carbonyl groups are formed. Because of these carbonyl groups, the glucosidic linkage in the chain becomes unstable in alkali. Pectin, in which such carbonyl groups (ester groups) are present in the molecule right from the start, can readily be degraded with 0.1 N sodium hydroxide, even with complete exclusion of oxygen, but only until the ester group is saponified. After that, the degradation stops. If one esterifies again, the degradation starts again. Because in an alkaline medium, chain scission and saponification of the ester groups along the chain always occur side by side, the number of alkali-unstable chain linkages, and therefore the rate of degradation, decreases under the influence of alkali to the same extent as the saponification of the methoxyl groups proceeds. Thus the saponification reaction behaves as an automatic brake in the alkaline degradation of pectin (see Figure 630-1).* Before each step the pectinic acid was reacted with diazomethane to regenerate the ester.

The chain scission can be formulate; in the following manner:

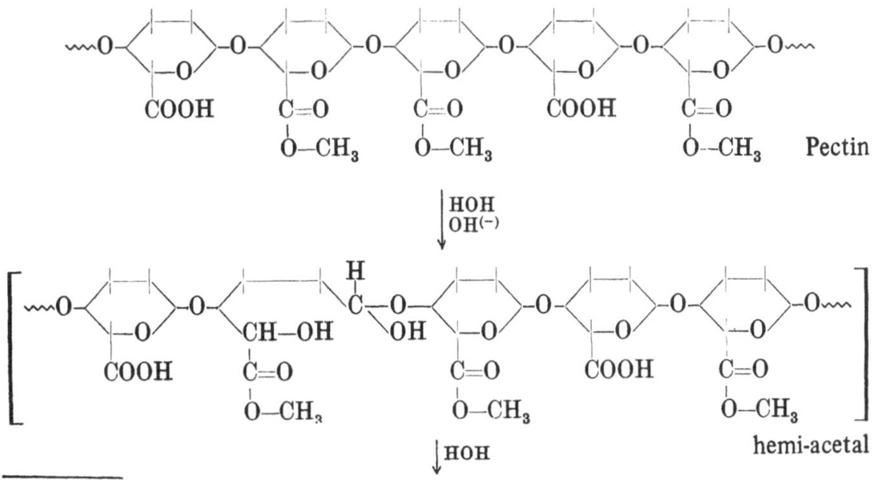

*This figure appears on Page 630.

In a similar manner one can also formulate the oxidative alkaline degradation of cellulose, which has practical importance because it is responsible for the loss in tensile strength of cotton fabrics on washing.

Enzymatic Degradation

A special form of hydrolytic degradation is degradation by enzymes. This process can be differentiated from acid or base hydrolysis, because enzymes usually degrade in a strongly selective manner, i.e., they split only certain well-defined bonds. For example, hydrochloric acid degrades polysaccharides as well as proteins and nucleic acids: (in the first case, glucosidic linkages are splits; in the second, peptide bonds; and in the third, phosphoric acid ester bonds). However, enzymes only attack one type of bond. In addition, among the enzymes which attack polysaccharides (glucosidases), one finds those which hydrolyze only α-1,4-linkages, others which hydrolyze only α-1,6-linkages, and still others which split only β-1,6-linkages, and still others which split only β-1,4-linkages. The degree of specificity of enzymes can be readily seen from the example of the amylases: α-amylase splits the amylose molecule randomly at any point along the chain, whereas β-amylase degrades the molecule beginning at the chain ends; neither α- nor β-amylase can degrade maltose to glucose (in spite of the fact that the glucosidic linkages in maltose can not be distinguished from the other glucosidic linkages in the amylose chain). On the other hand, the phosphorylases degrade starch in the presence of phosphate to glucose-1-phosphate. The proteolytic enzymes act on various compounds in a similarly specific manner. The most well-known are pepsin (the enzyme of the stomach, which is active in acid medium) and trypsin and chymotrypsin (pancreas enzymes, which are active in alkaline medium).

Enzymes are proteins with a special tertiary structure, and in some cases, with a catalytically active component called a prosthetic group or coenzyme. Their chemical nature can sometimes be distinguished with the aid of blocking reagents. Series of enzymes, for example, the amylases, can be obtained in crystalline form. Usually the hydrolytically active enzymes are named after the polymers which they degrade. Table 630 summarizes the polymers and the enzymes which degrade them.[*]

Enzymatic degradation, and in general, hydrolytic chain degradation, was, and still is, of great importance for the elucidation of the structure of polysaccharides, proteins, and nucleic acids. This applies especially to the stepwise degradation of these materials.

---

[*]This table appears on Page 630.

The procedure consists of carrying out the degradation under different reaction conditions (acid, alkaline, and enzymatic) to obtain low molecular weight chain segments with two or three structural units, whose constitution can readily be determined. In this manner one obtains information about the monomers, or structural units, of which the chain is built up. One also obtains information on the type of bonding between the individual structural units in the chain. Of especial importance is the elucidation of the chain structure of proteins where the chain usually consists of a large number of different amino acids. For the separation and identification of the structural products obtained in acid, alkaline, or enzymatic hydrolysis, one usually makes use of chromatographic methods (column chromatography or paper chromatography) and of electrophoresis. Through the interpretation of a large number of such protein hydrolysate analyses (made by automatic methods), it has been possible in a certain number of cases to determine the sequence of amino acids in the polypeptide chains of proteins; for example, with insulin (F. Sanger) and with the protein component of tobacco mosaic virus (Anderer and Schramm).

272 Reactions without Chain Scission

It is difficult to distinguish between the reactions which maintain the chain and those which degrade it, because many reactions of polymer chains which originally occur without chain scission, eventually result in chain degradation. One can separate these reactions into those which occur unintentionally, and those carried out on purpose in order to modify the properties of macromolecular compounds in certain well-defined ways.

To the first group belong (1) the aging reactions which, with many polymers, result in such modifications as yellowing and brittleness; and (2) those reactions which bring about decomposition phenomena at elevated temperatures and in a short time with some polymers, such as polyvinylchloride, polyvinylidene chloride, and polyacrylonitrile.

To the second group, i.e., the intentional reactions, belong the so-called polymer analogous transformations, which are carried out in order to demonstrate the macromolecular nature of certain compounds; also the technical processes for the preparations of certain plastics and elastomers, reactions which, in principle, are not different from polymer analogous transformations.

Polymer analogous transformations will be discussed in the section dealing with molecular weight (compare p. 332).

2721 Aging Phenomena

Aging phenomena are those processes responsible for the changes which occur in the course of weeks, months, and years in most polymers, especially plastics, at ordinary temperature under the influence of light, air, moisture, and CO_2. The results of these aging processes are usually the same: lowering of the mechanical strength, yellowing, etc. However, the individual reaction steps differ from one polymer to the other in most cases and are difficult to define. In some cases one deals with degradation reactions (however, in some cases cross-linking reactions are to blame). In many cases, especially with polymeric hydrocarbons (polyethylene, polypropylene, and natural rubber), one obtains hydroperoxide groups as the first step in a chain reaction brought about by light (autoxidation, compare p. 309).

The rate of aging is very different with different polymers. Polyvinylchloride, polyvinylidene chloride, polyacrylonitrile, polyacrylates, and polymethacrylates are those which are most resistant to aging and the influence of weathering. Those polymers least resistant to aging are natural rubber and synthetic rubber based on isoprene and butadiene, which are readily attacked by oxygen because of the double bond in the chain. Aging resulting from the influence of air was already discussed in section 2712 (p. 309).

The study of aging reactions is especially important for the discovery of suitable stabilizers, because the more clearly the reaction process is known, the simpler it is to prevent the reaction or to transform it in such a way that no damage results.

2722   The Degradation of Polyvinylchloride, Polyvinylidene Chloride, and Polyacrylonitrile at Elevated Temperature

Polyvinylchloride, polyvinylidene chloride, and polyacrylonitrile become discolored on heating to 80°C-160°C (even with exclusion of air). First they become yellow, then brown, and finally dark brown and black. As the color gets darker, the mechanical strength decreases. This degradation occurs most rapidly with polyvinylidene chloride and most slowly with polyacrylonitrile.

With polyacrylonitrile one assumes that cyclization of the chain occurs according to the following reaction scheme:

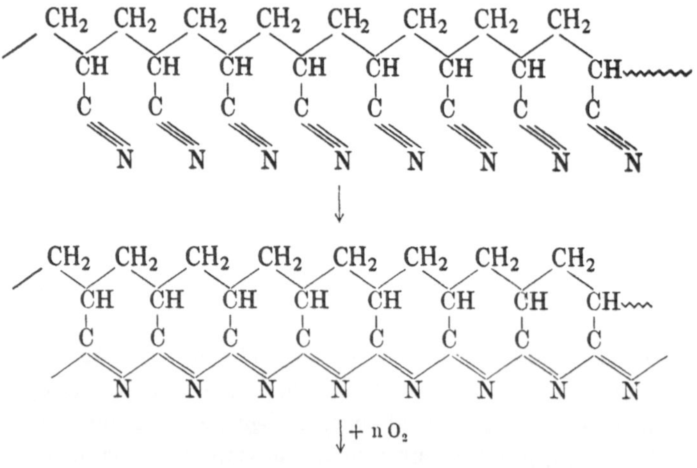

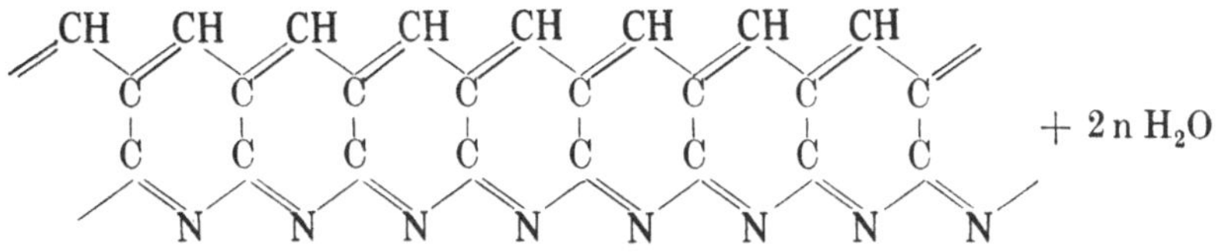

$$\begin{array}{cccccccc} CH & CH & CH & CH & CH & CH & CH & CH \\ | & | & | & | & | & | & | & | \\ C & C & C & C & C & C & C & C \\ | & | & | & | & | & | & | & | \\ C & C & C & C & C & C & C & C \\ | & | & | & | & | & | & | & | \\ N & N & N & N & N & N & N & N \end{array} \quad + 2\,n\,H_2O$$

Such compounds exhibit semiconductor properties to a certain degree (which however is not sufficient for technical use). Because of their ladder-like structure (ⅠⅠⅠⅠⅠⅠⅠⅠⅠⅠ) they are called "ladder polymers." It can not be demonstrated that the arrangement of the ring units in the chain of cyclized polyacrylonitrile as it is described by the formula shown above, is true for the whole polymer chain. Rather this formula represents a more or less long segment of the chain. As we know from similarly condensed hydrocarbons (acenes), as the number of condensed rings in the chain increases, the tendency of chains to mutually condense increases (compare p. 321).

The cyclization reaction of polyacrylonitrile occurs already below the glass temperature (softening temperature). Therefore polyacrylonitrile can not be machined by the methods used for thermoplastic materials (extrusion, injection molding, etc.). Polyacrylonitrile fibers can be manufactured only via solution of PAN in dimethylformamide (DMF). On the other hand, the cyclization reaction of PAN is used to produce a flame-resistant fabric ("Black Orlon," DuPont).

Polyvinylchloride and polyvinylidene chloride split off HCl more or less rapidly, depending on the temperature. The following equation summarizes this reaction, which is actually a chain reaction (Winkler, Frye and Horst, D. Braun):

$$\sim\!\!\sim\!CH_2\!-\!CH\!-\!CH_2\!-\!CH\!-\!CH_2\!-\!CH\!-\!CH_2\!-\!CH\!-\!CH_2\!-\!CH\!-\!CH_2\!-\!CH\!\sim\!\!\sim$$
$$\begin{array}{cccccc} \quad\; | & | & | & | & | & | \\ \;\; Cl & Cl & Cl & Cl & Cl & Cl \end{array}$$

$$\downarrow -4\,HCl$$

$$\sim\!\!\sim\!CH\!=\!CH\!-\!CH\!=\!CH\!-\!CH\!=\!CH\!-\!CH\!=\!CH\!-\!CH_2\!-\!CH\!-\!CH_2\!-\!CH\sim\!\!\sim$$
$$\begin{array}{cc} | & | \\ Cl & Cl \end{array}$$

Here also, the darkening is a result of the conjugated double bonds which are formed. The reaction is catalyzed by HCl and therefore is autocatalytic. Iron brings about a further increase in the rate. On further heat treatment of polyvinylchloride, in addition to the HCl split off, there is oxidative chain scission which finally leads to complete destruction of the material. Because the processing of polyvinyl-chloride is complicated by the splitting off of hydrochloric acid and therefore the usefulness of polyvinylchloride products decreased, one has to add stabilizers to the raw material. The compounds which have a stabilizing action are those which can bind the HCl which is formed. For example, one adds sodium carbonate to polyvinyl-chloride which has been produced by emulsion polymerization. Since the stabilizing action of sodium carbonate is not satisfactory in most cases, one has (and with success) tried to find better PVC stabilizers. The producer and processor of PVC nowadays has the choice of a number of compounds, depending on the end use of the PVC. These are usually organic compounds of zinc, tin, or cadmium. For example:

> Zincoctoate (zinc salt of 2-ethylhexanoic acid)
> Cd-caprylate
> Ba-Cd-laurate-mixture

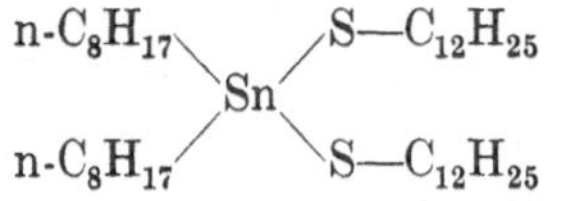

$$\begin{array}{ccc} \text{n-}C_8H_{17} & & S\text{—}C_{12}H_{25} \\ & \diagdown Sn \diagup & \\ \text{n-}C_8H_{17} & \diagup \quad \diagdown & S\text{—}C_{12}H_{25} \end{array}$$

> Di-n-octyl-tin-didodecylmercaptid

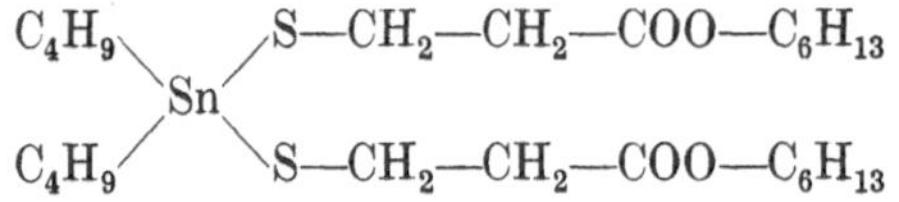

$$\begin{array}{ccc} C_4H_9 & & S\text{—}CH_2\text{—}CH_2\text{—}COO\text{—}C_6H_{13} \\ & \diagdown Sn \diagup & \\ C_4H_9 & \diagup \quad \diagdown & S\text{—}CH_2\text{—}CH_2\text{—}COO\text{—}C_6H_{13} \end{array}$$

> Dibutyl-tin-bis-thioglycolic acid hexyl ester

The mode of action of such compounds is not restricted to the binding of free HCl. One assumes that already in the polymer chain there is an exchange of labile chlorine atoms with $R-$, $R-S-$, or $R-\overset{\displaystyle \|}{\underset{\displaystyle O}{C}}-O-$ groups, so that HCl is never split off.

2723   Reactions Along the Polymer Chain

While one tries to prevent reactions which modify the polymer in an undesirable manner, there are other reactions which one carries out in order to modify the properties of the polymer in a certain well-defined way. The following reactions have been carried out on a technical scale:

cellulose + NaOH + CS_2 → cellulose xanthate

$$\downarrow \quad + H_2O$$

cellophane, cellulose fibers

cellulose + acetic anhydride → celluloseacetate (fiber)

cellulose + HNO_3 → cellulosenitrate (explosive)

cellulose + dimethylsulfate → methylcellulose (water soluble thickening agent)

cellulose + reactive dyestuffs → dyed fibers (dye component bound by covalent bonds)

polyethylene + Cl_2 + SO_2 → sulfochlorinated polyethylene (special rubber HYPALON)

polyisoprene or polybutadiene + sulfur → (vulcanized) rubber

polystyrene (cross-linked) + H_2SO_4 → sulfonated polystyrene (cation-exchanger)

polystyrene + $CH_2 = O$ + HCl ——→ chloromethyl polystyrene

$$\downarrow + NH_3$$

Aminomethylpolystyrene
(anion-exchanger)

polyvinylchloride + Cl_2 → chlorinated PVC (primary material for special fibers)

polyamides (e.g., nylon-6) + ethylene oxide → water-soluble ethoxy-derivatives

polyamide carboxylic acid $\overset{250°C}{→}$ polyimide (see p. 227).

polyvinyl acetate + KOH → polyvinyl alcohol

polyvinyl alcohol + aldehydes → polyvinylacetals

If one says that with these reactions the polymer chain remains unchanged, this means only that the reaction products are still macromolecular compounds with molecular weights between 50,000 and 500,000. The reactions do not all occur in a strictly polymer analogous manner, for, because of various side reactions, it is not always possible to prevent a certain chain degradation, especially under technical conditions.

If only a few of these reactions have achieved technical importance, this is because macromolecular compounds are not very satisfactory starting materials. They can not be distilled or recrystallized. The solutions have a high viscosity which makes stirring difficult; also heat exchange, the passage through pumps, and especially filtration become difficult, so that one has to work with relatively dilute solutions. However, the removal of the large amounts of solvents then becomes a considerable problem, especially when side products have to be removed, so that the polymer has to be precipitated. These difficulties bring about a price increase in the resulting products, and one therefore prefers through suitable synthesis, for example, copolymerization, to build the desired structures into the polymers rather than to modify them afterwards.

However, polymer modification may be the only way to obtain certain reaction products of considerable interest because of their special properties. This area has therefore become more important in spite of the above-mentioned technical difficulties. This scope of this area of investigation is unlimited, because polymer chains can be submitted to a host of different organic reactions. Particularly suitable starting materials are polymers which have reactive functional groups, such as polyacrylic acid, polyvinyl alcohol, cellulose, and maleicanhydride copolymers.

Not all the structural units have to take part in these reactions. Often it is not even possible to carry out the reaction in a quantitative manner. In many cases one wants only a small number of the structural units to react—for example, if one wants to produce graft-copolymers or cross-linked polymers. Such cross-linking reactions are especially important with elastomers, lacquers, and casting resins.

Two groups of reactions along the polymer chain are described in the following paragraphs in more detail: cyclization reactions and cross-linking reactions.

Cyclization reactions

Polymers with cyclic structural units in the chain can be synthesized by ring-forming polymerization (cyclopolymerization, see pp. 58), but they can be prepared also by cyclization reactions of preformed polymers.

One of the oldest examples is the reaction of polyvinyl alcohol with formaldehyde resulting in polyvinylformal:

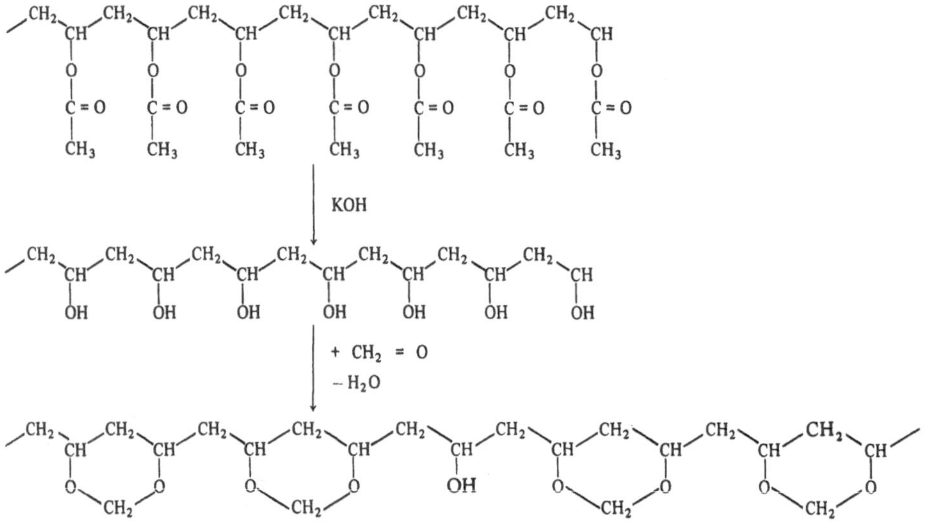

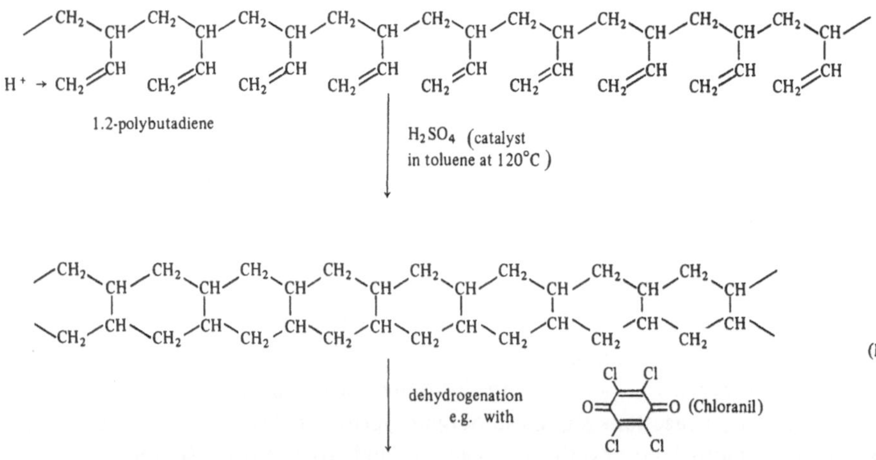

Another type of cyclization reaction is the cationic polymerization of 1,2-polybutadiene (I), which can be prepared by polymerization of butadiene with a complex of butyl lithium and tetramethylethylene diamine (for instance) as initiator

possible structure of aromatized cyclopolybutadiene   (III)

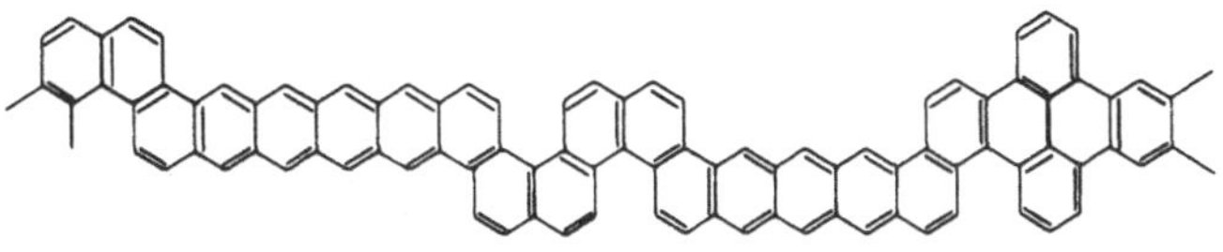

This reaction is carried out on a technical scale and leads to "Pluton" fabric (3 M), a black, inflammable (up to 1,500°C) material, which is insoluble and infusible. Therefore, the spinning process is made with the soluble intermediate ladder polymer (II). The structure of the final ladder polymer (III) has not been elucidated. Since acenes with more than six linearly condensated rings are unstable even at normal temperature (Clar and Boggiano) and form higher condensed systems with a more stable aromatic character, the ladder polymer (III) will probably have an irregular structure with different condensed ring systems with a beginning tendency to form graphitelike structures. The graphitizing process is completed at higher temperatures (> 1,000°C).

In the last few years many other fibers (e.g., cellulose, polyacrylonitrile, PVC, and PPO) have been treated at high temperatures to form carbon fibers (compare p. 317).

A third group of cyclization reactions leads to heterocyclic unit-containing polymers by intramolecular splitting-off of water at elevated temperatures:

$$\text{polyamino amides} \xrightarrow{250°C} \text{polybenzimidazoles ("Imidite", Narmco)}$$

$$\text{polyamide carboxylic acids} \xrightarrow{250°C} \text{polyimides ("H-film", "Kapton", "Vespel" Du Pont)}$$

$$\text{polyamino amide carbocylic acids} \rightarrow \text{polyimidazopyrrolones}$$

These reactions have been described previously on page 225 ff.

Cross-linking reactions

"Crosslinking" means covalent, or secondary valence-type linking between polymer chains (for structure and reaction schemes compare pp. 24-30; pp. 553-559, and pp. 288-301). As long as not all the polymer chains in the system are involved in cross-linking, i.e., if on the average only two, three, five, or ten polymer chains are linked through the cross-linking reaction, then these polymer bundles are still soluble. In these preliminary stages of cross-linking one can still speak of branched macromolecules. However, if the cross-linking reaction proceeds further, then these cross-linked polymer chain systems soon reach a critical size where insolubility in all solvents occurs. This is true for covalent cross-links. For secondary valence cross-links (e.g. H-bonds), it is characteristic that there are certain solvents, in which these cross-links are removed. Secondary valence cross-linked polymers are furthermore mostly thermoplastic.

Because of the long chains of the polymer molecules, quite small amounts of the cross-linking agent are usually sufficient to bring about total cross-linking of the system, i.e., involving all the polymer chains. This necessary amount is the smaller, the larger the molecular weight of the polymer (compare p. 25).

Because of the polydispersed character of most macromolecular compounds, the critical cross-linking concentration, where the polymer becomes insoluble, is usually only surpassed by the highest molecular weight components. In this stage, one part of the macromolecules (those with shorter chains) is soluble; however, the other parts (the molecules with the longer chains are insoluble. On further increasing the number of cross-linking points, the soluble parts become smaller and smaller. Thus, for example, with a copolymer of styrene and 0.008% divinylbenzene, approximately one third is still soluble in toluene, whereas the rest remains as a strongly swollen gel. With 0.01% divinylbenzene, only 27% can be dissolved, and with 0.1% divinylbenzene, only 5% of the cross-linked copolymer can be dissolved with toluene. This refers to a molecular weight of $M_w \approx 1,000,000$ (determined on the basis of an uncross-linked polystyrene prepared under similar conditions).

The reactions which lead to cross-linking of macromolecular compounds can be divided into three groups: (1) cross-linking through copolymerization; (2) cross-linking through radical combination; and (3) cross-linking through functional groups.

Before describing the different methods of cross-linking, we have to consider another point of general interest: Polymer molecules are random coils. Therefore, in general, cross-linking reactions can occur within a coil between two chain segments of the same polymer chain (*intra*molecular cross-linking), and they can occur between chains of different coils (*inter*molecular cross-linking). Figure 554 shows the two types of cross-links schematically.

Intramolecular cross-linking reactions alone do not lead to cross-linked polymer systems. They result from cross-linking reactions in extremely dilute polymer solutions and cause a coil contraction and thus a decrease of the viscosity number (compare p. 512). Exclusively intermolecularly cross-linked systems can be prepared by reactions of two polymers of the same type with different reactive functional groups, the group X being in one polymer, and the group Y in the other. By means of such reactions it was shown, by Vollmert and Stutz, that polymer coils can not penetrate one another (see pp. 494). Therefore, the functional groups in the interior of the X-coils are not attainable for the groups of the Y-coils.

In nearly all cross-linked polymers the two types of cross-links are mixed so that there is a homogeneous cross-link distribution in the whole system. For some remarkable differences in the mechanical behavior of homogeneously cross-linked and only intermolecular cross-linked systems, see p. 558.

Cross-linking Through Copolymerization

Copolymerization can be used to cross-link macromolecules by carrying out the polymerization in the presence of compounds which contain more than one polymerizable double bond in the macromolecule:

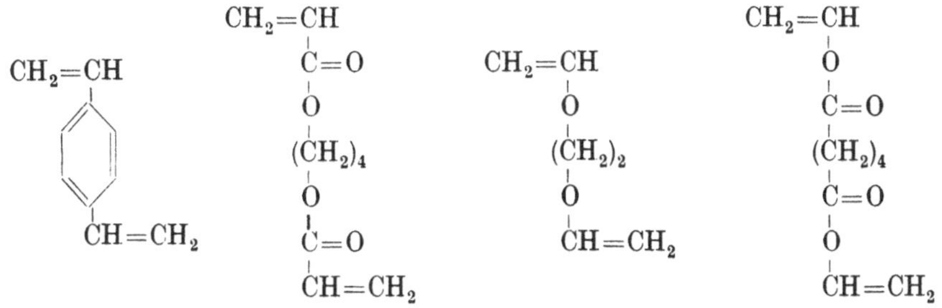

$$CH_2=CH$$

Divinylbenzene

$$CH_2=CH$$
$$C=O$$
$$O$$
$$(CH_2)_4$$
$$O$$
$$C=O$$
$$CH=CH_2$$

Butanediol-
diacrylate

$$CH_2=CH$$
$$O$$
$$(CH_2)_2$$
$$O$$
$$CH=CH_2$$

Glycoldivinyl-
ether

$$CH_2=CH$$
$$O$$
$$C=O$$
$$(CH_2)_4$$
$$C=O$$
$$O$$
$$CH=CH_2$$

Adipic acid
divinylester

$$CH_2=CH-CH_2-O-CH=CH_2$$
Allylvinyl ether

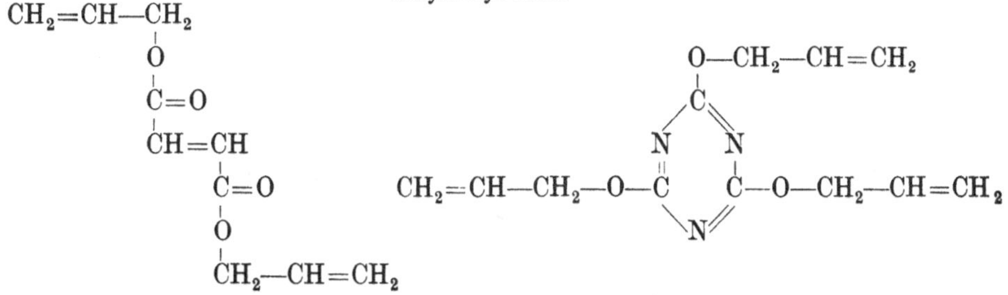

$$CH_2=CH-CH_2$$
$$O$$
$$C=O$$
$$CH=CH$$
$$C=O$$
$$O$$
$$CH_2-CH=CH_2$$
Diallylfumarate

$$CH_2=CH-CH_2-O-C$$

Triallylcyanurate

In this group of compounds with several polymerizabl_ double bonds in the molecule, one may include unsaturated polymers (for example, polybutadiene or unsaturated polyesters) such as are produced on a technical scale from glycols and maleic anhydride:

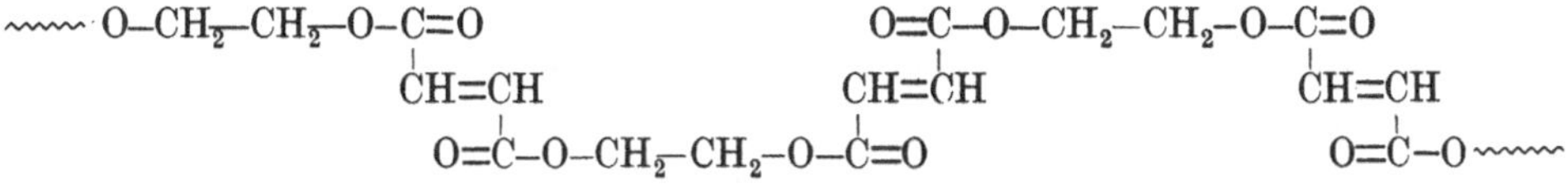

Fumaric Acid – Glycol-Polyester

If such molecules are incorporated into a polymer chain through one of their double bonds, then the other double bond can be incorporated into another polymer chain thus bringing about cross-linking (compare p. 26 and also pp. 299-301). Already the addition of 0.01% divinylbenzene in styrene polymerization is sufficient to bring about a cross-linked polymer (which however still swells strongly in benzene or toluene). Through variation of the concentration of the cross-linking agent it is possible to prepare cross-linked polymers of any desired degree of cross-linking. With dissolved unsaturated polyesters in styrene, one prepares casting resins on a large industrial scale, by placing these solutions in molds which have been filled with glass fibers and then polymerizing them in the mold.

With this first type of cross-linking through bifunctional or poly-functional monomers, the cross-linking usually occurs during copolymerization. This has the result that the polymers formed in this way can not be fabricated after polymerization, unless one degrades the resulting cross-linked polymer thermally or mechanically to such an extent that the cross-linked material forms branched molecules. With bulk and solution polymerization in the presence of bifunctional monomers, after a low degree of conversion, one obtains a gel which can no longer be stirred, so that one has to carry out the reaction very slowly or in relatively small cans (high surface/volume ratio) so that the heat of polymerization can also be removed without stirring. Only pearl polymerization and under suitable circumstances also emulsion polymerization can be carried out in the presence of bifunctional monomers without running into difficulty.

As a special type of cross-linking through copolymerization by means of di- or polyvinyl compounds, we have to consider the case in which these compounds are macromolecular substances. The best known example, unsaturated polyesters, was mentioned before, but many other unsaturated polymers can be prepared and copolymerized with a wide variety of monomers. Cross-linked graft-copolymers are generally formed by these reactions, some of which were described previously on pp. 299 ff).

Because of the closed character of polymer coils and because of the incompatibility of different polymers, the conversion of the double bonds present in the unsaturated primary polymers is not complete, even if this should be possible according to the copolymerization parameters of the system. As was confirmed by recent work (Vollmert and Mueller), the degree of conversion depends strongly on the molecular weight of the unsaturated polymer, because the double bonds become less available with increasing molecular weight. That means: the monomer in which the unsaturated polymer was dissolved prefers to polymerize in the spaces between the primary polymer coils rather than inside the coils. We do not understand this phenomenon completely, but it seems to relate to the fact that random coils do not penetrate one another if they can avoid it (compare pp. 494).

The minimum number of double bonds in an unsaturated polymer with cross-linking ability is two per macromolecule. In order to be sure that every polymer chain has two double bonds, one can prepare this type of unsaturated polymer through reaction of polymers with two reactive end groups with suitable monomers, such as reactive acrylic acid derivatives (acrylic acid chloride, isocyanato-ethyl-acrylate, butanediol-monoacrylate, acrylic acid amidomethylol ethers, etc.). With the aid of such polymers with unsaturated end groups one can synthesize cross-linked systems with distinct structures and properties, especially if different monomers are used for the unsaturated polymers on the one hand, and the cross-linking polymerization on the other hand (compare Figure 325).

No cross-linking occurs with divinyl monomers in which the distance between the vinyl groups is just so large that 6-membered rings are formed by chain propagation, as for example with acrylic anhydride (compare p. 58).

In every case, ring- (or loop-) forming reactions (=intramolecular crosslinking) and intermolecular cross-linking are competing reactions, as shown in Figure 326. In a first copolymerization step a copolymer with vinyl side groups is formed. These vinyl groups, fixed on the growing chains, can react in a second copolymerization step with the growing end of the same chain by forming a loop or with the growing end of another chain by forming an intermolecular cross-link.

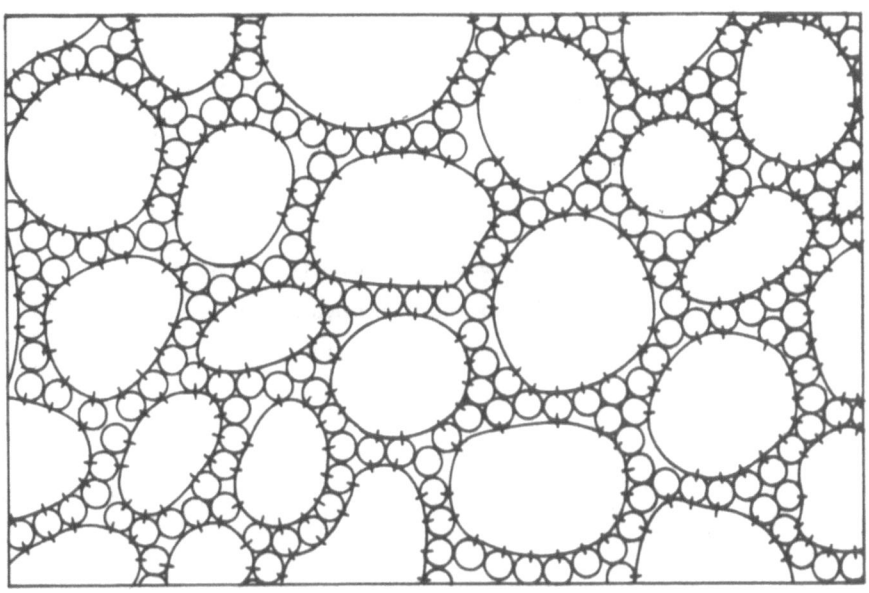

Examples:

A	B
Polystyrene	Polytetrahydrofuran
Polymethylmethacrylate	Polybutylacrylate
Polyacrylonitrile	Polyester (hexanediol-adipic acid)
Polybutylacrylate	Polystyrene
Polyvinylpyrrolidone	Polyacrylic acid

FIG. 325 — Cross-linked polymer-polymer-system with a special tertiary structure

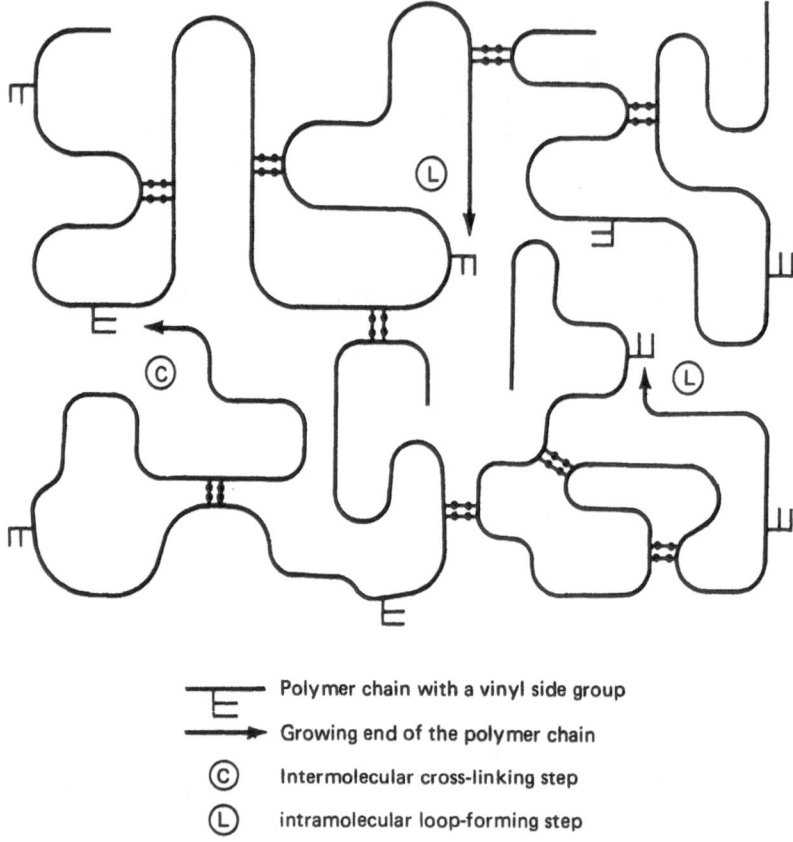

⊢ ⊐	Polymer chain with a vinyl side group
⟶	Growing end of the polymer chain
©	Intermolecular cross-linking step
Ⓛ	intramolecular loop-forming step

FIG. 326 – Diagram representing cross-linking copolymerization

Cross-linking Through Radical Combination

The second method for cross-linking is through radical combination. It does not have any large technical importance. It is most readily brought about through irradiation of the polymers by (X-rays, electrons, or radiation from radioactive isotopes). For example, one can cross-link polyethylene which thereby loses its plastic character to some extent and becomes harder and transparent.

Cross-linking Through Functional Groups

A prerequisite for this type of cross-linking is the presence of reactive groups in the polymer chain. Such polymer chains are easily obtained through copolymerization with monomers which contain the desired group, for example, acrylic acid, maleic anhydride, aminostyrene, glycidylacrylic ester, glycolmonoacrylate, and isocyanato-acrylic ester (compare p. 148); or through chemical reactions of polymers, for example, OH groups through saponification of vinyl acetate polymers or copolymers. If such polymers are reacted with suitable polyfunctional compounds such as glycols, bisepoxides, diisocyanates, diamines, and dicarboxylic acids, then these compounds react with the functional groups of polymer chains, and cross-linking is brought about. Thus with polymeric coatings one can apply solutions which have been mixed with a cross-linking agent, and which, on heating, bring about cross-linking and the coating becomes insoluble. This type of cross-linking is of greater importance with polymers which have been prepared through polycondensation (for example, with polyesters and with phenol-formaldehyde, or urea-formaldehyde resins). With polycondensation the cross-linking agents are usually trifunctional monomers (trifunctional in relation to the number of reactive groups taking part in the polycondensation). Examples are glycerol, cyanuric acid, melamine, trichlorosilane, phenol (reaction scheme on p. 25, p. 205, and pp. 222-224). From phthalic anhydride and glycerol, under addition of unsaturated fatty acids, and in many cases also styrene, one prepares the alkyd resins. From phenol and formaldehyde one prepares the phenol-formaldehyde resins, and from urea (or melamine) and formaldehyde, one forms the urea-formaldehyde (melamine-formaldehyde) resins. Initially the condensation is only carried to the point where the condensates are still soluble and still flow. For the preparation of molded products or plates, these precondensates are mixed with fillers such as lignin or wood flour, or one impregnates layers of textiles; then the whole is heated under a press, so that the condensation is carried out further, and one obtains resins with a cross-linked structure. In using such compounds as raw material for lacquers, the objects which are to be coated—for example, automobiles—are first coated with the condensate. Then the whole object is heated, and the coating is cross-linked to the final structure.

In this manner one also prepares products from polyester rubbers. One reacts the relatively free-flowing polyester with diisocyanate and then lets the reaction continue to the stage where the mass still can flow. Finally, the reaction of the isocyanate group of one chain with an active hydrogen of another chain, which causes the cross-linking, occurs in the final mold.

Cross-linking Through Secondary Valences

If one extends the concept of secondary valence cross-linking sufficiently, one may characterize all solid and rubbery polymers as cross-linked through secondary valence forces, because in every solid and liquid material the molecules are more or less tightly bound by secondary valence forces. However, there are certain groups, for example COOH- or $CONH_2$-groups, which, because of hydrogen bond formation, are especially tightly bound together. Thus if there are such groups in a nonpolar macromolecule, the chains will be more tightly held to each other at the places where these groups are to be found, and this resembles a regular cross-link (compare, Ionomers):

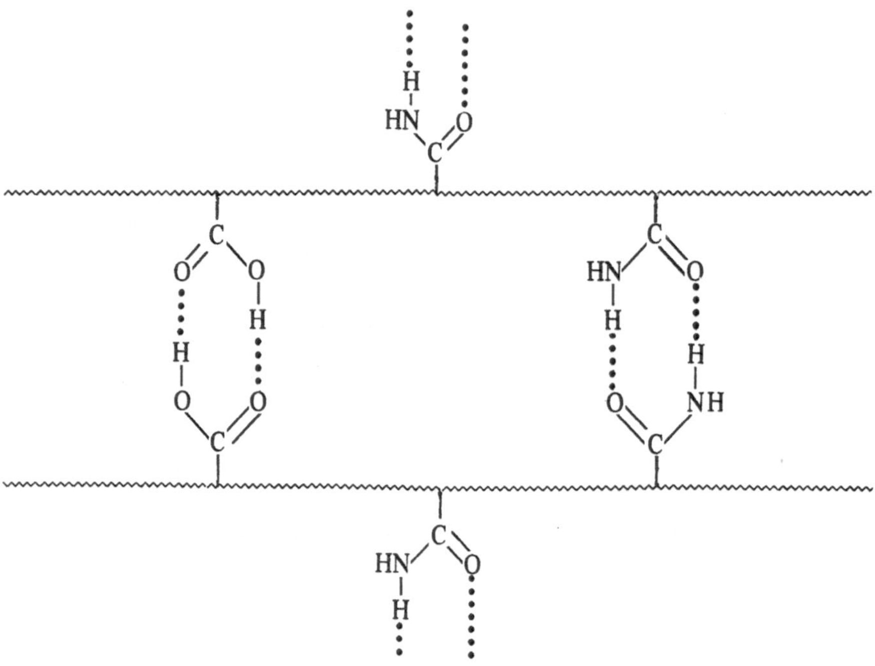

Such secondary valence cross-linking does not always have to be caused by hydrogen bonds. In principle all groups with high molar cohesion are capable of secondary valence cross-linking, especially highly polar groups, such as neutralized carboxyl groups. This type of secondary valence cross-linking is used industrially: in the copolymerization of ethylene with small amounts of acrylic acid (10%) and subsequent neutralization of a part of the carboxyl groups ("Surlyn A", Du Pont).

Secondary valence cross-linking distinguishes itself from covalent cross-linking particularly by the fact that such secondary valence cross-links are thermo-reversible, i.e., the cross-links disappear on heating, and on cooling they are formed again. With thermoplastic polymers this has considerable advantages. Covalent cross-linking destroys the thermoplastic processability (compare p. 564) of polymeric materials, or at least it is very much hindered even at small amounts of cross-linking, whereas with secondary valence cross-linking the processability is much less affected.

Secondary valence cross-links can be removed not only by heating, but also by treatment with suitable polar solvents. This is used in the preparation of elastic fibers (Lycra, Virene). Thus in polar solvents, such as dimethylformamide, the chains consisting of nonpolar polyether segments and strongly polar polyurea or polyurethane segments (compare 241) are soluble. After removal of the solvent during spinning, the resulting threads behave like vulcanized rubber.

Secondary valence cross-linking plays a highly important and decisive role in nature: all proteins are reversibly cross-linked through the CONH groups. By changing the pH value, or the salt concentration, one can create or remove cross-links. A particular role is played by hydrogen bonding between molecular chains in the formation of helices, and particularly in the formation of the double helix of the nucleic acids (compare p. 260).

Removal of Cross-linking

A cross-link is most simply removed when it has been brought about by secondary valence forces—for example, in the case of plasticized polyvinylchloride (compare pp. 518, 545), or with gelatins or pectins, which go over into solution on warming. With main valence cross-linked materials, an elimination of cross-linking can only be brought about without a corresponding change in the polymer chains, if these cross-links are different from the main chain links chemically, i.e., if the chain consists of carbon-carbon bonds, and the cross-links are ester bonds. In such cases one can break the ester bonds through hydrolysis, without attacking the polymer chains.

One is often less interested in removing the cross-linking, than in removing certain effects of the cross-linking. For example, polystyrene loses its thermoplasticity through cross-linking; however, through mechanical degradation (strong milling or grinding) it can obtain a thermoplastic character again. In the degradation of cross-linked polymers, depending on the degree of cross-linking, one can obtain more or less branched macromolecules. Systems where cross-linking occurs between polymer chains of different character, result in graft-copolymers on mechanical degradation, as shown in the following scheme:

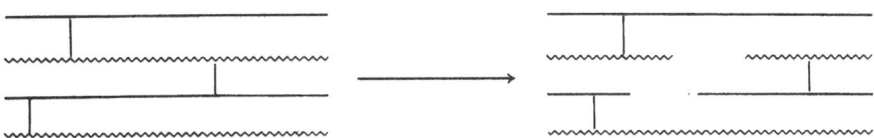

Vulcanization of Rubber

The cross-linking reaction which is carried out on the greatest technical scale is the vulcanization of natural and synthetic rubber. In the strictest sense, one understands under this reaction the treatment of polyisoprene, or butadiene copolymers, with sulfur, or sulfur compounds, in the presence of vulcanization accelerators such as, 2-mercaptobenzothiazole (I), tetramethylthiuram disulfide (II), or zinc dimethyldithiocarbamate (III):

In this reaction the double bonds present in the chains of the polyisoprene, or polybutadiene, react with sulfur with the formation of sulfur bonds between the polymer chains. A strictly schematic representation of cross-linking reactions occuring during the vulcanization is not possible, at least not of the reactions occuring during hot vulcanization where the rubber is milled with 2-4% sulfur and vulcanization accelerators, and is then heated after molding to 130°C-150°C. With this type of vulcanization one does not deal with a single, well-defined reaction, but with a series of reactions, including hydration and dehydration. One thing is certain, that the polymer chains of the vulcanized rubber are cross-linked and that bonds between the chains can be formed through single sulfur atoms, or through chains of sulfur atoms.

In addition to hot vulcanization, one can dip thin layers of rubber into solutions of S_2Cl_2 in carbon disulfide (cold vulcanization).

Since it is impossible to isolate well-defined reaction products in the vulcanization of rubber, one has to investigate model reactions involving the reaction of low molecular weight olefins with sulfur or sulfur compounds. Thus the following reactions bring about cross-linking through sulfur chains:

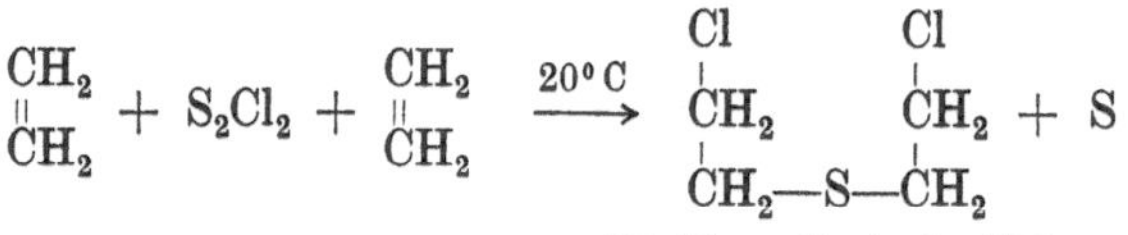

Dichlorodiethylsulfide

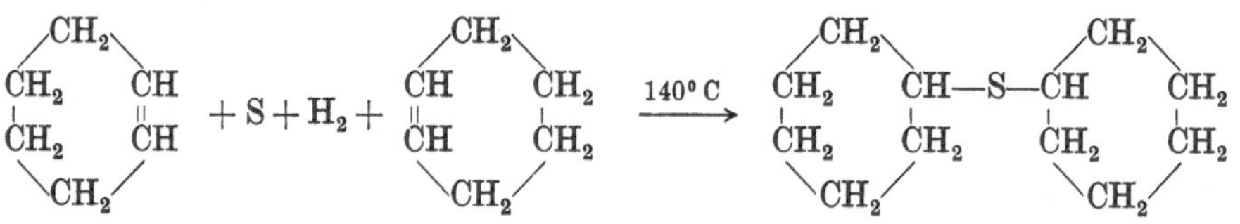

Cyclohexene   Dicyclohexylsulfide

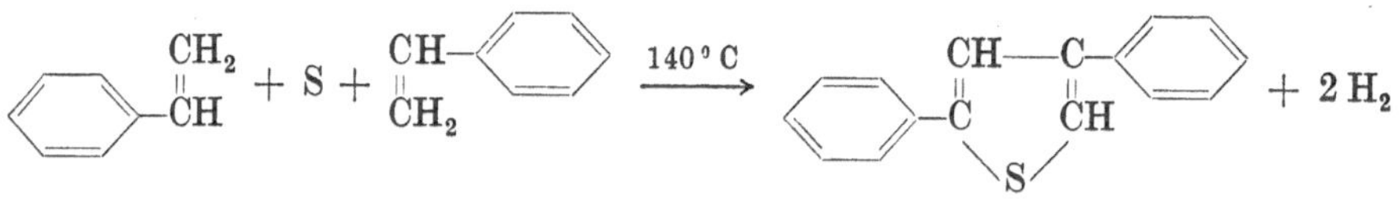

Styrene   Diphenylthiophene

The term vulcanization is used for all the reactions which lead to the cross-linking of elastic polymers (elastomers). Thus one can vulcanize unsaturated elastomers (from isoprene or butadiene), not only with sulfur, but also with peroxides, trinitrobenzene, dinitrosobenzene, and urea-and phenol-formaldehyde precondensates, whereby especially the last possibility is also of technical interest. In addition, there is a certain number of saturated rubber-elastic polymers which can not be cross-linked with sulfur, so that one has to look for other cross-linking agents and cross-linking methods (compare Table 565).

2724   Reactions at the Chain End

Because the end groups of macromolecular compounds are always present only in low concentration, one can only use highly reactive compounds for reaction with chain ends (for example, isocyanates and epoxides). The diisocyanates especially are used in order to increase the chain length of polymers having OH or COOH end groups (especially polyesters and polyethers):

$$HO \sim\!\!\sim\!\!\sim OH + O = C = N - R - N = C = O + HO \sim\!\!\sim\!\!\sim OH + O = C = N - R - N = C = O + \ldots$$

$$
\underset{\displaystyle HO\sim\!\!\sim\!\!\sim O - \overset{\overset{\displaystyle O}{\|}}{C} - NH - R - NH - \overset{\overset{\displaystyle O}{\|}}{C} - O \sim\!\!\sim\!\!\sim O - \overset{\overset{\displaystyle O}{\|}}{C} - NH - R - NH - \overset{\overset{\displaystyle O}{\|}}{C} - O\sim\!\!\sim\!\!\sim}{}
$$

$\sim\!\!\sim\!\!\sim$ = Polyether (or polyester) chain

According to these principles, one can prepare specialty rubbers based on polyesters, foams based on polyethers, and elastic fibers based on polyesters or polyethers (see p. 239).

Reactions of this type are reactions of polymer molecules with one another: a polyether chain reacts in a first step with a diisocyanate and then, in a second step, with another polyether chain. The rate of such reactions is in a certain range strongly dependent on the molecular weight. Polymer chains above a certain length (corresponding to a molecular weight of about 10,000 to 20,000) are arranged as random coils. Because polymer coils can not penetrate one another, the probability of reaction decreases with increasing molecular weight.

All condensation polymerizations, i.e., all polymer syntheses occurring by means of functional groups, are, at higher conversions, reactions of polymer molecules with other polymer molecules of the same type (with few exceptions). Therefore, it is understandable that it is nearly impossible to prepare polyesters or polyamides with high molecular weights of about 500,000 or more. This is possible only in the case of highly reactive functional groups (i.e., N=C=O-groups or carboxylic anhydride groups), especially if the resulting polymer coils are relatively open, as for example with polyimides prepared from pyromellitic anhydride and benzidine. Another important example of a reaction of polymer molecules with highly reactive end groups is, of course, the recombination of growing chains during radical polymerizations.

3. THE PROPERTIES OF THE INDIVIDUAL MACROMOLECULE

31 MOLECULAR WEIGHT

311 Polymer Analogous Conversion

Even before the start of Staudinger's work on macromolecular compounds, it was known that starch and cellulose must be polymers of glucose, since they yield glucose when hydrolyzed with acids. Also, the type of linkage between the glucose units, β-1,4,-glucosidic with cellulose and α-1,4-glucosidic with starch, was already exactly known. However, it was still unexplained why starch and cellulose have such completely different physical properties from the known crystallized oligomers, i.e., the di-, tri-, tetra-, up to the deca-saccharides. Since osmotic measurements had shown that in cellulose solutions, one has particles of the same order of size as colloidal particles, one usually assumed that the typical properties of cellulose (such as the formation of fibers and the high viscosity of cellulose solutions), were results of secondary valence aggregates of glucose chains, consisting of about 10 glucosidically linked units, to colloidal particles of the same character

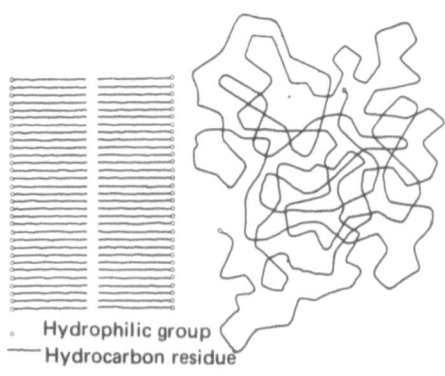

        Hydrophilic group
——— Hydrocarbon residue

FIG. 332 — Schematic representation of a micellar colloid (soap micelle), and a molecular colloid (random coil).

as soap micelles. The length of the glucosidic main valence chain (i.e., 10 glucose residues) was derived from X-ray structure analysis and the resulting dimensions of the elementary unit.

H. Staudinger disagreed with this view. He thought that the colloidal particles in solutions of cellulose, starch, and rubber and synthetic polymers (such as polystyrene and polyacrylates) must be identical with the molecules of these materials. He differentiated these new types of colloids from the particles of colloidal dimensions held together by secondary valence forces by calling them molecular colloids as opposed to micellar colloids (Figure 332). Table 333 presents a survey of different colloids according to the Staudinger classification:

TABLE 333. Classification of Colloids According to STAUDINGER

Dispersion Colloids	Molecular Colloids	Micellar Colloids
Oil-in-water- emulsion Graphite-in-oil dispersion Colloidal gold solution Rubber latex Plastics dispersions	Sphero-colloids: glycogen in water albumin in water Linear colloids: cellulose in cuprammonium soln. cellulose nitrate in acetone polymer solutions	Soap solution in water V_2O_5 – sol Azomethine dye solutions

The idea that there are molecules in which thousands of glucose residues (or isoprene residues, etc.) are linked through covalent bonds to long chain molecules (an idea which now has become so accepted that one does not think it peculiar any more) was so new and unreconcilable with then current concepts 50 years ago when it was first mentioned,[1] that it was generally opposed.

Staudinger's concept of the macromolecular structure of cellulose and other macromolecular compounds was proved by a series of reactions which he called a polymer analogous conversion. This series consisted of the conversion of a macromolecular compound to a derivative with completely different solubility properties, however, with the original degree of polymerization maintained. This degree of polymerization was measured before and after the conversion by means of careful molecular weight determinations:

$$\text{Cellulose} \quad \xrightarrow{\text{HNO}_3} \quad \text{Cellulose nitrate} \quad \xrightarrow{\text{OH}^-} \quad \text{Cellulose}$$
$$(\bar{P} = 1000) \qquad (\bar{P} = 1000) \qquad (\bar{P} = 1000)$$

---

[1] H. Staudinger, Ber. *59*, 3019 (1926).

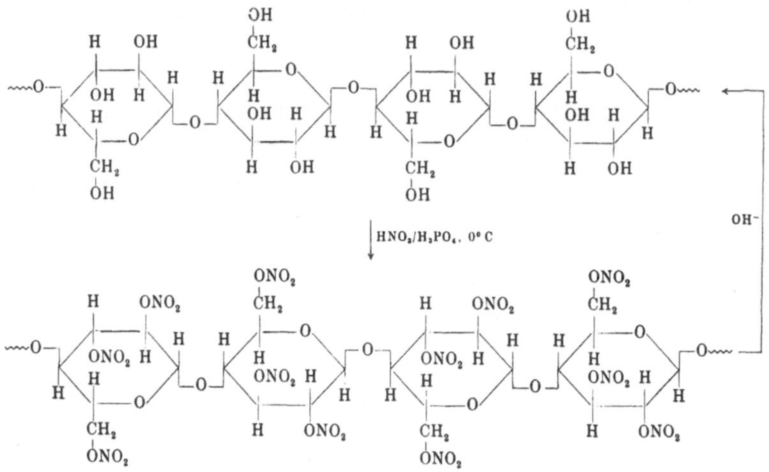

In a similar manner, many other polymer analogous conversions were carried out:

Cellulose	$\xrightarrow[\text{H}_2\text{SO}_4]{\substack{\text{acetic}\\\text{anhydride}}}$	Celluloseacetate $\xrightarrow{\text{OH}^-}$ Cellulose
Cellulose	$\xrightarrow{\substack{\text{Dimethyl-}\\\text{sulfate}}}$	Methylcellulose
Starch	$\xrightarrow{(\text{CH}_3\text{CO})_2\text{O}}$	Starch acetate $\xrightarrow{\text{OH}(-)}$ Starch
Glycogen	$\xrightarrow{(\text{CH}_3\text{CO})_2\text{O}}$	Glycogen acetate $\xrightarrow{\text{OH}(-)}$ Glycogen
Polyvinylalcohol	$\xrightarrow{(\text{CH}_3\text{CO})_2\text{O}}$	Polyvinylacetate $\xrightarrow{\text{OH}(-)}$ Polyvinylalcohol
Polymethyl-acrylate	$\xrightarrow{\text{OH}(-)}$	Polyacrylic acid $\xrightarrow{\text{CH}_2\text{N}_2}$ Polymethyl-acrylate
Polystyrene	$\xrightarrow{\text{Pd, H}_2}$	Polyvinyl-cyclohexane
Rubber	$\xrightarrow{\text{Pd, H}_2}$	Hydrorubber
Pectin	$\xrightarrow{\text{HNO}_3}$	Pectin nitrate
Pectinic acid	$\xrightarrow{\text{CH}_2\text{N}_2}$	Pectinic acid methylester
Acrylic acid ester	$\xrightarrow{\text{LiAlH}_4}$	Polyallylalcohol

Cellulose is soluble in aqueous copper tetrammine hydroxide solutions, and other aqueous salt, or complex salt, solutions; the cellulose esters (nitrate and acetate), are soluble in such organic solvents as acetone or chloroform. The fact that the colloidal particles of cellulose esters in acetone or chloroform have a particle weight that corresponds to the same degree of polymerization as the particle weights of the colloidal particles of the same cellulose compound in water before the esterification and after the saponification demonstrates that there must be a polymer radical with main valence linkages which maintains its size during these conversions. Thus it is completely impossible that the particle size (degree of polymerization or degree of aggregation) in water of a secondary valence aggregate, consisting of primary particles with hydroxyl groups, should be the same as that of an aggregate of the esterified primary particles in organic solvents. Thus, the secondary valence forces caused by the OH groups must be quite different from the secondary valence forces caused by the ester groups, which, of course, accounts for their completely different solubility properties. Similarly, one finds that with soap, a typical micellar colloid, the formation of micelles does not take place if one esterifies the fatty acids. Thus, the fatty acid esters, when dissolved in an organic solvent, form completely normal solutions without any colloidal character.

Not all macromolecular compounds permit polymer analogous reactions to take place with equal facility. However, since the macromolecular character of many natural and synthetic polymers has now been demonstrated, and one has learned more about the properties and the synthesis of macromolecular compounds, it is possible to say on the basis of a total property picture (rubber elasticity, fiber and film formation, swelling, and viscosity of the solution), and especially on the basis of the synthesis of the compound, whether or not it belongs into the class of macromolecules. Polymer analogous conversions are therefore carried out only seldom today as a means of demonstrating that a compound is macromolecular. However, such reactions are carried out quite frequently on a technical scale in order to change the properties of macromolecular compounds in a certain way (for example, the preparation of cellulose acetate from cellulose, polyvinylalcohol from polyvinylacetate, polyvinylacetals from polyvinylalcohol, preparation of Hypalon by sulfochlorination of polyethylene and preparation of polyacrylic acid salts by saponification of polyacrylonitrile or polyacrylic esters. Of course, such reactions carried out on a technical scale are usually no longer truly polymer analogous reactions, because one always finds more or less strong chain degradation, i.e., a splitting of the chain molecule into two or more smaller fragments. If one remembers that 0.01 mg of water is sufficient to degrade 1 g of cellulose of molecular weight of 1.8 million by splitting of glucosidic linkages to the point where the molecular weight is reduced to half its original value one understands that polymer analogous esterification can be successful only if one carefully excludes all possibility of chain degradtion. This has to be done more and more carefully, the larger the molecular weight before the reaction. Therefore, a successful polymer analogous conversion is a proof that a compound is macromolecular. However, the opposite is not true, i.e., one cannot say that a compound is not macromolecular because the molecular weight before the reaction is greater than the degree of polymerization of the derivative after the reaction. In such a case, one has to look for side reactions which can result in a decrease in molecular weight as a result of chain degradation.

312 Molecular Weight Determination

With low molecular weight compounds, each compound has its own molecular weight: urea always has a molecular weight of 60.06, indigo always has a molecular weight of 262.25, and the molecular weight of anilin is always 93.12. A compound with a molecular weight of 100 can never be urea, and a compound with a molecular weight of 60 can never be anilin. However, with macromolecular compounds the situation is completely different. Thus, polystyrene can have a molecular weight of 80,000, but it can also have a molecular weight of 5,000, 800,000, or even 8 million. Macromolecular substances do not have a certain definite and permanent molecular weight. Thus, there are a large number of celluloses, polystyrenes, polymethacrylates, etc., which differ in their properties only by having a different molecular weight. For example, polystyrene with a molecular weight below 10,000 is a brittle material without any mechanical strength, which can be easily rubbed into a fine powder, and which at a temperature of 50°C becomes an oily liquid. On the other hand, molded polystyrene of a molecular weight of 250,000, is a hard and strong glassy plastic, and polystyrene with a molecular weight over 1 million, is a compound which precipitates out of its solutions as a fibrous material and which at temperatures over 100°C turns into a tough rubbery material. Nearly all properties of macromolecular compounds change with the molecular weight, some only in a small region, up to a certain value of the molecular weight which is usually around 100,000. However other properties, such as the viscosity, change continuously over the entire region of molecular weights obtainable. This is also why the plastics industry must be able to control the molecular weight of the products that are manufactured.

It is therefore very important in characterizing a polymeric material to give the molecular weight as a part of its description. Thus, by the designation of "polystyrene, molecular weight 180,000," one limits the number of possible polystyrenes to a much smaller range. Unfortunately the characterization is still not complete, because macromolecules do not have a uniform length. Thus, a molecular weight of 180,000 can result from a number of different situations. All molecules could have molecular weights near 180,000 (for example, from 170,000 to 190,000) in which case one speaks of a narrow distribution; however, the molecular weights could also lie between 104 (the molecular weight of the monomer) and 5 million or more, in which case one has a broad distribution. The way in which the macromolecules of a product are distributed over the different molecular weights can be seen from the molecular weight distribution curve, which has its maximum at the molecular weight most frequently present. In special cases one can also find distribution curves with two or more maximums, such as is found with natural products, or on mixing two polymers with very different molecular weights. Instead of drawing a distribution curve, it is often sufficient to determine the molecular weight according to two methods which yield different averages such as $\overline{M}_w$ and $\overline{M}_n$. The further these values are separated from each other, i.e., the more the ratio $\overline{M}_w/\overline{M}_n$ is different from 1, the broader the distribution is (compare pp. 411 and 431).

In this chapter we will discuss the different methods of molecular weight determination, the determination of the distribution function by means of fractional precipitation, and finally, the molecular weight averages obtained by different methods.

Instead of the molecular weight, one often uses the degree of polymerization, which is defined as the average number of structural units in each macromolecule:

$$\overline{P} = \frac{\overline{M}_{polymer}}{M_{monomer}} \tag{1}$$

3121 The End Group Method

If one knows the composition of a macromolecular compound to the extent that one knows how many times a functional group, which is easily determined analytically, occurs in the molecule, one can then obtain the molecular weight of the compound through quantitative determination of this particular group. For example, if one has a polyester which has been prepared through polycondensation of a hydroxyacid, then one knows that this polyester has a carboxyl group for each molecule and that on titration 1 mole of sodium hydroxide is used per mole of polyester. To find the molecular weight of the polyester, one therefore only has to calculate how much polyester (in grams) will be neutralized by 1 mole of sodium hydroxide. Thus, if the titration of 1.5 g of polyester (E = 1.5 g) would use up 0.75 cm^3 of 0.1 N sodium hydroxide (=0.003 g of NaOH), then one would need $1.5 \cdot 40/0.003$ g of the polyester in order to use up 1 mole = 40 g of NaOH. The molecular weight of the polyester is $\overline{M} = 1.5 \cdot 40/0.003 = 20,000$. Usually with a polyester one gives the acid number (AN), which is defined as the number of milligrams of KOH used up on titration of 1 g of polyester. In the above numerical example, the acid number would therefore be AN = 2.8. The molecular weight of the polyester can be calculated accordingly $\overline{M} = 56 \cdot 1000/2.8 = 20,000$ or in general

$$\overline{M}_{polyester} = 56,000/AN. \tag{2}$$

If the polyester has been formed through condensation of dicarboxylic acids and glycols, this does not change anything as long as neither one of these components is used in excess. If, on the other hand, the dicarboxylic acid has been present in excess in the condensation, then the resulting polyester (after removal of the unreacted monomers) uses up more sodium hydroxide, for the same molecular weight, as a polyester which has been prepared with complete equivalence of the monomers. If the excess of dicarboxylic acid during synthesis is so great that with all molecules there are carboxyl groups at both ends, then obviously the consumption of sodium hydroxide is twice as large, or conversely, with the same amount of sodium hydroxide used up, the molecular weight is twice as large. Quite generally one can write (see the above numerical example):

$$\overline{M} = n \cdot q \cdot E/a, \tag{3}$$

where n is the number of groups which can be determined per macromolecule, q is the equivalent weight of the reagent used for the determination in grams corresponding to a particular group, E is the weight of the polymer in grams, and a is the consumption of the reagent in grams.

From the analytical determination it is also possible to calculate the fraction of the groups which have reacted, during the polycondensation, P, and one obtains the degree of polymerization, $\overline{P} = \overline{M}_P/M_M$ according to the following simple relation (for derivation see p. 220):

$$\overline{P} = 1/1 - p, \tag{4}$$

where the conversion p is defined by: $p =$ number of reacted functional groups/number of functional groups present.

For example, if one reacts (model example) four molecules of an ω-hydroxy acid to a polyester with $P = 4$,

$$HO{-}R{-}COOH + HO{-}R{-}COOH + HO{-}R{-}COOH + HO{-}R{-}COOH$$
$$\downarrow$$
$$HO{-}R{-}\underset{\underset{O}{\|}}{C}{-}O{-}R{-}\underset{\underset{O}{\|}}{C}{-}O{-}R{-}\underset{\underset{O}{\|}}{C}{-}O{-}R{-}COOH + 3\,H_2O$$

then of the total number of eight functional groups present, six have reacted, and two remain unreacted. Therefore the degree of polymerization $\overline{P} = 8/2 =$ the total number of functional groups present/the number of unreacted functional groups $= 4$ If one designates the total number of functional groups by N, then $p \cdot N$ is the number of reacted groups, and $(N - pN)$ the number of unreacted functional groups. For P one therefore has the relation:

$$\overline{P} = N/(N - pN) = N/N(1 - p) = 1/1 - p$$

The end-group method is useful with polymers which have end groups that can be determined precisely by some analytical reaction. In addition to polyamides and polyesters, where the end group method is used quite frequently, it can also be applied to polysaccharides and polyethers:

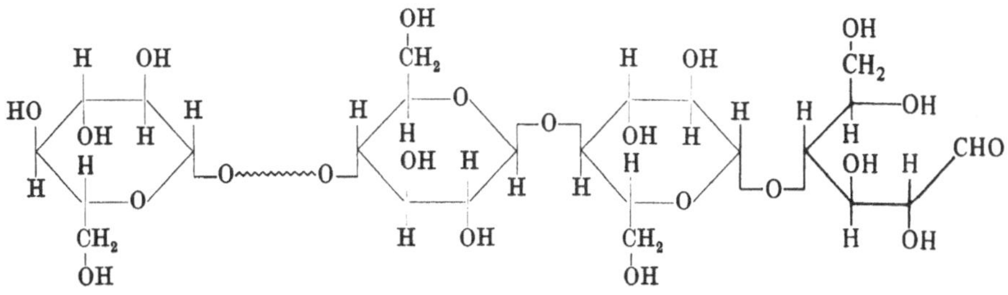

Partially hydrolyzed Cellulose

Determination of the aldehyde group by iodine titration, or determination of tri-methylglucose after complete methylation (too inaccurate with high molecular weights).

$$R—NH—CH_2—CH_2—O—CH_2—\overset{*}{C}H_2—O\sim\sim CH_2—CH_2—O—CH_2—CH_2—OH$$

Polyether from ethylene oxide (amine as initiator)

Determination of the OH end group by acetylation and saponification, or determination of the NHR end group.

$$HO—CH_2—O—CH_2—O—CH_2—O—CH_2\sim\sim O—CH_2—O—CH_2—OH$$

Polyoxymethylene (polymerized in water)

Determination of the hemi-acetal end groups by acetylation and saponification.

In many cases one can choose a method of synthesis which will yield a polymer with end groups that can be analytically determined through the choice of special initiators. For example, polymers with two bromobenzoyl end groups are obtained if one starts the polymerization of styrene with bromobenzoylperoxide. Even more exact is an end group determination if one uses initiators which contain carbon-14. Thus, for example, many polymerizations have been described using azoisobuty-ronitrile containing C^{14} as the initiator. In using this method it is a prerequisite that one definitely knows how many end groups of this type one has per macromolecule. If one does not know that, then one can still use the end-group method to determine how many end groups there are per macromolecule, but it is then necessary to determine the molecular weight by another method, such as the osmotic pressure method. If one knows the molecular weight, $\overline{M}_n$, and the number of end groups per gram, it is easy to calculate the number of end groups per molecule, and this gives information, for example, about the type of chain termination occurring during polymerization.

Also the calculation of the molecular weight with stoichiometric polymerization (for example, living polymers with Li- or Na-alkyls, compare p. 173) is an end-group method. One knows that for each macromolecule one or two molecules of sodium (depending on the initiator) are necessary. With sodium naphthalene as initiator, for example, the molecular weight of the polymer is simply given through the ratio $200/[n]$ where $[n]$ is the sodium concentration in moles per 100 g of polymer (compare p. 169 and 173).

A limitation for the application of the end-group method is the sensitivity of the analytical method. In the example on page 337, we used in the titration of 1.5 g of polyester with a molecular weight of 20,000, 0.75 cm^3 of n/10 NaOH. From this one sees that with a molecular weight of 20,000 the sensitivity of the method is no longer very great. Even by increasing the amount of polymer used or by using a more dilute NaOH solution for the titration, the sensitivity cannot be markedly increased. It is not the small concentration of the end groups which presents a problem, but the fact that the ratio of the end groups to any impurities present becomes smaller and smaller, the higher the molecular weight. For example, the smallest amounts of reducing agent suffice to make the iodine titration of polysaccharides completely useless for polymers of high molecular weight.

An important role can be played by end-group determinations in the characterization of branched polymers. Thus, if one determines the molecular weight osmotically, then one can determine the number of side chains per macromolecule, if one knows that each side chain carries an end group. Of course one does not obtain any information about the length and the frequency of the side chains, as can be seen in the following scheme. All three structures will yield the same result in an end-group determination:

```
1.    ⊗—A—A—A—A—A—A—A—A—A—A—A—A—A—A—A—A—⊗
              |           |           |           |
              A           A           A           A
              |           |           |           |
              ⊗           ⊗           ⊗           ⊗

2.  ⊗—A—A—A—A—A—A—A—A—⊗        3.  ⊗—A—A—A—A—A—A—A—A—⊗
        |   |   |   |                      |           |
        A   A   A   A                      A           A
        |   |   |   |                      |           |
        A   A   A   A        ⊗—A—A—A—A      A—A—A—A—⊗
        |   |   |   |                |      |
        ⊗   ⊗   ⊗   ⊗                A      A
                                     |      |
                                     ⊗      ⊗
```

By combination of end-group determination and osomotic molecular weight determination, it was possible to find out something about the structure of the polysaccharides starch and glycogen (polysaccharide in the liver).

It is possible to distinguish between types 1, 2, and 3, by means of comparative viscosity measurements.

With proteins it is possible to determine the amino acids present in the chain quantitatively by means of hydrolysis. If one assumes that with all molecules of the particular protein the ratio of the amino acids is identical, then it is possible to calculate a minimum molecular weight from the amount of that amino acid which is present in the smallest amount. This minimum molecular weight is actually found if the particular amino acid occurs only once in the molecule. If it occurs twice in the same molecule, and the molecular weight is twice as great as the minimum

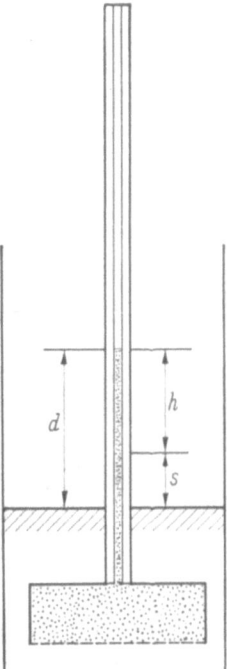

FIG. 341 — Basic principle of an osmotic pressure
measurement.

molecular weight, the number of times an amino acid occurs in the macromolecule
can only be determined by additional osmotic pressure measurments.

3122   The Osmotic Method

Figure 341 shows the principle of osmotic measurements. There is a capillary
with a cell, the bottom of which consists of a semipermeable membrane. The entire
assembly is placed into a vessel filled with solvent. In the cell there is a dilute
solution of a macromolecular compound whose molecules are so large that they
cannot pass through the membrane, whereas the smaller solvent molecules can pass
in both directions through the pores of the membrane.

The system has, just as every system, the desire to increase its entropy. This
becomes possible through the fact that the solvent entering into the cell through
the membrane dilutes the solution, (i.e., the increase in entropy corresponds to the
value of the entropy of dilution ΔS). Because of the corresponding volume increase,
the solution rises in the capillary, and the potential energy is increased by ΔH,
which is the work against the gravity of this volume increase in the capillary. The
dilution process is then over, and the system is in equilibrium when $\Delta H = T \cdot \Delta S$

($\Delta F = \Delta H - T \cdot \Delta S = 0$).[2] The hydrostatic pressure of the column of liquid of height h is now equal to the osmotic pressure in the cell. The height s corresponds to the capillary rise caused by the surface tension, and this has to be measured separately and then subtracted from the total height d in order to obtain the height corresponding to the osmotic pressure h. Usually one gives the osmotic pressure in atmospheres. Then one has to convert the height h into Torr. Since the hydrostatic pressure of a column of liquid is inversely proportional to the density of the solution, one can write:

$$\frac{h_{Hg}}{h_{solution}} = \frac{\rho_{solution}}{\rho_{Hg}} \quad \text{and} \quad h_{Hg} = \frac{h_{soln} \cdot \rho_{soln}}{\rho_{Hg}} \quad [\text{Torr}].$$

Since 760 Torr = 1 atm,

$$\frac{h \cdot \rho_{soln'}}{\rho_{Hg}} \; [\text{Torr}] = \frac{h \cdot \rho_{soln}}{\rho_{Hg} \cdot 760} \quad [\text{atm}]$$

Then from h in mm, one obtains the osmotic pressure p in atm:

$$p = \frac{h \cdot \rho_{soln}}{10330} \quad [\text{atm}] . \tag{6}$$

For dilute solutions, just as for ideal gases, van t'Hoff's law holds:

$$p \cdot v = n \cdot R \cdot T., \tag{7}$$

where p = osmotic pressure in atm, v = volume of the solution, T = absolute temperature,

R = gas constant = 82.06 $\left[\dfrac{cm^3 \cdot atm}{deg. \; mole}\right]$, and n = number of moles of dissolved substance.

If in a volume v, 1 mole, i.e., M grams of macromolecular compound are dissolved, then the number of moles $n = 1$. If m grams of substance are dissolved in the volume v, then $n = m/M$, so that Equation (7) becomes:

$$pv = \frac{m}{M} \cdot RT$$

$$p = \frac{m}{v} \cdot \frac{RT}{M} ,$$

---

[2] ΔF is a measure of the ability of the system to do work and is called the free energy.

where m/v is the mass per unit volume, i.e., the concentration of the solution, $m/v = c$ [g/cm^3], and one then obtains for p:

$$p = c \cdot \frac{R \cdot T}{M} \tag{8}$$

and for the corresponding molecular weight $\overline{M}$:

$$\overline{M}_n = \frac{RT}{p/c} . \tag{9}$$

As will be discussed later (p. 429), osmotic measurements yield the number average $\overline{M}_n$. The van t'Hoff equation is an ideal case, which holds for infinitely dilute solutions. In order to account for the deviations which one finds with practical concentrations, one can write the equation in the form of a power series:

$$p/c = R \cdot T/\overline{M}_n + B \cdot c + C \cdot c^2 , \tag{10}$$

where B is called the second virial coefficient and p/c the reduced osmotic pressure. Equation (10), if one neglects the third member, gives a straight line, and this is obtained if one plots p/c versus c. One then obtains the expression $RT/\overline{M}_n$ as the intercept on the ordinate, and the second virial coefficient from the tangent of the straight line. This method is always used to calculate osmotic molecular weight in practice. Therefore, for an osmotic molecular weight determination one needs a series of four to six osmotic measurements at different concentration. The resulting values of p/c are plotted against c, and by extrapolation to concentration O, one obtains the value $\lim_{c \to o} p/c = RT/\overline{M}_n$. The second virial coefficient, which is the larger the more the solution deviates from ideal behavior, is with macromolecular solutions an important characteristic for the interaction between the dissolved molecules and the solvent (compare p. 530). The greater B, the stronger is the swelling tendency (expansion) of the coiled macromolecules caused by the solvation (see Figure 518). Therefore, the greater B, the better the solvent. B is a thermodynamic quantity for the solvent power of different solvents for a macromolecular compound (G. V. Schulz).

Limits of Osmotic Molecular Weight Determination

Figure 344 presents a series of examples for osmotic molecular weight determination. As can be seen, one does not always obtain a straight line, i.e., the deviation from ideal behavior is not always that described by a single coefficient, B. The slope of the p/c versus c curve is for one and the same polymer quite different, depending on the solvent. This is a demonstration of the fact that the value of B in

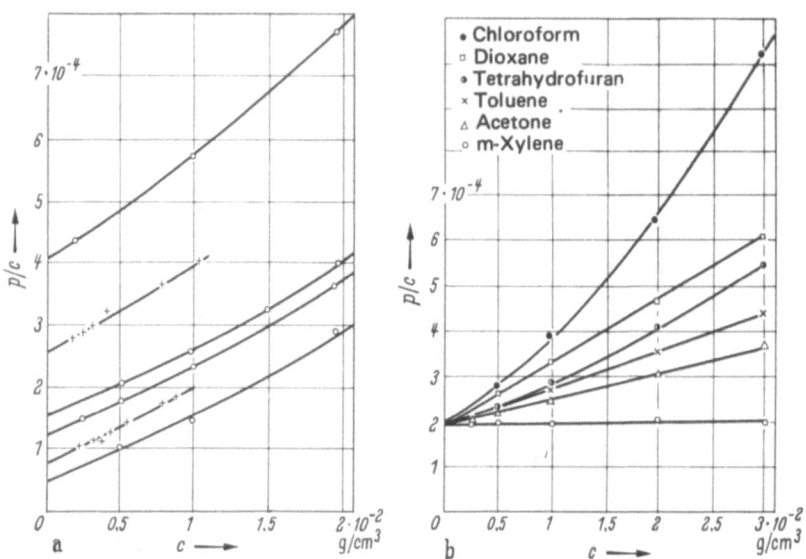

a) p/c vs. c curves for polystyrene of different molecular weights in toluene (G.V. SCHULZ and C.E.H. BAWN).

b) p/c vs. c curves for polymethylmethacrylate (M = 128,000) in different solvents (G.V. SCHULZ and H. DOLL)

FIG. 344(a and b) – Examples of osmotic molecular weight determinations.

Equation (10) mainly describes the influence of the solvent on the behavior of the polymer molecule. It is always useful to carry out osmotic molecular weight determinations with different solvents, because faulty measurements can be recognized by the fact that extrapolation of the curves would not lead to the same intercept on the ordinate.

In addition to experimental reasons for this, such as leakage of the apparatus and unsuitable membranes, it can also be due to association phenomena, i.e., combination of several macromolecules to larger particles.

With increasing molecular weight the height d, or h, in the capillary becomes smaller, and the accuracy of the measurement therefore becomes less. The accuracy can be improved if one measures at higher concentrations, however, then the point measured is further from the ordinate and the extrapolation becomes uncertain. In such cases it is especially important to carry out measurements in several solvents.

The usefulness of the osmotic method is limited with very high molecular weights by the poor accuracy (however, molecular weights of a million can still be determined). The limitation with lower molecular weights is that one can not find really semipermeable membranes, i.e., the small molecules of the polymer can diffuse through the membranes. In general, it is good before making a definite selection of the material, to test it out with known molecular weight compounds. For example, by means of anionic polymerization, relatively homogeneous polystyrene with well-defined molecular weight can be prepared, which is highly suitable as a test material (compare pp. 164-175). If the membranes are too tight, i.e., have too small pores, the measurements take a long time. On the other hand, if they are very large, the low molecular weight fractions of the polymer can diffuse through them, and one obtains too high a molecular weight. Therefore, it does not make much sense to use the osmotic method for unfractionated polymers with large amounts of low molecular components. Such products are usually fractionated before the molecular weight is determined.

The importance of fractionation can be recognized from Figure 345, which shows osmotic measurements of an industrially prepared polystyrene (by bulk polymerization), using a series of membranes of different porosity. Next to the p/c scale the figure shows corresponding molecular weights, and it can be seen that for one and the same unfractionated polystyrene, molecular weights between 7,000 and 200,000 were obtained, depending on the membrane used. Therefore, one should not attempt to carry out osmotic pressure measurements with such unfractionated products. Instead it is better to fractionate first, or to carry out molecular weight determinations according to a different method.

The osmotic molecular weight determination is characterized by the fact that (1)

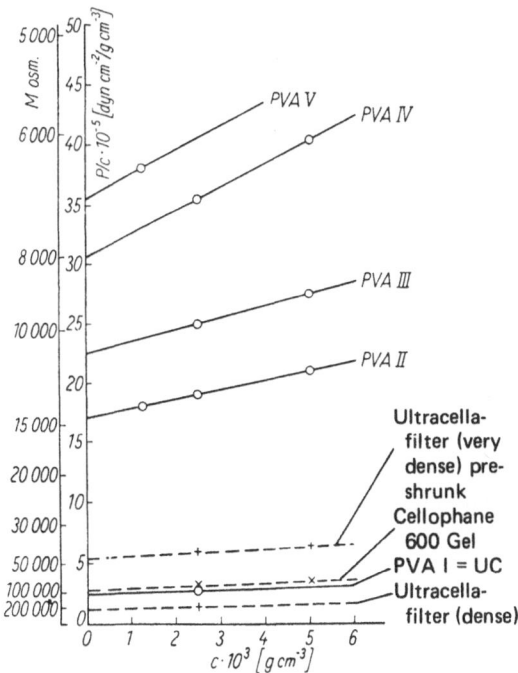

FIG. 345 — p/c vs. c curves of unfractionated polystyrene in toluene using different membranes (G. MEYERHOFF)

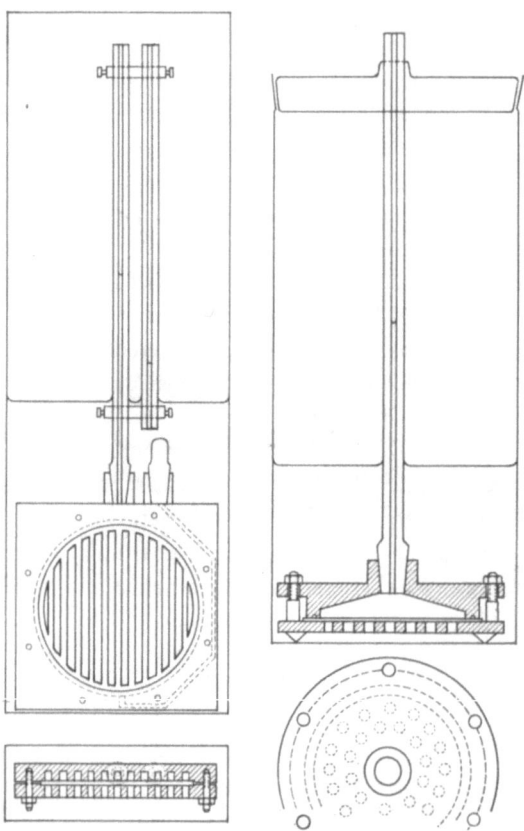

FIG. 346 — HELLFRITZ-osmometer
(left), and SCHULZ-osmometer (right).

it is based on a clear theoretical basis without any particular assumptions, and (2) it is the only absolute method which is applicable over a wide molecular weight range, and which yields stoichiometrically useful averages ($\overline{M}_n$-averages, compare p. 425). (3) Osmotic measurements can be carried out in any laboratory with simple apparatus and without any special physical expertise. (This does not mean, however, that careful work and clean equipment are unnecessary!)

Osmometers

Over the years numerous types of osmometers have been developed and used. They differ usually only in the details of their technical construction, and all are based on two types of cells, namely the cell with a horizontal membrane (G. V. Schulz) and the cell with a vertical membrane (R. M. Fuoss and D. J. Mead). (see Figure 346).

One can determine the osmotic pressure by measuring the height difference in two capillaries, one of which is connected to the osmotic cell and the other to the surrounding solvent. One can either wait for attainment of an equilibrium height difference (static method), or one measures the counter-pressure which is necessary to prevent the diffusion of solvent through the membrane and the corresponding volume change of the solution (dynamic method). The dynamic method is now most commonly used because the time of measurement is much shorter than with the equilibrium method.

The osmometers shown in Figure 346 have the advantage that they require only very simple apparatus and that they are easily constructed in any workshop and do not need any customer service. For this reason, and also because their method of operation is easily understood, they are quite useful in laboratory courses in physical polymer chemistry.

For research purposes one uses electrical and automatically recording dynamic osmometers (for example Mechrolab- Shell-, or Stabin-osmometer) where equilibrium is reached in 15 to 30 minutes at the utmost (Figure 347). For molecular weights under $\overline{M}$ = 20,000, highly sensitive vapor pressure osmometers are used (no membrane problems).

Figure 347 shows on the left side a modern vapor pressure osmometer and on the right side a modern membrane osmometer.

Particularly exact measurements can be carried out with an osmotic balance, which has been described by Jullander.

Membranes are usually films of cellulose nitrate for aqueous measurements, and films of cellulose, or polyvinylalcohol, for measurements in organic solvents.

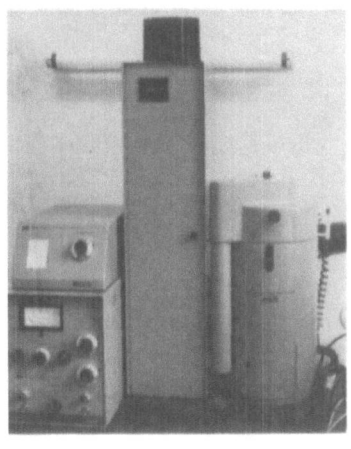

(a)   (b)

a) Vapor-pressure osmometer (PERKIN-ELMER CO.)

b) Membrane osmometer (MECHROLAB CO.)

FIG. 347.

3123   The Light-Scattering Method

For Particle Diameters smaller than λ/20

If a beam of light rays passes through a colloidal solution, then it is possible to see the light beam from the sides (the Tyndall Effect). This results from the fact that a part of the entering light is scattered by the colloidal particles in all directions. With colloidal particles which are smaller than $\lambda/20$, the intensity of the scattered light in all directions is the same, if one uses for the primary light a polarized light which vibrates vertically to the plane of observation (spherically symmetrical scattering function).

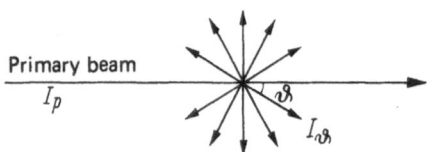

FIG. 348 — Spherically symmetrical scattering function for particles with diameter $< \lambda/20$.

Since with macromolecular solutions the colloidal particles consist of small gel particles, i.e., the coiled chain molecules which have become solvated with solvent to the extent that the entire gel particle consists of 90 to 99% of solvent, (depending on the average density of the coil), the refractive index of the colloidal particles is in such cases only slightly different from the refractive index of the solvent. This gives rise to simple physical laws for the scattering of the light. According to P. Debye, the following power function applies:

$$\frac{K \cdot c}{i_0} = \frac{1}{\overline{M}} + \frac{2 \cdot B \cdot c}{RT} + \frac{3 \cdot C \cdot c^2}{RT} + \cdots. \tag{11}$$

where K equals

$$K = \frac{4\pi^2 n_0^2}{N_A \lambda_0^4} \cdot \left(\frac{n - n_0}{c}\right)^2 \tag{12}[3]$$

$(n-n_0)/c$ = refractive index increment; n = refractive index of the solution; n_0 = refractive index of the solvent; c = concentration in g/cm^3; λ_0 = wave length of the incident light (primary light) in vacuum; B = identical with the 2. virial coefficient of osmotic pressure (compare Equation 10); i_0, that is, i_Θ = reduced scattering intensity = absolute intensity of the scattered light at an observation angle[4] θ under

---

[3]If one uses unpolarized light, one has to multiply Equation (12) by $(1 + \cos^2 \theta)/2$.

[4]For systems with particles $<\lambda/20$, any observation angle may be selected since the scattering function is spherically symmetrical ($i_\theta = i_0$). However, for experimental reasons, to avoid interference of the primary beam, one usually selects values of θ not very different from $90°$.

standard conditions, i.e., with a scattering volume $v_{sc} = 1$ cm^3, a distance $R = 1$ cm between the observer and the scattering volume, an incident scattering intensity $I_p = 1$ cm^{-1}. From the scattering intensity I under any conditions, one can calculate i_o by the following relation: $i_o = IR^2/I_p v_{sc}$.

One determines the intensity of the scattered light at different concentrations and plots the resulting values of K_c/i_o as a function of c (Figure 349). Just as with the osmotic measurements according to Equation 10, $\lim/c\rightarrow o$ $p/c = RT/M_n$, one obtains here according to Equation 11 a value $\lim/c\rightarrow o$ $K_c/i_o = 1/\overline{M}_w$. Independent of osmotic measurements, one can determine the second virial coefficient B as tan δ of the slope of the line $Kc/i_o - c$ from the light-scattering measurements, provided the 3. virial coefficient C can be neglected.

In this simple way, i.e., through the measurement of scattered light intensity at a *single* angle θ, one obtains the molecular weight, but only for particle diameters smaller than $\lambda/20$, which is only true with glycogen, with a number of pro-

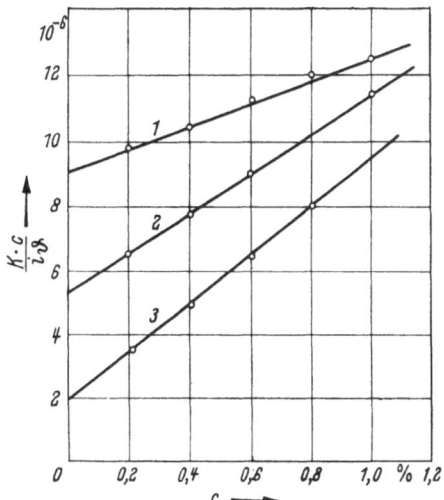

1) $\overline{M} = 108,000$ polystyrene in methylethylketone
2) $\overline{M} = 182,000$ polystyrene in methylethylketone
3) $\overline{M} = 500,000$ polystyrene in methylethylketone

FIG. 349 — Example of the determination of the molecular weight and the second virial coefficient by light scattering measurements according to Equ. (11). (Particle diameter $< \lambda/20$. Measurements of DOTY, ZIMM and MARK).

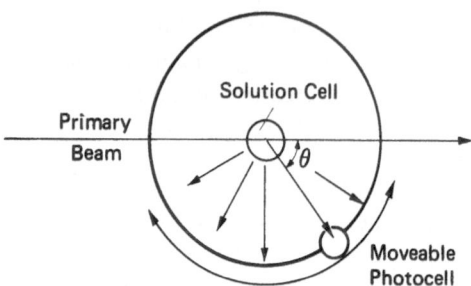

FIG. 350-1 — Designation of the observation angle θ.

teins, and with linear polymers in poor solvents at a relatively low molecular weight. In general, however, macromolecules with a linear chain structure, in good solvents, and with molecular weights between 10^5 and 10^7, have diameters between 200 and 3,000 Å. For light-scattering measurements, one generally uses mercury arcs (λ_0 = 4,000-5,000 Å), and therefore the particle diameter is larger than $\lambda/20$. Therefore, the molecular weight must be determined by measuring the dependence of the scattered light on the scattering angle.

For Particle Diameters larger than $\lambda/20$

With macromolecules where the coil diameter is larger than $\lambda/20$, the scattering function is no longer spherically symmetrical. Instead one finds with larger particles a diminution of the scattered light by interference, which is different with different observation angles. This results from the fact that for larger particles, the different scattering centers within a particle are so far from one another that the resulting scattered rays have a path difference of the order of 0 to $\lambda/2$. Since the physically separate scattering centers within a macromolecule are activated by one and the same wavefront of the primary light, the scattered radiation resulting from the centers is coherent and therefore capable of interference. The degree of diminution resulting from interference depends on the path difference of the interfering rays, and this path difference in turn depends on the angle θ. Therefore, the intensity of the scattered light differs depending on the observation angle θ. The reduced scattering intensity i_θ is diminished the more, the larger the angle θ (compare also

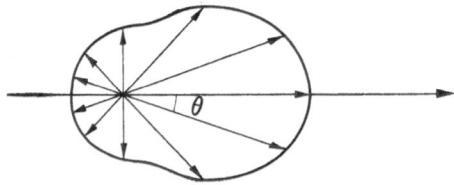

FIG. 350-2 — Diagram representing the scattering intensity i_θ at different observation angles θ

p. 457). The factor by which the scattering intensity i_o (for a perpendicularly polarized incident beam, undiminished by interference) is diminished by interference at the angle θ is called the scattering function, P_θ (where i_o equals the scattered light in the immediate vicinity of a primary beam, i.e., with $\theta \to 0$). Therefore, for the scattering intensity i_θ reduced to standard conditions and measured at the angle θ, one can write

$$i_\theta = i_o \cdot P_\theta. \tag{13}$$

If one then substitutes this in the Debye Equation (11), one obtains:

$$\frac{Kc}{i_\theta} = \frac{1}{M \cdot P_\theta} + \frac{2\,Bc}{P_\theta \cdot RT} + \;\cdots \;,$$

and according to Zimm the following approximation can be written:

$$\frac{Kc}{i_\theta} = \frac{1}{\overline{M} \circ P_\theta} + \; 2\,Bc \; + \dots \tag{14}$$

The scattering function P_θ, for random coils, as they are usually found in macromolecular solutions, is dependent on the scattering angle θ, as well as on the wave length λ, and on the average coil diameter $\sqrt{\overline{h}^2}$ [$\sqrt{\overline{h}^2}$ = mean end-to-end distance of a coiled polymer chain (compare p.441)]. One can therefore write the following approximation according to Zimm:

$$1/P_\theta = 1 + (8\pi^2/9\lambda^2) \cdot \overline{h}^2 \cdot \sin^2\theta/2.* \tag{15}$$

If one substitutes this in Equation (14), one obtains:

$$Kc/i_\theta = 1/M + (1/M) \cdot (8\pi^2/9\lambda^2) \cdot \overline{h}^2 \cdot \sin^2\theta/2 + 2\,Bc/RT \tag{16}$$

or, if one substitutes for the expression $[(1/M)\cdot(8\pi^2/9\lambda^2)\cdot\overline{h}^2]$ the term S, one obtains:

$$Kc/i_\theta = 1/M + S \cdot \sin^2\theta/2 + 2Bc/RT \tag{17}$$

This equation contains a linear dependence of the factor Kc/i_θ on the concentration c (for a constant angle of observation). Kc/i_θ is also linearly dependent on $\sin^2\theta/2$ (if the concentration c is held constant). Numerous light-scattering measurements have demonstrated the applicability of Equation (16).

In order to determine the molecular weight according to this equation, one has to carry out light-scattering measurements at different concentrations and over an as large as possible angular range. The results obtained are plotted graphically as Kc/i_θ vs. c, and Kc/i_θ vs. θ. As an example, Figure 352a,b shows measurements of polystyrene in methylethylketone. In both cases (Kc/i_θ vs. c at constant θ and Kc/i_θ vs. $\sin^2\theta/2$ at constant c), the measurements yield a series of straight lines

*$\lambda = \lambda_o/n_o$. (Compare p. 348.)

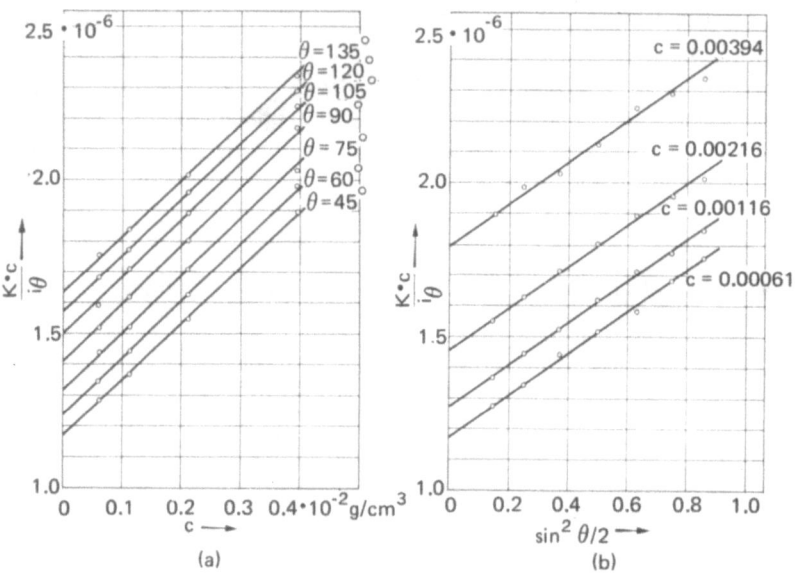

FIG. 352(a and b) — a) Kc/i_θ as a function of the concentration c at constant observation angles θ. b) Kc/i_θ as a function of the observation angle θ (i.e. $\sin^2 \theta/2$) at constant concentrations c. (Polystyrene, $\overline{M}$ = 940,000 in methylethylketone. Measurements of HUSEMANN and STUART)

which can be extrapolated to the ordinate and therefore permit the determination of Kc/i_θ at c→0, and at $\sin^2 \theta/2$→0.

Therefore in Equation (17), in one case the term 2 Bc = 0, and in the other case the term $S \cdot \sin^2\theta/2 = 0$. One obtains at c→0:

$$Kc/i_\theta = S \cdot \sin^2\theta/2 + 1/M, \tag{18}$$

and for $\sin^2\theta/2$→0:

$$Kc/i_\theta = \frac{2\ Bc}{RT} + 1/M. \tag{19}$$

If one plots the extrapolation values from Figure 352a for lim/c→0 Kc/i_θ at 45°, 60°, 75°, and 90° versus the $\sin^2\theta/2$-values corresponding to those θ-values, one obtains straight lines according to Equation (18) (Figure 353a). In a similar manner, one can obtain by extrapolation in Figure 352b the Kc/i_θ-values for the concentrations 0.001, 0.002, and 0.004 for $\sin^2\theta/2 = 0$. If these extrapolation values are plotted as a function of concentration, one obtains the straight line corresponding to Equation (19) (Figure 353b). Both straight lines (Figure 353a and Figure 353b) yield 1/M as the intercept at the ordinate. From Figure 353a one can obtain from the slope of the straight line the value $\tan\delta = S = (1/M)\ (8\pi^2/9\lambda^2)\cdot\overline{h}^2$ [compare

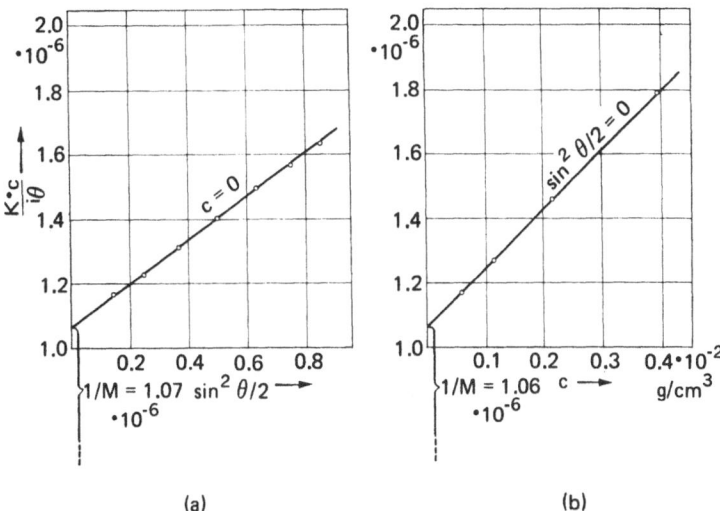

(a)   (b)

FIG. 353 — a) Extrapolation of the limiting values of Kc/i_θ [$c \to 0$] to $\sin^2 \theta/2 \to 0$ b) Extrapolation of the limiting values of Kc/i_θ [$\sin^2 \theta/2 \to 0$] to $c \to 0$

Equation (17)], and from this calculate the coil diameter $\sqrt{\overline{h^2}}$.[5] The tangent of the slope of the Kc/i_θ versus c line **(Figure 353b)** is equal to 2B. By using this method of plotting (introduced by B. Zimm), one can obtain from light-scattering measurements at different observation angles θ and different concentrations c, in addition to the molecular weight, the coil diameter $\sqrt{\overline{h^2}}$ and the 2. virial coefficient B, which in general corresponds well with B obtained from osmotic measurements.

In general, the extrapolation according to Figures 352a and 352b is not carried out separately, but in a single diagram by plotting Kc/i_θ versus $(\sin^2\theta/2) + 100c$. Figure 354 shows such a Zimm diagram for the same experimental values which have been plotted in Figures 352 and 353. The lines of low slope show the dependence of Kc/i_θ-values on $\sin^2\theta + 100c$ for each constant concentration c. If one extrapolates the line to the intercept with each correspondent 100c value of the abscissa ($\sin^2\theta/2 = 0$), one obtains a series of points which together form the extrapolation of line $Kc/i_\theta = f(c)$ at $\theta = 0$. In a similar manner, the straight lines of high slope show the dependence of Kc/i_θ on $\sin^2\theta/2 + 100c$ at constant scattering angles θ. If one extrapolates these lines to the intercept with the corresponding abscissa value $\sin^2\theta/2$ (c=0), one obtains the points for the extrapolation $Kc/i_\theta = f(\sin^2\theta/2)$ at c=0. The two extrapolations must intercept the ordinate at the same point and this is a criterion for the validity of Equation 16.

In addition to the Zimm extrapolation procedure, there is a simpler method

---

[5]A more accurate method for $\sqrt{\overline{h^2}}$ is shown on pp. 456-463. The significance of this parameter is discussed on p. 441 ff.

which in principle is not very different. This consists of interpreting the light-scattering measurements with the aid of the dissymmetry Z:

$$Z = \frac{i_{45°}}{i_{135°}} = \frac{P_{45°}}{P_{135°}} \tag{20}$$

For this purpose one needs only to measure at two different angles, for example, at 45° and 135°. In order to determine the molecular weight from such measurements, one has to know the form of the molecule, i.e., coil, rod, or isotropic sphere (compare p. 447). With the extrapolation procedure for the determination of M, on the other hand, the particle form is of no importance. Only the value of S (compare Equation 17) changes with the particle form, however, not the linear dependence of the reciprocal scattering intensity Kc/i_θ on $\sin^2\theta/2$.

Measurement Technique

In order to obtain a Zimm diagram, one has to determine the following experimental values (all measurements with the temperature as constant as possible): the refractive index of the solvent, n_o; the difference of the refractive index between solution and solvent (n-n_o) (compare Equation 12); the polymer concentration; and the absolute intensity of the radially scattered light i_θ at different angles θ.

The refractive index n_o of the solvent can be measured in the usual refractometers in so far as the value is not already known from the literature. The refractive index difference (n-n_o), and therefore the refractive increment (n-n_o)/c, is usually only very small (of the order of 0.1 cm^3/g) and therefore one has to determine (n-n_o) very accurately.

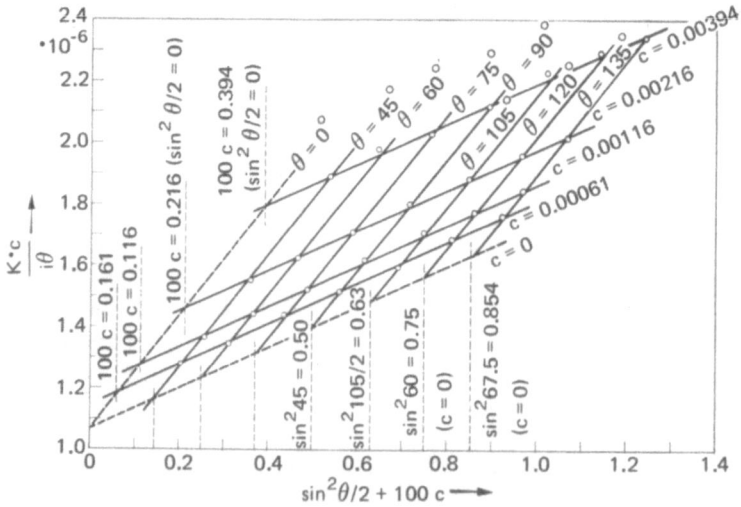

FIG. 354 – ZIMM-plot for polystyrene (M = 940,000) in methylethylketone (measurements of HUSEMANN and STUART).

A differential refractometer constructed especially for this purpose has recently been described. This permits the determination of $(n-n_0)$ to very high accuracy.

The reduced scattering intensity i_θ is the intensity measured at the angle θ of the radially scattered light from a scattering volume of 1 cm^3, a primary light intensity $I_p = 1$ cm^{-1}, and a distance of observation of 1 cm. Since the scattering volume, and the distance of the scattering center to the photometer can be determined only with difficulty, one compares the relative values of I_θ, measured in any instrument, with the scattering value of a standard substance (for example, benzene) measured

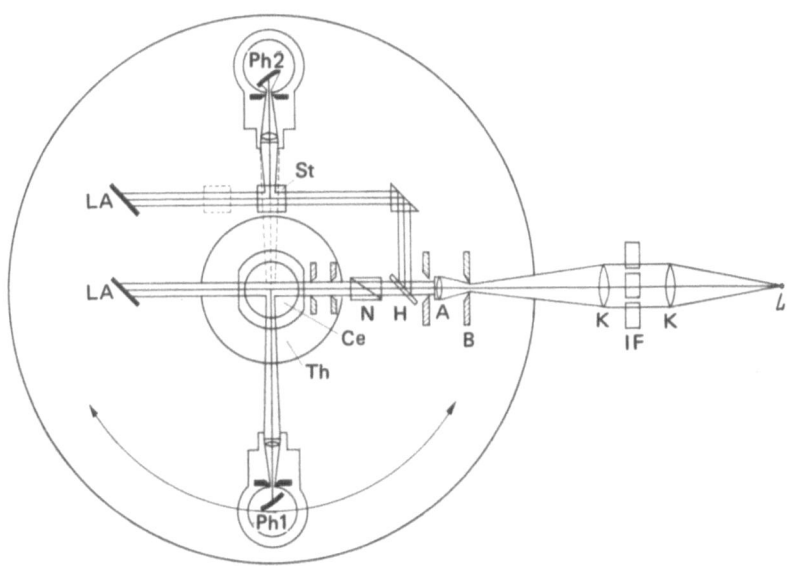

Ph 1	moveable photocell
Ph 2	reference photocell
St	calibrated turbidity standard (translucent glass)
Ce	solution cell
Th	thermostat
N	polarizer
H	lightbeam splitter
A	achromatic system
B	slit
K	condensing lenses
IF	interference filter
L	light source (high pressure mercury lamp)
LA	light absorber (beam stop)

FIG. 355 – Light scattering photometer designed by H.J. CANTOW.

 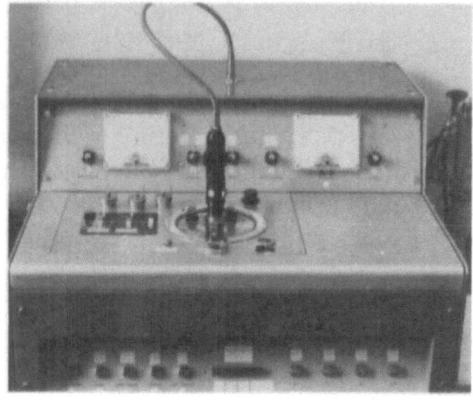

(a) (b)

FIG. 356 — a) Differential refractometer b) Light scattering photometer (SOFICA)

under the same conditions. The i_θ-values of the standard substance are determined by very careful measurements, and they can be found in the literature. (For example, with benzene $i_{\theta = 90°} = 48.4 \cdot 10^{-6}$ cm^{-1} at $\lambda = 4358$ Å using unpolarized light). One then obtains the absolute values of i_θ from the measured relative values I_θ of the solution to be measured and I_θ (benzene) of the standard substance, and from the absolute value i_θ (benzene) of the standard substance (from literature) in the following way:

$$i_\theta/i_{\theta \text{ (benzene)}} = I_\theta/I_{\theta \text{ (benzene)}}$$ (21)

In practice one proceeds as follows: a glass cylinder of constant turbidity is calibrated by means of benzene and this glass cylinder is then used as a standard.

The determination of the absolute scattering intensity may be avoided by measuring the extinction, i.e., the diminution of the primary beam in passing through the macromolecular solution. The interpretation of the measurements, however, is less certain and accurate than the Zimm extrapolation procedure.

The determination of the scattering intensity as a function of the observation angle θ is carried out with a light-scattering photometer constructed especially for this purpose. Figure 355 shows the schematic diagram of an apparatus developed by H. J. Cantow. Figure 356 shows a Sofica light-scattering photometer based on similar principles (see Figure 356b). Figure 356a shows a highly sensitive differential diffractometer which is absolutely essential for molecular weight determinations by light scattering.

If one carries out the calibration of the standard only at a *single* angle θ (for example, 90°), one has to take into account that the scattering volume V_S changes with the observation angle θ. It is smallest at $\theta = 90°$ and becomes larger on going

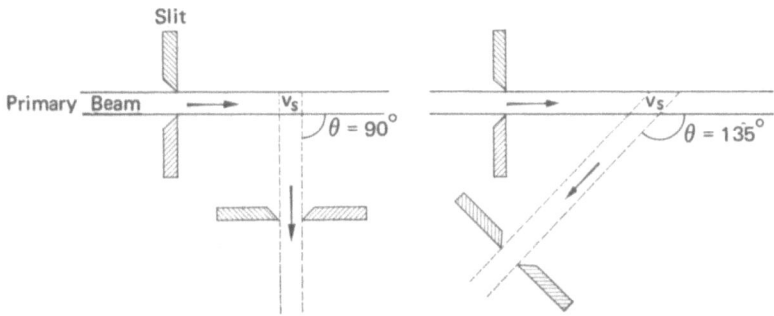

FIG. 357 — Diagram to demonstrate the effective scattering volume v_s (after G. MEYERHOFF)

to larger and smaller angles (Figure 357). It can be derived rather simply that:

$$V_{S, 90°} = V_{S, \theta} \cdot \sin\theta. \tag{22}$$

The cost of the apparatus for molecular weight determinations by light scattering is considerably higher than that for simple osmotic measurements equipment (Figure 346). In spite of that fact, the method has become important and is now one of the most valuable means for the characterization of macro-molecular compounds. Like the osmotic pressure, this is an absolute method, i.e., it is possible not only to observe changes in the molecular weight, but also to derive a direct and quantitative relation between the measured intensity of the scattered light i_θ, or rather, the angle-and wavelength dependence of i_θ, and the molecular weight, i.e., the reciprocal 1/M (Equation 16). Unlike with osmotic pressure measurements, however, one does not obtain the stoichiometrically useful average $\overline{M}_n$ (number-average), but one obtains $\overline{M}_w$, the weight-average. Especially with unfractionated products this differs considerably from the stoichiometric number-average molecular weight (with polymeric products having a normal distribution[6] $\overline{M}_w/\overline{M}_n$ = 2). Furthermore, with light scattering, the measured effect and therefore the accuracy, increases with increased molecular weight, in contrast with the osmotic method, where the accuracy diminishes with increasing molecular weight. Thus molecular weights over 1 million can also be determined accurately by the light-scattering method.

A special advantage of the light-scattering method lies in the fact that from a series of measurements one obtains not only the molecular weight and the 2. virial coefficient, but also the particle size. Thus for chain molecules the coil "diameter" $\sqrt{\overline{h}^2}$ (compare p. 441) can be obtained, which is not possible by any other direct measurement.

---

[6]Compare pp. 411-425, especially p. 421.

The greatest problem in the experimental determination of the scattered light consists in the purification of the solutions. These have to be carefully cleaned and purified from all dust before the measurement because the presence of dust particles prevents meaningful measurements. The solutions can be cleaned by passing through filters of small pore size (glass filters G 5 or membrane filters). They can also be cleaned up by centrifuging in a high speed centrifuge. In the latter case one has to watch that, especially with very high molecular weight products, parts of the polymer do not sediment together with the dust particles. The most accurate procedure is to measure the solution in a cell which can be introduced directly into the centrifuge. This avoids the introduction of new impurities on transferring the solution from the centrifuge glass into the measurement cells.

3124   Determination of the Molecular Weight Through Sedimentation in the Ultracentrifuge and Diffusion Measurements

Theoretical Principles

In vacuum a body falls under the influence of gravity with a constantly increasing velocity (free fall). In a real medium, however, there is a force opposed to the force of gravity, a frictional force, which is proportional to the velocity of fall. Therefore, at a certain velocity u, a situation arises where the frictional force, k_F, becomes equal to k_S, the force of gravity minus the buoyancy, and therefore compensates the effect of gravity:

$$k_F = k_S. \tag{23}$$

The body continues therefore from this point on by falling without the influence of other forces, i.e., with a velocity u which remains constant. The sedimentation velocity is dependent on the particle size and increases with increasing particle size. This arises from the fact that the force of gravity is proportional to the size of the particles, i.e., to r^3, whereas the frictional force (according to Stoke's law) is proportional to r, i.e., $k_F \propto r$. In going from smaller to larger particles, the gravitational force therefore increases faster than the frictional force k_F, and the result is that k_F with large particles only reaches the value of k_S at a higher sedimentation velocity than is the case with smaller particles under otherwise identical conditions. This is so with spheres and with all particles which have a compact form, such as ellipsoids and cylinders, where the axial ratio does not differ too much from 1.

However, this is not the case with long rod-like particles, threads, thin wires, or chains. With such shapes both the gravitational force and the frictional force are proportional to the length (because both the weight $\rho \cdot \pi r^2 h$ and also the surface $2\pi r h$ are proportional to the length; r, in comparison to the length h, is so small, that it does not make any difference whether it increases with the square or the first power). This means that the sedimentation velocity is essentially independent of the length, regardless of the form which the chain molecules assume, i.e., whether they are stretched out or coiled, as long as the liquid in which they are sedimenting can penetrate unhindered throughout all parts of the coil.

Thus, if the macromolecules, as was formerly often assumed, are long rods or coils freely penetrated by the solvent, one would not expect any dependence of the sedimentation velocity on the molecular weight and, therefore, molecular weight determination in the ultracentrifuge through measurement of the sedimentation velocity would not be possible.

However, it was found that in the ultracentrifuge one can determine the molecular weight not only of compact macromolecules (such as, glycogen and the so-called globular proteins) but also of chain molecules, such as are found with macromolecules of linear structure (for example, cellulose, rubber, polystyrene, polyvinylchloride, and most other thermoplastic materials). This is an obvious proof that the molecules of polymers with chain structure can not be present in solution as stretched out chains or open, freely drained coils. Instead they have the character of gel particles,[7] whose structure is formed by the coiled-up chains of the macromolecules. In the spaces within the coil, the solvent is held[8] as in a highly porous sponge. During sedimentation in a centrifugal field the entire structure (coil and included solvent) moves as one. In the calculation, however, the solvent carried with the macromolecule does not have to be taken into account. One can therefore work with the density of the macromolecular compound, i.e., the density of the matrix substance, and not with the average density of the entire sedimenting particle, which of course is not known[9]. As a model for sedimenting coils, one can assume hollow spheres with a thin wall, whose interior is filled with the same solvent in which the sedimentation velocity is measured. One, therefore, obtains as the molecular weight a value for the chain substance of the coil, and not the particle weight of the entire gel structure. Since the solvent carried with the particle has the same, or almost the same, density as the solvent on the outside of the particle, its weight is compensated by its buoyancy. Therefore, for the force k_S which causes the sedimentation of the particle, only the mass m of the chain matrix (mass of the chain molecule) has to be taken into account. One has to subtract from m the buoyancy, i.e., the weight of the volume of liquid which is displaced by the mass of chain material. One can write:

$$k_S = (m - v_m \cdot \rho_{solv}) \cdot b, \tag{24}$$

where k_s is the force of gravity less the buoyancy, m is the mass of a single macromolecule (mass of a chain molecule) b is the centrifugal acceleration, v_m is the volume of the chain substance of a single coil (not the volume of the entire coil), ρ_{solv} is the density of the solvent, and ρ_m is the density of the chain

---

[7]The coil structure of the macromolecule is treated in greater detail in the chapter on molecular structure (compare pp. 435-483).

[8]The solvent in the interior of the coil is in constant exchange through diffusion with the surrounding solvent. "Held liquid" or "included liquid," therefore, does not mean anything beyond saying that the solvent in the interior of the coil is entrained during sedimentation.

[9]The equivalent density ρ_{equ} of the gel particle (coil) can be determined by means of viscosity measurements (compare pp. 450-454).

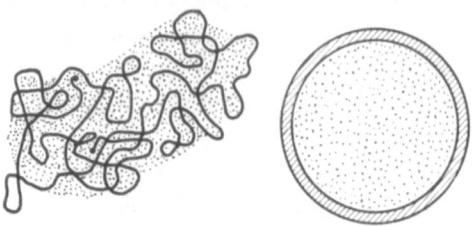

FIG. 360 — Statistical coil and equivalent sphere (= sphere, which would behave during sedimentation just like the coil)

substance (= the density of the polymer in solution).

The volume v_m belonging to a mass m is not known. It can be replaced by the mass m and the density of the chain substance (in the dissolved state), by writing $\rho_m = m/v_m$, and $v_m = m/\rho_m$ and one then obtains:

$$k_S = b\left(m - m\,\frac{\rho_{solv}}{\rho_m}\right) = m \cdot b \left(1 - \frac{\rho_{solv}}{\rho_m}\right) = m \cdot b\,(1 - V\,\rho_{solv}). \qquad (25)$$

$V = 1/\rho_m = v_m/m$: specific volume of the dissolved macromolecular compound (this is not exactly identical with the reciprocal density of the solid substance and therefore has to be determined in solution).

For the force of friction k_F, which opposes the force of gravity, the radius of the entire particle (coil and included solvent) is determining. According to Stoke's law for spherical particles, one can write:

$$k_F = 6\pi\,r\,\eta\,u = F \cdot u \qquad (26)$$

If one sets k_S and k_F equal, one obtains:

$$m \cdot b\,(1 - V\rho_{solv}) = 6\,\pi\,r\,\eta\,u$$

and since $m \cdot N_A = M$:

$$\overline{M} = \frac{6\,\pi\,r\,\eta\,u \cdot N_A}{b(1 - V \cdot \rho_{solv})} \qquad (26a)$$

where $F = 6\pi r\eta$ is the frictional factor, r is the radius of the sedimental particles, η is the viscosity of the solvent in which the sedimentation is determined, and u is the sedimentation velocity.

With compact spherical particles, not solvated in the interior, one can again replace the radius r by the specific volume V,[10] and one can determine the molecular weight $\overline{M}$ according to the following equation [which is obtained by setting k_S and k_F equal and replacing in Equation (26a) r by the expression $(3mV/4\pi)^{1/3}$] directly from the measured sedimentation constant s without any additional measurements:

$$\overline{M} = \frac{9\,\pi\,N_A}{2\,V}\left(\frac{4}{2}\cdot\frac{\eta\,s\,V}{1-V\rho_{solv}}\right)^{3/2} \tag{27}$$

s is the sedimentation constant = sedimentation velocity/centrifugal acceleration [sec].

With coils this cannot be done, because first, the diameter responsible for the frictional force is much larger than the diameter which would be obtained from the pure chain substance alone, and second, the coils do not have a spherical shape, but are irregular, elongated ellipsoids (compare pp. 445 and 505). One therefore has to determine experimentally the frictional factor F (corresponding to $6\pi r\eta$ with spheres). Usually one does this by diffusion measurements, which are not simple to carry out, but which have the advantage that the phenomenon of diffusion is very similar to that of sedimentation, so that one can expect the frictional factor F determined from diffusion to be identical with the frictional factor F which is characteristic for sedimentation. By means of measurements of compounds with known molecular weight, this has been experimentally confirmed. The relationship between the experimentally determined diffusion constant D and the frictional factor F is given by the Nernst diffusion equation[11] :

$$D = R \cdot T/N_A \cdot F. \tag{28}$$

If in Equation (26) one replaces r by $1/2 d_{equ(D)}$, $= RT/6\pi \cdot D \cdot \eta \cdot N_A$ then one obtains, after setting k_s and k_F equal, Equation (30), from which one can now calculate the molecular weight $\overline{M} = m N_A$ after measuring the sedimentation velocity, u, in the ultracentrifuge, and the diffusion constant, D, in a diffusion cell.

In a similar manner one can determine by means of viscosity measurements the equivalent particle diameter $d_{equ,[\eta]}$ by substituting the experimentally determined viscosity number into the Einstein equation for spherical particles. $d_{equ,[\eta]}$ does not correspond exactly to $d_{equ,D} = d_{equ,sed.}$, and one therefore has to add a

---

[10] $V = 1/\rho_m$; $\rho_m = m/v_m = m/(4/3)\pi r^3$

[11] For spheres the diffusion equation is $D = RT/6\pi r\eta NA$ (the derivation can be found in the standard textbooks of physical chemistry). If one sets the experimentally determined D values into this equation, one obtains a numerical value for the radius r, or the corresponding diameter d, which is designated as the diameter of an equivalent sphere: $d_{equ,(D)} = 2RT/6\pi D\eta N_A$. This is the diameter of spherical particles which in diffusion measurements would give the same results as the measured particles with any particular shape (for example, coils with an ellipsoidal shape).

factor in order to reduce $d_{equ,[\eta]}$ to d_{equ,s_o}, if instead of the diffusion constant one wants to use the intrinsic viscosity which is determined much more readily. With polymethylmethacrylate, for example, $d_{equ,s} = 0.86 \cdot d_{equ,[\eta]}$ (compare p. 466). This factor has the same order of magnitude for many polymer solvent systems, but is not exactly equal for all polymers. Stronger deviations are, for example, found for cellulose nitrate in acetone ($d_{equ,s_o}/d_{equ,[\eta]} \approx 0.7$). According to the Einstein equation one can write (for derivation see p. 464):

$$d_{equ,[\eta]} = (6\,M[\eta]/2.5\pi N_A)^{1/3}. \tag{31}$$

By substituting for r in Equation (26a) or (27) $1/2\ d_{equ,[\eta]}$ according to (31), one obtains:

$$\overline{M}_u = [3\pi\eta u \cdot 0.86/1 \cdot V\rho_{solv})b]^{3/2} \cdot [6[\eta]/2.5\pi]^{1/2} \cdot N_A. \tag{32}$$

If one substitutes the Nernst diffusion equation into Equation (26), one obtains:

$$k_F = (RT/N_A D) \cdot u. \tag{29}$$

Constant sedimentation velocity is reached after the frictional force k_F (which increases with increasing sedimentation velocity) has become so great that it has become equal to k_S (the gravitational force less the buoyancy):

$$mb(1 - V\rho_{solv}) = (RT/DN_A) \cdot u. \tag{30}$$

Usually instead of calculating with the sedimentation velocity u, one works with the so-called sedimentation constant, s:

$$s = u/b = \text{(sedimentation velocity/centrifugal acceleration) [sec]} \tag{33}[12]$$

s = sedimentation constant, measured in [Svedberg]: 1 [Svedberg] = 10^{-13} [sec].

If one further takes into account that the mass of the single molecule, m, multiplied by Avogadro's number gives the molecular weight, M, one can write:

$$\overline{M} = RT\ s_o/D_o\ (1 \cdot V\rho_{solv}). \quad \text{(Svedberg Equation)} \tag{34}$$

The average which is obtained by measurement of the sedimentation velocity and of the diffusion constant is approximately an $\overline{M}_w$-average (compare p. 425). s_o and D_o are the extrapolated values of s and D at c = o.

---

[12]The expression "sedimentation constant" is not very meaningful, because s is not really a constant. It would be better to call s, or rather s_o [obtained after extrapolating s to infinite dilution (c→0)], a reduced sedimentation velocity. However, since s not only depends on the centrifugal acceleration, but also on the density, ρ_{solv}, and the viscosity, η, of the solvent, and on the density of the dissolved polymer 1/V, one can obtain more meaningful s values, which now depend only on the molecular size, if one eliminates these parameters and, as with the viscosity number $[\eta]$, defines a sedimentation number [s]:

$$[s] = s_o\eta/(1 \cdot V\rho_{solv}) = \text{sedimentation number.}$$

[s] now depends only on the molecular weight and the particle diameter, i.e., parameters which are characteristic for the polymer molecule [compare Equation (26a)]. A practical example for the utility of the sedimentation number can be seen in Figure 472

The Measurement of V, s, and D

In order to calculate the molecular weight according to Equation (34), one has to determine experimentally the following parameters: the specific volume V, the density of the solvent ρ_{solv}, the diffusion constant D, and the sedimentation constant s.

The specific volume V is determined by accurate measurements of the density of the solvent and of the solution in a pycnometer.

If v_p is the volume and m_p the mass of the polymer in solution, v_s the volume and m_s the mass of the solution, v_{solv} the volume, m_{solv} the mass, and ρ_{solv} the density of the solvent, one can write:

$$V = \frac{v_p}{m_p} \; ; \quad v_p = v_s - v_{solv}; \; v_{solv} = \frac{m_{solv}}{\rho_{solv}} \text{ and } m_{solv} = m_s - m_p$$

$$V = \frac{1}{m_p} \left[v_s - \frac{1}{\rho_{solv}} (m_s - m_p) \right] = \frac{1}{c_v} - \frac{1}{\rho_{solv}} \left(\frac{1}{c_g} - 1 \right)$$

where c_v is the volume concentration and c_g the weight concentration in $[g/cm^3]$ and $[g/g]$, respectively. The values of V determined in this way are not exactly identical with the reciprocal density $1/\rho$ of the solid substances.

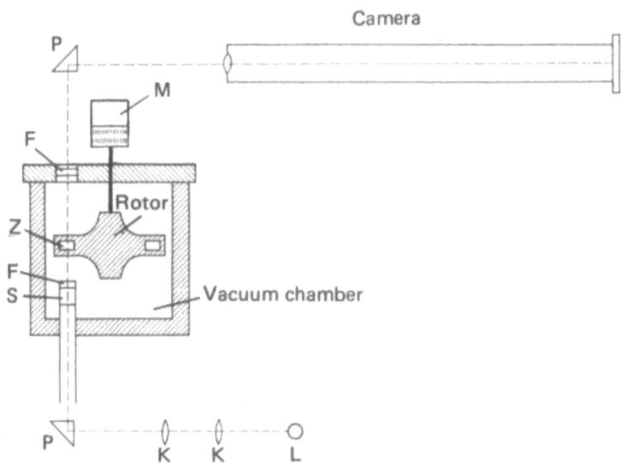

L   light source (high pressure mercury lamp)
K   condensor lense
P   reflecting prism
S   scale
F   quartz window
Z   solution cell with quartz windows
M   electromotor

FIG. 363 — Schematic representation of an ultracentrifuge measurement (after G. MEYERHOFF).

The sedimentation constant is obtained, according to its definition (Equation 33), from the sedimentation velocity, which is measured in the ultracentrifuge (see Figure 363). The dilute solution of the polymer is in a cell placed into the rotor of the centrifuge. The cell is fitted with two quartz windows which permit photographing the solution through a window in the vacuum-chamber. During rotation of the centrifuge, the macromolecules slowly sediment with constant velocity under the influence of the centrifugal force toward the bottom of the cell (order of magnitude: 2mm/h). Depending on the homogeneity of the dissolved macromolecules, one observes at the upper edge of the cell a more or less sharp boundary against the pure solvent, which, during the course of several hours (depending on the particle size and the centrifugal acceleration), moves toward the bottom of the cell. The velocity with which this boundary (meniscus) moves is the sedimentation velocity. The boundary between the solution and the solvent layer above it can never be completely sharp, because first of all the separation is smeared out through rediffusion of the sedimenting particles (an effect which is the larger, the smaller the particles), and secondly, because the boundary for nonhomogeneous systems (i.e., most macromolecular substances) is spread out even further because of the fact that the components with low molecular weight stay behind during sedimentation while the components with high molecular weight sediment more rapidly (compare Figure 364).

The simplest method of observation consists in photographing the rotating cell in suitable intervals with diffuse ultraviolet light. If one chooses the wavelength of the ultraviolet light in such a way that the light is absorbed much more strongly by the solution than by the solvent, then it is possible to recognize the sedimentation boundary by the different extent of darkening of a film placed under the cell. If the substance being investigated consists of several components with different (but within one component uniform) particles, then these components have a different

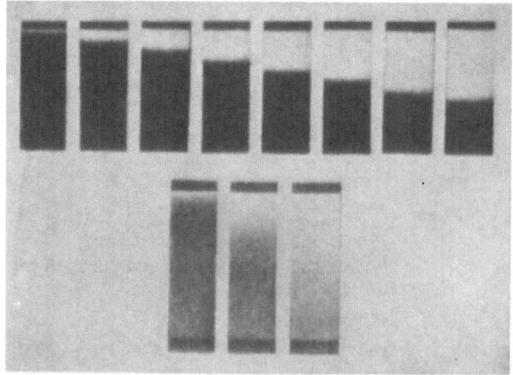

FIG. 364 — Sedimentation photographs of
a monodisperse system (top) and a polydisperse
system (bottom) (from MEYER and MARK).

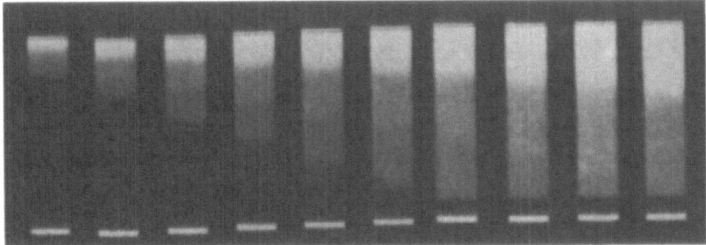

FIG. 365 – Sedimentation photographs of an oligodisperse system (after MEYER and MARK.)

sedimentation velocity depending on their particle size, and one obtains a picture of stepwise decreasing blackening. Figure 365 shows such photographs of protein solutions. This simple method of observation and measurement is however not sufficiently exact, especially in dispersed systems.

For precise measurements one has used first Lamm's scale method, which essentially consists of photographing a uniformly illuminated microscale in suitable time intervals obliquely through the cell. Since the refractive index of the solution depends on the concentration (the concentration gradient dc/dx is proportional to the gradient of the refractive index dn/dc), the strong concentration gradient in the boundary layer produces a distortion of the scale photographed through the boundary. This manifests itself by compression of the marks on the scale corresponding to the boundary region and a spreading out of the marks immediately next to this region. The magnitude, Z, of this distortion is proportional to the concentration change in the cell. The distance x from the axis of rotation where the Z value reaches its maximum is an exact definition of the boundary between the solution and the supernatant solvent. The determination of the distortion is obtained by comparing the distorted scale with one photograph without any distortion through the pure solvent. One can also project a picture of the scales one on top of the other for this purpose (Figure 366).

If one remembers that for a single molecular weight determination an entire series of such photographs has to be evaluated (with different concentrations and photographed at different times), one can form an idea about the time needed for such experiments. One therefore usually prefers the Schlieren method of Philpot and Swensson, which, through a cleverly selected optical system, makes it possible to observe the boundary as a maximum of the concentration gradient directly on the photographic film.

Whereas with monodisperse systems the sharpness of the maximum of the concentration gradient and therefore the height of the maximum of the distortion curve (Figure 366) remains unchanged during the entire sedimentation run, with polydisperse systems this maximum becomes more and more flat during the sedimentation because the larger particles in the system sediment more rapidly than

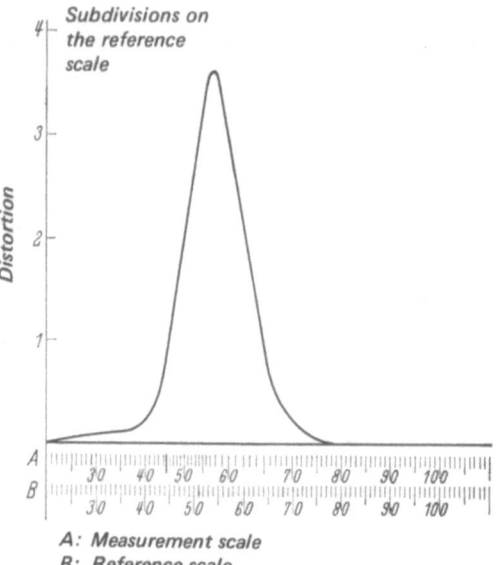

A: Measurement scale
B: Reference scale

FIG. 366 — Evaluation of sedimentation
photograph by the LAMM Scale Method.

the smaller ones. With oligodisperse systems one obtains curves with several
maximums, which lie fairly close to each other during a short period of sedimenta-
more corresponding to the broadening of the single maximum with polydisperse
more corresponding to the broadening of the single maximum with polydispersed
materials. One can therefore see immediately on sedimentation in the ultracentrifuge whether a macromolecular compound is monodisperse, oligodisperse, or
polydisperse. The velocity and the degree of flattening of the curves found with
polydisperse materials are a rough estimate of the degree of inhomogeneity.

The velocity with which the distance x, between the rotor axis and the
maximum of the concentration gradient, increases with time t, is the sedimentation
velocity, u, from which one can calculate the sedimentation constant, s according
to Equation (33) (see Figure 367-1):

$$s = \frac{u}{\omega^2 \cdot x} = \frac{x_2 - x_1}{t_2 - t_1} \cdot \frac{2}{(x_1 + x_2) \omega^2} \tag{35}$$

where ω is the angular velocity, u is the sedimentation velocity, and x_1 is the
distance of the boundary between solution and solvent (maximum of the concentration gradient) from the center of rotation at time t_1.

Since the distances x_1 and x_2 are only slightly different, one can express x
through the average distance $(x_1 + x_2)/2$. As with all molecular weight determinations, the measurements must be carried out at different concentrations, so that by

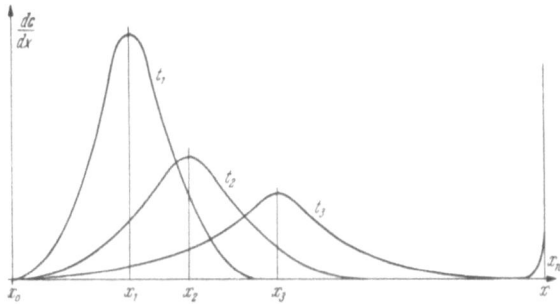

FIG. 367-1 — Schematic sedimentation diagram for a polydisperse system.

extrapolation of the sedimentation velocity to infinite dilution, one can graphically obtain s_o. In order to obtain the extrapolated curves essentially as straight lines, it is best to plot the reciprocal sedimentation constant $1/s$ as a function of concentration. The following equation holds:

$$1/s = 1/s_o + k_s \cdot c. \qquad (36)$$

s_o is called the reduced sedimentation velocity. The sedimentation number [s] is given by the relation (compare p. 362):

$$\frac{s_o \, \eta}{1 - V \, \rho_{solv}} = [s] \ ,$$

where η is the viscosity of the solvent, ρ_{solv} is the density of the solvent, and V is the specific volume of the polymer chain.

Also the determination of the diffusion constant consists of measuring the concentration gradient at different times t, and one can therefore make use of the same observation method as is used with sedimentation. Figure 367-2 shows the

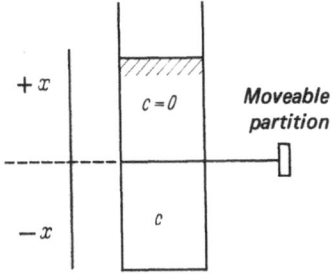

FIG. 367-2 — Schematic representation of a diffusion cell.

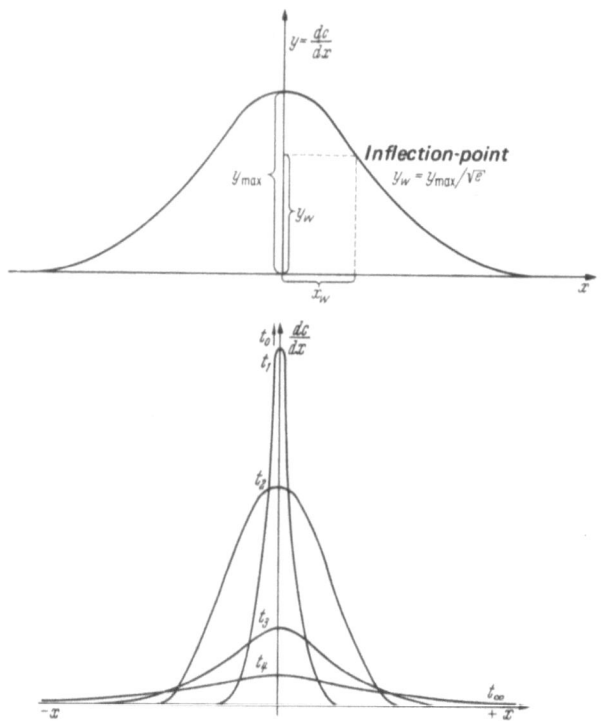

FIG. 368 — Interpretation of the diffusion diagram (after J. HENGSTENBERG)

schematic diagram for the experimental arrangement: on opening the barrier, the macromolecules from the lower solution layer begin to diffuse into the pure solvent. Thus the concentration above the barrier increases in the same measure as the concentration below the barrier decreases. The concentration gradient remains greatest close to the boundary and decreases with increasing distance up and down from the boundary in the form of a Gaussian curve. The bell shaped curves become flattened as the time of measurement t increases (Figure 368).

As a measure for the continuing diffusion, one chooses the distance x_i of the inflection point of the diffusion curve from the original boundary between the solvent and solution. This distance, divided by the time t, which has passed since the beginning of the diffusion experiment, is a sort of mean diffusion velocity. It is inversely proportional to the corresponding distance x_i, i.e., the mean diffusion velocity defined as x_i/t decreases in the same ratio as the distance x_i increases:

$$x_i/t = \text{constant} \cdot 1/x_i \tag{37}$$

Equation (37) is at the same time the definition of the diffusion constant thus the proportionality constant of Equation (37) is equal to 2D:

$$\frac{x_i}{t} = 2D \frac{1}{x_i} \quad \text{or } D = \frac{x_i^2}{2t} \qquad (38)$$

The inflection point of the curves can be determined mathematically from the y value of the maximum of the curve, y_{max}. For a Gaussian error curve the following holds: $y_i = y_{max}/e$. In most cases the diffusion curves differ from Gaussian curves more or less strongly, and one has to make use of the integrals of the areas under the diffusion curves for the definition of the inflection point. According to the so-called moment method, x^2 is defined as follows:

$$x_i^2 = \frac{\text{second moment of the diffusion curve}}{\text{area under the diffusion curve}} = \frac{\int_{-\infty}^{+\infty} x^2 y \, dx}{\int_{-\infty}^{+\infty} y \, dx} \cdot \qquad (39)$$

The values of the diffusion constants obtained according to Equations (38) and (39), are called D_m values. Further, one usually finds also the D_A values which are determined according to the so-called area method and which are only slightly different from the D_m values.

Thus from each diffusion curve obtained at different times, t, one can calculate a value for the diffusion constant D. In order to recognize an accidental error, the average value of D is obtained by plotting the product $D \cdot t$ as a function of t and determining the value of D as the tangent of the slope of the line $Dt = D \cdot t$.

Determination of the 2. Virial Coefficient B by Diffusion and Sedimentation Measurements

Just as with the sedimentation constant, one also has to determine the diffusion constant at different concentrations, c. The D values are usually plotted as a function of c and the resulting straight lines, which usually have a low slope, can be extrapolated easily to the value of the diffusion constant at infinite dilution, D_0. One can write:

$$D = D_0 + D_0 \cdot k_D \cdot c. \qquad (40)$$

(With regard to the molecular weight, D_0 is just as little a constant as s_0. The value $(D_0 \cdot \eta)$, which is independent of the viscosity of the solvent and depends only on the molecular size, is called the diffusion number $[D] = D_0 \cdot \eta$).

The constant k_D, like the constant k_S in Equation (36), is not identical with the 2. virial coefficient B of the osmotic pressure as it appears in the corresponding equations $P/c = f(c)$ and $K \cdot c/i_\theta = f(c)$. However, the values k_D and k_S [Equations (40) and (36)] are related to the 2. virial coefficient according to the following equation (G. V. Schulz and G. Meyerhoff):

$$B = \frac{R \cdot T}{2 M} \cdot (k_D + k_S). \qquad (41)$$

Thus it is possible to determine the important parameter B, which is a measure for the degree of the solvation of the macromolecules, i.e., the degree of interaction between the solvent and the macromolecule, by three completely independent methods.

Determination of the Molecular Weight from Sedimentation Equilibrium

The sedimentation process in the ultracentrifuge, if it is continued sufficiently long, leads to an equilibrium between sedimentation and diffusion. In this case there is a constant concentration gradient in the solution, which corresponds entirely to the pressure gradient in the atmosphere of the earth, and can therefore be described by the following law:

$$\ln \frac{c_2}{c_1} = \frac{M(1 - V\rho_{solv}) \cdot b}{RT} (x_2 - x_1) \qquad (42)$$

and

$$\overline{M} = \frac{2R\,T \ln (c_2/c_1)}{\omega^2(x_2^2 - x_1^2)(1 - V\rho_{solv})}, \qquad (43)$$

where c_1 and c_2 are the concentrations at the distances x_1 and x_2, respectively, from the center of rotation. b is given by $\omega^2 x$ and x by $(x_1 + x_2)/2$ (ω is the angular velocity). The concentration ratio c_2/c_1 can then be determined, for example, by photometric determination of the degree of blackening of a photographic plate.

The determination of the sedimentation equilibrium requires a long time of centrifugation (up to several days or weeks) and is therefore carried out only very seldom with polymer materials having molecular weights below 10^6. Furthermore, one obtains for $\overline{M}$ the so-called $\overline{M}_z$ value, which is different from $\overline{M}_n$ and also from $\overline{M}_w$. With a "normal-distribution" (compare pp. 411 ff. and p. 419) the relationship between these averages is given by $\overline{M}_n : \overline{M}_w : \overline{M}_z = 1:2:3$.

The Ultracentrifuge

The first ultracentrifuge was constructed by T. Svedberg and was considered a sort of miracle. Today one can obtain commercial ultracentrifuges with complete optical systems just as any other laboratory equipment, even though they will probably never became very inexpensive. Figure (371) shows a modern ultra-centrifuge.

Diffusion Cells

The greatest difficulty in determination of the diffusion constant lies in the formation of a sharp boundary between the solution and the solvent at the beginning of the diffusion experiment. Figure (372) shows the Meyerhoff cell. With this type of cell, the difficult insulation of the sliding barrier against the water

FIG. 371 — SPINCO– Ultracentrifuge. (In the open chamber at left the rotor can be seen; at the bottom on the right is the high vacuum system).

of the thermostat is circumvented by the fact that the barrier is not removed from the outside, but in a solvent cell communicating with the solution cell. The measurements (sedimentation measurements as well as diffusion measurements) are carried out at carefully controlled constant temperatures.

Significance of the Ultracentrifuge

The sedimentation method, which in every respect (technique of measurement, apparatus, time, and experimental and scientific practice and experience) requires a lot of effort, is characterized by great dependability and the occurrence of few errors. Requirements for the purity and cleanliness of the solutions are not as critical as with light-scattering measurements. The ultracentrifuge is applicable over the entire range of molecular weights occurring with macromolecular substances. The ultracentrifuge is of special importance in the examination of proteins and nucleic acids because with proteins and nucleic acids one mostly has to work with monodisperse and oligodisperse systems. (Oligodisperse systems consist of two or more components with different, but homogeneous, particle weights.) The ultracentrifuge produces photographs with clearly distinguishable degrees of blackening (compare Figure 365) or, (with refractometric measurements with the aid of the scale method), sedimentation curves with two or more maximums, so that through a *single* measurement one can determine the molecular weights of all the components (Figure 373).

Oligodisperse and monodisperse systems do not exist other than in natural polymer systems. Synthetic polymers, and also many natural polymers, are polydisperse, i.e., within certain boundaries each compound contains all molecular weights. During sedimentation runs the high molecular weight components sediment rapidly and the low ones sediment slowly, so that the boundary between the solvent and the solution in the ultracentrifuge cell is gradually spread out. This has the result that also the sedimentation curve becomes broad and flat. Even though the quantitative interpretation of the sedimentation curve for the calculation of the molecular weight distribution curve is possible in principle, and has in fact also been carried out, it is not quite satisfactory as far as accuracy is concerned. However, one can obtain from the form of the sedimentation curve and the steepness of the maximum a preliminary idea about the homogeneity of the product. If the product consists of a mixture of polymers of different molecular weight, one obtains sedimentation curves with several maximums, which, of course, are less sharp than is found with oligodisperse systems. The ultracentrifuge is not only of great importance as an instrument for molecular weight determinations, but it also has

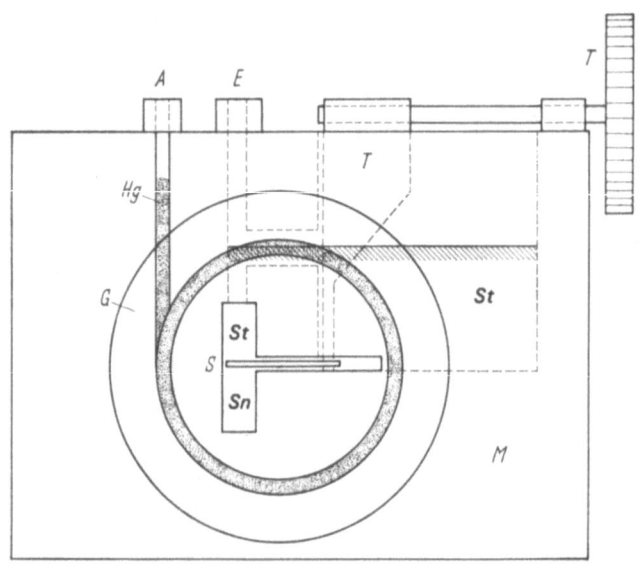

E   opening for introduction of solution and solvent

Hg   mercury seal for the cell

T   turn-screw to move the barrier between solution and solvent

S   sliding barrier

St   solvent in the upper half of the cell

Sn   solution in the lower part of the cell

G   glass plate on both sides of the cell chamber

M   metal block

FIG. 372 – Diffusion cell (G. MEYEROFF)

given important results in the examination of the shape (secondary structure) of polymer chains (compare pp. 468-481).

Influence of Solvent and Coil Density

With chain-like polymer molecules, the sedimentation velocity depends also on the solvent. In strongly solvatizing solvents (in which there occurs a large increase in viscosity) the sedimentation velocity is smaller than in poor solvents (in which the same product gives rise to a smaller increase in viscosity). This results from the fact that in good solvents the coils have a smaller density and, therefore, a larger volume and a larger particle radius, than in a poor solvent. This is seen immediately if one does not express the frictional factor F, i.e., the frictional force k_F, by the experimentally determined diffusion constant D, but instead by means of Stoke's law for spheres [Equation (26)], and then sets k_F equal to k_S:

$$6\pi r \eta u = mb\,(1 - V\rho_{solv}) \tag{44}$$

or, (because $m \cdot N_A = \overline{M}$ and $u/b = s$):

$$N_A\,6\pi r \eta s = \overline{M}\,(1 - V\rho_{solv})$$

and

$$s = \frac{\overline{M}(1 - V\rho_{solv})}{6\pi r \eta\, N_A} \tag{45}$$

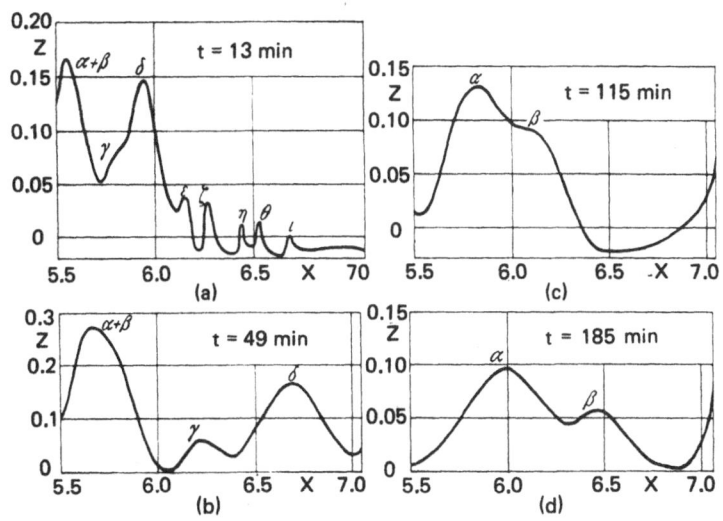

β	Palmer's lactoglobulin	Z	distortion of the scale
γ	normal lactoglobulin	X	distance from the rotor axis
δ, ϵ	normal casein components	α	lactalbumin

FIG. 373 — Sedimentation diagram of dialyzed skimmed milk (an oligodisperse system). (A.S. McFARLANE)

It can be seen that the sedimentation velocity is inversely proportional to the radius, i.e., the larger the sedimenting coil, (for a given molecular weight), the lower the sedimentation velocity and also the lower the accuracy of the measurements. One should therefore try to use poor solvents for sedimentation measurements in which, for a given molecular weight, s is largest.

Poor solvents should be preferred for sedimentation experiments for another reason: the density of the coil, and therefore the ratio of the particle weight to the particle diameter, is dependent not only on the solvent, but also on the molecular weight itself. Thus, one finds that the coil density decreases with increasing molecular weight, i.e., the coil radius increases more rapidly than one can account for by the increase of $\overline{M}$. This leads to the fact that the change in the sedimentation constant with the molecular weight is less with coils than with systems with a particle density which is independent of molecular weight. As will be discussed more thoroughly later on, for particles with constant density: $s \propto \overline{M}^{2/3}$; for particles with decreasing density: $s \propto \overline{M}^{(2-a)/3}$ where a is the exponent of the coil density function $\rho_{coil} \propto \overline{M}^{-a}$ and also the exponent of the viscosity function: $[\eta] \propto \overline{M}^a$ (compare p. 469). In good solvents a is larger than in poor solvents, however, the greater a the smaller is the change of s with the molecular weight. Thus, in order to obtain the largest possible difference in the sedimentation velocity of different products with given molecular weights $M_1 \neq M_2 \neq M_3$, etc., one selects a poor solvent, because, in such a solvent, a is smaller (close to 0.5), and the molecular weight determination by sedimentation is therefore more sensitive.

The same phenomenon, i.e., greater coil density with small particles and smaller coil density with large particles, leads to the result that with polydisperse products the larger particles sediment more slowly and the smaller particles more rapidly, as if they would have all the same density. This means that the maximum of the concentration gradient in the boundary layer is sharper than it would be with particles of the same density. This is because the decrease in the coil density with increasing molecular weight works against the spreading out of the sedimentation boundary layer caused by polydispersity. This effect is observed most strongly with good solvents, so that with strongly polydisperse products one should use a good solvent.

3125 Molecular Weight Determination by Viscometry

Of all the methods used for the determination of the molecular weight of macromolecular compounds, the viscosity method introduced by H. Staudinger is the one most commonly used in day-to-day research and development. Although by means of viscosity measurements alone, it is only possible to observe *changes* in the molecular weight, in many cases this is already sufficient. Secondly, if one needs absolute values for $\overline{M}$, these can be determined with the aid of a calibration curve for $[\eta]$ versus $\overline{M}$. Such calibration curves can be found in the literature for all important polymers.

Theoretical Principles

As has been theoretically derived by A. Einstein, the relative viscosity η_{rel} of a solution (or dispersion) of spherical, colloidal particles depends only on the volume fraction φ of the dissolved (or dispersed) phase:

$$\eta_{rel} = 2 \cdot 5 \cdot \varphi + 1. \tag{46}$$

The relative viscosity expresses the number of times which the viscosity of the solution, η is greater than the viscosity η_0 of the solvent in which the macromolecular substance is dissolved:

$$\eta = \eta_{rel} \cdot \eta_0 \text{ or } \eta_{rel} = \eta/\eta_0. \tag{47}$$

The volume fraction φ is the volume occupied by the dissolved particles in 1 cm^3 of solution:

$$\varphi = \text{volume of the dissolved macromolecules/volume of the solution.} \tag{48}$$

If the mass of the dissolved macromolecular compound is m and the density of the empty molecular coils of this compound is $\overline{\rho}_{equ,[\eta]}$,[13] then the volume which m grams of the compound occupy in solution in the form of statistical coils is given by: $m/\overline{\rho}_{equ,[\eta]}$, and φ is then:

$$\varphi = (m/\overline{\rho}_{equ,[\eta]})/v_S,$$

where v_S is the volume of the total solution. Since m/v_S in grams per cm^3 is usually called the concentration, c, of the solution one can also write:

$$\varphi = c/\overline{\rho}_{equ,[\eta]}. \tag{49}$$

If one substitutes this in Equation (46), one obtains the Einstein viscosity law in the following form:

$$\eta_{rel} = (2 \cdot 5 \, c/\overline{\rho}_{equ,[\eta]}) + 1 \tag{50}$$

or[14]

$$\frac{\eta_{rel} - 1}{c} = 2.5 \, \frac{1}{\overline{\rho}_{equ}, [\eta]} = \eta_{sp}/c \tag{51}$$

---

[13] ρ_{equ} is the density of equivalent spheres, which in terms of their viscosity-increasing function, behave exactly like the dissolved macromolecules whose form cannot be exactly determined. A more detailed discussion can be found in the chapter on "Molecular Shape" (compare pp. 435-438) also in the section "The Macromolecular Solution (p. 488-510). $\rho_{equ,[\eta]}$ is not the density of the chain substance, but the much lower density of the entire coil, that is, the empty coil without the included solvent. For the definition of the coil volume, compare Figures 493 and 509.

[14] Instead of $\eta_{rel}-1$ one can write $(\eta-\eta_0)/\eta_0 = \eta_{sp}$ specific viscosity. This form is usually found in the German and American Literature. $\eta - \eta_0$ is the viscosity increase brought about in the solvent by the dissolution of the macromolecular compound, and $(\eta - \eta_0)/\eta_0 = \eta_{sp}$ is therefore the relative viscosity increase.

Since this equation applies only if the dissolved particles do not interfere with one another (with macromolecular solutions this is only true at large dilutions), one has to plot the experimentally determined $(\eta_{rel}-1)/c$ values versus c and to extrapolate to c=0. Because the viscosity of macromolecular solutions, especially if we are dealing with solutions of polymers with very high molecular weight, can depend on the flow gradient G [i.e., on the flow (shear) gradient in the instrument of measurement], one has in such cases to determine η_{sp}/c also at different flow gradients G and to extrapolate to G=0. The limiting value of $[\eta]$ obtained in this way, is called the viscosity number,[15] or Staudinger-Index (=intrinsic viscosity in the older literature):

$$\frac{\eta_{sp}}{c} = \frac{\eta_{rel}-1}{c} = [\eta] = 2.5 \quad \frac{1}{\bar{\rho}_{equ,[\eta]}} . \tag{52}$$
$$(c \to 0) \ (c \to 0)$$
$$(G \to 0) \ (G \to 0)$$

From this it can be seen that the viscosity increase, which occurs on dissolving a macromolecular compound, is inversely proportional to the average particle density. In general the density is a constant of the material, which usually is independent of the size of a certain object and therefore with dispersions of the usual solid compounds (rubber latex, plastics dispersions, and gold sols, etc.) one cannot find any dependence of the viscosity on the particle size. This, however, is not the case with the random coils present in solutions of macromolecular compounds having a chain structure. As will be discussed in greater detail in the discussion of statistical coils (compare pp. 441-450), it is a theoretically clearly founded characteristic quality of coils that their average density decreases with increasing molecular weight. With ideal coils this occurs according to the square-root law of Werner Kuhn:

$$\bar{\rho} \ coil = K_\rho \cdot \bar{M}^{-0.5} \tag{53}$$

---

[15]The extrapolation to G = 0 can be avoided in many cases because the effect is usually within the error of the measurements, if one uses a suitable viscometer. In doubtful cases, however, one should first carry out a measurement with the highest molecular weight sample before one neglects G entirely. With $[\eta]$ values over 300 cm^3/g, one has to count on a considerable influence of the velocity (shear) gradient on the η_{sp}/c-values.

$[\eta]$ has the dimensions of a reciprocal density. Since the density is usually given in grams per cm^3, one should always use grams per cm^3 for the concentration unit in giving the value of η_{sp}/c, and this in turn will yield an $[\eta]$ value in the correct dimension of cm^3/g. Unfortunately at the beginning this was not taken into account, and therefore one always has to check viscosity values in the literature as to their dimension, before it is possible to compare them with each other. In the older literature one usually finds c in grams per liter or in grams per/100 cm^3. Instead of $[\eta]$, one often finds in older German papers the designations $Z\eta$, and instead of viscosity number one finds limiting viscosity or intrinsic viscosity. Since, as is shown by Equation (52), $[\eta]$ is not a real viscosity, but only a relative viscosity increase proportional to the reciprocal coil density, the designation "viscosity number" is more correct.

Therefore, one has to expect from the Einstein law that also the degree of viscosity increase $(\eta_{rel} - 1)/c = \eta_{sp}/c$ is a function of the molecular weight with such systems. Thus if one substitutes Equation (53) in Equation (52), one obtains:

$$[\eta] = 2.5/K_\rho \cdot \overline{M}^{-0.5} = K_{[\eta]} \cdot \overline{M}^{0.5}. \qquad (54)$$

This is the Kuhn viscosity law for solutions with ideal statistical coils. This equation, which has been derived on purely theoretical grounds, can also be confirmed experimentally. Figure 377 shows as an example the plot of $[\eta]$ versus $\overline{M}$ on a double logarithmic paper for polyisobutylene in benzene at 24°C. In this logarithmic form, $\log [\eta] = \log K + 0.5 \log \overline{M}$, Equation (54) is the equation of a straight line, and the exponent 0.5 is found as the tangent of the slope. Actually the exponent is only equal to 0.5 at a certain well-defined temperature, the so-called θ-temperature, or Flory-temperature, whereas at higher temperatures one finds higher exponents (usually 0.7). The exponent is furthermore dependent on the type of solvent and the structure of the polymer. This means that with real systems, because of the energetic interaction between the solvent and the polymer chains, the coil density does not decrease according to Equation (53) exactly with $\overline{M}^{-0.5}$ but decreases quite generally with $\overline{M}^{-a}$. a usually lies between 0.65 and 0.75, so that the equation for real systems [corresponding to the ideal Equation (53)] is:

$$\overline{\rho}_{equ,[\eta]} = const. \cdot \overline{M}^{-a}. \qquad (55)$$

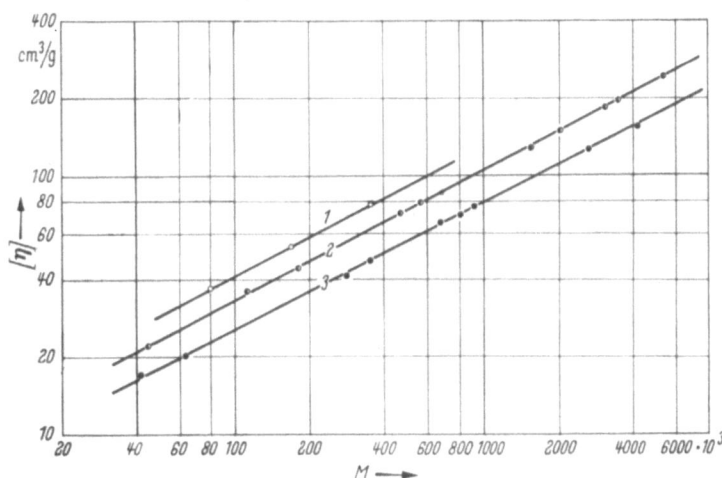

1   cellulose caprylate in γ-phenylpropylalcohol at 48°C

2   polyisobutylene in benzene at 24°C

3   polystyrene in cyclohexane at 34°C

FIG. 377 — $[\eta]$ — M-relation for three polymers under θ-conditions; $a = 0.5$ (P.J. FLORY).

Correspondingly, one obtains from the Einstein law, by substituting Equation (55) in Equation (51) the following general viscosity equation for macromolecular solutions:

$$[\eta] = K_{[\eta]} \cdot \overline{M}^a \cdot \tag{56}[16]$$

In certain cases a can be equal to O. This is always found if the dissolved macromolecules are not in the form of coils, but are compact, non- or only poorly-solvated particles with constant density. Macromolecular compounds of this type are, for example, glycogen (liver polysaccharide) and the so-called globular proteins.

Furthermore there are polymers (especially cellulose, pectins, and other poly-saccharides) whose chains are relatively stiff, and which therefore in solution form especially loose coils whose density decreases with increasing molecular weight according to the relation $\overline{\rho}_{coil} = K \cdot M^{-0.9 \text{ to } 1.2}$. In the first case ($a = 0$), the viscosity number $[\eta]$ is relatively small and independent of the molecular weight, whereas in the second case ($a \cong 1$), $[\eta]$ is relatively high and almost proportional to the molecular weight. It is entirely possible, (although for reasons of experimental accuracy it can be demonstrated only with difficulty), that with a certain polymer-solvent system a might be exactly equal to 1. However, one should avoid drawing from this fact any conclusions about a special molecular structure, or molecular form, which might account for this.

The $[\eta]$ – $\overline{M}$ Calibration Curve

Neither the constants $K_{[\eta]}$ in Equation (56) nor the exponents a can be derived theoretically. They have to be determined experimentally. One proceeds as follows: The polymer is separated into as many fractions as possible with the greatest possible homogeneity, and the molecular weight and the viscosity number are determined for each fraction. For the molecular weight determination, provided the fractions are sufficiently homogeneous, any absolute method is suitable. In practice, one usually gives some preference to such absolute methods which give averages close to $\overline{M}_w$ (i.e., sedimentation in the ultracentrifuge or light-scattering) because Equation (56) also gives approximately $\overline{M}_w$ averages, and therefore differences in the homogeneity of the fractions have the least influence on the position on the points in the $[\eta]$ – $\overline{M}$ diagram (for further discussion see section on molecular weight averages, p. 425).

---

[16]The correct relationship between $[\eta]$ and $\overline{M}$ with macromolecular solutions was first recognized by H. Staudinger and demonstrated experimentally with solutions of cellulose and cellulose derivatives, ($a \cong 1$). W. Kuhn has derived Equation (56) on theoretical grounds for the statistical coil by means of the random flight equation. H. Mark was the first to formulate this equation in the above (power) form and to demonstrate its validity (at the same time as R. Houwink) by means of empirical values. One would really have to call this equation, if one were to use the names of all the scientists involved, the Einstein-Staudinger-Kuhn-Mark-Houwink equation, because in reality this is simply the Einstein viscosity law for spherical colloids transferred to particles with size dependent particle density (random flight particles = statistical coils; for the derivation see p. 454).

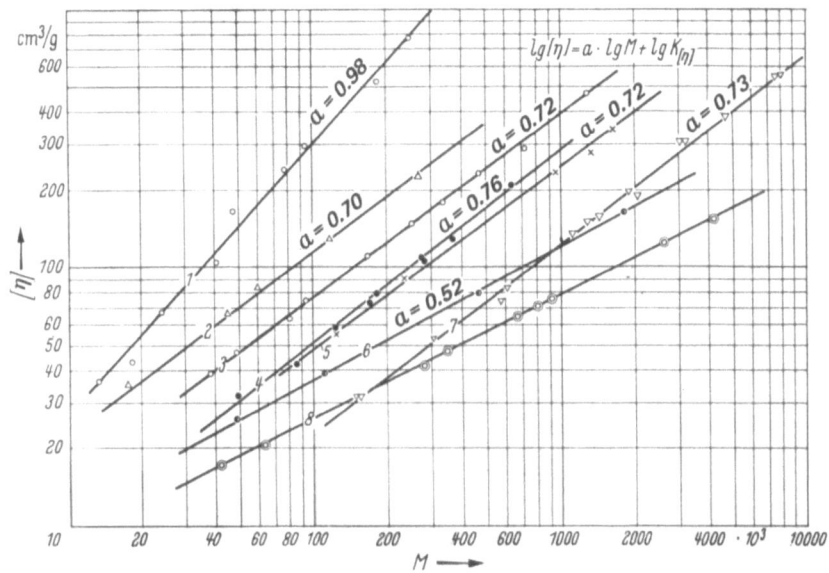

1 cellulose nitrate (acetone, 20°C)

2 polycarbonate (tetrahydrofuran, 20°C)

3 polyisobutylene (cyclohexane, 30°C)

4 isotactic polypropylene
   (α-chloronaphthalene, 145°C)

5 polystyrene (toluene, 22°C)

6 polyisobutylene (benzene, 25°C)

7 polymethylmethacrylate (acetone, 20°C)

8 polystyrene (cyclohexane, 34°C)

FIG. 379 — $[\eta]$ — $\overline{M}$ curves on a log-log plot (measurements of G.V. SCHULZ, MEYERHOFF, FLORY, ZIMM).

The experimentally determined $[\eta]$ and $\overline{M}$ values of the fractions are plotted on a double logarithmic coordinate system, and one obtains a straight line which according to Equations (56), (57), or (58),

$$\log [\eta] = a \cdot \log \overline{M} + \log K_{[\eta]} \qquad (57)$$

$$\ln [\eta] = a \cdot \ln \overline{M} + \ln K_{[\eta]} \qquad (58)$$

gives the exponent a as the tangent of the slope and the constant $K_{[\eta]}$ as the intercept on the ordinate.

Figure 379 shows for a series of polymers such experimentally obtained $[\eta]$ versus $\overline{M}$ calibration curves, and in Table 380 the corresponding a and $\overline{M}$ values[17] are listed for a large number of polymers.

---

[17]Higher $K_{[\eta]}$ constants do not automatically mean higher viscosity numbers, because $[\eta]$ depends not only on K, but also on the exponent a which determines both the change in the viscosity number with the molecular weight and the absolute value of $[\eta]$. Even though, for example, the $K_{[\eta]}$ value for polyisobutylene in cyclohexane is smaller than in benzene, the

TABLE 380. The constants $K_{[\eta]}$ and *a* for a selection of polymer/solvent systems. (For a very comprehensive list refer to BRANDRUP-IMMERGUT, *Polymer Handbook*, Interscience, 1966)

Polymer	Solvent	Temperature °C	$K_{[\eta]}^*$	a
Natural rubber	Toluene	25	5.0 10^{-2}	0.67
Pectic acid	0.155 n NaCl-soln.	25	0.014 10^{-2}	1.34
Cellulosenitrate	Acetone	20	0.28 10^{-2}	1.0
Celluloseacetate	Acetone	25	0.25 10^{-2}	1.0
Polyethylene (low pressure)	Tetrahydronaphthalene	120	2.36 10^{-2}	0.78
	α-Chloronaphthalene	125	4.3 10^{-2}	0.67
Polyethylene (high pressure)	Decalin	70	0.38 10^{-2}	0.74
	Xylene	75	1.35 10^{-2}	0.63
Polyisobutylene	Benzene	24	8.3 10^{-2}	0.50
Polyisobutylene	Benzene	60	2.6 10^{-2}	0.66
Polyisobutylene	Toluene	20	2.6 10^{-2}	0.64
Polyisobutylene	Cyclohexane	30	2.6 10^{-2}	0.70
Neoprene (Poly-chlorobutadiene)	Toluene	25	5.0 10^{-2}	0.62
SBR (synthetic rubber)	Toluene	25	5.2 10^{-2}	0.67
Polystyrene	Benzene	25	1.03 10^{-2}	0.74
Polystyrene	Methylethylketone	25	3.9 10^{-2}	0.58
Polymethylmethacrylate	Benzene	20	0.835 10^{-2}	0.76
Polymethylmethacrylate	Acetone	20	0.390 10^{-2}	0.73
Polymethylmethacrylate	Chloroform	20	0.485 10^{-2}	0.80
Polyvinylacetate	Acetone	25	1.76 10^{-2}	0.68
Polyvinylpyrrolidone	Water	25	1.40 10^{-2}	0.70
Polycaprolactam	m-Cresol	20	0.73 10^{-2}	1.0
Polycaprolactam	H_2SO_4	25	2.9 10^{-2}	0.78
Polycarbonate (4,4'-Dihydroxy-2,2-diphenylpropane)	Tetrahydrofuran	20	3.99 10^{-2}	0.70
	Methylenechloride	20	1.11 10^{-2}	0.82
N-Butyl-nylon 1	Tetrahydrofuran	20	0.045 10^{-2}	1.18

*These constants apply for viscosities in cm^3/g

---

viscosity number in the better solvent cyclohexane is approximately five times as large as in the poorer solvent benzene. For polyisobutylene of $\overline{M} = 10^6$, in cyclohexane: $[\eta] = 2.6 \cdot 10^{-2} (10^6)^{0.7} = 410 \, cm^3/g$; and in benzene: $[\eta] = 8.5 \cdot 10^{-2} (10^6)^{0.5} = 85 \, cm^3/g$. For the definition of $K_{[\eta]}$ in terms of the length of the statistical chain element, compare p. 513.

Measurement of the Viscosity

The viscosity measurements used for the determination of the viscosity number can be carried out simply and rapidly. According to the Hagen-Poiseuille law:

$$\eta = \pi/8 \cdot \Delta p/l \cdot r^4/v \cdot t, \tag{59}$$

where $\Delta p = \rho \cdot g \cdot h$ is the pressure difference; l is the length of the capillary; and r is the radius of the capillary. Equation (59) states that with a constant flow volume and constant apparatus dimensions, the viscosity is proportional to the product of the flow time, t, through the capillary, and the density of the solution, or the solvent, respectively. One therefore needs for a determination of the relative and the specific viscosity $\eta_{rel} = \eta/\eta_0$ and $\eta_{sp} = (\eta_{rel}-1)/c$, not the absolute η values, but since the density of the solution and of the solvent are only slightly different in

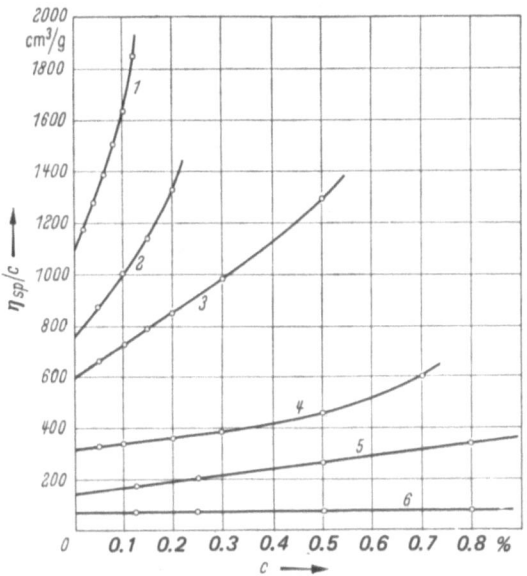

1 cellulose $\bar{M}$ = 1,250,000 in $[Cu(NH_3)_4](OH)_2$ solution

2 polyisobutylene $\bar{M}$ = 9,500,000 in toluene

3 polyisobutylene $\bar{M}$ = 1,710,000 in cyclohexane

4 polyisobutylene $\bar{M}$ = 2,459,000 in toluene

5 polystyrene $\bar{M}$ = 560,000 in toluene

6 polystyrene $\bar{M}$ = 312,000 in methylethylketone

FIG. 381 — η_{sp}/c vs. c curves of different polymers at 20°C (from STAUDINGER, SCHULZ, HUSEMANN)

most cases, it is quite sufficient to determine the flow time t for different concentrations:

$$\eta_{sp} = \eta_{rel} - 1 = (\eta - \eta_0)/\eta_0 = (t \cdot \rho - t_0 \cdot \rho_0)/t_0 \cdot \rho_0 \cong (t - t_0)/t_0 \qquad (60)$$

The η_{sp}/c determined in this way are then plotted against c. By extrapolation of the curves to c = 0, one obtains as the intercept at the ordinate the value of the viscosity number $\eta_{sp}/c = [\eta]$ (Figure 381).

With the aid of the $[\eta]$ values determined in this way, one can by means of $[\eta]$ versus $\overline{M}$ calibration curves (as shown in Figure 379 for a number of polymers) obtain the corresponding molecular weights. One always has to take care however, that $[\eta]$ measurements are carried out at the temperatures and in the solvents for which the calibration curves have been determined.

The Dependence of η_{sp}/c Values on the Concentration

The slope of the η_{sp}/c vs. c curve can be described in many cases through an empirical equation established by Huggins:

$$\eta_{sp}/c = [\eta] + k_H \cdot [\eta]^2 \cdot c. \qquad (61)$$

The equation is a power series which has been ended after the second term, and corresponds, if one expresses $k_H \cdot [\eta]^2$ as a single constant, in its form exactly to the power series developed generally for gases, which has already been used to describe the slope of the p/c versus c curves from osmotic pressure measurements [Equation (10)]:

$$p/c = RT/M + Bc. \qquad (62)$$

The 2. virial coefficient of the osmotic pressure, B, and the Huggins coefficient $k_H [\eta]^2$ depend on the solvent and represent a measure for the strength of the solvent, i.e., for the degree of solvation (Compare also Chapter 4, p. 517).

The Huggins equation is not so general that it can replace the extrapolation. First of all the constant k_H is not exactly equal for each polymer-solvent system (k_H has values between 0.2 and 0.6), but there are also cases where the η_{sp}/c vs. c curve is not a straight line, but curves upward (especially with compounds having especially high molecular weight and in general with all solutions having high viscosity number). High viscosity numbers are not always caused by a high molecular weight; they can also result from a high value of the constants $K_{[\eta]}$ and the exponent a of the viscosity Equation (56), i.e., the result of loose coiling of the molecules due to stiff chains and high A_m values (compare pp. 441-444).

In such cases it has often been useful to employ the equation of G.V. Schulz and Blaschke.[18]

---

[18]Schulz-Blaschke: $\eta_{sp}/c = [\eta](1 + k_{SB} \cdot \eta_{sp}) = [\eta](1 + k_{SB} \cdot \dfrac{\eta_{sp}}{c} \cdot C).$

Huggins: $\eta_{sp}/c = [\eta](1 + k_H[\eta] \cdot C) = [\eta](1 + k_H \cdot \dfrac{\eta_{sp}}{c} \cdot C).$

$$[\eta] = \frac{\eta_{sp}/c}{1 + k_{SB}\,\eta_{sp}} \quad \text{or} \quad \frac{\eta_{sp}}{c} = k_{SB}\cdot[\eta]\cdot\eta_{sp} + [\eta], \qquad (63)$$

where η_{sp}/c is plotted against η_{sp}, and k_{SB} in most cases has a value of 0.27. One then obtains straight lines with a slope $k_{SB}\cdot[\eta]$, and $[\eta]$ as the intercept at the ordinate. Figure 383 shows the advantage of this procedure with cellulose derivatives as an example.

Macromolecular solutions do not always behave like Newtonian liquids, and one therefore has to count on the fact that the viscosity depends on the velocity gradient and, therefore, on the experimental conditions. If one doesn't know the behavior of the polymer in this connection, one has to carry out measurements at different velocity gradients G and then if necessary extrapolate to G = 0. This is necessary, for example, with cellulose derivatives.

Viscometers

Since molecular weight determinations by viscometry do not require absolute η values, one can carry out the measurements in simple Ostwald viscometers (Figure 384a). Because of the non-Newtonian behavior of most macromolecular solutions

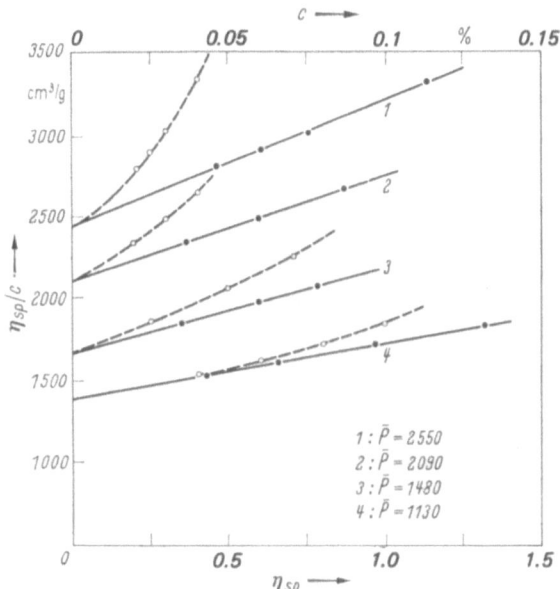

FIG. 383 — η_{sp}/c vs. η_{sp} curves (———) for cellulose-nitrate in acetone, and the corresponding η_{sp}/c vs. c curves (– – – –) (M. MARX and G. V. SCHULZ)

at high velocity gradients in the capillary, one usually chooses the viscometer dimensions in such a way that the velocity gradient is the smallest possible. According to G. V. Schulz the capillary diameter should be selected in such a way that for a difference in level of 10 cm, the flow time should be of the order of 100-150 sec. By a special arrangement of the bulbs in the Desreux instrument (Figure 384b), one achieves a relatively small difference h between the liquid levels, so that one can therefore choose a diameter which is quite large. This gives rise to an especially low velocity gradient.

Since the viscosity is strongly temperature dependent, the viscometer has to be kept in a thermostat during the measurements. The usual temperature of measurements is 25°C.

In place of the Ostwald viscometer, one can also use different other viscometers: the falling ball viscometer and especially rotational viscometers, such as the Couette viscometer. These are usually considerably more expensive. Rotational viscometers with cylinders of different size and variable rotation velocities are especially convenient if one wants to carry out measurements at different gradients. By variation of the rotation speed and the distance between the cylinders, one can vary the velocity gradient G over a large range in a well-defined way.

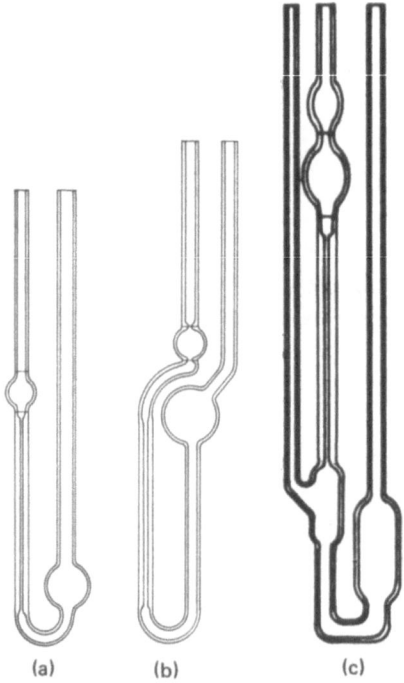

(a)   (b)   (c)

FIG. 384 — a) OSTWALD — Viscometer b) DESREUX-BISCHOFF — Viscometer c) UBBELOHDE — Viscometer.

313 Molecular Weight Distribution

3131 The Distribution Curve

Because of the polydisperse character of most macromolecular compounds, the molecular weight, determined in one way or another, yields an average. Even if one knows whether it is a number-average or a weight-average (see also p. 411 and p. 426), this number does not say very much about the state of the polydisperse system: i.e., whether a polymer with an average molecular weight of 100,000 corresponds to molecules between 80,000 and 120,000 (a narrow molecular weight distribution, so that it can be regarded practically as uniform), or whether such a polymer consists of molecules varying in molecular weight between 500 and 10 million (corresponding to a very broad distribution.) Whether one or the other is the case can be determined by carrying out an osmotic *and* a light-scattering, or sedimentation, molecular weight determination. This yields one $\overline{M}_n$-and one $\overline{M}_w$-value, and the quotient $\overline{M}_w/\overline{M}_n$ provides a measure of the breadth of the distribution. The larger $\overline{M}_w$ in relation to $\overline{M}_n$, i.e., the more the ratio $\overline{M}_w/\overline{M}_n$ is different from 1, the more inhomogeneous is the material. However, even then one still doesn't know how the broad or narrow distribution is obtained, i.e., whether the components with medium molecular weight are present in much greater amounts than the low and high molecular weight components, but where these might have extremely low and extremely high molecular weights. However, there could also be several different molecular weights appearing in relatively large amounts, which then leads to distribution curves with several maximums. About questions such as these, one obtains information only if one separates the polymer sample into a series of fractions and determines their molecular weight and the amount in which they are present. The larger the number of fractions, the more accurate the determination of the molecular weight distribution. The extreme theoretical case, where the polymer is divided evenly into completely homogeneous fractions of degree of polymerization $P = 1, P = 2, P = 3$, etc. to $P = 10^6$ or 10^7, can not even be approximated. (The degree of polymerization P, is defined as the number of structural units in a chain-like macromolecule). Instead, one obtains (independent of the method used) fractions which are always more uniform than the original mixture from which they are separated, but which, even in the most careful fractionation, still overlap to the extent of a hundred or more structural units. For each fraction, one determines the molecular weight and the actual weight of the fraction, and then obtains the type of result which is represented in Table 386 for a hypothetical polymer.

One can represent such a result in the form of a step curve, such that the amount m_F with which each fraction is represented, is plotted against the degree of polymerization (Figure 386). This representation shows that the percentage of molecules with a degree of polymerization between 2,500 and 3,500 is 9%, that between 4,500 and 5,500 is 20%, etc., where however the boundaries between the two fractions can not be determined as accurately as the step curve seems to imply.

TABLE 386. A simplified example of the results obtained in a typical fractionation experiment.

Fraction No.	Amount in % m_F	Cumulative amount in % Σm_F	Degree of Polymerization $\bar{P}$
12	1	1	1000
11	4	5	2000
10	9	14	3000
9	15	29	4000
8	20	49	5000
7	17.5	66.5	6000
6	13.5	80	7000
5	9	89	8000
4	5.5	94.5	9000
3	3	97.5	10000
2	2	99.5	11000
1	0.5	100	12000

However, from an overall point of view, the result of the fractionation can be read from such a curve as long as the width of the steps is not too different. Unfortunately, this last condition is especially very difficult to fulfill in practice. Usually the differences between the molecular weights of the neighboring fractions are quite different. The effect of this irregularity on a curve such as the one shown in Figure 386 can be understood if one tries to represent the values in Table 387 in the same form as was done in Figure 386 for the values from Table 386. A glance at the m_F values in Table 387 shows that the steps next to the third fraction show a maximum right next to a minimum. This, however, leads to a completely erroneous picture of the molecular weight distribution of this polymer. Thus the m_F value of the second fraction is not higher than that of the third fraction because the amount of molecules with a degree of polymerization 1,450 is greater than that with a degree of polymerization of 1,300 (in fact, just the opposite is true!); but it is higher because fraction 2 covers a much wider range in the degree of polymeriza-

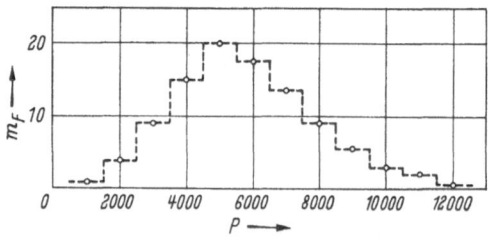

FIG. 386 — Distribution step-diagram, which indicates the fraction (in %) of the total product represented by the successive polymer fractions.

tion than fraction 3. From this one sees that the m_F values 26.5 and 9.9, just as all the other m_F values, cannot be readily represented by a length such as the height of the step, but only by an area (compare Figure 397: m_F is the area under the distribution curve of fraction F with an average degree of polymerization $\bar{P}$). As will be explained more fully below, these conditions can be met if one plots first the sum of the fractions (Σm_F) against the degree of polymerization. Through graphical differentiation of the sum-curve thus obtained, one readily finds the corresponding distribution curve, which in principle corresponds with the curve shown in **Figure 386** except that now the steps have the narrowest possible width, i.e., the width 1. Figure 388-1 shows the integral curve for the hypothetical polymer in Table 386.

TABLE 387. Separation of a polystyrene into 8 and into 12 fractions.

Fraction No.	Weight of the Fraction m_F	Degree of Polymerization $\bar{P}$	Fraction No.	Weight of the Fraction m_F	Degree of Polymerization $\bar{P}$
8	3.4	170	12	2.5	153
7	3.7	360	11	2.0	335
6	7.3	430	10	8.3	400
5	16.8	680	9	5.6	565
4	24.9	900	8	5.2	620
3	9.9	1300	7	13.3	710
2	26.5	1450	6	7.5	910
1	7.5	2250	5	12.8	990
			4	13.6	1280
			3	5.5	1390
			2	19.6	1760
			1	4.1	2480

It is not correct to draw a curve directly through the points for the sum values, Σm_F. Instead one has to draw the curve in the way shown in **Figure 388-1**. This takes into account the fact that the first fraction does not have only molecules up to a degree of polymerization of 1,000, but also molecules with a higher degree of polymerization. Similarly, fraction 2 does not only have molecules up to P = 2,000, but also such of 3,000 and 4,000. The integral curve yields the result that the fraction of molecules up to a degree of polymerization of 3,000, is equal to 10%,[19] and that the fraction up to a degree of polymerization P = 5,000 equals 45%, etc. Σm_F is the weight sum of the fractions (for example, the fractions from 1 to 4). Σm_P is the weight sum of the molecules of a product with a degree of

---

[19]Instead of percent (fraction of 100 g of substance), one often finds the fraction of 1 g of substance. In this example, then for P = 1,000, the fraction $m_P = 4.3 \cdot 10^{-5}$ (as grams in 1 g of substance).

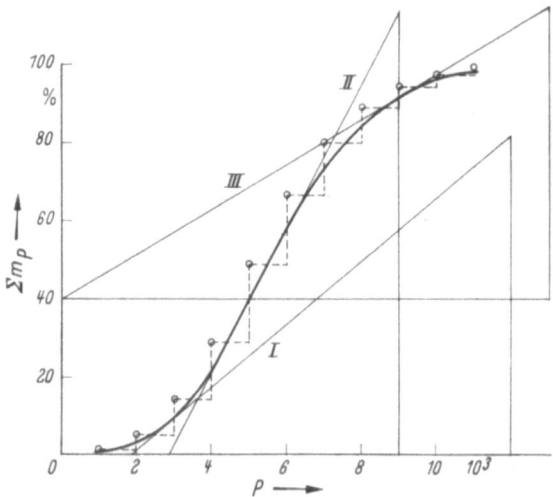

FIG. 388-1 — Integral distribution curve.

polymerization up to P (for example, from P = 1 to P = 5,000). The differential distribution-curve is obtained most simply by graphical differentiation of the integral curve by determining the tangent at a large number of points on the integral curve. With the tangent I (in Figure 388-1) for example, one obtains $82/10{,}000 = 8.2 \cdot 10^{-3}$ or with tangent II, the inflection point tangent (corresponding to the maximum of the differential curve), the tangent of the angle of inclination is $114/6000 = 18.8 \cdot 10^{-3}$. The result of this type of differentiation is shown in Figure 388-2.

Such a differential distribution curve gives information as to the size of the components with a certain polymerization degree P as a fraction of the total

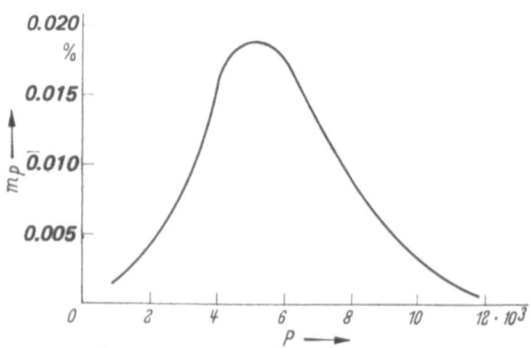

FIG. 388-2 — Differential distribution curve.

product, and, it therefore provides an accurate description of the polymer with regard to the size of its molecules.

With this type of representation it is more suitable to plot the degree of polymerization rather than the molecular weight, because the degree of polymerization from one molecule to the next always changes by the amount 1. This means that the subdivision of the abscissa is the same for all polymers. On the other hand, if one would plot the molecular weight, one would have to let the abscissa values increase by the molecular weight of the structural unit (for example, with polymethylmethacrylate by 100, for polystyrene by 104, etc.). Values between these would be forbidden. However, even then the subdivision would not be exact because the absolute values of the individual molecular weights would be exact only if one would know exactly the nature of the end groups, i.e., the mechanism of the initiation and termination as well as the degree of branching (chain transfer reactions). On the other hand, if one plots the degree of polymerization, one does not have to worry about these questions and the percentage scale obtained by differentiation of the integral curve represents exactly the fraction of each individual degree of polymerization.

3132 Determination of the Distribution Curve by Means of Fractional Precipitation

The molecules of a compound in equilibrium with its solvent have two possibilities. They can combine with solvent molecules through secondary valence forces until they are solvatized, and they can then, surrounded by their solvation shell, move freely in the solvent. The second possibility is that the dissolved molecules again lose their solvation shell and combine with each other. In this latter case one first obtains aggregates of several molecules, and if the process continues, finally macroscopic particles, i.e., the dissolved substance precipitates. Under constant conditions (temperature and concentration) this results in an equilibrium such that in a certain unit time as many molecules go into solution as are precipitated out of the solution.

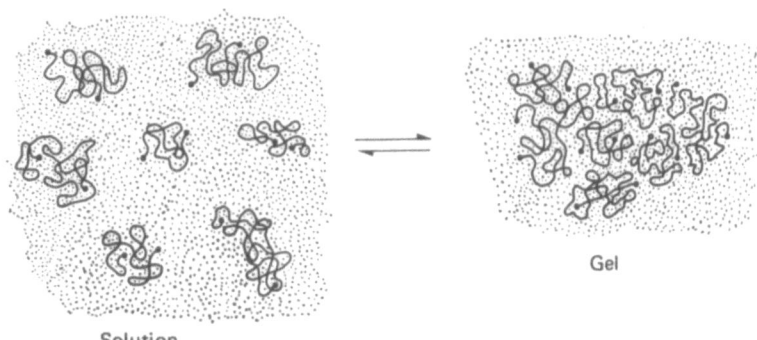

Gel

Solution

FIG. 389 — Schematic representation of the equilibrium between gel and solution.

These processes are in principle identical with macromolecular substances and with low molecular weight substances. If one adds to the solution of the compound a different solvent in which the dissolved substances are insoluble (a precipitant), the dissolved substance precipitates after a certain time. With low molecular weight compounds this usually occurs in the form of crystals; with macromolecular substances, in general, this occurs in the form of concentrated solutions or gels. In the gels, the macromolecules are no longer present as single molecules, but in the form of aggregates. In contrast to the crystallites of the low molecular weight compounds, such aggregates usually contain almost the entire solvatizing liquid of the individual molecules as a swelling liquid. This results from the fact that the individual macromolecules do not form aggregates by joining each other along their full length, but are combined only at a few places along the polymer chain, i.e., the coil retains its character as a gel particle (coil + bound solvent) even after it becomes an aggregate of several molecules. One can consider this process of gel formation, which occurs in the precipitation of macromolecules from a solution through slow addition of precipitant as a secondary valence cross-linking process, in contrast to gel formation through functional groups and bifunctional cross-linking agents. The density of the cross-links is determined here by the concentration of the precipitant. The more precipitant, the closer to each other the points of cross-linking. As is shown in greater detail on pages 25, 491, and 544, the critical linkage concentration at which gelation occurs is the smaller, the higher the molecular weight. This means that with a higher molecular weight, one obtains gelation at a lower concentration of cross-linking or adhesion points. Therefore, if one increases the concentration of cross-linking points by slow addition of precipitant, the critical concentration will first be reached by the components of the polymer with the highest molecular weight, which are then separated from solution as a gel phase and removed from the solution as the first fraction. Only on further addition of precipitant and the corresponding increase in the cross-link concentration will components with a lower molecular weight aggregate to a gel, and one then obtains the second, third, fourth, etc., fraction of the polymer, each with an increasingly lower molecular weight.

Experimental Procedure for Fractional Precipitation

One proceeds as follows: one takes a dilute (0.5-0.1%) solution of the polymer to be fractionated and then adds the precipitating agent under strong stirring. In most cases methanol is a suitable precipitant: it is miscible in all proportions both with water and with organic solvents, and almost all macromolecular compounds are insoluble in methanol. Since the aggregation to the gel phase occurs through secondary valence bonds which can be easily split by an increase in temperature, the separation of the gel is temperature dependent, and one always has to carry out the fractionation in a thermostat. In order to obtain homogeneous precipitation, one adds precipitant until the first onset of turbidity, then one increases the temperature enough so that this turbidity just disappears again, and finally one places the solution in a thermostat and decreases the temperature slowly so that precipitation does not occur through increased precipitant concentration, but through slow and uniform cooling of the solution. This is necessary to allow an equilibrium to establish itself between the sol and the gel phase (to be further discussed in the following sections). If one were to use the precipitate obtained directly by addition of the precipitant, this fraction would contain considerable amounts of lower molecular weight material carried down by adsorption. The separation of the gel phase is carried out through decantation of the supernatant. One can accelerate the separation considerably by centrifuging the solution.

The Successive Precipitation of Fractions

After separation of the first fraction, one adds sufficient precipitant to the remaining solution so that the second fraction separates out. This fraction is also separated from the remaining solution, which is then again reacted with methanol, etc. In this way one obtains fractions with decreasing molecular weight. The amount of each fraction separated out is the smaller, and their number the larger, the less precipitant one adds from one time to the next.

Through the addition of precipitant the volume of the solution becomes larger and larger from fraction to fraction. In many cases one can improve the situation by choosing a solvent which is more volatile than the precipitant. This permits

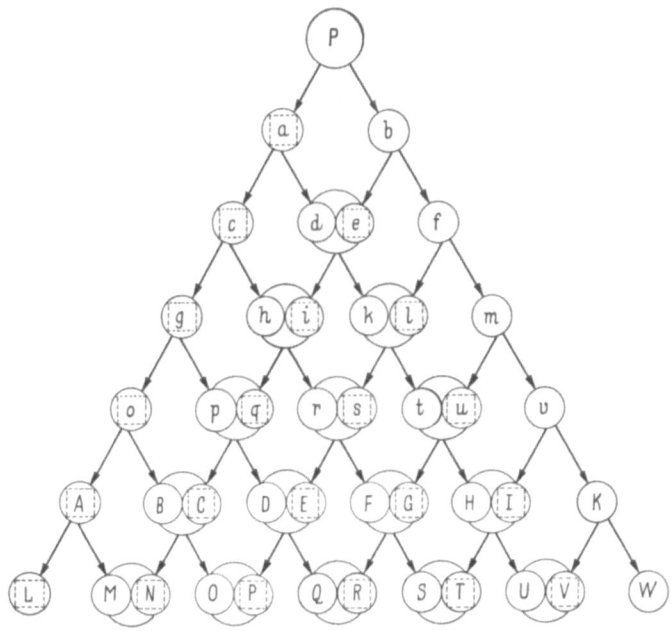

FIG. 391 – The triangular fractionation scheme.

increasing the precipitant concentration through evaporation of the solvent, so that the total volume becomes smaller from fraction to fraction, or at least remains the same.

The sharpness of the fractions can be improved by redissolving the fraction after each separation and reprecipitating it. The supernatant is in each case combined with the total solution.

The Triangular Fractionation (Meffroy-Biget)

Instead of successive separations with the degree of polymerization becoming smaller from fraction to fraction, one can also carry out fractionational precipitation by adding sufficient precipitant to the polymer solution at the first addition to bring about precipitation of approximately half the polymer with the other half remaining in the solution. After separation of the gel this is again brought into solution and one then has two solutions (a and b) which are treated in a way similar to that in which the original solution P was treated (compare Figure 391): To each solution one adds sufficient precipitant to bring about precipitation of approximately half the material, and the other half remains in solution. Thus from solution a one obtains the gel c and the solution d, and from the solution b, the gel e and the solution f. The solution d and the gel e are combined to give the solution d/e, and gel c is again dissolved so that one now has three solutions. Each of the three solutions one reacts with sufficient precipitant so that approximately one half of the dissolved polymer fraction precipitates. Solution c gives the gel g and the solution h; solution d/e results in the gel i and the solution k; and from solution f one obtains gel l and the solution m. One can continue this game in this way until one feels that enough fractions have been obtained. The larger the number of fractions, the more accurate the fractionation but the greater, of course, the effort.

In most cases the separation of the solutions into 50:50 portions becomes difficult after six or eight steps, either because the major portion remains in solution, or, if one adds only a small additional amount of precipitant, because the largest part precipitates. In the end one always obtains the same number of precipitated and dissolved fractions. One can then precipitate the solutions separately, or one can mix them with the corresponding gels and precipitate them together.

This triangular situation has certain advantages over successive fractionation: (1) The separation is sharper because one always fractionates fractions which become more homogeneous from one step to the next. (2) It is more rapid because one can treat several batches at the same time either allowing them to separate, or adding new precipitant. (3) It has been possible in an elegant fashion to avoid the usual great increase in the volume of the solutions.

Each fraction is weighed, and its molecular weight is determined. The experimental results are then interpreted in the way which has previously been described (compare Table 387 and Figures 388 and 401).

Theoretical Treatment of Fractional Precipitation (according to G. V. Schulz)

Theoretically one can treat fractional precipitation as a distribution into two nonmiscible phases: a gel phase (') and sol phase ("). In going from the sol phase to the gel phase, the internal energy of the system is decreased and heat is given off (in the amount ΔH, when 1 mole of material changes phase). At the same time the entropy of the system is decreased by the amount $T \cdot \Delta S$. The transfer only occurs as long as the value ΔH is larger than $T \cdot \Delta S$, i.e., as long as $\Delta F = \Delta H - T \cdot \Delta S$ is negative. ΔF, the free energy of the system, is a measure for the driving force for phase transfer and corresponds to the osmotic work which a system does when 1 mole of a polymer fraction with the degree of polymerization P goes over from the sol phase (") to the gel phase ('), or which has to be used in order to achieve a change from the gel phase to the sol phase. If this process occurs in a reversible manner, i.e., in the immediate neighborhood of the equilibrium concentrations c_P' and c_P'', then one can write:

$$\Delta F = \Delta H - T \cdot \Delta S = A = RT \cdot \ln c_P'/c_P'' \quad \text{or} \quad c_P'/c_P'' = e^{A/RT}. \qquad (64)^{20}$$

If one assumes that the work A with a polymer can be additively constructed from the ϵ values for the structural units (Brönsted), then $A = P\epsilon$ (P = number of structural units in the polymer chain = degree of polymerization), and one can write:

$$c_P'/c_P'' = e^{P \cdot \epsilon/RT}. \qquad (65)$$

If one multiplies both sides of the equation with v'/v'' then, because $v \cdot c = m$, the concentration ratio c_P'/c_P'' has been replaced by the mass ratio m_P'/m_P'' (v' is the volume of the gel phase, and m' is the mass of the gel phase):

$$\frac{m_P'}{m_P''} = \frac{v'}{v''} e^{P\epsilon/RT} \quad \text{or} \quad \theta = \frac{1}{\varphi} \cdot e^{P\epsilon/RT}, \qquad (66)$$

where θ is the ratio m_P'/m_P'' according to which the mass m_P of the molecules with a degree of polymerization P, is distributed over both phases:

$$\theta = m_P'/m_P'', \qquad (67)$$

and where the volume ratio of the two phases, v'/v'', is represented by $1/\varphi$:

$$1/\varphi = v'/v''. \qquad (68)$$

m_P is the percentage of the molecules with degree of polymerization P in the total product; m_P' is the percentage of molecules with the degree of polymerization P (also based on the total product) present in the gel phase; and m_P'' corresponds to

---

[20] c_P'/c_P'' corresponds to the Nernst distribution coefficient (for example, for the removal of a compound from solution by ether).

the percentage of molecules with a polymerization degree P remaining in the sol phase. The sum of m_P' and m_P'' therefore equals m_P:

$$m_P = m_P' + m_P'' \qquad (69)$$

and

$$m_P' = m_P \; \frac{m_P'/m_P''}{m_P'/m_P'' + m_P''/m_P''} = m_P \cdot \frac{\theta}{1+\theta} \cdot \qquad (70)^{21}$$

Accordingly:

$$m_P'' = m_P \cdot \frac{1}{1+\theta} \qquad (71)$$

If the total distribution $m_p = f(P)$ is known, one can calculate the distribution curves of the fractions with the help of Equations (66) and (70/71). For this purpose one has to arbitrarily determine the volume ratio of the phases ($\varphi = v''/v'$) and for each fraction again arbitrarily set a value of ϵ. The values φ and ϵ are determined by the experimental conditions under which the fractions are separated.

Figure 395 shows the influence of φ on the fractionation. One obtains the curves by substituting in Equation (66) increasing values of P at various constant values φ = 1, φ = 3, φ = 10, φ = 100, etc. (assuming for simplicity that m_P = 1; R = 1.986 cal/degree; and T = 300°K). One sees that for φ = 1 more than 70% of the fraction of degree of polymerization 100 is already in the gel phase (m_P' = 70%) and therefore practically no separation occurs. Also with φ = 10, the gel phase still contains too large a proportion of molecules of low degree of polymerization, and only when φ is greater than 100, can one speak of a practically useful separation. In practice this means that one has to achieve a sufficient high dilution of the solution to be fractionated: the larger the dilution, the larger the volume of the sol phase versus the gel phase, and therefore the larger the ratio of $\varphi = v''/v'$.

As can be seen from Equation (64), the concentration c_p' of the polymer molecules with degree of polymerization P in the gel phase is the larger, the larger the work A which the system can carry out in the transfer of macromolecules from the sol phase to the gel phase, i.e., the larger the driving force of the process. Since the driving force is $A = P \cdot \epsilon$, one always finds (for a given ϵ) that in the gel phase the

---

[21] $\dfrac{m_P'}{m_P''} = \dfrac{m_P}{m_P' + m_P''} = 1; \quad m_P' = m_P \; \dfrac{m_P'}{m_P' + m_P''} = m_P \; \dfrac{m_P'/m_P''}{m_P'/m_P'' + m_P''/m_P''}$

Through multiplication of numerator and denominator by $1/m_P''$ one obtains the result that on the right side of the equation, only the ratio $m_P'/m_P'' = \theta$ appears, which can be calculated according to Equation (66).

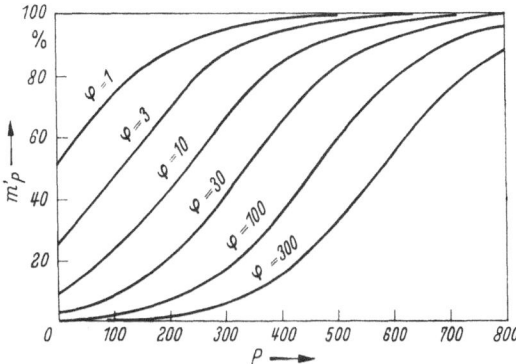

FIG. 395 — Influence of the ratio φ of the phase volumes on the separation effectiveness of the fractionation (G. V. SCHULZ)

molecules with the highest degrees of polymerization are present in greater concentration than those with low degrees of polymerization.

The value ϵ can be arbitrarily fixed in the experiment through the amount of the added precipitant, because the more precipitant is added to the volume of solution, the larger becomes ϵ (= the work required for phase transfer per structural unit). The stepwise additions of precipitant therefore correspond to a stepwise increase in ϵ. The greater ϵ becomes, the smaller the macromolecules which can reach a definite value of $A = P \cdot \epsilon$ and thus a certain equilibrium-concentration c'_p in the gel phase.

TABLE 395. The separation of the first fraction of a polymer carried out mathematically with the aid of Equations (66) and (70, 71).

$$R = 1.986 \text{ cal/deg.} \qquad T = 300^\circ K \qquad \epsilon_I = 0.3 \text{ cal}$$
$$\varphi = 100 \qquad \qquad \log e = 0.4343$$

P	m_P*	$\dfrac{P \cdot \epsilon_I}{RT}$	$\log \theta_I$**	θ_I	$\dfrac{\theta_I}{1+\theta_I}$	$m'_{P(I)}$	$\dfrac{1}{1+\theta_I}$	$m''_{P(I)}$
2000	4.25 10^{-3}	1.01	0.437−2	0.0274	0.0266	0.113 10^{-3}	0.975	4.15 10^{-3}
4000	16.0 10^{-3}	2.02	0.875−2	0.0750	0.070	1.12 10^{-3}	0.930	14.9 10^{-3}
5000	18.8 10^{-3}	2.52	0.090−1	0.123	0.110	2.06 10^{-3}	0.890	16.7 10^{-3}
6000	17.5 10^{-3}	3.03	0.315−1	0.206	0.165	2.90 10^{-3}	0.830	14.5 10^{-3}
7000	13.1 10^{-3}	3.54	0.530−1	0.339	0.253	3.32 10^{-3}	0.746	9.80 10^{-3}
8000	9.0 10^{-3}	4.04	0.750−1	0.562	0.360	3.23 10^{-3}	0.641	5.76 10^{-3}
9000	5.75 10^{-3}	4.54	0.970−1	0.933	0.482	2.78 10^{-3}	0.518	2.98 10^{-3}
10000	3.06 10^{-3}	5.04	0.190	1.55	0.610	1.86 10^{-3}	0.392	1.20 10^{-3}
11000	1.55 10^{-3}	5.54	0.450	2.515	0.715	1.11 10^{-3}	0.285	0.44 10^{-3}

*Taken from the toal distribution curve (Fig. 388-2)

**$\lg \theta_I = \dfrac{P \epsilon_I}{RT} \log e - \log \varphi$ [According to Equ. (66)]

Calculated Fractionation

In order to describe quantitatively a fractional precipitation by means of Equations (66) and (70/71), one only has to increase the transfer energy ϵ stepwise by suitable amounts (i.e., corresponding to practical conditions), in a manner similar to the stepwise addition of precipitant. At the same time one has to calculate the m_P' and m_P'' values, for the molecular range covered by the total distribution curve, by substituting increasing P values in Equations (70), and (71), and (66), respectively. The numerical values for m_P are given by the total distribution curve which must be known. One can take an experimentally determined distribution curve, or one calculated according to Schulz and Flory, or a completely arbitrary one, as the one shown in Figure 388-2, for example. One then obtains, according to Equation (70), a series of m_P' values (gel phase) for different degrees of polymerization P, which are plotted in a diagram corresponding to Figure 388-2 (compare curve 1 in Figure 398). These characterize the amount of molecules of degree of polymerization P present in the gel phase and expressed in percent of the total material. The m_P'' values (sol phase) are calculated according to Equation (71) (for the same ϵ value) and represent the fraction of the polymer which, after separation of Fraction I still remains in solution.

$$\text{Gel phase of Fraction I: } m_{P(I)}' = m_P \; \frac{\theta_I}{1 + \theta_I} \quad \text{(Curve I in Figure 398)}$$

$$\text{Sol phase of Fraction I: } m_{P(I)}'' = m_P \; \frac{1}{1 + \theta_I} \quad \text{(Curve RI in Figure 398)}$$

Table 395 represents such a calculation and Figure 398 the graphical representation of the results. As the total distribution curve, m_p, the arbitrarily constructed curve shown in Figure 388-2 was selected.

The second fraction is obtained by adding to the remaining solution some more precipitant (increase of ϵ). In place of the total distribution curve one now uses the RI-curve (= distribution curve of the remaining product after removing the first fraction). Instead of m_p one now has to substitute in Equation (70) the corresponding value $m_{P(I)}''$. Therefore for Fraction II the following holds:

$$\text{Gel phase of Fraction II: } m_{P(II)}' = m_{P(I)}'' \cdot \frac{\theta_{II}}{1 + \theta_{II}} = m_p \cdot \frac{1}{1 + \theta_I} \cdot \frac{\theta_{II}}{1 + \theta_{II}} .$$

$$\text{Sol phase of Fraction II: } m_{P(II)}'' = m_{P(I)}'' \cdot \frac{1}{1 + \theta_{II}} = m_p \cdot \frac{1}{1 + \theta_I} \cdot \frac{1}{1 + \theta_{II}} .$$

In the same manner one can calculate the gel and sol curves also for the third, fourth, and other fractions. One obtains for each fraction (i.e., for each ϵ) a distribution curve, such as is shown in Figure 397 in the case where $m_p = f(P)$ is a normal distribution.

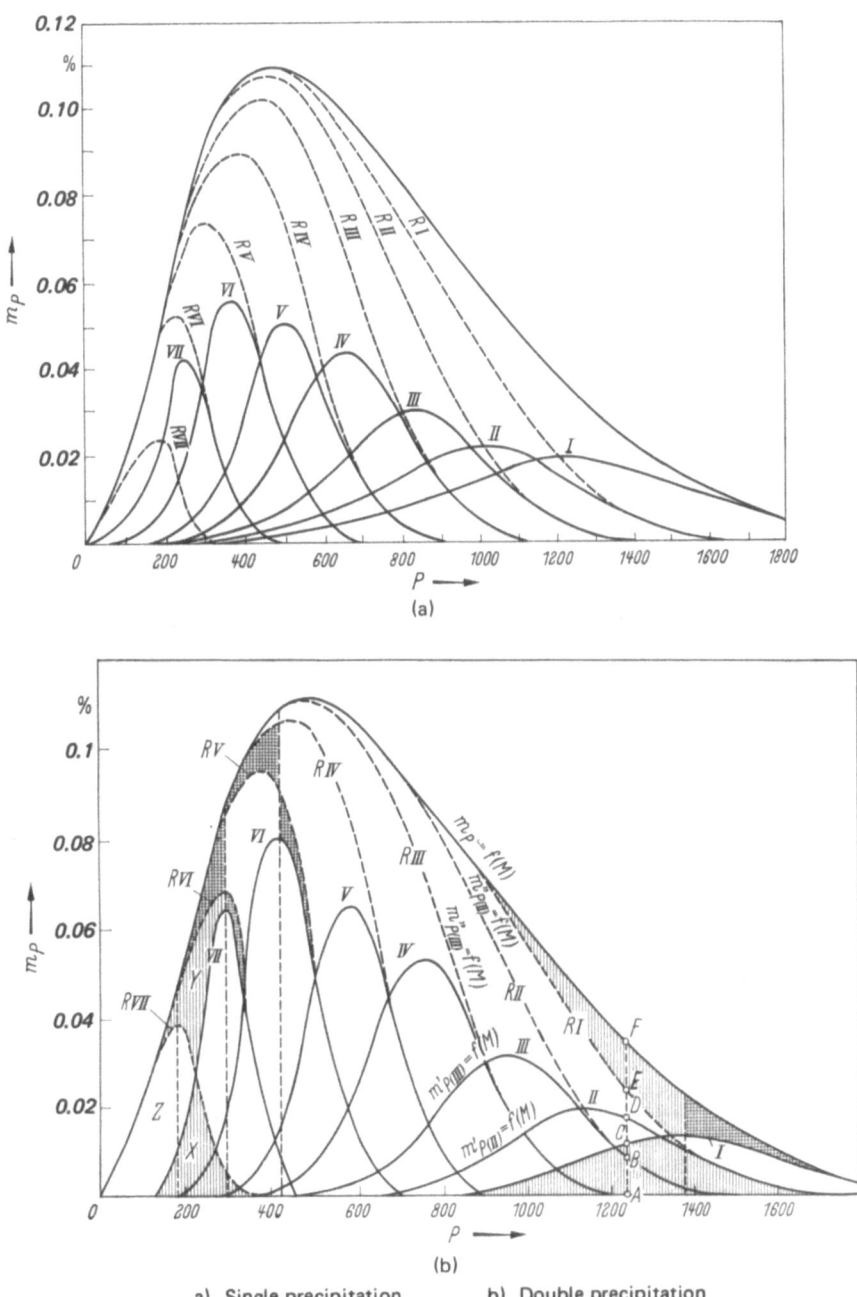

a) Single precipitation   b) Double precipitation

FIG. 397(a and b) — Separation of a polymer with a Schulz-Flory-distribution into 8 fractions by a fractional precipitation carried out mathematically according to Equations (66) and (70/71) (after G.V. SCHULZ).

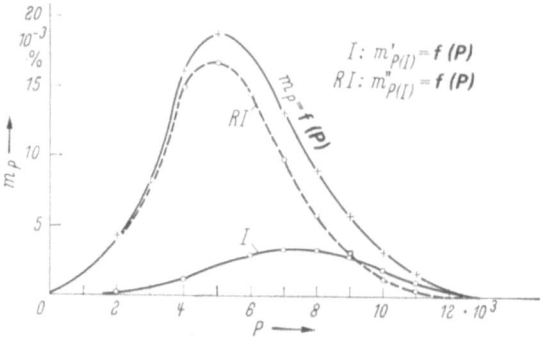

FIG. 398 — Separation of the first fraction (Fraction I) from a polymer with an arbitrarily selected distribution curve $m_P = f(P)$, $\epsilon_I = 0.3$ and $\varphi = 100$.

Figure 397a shows a fractionation with a single precipitation, Figure 397b shows one where each fraction is reprecipitated after dissolution.

One recognizes that with fractional precipitation one cannot speak of a really sharp boundary between the various fractions. The separation does become better through reprecipitation, even though there is still some overlap between the fractions, especially those of high molecular weight. Thus a representation according to Figure 386 is quite far removed from reality. For example, if one takes the degree of polymerization 1,230, one finds in Figure 397b that the molecules at this degree of polymerization are present in the total product to the extent of 0.035% ($m_{P=1,230} = AF = 0.035\%$). A part of this, i.e., $m'_{P(I)} = AC \cong 0.011\%$, precipitates already with Fraction I. The remainder, $m''_{P(I)} = CF = AE^{22} \cong 0.024\%$ distributes itself over the second and the third fraction, and actually the greater part, $m'_{P(II)} = AD \cong 0.017\%$, precipitates in Fraction II. The remainder, i.e., $DE = m''_{P(II)} = AB \cong 0.007\%$, precipitates only in Fraction III, without any remainder. Therefore, $AB = m''_{P(II)} = m'_{P(III)}$.

For the experimental determination of the molecular weight distribution of a macromolecular substance, it is important to obtain a correct distribution curve through fractional precipitation, in spite of the completely different width of the distribution within any one fraction of high and low molecular weight. With the aid of Figure 397 one can study this question and determine how much an experimentally determined distribution curve differs from the actual distribution (which in the case of Figure 397 is known). With an experimental fractional precipitation, one determines the weight of each fraction and its average molecular weight and its

---

[22]In order to understand Figure 397b, one has to clearly understand the following point: CF is the residue which remains in solution of the total amount m_P of degree of polymerization 1,230 ($m_{P=1,230} = AF$), when fraction I ($m_{P=1,230(I)} = AC$) has been removed. According to Equation (71) this residue is described by the corresponding value of the $m_{P(I)}$-function, shown in Figure 397b as the dashed curve RI. The value of this function for $P = 1,230$ is given by the distance AE ($m''_{P=1,230} = AE$). Therefore, CF = AE and, of course, AC = EF.

degree of polymerization. With the calculated example in Figure 397 one can take for the average degree of polymerization, $\overline{P}$, of the individual fractions simply the degree of polymerization at the maximum of the individual curves.[23] The weight of the fraction is given by the area under the curves. By means of planimetry of the areas under the curve in Figure 397b, one obtains the following "experimental" results:

TABLE 399. Weight m_F of the individual fractions in Fig. 397b (in %)

	VIII	VII	VI	V	IV	III	II	I
Degree of Polymerization . .	180	300	420	585	755	955	1140	1370
Weight m_F in %	6.4	9.0	17.2	17.6	18.5	13.5	9.7	8.1

Construction of the Distribution Curve from the Mathematical Results

As discussed earlier, it is possible to construct the distribution curve from such results by first drawing the integral curve and then differentiating it. The integral curve shows the content, for example, the percent of polymer up to a degree of polymerization of 180, 300, 420, etc. present in the total product. Figure 397b shows that the fraction of molecules up to a degree of polymerization of 180 (= area under the total distribution curve up to P = 180) is approximately half as large as the amount $m_{F(VIII)}$ of the lowest fraction. (Fraction VIII is obtained by complete precipitation or evaporation of the entire remaining solution and therefore contains everything that has not precipitated with the seven earlier fractions.) Therefore, in constructing the integral distribution curve above $\overline{P}$ = 180, one does not plot the value $m_{F(VIII)}$ = 6.4%; rather, 3.2% = $\sum\limits_{P=1}^{P=180} m_P$ is plotted.

The corresponding integral value for fraction VII (P = 300) is found, if one sees that the horizontally shaded areas X and Y have the same area and that therefore the nonoverlapping residue Z of Fraction VIII and the area Y together equal the total area under the distribution curve of Fraction VIII. [The amount of polymer which still remains in solution after separation of Fractions I to VI is represented by the area under the curve R VI. After separation of Fraction VII there still remains some material which is represented by the area Z + Y. This remainder is, however, identical with Fraction VIII (= R VII), which is obtained by evaporation of the

---

[23]There are different types of averages for $\overline{P}$ or $\overline{M}$. (These are treated in greater detail in the next section of this chapter, p. 425). Depending on the form of the distribution curve, the maximum degree of polymerization can represent different averages. With symmetrical distribution (Gaussian curves) the polymerization degree at the maximum corresponds to the *weight*-average ($P_{max} = \overline{P}_w$), whereas the number-average is smaller ($\overline{P}_n < P_{max}$). With the Schulz-Flory distribution [also called the most probable distribution or normal-distribution (compare p. 411)] which is found with many polymers prepared by polymerization or polycondensation, $P_{max} = \overline{P}_n$ and $\overline{P}_w = 2\overline{P}_n$. Since the distribution curves of the different fractions are quite symmetrical the above example must describe weight-averages ($\overline{P}_w$).

remaining solution. Therefore the area $Z + Y$ is equal to the area $Z + X$, which means that $X = Y$.] If one also adds half of the area under the distribution curve of Fraction VII to $Y + Z$ one obtains an area almost equal to the area under the total distribution curve up to $P = 300$. [These areas would be exactly identical if the small double-dashed areas in Figure 397b would have the same area, because only then the amount of error to the left of $P = 300$, 420, etc., would be compensated by an extra area to the right. This is, however, only the case with the middle fractions. With the lower fractions the minus error, and with the high fractions the plus error, outweights this. The error becomes especially important with the highest fractions, because with them the right area does not have a counter part, and its entire content therefore becomes a plus error.] One therefore obtains the value which one has to plot above $P = 300$ on the integral curve if one adds half the value of Fraction VII to the total value of Fraction VIII: $6.4\% + 4.5\% = 10.9\%$. The same considerations can be applied to all further fractions; therefore, the integral curve point for Fraction VI ($P = 420$) is $\sum\limits_{P=1}^{P=420} m_P = 6.4 + 9 + 17.2/2 = 24.0\%$, and so on. Instead of first calculating the values for the integral curve, one can also, as was already done in Figure 388, plot the sum of the m_F values directly (m_F = weight of a fraction in percent of the total polymer) up to each respective degree of polymerization. These values are then linked by a step curve, and the integral curve is obtained by drawing it through the half-value of the vertical steps. The result of the fractional precipitation described in Table 399 is plotted in this manner in Figure 401.

One sees that the "experimental" curve corresponds quite well with the known original distribution curve in Figure 397. As was to be expected (compare preceding paragraph), with the lower fractions one finds small deviation downward, and with the high fractions a somewhat larger deviation upward, which has to be taken into account. If one repeats the same procedure that was carried out in Figure 401 for the fractions in Figure 397b also for the fractions in Figure 397a, and then compares the resulting distribution curves, one sees that the distribution curve depends relatively little on whether the precipitation is carried out once or whether a double precipitation is made. This is in spite of the fact that the fractions become considerably sharper through repetition of the precipitation. A sharp fractionation is still advantageous, because the differences in the molecular weight averages begin to disappear, and it then becomes unimportant which method of molecular weight determination one uses. With broad fractions, however, the molecular weight averages $\overline{M}_n$ and $\overline{M}_w$ are more or less different from each other. (Compare footnote 23.) Further, one has to consider that the separation into two liquid phases is a borderline case which is not at all possible with all polymers. For example, with very high fractions one can find that considerable amounts over and above the equilibrium situation are precipitated because lower molecular weight material falls out adsorbed on the high molecular weight material. This is an effect which can be considerably reduced by repetition of the precipitations. Finally, one does not

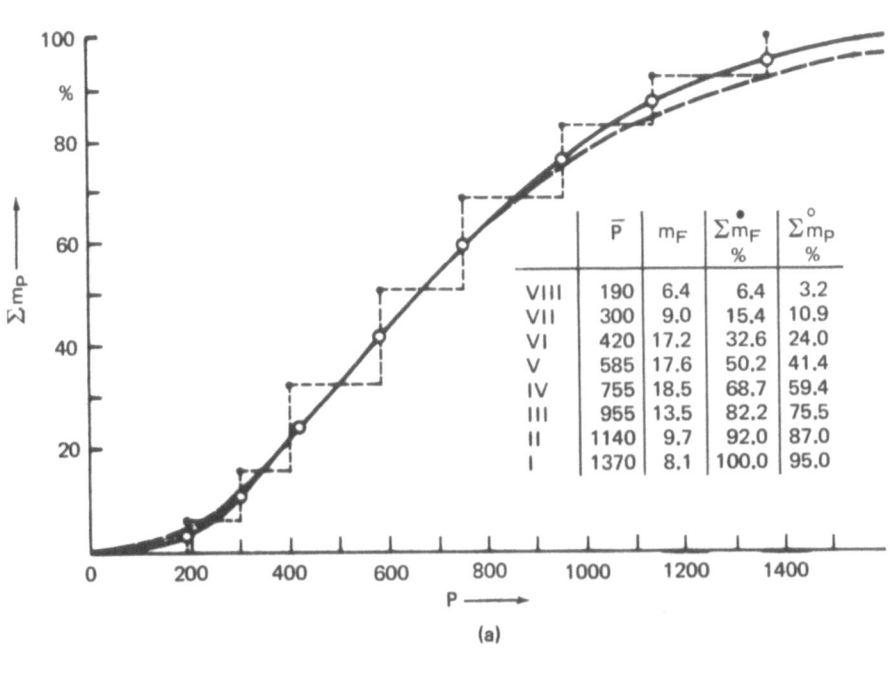

	$\bar{P}$	m_F	Σm_F %	Σm_P %
VIII	190	6.4	6.4	3.2
VII	300	9.0	15.4	10.9
VI	420	17.2	32.6	24.0
V	585	17.6	50.2	41.4
IV	755	18.5	68.7	59.4
III	955	13.5	82.2	75.5
II	1140	9.7	92.0	87.0
I	1370	8.1	100.0	95.0

(a)

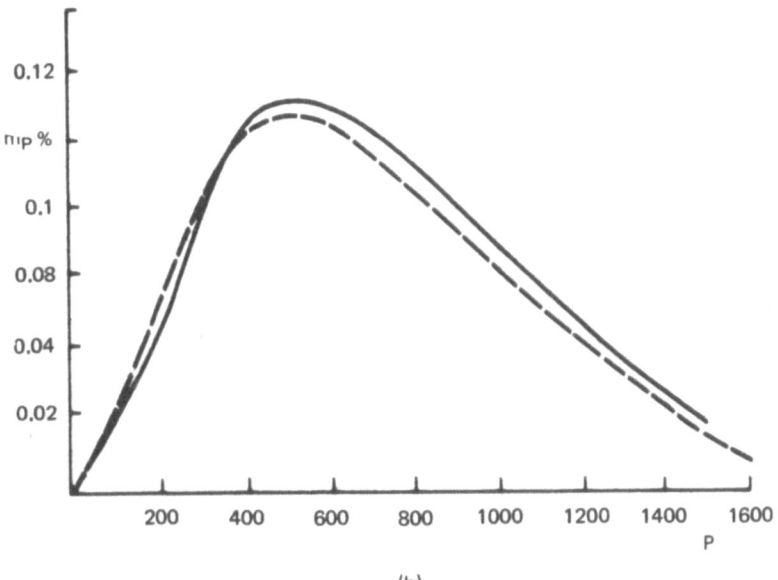

(b)

FIG. 401(a and b) — a) Integral curve based on the fractions in Fig. 397 and Table 399 (—————) Theoretical curve ("normal" distribution) (— — — —) b) Plot of the distribution curve obtained via the integral curve based on the calculated fractions [from Equs. (66) and (70/71)].

only carry out fractional precipitation in order to obtain distribution curve, but often also in order to determine the dependence of the physical properties (intrinsic viscosity, sedimentation constant, etc.) on the molecular weight. In such cases one has to be particularly careful to obtain sharpness of fractionation.

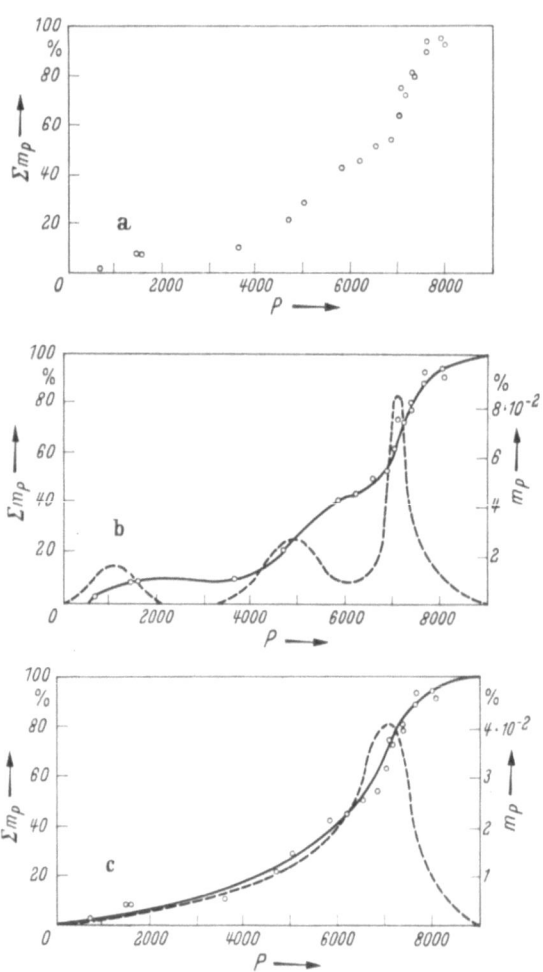

a) experimental results (measurements of SCHULZ
and MARX)

b) and c) possible interpretations

FIG. 402(a-c) — Distribution curve (——— = integral; —— —— = differential) of a nitrated cotton cellulose obtained by fractional precipitation of acetone solutions with water.

With the mathematical fractionation, the phase ratio φ is held constant, but in practice this is not always possible. However, since the calculation shows (compare Figure 395) how especially important the phase ratio is for a successful fractionation, it can easily be seen that the non-constancy of the φ value can give rise to complications in fractional precipitation. In extreme cases it can happen that the first fraction precipitated out has a lower molecular weight than the second. Knowing about such error possibilities, it is best to look at the results of fractionation experiments with a certain amount of reservation, and if possible, to carry out a precipitation fractionation with different solvent-precipitant mixtures.

In Figure 402a one sees the fractionation results for a cotton cellulose. Figure 402b shows the resulting distribution curve as it was obtained from the literature, and Figure 402c shows that the experimental values can also be interpreted in a different manner. What can be said with certainty is only that the molecular weight distribution of natural cotton cellulose is considerably different from that found with synthetic polymers, which usually have a normal distribution. The asymmetry of the curves has a mirror image (compare Figure 421 or 424).

3133   Other Methods of Fractionation

In view of the large amount of effort and time required for fractional precipitation, there have been a number of attempts to fractionate by simpler methods. The following methods have been used:

Fractionation in Columns

The Baker-Williams Procedure

According to this procedure the separation of polymer fractions of different molecular weight by frequently repeated fractional dissolution and precipitation is carried out in a tube filled with small glass beads (Figure 404). The precipitation is brought about by a linear temperature gradient along the column (at the upper

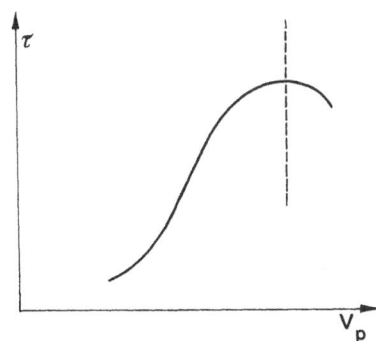

FIG. 403 — Increase in the turbidity τ of a polymer solution on the addition of increasing amounts of precipitant (v_p = volume of precipitant). (Schematic representation according to G.V. SCHULZ).

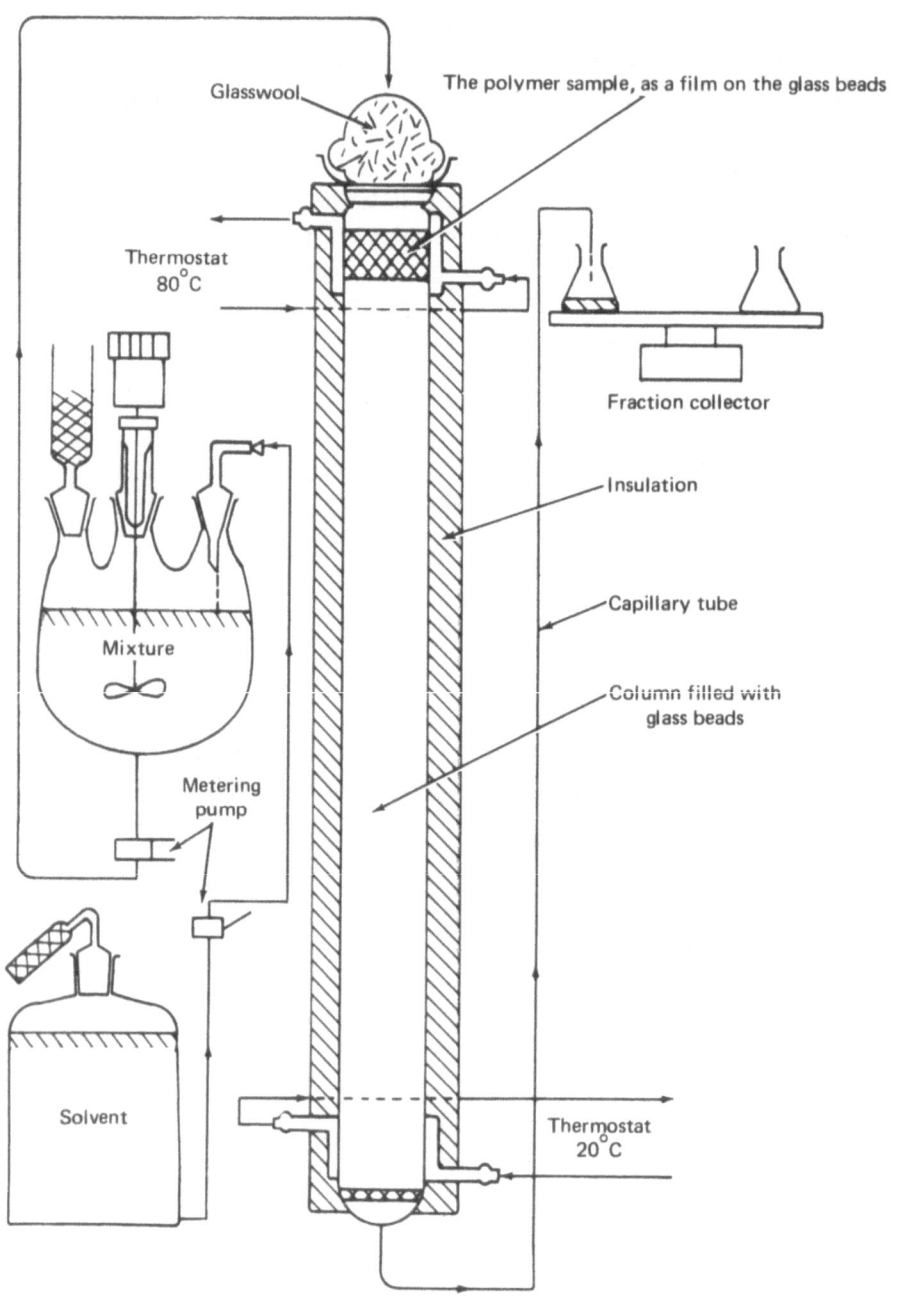

FIG. 404 —   BAKER-WILLIAMS fractionation procedure

end of the column 60°C-80°C; at the lower end 10°C-20°C). Dissolution is caused by a time-dependent concentration gradient of the solvent-precipitant mixture which is passed through the column from top to bottom.

At the beginning of the fractionation the column is filled with a poor solvent (precipitant). Usually the polymer which has to be fractionated is first dissolved, and after the addition of small glass beads, the solvent is evaporated so that the polymer remains on the glass beads in the form of a thin film. The coated glass beads are then placed at the top of the column, and then one begins to add the solvent-precipitant mixture (for example, toluene-methanol). In the course of hours (sometimes, days), the solvent concentration is increased from 0 to 100%, at first more rapidly and later on more slowly, because the solubility differences become smaller with increasing molecular weight.

At first only the low molecular weight components are dissolved out of the polymer film and travel towards the lower end of the column, carried along by the solvent mixture. Because of the drop in temperature, the polymer precipitates again as it moves down the column, first the high molecular weight components, and only further down (at still lower temperature), the very low molecular weight components (assuming that they are not soluble even in the cold precipitant). The succeeding, more solvent-rich, solvent-precipitant mixture redissolves the precipitated polymer, and now the fractions with lowest molecular weight dissolve first and those with high molecular weight last. Through these two processes, (1) the precipitation caused by lowering of the temperature, and (2) the redissolution caused by the succeeding, more solvent-rich, solvent-precipitant mixture coming from the top, the components with lower molecular weight move more rapidly down the column (they are precipitated last and dissolve first) than the components with higher molecular weight. Thus, the components with the lowest molecular weight leave the column first, and the components with higher molecular weight follow later. The interval between higher and lower molecular weight components will be the larger, the slower the concentration of the good solvent in the extraction mixture is increased, and the flatter the temperature gradient for a given temperature difference between top and bottom of the column, i.e., the longer the column. Of course, this also leads to an increase in the material and time used for the procedure. However, since it is possible to automatize the fractionation procedure by means of pumps which meter out the solvent/non-solvent mixture, and an automatic fraction collector followed by turbidity titration (see Figure 403 and p. 409), column fractionation (determination of the distribution curve) represents a considerable improvement over stepwise fractional precipitation.

In determining the dimensions of the column and the number of fractions that should be collected, one has to take into consideration that the molecular weight of each fraction must be determined separately (usually by viscometry). For this molecular weight determination one needs a certain minimum amount of polymeric material, and therefore the number of useful fractions depends on the amount of polymer that one uses as starting material. The η_{sp}/c vs. c curves are determined after concentrating the fractions by evaporation of solvent to a concentration of approximately 0.5%. Often it is sufficient to measure the viscosity of all the fractions at a certain fixed concentration for each fraction and to take only the η_{sp}/c value, rather than the viscosity number $[\eta]$, as a measure for the size of the molecules.

One disadvantage is that one has to start with a dry polymer film, so that at the beginning of the fractionation the fractional dissolution of the polymer begins at a low degree of swelling, i.e., a high gel density. As a result it is possible that low molecular weight components are retained longer than they should be, because they are held back for steric reasons and simply cannot come out of the slightly swollen polymer film. If one is not concerned particularly with a separation in the region of low molecular weight components, it is better to start with a fractional precipitation at the top of the column by adding a solution of the polymer and subsequently adding at short intervals pure nonsolvent into the upper part of the column so that precipitation occurs. (The upper portion of the column, where precipitation takes place, has to be stirred in this procedure.) The subsequent course of the fractionation is the same as with the normal procedure described above.

It does not always seem quite certain that, on cooling a very finely divided polymer gel in the presence of the relatively large glass pearls or beads, the polymer is immediately fixed to the glass bead surface. However, the sharpness of the fractionation would be very much decreased if, during the passage of the polymer from the top to the bottom of the column, the precipitating fractions could be partly flushed through the column without being absorbed on the glass bead surface. From this point of view it might be better to use a tight layer of glass wool in the column rather than glass beads. Possibly this effect influences certain fractionations which do not show a better separation according to the Baker-Williams procedure than without the temperature gradient.

If a sufficiently large Baker-Williams column is used, the method can also be applied to the preparative separation of a polymer into more or less homogeneous fractions.

Gel Permeation Chromatography (GPC)

Gel permeation chromatography is currently the most important technique for the determination of molecular weight distributions. It is based on the diffusion of the polymer molecules into the capillary interstices of a cross-linked, swollen polymer gel, which is used for the filling of the column, and which permits a particularly sharp separation into fractions of different molecular weights and therefore allows a particularly exact determination of the distribution curve. In this procedure a certain amount of a dilute solution of the polymer to be fractionated is allowed to stream slowly through a column filled with gel beads, followed by pure solvent. The desired number of fractions are collected and characterized. In contrast to the Baker-Williams column, in a GPC column, one first obtains the components with the *highest* molecular weight and those with the lowest molecular weight come at the end. This is because the gel beads used for GPC are full of capillaries (compare Figure 407) and the capillary diameters differ from each other (10-10^5 Å). Therefore, certain capillaries, or pores, are accessible only to very small polymer molecules contained in the passing solution, whereas other capillaries are accessible to macromolecules of medium molecular weight, and still others (relatively few) are accessible to very high molecular weight polymer molecules. If $v_{(P = 100)}$ is the total volume of the pores which are just accessible only to macromolecules up to a degree of polymerization $P = 100$ and $v_{(P = 10)}$ is the total capillary volume which is still accessible to macromolecules up to a degree of polymerization $P = 10$, then it is clear that v_P is the larger, the smaller P:

$$v_{P = 10} \quad > \quad v_{P = 100} \quad > \quad v_{P = 1,000} \quad > \quad v_{P = 10,000}.$$

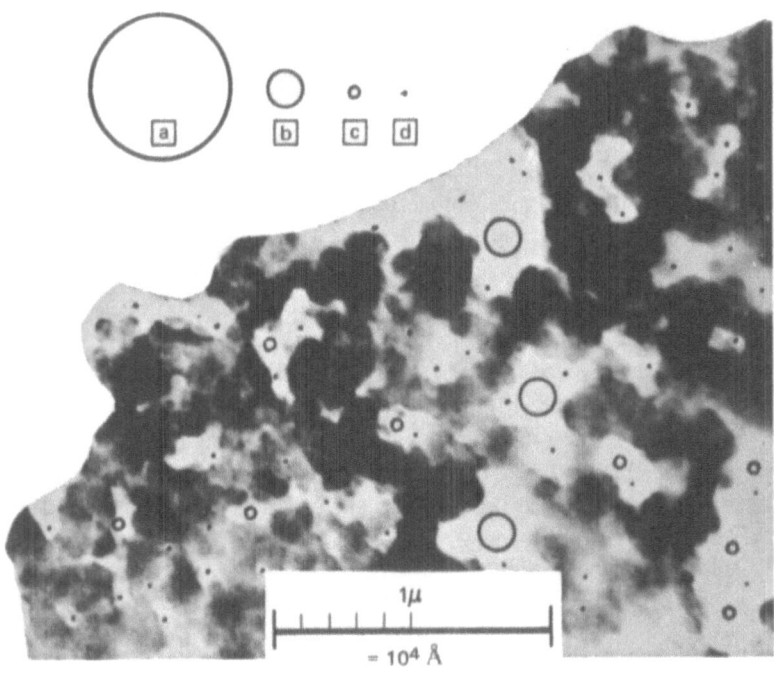

dark: polystyrene

a: $M \approx 6 \cdot 10^7$

b: $M \approx 5 \cdot 10^6$

c: $M \approx 5 \cdot 10^5$

d: $M \approx 5 \cdot 10^4$

light: pores

(compare random coil diameter $d_{equ[s]}$ of polymethylmeth-acrylate fractions of table 466).

FIG. 407 — Electron micrograph of a highly crosslinked polystyrene-gel, prepared by polymerization of 30 % styrene, 10 % divinylbenzene in the presence of 60 % of a diluent (mixture of diethylbenzene and isoamylalcohol 20/80) (K.H. ALTGELT and J.C. MOORE).

Thus molecules with P = 10 can enter into almost all the capillaries in the gel (those with small and those with large diameters), whereas molecules with P = 10,000 can only enter into very few capillaries, i.e., those whose diameter is large enough to permit the molecules with P = 10,000 to enter. If one considers that for a macromolecule in the solution streaming past a gel bead, the probability of entering a capillary, and therefore also the residence time in the capillary, is the larger, the larger the accessible capillary volume, then it is easy to see that the macromolecules will spend a longer time in the capillaries, the smaller the molecules are. If one further considers that the macromolecules, as long as they are within a capillary in one of the gel particles, are inaccessible to the solution (or solvent) streaming past, then this leads to the result that the solution leaving the column, first only contains such macromolecules which because of their large size have not had occasion, or only seldom, to enter into the pores of the gel. The smaller macromolecules on the other hand, which during their passage through the column have often been able to enter into the capillary spaces, will leave the column much later.

This procedure was first used for the separation of peptide mixtures, as they are obtained in the degradation of proteins or in protein syntheses, in aqueous solutions. The gels used in aqueous solutions are the so-called Sephadex-gels (Pharmacia Co., Sweden). For the fractionation of synthetic polymers, such as polystyrene, polyethylene, and polyacrylates, the gels are usually specially prepared, strongly cross-linked polystyrene beads prepared by pearl polymerization. The pores in these gel beads generate themselves if the pearl polymerization is carried out in the presence of divinylbenzene and poor solvents. Silica gels have also been used for the GPC separation of vinyl polymers.

The GPC procedure can be carried out in a fully automatic manner (Waters instrument) by determining the concentration of the solution leaving the column with a differential refractometer and by measuring the molecular weight through the intensity of the scattered light. By means of a calculator the measurement results are transferred to a recorder, which then immediately gives the distribution curve. (See Figure 409).

Fractional Dissolution

This is really the reverse of fractional precipitation. The entire polymeric product exists as a swollen gel, from which each fraction moves into the solution phase. Since the molecules in the gel phase are only slightly mobile, the phase transition is considerably restricted and the possibilities for a sharp separation are relatively slight. In spite of this, it has been possible to make this method useful in practice by using the polymer in the form of a very thin film, which gives rise to the best possible contact between the solvent-precipitant mixture and the polymer. The polymer is first dissolved and with the dilute solution one coats a sheet of metal, or a porous sheet such as filter paper, where the polymer precipitates on evaporation of the solvent. This leaves behind a thin film of the polymer. If one then treats the impregnated sheet with the solvent-precipitant mixtures of increasing solvent content, first the low molecular weight fractions go into solution, and then, as the solvent content of the mixtures increases, one also obtains higher and higher molecular weight components. This process can also be carried out in a column filled with a granular support similar to chromatography. (Desreux method)

Turbidity Titration

If one adds a precipitant to a very dilute solution of a polymer slowly, and under strong mixing, then the polymer precipitates in the form of very fine gel particles, the amount of which can be measured from the increasing turbidity of the solution. The turbidity appears the earlier, the higher the molecular weight of the polymer, especially the higher the molecular weight of the highest fraction. The greater the amount of these components, the faster the turbidity increases at the beginning of the titration. With a benzene solution of carefully prepared living polystyrene, for example, which is characterized by a very narrow molecular weight distribution, one finds that on addition of methanol there is at first no turbidity at all, but then from a certain amount of methanol on, there is a very steep increase in turbidity. With polymers prepared through radical polymerization, the turbidity appears much sooner and increases in a constant fashion, which results from the fact that

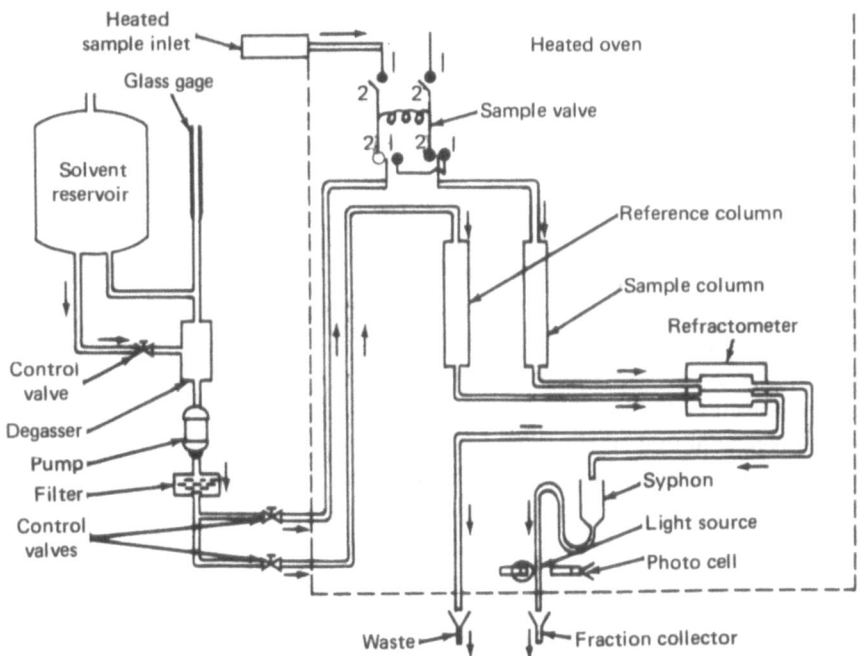

FIG. 409 – WATERS– apparatus for gel permeation chromatography (GPC) (after L.E. MALEY).

such products have relatively larger components, with molecular weights lying very much above the average. In general, the turbidity is proportional to the amount of the precipitated polymer, whereas the amount of precipitant is proportional to the reciprocal value of the molecular weight (the greater the amount of the precipitant required, the smaller the molecular weight of the polymeric component which precipitates at this precipitant concentration, and vice versa). Therefore, if one plots the turbidity τ versus the amount of added precipitant, one obtains a reverse integral distribution curve which one can calibrate by means of polymers whose distribution curve is known. But even without such a calibration the turbidity titration method is often used as a rapid and easy relative method in order to compare different polymeric products with regard to their molecular weight distribution curve (for example, in industrial laboratories for production control purposes).

Because of the dilution which occurs with the addition of precipitant, the turbidity titration curve results from two effects: one is the increase of turbidity resulting from the constantly precipitating polymer, and the second is the decrease of the turbidity at the same time through dilution of the medium with the precipitant. The second effect is at first much smaller than the first. Only after the entire polymer has precipitated, does the dilution then lead to a diminution of the turbidity (see Figure 403).

For this reason, and because measurements become less accurate with strong turbidity, the turbidity titration method gives a correct representation of the distribution only in the high molecular region of the distribution curve, whereas in the low molecular zone of the distribution only very major differences can be recognized.

Sedimentation in the Ultracentrifuge

If one centrifuges a solution of a polymer with homogeneously long molecules, one obtains on sedimentation a relatively sharp boundary between the solution and the supernatant solvent, with the concentration of the solution in the entire solvent region remaining the same. On the other hand, with polydisperse systems, one finds a completely different situation because the components with high molecular weight sediment much more rapidly than the small macromolecules. This leads to a concentration gradient at the boundary, which with longer times of sedimentation becomes more and more flat. (One can picture this process, if one thinks of the start of a race with a large group of cyclists. At first, they form a thick pack, but then slowly they spread out over a large region because of their different velocities.) This process occurs the faster, the less homogeneous the system. Several times construction of an integral distribution curve from this flattening-out process has been attempted. This theoretically clear and elegant determination of the molecular weight distribution has, however, found only relatively little application.

The Ratio $\overline{M}_w/\overline{M}_n$

If one is less interested in the course of the distribution curve than in the width of the distribution, then one can be satisfied with two molecular weight determinations, i.e., an osmotic determination giving the average $\overline{M}_n$ and a sedimentation or light-scattering measurement which gives essentially an $\overline{M}_w$ value. As will be explained more clearly in the section after the next, these two averages are the further apart, the broader the distribution. (Compare p. 431 and Table 431; $\overline{M}_w/\overline{M}_n$ with normal distribution: see pp. 419 and 345). The ratio $\overline{M}_w/\overline{M}_n = 1$ means that the product consists of completely homogeneous macromolecules (sometimes one finds the value $(\overline{M}_w/\overline{M}_n) - 1$ defined as the *inhomogeneity*). The further the ratio $\overline{M}_w/\overline{M}_n$ differs from 1, the less homogeneous the product. For living polystyrene (Scott-Szwarc polystyrene), $\overline{M}_w/\overline{M}_n$ lies between 1.05 and 1.2. On the other hand for a technical bulk-polymer, $\overline{M}_w/\overline{M}_n$ may be approximately 10. For polymers which have been polymerized by radical initiation at constant temperature and not to very high conversion (less than 80%), the ratio $\overline{M}_w/\overline{M}_n$ lies between 2 and 3 (with normal distribution $\overline{M}_w/\overline{M}_n$ is 1.5 and 2, respectively). Instead of ultracentrifuge or light-scattering measurements one can, if the calibration curve of $[\eta]$ versus $\overline{M}$ is known, determine $\overline{M}_w$ simply through viscosity measurements. For the determination of $\overline{M}_n$ one uses osmotic measurements. This means that the measurement results, especially with high molecular weight products, are relatively inaccurate. With relatively homogeneous polymers, for example, living polystyrene, where $\overline{M}_n$ is only slightly different from $\overline{M}_w$, the $\overline{M}_w/\overline{M}_n$ values are meaningful only if $\overline{M}_n$ has been calculated from the experimentally determined distribution curve according to Equation (117), p. 426.

3134 Theoretical Derivation of the Distribution Function
(Schulz-Flory Distribution)

If the mechanism of a polymer synthesis reaction is known, then in principle it is possible to calculate the size distribution of the resulting macromolecules. The distribution function for polymers formed through addition polymerization was first derived by G. V. Schulz, and that for polymers formed by condensation polymerization (polycondensation) was derived in different ways by P. Flory and G. V. Schulz. One calls these theoretically derived distribution functions: normal distribution, or most probable distribution, or Schulz-Flory distribution.

Normal Distribution for Addition Polymerization

As has already been discussed in detail in the section on the synthesis of macromolecules, polymerization is a chain reaction in which the number of monomer additions following a single initiation reaction (that is, the kinetic chain length) is proportional to the chain length (described by the degree of polymerization P) as long as no chain transfer takes place. With radical polymerization, chain termination often occurs through recombination, in which case the degree of polymerization is twice the kinetic chain length. On the other hand, if termination occurs through disproportionation instead of recombination, or if there is a reaction of the active chain end with some other added small molecules (such as water in anionic polymerization), then the kinetic chain length is equal to the degree of polymerization. The length of the kinetic chain is given by the ratio of the propagation rate and the termination rate:

$$\bar{P} \sim L_{kin} = r_p/r_t. \tag{72}$$

If these rates are equal, the fraction of addition (propagation) reactions and the fraction of termination reactions are equal, i.e., 0.5 (50%); or, expressed differently, the probability w_p that a chain radical adds another monomer is equal to the probability w_t that the chain is terminated, i.e., each is 0.5. This conclusion is justified if one assumes that one of these reactions must take place, and, accordingly, that the total probability of these alternatives has to be equal to 1. In reality with polymerization the propagation rate r_p is approximately a thousand times greater than the rate r_t of the termination reaction. The probability w_p of a monomer addition to a radical chain end therefore has the order of magnitude 0.999, and the probability w_t of a chain termination is 0.001. The probability w_p of the growth reaction will be called α in the following text, where α is only slightly smaller than 1, i.e., $1 > \alpha > 0.99$:

$$w_p = \alpha \tag{73}$$

and

$$w_t = 1 - \alpha. \tag{74}$$

Because of the ratio of the reaction rates r_p and r_t is equal to the ratio of the fractions with which the growth and the termination reactions are represented in the totality of reactions, and, therefore, is equal to the ratio of the probability of the two reactions, one can express the average kinetic chain length L_{kin}, or the average degree of polymerization, $\bar{P}_n$, also in terms of α [compare Equation (97)]:

$$\bar{P}_n = L_{kin} = r_p/r_t = w_p/w_t = \alpha/(1-\alpha) \tag{74a}[24]$$

Because α usually has values close to 1, we can write:

$$\bar{P}_n = \alpha/(1-\alpha) \cong 1/(1-\alpha). \tag{74b}$$

Therefore, the numerical value of α is given by the average degree of polymerization:

$$\alpha = \bar{P}_n/(\bar{P}_n + 1) \cong (\bar{P}_n - 1)/\bar{P}_n. \tag{74c}$$

If α is the probability that a monomer is added, then the probability that this event occurs twice, one after the other, i.e., that also a second monomer is added to the same chain, is given by α^2, and the probability that after the second monomer

---

[24]With the understanding that disproportionation is the only termination reaction. The influence of recombination will be discussed later (p. 415).

without interruption a third and a fourth monomer adds is given by α^3 and α^4.[25]

In general, the probability w_p that P monomers are linked into one chain, i.e., that a polymer radical consists of P structural units, is given by

$$w_P = \alpha^P. \tag{75}[26]$$

Such a polymer is however still capable of growth. For a terminated polymer, there has to be for each polymer molecule the event of chain termination, whose probability, as discussed above, is given by $1-\alpha$. The probability, w_P, that a polymer with a degree of polymerization P exists, is therefore:

$$w_P = \alpha^P \cdot (1 - \alpha). \tag{76}$$

The probability of an event occurring is always at the same time the fraction that this event represents of the total number of possible events (1/6 is, for example, the fraction of sixes, or of any other eye number, of a given, sufficiently high number of dice throws). Thus w_P is the fraction of macromolecules with a degree of polymerization P among the total number of macromolecules which have been formed. If one designates the total number of macromolecules of all degrees of polymerization with N, then the number N_P of macromolecules with degree of polymerization P contained in the total product is given by:

$$N_P = N\alpha^P \cdot (1 - \alpha). \tag{77}$$

With the number fraction (N_P/N) given by Equation (77) of the molecules with the degree of polymerization P, one cannot in practice do very much. One needs the weight fraction, and, in order to obtain this, one has to introduce the total number n of reaction steps required for the formation of N macromolecules. [27]

---

[25]It is easier to demonstrate this using a die as an example. The probability to throw a six (that is, that of six different, but equally possible events, a certain one occurs) is 1/6. This holds for every single throw, however many one makes. However, if one asks after the probability to throw six twice in a row, then this is $1/6 \cdot 1/6$, because this event (twice six, one after the other) is only one out of 36 possibilities: 6-6, 6-5, 6-4, etc. 5-6, 5-5, 5-4, etc. 4-6, 4-5, 4-4, etc., etc.

[26]Sometimes one finds instead α^{P-1}. Whether P or P − 1, depends on how one defines the initiation reaction: $R\cdot + M$, or $R - M\cdot + M$. Since P in general is much greater than 1, one can consider P and P − 1 as equivalent.

[27] Instead, one can substitute in Equation (77) for the number N of macromolecules, the number of moles $x = 1/\overline{M}_n$ present in 1 g of the polymer. This gives:
$$x_P = x \cdot \alpha^P (1 - \alpha) = (1/\overline{M}_n)\alpha^P(1 - \alpha) = m_P/M_P$$
where x_p is the number of moles, m_p the mass of molecules, and M_p the molecular weight of the molecules with degree of polymerization P present in 1 g. $\overline{M}_n$ is the average molecular weight. Since the molecular weight $M_P = P \cdot M_{mon}$, and correspondingly $\overline{M}_n = \overline{P}_n \cdot M_{mon}$ (M_{mon} = molecular weight of the monomer, i.e., of the structural unit), one obtains:
$$m_P/P \cdot M_{mon} = (1/\overline{P}_n \cdot M_{mon}) \cdot \alpha^P(1 - \alpha)$$
and after multiplying by M_{mon}:
$$m_P = (P/\overline{P}_n)\alpha^P(1 - \alpha).$$
If one replaces $\overline{P}_n$ according to Equation (74b) by $\alpha/(1 - \alpha) \cong 1/(1 - \alpha)$ one obtains:
$$m_P = P \cdot \alpha^P(1 - \alpha)^2,$$ which is identical with Equation (83).

Since $(1 - \alpha)$ is the fraction with which the termination steps are represented in the totality n of all reaction steps, then $n \cdot (1 - \alpha)$ is the number of termination reactions which have occurred in the formation of N macromolecules. Since each termination reaction signifies the formation of a polymer molecule, the number of termination reactions is equal to the number of macromolecules:

$$N = n(1 - \alpha). \tag{78}$$

If one substitutes this in Equation (77), one obtains:

$$N_P = n\alpha^P(1 - \alpha)^2. \tag{79}[28]$$

The number of termination reactions is in general very small in comparison with the addition reactions (growth steps). One can neglect the difference between the number of total reactions n and the addition reactions $n\alpha$, so that n is practically identical with the number of structural units which are involved in the formation of N macromolecules. If m_{Mon} is the mass of a structural unit, then the total weight G of N macromolecules is given by:

$$G = n \cdot m_{Mon}. \tag{80}$$

N_P is the number of macromolecules with a degree of polymerization P, and $m_{Mon} \cdot P$ is the mass of one of these macromolecules. Therefore, the mass G_P of all macromolecules with degree of polymerization P is given by:

$$G_P = N_P \cdot m_{Mon} \cdot P, \tag{81}$$

and the weight fraction m_P of macromolecules with degree of polymerization P is therefore:

$$\frac{G_P}{G} = m_P = \frac{N_P \cdot P \cdot m_{Mon}}{n \cdot m_{Mon}} = \frac{N_P \cdot P}{n}. \tag{82}$$

If one replaces N_P in this equation by $n\alpha^P(1 - \alpha)^2$ according to Equation (79), one obtains for the weight fraction m_P (which gives the number of grams with which a molecule of degree of polymerization P is represented in the substance):

$$m_P = P\,\alpha^P\,(1-\alpha)^2 \tag{83}$$

or in percent:

$$m_P = 100\,P\,\alpha^P\,(1-\alpha)^2. \tag{84}$$

---

[28] With G. V. Schulz, Equation (79) has the form $N_P = n\alpha^P(1/\Sigma P\alpha^P)$. The expression for the summation represents an infinite series: $\Sigma P\alpha^P = \alpha + 2\alpha^2 + 3\alpha^3 + 4\alpha^4 + \ldots \ldots$ Since $\alpha < 1$, the series has the value $1/(1 - \alpha)^2$, and one sees that the Schulz expression is identical with Equation (79). If one continues the derivation continously, one obtains instead of the sum the integral $\int P\alpha^P dP$, and instead of $1/(1 - \alpha)^2$ the equivalent expression $1/\ln^2\alpha$.

If this derivation is carried out slightly differently, one obtains Equation (83) in the following form:

$$m_P = P \alpha^P \ln^2 \alpha. \tag{85}$$

Since the α values, in general, lie around 0.99 to 0.9999, the expressions $\ln^2 \alpha$ and $(1 - \alpha)^2$ yield equal numbers, and therefore the functions (83) and (85) are also equal.

Distribution when Termination is by Recombination

With radical polymerization, chain termination occurs at lower temperatures predominantly through recombination of two growing chain ends:

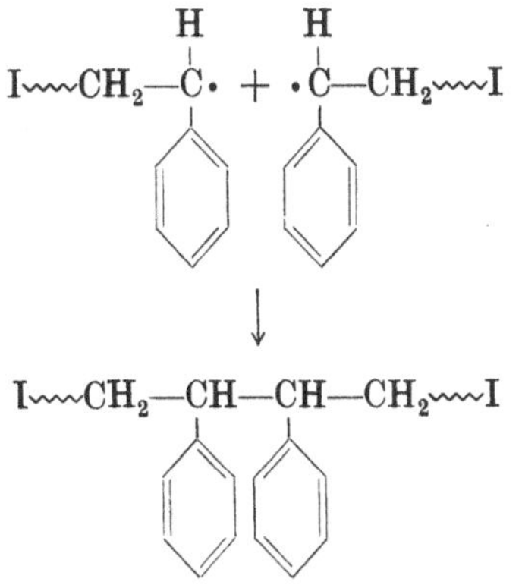

For the production of a chain of definite length, for example, degree of polymerization $P = 1,000$, there are the following possibilities. First, an initiator radical which has just added a monomer (i.e., a chain radical with the degree of polymerization $x = 1$) recombines with a chain radical with a degree of polymerization $y = 999$. Second, a chain radical with two monomer residues ($x = 2$) can combine with a chain radical with the degree of polymerization of $y = 998$. Third, a chain radical $x = 3$ finds a chain radical $y = 997$, etc., up to $(x = 500) + (y = 500)$. In general, one obtains a polymer with P structural residues through a combination of the following chain pairs: $1 + (P-1)$, $2 + (P-2)$, $3 + (P-3)$ ... up to $P/2 + [P-(P/2)]$. Thus altogether there are $P/2$ different combinations possible by which a polymer with the degree of polymerization P can be formed by recombination of two chain radicals (compare footnote 31). What is now the probability that of these $P/2$ different combinations a certain definite combination of two radicals with $P = x$ and $P = y$ is formed? It can be shown with the aid of two dice and the resulting combination of numbers, that this probability is twice as large as the product of the single probabilities w_x and w_y for the chain radicals x and y. With a pair of dice there are 36 number combinations. If all these number combinations occur equally frequently, then the probability that of these 36 possible combinations, one particular one occurs is 1/36, i.e., 1/6 times 1/6. We must remember, however, that combinations with different numbers always can occur twice, for example, 6-1, and 1-6, 6-2, and 2-6, etc., whereas the combinations with two equal numbers 6-6, 5-5, etc., can occur only once. Therefore, the occurrence of any definite combination of two different numbers is $2 \cdot 1/6 \cdot 1/6 = 1/18$, or, in general, the probability of combination w_{x+y} of two radicals[29] of different size with the single probabilities w_x and w_y is:

$$w_{x+y} = 2 \cdot w_x \cdot w_y. \tag{86}$$

The probability for the combination of two equal numbers is given by $1/6 \cdot 1/6 = 1/36$, or, in general, for the combination of two radicals of the same length:

$$w_{x+x} = (w_x)^2. \tag{87}$$

When two chain radicals combine during polymerization, one obtains from the two single radicals with degrees of polymerization x and y, a final macromolecule with the degree of polymerization $x + y = P$, and one can therefore replace y by $P-x$. The proportions in which the radicals x and $P-x$ are represented in the polymerizing system, and therefore the single probabilities w_x and w_{P-x} are, according to Equation (76), given by $w_x = \alpha^x(1 - \alpha)$ and $w_{P-x} = \alpha^{P-x}(1 - \alpha)$. By substitution in Equation (86) and (87), one obtains for the combination of radicals of unequal size:

$$w_{x + (P-x)} = 2\,\alpha^x\,(1-\alpha)^2\,\alpha^{P-x} = 2\,\alpha^P\,(1-\alpha)^2, \tag{88}$$

and for the combination of equally long radicals, for example, x = y = P/2:

$$w_{(P/2 + P/2)} = \alpha^{P/2}\,(1-\alpha)^2\,\alpha^{P/2} = \alpha^P\,(1-\alpha)^2. \tag{89}$$

Equation (88) describes the probability that a certain chain radical with a degree of polymerization x reacts with another chain radical of degree of polymerization $P-x$, and that in this reaction a macromolecule of degree of polymerization $x + (P - x) = P$ is formed. However, macromolecules with degree of polymerization P are not formed only by this one combination of chain radicals. As was seen

---

[29]The comparison with a pair of dice is permitted in discussing the combination of two chain radicals, because the number of chain radicals with the same degree of polymerization is very large. In the following model, we have four radical pairs with degrees of polymerization; $P = a$, $P = b$, $P = c$, and $P = d$, i.e., a total of eight radicals. The number fraction of each degree of polymerization is therefore 2/8. This is also the probability of finding among these eight radicals one with a certain degree of polymerization, for example, $P = c$. By substitution into Equation (88) and (89), one finds for the probability w_{x+y} that two particular different radicals combine $w_{x+y} = 2 \cdot w_a \cdot w_b = 2 \cdot (2/8) \cdot (2/8) = 1/8$. For the probability w_{x+x} that two equally large radicals combine, one finds $w_{x+x} = (w_a)^2 = (2/8)^2 = 1/16$. The actual probabilities for combination are, however, 1/7 and 1/28 respectively which can be seen immediately if one considers the possibilities for combination: there are six different combinations possible of unequal chains and for each exist four possibilities. For example, there are four ways possible for the combination of radicals of the kind a with those of the kind b: 1-3, 1-4, 2-3, 2-4. The same is true for the combinations of a with c, a with d, b with c, b with d, and c with d. Finally,

above, they can be formed as well by a series of altogether P/2 different combinations: $1 + (P - 1)$, $2 + (P - 2)$, $3 + (P - 3)$, etc., up to $P/2 + (P - P/2)$. In order to obtain the probability w_P for the formation of a macromolecule with a well-defined degree of polymerization P by one of these combinations [each one of which has the probability given by Equation (88)], one has to add the probabilities of all possible combinations from $1 = (P - 1)$ up to $(P/2) + (P/2)$.[30] This means that one simply has to multiply Equation (88) by P/2, because the individual probabilities for all combinations which lead to a macromolecule of the size P are

---

there is also the possibility of a combination a-a, b-b, c-c, and d-d. Therefore, there are a total of $6 \cdot 4 + 4 = 28$ possibilities of combination. Corresponding to the four possibilities of combination of unequally long radicals, the probability that one of the kind a combines with one of the kind b is given by $4/28 = 1/7$. The probability of a combination of two equally long radicals, for example, a with a, is given by $1/28$, and is therefore smaller by a factor of 4 [not by a factor of 2, as demanded by Equations (86) and (87)], than the combination of radicals of unequal length. If one now increases the number of radicals of each kind to four, then one has altogether 120 possibilities of combination, i.e., $6 \cdot 16$ for unequal radicals and 4 times 6 for equal radicals. The probability that a certain radical of the kind a reacts with one of the kind b is therefore $16/120 = 1/7.5$, and the probability for the combination of radicals of equal kind, for example, b with b, is $6/120 = 1/20$. With 8 radicals of the same kind, the corresponding probabilities are $64/496 = 1/7.75$ and $28/496 = 1/17.7$. As the number of radicals of the same length increases, the probabilities approach more and more the limiting values 1/8 and 1/16, as one has to expect in analogy to the dice example.

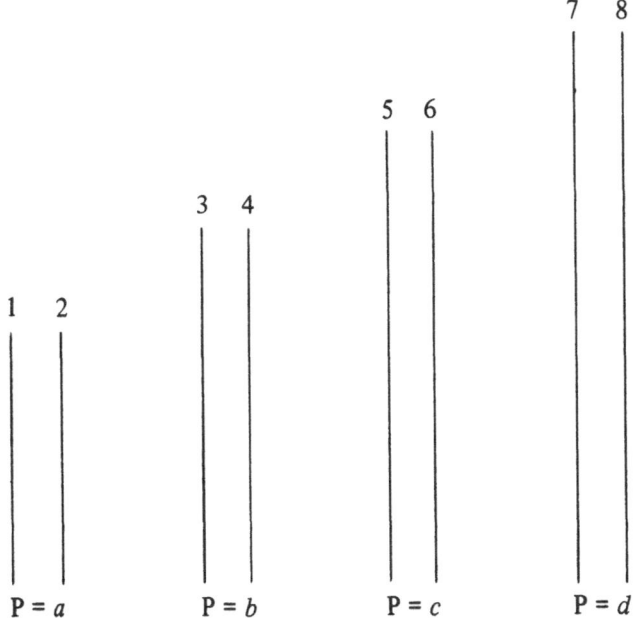

[30]If the probability of throwing two sixes (6-6) with one pair of dice is 1/36, then if one uses two pairs of dice, the probability of throwing two sixes with one pair or the other is twice as large, and with three pairs of dice, three times as large, etc.

equally large.[31] Thus one obtains:

$$w_P = P\alpha^P(1 - \alpha)^2, \tag{90}$$

where w_P is the probability for the creation of a macromolecule of length P through recombination. w_P is also equal to the fraction of macromolecules of length P in a polymeric product, where the chain termination reaction occurs only through recombination.

If one now would like to know the number N_P of macromolecules with a degree of polymerization P, which are formed from N chain radicals through recombination, one has to take into account that on recombination of two chain radicals to a macromolecule the number of macromolecules formed is only equal to half of that obtained without recombination. According to Equation (77) one can write:

$$N_P = \frac{1}{2} N P \alpha^P (1-\alpha)^2. \tag{91}$$

Since $N = n(1 - \alpha)$, [n is the number of monomer residues in N macromolecules; $(1 - \alpha)$ is the fraction of termination reactions, compare Equation (78)], one obtains:

$$N_P = \frac{1}{2} n P \alpha^P (1-\alpha)^3. \tag{92}$$

---

[31]Actually, one should first have to multiply only by $P/2 - 1$, and then, according to Equation (89), add $\alpha^P(1 - \alpha)^2$, because the last combination of the series, $P/2 + P/2$, has only half the probability of a combination of two unequally long radicals, i.e.,:

$$w_P = 2(P/2 - 1) \alpha^P(1 - \alpha)^2 + \alpha^P(1 - \alpha)^2 = (P - 2)\alpha^P(1 - \alpha)^2.$$

But, since P on the average is much larger than 1, one can without making a great error use $(P - 2) \cong P$, so that Equation (90a) becomes Equation (90). A more important limitation for Equation (90) stems from the fact that the number of combinations leading to a macromolecule with a degree of polymerization P is only equal to $P/2$ up to a degree of polymerization $P = \frac{1}{2}P_{max}$. For example, if the highest degree of polymerization of a chain radical that can occur in a polymerizing medium is $P_{R,max} = 500$, then the largest degree of polymerization that can occur through recombination must be $P_{max} = 1,000$. For the formation of such a macromolecule one does not have $P/2 = 500$ possibilities of combination, but only a single one: the reaction of two radicals of equal length each of a degree of polymerization of $P_R = 500$. All other combinations would lead to smaller macromolecules. Using a simple model system; for example, $P_{max} = 20$ and $P_{R,max} = 10$, one can easily see that for $P = 20$ there is only a single combination possibility, i.e., $(1/20)P$. for $P = 18$, there are two possibilities, i.e., $(1/9)P$; for $P = 16$, three, i.e., $(1/5.3)P$; for $P = 14$, four possibilities, i.e., $(1/3.5)P$, etc. Only from $P = 10 = \frac{1}{2}P_{max}$ downward is the number of combination possibilities equal to $\frac{1}{2}P$. Thus for $P = 10$ there are five possibilities $(= \frac{1}{2}P)$, for $P = 8$, four possibilities $(= \frac{1}{2}P)$, for $P = 6$, three possibilities $(= \frac{1}{2}P)$, etc. The constant lowering of the number of combinations with the higher degree of polymerization (over $\frac{1}{2}P_{max}$) below the value of $\frac{1}{2}P$ leads to the result that Equation (90) and the resulting distribution function, Equation (94), give too high a value for the fraction of the macromolecules with molecular weights greater than $\frac{1}{2}P_{max}$. The error is the greater, the higher the value of P. This effect is not so very important because the normal distribution runs out to a long and flat end toward the high molecular weights (Figure 419), so that the critical boundary $\frac{1}{2}P_{max}$ lies considerably to the right of the curve maximum, and therefore the limitation of the applicability of Equation (90) does not influence the overall character of the distribution function too strongly. However, one should consider it in comparing experimental distribution curves with the theoretical function (compare Figure 421).

According to Equations (79) through (82), one can write:

$$\frac{N_P}{n} = \frac{1}{2}\frac{m_P}{P} = \frac{1}{2} P\alpha^P (1-\alpha)^3. \tag{93}$$

The weight fraction m_P of the molecules with a degree of polymerization P in a certain product, is given by Equation (94):

$$m_P = \frac{1}{2} P^2 \alpha^P (1-\alpha)^3 \cong \frac{1}{2} P^2 \alpha^P (\ln \alpha)^3 \tag{94}$$

or expressed in percent:

$$m_P = \frac{1}{2} P^2 \alpha^P (1-\alpha)^3 \cdot 100. \tag{95}$$

Figure 419 shows the graphical representation of Equation (95) for different average degrees of polymerization, i.e., for different α values. Figure 420 shows that the distribution becomes more homogeneous through recombination. In Figure 421, finally, the experimentally determined distribution curve (through fractional precipitation) of polystyrene is compared with the theoretical distribution according to Equation (95). The curves do not check completely, but one sees that the character of the molecular distribution is represented properly by Equation (95).

Normal Distribution and Molecular Weight Averages

Averages with Simple Chains (without Recombination)

After N polymer chains of different lengths have been formed from a total of n monomers, then a single chain consists of an average of n/N structural units, i.e., the average degree of polymerization $\bar{P}_n$ is:

$$\bar{P}_n = n/N. \tag{96}$$

Since according to Equation (78) $N = n \cdot (1 - \alpha)$, one can write for $\bar{P}_n$:

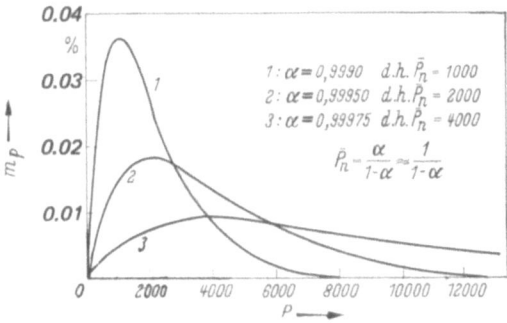

FIG. 419 – Normal distribution according to Equ. (84) for different α-values (G.V. SCHULZ)

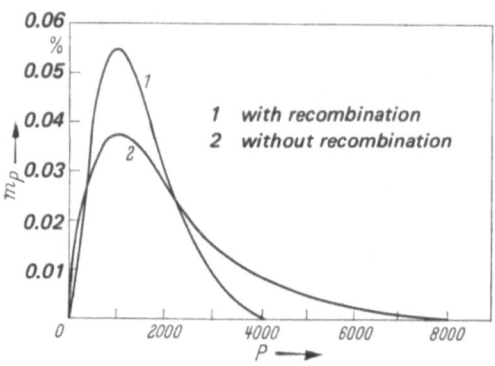

FIG. 420 — Normal distribution with and without recombination according to Equs. (84), resp. (95), for the same α (G.V. SCHULZ).

$$\bar{P}_n = \frac{n}{n(1-\alpha)} = \frac{1}{1-\alpha}.\qquad(97)$$

Since the total number N of the macromolecules of a certain product is equal to the sum of the macromolecules of degree of polymerization, $P = 1, P = 2, P = 3$, etc., up to $P = \infty$,[32] i.e., $N = \Sigma N_P$, and since according to Equation (82) $m_P = N_P \cdot P/n$ and $n_P = n \cdot m_P/P$, one can write for the total number N of the macromolecules:

$$N = \sum_1^\infty n \cdot m_P/P = n \cdot \sum_1^\infty m_P/P.\qquad(98)$$

If one substitutes this in Equation (96), one can write for the degree of polymerization $\bar{P}_n$:

$$\bar{P}_n = n/(n \cdot \Sigma m_P/P) = 1/(\Sigma m_P/P).\qquad(99)$$

This is the general equation defining the so-called number average (index n), which is identical with the average degree of polymerization and which, for example, is determined by osmotic measurements. As will be shown in greater detail in the following section, in addition to the number average $\bar{P}_n$, there is also the so-called weight average, $\bar{P}_w$, which one obtains by light-scattering or sedimentation measurements:

$$\bar{P}_w = \sum_1^\infty m_P \cdot P.\qquad(100)$$

If one now goes the opposite way and substitutes in Equation (100) for m_P according to Equation (83) $P \cdot \alpha^P(1-\alpha)^2$, or according to Equation (85) $P\alpha^P \ln{}^2\alpha$, one obtains:

$$\bar{P}_w = \sum_1^\infty P^2 \alpha^P(1-\alpha)^2 = (1-\alpha)^2 \sum_1^\infty P^2 \alpha^P.\qquad(101)$$

---

[32] ∞ does not really mean infinity here, but only any large number: 1,000, or 10,000, or 100,000.

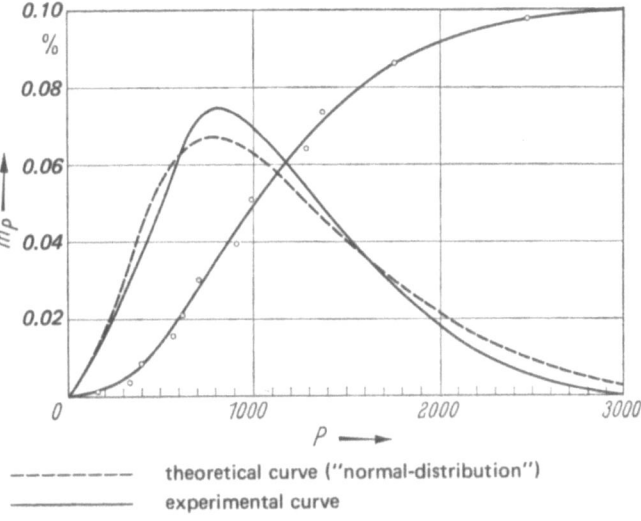

--------- theoretical curve ("normal-distribution")
――――――― experimental curve

FIG. 421 — Comparison of a distribution curve obtained by fractional precipitation of a polystyrene sample with the "normal distribution" based on Equ. (95) (G.V. SCHULZ).

The value of the infinite series, which is given by the expression for the summation, is for $\alpha < 1$:

$$\sum_{1}^{\infty} P^2 \alpha^P = \alpha + 4\alpha^2 + 9\alpha^3 + 16\alpha^4 + \ldots\ldots$$

$$= \alpha(1 + \alpha)/(1 - \alpha)^3 = 2\alpha/(1 - \alpha)^3 \cong 2/(1 - \alpha)^3 \qquad (102)[33]$$

and for $\bar{P}_w$ one can therefore write:

$$\bar{P}_w = \frac{2}{1-\alpha}. \qquad (103)$$

If one compares Equation (103) with Equation (97), one sees that for the normal distribution, the weight average $\bar{P}_w$ is twice as large as the average degree of polymerization $\bar{P}_n$ (the number average):

$$\bar{P}_w = 2\,\bar{P}_n. \qquad (104)$$

Averages for Polymer Chains Formed Through Recombination

If one substitutes the distribution function corresponding to recombination, [Equation (94)], into the equation defining $\bar{P}_n$ and $\bar{P}_w$ [Equations (99) and (100)], one obtains the following results:

---

[33]Since α is close to 1, $(1 > \alpha > 0.99)$, $\alpha^2 \cong \alpha$, and $\alpha + \alpha^2 \cong 2\alpha \cong 2$.

$$\bar{P}_n = \frac{1}{\tfrac{1}{2}(1-\alpha)^3 \sum P^2 \alpha^P/P} = \frac{2}{1-\alpha}, \qquad (105)^{34}$$

$$\bar{P}_w = \frac{1}{2}(1-\alpha)^3 \sum P^2 \alpha^P \cdot P = \frac{3}{1-\alpha}. \qquad (106)^{34}$$

As is to be expected, the recombination of two chain radicals leads to a doubling of the average degree of polymerization. The ratio of $\bar{P}_w$ to $\bar{P}_n$ is no longer 2:1 but 3:2, which follows from the fact that through recombination the distribution is more homogeneous (see Figure 420). The quotient $\bar{P}_w/\bar{P}_n$, which one uses as a measure of the homogeneity of the polymer, is equal to 2 for chain termination through disproportionation, and equal to 1.5 for chain termination through recombination (when a normal distribution holds).

Average Degree of Polymerization and Most Frequent (Mean) Degree of Polymerization

The average degree of polymerization $\bar{P}_n$ for the normal distribution is at the same time that degree of polymerization which occurs most frequently in a polydisperse product, i.e., the degree of polymerization P_{max} occurring at the maximum of the distribution curve. This can be seen immediately if one determines the maximum of the m_P–P function, differentiating Equation (84) according to P, and then setting the differential quotient equal to 0:

$$\frac{d}{dP}(1-\alpha)^2 P \alpha^P = (1-\alpha)^2 P \alpha^P \ln \alpha + \alpha^P (1-\alpha)^2 = 0$$

$$(1-\alpha)^2 P_{\text{Max}} \alpha^P \ln \alpha = -\alpha^P (1-\alpha)^2,$$

$$P_{\text{Max}} = \frac{-\alpha^P (1-\alpha)^2}{(1-\alpha)^2 \alpha^P \ln \alpha},$$

$$P_{\text{Max}} = -\frac{1}{\ln \alpha}. \qquad (107)$$

Or since $-\ln\alpha = 1 - \alpha \ (1 > \alpha > 0.99)$:

---

34 1 $$\bar{P}_n = \frac{1}{\sum\limits_1^\infty \dfrac{m_P}{P}} = \frac{1}{\tfrac{1}{2}(1-\alpha)^3 \cdot \sum\limits_1^\infty P^2 \alpha^P/P} = \frac{1}{\tfrac{1}{2}(1-\alpha)^3 \cdot 1/(1-\alpha)^2}$$

$$= \frac{1}{\tfrac{1}{2}(1-\alpha)} = \frac{2}{1-\alpha},$$

because: $\sum P \cdot \alpha^P = \alpha + 2\alpha^3 + 3\alpha^3 + 4\alpha^4 + \cdots = 1/(1-\alpha)^2$

$$\bar{P}_w = \sum_1^\infty m_P P = \frac{1}{2}(1-\alpha)^3 \cdot \sum_{P=1}^{P=\infty} P^3 \cdot \alpha^P = \frac{1}{2}(1-\alpha)^3 \cdot 6/(1-\alpha)^4 = 3/(1-\alpha),$$

because: $\sum P^3 \cdot \alpha^P = \alpha + 8\alpha^2 + 27\alpha^3 + 64\alpha^4 + \cdots$

$$= \frac{\alpha(1+4\alpha+\alpha^2)}{(1-\alpha)^4} \approx \frac{6\alpha}{(1-\alpha)^4} \approx \frac{6}{(1-\alpha)^4}.$$

$$P_{\text{Max}} = \frac{1}{1 - \alpha} \cdot \tag{108}$$

If one compares Equation (108) with Equation (97), one sees that for a normal distribution the average degree of polymerization $\bar{P}_n$ coincides with the most frequent degree of polymerization P_{max}:

$$\bar{P}_n = P_{\text{Max}} = \frac{1}{1 - \alpha} \cdot \tag{109}$$

In the same manner one obtains through differentiation of Equation (94) for the case of chain termination through recombination:

$$\bar{P}_n = P_{\text{Max}} = \frac{2}{1 - \alpha} \quad \text{(with recombination)}. \tag{110}$$

With the aid of the derived Equations (99) and (100), or respectively, (105) and (106), one can express α and $(1 - \alpha)$ through the averages $\bar{P}_n$ or $\bar{P}_w$. The distribution curves can then be calculated according to the following equations:

1. Without recombination:

$$\left(1 - \alpha = \frac{1}{\bar{P}_n} = \frac{2}{\bar{P}_w}\right)$$

$$m_P = P\,(1 - \alpha)^2\,\alpha^P = \frac{P}{\bar{P}_n^2}\left(1 - \frac{1}{\bar{P}_n}\right)^P = \frac{4P}{\bar{P}_w^2}\left(1 - \frac{2}{\bar{P}_w}\right)^P. \tag{111}$$

2. With recombination:

$$\left(1 - \alpha = \frac{2}{P_n} = \frac{3}{P_w}\right)$$

$$m_P = \frac{1}{2}\,P^2\,\alpha^P\,(1 - \alpha)^3 = \frac{4P}{\bar{P}_n^3}\left(1 - \frac{2}{\bar{P}_n}\right)^3 = \frac{27\,P^2}{2\,\bar{P}_w^3}\left(1 - \frac{3}{\bar{P}_w}\right)^P. \tag{112}$$

Normal Distribution for Polycondensation

As a typical example of a polycondensation, one can use the formation of a linear polyester from an ω-hydroxy acid:

$$n\,\text{HO—R—COOH} \xrightarrow{\;-(n-1)\,\text{H}_2\text{O}\;} \text{HO}\text{\textsf{—[}}\text{R—COO}\text{\textsf{—]}}_n\text{H}$$

For example, if a single carboxyl group out of 10 hydroxy acid molecules has been esterified, then the probability, p, that the carboxyl group forms part of an ester group, can be given for each of the present carboxyl groups by the expression p = 1/10. And the probability that a carboxyl group is a part of the unesterified groups is given by 9/10, or in general $(1 - p)$. If out of the 10 carboxyl groups, two are esterified (which can be determined through titration), then the probability p that a group belongs to the esterified carboxyl groups is twice as large, etc. In every case the probability of being esterified is equal to the fraction p of the esterified carboxyl groups, or in general to the fraction p of the converted functional groups. This fraction p is identical with the probability that a conversion of a functional group occurs. With respect to the reaction of two monomers this conversion step is identical with the formation of a dimer.

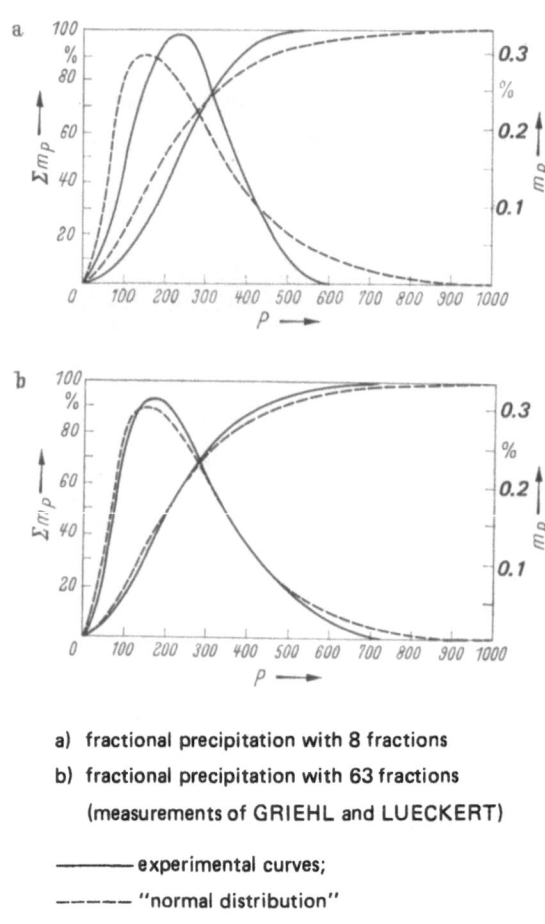

a) fractional precipitation with 8 fractions

b) fractional precipitation with 63 fractions

(measurements of GRIEHL and LUECKERT)

——— experimental curves;

– – – – "normal distribution"

FIG. 424(a and b) – Experimental confirmation of the SCHULZ-FLORY-distribution with Nylon-6.

Exactly as with polymerization the probability that an event such as esterification occurs twice in a row and that, therefore, a chain with a minimum[35] of three structural units occurs, is given by p·p. The probability that three esterifications

---

[35] Esterification can occur not only between two monomers, but also between monomers and oligomers or between several oligomers.

occur one after the other, is p^3, etc. (see footnote 25, p. 413). In general, the probability w_P that a chain with P links (and $P - 1$ ester links) is formed, is given by p^{P-1}. This is exactly the probability that the chain contains a minimum of P structural units. If one sets the condition that the chain may only be so long and no longer, then the carboxyl group at the chain end must remain free and is not permitted to react with the OH end of another chain. The probability that a carboxyl group remains unesterified is, as given above, $1 - p$. Therefore the probability w_P that a polymer chain is formed, with no more and no less then P structural units, is given by:

$$w_P = p^{P-1}(1 - p),\qquad (113)$$

where p is the probability that a particular COOH group is esterified. p is also the fraction of the esterified carboxyl groups of a certain polymeric product. If therefore in a sample a total of n carboxyl groups are present, then the number of esterified carboxyl groups is n times p. Similarly w_P is the probability that a polyester chain contains P structural units. It is also equal to the fraction of chains with P monomer residues present in the total number of polymer chains. If therefore a polyester consists of N polymer molecules with all possible degrees of polymerization, then the number N_P of molecules with the degree of polymerization P, is given by:

$$N_P = N \cdot p^{P-1} \cdot (1-p).\qquad (114)$$

In the same way as with addition polymers, (compare p. 414), one then obtains the weight fraction m_P of macromolecules with the degree of polymerization P in the total mass of the polyester:

$$m_P = P\, p^{P-1}\, (1-p)^2.\qquad (115)$$

As can be seen from comparison of Equations (115) and (83), the distribution function with polymers prepared by polycondensation is the same as with polymers formed through addition polymerization. In place of the fraction α of the addition reactions, with polycondensation one uses the fraction p of the converted functional groups. As with addition polymerization, the distribution function is the broader, the higher the average molecular weight (compare Figure 419).

Equation (115) has been checked with many polymers obtained by polycondensation and has always been found to apply. Figure 424 shows as an example a distribution curve determined through fractional precipitation for a nylon-6 product in comparison with calculated curves according to Equation (115). One can see that the experimental distribution corresponds better and better with the theoretical calculated one, the more careful the fractionation has been carried out.

3135 Molecular Weight Averages

The only completely satisfactory description of the molecular weight (i.e., the degree of polymerization) of a macromolecular compound is the distribution curve

$[m_p = f(P)]$ as determined through fractionation. However, in view of the great effort involved in fractionation, one is often satisfied with an average value for the molecular weight. In addition to the fact that this is an incomplete description of the molecular weight, one also has to take into account here that there are several molecular weight averages which differ considerably from each other (as has been pointed out previously in the discussion of the normal distribution).

For example, if one says that a polystyrene has an osmotically determined molecular weight of $\overline{M}_n$ = 150,000, this means that this polydisperse product contains perhaps a mixture of molecules with M = 1,000 to M = 10,000,000, which would behave in the osmometer just like a homogeneous polystyrene existing only of molecules of M = 150,000. It should be understood, however, that this mixture of molecules does not behave like a homogeneous product in every other respect. In addition to the osmometric behavior, these two products would be alike only in so far as their chemical reactions are concerned, and possibly also with regard to their mechanical strength. However, as regards other properties (for example, the increase in viscosity to which it gives rise in a solvent, or the light-scattering intensity, or the sedimentation velocity in the ultracentrifuge), it behaves like a homogeneous polystyrene of M>150,000 (if a normal distribution is present), similar, for example, to a homogeneous polystyrene of M = 225,000 (assuming chain termination by recombination).

Definition of the Molecular Weight Averages

All averages can be described by the following equation developed by G. Meyerhoff [Equation (116)] where c_i is the weight fraction of the molecules with the molecular weight M_i:

$$\overline{M}_\beta = \frac{\sum c_i \cdot M_i^\beta}{\sum c_i \cdot M_i^{\beta-1}}. \tag{116}$$

(for example, c_i can be the fraction of molecules with the molecular weight M_i, then the sum $c_1 + c_2 + c_3 + c_4 + \ldots = \sum c_i$ has the value 1; c_i can also be the percentage of molecules with molecular weight M_i, then $\sum c_i$ is equal to 100.)

If one sets $\beta = 0$, one obtains the equations defining the number averages $\overline{M}_n$ and $\overline{P}_n$.

$$\overline{M}_{(\beta\,=\,0)} = \overline{M}_n = \frac{c_1 + c_2 + c_3 + c_4 + \cdots}{c_1/M_1 + c_2/M_2 + c_3/M_3 + c_4/M_4 + \cdots} = \frac{\sum c_i}{\sum c_i/M_i}, \tag{117}$$

$$\overline{P}_n = \frac{\overline{M}_n}{M_{\mathrm{Mon}}} = \frac{c_1 + c_2 + c_3 + \cdots}{c_1/P_1 + c_2/P_2 + c_3/P_3 + \cdots} = \frac{\sum c_i}{\sum c_i/P_i}. \tag{117a}$$

With $\beta = 1$, Equation (116) gives the weight averages $\overline{M}_w$ and $\overline{P}_w$:

$$\overline{M}_{(\beta\,=\,1)} = \overline{M}_w = \frac{c_1 M_1 + c_2 M_2 + c_3 M_3 + c_4 M_4 + \cdots}{c_1 + c_2 + c_3 + c_4 + \cdots} = \frac{\sum c_i \cdot M_i}{\sum c_i}, \tag{118}$$

$$\overline{P}_w = \frac{\overline{M}_w}{M_{\mathrm{Mon}}} = \frac{c_1 P_1 + c_2 P_2 + c_3 P_3 + \cdots}{c_1 + c_2 + c_3 + \cdots} = \frac{\sum c_i P_i}{\sum c_i}. \tag{118a}$$

Then there is also the Z average, which one obtains in molecular weight determinations from the sedimentation equilibrium in the ultracentrifuge. It can be derived from Equation (116), if one sets $\beta = 2$. In practice, this only plays a relatively unimportant role, however.

Number Averages and Weight Averages

If one wants to make clear where Equations (117) and (118), and the terms "number average" and "weight average" come from, one can use a simple model example:

TABLE 427. Model example to illustrate the formation of averages

Fraction	m Weight of a single sphere g	$c = n \cdot m$ Weight of one fraction g	$n = c/m$ Number of spheres in one fraction
I	6.25	25	$25/6.25 = 4$
II	10	50	$50/10 = 5$
III	25	25	$25/25 = 1$

In order to obtain the average weight of a ball, one must (as can be seen immediately from the model example) divide the total weight of the system by the total number of balls (here $\overline{m}_n = 100/10$); or in general:

$$\overline{m}_n = \text{total weight of particles/total number of particles.} \qquad (119)$$

The total weight consists of the weights of the individual fractions:
Total weight $= c_1 + c_2 + c_3 = \Sigma c_i = 25 + 50 + 25 = 100$.

The total number of particles is the sum of the number of particles of the individual fractions $n_i = c_i/m_i$ (Table 427, column 4). Here in our model example: $c_1/m_1 + c_2/m_2 + c_3/m_3 = \Sigma c_i/m_i = 4 + 5 + 1 = 10$. The average particle weight $\overline{m}_n$ (number average) according to Equation (119) is:

$$\overline{m}_n = (25 + 50 + 25)/(4 + 5 + 1) = 10, \text{ or, in general:}$$

$$\overline{m}_n = \Sigma c_i/(\Sigma c_i/m_i), \text{ and after multiplication with } N_A: \overline{M}_n = \Sigma c_i/(\Sigma c_i/M_i).$$

One can therefore assume that the number average is formed in such a way that first all the particles of a polydisperse system are melted together into a homogeneous mass, which is then divided up again into particles of the same size, so that the number of particles in the original polydisperse system and the number of particles in the new monodisperse system are the same.

This monodisperse new system with the same number of particles as the old polydisperse system is not the only possibility of replacing it. Instead of breaking up the molten mass into 10 balls of 10 g each, one can also break it up into 9 balls of 11.1 g, or 8 balls of 12.5 g, or 7 balls of 14.3 g, etc. This at first sounds quite meaningless, since all these monodisperse systems no longer have any relation to the original polydisperse system (Table 427). However, as will be seen immediately, the system "8 balls of 12.5 g each" represents a special case because 12.5 g (actually 12.8 g), is the weight average $\overline{m}_w$, which means that this particular system of 8 balls of 12.5 g, when converted to molecular dimensions, gives the same result in sedimentation, or light-scattering, or in its viscosity increase, as the original polydisperse system (Table 427).

One can construct the weight-average molecular weight $\overline{M}_w$, if one allows the individual fractions to take part in the formation of the sum of the weights, not according to the ratio of the effectively present number of particles, but in the ratio of the weight fractions c_1, c_2, and c_3. Thus one proceeds as if the system does not consist of four balls of 6.25 g, 5 balls of 10 g each and 1 ball of 25 g, but as if the system consists of 25 balls of 6.25 g each, 50 balls of 10 g each, and 25 balls of 25 g each.

The weight fractions of the individual fractions are then no longer $n_1 \cdot m_1 = c_1$, $n_2 \cdot m_2 = c_2$, and $n_3 \cdot m_3 = c_3$, but $c_1 \cdot m_1$, $c_2 \cdot m_2$, and $c_3 \cdot m_3$:

25 balls of 6.25 g each give a weight fraction of Fraction I equal to $c_1 \cdot m_1 =$ $25 \cdot 6.25 = 156$ g

50 balls of 10 g each give a weight fraction of Fraction II equal to $c_2 \cdot m_2 =$ $50 \cdot 10 = 500$ g

25 balls of 25 g each give a weight fraction of Fraction III equal to $c_3 \cdot m_3 =$ $25 \cdot 25 = 625$ g

Total weight $= 1281$ g

If from this one again forms the arithmetic mean according to Equation (119), one obtains (since the total number of particles is $25 + 50 + 25 = 100$):

$$\overline{m}_w = (c_1 m_1 + c_2 m_2 + c_3 m_3)/(c_1 + c_2 + c_3) = \Sigma c_i m_i / \Sigma c_i = 1281/100 = 12.8,$$

and after multiplication of both sides with Avogadro's number N_A

$$\overline{M}_w = (c_1 M_1 + c_2 M_2 + c_3 M_3 + \ldots)/(c_1 + c_2 + c_3 + \ldots) = \Sigma c_i M_i / \Sigma c_i.$$

In place of the number fractions, we have now used the weight fractions of the different fractions. For the chemist, the weight average $\overline{M}_w$ is only of minor importance, since one cannot use it for stoichiometric calculations. For chemical transformations, one needs a molecular weight average, which allows calculations of the number of end groups capable of reaction, and this is the number average $\overline{M}_n$.

Table 429 shows the different methods used for the determination of molecular weights and the corresponding averages.

With the osmotic and the viscometric methods the expected average can be derived easily from the equations on which the calculation of the molecular weight is based, and one can confirm that the osmotic measurements give a number average, and the viscometric measurements a weight-average molecular weight.

TABLE 429. Molecular weight determination methods and the resulting averages.

Determination Method	Average	β
End-group determination	M_n	0
Osmotic pressure	M_n	0
Light scattering	M_w	1
Sedimentation (ultracentrifuge)	$\approx M_w$	≈ 0.9
Sedimentation equilibrium (ultra-centrifuge)	M_z	2
Viscosity	$\approx M_w$	$0.8 - 1$

Osmotic Molecular Weight Average

The molecular weight is calculated from the osmotic pressure according to the equation $\overline{M} = RTc/p$ (compare p. 343). The osmotic pressure p consists additively of the partial pressures $p_1, p_2, p_3 \ldots$ for the molecules of different degrees of polymerization $1, 2, 3, \ldots$ etc.:

$$p = p_1 + p_2 + p_3 + \cdots = \frac{R \cdot T \cdot c_1}{M_1} + \frac{R \cdot T \cdot c_2}{M_2} + \frac{R \cdot T \cdot c_3}{M_3} + \cdots$$

The same holds for the total concentration c:

$$c = c_i + c_2 + c_3 + \ldots\ldots$$

For the osmotic molecular weight one can then write:

$$\left.\begin{aligned} M_{osmotic} &= \frac{R \cdot T(c_1 + c_2 + c_3 + \cdots)}{p_1 + p_2 + p_3 + \cdots} \\ &= \frac{RT(c_1 + c_2 + c_3 + \cdots)}{RT(c_1/M_1 + c_2/M_2 + c_3/M_3 + \cdots)} = \frac{\sum c_i}{\sum c_i/M_i} = \overline{M}_n \end{aligned}\right\} \quad (120)$$

Viscometric Molecular Weight Average

When the exponent a in the viscosity equation $[\eta] = K \cdot M^a$ has the value $a = 1$, the viscometric molecular weight (compare p. 378) is $\overline{M}_{[\eta]} = [\eta]/K = (\eta_{rel} - 1)/Kc$. Also here the $(\eta_{rel} - 1)$ value can be constructed additively from the $(\eta_{rel} - 1)$ values of the individual fractions:

$$(\eta_{rel} - 1) = (\eta_{rel} - 1)_1 + (\eta_{rel} - 1)_2 + (\eta_{rel} - 1)_3 + \cdots$$
$$= K M_1 c_1 + K M_2 c_2 + K M_3 c_3 + \cdots$$

For the molecular weight one can then write:

$$\overline{M}_{visc} = \frac{K(M_1 c_1 + M_2 c_2 + M_3 c_3 + \cdots)}{K(c_1 + c_2 + c_3 + \cdots)} = \frac{\sum c_i M_i}{\sum c_i} = \overline{M}_w. \quad (121)$$

If a is not equal to 1, the equation is: $\overline{M}^a = (\eta_{rel} - 1)/Kc$, or:

$$\overline{M}^a = \frac{K \cdot (M_1^a \cdot c_1 + M_2^a \cdot c_2 + \cdots)}{K \cdot (c_1 + c_2 + \cdots)} = \frac{\Sigma c_i \cdot M_i^a}{\Sigma c_i}.$$

For $\overline{M}_{[\eta]}$ one can then write:

$$\overline{M}_{[\eta]} = \left(\frac{\Sigma c_i \cdot M_i^a}{\Sigma c_i}\right)^{1/a} \qquad (122)$$

As $a \to 1, \overline{M}_{[\eta]} \to \overline{M}_w$.

These examples have shown that a monodisperse system with an average molecular weight $\overline{M}$ which has *all* the properties of a polydisperse polymeric product does not exist. If one considers the osmotic behavior, then the equivalent monodisperse system has the molecular weight $\overline{M}_n$. However, if one considers light-scattering, then the equivalent average is $\overline{M}_w$, with the viscosity, $\overline{M}_{\beta=1}$ to $\overline{M}_{\beta=0.8}$, and with sedimentation, $\overline{M}_{\beta=0.9}$ or $\overline{M}_z$. It is not always possible for each average to give a closed expression as is the case with the osmotic and viscometric averages [compare Equation (120) and (121)], because the β-value itself can be a function of the inhomogeneity of the product, as for example with sedimentation in the ultracentrifuge. In this case the β-value can be expressed by means of a relation between the exponent a of the viscosity equation and the exponent β for Equation (116). This will be described in greater detail in the next section.

Molecular Weight Averages and the Viscosity Number [η]

In order to determine the dependence of the viscosity number on the molecular weight, one usually fractionates the product into a series of fractions of decreasing molecular weight by means of fractional precipitation. One then determines the molecular weight and viscosity number of each fraction. If the fractions are completely homogeneous, it becomes unimportant which molecular weight determination method one uses to obtain the molecular weight. However, and this is always the case in practice, if the fractions still have a certain inhomogeneity, then the method of molecular weight determination has to be taken into consideration. This can easily be shown by the following example.

Two homogeneous fractions of molecular weight $M_1 = 1,000,000$ and $M_2 = 100,000$ are mixed in different proportions. In this way one obtains mixtures of different, but well-defined, inhomogeneity. According to Equations (121) or (122), one can calculate for each mixture the viscometric molecular weight, and, according to $[\eta] = K \cdot M^a$, the $[\eta]$ value, if one arbitrarily sets values for K and a. Furthermore, according to Equation (116), one can calculate any other desired average for these fractions.

Assuming that one can find a solvent for which the Staudinger equation applies (compare p. 378), then $a = 1$. For $K_{[\eta]}$, we arbitrarily assume the value $4 \cdot 10^{-4}$. Then one can calculate for the mixtures 30/70, 50/50, and 80/20, the $[\eta]$ values shown in Table 431. ($[\eta] = K_{[\eta]} \cdot \overline{M}_{[\eta]}$). Since $a = 1$, the $\overline{M}_{[\eta]}$ values of the mixtures are identical with the $\overline{M}_w$ values.

TABLE 431. Calculated $[\eta]$-values for well-defined mixtures of two homogeneous fractions $M_1 = 10^6$ and $M_2 = 10^5$ ($a = 1$ and $K_{[\eta]} = 4 \cdot 10^{-4}$) and the corresponding $\overline{M}_w$ and $\overline{M}_n$ of the mixtures

c_1	c_2	$\overline{M}_{[\eta]}$ *	$[\eta]$	$\overline{M}_{\beta=1}$ $(=\overline{M}_w)$	$\overline{M}_{\beta=0}$ $(=\overline{M}_n)$
0	100	100000	40	100000	100000
30	70	370000	148	370000	137000
50	50	550000	220	550000	182000
80	20	820000	328	820000	358000
100	0	1000000	400	1000000	1000000

*from Equ. (122), a = 1.

As the number in the table show, the $\overline{M}_n$ values differ more and more from the corresponding $\overline{M}_w$ values, the more the composition of the mixtures approaches the ratio of 50/50, i.e., the less homogeneous the mixtures are. Conversely, the quotient $\overline{M}_w/\overline{M}_n$ can be taken as a measure for the degree of homogeneity (or inhomogeneity) of a polydisperse mixture.

If one plots as usual the $[\eta]$ values against the molecular weight $\overline{M}_{[\eta]}$ (Figure 432-1), then the $[\eta]$ values of all mixtures must lie, of course, on a straight line, even though these mixtures have completely different homogeneities. The same straight line would be found for homogeneous fractions, if one plots $[\eta]$ against $\overline{M}_{[\eta]}$ (here equal to $\overline{M}_w$). However this would be quite different, if one plots $[\eta]$ against $\overline{M}_n$ (equals $\overline{M}_{\beta=0}$), because now the $[\eta]$ values of the mixtures lie outside the straight line, and the more inhomogeneous the mixture, the further the points will lie away from the straight line.

This result may be expressed in a more general fashion. Always when the viscometric average $\overline{M}_{[\eta]}$ coincides with the average $\overline{M}_\beta$ of the absolute method used for the determination of the molecular weight, then the $[\eta]$ versus $\overline{M}$ points in the log-log diagram, independent of the homogeneity of the products, will lie on the straight line which applies for homogeneous fractions. In practice, this leads to the conclusion that for the calibration of the $[\eta]$ $-\overline{M}$ curves, one should use absolute methods whose averages correspond as much as possible with the $\overline{M}_{[\eta]}$ values, because then the measured values will be independent of the homogeneity of the fractions. Since the $\overline{M}_{[\eta]}$ averages [at least in good solvents (high a values)] lie close to the $\overline{M}_w$ averages, the $\overline{M}_w$ methods (ultracentrifuge, light-scattering, etc.) are therefore more suitable for the determination of the $[\eta]$ versus $\overline{M}$ function than the osmotic measurements which give $\overline{M}_n$ values.

Figure 432-2 shows the $[\eta]$ $-\overline{M}$ lines for polymethylmethacrylate in acetone, one for osmotic measurements, the other for sedimentation measurements in the ultracentrifuge.

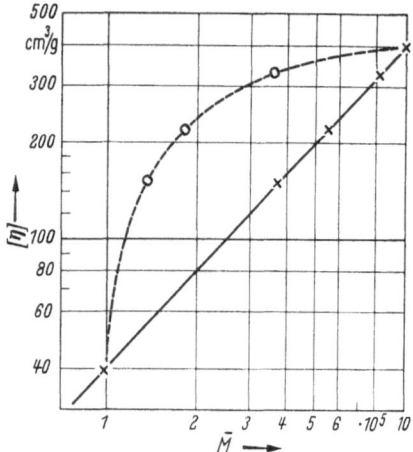

FIG. 432-1 — Log-log plot of $[\eta]$ vs. M_w (+) and $[\eta]$ vs. $\overline{M}_n$ (o).

One can see that the osmotic measurements all lie somewhat above the straight line obtained from the sedimentation measurements obviously because the fractions are not completely homogeneous (compare also Figure 432-2), Also that the $\overline{M}_n$ values often scatter much more in the $[\eta]$ versus $\overline{M}$ diagram than the $\overline{M}_w$ values can be readily understood on the basis of Figure 432-1. The explanation lies in the fact

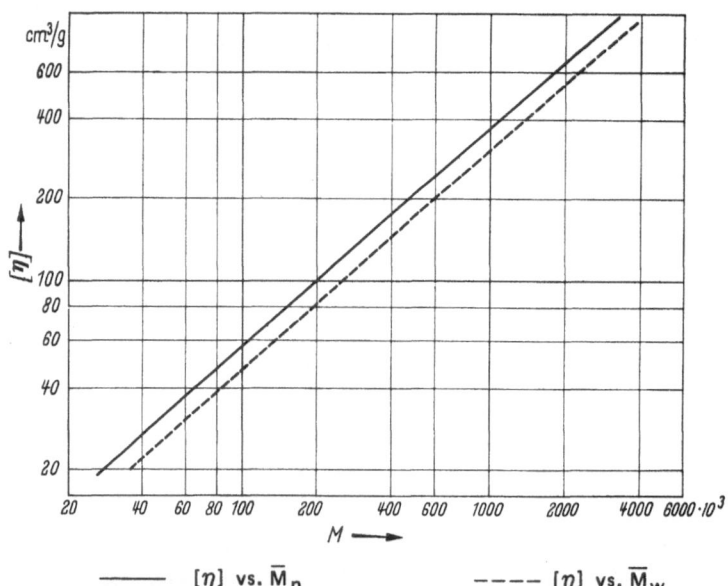

——— $[\eta]$ vs. $\overline{M}_n$ ---- $[\eta]$ vs. $\overline{M}_w$

FIG. 432-2. — Experimentally determined $[\eta]$ — M lines for polymethylmethacrylate in chloroform (G.V. SCHULZ and G. MEYERHOFF).

that the homogeneity of the fractions differs from fraction to fraction, and therefore the osmotic values are more or less farther away from the straight line (which applies for homogeneous fractions).

The Influence of the Exponent a in the Viscosity Equation

The viscometric average only coincides with the weight average $\overline{M}_w$ when $a = 1$. If a is smaller than 1 (in good solvents, a has a value of ~ 0.7 with most polymers), then the viscometric average is smaller than $\overline{M}_{\beta=1} = \overline{M}_w$. In order to have the $[\eta]$ values lie on the straight line in the $[\eta] - \overline{M}$ diagram independent of the homogeneity of the fractions, one needs a molecular weight determination method where the averages coincide with the corresponding viscometric averages (which depend on a). Thus one needs a method which gives $\overline{M}_{\beta=0.8}$ averages when $\overline{M}_{[\eta]} = \overline{M}_{\beta=0.8}$, and so on.

In order to obtain the dependence of the β values on the exponent a one can again make use of a "synthetic" mixture of two homogeneous fractions and calculate, according to Equation (122), which averages one obtains with different a values. According to Equation (116), one can then further determine which β value belongs to the different average values obtained. In Figure 434-1 the dependence of the calculated [according to Equation (122)] viscometric average $\overline{M}_{[\eta]}$[36] on a is shown for a 1:1 mixture of fractions of $M_1 = 10^6$ and $M_2 = 10^5$.

Figure 434-2 shows for the same mixture the dependence between β and the average value M_β calculated according to Equation (116).[37]

With the help of these curves one can now represent β as a function of a. First, for a certain exponent a, one finds the corresponding average $\overline{M}_{[\eta]}$ in Figure 434-1. Then, using Figure 434-2, one obtains the value of β corresponding to this molecular weight. In Figure 435 we find the resulting function $\beta = f(a)$. From the $\beta - a$ curve, one determines the value of β which permits calculation of the same average [according to Equation (116)], as is obtained [according to Equation (122)] with the corresponding values of a. This is the value β must have, with an absolute molecular weight method, if one wants the viscosity function $[\eta] = K \cdot M^a$ to be a straight line for inhomogeneous products and independent of the degree of inhomogeneity of these products. For example, for polymethylmethacrylate in

---

[36] For $a = 0.6$, for example:

$$\overline{M}_{[\eta]} = \left(\frac{50(10^5)^{0,6} + 50(10^6)^{0,6}}{50 + 50} \right)^{1/0,6} = \left(\frac{50(10^3 + 10^{3,6})}{100} \right)^{1,666} = 455{,}330$$

[37] For $\beta = 0.6$, one has

$$\overline{M}_\beta = \frac{50(10^5)^{0,6} + 50(10^6)^{0,6}}{50(10^5)^{-0,4} + 50(10^6)^{-0,4}} = \frac{10^3 + 10^{3,6}}{10^{-2} + 10^{-2,4}} = 357{,}500.$$

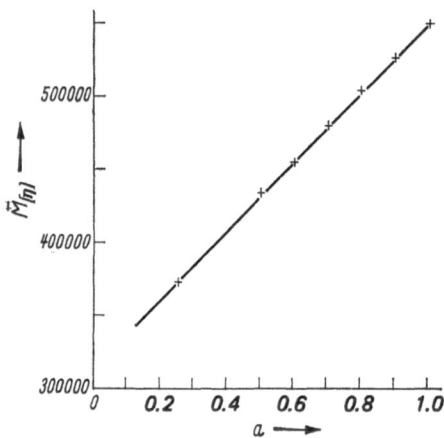

FIG. 434-1 — $M_{[\eta]}$ as a function of
the exponent a for a 1:1 mixture of
two homogeneous fractions with $M_1 =$
10^6 and $M_2 = 10^5$

acetone, the equation $[\eta] = K \cdot M^{0.73}$ holds. This means that the β value, for which
this requirement holds, is approximately $\beta = 0.9$ according to Figure 435. If one
now observes that the molecular weights which are obtained by sedimentation
measurements in the ultracentrifuge lie always exactly on the straight line in the
double-logarithmic $[\eta] - M$ diagram, whether the products are fractions or not
(measurements of G. Meyerhoff), then one can deduce from this fact that the
ultracentrifuge must in this case yield $\overline{M}_{\beta=0.9}$ values.

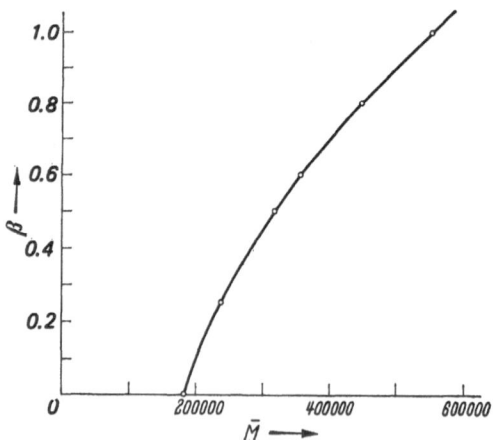

FIG 434-2 — β as a function of the average molecular weight of a 1:1 mixture of
two homogeneous fractions with $M_1 = 10^6$
and $M_2 = 10^5$

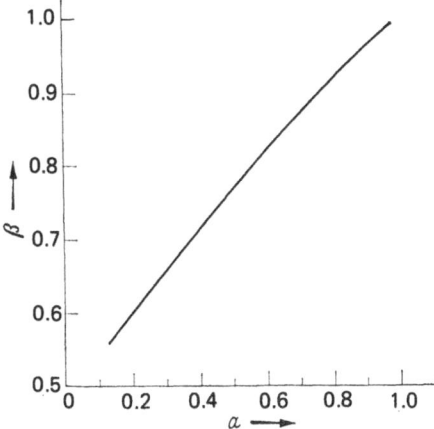

FIG. 435 — Dependence of the β-value on the exponent a of the viscosity equation.

Knowledge of the function β = f(a) has no particular practical significance, especially because one does not have a wide selection of absolute methods for molecular weight determination with β values between 0.5 and 1. Thus even if one knows that for polyisobutylene in benzene at 24°C, for example, the exponent a of the viscosity equation $[\eta]$ = K·M^a is equal to 0.5, and therefore for the exact determination of the $[\eta]-\overline{M}$ curves, one needs a molecular weight determination method which, according to Figure 435, would give $\overline{M}_{\beta=0.77}$ values, then this is not particularly useful because one just does not have such a method. However, knowledge of the fraction is of theoretical interest, because it relates the viscometric averages [according to Equation (122)] with the $\overline{M}_w - \overline{M}_n$ scale [Equation (116)].

As Figure 435 shows, the β value changes only relatively slowly with a. Even for the (for practical purposes) very low value of a = 0.5, β is still close to 0.8, so that even using θ solvents (where a = 0.5), one does not expect important deviations for the $[\eta]-\overline{M}$ function when the molecular weight is determined by light-scattering ($\beta \cong 1$), or sedimentation ($\beta \cong 0.9$) even if the fractions are not very sharp.

32 THE FORM OF THE MACROMOLECULE (MOLECULAR SHAPE)

The shape of a molecule is determined by the spatial arrangement of the atoms. In some cases, for example where there are stiff rings, the shape of the molecule is determined by the constitution of the bonds. With most organic compounds, however, the atoms have so much freedom of movement that several spatial arrangements are possible. Only then, when the different forms are prevented from going from one form to another by a potential barrier, one can isolate them as stereoisomers (*cis-*, *trans-* isomerism). In general however, the kinetic energy of the heat motion at room temperature is sufficient to bring about a continuous rapid movement, or change, from one form into another so that the isolation of definite individual forms is not possible. Naturally, the many different shapes of a molecule are limited by its constitution, and one can make use of molecular models to obtain some information about this.

Macromolecules have a chain structure. The constitution of the chain can differ according to the type of structural units (compare Table 12). However these differences will not bring about a principally different shape of the macromolecule.

For the time being, we will not consider all the special types of structures which can make the problem of the molecular shape more difficult, such as chain branching, cross-linking, stiff substituents which prevent free rotation, etc. The following considerations refer to an ideal macromolecule with strictly linear structure.

Furthermore, the macromolecule should not be restricted in its mobility through outside influences, i.e., the macromolecules should not hinder each other through secondary valence forces, as is the case in the solid state (i.e., in the absence of solvents). Furthermore, the solvent should not have any energetic influence (heat of solution, $\Delta H = O$). Thus we are dealing with the shape of the macromolecule in an extremely dilute solution in an indifferent solvent.

The phenomenon that the paraffin chain can assume different shapes as a result of rotation around the carbon-carbon axis is due to the fact that the valence angle differs from $180°$, i.e., it is approximately $110°$. This can best be seen with the help of molecular models. Figure 436 shows two different forms (the chair and boat

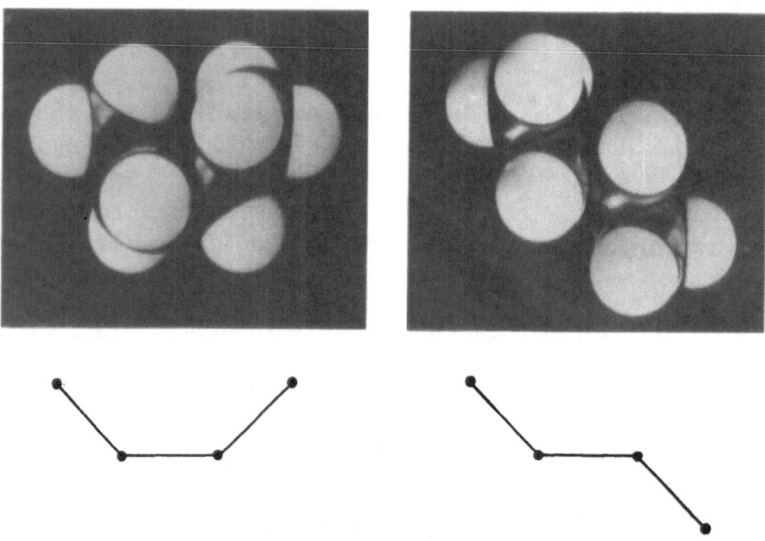

FIG. 436 — Butane molecule, boat-form and chair-form

form) of a molecule of butane, and below the models one sees the corresponding schematic line diagrams, which will be used frequently in the following text. For the valence angle in the line diagram we have used simply an angle of 120°.

Rotation around the C–C axis is not completely free. This is because not every arrangement of two carbon atoms relative to one another has the same energy. For example, if one looks at a molecule of ethane along the direction of the carbon-carbon axis, ("from above"), there are two preferred positions: one (position 1 in Figure 437), in which the hydrogen atoms cover each other, and the second (position 2 in Figure 437), where the hydrogen atoms are in the so-called gauche position.

Position 1 has an energy which is 3,000 calories above that for position 2. For a rotation of the carbon atoms around their axis from one gauche position to the next (rotation through 120°), the energy barrier of 3,000 cals (position 1) must be overcome. This amount is so small, however, that even at a temperature of 20°C, most of the molecules can undergo this rotation. As Table 438 shows, the energy barrier is smaller with longer chains, especially those with double bonds.

The height of the energy barrier between the gauche positions determines how often, at a given temperature, there is a rotation around the carbon-carbon axis, i.e., how mobile the carbon-carbon chain is. The shape of the chain, however, does not depend on the potential energy barrier, because the three possible positions 1 and the three possible positions 2 are completely equivalent energetically; therefore the

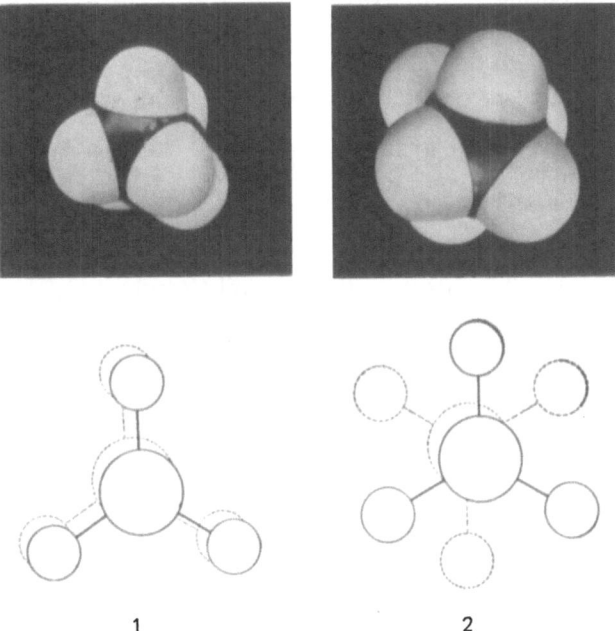

1   2

FIG. 437 – Ethane molecule, seen along the C – C axis.

different chain forms, which result from rotation from position 2 to position 2 or from position 1 to position 1 are not different from each other in energy content.

Of course, there will also be transitions from 1 to 2 and from 2 to 1, which correspond to a change in the potential energy of the chains, i.e., the chain atoms. However, at constant temperatures the transitions from 1 to 2 ($-\Delta H$) and the transitions from 2 to 1 ($+\Delta H$) are equally frequent, so that the internal energy of a system consisting of many macromolecules is not affected by them. Therefore, the type of shape changes which a chain molecule undergoes is determined only by the condition that the entropy of a system consisting of many chains, or chain links, may not become smaller (ΔS must be positive):

$$\Delta F = -T \cdot \Delta S \qquad (\Delta H = 0).$$

Changes of shape which would bring about an entropy decrease of the entire system (positive ΔF) are unlikely, whereas those which bring about an entropy increase will be favored.

TABLE 438. Opposition to rotation of the chain around the C–C bond (according to H.A. STUART)

Compound		Hindrance potential in cal/mole
Ethane	$CH_3—CH_3$	2700
Propylene	$CH_2=CH—CH_3$	2100
Butylene	$CH_3—CH_2—CH=CH_2$	<800
Dimethyl-acetylene	$CH_3—CH=CH—CH_3$	<800
	$CH_3—C\equiv C—CH_3$	<500

321 The Ideal Statistical Coil (Random Coil)

Which form a certain macromolecule will assume if it is completely free of outside influences can be answered in a simple way: The chain molecule will always try to assume a condition of maximal possible entropy, i.e., it will assume the most probable shape, which is the most irregular shape: the one for which there are the largest number of possible ways of attaining it. Figure 439 shows schematically (the valence angle, which in reality lies around 110°, was for easier representation arbitrarily set at 120°) the different possibilities of forming a carbon chain with eight atoms. Up to three carbon atoms (double shading) one cannot have different chain forms. However, whether the next carbon atom (single shading) adds in position 4a or 4b (with spatial models, there are, of course, three possibilities) is simply a matter of accident (within the limitations of the valence angle and the gauche position of the hydrogen atoms). If one extends the chain by another atom (white), then the following positions become possibilities: 5a', 5a'', 5b', and 5b'',. If one now considers the different forms, one sees that the fraction of regular forms (stretched out zig-zag chain, circles, i.e., spirals, symmetrical forms which on continuation of the same sequence with longer chains lead to regular—for example, snake-like—foldings) decreases rapidly from step to step—with a spatial model, of course, even faster than with a planar model. This does not say anything other than that the probability for the formation of regular forms is the smaller, the larger the chain molecule.

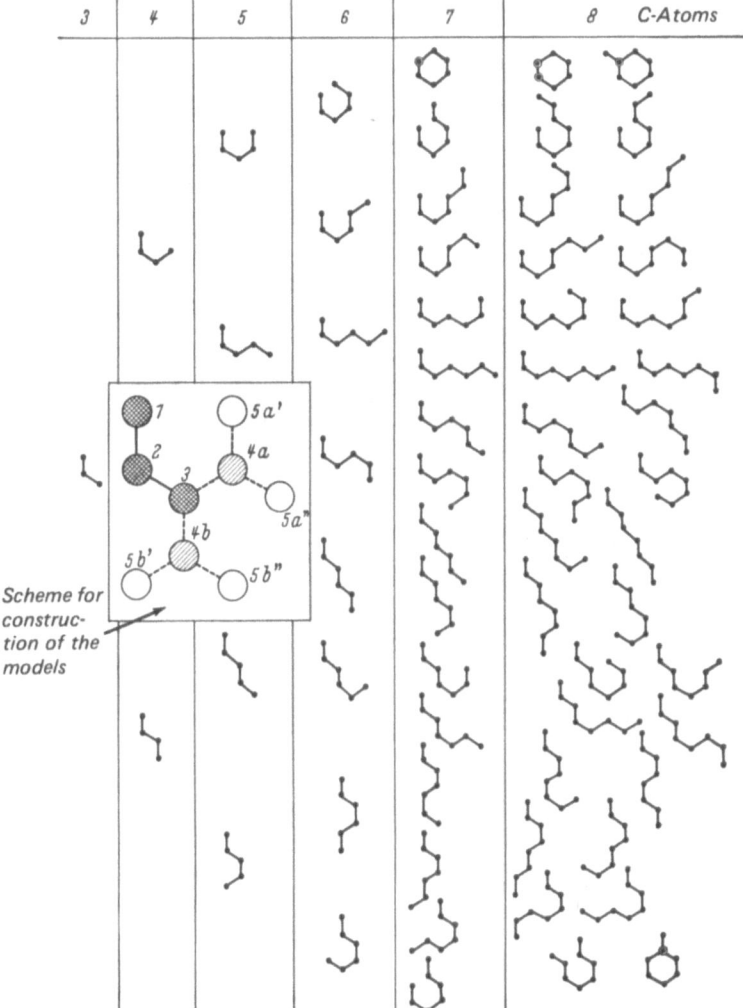

FIG. 439 — Possible ways of constructing the different conformations of a C_8-chain.

As Figure 439 shows, the number of molecules which arise as a result of the addition of a further atom (they do not have all different forms) grows according to 2^{n-3}, where n is the number of the carbon atoms forming the chain of the macromolecule.

With 4,000 atoms (consider, for example, a polyethylene with the degree of polymerization 2,000), one obtains roughly $2^{4,000} = 10^{1,200}$ chain molecule models, among which the fully stretched chain is represented only a single time. All other models are more or less strongly coiled. Even if the entire solar system would consist of polyethylene (of P = 2,000), one would still not have much expectation of finding molecules in their fully stretched chain form.

If one wants to picture how the most likely shape of a long chain molecule looks, one can make up a spatial, or even simpler, planar model by throwing a die. If one starts with a chain of three atoms, depending on the number thrown, the next atom is added in the a- or b-position (1, 2, and 3 = a; 4, 5, and 6 = b). In carrying out such a model experiment, one finds that longer, straight chain segments, which would occur if one alternatively throws always a and b for a longer time, hardly ever occurs. Actually any kind of regularity in the sequence of a and b is quite rare. As the result of such an experiment, with a sufficiently large number of throws, one always obtains a statistical coil, i.e., its two dimensional planar analogue, as shown in Figure 440).

Figure 440 shows such an experiment for 1,000 throws. The statistical coil is the most probable shape of a long chain molecule, and the form of the macromolecules therefore leads us into a discussion of the properties of a statistical coil.

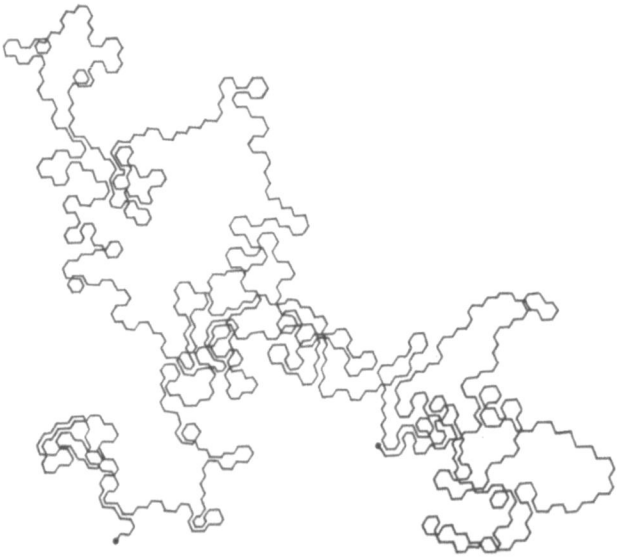

FIG. 440 — Statistical coil model constructed by throwing dice

3211   The Quantitative Description of the Statistical Coil with the Aid of Random Flight Statistics

W. Kuhn (1934) and H. Mark (1935) first recognized the coiling of macromolecules. Kuhn had the idea to describe the statistical coil with the aid of random flight statistics. The random flight problem is well known to the chemist through Brownian motion (compare Figure 441). Thus, if a small particle, from one point of reversal of its irregular zig-zag motion to the next covers on the average a distance Δx and has, after Z deflections from its direction of motion, reached from A to the point B, then the distance x between A and B is given by:

$$\overline{x^2} = Z\,(\overline{\Delta x})^2. \tag{123}$$

For the application of this equation to statistical coils there is only one obstacle: the fixed valence angle (with carbon-carbon and carbon-oxygen approximately 110°) given by the tetrahedral structure of the C atoms leads to the result that the direction of the next carbon-carbon axis is predetermined by the position of the preceding carbon-carbon axis on the surface of a cone, i.e., it is limited to three energetically preferred directions in this surface (See Figures 437 and 442). Thus one first has to represent the coil by a segment model, where enough structural units are combined into a chain segment of the length A_m, such that the condition of completely free rotation in all directions is fulfilled. How many carbon atoms or structural units are required before this is the case cannot be determined from the size of the valence angle, because one cannot know how large the number of the permitted spatial directions must be (after 1 carbon atom there are 3, after 2 carbon atoms 3^2 after n carbon atoms 3^n, etc.) before the mobility can be considered as completely unhindered in the sense of random flight statistics. This determination results, however, directly from the random flight statistics equation,

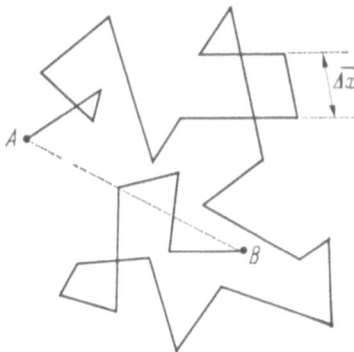

FIG. 441 — Zig-zag path of a gas molecule (Brownian motion).

if one applies it to a certain polymer and knows how large the fully stretched chain is (L_{max}) and the average end-to-end distance ($\sqrt{\overline{h^2}}$) of the coiled chain:

$$\overline{h^2} = Z \cdot A_m^2 \quad \text{or} \quad \sqrt{\overline{h^2}} = A_m \cdot \sqrt{Z}. \tag{124}$$

where $\overline{h}$ is the average distance between chain ends, A_m is the length of the statistical chain segment, and Z is the number of chain segments.

With Brownian motion [compare Equation (123)] the corresponding parameters are $\overline{x}$ the average distance between A and B; $\overline{\Delta x}$, the average length of the distance

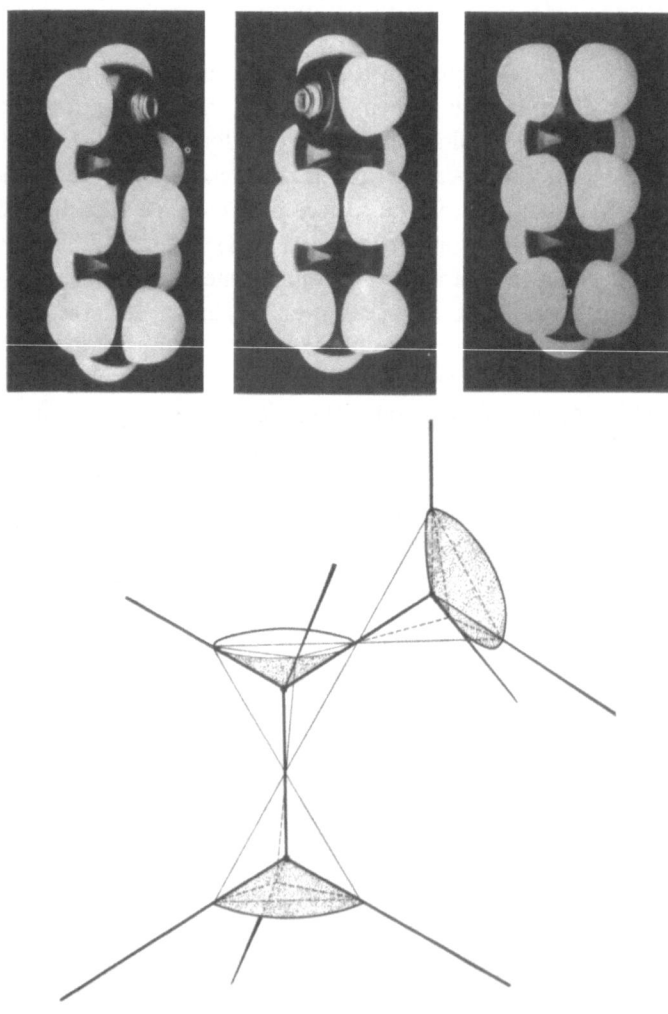

FIG. 442 —The significance of the valence angle for the form of the C–C -chains.

between two changes in direction; and Z; the number of changes in direction, i.e., collisions, along the path from A to B.

Since $Z \cdot A_m$ is nothing other than the maximum chain length L_{max},

$$L_{max} = Z \cdot A_m, \qquad (125)$$

one can replace Z by L_{max}/A_m and obtain:

$$\overline{h^2} = L_{max} \cdot A_m \quad \text{and} \quad \sqrt{\overline{h^2}} = \sqrt{A_m} \cdot \sqrt{L_{max}}. \qquad (126)$$

Figure 443 demonstrates the transition between the valence angle coil and the segment coil (segment model) according to Equations (124) and (126). Diagram 1 in Figure 443 represents, in the same way as Figure 439, a valence angle coil (valence angle in the planar models is 120° instead of 110° with actual molecules). The length 1 of a structural unit in the model is 5 mm. With "degree of polymerization" = 22, one therefore has L_{max} = 110 mm. The distance of the chain ends $\sqrt{h^2}$ is 28 mm. According to Equation (126), A_m is given by $A_m = h^2/L_{max} = 784/110$ mm. A_m is therefore approximately 7 mm. The number of chain segments of length A_m = 7 mm is $Z = L_{max}/A_m = 110/7 \cong 15$. In diagrams 2 to 7 (in Figure 443), several segment models are shown which obey these conditions, h^2 = 28 mm, A_m = 7 mm,

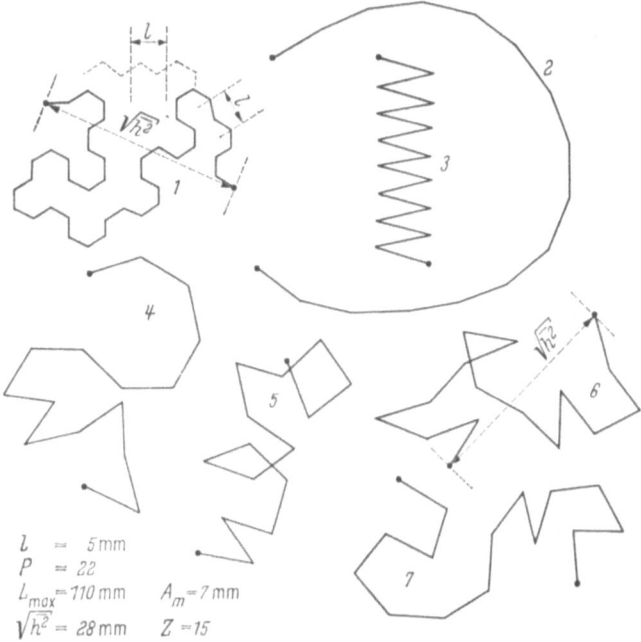

$$l \quad = \quad 5\,mm$$
$$P \quad = \quad 22$$
$$L_{max} = 110\,mm \qquad A_m = 7\,mm$$
$$\sqrt{h^2} = 28\,mm \qquad Z = 15$$

FIG. 443 — Replacement of the valence-angle-coil by a segment model (random flight model).

and Z= 15. Equation (126) does not give any information about the behavior of the line in these drawings. However, the basic condition for the validity of Equations (124) and (126) is that the direction of each successive segment is completely determined by accident and therefore the models number 2 and number 3 must be eliminated as possible models for real chain forms. These forms are just as improbable as the maximally stretched chain with $\sqrt{\overline{h^2}} = L_{max}$. For the direction of each successive segment, one must assume that no direction is preferred to any other. This implies that flat and sharp angles and the actual deviation from 90° are on the average equally probable. Furthermore, turns in a clockwise and a counter-clockwise manner must occur equally often in the model.

The end-to-end distance of the chain ends does not have any particular relation to the maximal length of the chain for any individual coil, and it is also not related to the size of the coil. It is not impossible that the chain ends with a very large coil might find themselves next to each other, whereas with a much smaller coil they might have a relatively large distance. This is very improbable, however, as can be readily understood from the above model. Thus if in diagram 1 one assumes the dashed curve instead of the heavy curve, then one obtains for $\sqrt{\overline{h^2}}$ the value of 6 mm instead of 28 mm. For the values of A_m and Z according to Equation (126), one obtains A_m = 0.325 mm and Z = 340. If one then tries to draw a model with these values according to diagrams 5, 6, or 7, where sharp and flat angles occur equally often, one obtains considerably tighter, and, with respect to the maximum diameter, considerably smaller coils. Coils which in their density and size correspond to diagram 1 can only be drawn under strong violation of the conditions for validity of Equation (124), and are therefore completely improbable and deviate considerably from the average. Segment models with $\sqrt{\overline{h^2}}$ = 6 mm, A_m = 0.32, and Z = 340 have obviously nothing to do with valence angle coil of diagram 1, because A_m is smaller than the length 1 of the structural unit.

One sees from this that $\sqrt{\overline{h^2}}$ as it is defined by Equation (124)—even though in the individual case it bears no relation to the coil size—can with a sufficiently large number of coils be taken as a measure for the average coil diameter. Furthermore, the random flight equations (124), or (126), if they are assumed to be valid, always lead to a certain coil size and coil density: If for a given length L_{max} the end-to-end distance $\sqrt{\overline{h^2}}$ is small, A_m is also small and the coil must be relatively tight. If $\sqrt{\overline{h^2}}$ for the same L_{max} is larger, than A_m must also be larger and the coil less tight, so that the end-to-end distance and the maximum coil size always remains in the same relationship to each other.

Since the carbon-carbon distance with polymer chains has been very exactly determined through X-ray measurements, it is possible to calculate L_{max} from the degree of polymerization. $\sqrt{\overline{h^2}}$ can be determined from light-scattering, so that one can then calculate A_m according to Equation (126). Polymers with flexible chains (most vinyl polymers) have small A_m values, and those with stiff chains (for example, cellulose derivatives) have large A_m values.

3212 The Dependence of the Coil Diameter $\sqrt{\overline{h^2}}$ and the Coil Density on the Molecular Weight M

For the experimental examination of the question whether macromolecules in solution actually behave as statistical coils, one studies the change of $\sqrt{\overline{h^2}}$ as a function of $\overline{M}$. Since the segment length A_m for a certain polymer is constant in a particular solvent, and the maximum chain length L_{max} is proportional to the degree of polymerization and therefore to the molecular weight, it follows from Equation (126) that:

$$\sqrt{\overline{h^2}} = \text{const.} \cdot \sqrt{\overline{M}}. \qquad \text{(Kuhn's square-root law).} \qquad (127)$$

With polymers of pure C–C chains, the monomer length $l_M = 2.52$ Å was determined by X-rays, so that $L_{max} = 2.52 \cdot \overline{P}$. Then one can write for the chain end-to-end distance $\sqrt{\overline{h^2}} = \sqrt{A_m \cdot 2.52} \cdot \sqrt{\overline{P}}$, or in general:

$$\sqrt{\overline{h^2}} = \sqrt{A_m \cdot l_{Mon}} \cdot \sqrt{\overline{P}} = \sqrt{\frac{l_{Mon}}{M_{Mon}} A_m} \cdot \sqrt{\overline{M}_{Pol}}. \qquad (127a)$$

Between compact spheres of the same material and statistical coils there is therefore a principal difference: with ordinary spheres the diameter changes as a function of $m^{1/3}$, but with statistical coils as a function of $\overline{M}^{1/2}$.

From this one can derive a property of the statistical coil which determines the entire behavior of the macromolecular solution, namely the dependence of the average coil density on the molecular weight: the density of an average sphere is $\rho_{sphere} = m/(1/6\pi d^3)$. Quite similarly one can define the average coil density, where m is only the mass of the pure chain molecule, i.e., which does not contain the mass of the solvent taken up by the coil. Of course, one could also take into the calculation the mass of the solvent. However, this would not be easy to do, since the total mass of the coil is not determined very easily. If N_A is the number of molecules in 1 mole (i.e., Avogadro's number) then one can write for the mass of a single chain molecule: $m = M/N_A$ and for the average coil density one obtains:

$$\overline{\rho}_{coil} = \frac{\overline{M}}{\frac{1}{6}\pi d^3 N_A}. \qquad (128)$$

Since one can not expect a coil to be completely spherical, (a sphere is a very improbable form, which in nature is found so frequently only because it is the result of surface tension; but coils have no, or only extremely small, surface tension), the equivalent coil diameter d is not equal to $\sqrt{\overline{h^2}}$, because it is not the diameter of a sphere, but it is proportional to $\sqrt{\overline{h^2}}$ and one can write:

$$\overline{\rho}_{coil} = \text{const.} \cdot \frac{\overline{M}}{(\sqrt{\overline{h^2}})^3}. \qquad (129)$$

If one takes into account that $\sqrt{\overline{h^2}}$ is proportional to $M^{0.5}$, one writes for the coil density:

$$\bar{\rho}_{coil} = const. \cdot \frac{\overline{M}}{(\sqrt{\overline{M}})^3} = const. \cdot \frac{1}{\sqrt{\overline{M}}} . \tag{130}$$

With compact particles the density is independent of the particle size (if one does not take into consideration "particles" of the size of stars!). However, with coils of the same material the density is a function of the molecular weight: the average coil density is inversely proportional to the square root of the molecular weight. The coils of one and the same polymer are therefore the looser, the greater the molecular weight of the polymer and vice versa.

Because the form of statistical coils is irregular, one cannot give an absolute value for the coil density (compare pp. 505-510). However, by means of viscosity and sedimentation measurements with dilute polymer solutions, one can determine how large the diameter and density of a sphere would have to be in order to give the same behavior with regard to viscosity or sedimentation as the actual coils present in the solution. In this sense one talks about an equivalent sphere, and one can also define average coil density:

$$\bar{\rho}_{equ} = \frac{\overline{M}}{(1/6)\, \pi\, d^3_{equ}\, N_A} \tag{131}$$

which can be determined very easily by means of viscosity measurements. Because of the elongated form of the coil it is not immediately possible to compare the diameter of the equivalent sphere d_{equ} with an analogous parameter of the real coil. However, the average coil density $\bar{\rho}$ is a term which can be transferred without any limitations to the coils actually present. The equivalent sphere density more closely corresponds to the actual average coil density, the more the equivalent sphere corresponds to a sphere of the same volume as the coil. As we will see later on, the experimentally determined diameter of a sphere equivalent to the coil with regard to its sedimentation velocity or viscosity increase (d_{equ,s_0}, $d_{equ[\eta]}$) is usually larger than the diameter of a sphere equal in volume to the coil ($d_{equ(vol)}$). Therefore, the real average coil density is usually somewhat larger than the experimentally determined equivalent coil density, which, for example, for a polymethylmethacrylate, or polystyrene of $M = 1,000,000$ is approximately 0.02 g/cm^3, and for a cellulose molecule of the same molecular weight, approximately 0.001 g/cm^3. From this one can see that the chain molecule itself occupies only a very small part of the coil volume.

In giving the coil density, one also knows the degree of coiling in a quantitative manner. If one would like to define in detail a particular degree of coiling, one can make use of the quotient $L_{max}/(\sqrt{\overline{h^2}})^3$ which is proportional to the coil density. However, it is completely wrong to define the degree of coiling, Q, by the quotient

$L_{max}/(\sqrt{\overline{h^2}})$, as is often done in the literature. This value of Q is not only unnecessary, but also misleading, because it increases proportionally with the square root of the molecular weight ($Q = \text{const.} \cdot M^{0.5}$), as one can easily see if one replaces L_{max} by $K \cdot \overline{M}$ and $\sqrt{\overline{h^2}}$ by $K \cdot \sqrt{M}$. This means that the degree of coiling defined in this manner would paradoxically become larger for a certain polymer, the looser the coil, because the density of the coil decreases with increasing $\overline{M}$. The definition $Q = L_{max}/\sqrt{\overline{h^2}}$ would apply to a chain which is linearly compressed, for example, if it has a spiral form, or some otherwise folded chain, whose length changes only in *one* dimension, like a concertina. This would not apply to a statistical coil whose chain is distributed throughout a well-defined *volume*.

Each quantitative value for the coil density is from all points of view an average value. First, the volume of a coil is not filled in a homogeneous manner by the polymer chains, but the density decreases in going outward and the statistical coil has a density distribution curve similar to a Gaussian curve. Second, the density and the density distribution of the coil change constantly because of the Brownian movement, and the average coil density is a value around which, over a sufficiently long period of time, the actual density of the coil fluctuates. Third, the density is different from coil to coil, and one deals with an average which refers to a large number of different coils. The density of the individual coils is different because macromolecular compounds are usually polydisperse, i.e., they consist of a mixture of molecules of different size. Since the coil density according to Equation (130) depends on the molecular weight, the density of the individual coils cannot be the same.

3213 The Geometric Shape of the Coils

With the observation that in general linear macromolecules are not present in solution as stretched-out chains, but as irregular and, depending on the chain structure, more or less dense coils, the question of molecular shape really becomes a question of the shape of the coils. With the shape of the coils, the picture is exactly the same as with the coil density: it is not possible to ascribe one particular shape to an individual coil. Within the limits of the chain structure, it can have any desired shape. However, it will not maintain a particular shape over a longer period of time—because of the heat movement there is a constant change of shape determined by the flexibility of the atoms in the chain and the temperature. However, there is a most probable form which can be most frequently found with a sufficiently large number of coils, or, if we look at the single coil: over a sufficiently long period of time, there will be a shape which this coil will prefer to assume most frequently. As model considerations and model experiments have shown, the most probable form can be best represented as a bean-like (irregularly ellipsoidal) structure, because for such an irregular overall shape (smaller irregularities are not at all considered here), there is the largest number of possible formations. By means of a large number of experiments with wire models fashioned according to the results of throwing dice, H. Kuhn determined for statistical coils the following axial ratios: $\overline{H}_1 : \overline{H}_2 : \overline{H}_3 = 1.36 : 0.78 : 0.50$. For the ratio of the maximal length of the coil, $\overline{H}_1$, to the root-mean-square end-to-end distance $\sqrt{\overline{h^2}}$ the wire models give the value of 1.25:

$$\overline{H}_1 = 1.25 \cdot \sqrt{\overline{h^2}}$$

3214   The Problem of Solvent Streaming through the Coil (Draining)

The low density of the statistical coil, whose volume is only filled to a small percentage by the polymer chain, leads to the assumption that the coil, or at least a part of the coil, is more or less freely streamed through (or drained) by the solvent. This assumption, and especially the conclusion which can be derived from this as far as the viscosity of macromolecular solutions is concerned, has been examined by Debye and Bueche, by Kirkwood and Riseman, and by Flory and Fox, and their theoretical work has been discussed extensively in the polymer literature. Up to now, there has been no experimental confirmation for total, or even partial, draining of the coils. Instead, the experimental study of sedimentation of dissolved macromolecular compounds in the ultracentrifuge has shown that these coils behave as closed undrained ellipsoids (gel particles). Also the linear relationship between $[\eta]$ and M in the double logarithmic plot for all macromolecular compounds (in the range of about $M = 10^4$ to $M = 10^7$) can not be reconciled with a partially drained coil (compare p. 470).

322   The Real Statistical Coil

3221   Space Filling and Energetic Interaction

The statistical (random) coil, as it is described by random flight statistics, is an ideal structure from which real coils, depending on the constitution of the chain and on the solvent and the temperature (even considering only extremely dilute solutions), deviate more or less. Among the random flight conditions is the requirement that the line can cross its own path as often as it wants to, and that the individual segments of the random flight path do not influence each other in any way. The actual chain molecule can obviously not fulfill these requirements exactly, because (1) the chain has a certain volume of its own and therefore a particular point in space which is already occupied by the chain is no longer accessible to another chain segment (restricted volume effect), and (2) the different parts of the chain molecule (chain segments) are partners in the solvation equilibrium (energetic interaction). According to the constitution of the chain and the type of solvent, the chain segments may have the desire to come together in the form of intramolecular aggregates and to avoid interaction with the solvent (so that the coil assumes a more dense form). However, the opposite case can also occur, that the structural units of the chain prefer combination with the solvent molecules (i.e., if the solvated state of the chain is an energy-poorer state), and therefore the individual chain segments will try to avoid close contact with each other and the coil will assume a larger shape. With a given polymer, in the first case, where there is a tendency for the coils to become denser and also to come closer together, one speaks of a *poor* solvent, and in the second case, where the coils are enlarged because of the strong solvation tendency, one speaks of *good* solvent. The chain-stiffening effect of good solvents can be confirmed by ring-closing reactions, which occur more rapidly in poor solvents than in good solvents.

The tendency to associate through secondary valence forces, which occur through the interaction of different segments in one and the same coil, and the enlargement of the coil through the chain-stiffening action of solvation (secondary valence forces: chain to solvent), are always in equilibrium with each other. With good solvents the coil-expanding effect is predominating, with poor solvents the contracting effect. There are solvents which for a certain polymer at a certain temperature show a balance between these two effects, so that the coil appears to be without the influence of other forces (i.e., this is a pseudo-ideal solvent or θ solvent). With increasing temperature, one usually finds that solvation becomes predominant, whereas with lowering of the temperature the attraction between the chain segments becomes more important. In principle, therefore, for each solvent for a definite polymer, there is one temperature where the two forces (solvation and association forces) are of the same magnitude and therefore their effects cancel out (θ temperature or Flory temperature). In θ solvents, or generally at the θ temperature, the coil is therefore without the influence of outside forces, and the laws of the ideal statistical coil will hold: $\bar{p}_{\text{coil}} = \text{const.} \cdot M^{-0.5}$ and $\sqrt{\overline{h^2}} = \text{const.} \cdot M^{0.5}$ [Equations (130) and (127)]. As will be discussed again further on, this has the result that the exponent a of the $[\eta]$-M function is equal to 0.5 (compare p. 456). Furthermore, the second virial coefficient of the osmotic pressure is zero (compare pp. 343 and 535).

Good solvents can be easily recognized by the fact that through the dissolution of a polymer in the liquid, the viscosity is increased very strongly, whereas poor solvents show a lower viscosity increase. An especially strong loosening-up of the coil (perhaps even maximum stretching of the chain molecules) is found with polyelectrolytes, i.e., with solutions of salts of polymeric acids or bases (for example, polyacrylic acid or polyuronic acid), because the groups of the same charge along the chain repulse each other electrostatically, and this leads to an energetic preference of a stretched-out chain form.

A stiffening of linear chain molecules can also be brought about by the primary chain structure (1) by inflexible structural units such as aromatic or heterocyclic ring units (compare pp. 225-230) and (2) by steric hindrance caused by large side groups in the chain, which prevent free rotation and therefore only allow certain positions of the chain atoms with respect to each other. Such real restriction in the free rotation must be distinguished from a simple increase in the difficulty of free rotation, such as results from the increase in the potential energy barrier which separates the three stable positions of neighboring chain atoms. Since in this case all positions have the same probability, the density of the coil is not influenced; only the flexibility of the coil is, in the sense of the velocity of transition of one chain formation into another. This flexibility of the chains, i.e., the movement of the chain segments within the coil, which can also be described as *intra*molecular Brownian motion (in contrast to *inter*molecular Brownian motion, which refers to the movement of the entire coils with respect to one another) is of greatest importance for the physical properties, especially the mechanical behavior of the polymer. Thus, for example, for rubber elasticity, a degree of chain flexibility is absolutely required. Polymers with restricted free rotation are hard and often brittle.

The deviation of the real coil from the random flight path because of excluded volume and energetic interaction leads not only to the result that the density of the real coil deviates to lower values than the density of the statistical coil. One also has to expect that the dependence of the coil diameter and the coil density on the molecular weight is more or less different from that of the ideal function [Equation (130)]. For example, one should assume that the exponent may be larger or smaller than 0.5:

$$\rho_{coil} = const. \cdot M^{-a}, \tag{132}$$

or that the exponent a depends on the molecular weight. This will be treated more extensively in the following section and in discussing macromolecular solutions.

3222   Experimental Determination of Coil Properties

Determination of the Coil Density from the Viscosity Number

As can be seen further on in the discussion of the behavior of the coil in sedimentation in the ultracentrifuge, the solvent sucked up by the coil is held so tightly that the entire structure consisting of the chain molecule and the included liquid sediments as a complete unit, and thus behaves like a small gel particle. One can describe the viscosity of such a solution by the Einstein law for a solution of spherical colloidal particles (compare also p. 375 and pp. 501-510):

$$\eta_{rel} = \eta/\eta_0 = 2.5\varphi + 1, \tag{133}$$

where η is the viscosity of the solution; η_0 is the viscosity of the solvent; and φ is the volume fraction of the dissolved phase (= volume of the dissolved phase/volume of the total solution = v_m/v_{soln}).

If one describes the volume of the dissolved phase v_m by the mass and the density of the dissolved phase, i.e., $v_m = m/\rho$ one obtains $\varphi = m/\rho \cdot v_{soln}$, or $\varphi = (1/\rho) \cdot c$, because m/v_{soln} is nothing other than the concentration defined as the mass dissolved in 1 cm^3 of solution, if one measures the volume of the solution v_{soln} in cm^3. One can also describe the Einstein law as follows:

$$\eta_{rel} - 1 = 2.5 \frac{1}{\varrho} \cdot c,$$

or if one replaces $\eta_{rel} - 1$ by $\eta_{sp} = (\eta - \eta_0)/\eta_0$ and calls the $\lim_{c \to o} \eta_{sp}/c$ the viscosity number $[\eta]$, one obtains:

$$[\eta] = 2.5/\rho. \tag{134}$$

$\eta_{sp}/c = [(\eta-\eta_0)/\eta_0]/c$ is the relative viscosity increase for a concentration c = 1 (that is, with reference to the viscosity of the solvent η_0 as unity). If one measures the concentration in grams per cm^3 of solution, one obtains the viscosity number $[\eta]$ in cm^3 per gram.

Since one does not know the mass of the gel coils (chain molecules + solvent carried along), but only the mass of the chain substance (weight of the polymer in preparing the solution), one also has to refer the concentration c to this mass, and the density ρ is therefore the density of the empty coil without the included solvent. The mass of the polymeric chain molecule is not distributed homogeneously throughout the coil volume. Furthermore, all coils do not have the same density. Therefore, the coil density is a complicated average value. We should also remember that the coils are not completely spherical, but that they have the form of irregular elongated ellipsoids. The Einstein law, however, was theoretically derived for spherical colloidal particles, and if one substitutes a measured viscosity number into it, it gives the density of spherical colloidal particles which would have the same viscosity number as the actually measured system. In other words, the value for the average coil density which one obtains from the Einstein viscosity law is the density of an equivalent sphere producing the same viscosity increase as the coil:

$$[\eta] = 2.5/\bar{\rho}_{equ}. \tag{135}$$

TABLE 451. Viscometrically determined coil-densities

Polymer	Solvent	$\bar{M}_w$	$[\eta]_{20^0}$ cm^3/g	$\bar{\rho}_{equ}[\eta]$ g/cm^3	$d_{equ}[\eta]$ Å
Polymethyl- methacrylate	Acetone	10^6	110	0.0227	520
Cellulosenitrate. . . .	Acetone	10^6	2800	0.00090	1520
Polyisobutylene . . .	Cyclohexane	10^6	410	0.0061	810
Polyisobutylene . . .	Benzene	10^6	83	0.03	470
Polystyrene	Benzene	10^6	290	0.0086	715
Polystyrene	Methylethylketone	10^6	80	0.031	465
Glycogenacetate . . .	Chloroform	10^6	10	0.025	233

If the deviations from the spherical shape are not considerable, one can consider the equivalent sphere as a sphere, which would result if one would take the coil and mold it into a sphere of the same volume as the coil. The closer this is to reality, i.e., the more the equivalent sphere coincides with a sphere of the same volume as the coil, the closer the equivalent coil density corresponds to the real average coil density, i.e., the density which results if one distributes the mass of polymer corresponding to a concentration c uniformly over the total volume required by the coils. As becomes apparent by comparison of sedimentation, viscosity, and light-scattering, the diameters of the equivalent sphere obtained by the different experimental methods do not coincide with one another and are usually somewhat larger than the diameter of a sphere equivalent in volume to the real coil, so that the $\bar{\rho}_{equ[\eta]}$ values are usually lower than the actual density of the coils (compare pp. 505-510).

This fact shows clearly that the attempt to describe the random coil by means of an equivalent sphere is not completely satisfactory. The same holds true for other models, which have been developed by different authors. All of this can describe the experimentally observed phenomena approximately but not exactly. In this book I have preferred the model of the equivalent sphere because of its clearness and relative simplicity. A more detailed quantitative description which considers the Gaussian distribution of the density of the random coil is given, for example, in theoretical papers by Flory and Fox or Stockmeyer, Zimm, and Fixmann, or more generally in some monographs on polymer physics.

In the discussion of the viscometric molecular weight determination, how one can obtain the viscosity number $[\eta]$ in a simple way from a series of viscosity measurements by extrapolation was described (compare p. 376). Table 451 gives $[\eta]$ values for a number of polymers and the corresponding [according to Equation (135)] calculated $\bar{\rho}_{equ[\eta]}$ values.

As the values show, the average coil density with all polymers, except for glycogenacetate, is considerably smaller in comparison to the density of normal solid or liquid substances. This means that the statistical coils are, therefore, rather loose structures whose volume is only filled to a few percent, or even fractions of a percent, by the polymer chain. All other material is solvent which has been taken up by the coil. One finds especially loose and extended coils with cellulose nitrate and with polyamidecarboxylic acids. Apparently these molecules are relatively stiff, and the values for A_m, and therefore also for $\sqrt{\bar{h^2}}$ [Equation (126)], are therefore rather high. One finds the opposite extreme of a very tight coil with glycogen, which is a highly branched water-soluble polysaccharide. With the values for polyisobutylene and polystyrene, one clearly sees the influence of the solvent on the flexibility or stiffness of the chain: in cyclohexane for polyisobutylene, and benzene for polystyrene, the coils are more extended because of solvation than in benzene and methylethylketone respectively. The better the solvent, the lower the coil density.

The coil density is also influenced by the temperature of the solution. In general, the coil density decreases with increasing temperature. Thus, the same solvent can be, for a given polymer, better or poorer depending on the temperature. This has an important technical use in the improvement of lubricating oils, whose viscosity decreases rapidly with temperature. Thus, if one dissolves a suitable oil-soluble polymer (usually complicated copolymers of acrylic esters or methacrylic esters) in such an oil at low temperatures ($0°C$-$20°C$), in view of the relatively high coil density, the viscosity increase is only very small. On the other hand, at higher temperatures ($80°C$-$100°C$), where the viscosity of the pure oil is undesirably low, the coils of the dissolved polymer have—because of the better solvation—a relatively low coil density. Therefore, the increase in viscosity is considerable, so that the strong viscosity decrease of the pure oil caused by raising the temperature is partly balanced by the viscosity increase caused by the polymer. Therefore, the viscosity-temperature curve is more flat.

The coil density may be determined without knowing the molecular weight. It is obtained very simply from Equation (135) as the average density of the dissolved phase. If one knows the molecular weight, one can calculate from the equivalent density or from the viscosity number (which, of course, is nothing other than 2.5 times the reciprocal coil density) the diameter of the equivalent sphere, $d_{equ[\eta]}$, ([η] measured in [cm^3/g]):

$$\bar{\rho}_{equ\,[\eta]} = \frac{\bar{M}}{N_A} \cdot \frac{1}{1/6\,\pi\,d^3_{equ\,[\eta]}} \tag{136}$$

and

$$d_{equ[\eta]} = \left(\frac{\bar{M} \cdot [\eta]}{2,5 \cdot 1/6 \cdot \pi \cdot N_A}\right)^{\frac{1}{3}} cm = 1,08\sqrt[3]{\bar{M} \cdot [\eta]}\ \text{Å}. \tag{137}$$

One can picture the low density of the coils in an especially simple way, if one considers the swelling of slightly cross-linked polymers. Figure 453 shows samples of polystyrene that have been cross-linked with 0.05% divinylbenzene, in both the solid and the swollen state. The coil density (density of the swollen phase), is here simply the "empty" density of the gel, which is calculated from the mass of the solid polymer and the volume of the swollen gel, because this gel is nothing other than a (depending on the degree of cross-linking) more or less tight packing of the

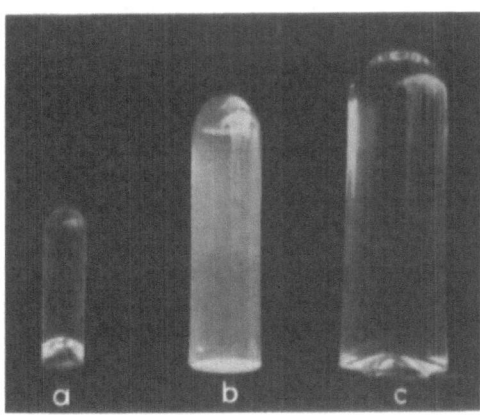

a) without swelling agent

b) after maximum swelling in methylethylketone

c) after maximum swelling in benzene

FIG. 453(a-c) — Polystyrene crosslinked with 0.05% divinylbenzene.

individual solvatized gel particles to a macroscopic gel. The swelling is limited because the individual coils are tied to each other through the cross-links.

One sees that the polystyrene samples are distinctly more swollen in benzene than in methylethylketone. This results from the fact that the coil density in benzene is lower than in the poorer solvent methylethylketone (compare Table 451). In principle one can determine the coil density from swelling measurements:

ρ_{coil} = mass of the cross-linked polymer/volume of the swollen sample.

However, one has to remember that the density of the cross-linked coils is higher than that of the uncross-linked coils.

The Dependence of the Coil Density on the Molecular Weight

The molecular weight dependence of the coil density is closely related to the nature of the statistical coil. For ideal statistical coils, the random flight equation yields a square-root dependence:

$$\rho_{coil} = \text{const.} \cdot \overline{M}^{-0.5}, \tag{138}$$

which cannot be expected to hold for real coil systems. Since the form of the energetic interaction is different, depending on the constitution of the polymers and the type of solvent, one should expect different $\overline{\rho}$-$\overline{M}$ functions from case to case. Since according to the Einstein viscosity law the viscosity number $[\eta]$ is

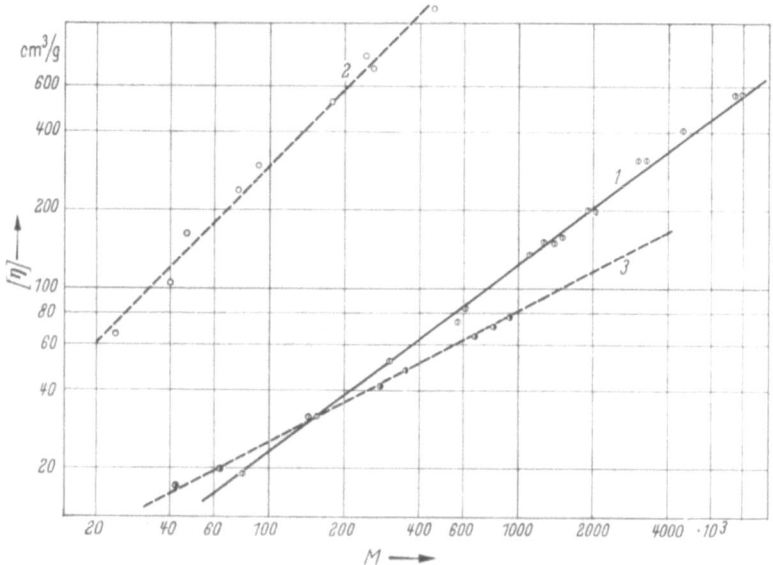

tan α (1) = 0.73;   tan α (2) = 1.0;   tan α (3) = 0.5

FIG. 454 − $[\eta]$ vs. M plot for 1) polymethylmethacrylate in acetone; 2) cellulose nitrate in acetone; 3) polystyrene in cyclohexane (according to SCHULZ and MEYERHOFF)

inversely proportional to the particle density, the change of the viscosity number with the molecular weight can be considered as an immediate result of the change of the density of the coil with the molecular weight. The viscosity function $[\eta]$ = f(M) is therefore at the same time a quantitative expression for the density function, because one can replace the viscosity number by "const. $1/\bar{\rho}_{[\eta]}$." Through many experimental investigations it has been known for some time that the viscosity number increases with the molecular weight according to the power law:

$$[\eta] = \text{const.} \cdot \overline{M}^a. \tag{139}$$

If one replaces $[\eta]$ by "const./$\bar{\rho}_{equ[\eta]}$" according to Equation (135), then one can write:

$$1/\bar{\rho}_{equ[\eta]} = \text{const.} \cdot \overline{M}^a$$

or: $\tag{141}$

$$\bar{\rho}_{equ[\eta]} = \text{const.} \cdot \overline{M}^{-a}.$$

Thus solutions of real coils deviate from the ideal law [Equation (127)] theoretically derived from the Einstein law and the random flight equation, only because the exponent is not always equal to 0.5, but usually to 0.6, or 0.7, or in general a. By determination of the molecular weight $\overline{M}$ and the viscosity number $[\eta]$, with a series of fractions with decreasing molecular weight, one can determine

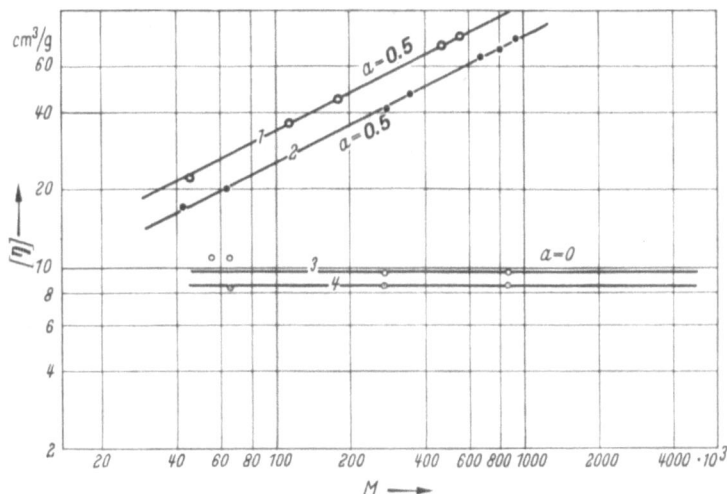

FIG. 455 — $[\eta]$ vs. $\overline{M}$ plots for 1) polyisobutylene in benzene at 24°C (FLORY) tan α = 0.5; 2) polystyrene in cyclohexane at 34°C (H.J. CANTOW) tan α = 0.5 3) glycogen acetate in chloroform at 20°C (STAUDINGER and HUSEMANN) tan α = 0;4) glycogen in formamide at 20°C (STAUDINGER and HUSEMANN) tan α = 0.

the exponent a by means of a double logarithmic plot of $[\eta]$ versus $\overline{M}$, such as is shown in Figure 454 for a number of polymers. Through transformation into logarithms Equation (139) becomes the equation of a straight line: $\log[\eta] = a \cdot \log\overline{M} + K$, so that one can obtain the constant a as the tangent of the slope from this diagram.

One can see that the exponent a for the same polymer and for the same solvent is constant over a large molecular weight region, i.e., from 50,000 to several millions. However, the value of the exponent a varies from one polymer to another and from one solvent to another. (Compare the a values in Table 380, also the $[\eta]$ versus $\overline{M}$ curves in Figure 379). With the especially loosely coiled (stiff) cellulose molecule, the coil density decreases very rapidly with increasing molecular weight ($a = 1$), and with the very tightly coiled glycogen, the coil density does not change at all with the molecular weight ($a = 0$). Figures 520 and 521 show the influence of the solvent: in the good solvent, cyclohexane, the coil density of polyisobutylene decreases wih increasing molecular weight more rapidly than in the poor solvent, benzene. The rate at which the density changes with the molecular weight also depends on the temperature of the solution. Thus the same solvent can be a poor solvent at 20°C, and a good solvent at 100°C. This means that at 20°C, the coil density (and therefore the viscosity number) changes only slightly with molecular weight or not at all. On the other hand at 100°C, the coil density decreases rapidly, and the viscosity therefore increases rapidly. Thus, it is possible to find for many solutions a temperature at which the exponent a has the value 0.5, where the coils behave like ideal coils (Flory temperature or θ temperature). This state is characterized by the fact that the tendency for solvation and the tendency for association of the chain segments are exactly balanced.

There are no confirmed cases known in which no constant value was found for the exponent a in a molecular weight range where the polymer chains have the conformation of random coils. With molecular weights below 10,000 this condition is satisfied less the more the molecular weight drops below 10,000. Oligomers with 10 or 20 structural units often have a rodlike shape (depending on the primary structure of the chain) and for rodlike particles the viscosity equation is $[\eta] = K_{[\eta]} \cdot M^2$, so that we have to expect a stronger upgrade of the logarithmic viscosity function with lower molecular weights. Therefore, the logarithmic $[\eta] - M$ plot can not be linear in the whole molecular region from 0 to ∞. However, linearity has been found for the most important range between $\overline{M} = 10^4$ to 10^7.

Determination of the Average End-to-End Distance $\sqrt{\overline{h^2}}$ and the Particle Shape by Light-Scattering[38]

If a light ray passes through a colloidal solution, a part of the incident light is scattered diffusely in all directions. If the colloidal particles are small in relation to the wave length of the incident light (diameter smaller than $\lambda/20$), then the intensity of the scattered light is equally great in all directions. But if the scattering

---

[38]Compare also section 3123.

particle (in macromolecular solutions, therefore, the coil) is larger than $\lambda/20$, then interference phenomena occur, because the scattering centers in one and the same molecule (coil), excited by a single primary light ray, are now so far from one another, that the scattered rays resulting from these centers suffer more or less large path differences, depending on the direction of the scattering. The larger the path difference (from 0 to $\lambda/2$), the greater their interference and the resulting diminution of the scattered light. As can be seen from Figure 457, the path difference is the larger, the larger the observation angle θ, so that the intensity of the scattered light always becomes smaller with increasing θ. The factor by which the intensity of the scattered light, i_o, which has not been decreased by interference (i.e., in the immediate neighborhood of the primary ray) becomes smaller at different angles, θ, is called the scattering-function P_θ. At infinite dilution and with polarized light perpendicular to the plane of observation, one can write:

$$i_\vartheta = P_\vartheta \cdot i_o. \qquad (142)$$

The diminution factor P_θ depends on the observation angle θ in a characteristic manner which depends on the particle shape (coil, sphere, or rod) as well as on the wave length and the particle size (particle size: with spheres d, with coils $\sqrt{\overline{h^2}}$ and with rods l).

For coils one can write:

$$P_\vartheta = \frac{2}{x^2}\left(e^{-x} - 1 + x\right)$$

$$x = \frac{8}{3}\,\pi^2 \sin^2 \vartheta/2\,\frac{\overline{h^2}}{\lambda^2}\,,^{*}$$

or approximately for small values of x:

$$\frac{1}{P_\vartheta} = 1 + \frac{8\pi^2}{9}\frac{\overline{h^2}}{\lambda^2}\sin^2 \vartheta/2 + \cdots. \qquad (143)$$

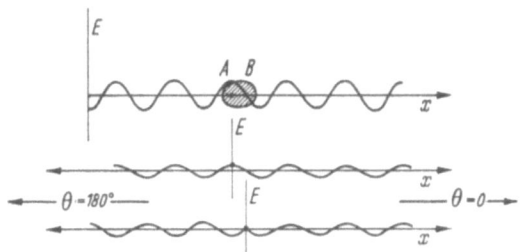

FIG. 457 — Diagram illustrating the reduction of the scattering intensity through interference with particles larger than $\lambda/20$.

$^{*}\overline{\lambda} = \lambda_o/n_o.$ See p. 351.

For spheres:

$$P_\vartheta = \left[\frac{3}{y^3} (\sin y - y \cdot \cos y) \right]^2$$

$$y = 2\pi \frac{d}{\lambda} \sin \vartheta/2,$$

or approximately for small values of y:

$$\frac{1}{P_\vartheta} = 1 + \frac{4\pi^2}{5} \left(\frac{d}{\lambda} \right)^2 \sin^2 \vartheta/2 + \cdots. \tag{144}$$

For rods:

$$P_\vartheta = \frac{1}{y} \int_0^{2y} \frac{\sin y}{y} \, dy - \left(\frac{\sin y}{y} \right)^2$$

$$y = 2\pi \frac{l}{\lambda} \sin \vartheta/2$$

or approximately for small values of y:

$$\frac{1}{P_\vartheta} = 1 + \frac{4\pi^2}{9} \left(\frac{l}{\lambda} \right)^2 \sin^2 \vartheta/2 + \ldots. \tag{145}$$

If one inserts in these theoretically derived equations (Debye, Neugebauer, and Zimm) for a certain angle θ (for the definition, compare Figure 350-1) different values of d/λ, or $\sqrt{\overline{h^2}}/\lambda$, or l/λ, one obtains the curves shown in Figure 459. The corresponding curves can be calculated for any desired angle. If one plots against the increasing values of D/λ ($D = d$, or $\sqrt{\overline{h^2}}$, or l) the ratio $P_{45°}/P_{135°} = Z$ [available through substitution of $\theta = 45°$ and $\theta = 135°$ in Equations (143) to (145)], one obtains the curves shown in Figure 460-1; or if one plots $Z = P_{45°}/P_{135°}$ versus $P_{90°}$, one obtains the curves shown in Figure 460-2. Since the ratio $P_{45°}/P_{135°} = Z$ is easily determined experimentally, one can with the aid of Figure 360-1 or 360-2, together with Figure 459, determine the corresponding value of $\sqrt{\overline{h^2}}$. The ratio of the scattering-functions $P_{45°}/P_{135°}$ is easily accessible, because it is identical with $i_{45°}/i_{135°}$, as seen from Equation (142):

$$i_\vartheta = P_\vartheta i_0 \text{ and therefore } \frac{i_{45}}{i_{35}} = \frac{P_{45}}{P_{135}} = Z. \tag{146}$$

Z is called the *dissymmetry*. It can be determined by measuring the intensity of the scattered light at $\theta = 45°$ and $\theta = 135°$ in a light-scattering apparatus. Of course, instead of 45° and 135° one could also select some other pair of angles between θ = 0 and 180°. The angles 45° and 135° are usually preferred for experimental reasons. The dissymmetry, which is the result of interference, becomes larger, the larger the difference of the angle, and therefore it is greatest with the pair of angles θ = 0 and θ = 180°. However, with these angles it is for technical reasons impossible to determine the scattering intensity i_θ and one has therefore selected the angles 45° and 135°, which differ from each other by 90°, and where the scattering intensity can still be easily measured and the dissymmetry effect almost appears at its maximum.

If one now wants with the aid of the measured dissymmetry Z to determine the $\sqrt{\overline{h^2}}$ values from the Z versus D/λ curves (Figure 460-1), one has to make the assumption that the molecules are coiled, because with the curves for spheres, or rods, one would, of course, obtain completely different values. In general, the coil form of the chain molecules can be regarded as confirmed. However, in doubtful cases, a definite distinction between the different shapes the particle might have and which calibration curve one has to use can be made with the aid of light-scattering measurements. This is done by determining the shape of the curve 1/$P_{90°}$ versus Z experimentally with the aid of the Zimm diagram and comparing this with the corresponding theoretical curves in Figure 460-2.

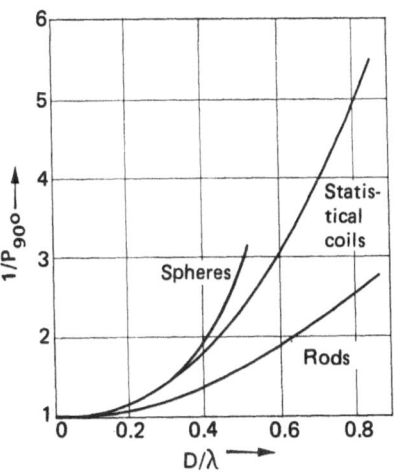

for coils: $D = \sqrt{\overline{h^2}}$;

for spheres: $D = d$;

for rods: $D = 1$

FIG. 459 — Scattering function 1/$P_{90°}$ as a function of the particle dimension D.

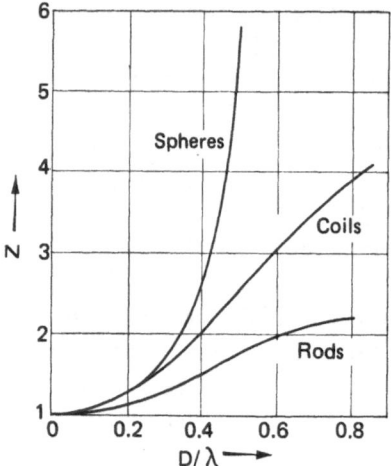

FIG. 460-1 – Dissymmetry $Z = P_{45}/P_{135}$ as a function of D/λ.

Experimental Selection of the Correct Z versus D/λ Curve

In the discussion of molecular weight determinations from the scattering intensity i_θ, it was shown how one can determine not only the molecular weight, but also $\sqrt{\overline{h^2}}$ without knowledge of the scattering function P_θ by using the Zimm extrapolation procedure. Thus, if one replaces in the light-scattering equation for the determination of the molecular weight (Debye-Zimm):

$$K \frac{c}{i_\theta} = \frac{1}{MP_\theta} + \frac{2Bc}{RT} + \cdots, \tag{147}$$

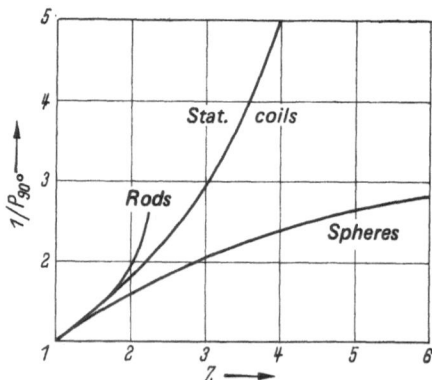

FIG. 460-2 — $1/P_{90°}$ as a function of the dissymmetry Z.

the scattering function P_ϑ by the theoretically derived series Equation (143):

$$\frac{1}{P_\vartheta} = 1 + \frac{8\pi^2}{9}\frac{\overline{h^2}}{\lambda^2}\sin^2\vartheta/2 + \cdots ,$$

one obtains the following equation:

$$K\frac{c}{i_\vartheta} = \frac{1}{M}\left(1 + \frac{8\pi^2}{9}\frac{\overline{h^2}}{\lambda^2}\sin^2\vartheta/2\right) + \frac{2Bc}{RT}. \tag{148}$$

If one now, for different angles θ, measures the scattering intensity $K\cdot c/i_\theta$ for different concentrations c, one can plot $K\cdot c/i_\theta$ versus c and extrapolate the line to $c = 0$. If one then plots the limiting values Kc/i_θ $(c = 0)$ versus $\sin^2\theta/2$, one obtains according to

$$K\frac{c}{i_\vartheta}\underset{(c\to 0)}{} = \frac{1}{M} + S\cdot\sin^2\vartheta/2 \tag{149}$$

TABLE 461. Comparison of an experimental $1/P_{90°}$ vs. Z relation with the corresponding theoretical values for coils, spheres, and rods.
(Measurements of H.J. CANTOW)

Molecular weight	Experimental values			Theoretical values calculated from Eqs. (143) − (145)		
	S Slope of the c→0 curve in the Zimm-plot	$1/P_{90°}$ according to Equ.(151) from S	$Z = \dfrac{i_{45}}{i_{135}}$	$1/P_{90°}$ for coils	$1/P_{90°}$ for spheres	$1/P_{90°}$ for rods
$7.7\ \cdot 10^6$	$0.74\cdot 10^{-6}$	3.85	3.69	4.24	2.20	10
$5.7\ \cdot 10^6$	$0.65\cdot 10^{-6}$	2.85	3.06	2.98	2.08	6
$3.0\ \cdot 10^6$	$0.54\cdot 10^{-6}$	1.81	2.01	1.84	1.61	4
$1.77\cdot 10^6$	$0.60\cdot 10^{-6}$	1.53	1.69	1.52	1.40	2.5
$1.47\cdot 10^6$	$0.51\cdot 10^{-6}$	1.30	1.47	1.34	1.30	1.34
$1.25\cdot 10^6$	$0.50\cdot 10^{-6}$	1.32	1.41	1.29	1.28	1.29

a straight line with a slope $\tan \partial = S = 8\pi^2\overline{h^2}/9M\lambda^2$ and the intercept at the ordinate, $1/M$. In the Zimm diagram both operations are combined by plotting the Kc/i_θ values obtained at different angles θ and different concentrations c against $\sin^2\theta/2 + 100c$ (compare p. 354). From the slope S, one can calculate the scattering function P_θ: If one (at c→0) combines Equations (147) and (149), one obtains:

$$K\frac{c}{i_\vartheta} = \frac{1}{M}\frac{1}{P_\vartheta} = \frac{1}{M} + S\cdot\sin^2\vartheta/2 \tag{150}$$

and from that:

$$\frac{1}{P_\vartheta} = 1 + \overline{M}\cdot S\cdot\sin^2\vartheta/2 \tag{151}$$

and for $\theta = 90°$:

$$\frac{1}{P_{90°}} = 1 + M \cdot S \cdot 0.5.$$

Since in order to obtain the Zimm diagram and to determine the slope S of the straight line, one has to measure i_θ for different angles, one therefore also knows $i_{45°}/i_{135°} = Z$. By experimental determination of Z and $P_{90°}$ for a series of fractions of compounds with increasing molecular weight [from the slope S of the Zimm diagram according to Equation (151)], one obtains an experimental $P_{90°}$ versus Z curve which can be compared with the curves obtained for the different particle shapes from theoretical calculations [Equations (143)-(145), Figure 460]. This has been presented in Table 461, for a series of fractions of polymethylmethacrylate. One clearly sees that the scattering function of polymethylmethacrylate corresponds best to that of a coil. The same holds true for the majority of all polymers. One can therefore use the theoretical curves, Z versus $\sqrt{\bar{h^2}}/\lambda$, or $Z - 1/P_{90°}$, and $1/P_{90°} - \sqrt{\bar{h^2}}/\lambda$ as calibration curves, if from the dissymmetry $i_{45°}/i_{135°} = Z$ measured at the angles $\theta = 45°$ and $135°$ one wants to calculate the ratio $\sqrt{\bar{h^2}}/\lambda$ and from it, by multiplication with the wave length λ, the average coil end-to-end distance $\sqrt{\bar{h^2}}$.

TABLE 462. Sample data for the determination of the mean end-to-end distance $\sqrt{\bar{h^2}}$ of polymethylmethacrylate in acetone from the dissymmetry Z (measurements of H.J. CANTOW)

Molecular weight $\overline{M}$	Degree of Polymerization $\overline{P}$	Length of the Polymer chain L_{max} A	Z	$\lambda = \frac{\lambda_0}{n}$ * A	$\sqrt{\bar{h^2}}/\lambda$ from Z according to Fig. 460-1	$\sqrt{\bar{h^2}}$ A
$5.73 \cdot 10^6$	57300	144000	3.06	4000	0.6	2400
$3.01 \cdot 10^6$	30100	75000	2.01	4000	0.4	1600
$1.09 \cdot 10^6$	10900	27500	1.35	4000	0.23	920
$0.536 \cdot 10^6$	5360	13500	1.15	4000	0.15	600

*λ_0 = wavelength of the light in vacuum

λ = wavelength of the light in the solvent used

n = refractive index of the solvent (for acetone, n = 1.36)

Table 462 gives a numerical example for the determination of the average chain end-to-end distance $\sqrt{\bar{h^2}}$ from the dissymmetry Z. The correct interpretation of the significance of this parameter is possible only by comparison with the maximum length of a chain molecule, which can be calculated from the degree of polymerization $P = M_{Pol}/M_{Mon}$ and the length of the monomeric unit l_M. Since with methylmethacrylate, the molecular weight of the monomeric residue $M_{Mon} = 100$, the maximum length of the macromolecule $L_{max} = P \cdot 2.52 = \overline{M}/100 \cdot 2.52$ Å. With the highest molecular weight (5 million), the length of the molecular chain is

60 times as large as the coil diameter, and for the smallest molecular weight (500,000) approximately 20 times as long. This ratio $L_{max}/\sqrt{\overline{h^2}}$ = 60, and 20, respectively, is often designated as the degree of coiling. One should not be confused by the increase in the "degree of coiling" with increasing molecular weight: in reality the coil becomes less tight as the molecular weight increases (compare p. 497). This can easily be illustrated by models: with an average thickness of the polymer chain of 8 Å [obtained from the molecular weight and L^{39}_{max}] and for $\overline{M}_w$ = $5\cdot10^6$, the ratio of the length to the thickness of the wire model is 126,000/8 = 15,800/1. For $\overline{M}_w$ = $1\cdot10^6$, the ratio is 25,200/8 = 3,150/1. If the coil becomes a sphere, and the diameter of the wire is 0.2 mm, then the small coil requires 63 cm of wire in a sphere of $1/6\pi(1.34)^3 \cong 1.2$ cm^3; with the large coil, one requires five times as long a wire: (316 cm) in a sphere of $1/6\pi(3.4)^3 \cong 20$ cm^3 (i.e., a 17 times larger volume). Therefore, the large molecule is only coiled about one-third as tightly as the smaller one.

In a similar way by measurement of the dissymmetry (measurement of the scattering intensity at different angles θ), one can determine the average chain end-to-end distance by measurement of the scattering intensity at different wave lengths. Since the two methods are independent of each other, one can use one to control the other.

The Equivalent Sphere

Coil dimensions can be obtained not only by light-scattering measurements but also by viscosity measurements, sedimentation measurements, or diffusion measurements. However, the molecular weight must be known. Viscosity measurements give the viscosity number, $[\eta]$. (For the determination of $[\eta]$,

TABLE 463. Coil dimensions for a wire coil model (wire diameter = 0.2 mm)

	Polymethylmethacrylate coil				Wire model coil			
M	Chain thickness [Å]	Chain length L_{max} [Å]	$\sqrt{\overline{h^2}}$ [Å]	$d_{equ[\eta]}$* [Å]	Wire diameter [mm]	Wire length [cm]	Coil diameter [cm]	Coil volume [cm^3]
$5\cdot10^6$	8	126000	2320	1370	0.2	316	3.4	21
$1\cdot10^6$	8	25200	910	537	0.2	63	1.34	1.26

The $\sqrt{\overline{h^2}}$ – values for the M – values (round numbers) are taken from the $\sqrt{\overline{h^2}}$ – $\overline{M}$ – curve in Fig. 467.

$d_{equ[\eta]}$ = diameter of the equivalent coil: $d_{equ[\eta]}$ = 0.59 $\cdot$ $\sqrt{\overline{h^2}}$ (see Table 466).

---

[39]The chain volume is $v = m/\rho$. With polymethylmethacrylate $(\rho \cong 1.25)$, $v = M/N_A \cdot 1.25 = 5\cdot10^6/7.5\cdot10^{23} = 0.66\cdot10^{-17}$ cm^3 = $6.6\cdot10^6$ Å^3. In order to obtain an average chain diameter, one has to consider the chain as a long cylinder of height $h = L_{max}$: $v = \pi r^2 h = \pi r^2 L_{max}$. Therefore, $\overline{r}_{chain} = (v/\pi L_{max})^{1/2} = (6.6\cdot10^6/\pi\cdot1.26\cdot10^5)^{1/2} = (16.6)^{1/2} \cong 4$Å.

compare pp. 381-384). According to the Einstein viscosity law:

$$[\eta] = 2.5/\bar{\rho}_{equ[\eta]}, \tag{152}$$

where $\bar{\rho}_{equ}$ is the average coil density (derivation compare pp. 375 and 450). If one regards the coil as a sphere with the volume v and diameter $d_{equ[\eta]}$, then for the density one can write:

$$\bar{\rho}_{equ[\eta]} = \frac{M/N_A}{v} = \frac{M}{1/6\,\pi d^3_{equ[\eta]}\,N_A} \;[\text{g/cm}^3] \tag{153}$$

If $\bar{\rho}_{equ[\eta]}$ in Equation (152) is substituted by Equation (153), we obtain the diameter of the equivalent sphere:

$$d_{equ[\eta]} = \left(\frac{6 \cdot M \cdot [\eta]}{2,5\,\pi\,N_A}\right)^{\frac{1}{3}} [\text{cm}] = 1,08 \cdot \sqrt[3]{M \cdot [\eta]} \;[\text{Å}] . \tag{154}$$

One obtains from this equation the diameter of spherical coils with molecular weight M, which in solution have the viscosity number $[\eta]$. However, since the coils are not spherical, one speaks of the diameter of an equivalent sphere, i.e., the diameter of spherical particles which are equivalent in terms of the viscosity number to the actual particles of any desired shape. Equivalent spheres are ideal particles which in solution would give the same relative viscosity increase $[\eta]$ as the effectively measured coils with irregular forms. In the same way the viscometrically determined value for the coil density must be an equivalent value (compare p. 445).

In a similar manner, by means of sedimentation measurements, one can also determine the diameter of equivalent coils d_{equ,s_0} by substituting the measured values of s_0 in Stoke's law. One thus determines how large the sphere would have to be in order to yield the value s_0 determined in the centrifuge. Such spheres are equivalent to the coils actually present in solution in terms of their sedimentation velocity. As has been shown on page 358, the constant sedimentation velocity u, is given through the equivalence between the frictional force k_F and the gravitational force (reduced by the buoyancy) k_S:

$$k_F = k_S$$

or:

$$6\,\pi\,r\,\eta\,u = m\,b\,(1 - V\,\varrho_{\text{solv}}).$$

For the coil radius r one therefore obtains:

$$r = \frac{m\,b\,(1 - V\,\varrho_{\text{solv}})}{6\,\pi\,\eta\,u}$$

If one considers that $m \cdot N_A = M$ and $u/b = s_0$, then:

$$d_{equ\,s_0} = \frac{2\,\bar{M}\,(1 - V\,\varrho_{\text{solv}})}{6\,\pi\,\eta\,s_0\,N_A} \tag{155}$$

where b is centrifugal acceleration; m, mass of the single particle; r, radius of the single particle; $\rho_{solv.}$, density of the solvent; s_0, sedimentation constant = sedimentation velocity/centrifugal acceleration; η, viscosity of the solvent; V, specific volume of the dissolved macromolecular substance; and s_0, sedimentation constant at infinite dilution measured in Svedberg; 1 svedberg = 10^{-13} seconds. (For the determination of s_0 compare pp. 358-367).

Finally, one can also calculate d_{equ,D_0} (by means of the Nernst equation) from experimentally determined diffusion constants:

$$D_0 = \frac{RT}{6\pi r \eta N_A}$$

and

$$d_{equ\,D_0} = \frac{2RT}{6\pi \eta N_A D_0}. \tag{156}$$

The particle diameter determined in this way is equal to the d_{equ,s_0} determined from sedimentation measurements.

The particle diameters determined viscometrically and by sedimentation measurements are not identical with one another and with the optically determined $\sqrt{\overline{h^2}}$ because of the elongated form of the coils. However, these values are (at least within a polymer homologous series, i.e., a series of fractions of a polymer with increasing molecular weights) directly proportional to each other:

$$d_{equ\,s_0} \sim d_{equ\,[\eta]} \sim \sqrt{\overline{h_{opt}^2}}. \tag{157}$$

Table 466 shows the coil dimensions of two carefully measured polymers. One sees that $d_{equ,s_0} < d_{equ[\eta]} < \sqrt{\overline{h_{opt}^2}}$.

The ratio $\sqrt{\overline{h^2}}/d_{equ,s_0}$ is a measure for the deviation of the coil from a spherical shape (compare also p. 479).

Dependence of the Average Chain End-to-End Distance $\sqrt{\overline{h_{opt}^2}}$ on the Molecular Weight

For the diameter d of a sphere of mass m, the density ρ and the volume $v = m/\rho$ = $1/6\pi d^3$, one can write in general:

$$d = \left(\frac{v}{(1/6)\pi}\right)^{\frac{1}{3}} = \left(\frac{m/\varrho}{(1/6)\pi}\right)^{\frac{1}{3}} = \left(\frac{6}{\pi}\right)^{\frac{1}{3}} \cdot \left(\frac{m}{\varrho}\right)^{\frac{1}{3}} = \text{const.} \left(\frac{m}{\varrho}\right)^{\frac{1}{3}}. \tag{158}$$

With normal spheres with constant density (i.e., independent of the size), the diameter is proportional to $m^{1/3}$:

$$d = \text{const.} \cdot m^{1/3}. \tag{159}$$

On the other hand, with the coils the density is not constant, but is a function of the molecular weight. As can be derived from the random flight statistics with ideal coils, one can write according to the Kuhn square-root law (compare p. 445):

$$\bar{\rho}_{coil} = \text{const.} \cdot M^{-0.5}. \tag{160}$$

With real coils we have seen that the exponent of the density function is not always exactly 0.5, but, as determined from viscosity measurements, is usually somewhat higher. One can write in general:

$$\bar{\rho}_{coil} = const. \cdot \overline{M}^{-a}, \qquad (160a)$$

where the exponent a can have different values from material to material and solvent to solvent. If one takes this dependence of the coil density on the molecular weight into consideration, by substituting Equation (160a) in Equation (158), one obtains for the dependence of the coil diameter on the molecular weight the following exponential equation:

$$\left. \begin{array}{l} const.\, d_{equ} = \sqrt{\bar{h}^2} = const. \left(\dfrac{M}{\bar{\rho}_{coil}} \right)^{\frac{1}{3}} = const. \left(\dfrac{M}{M^{-a}} \right)^{\frac{1}{3}} \\[3mm] = const.\, M^{\frac{1}{3}} \cdot M^{a/3} = const.\, M^{(1+a)/3} \end{array} \right\} \qquad (161)$$

If one designates the exponent of the $\sqrt{\bar{h}^2}$ versus $\overline{M}$-function as b, one obtains:

$$b = \frac{1+a}{3}. \qquad (162)$$

Therefore, if one knows the $[\eta]$ versus $\overline{M}$ function, i.e., the dependence of the viscosity number on the molecular weight, and therefore, the dependence of the coil density on the molecular weight, one can calculate the change in the average

TABLE 466. Coil dimensions of polymethylmethacrylate and cellulose nitrate (from measurements of MEYERHOFF, SCHULZ and CANTOW)

M	$\sqrt{\bar{h}_{opt}^2}$ Å	$[\eta]$ cm³/g	$d_{equ[\eta]}$* Å	s_0 in Svedb.	$d_{equ\,s_0}$* Å	$\dfrac{\sqrt{\bar{h}^2}}{d_{equ\,s_0}}$	$\dfrac{d_{equ[\eta]}}{\sqrt{\bar{h}^2}}$**
			Polymethylmethacrylate				
20 000			56.7	8	50.5		0.89
500 000	610	70	353	32	315	1.94	0.58
1 000 000	910	120	533	43	466	1.95	0.58
2 000 000	1370	202	798	58	691	1.98	0.58
3 000 000	1720	275	1012	69	867	1.98	0.59
5 000 000	2320	405	1367	86	1170	1.98	0.59
7 000 000	2820	520	1660	100	1415	1.99	0.59
			Cellulosenitrate				
20 000	240	60	115	8	75.7	3.2	0.48
100 000	700	300	330	13	229	3.0	0.47
1 000 000	3200	2800	1500	27	1120	3.2	0.47

*Because of the elongated form of the coils, the equivalent sphere is larger than a sphere with the same volume as the coil. Compare p. 507. The $[\eta]$ – and s_o– values used here are taken from the curves in Figs. 454, 467 and 472.

**This ratio is used as a form factor in some equations (p. 512).

chain end-to-end distance $\sqrt{\overline{h^2}}$ with molecular weight by means of Equation (161). For example, if the exponent a of the viscosity equation was found experimentally to be equal to 1, the exponent b of the $\sqrt{\overline{h^2}}$ versus $\overline{M}$ function is $(1 + a)/3 = 2/3 = 0.666$. Thus, if one can write for viscosity $[\eta]$ = const.$\cdot\overline{M}^1$ one obtains for the coil diameter $\sqrt{\overline{h^2}}$ = const.$\cdot\overline{M}^{0.66}$. If $[\eta]$ = const.$\cdot\overline{M}^{0.8}$, then $\sqrt{\overline{h^2}}$ = const.$\cdot\overline{M}^{0.6}$, (because b $= (1 + a)/3 = 1.8/3$, etc. This is so because both the viscosity and the coil diameter are determined by the coil density. Thus if the density of the coil changes with $\overline{M}^{-a}$ (and $[\eta]$ correspondingly with $\overline{M}^a$), then $\sqrt{\overline{h^2}}$ has to change with $M^{(1+a)/3}$ [Equation (161)]. This reaction has so far been found true with all polymer systems which have been measured with sufficient accuracy and care. As examples, we see in Figure 467 the $\sqrt{\overline{h^2}}$ functions of cellulose nitrate and polymethylmethacrylate on a double logarithmic plot. As can be easily checked, the slope of the lines for the cellulose is $b = 0.67$, and for polymethylmethacrylate $b = 0.58$, i.e., the experimentally determined functions are for cellulose:

$$\sqrt{\overline{h^2}} = \text{const.}\cdot\overline{M}^{0.67}$$

and for polymethylmethacrylate

$$\sqrt{\overline{h^2}} = \text{const.}\cdot\overline{M}^{0.58}.$$

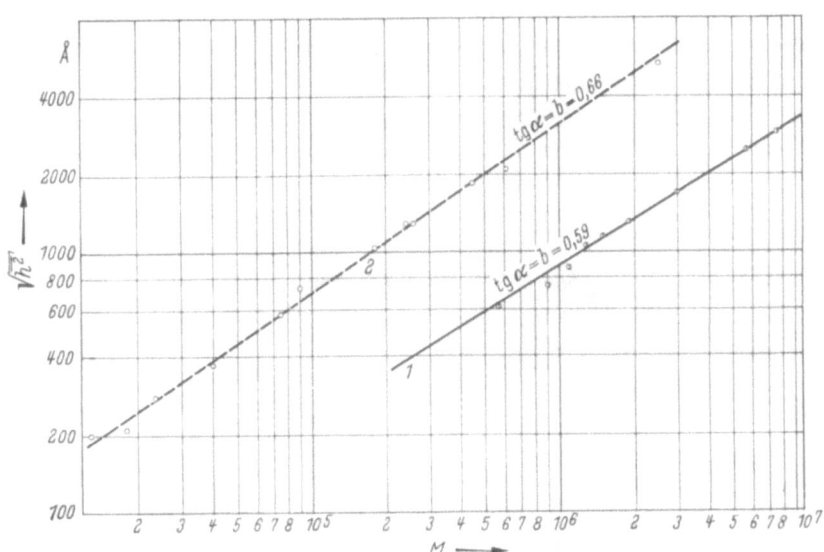

1) polymethylmethacrylate; $\tan \alpha = 0.59$

2) cellulose nitrate; $\tan \alpha = 0.66$

(measurements of SCHULZ, MEYERHOFF, and CANTOW)

FIG. 467 – Experimentally determined dependence of the average end-to-end distance $\sqrt{\overline{h^2}}$ on the molecular weight.

Since for cellulose $a = 1$ and for polymethylmethacrylate $a = 0.73$ (found experimentally), the exponent b for cellulose, calculated according to Equation (162) is:

$$b = (1 + a)/3 = 2/3 = 0.666$$

and for polymethylmethacrylate:

$$b = (1 + a)/3 = 1.73/3 = 0.58$$

Thus the experimentally determined exponents of the $\sqrt{\overline{h^2}}-\overline{M}$ functions agree exactly with the theoretical ones calculated according to Equation (161) from the density or the viscosity functions.

It is therefore not meaningful to plot $\sqrt{\overline{h^2}}$ versus $\sqrt{\overline{M}}$, as usually found in the older literature, because such a dependence $\sqrt{\overline{h^2}} \propto M^{0.5}$ (Kuhn's square-root law) is to be expected only when the exponent of the $[\eta]$ versus M function $a = 0.5$, i.e., in θ solvents. However, this will not hold when $a = 0.7$, or 0.8, or even 1. If the root law $\sqrt{\overline{h^2}} = $ const. $\cdot M^{0.5}$ has, in spite of this, always been confirmed experimentally in earlier years, this is only because measurements have been too inaccurate to show up the deviations. Only through experimental refinement of the molecular weight measurements (ultracentrifugation and light-scattering) has it been possible to confirm Equation (162).

Dependence of the Sedimentation—and Diffusion Velocity on the Molecular Weight

Investigations of sedimentation velocity are of special significance for the determination of coil properties, because in sedimentation one expects a basically different behavior from freely drained and undrained coils. Thus from sedimentation experiments one can resolve the question of the draining behavior of the coil (i.e., the passage of solvent through the macromolecular coil during sedimentation).

The Behavior of Undrained Coils in Sedimentation

According to Stoke's law (compare pp. 360 and 464), one can write for the sedimentation velocity of spheres:

$$s_0 = \frac{2 \, \overline{M} \, (1 - V \varrho_{solv})}{6 \, \pi \, \eta \, d_{equ \, s_0} \cdot N_A} . \tag{163}$$

If one puts together all the parameters which are constant for a given system, one can write:

$$s_0 = \text{const.} \cdot \overline{M}/d_{equ, s_0}, \tag{164}$$

or, if one replaces d_{equ, s_0} according to Equation (158) with const. $\cdot (\overline{M}/\rho)^{1/3}$:

$$s_0 = \text{Konst.} \frac{M}{(M/\rho_{coil})^{\frac{1}{3}}} = \text{const.} \; M^{\frac{2}{3}} \cdot \varrho^{\frac{1}{3}}_{coil} \tag{165}$$

If the particle density is independent of the particle weight, one can write:

$$s_0 = \text{const. } M^{\frac{2}{3}}.$$

This corresponds to Stoke's law, valid for simple spheres of the same material. With ideal statistical coils $\overline{\rho}_{coil} = \text{const.} \cdot \overline{M}^{-0.5}$. Thus, for this ideal case, one again finds a square-root law:

$$s_0 = \text{const. } M^{\frac{2}{3}} \cdot (M^{-0.5})^{\frac{1}{3}} = \text{const. } M^{0,5}. \tag{166}$$

If the density change does not occur with $\overline{M}^{-0.5}$, but more generally with $\overline{M}^{-a}$, then by replacement of ρ_{coil} in Equation (165) by "$\text{const.} \cdot \overline{M}^{-a}$" according to Equation (141), p. 455,) one obtains the following dependence of the sedimentation constant on the molecular weight:

$$s_0 = \text{const. } M^{\frac{2}{3}} (M^{-a})^{\frac{1}{3}} = \text{const } M^{(2-a)/3} \tag{167}$$

This is the s_0 versus $\overline{M}$ function, which one should expect for undrained, real statistical coils, whose density changes with $\overline{M}^{-a}$ (instead of $\overline{M}^{-0.5}$ for the ideal case). If one designates the exponent of the $s_0 - \overline{M}$ function with u, one can write:

$$u = (2 - a)/3 \tag{168}$$

The Behavior of Freely Drained Coils

The behavior of freely drained coils in sedimentation is the same as that of long thin threads or wires. The sedimentation velocity is in general determined by the equilibrium between the gravitational force (minus the buoyancy) and the frictional force. The latter always increases in proportion to the sedimentation velocity:

$$k_{friction} = k_{gravity}.$$

In the case of long, wire-like chains, both the gravitational force and the frictional force are directly proportional to the length of the chains because all parameters such as surface volume, and mass are proportional to the length l with such chains (compare p. 358):

$$k_{gravity} = \text{const.} \cdot l \tag{169}$$

and

$$k_{friction} = \text{const.} \cdot l \cdot s_0. \tag{170}$$

The sedimentation velocity, i.e., the sedimentation constant s_0, which is proportional to the sedimentation velocity, is therefore constant, independent of the length of the chains, as long as the condition remains true that the diameter is very small compared to the length:

$$s_0 \cdot l = \text{const.} \cdot l \text{ and } s_0 = \text{const.} \tag{171}$$

With freely drained coils, therefore, no change of the sedimentation velocity with molecular weight can be observed.

Sedimentation Velocity of Partially Drained Coils

If in a simplified way one regards the partially drained coils as a structure with an undrained center and a freely drained outer layer, one can divide the frictional force into one part corresponding to the undrained center, and a second part corresponding to the freely drained outer layer (k_{Fi} and k_{Fou}). The boundary zones thus behave exactly as freely drained coils or long wires, i.e., k_{Fou} is directly proportional to the chain length and therefore also to the mass m_{ou} present in the outer layer: $k_{Fou} = \text{const.} \cdot s_0 \cdot m_{ou}$. For the undrained center of the coil, one can count with Stoke's law, i.e., that the frictional forces $k_{Fi} = 6\pi r_i \eta s_u$, or, in general: $k_{Fi} = \text{const.} \cdot s_0 \cdot d_i$.

Since d_i according to Equation (158) is proportional to $(m_i/\rho_i)^{1/3}$, one can write: $k_{Fi} \backsim s_0 \cdot (m_i/\rho_i)^{1/3}$. If one then restricts (for simplicity) the number of possible cases to the one where the degree of draining is the same with all fractions, i.e., $m_i \backsim M$ and $m_o \backsim M$, and if one assumes that also ρ_i again [according to Equation (141)] decreases with M^{-a}, then one obtains:

$$k_{Fi} \sim s_0 \left(\frac{m_i}{\varrho_i}\right)^{\frac{1}{3}} \sim \left(\frac{M}{M^{-a}}\right)^{\frac{1}{3}} \cdot s_0 \sim M^{(1+a)/3} \cdot s_0 \sim M^b \cdot s_0 \qquad (172)$$

and

$$k_{Fou} \backsim M \cdot s_0. \qquad (173)$$

The force of gravity k_G is of course proportional to the molecular weight therefore, with partially drained coils the dependence of the sedimentation constant on the molecular weight is given by the following equations:

$$k_{Fi} + k_{Fou} = s_0 (K\,M^b + K'\,M)$$

$$k_G = \text{const. } M$$

$$s_0 (K\,M^b + K'\,M) = \text{const. } M \quad \text{and} \quad s_0 \sim \frac{M}{K\,M^b + K'\,M}^{40} \qquad (174)$$

$$s_0 \sim \frac{M^{1-b}}{K'' + M^{1-b}} \sim \frac{M^u}{K'' + M^u}.$$

$$40 \quad \frac{M}{K \cdot M^b + K' \cdot M} = \frac{M}{K \cdot M^b + K' \cdot M^b \cdot M^{1-b}} = \frac{M}{M^b (K + K' \cdot M^{1-b})}$$

$$= \frac{M^{1-b}}{K + K' \cdot M^{1-b}} = \frac{M^{1-b}}{K'\left(\dfrac{K}{K'} + M^{1-b}\right)}.$$

Since $b = \dfrac{1+a}{3}$, $\ 1-b = \dfrac{2-a}{3} = u$ [compare p. 469. Equations (167) and (168)].

If the constant K'' is very small in comparison with M^u, then even with partially drained coils the sedimentation constant is independent of the molecular weight. The assumption $K'' \ll M^u$ is fulfilled the sooner, the larger M. This means that the logarithmic $s_0 - \overline{M}$ curve is not a straight line in this case, but a curve with a steadily decreasing slope, which finally, for high molecular weights, asymptotically approaches a constant limiting value. Because with decreasing coil density the freely drained portion becomes larger with increasing molecular weight, one must expect an even stronger curvature than the one predicted by Equation (174).

Experimental Determination of the s_0 versus M, and D_0 versus M curves

The sedimentation velocity of macromolecules can be determined with the ultracentrifuge. As has been discussed previously in reference to the methods for molecular weight determination, it is common to refer to the reduced sedimentation velocity s_0 introduced by Svedberg instead of the sedimentation velocity. s is the quotient (sedimentation velocity)/(centrifugal acceleration), and s_0 is the corresponding value extrapolated to concentration zero.

In many older investigations concerning the dependence of the sedimentation velocity on the molecular weight, s_0 was usually plotted versus $\sqrt{M}$ in order to check the validity of the square-root law [Equation (166)]. Actually a straight line was found in almost every case. This shows at least that the exponent u of the general equation $s_0 = \text{const.} \cdot M^u$ cannot be very different from 0.5. Careful measurements over a broader molecular weight region which have been carried out—first by G. Meyerhoff with polymethylmethacrylate—more recently, have shown that the exponent u is not equal to 0.5, but agrees almost exactly with the theoretically predicted value which can be calculated according to $u = (2 - a)/3$ [compare Equation (167)] from the measured exponent a of the $[\eta] - \overline{M}$ function for the particular polymer. Thus, the decrease of the coil density with increasing molecular weight, as has been assumed in the derivation of the exponential relation Equations (163)-(168) is actually just as important for the slope of the logarithmic s_0 versus $\overline{M}$ curve, as for the slope of the $[\eta]$ versus $\overline{M}$, and the $\sqrt{\overline{h^2}}$ versus $\overline{M}$ curves. Figure 472 shows this for three examples: polymethylmethacrylate, cellulose nitrate, and polystyrene. For polymethylmethacrylate an exponent $a = 0.73$ is found by means of viscosity measurements. The exponent to be expected, $u = (2 - 0.73)/3 = 0.43$ checks well with the experimentally determined u from the figure. The same applies to cellulose nitrate: $a_{exp} = 0.98$; $u_{theor.} = (2 - 0.98)/3 = 0.34$; $u_{exp.} = 0.32$: For polystyrene: $a_{exp.} = 0.5$; $u_{theor.} = (2 - 0.5)/3 = 0.5$; $u_{exp.} = 0.5$ (compare Figure 472).

One can see how the deviations of the viscosity and density functions (Figure 454) from the ideal curves correspond to the deviations of the $s_0 - \overline{M}$ functions from the ideal exponent 0.5. This is to be expected for an undrained statistical coil. The straight lines show no curvature even for very high molecular weights, so that there is no experimental verification for even a small amount of partial draining.

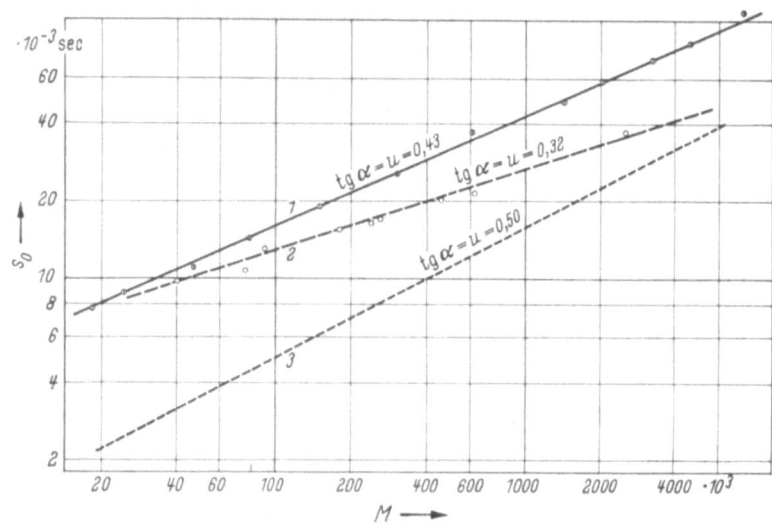

1) polymethylmethacrylate in acetone at 20°C

2) cellulose nitrate in acetone at 20°C

3) polystyrene in cyclohexane at 35°C

FIG. 472 — Log-log plot of s_0' vs. M for polymethylmethacrylate, cellulose nitrate, and polystyrene. (Measurements of SCHULZ, MEYERHOFF, and CANTOW)

The fact that the s_0-M curves for polymethylmethacrylate and cellulose nitrate intersect at approximately $\overline{M} = 20{,}000$ (Figure 472) may appear surprising at first, since the coil density of both materials at $\overline{M} = 20{,}000$ is still completely different (0.57 and 0.061) as the $[\eta]$ versus $\overline{M}$ lines in Figure 454 show. However, one has to consider that s_0 depends not only on the coil density, but also on the specific volume, which for polymethylmethacrylate $\cong 0.8$, and for cellulose nitrate $\cong 0.57$ cm^3/g. Furthermore, because of the nonspherical form of the coils, the viscometric coil density $\bar{\rho}_{equ,[\eta]}$ is different from the coil density $\bar{\rho}_{equ,s_0}$ effective in sedimentation. For polymethylmethacrylate $\rho_{equ,s_0} \cong 1.5 \cdot \rho_{equ,[\eta]}$ whereas for cellulose nitrate $\rho_{equ,s_0} \cong 3 \cdot \rho_{equ,[\eta]}$ (compare pp. 505-509). Thus one must have a crossing-over of the two s_0 versus M lines at $M = 20{,}000$.

It may also appear surprising that the sedimentation velocity of the more tightly coiled polystyrene molecules (θ system!) is smaller than that of the much more loosely coiled cellulose nitrate and polymethylmethacrylate molecules. This is because the viscosity of cyclohexane is approximately three times that of acetone.

In this instance one clearly recognizes that instead of using the reduced sedimentation velocity $s_0 = u/b$, it is more useful to employ the sedimentation numbers $[s] = u \cdot \eta / b(1 - V\rho_{solv.})$, where not only the influence of the centrifugal acceleration b has been eliminated, but also the influence of the other system parameters η, ρ_{solv} (= viscosity and density of the solvent respectively) and V (= specific volume of the polymer in solution) have been eliminated.

The dependence of the diffusion coefficient D_0 on the molecular weight can be obtained from the Nernst equation by replacing d_{equ,s_0} according to Equation (161) through const.$\cdot M^b$:

$$
\left.
\begin{aligned}
D_0 &= \frac{2RT}{6\,d_{equ\,s_0}\,N_A} = \text{const.}\ \frac{1}{d_{equ\,s_0}} = \text{const.}\ \frac{1}{M^b} \\
&= \text{const.}\ M^{-b} = \text{const.}\ M^{-(1+a)/3}.
\end{aligned}
\right\}
\tag{175}
$$

If one expresses d_{equ,s_0} by the volume $v = m/\rho = (1/6)\pi d^3$, one again recognizes clearly the influence of the coil density $\bar{\rho}_{coil}$:

$$
D_0 = 2\,RT\,/\,6\,N_A \left(\frac{m/\bar{\rho}}{(1/6)\,\pi} \right)^{1/3} = \text{const.}\cdot \bar{M}^{-1/3}\cdot \bar{\rho}^{1/3}_{coil}
$$

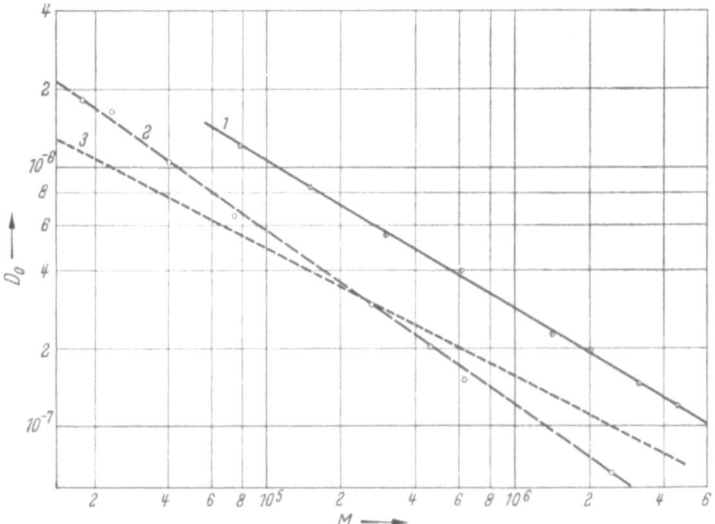

1) polymethylmethacrylate: $a_{exper} = 0.73;\ b_{calc} = 0.58;\ b_{exper} = 0.59$ (comp. Fig. 467, $d_{exper} = -0.58$

2) cellulose nitrate: $a_{exper} = 0.98;\ b_{calc} = 0.66;\ b_{exper} = 0.66$ (comp. Fig. 467, $d_{exper} = -0.67$

3) polystyrene: $a_{exper} = 0.50;\ b_{calc} = 0.50;\ d_{calc} = -0.50;\ d_{exper} = -0.50$

FIG. 473 — Log-log plot of D_0 vs. M for polymethylmethacrylate, cellulose nitrate, and polystyrene. (Measurements of SCHULZ, MEYERHOFF, and CANTOW).

and according to $\bar{\rho} \propto \bar{M}^{-a}$:

$$D_o \propto \bar{M}^{-1/3} \cdot \bar{M}^{-a/3} = \bar{M}^{-(1+a)/3} = M^{-b} = M^d. \qquad (176)$$

Thus, one must expect a negative exponent for the D_o versus $\bar{M}$ function, which has the same numerical value as the exponent b of the $\sqrt{\bar{h}^2}-\bar{M}$ function. Figure 473 shows the experimental curves for polymethylmethacrylate, cellulose nitrate, and polystyrene, which have a slope exactly equal to that to be expected according to $D_o = \text{const.} \cdot \bar{M}^{-b}$.

The Exponent Diagram

One can explain the coincidence of the experimentally determined exponents with the values calculated from a, only if one assumes that the macromolecules in solution are present as undrained random coils in which the solvent is held fast as in a gel, and therefore on sedimentation they behave like small closed spheres whose density, according to their coil character, decreases with increasing molecular weight. The decrease in density, does not, in general, occur exactly according to the exponent $\bar{M}^{-0.5}$ to be expected for ideal random coils, according to random flight statistics. Instead, depending on the solvent, temperature, and type of chain, the density decrease is proportional to a different exponent, which usually lies between 0.5 and 0.75 but which for certain polymers may also reach values of 1.0 (cellulose and cellulose derivatives) and 1.2 (pectin). On the other hand, it can also have values near 0 (glycogen and some proteins), and there are even systems theoretically imaginable where the exponent a has negative values, i.e., where the viscosity decreases with increasing molecular weight, assuming that the $[\eta]$ versus $\bar{M}$ relationship holds. Such polymer-solvent systems are difficult to realize, because

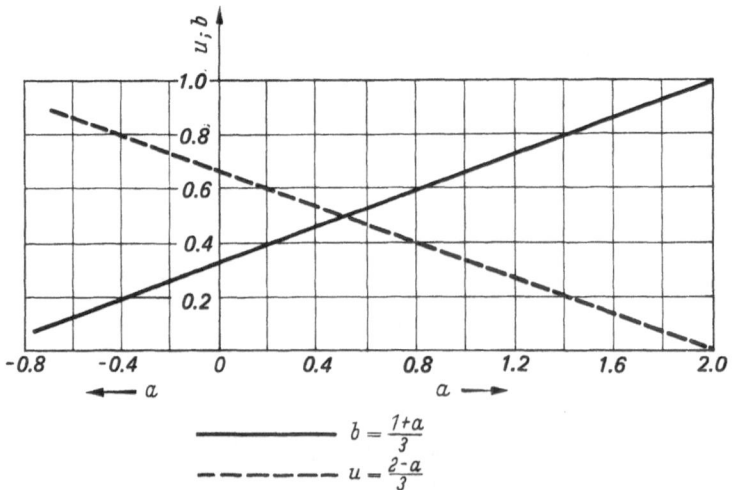

FIG. 474 — Exponent Diagram.

they assume a poor solvation of the chain, which would usually result in gel formation.

The deviation of the exponents b and u of the functions $\sqrt{\overline{h^2}}_{opt} \sim \overline{M}^b$, and $s_0 \sim \overline{M}^u$, from the ideal value 0.5, which is to be expected as a result of the change in density of the coils according to $\bar{\rho}_{equ} \sim \overline{M}^{-a}$ (instead of $M^{-0.5}$), can be represented in a simple manner with a diagram, by plotting u and b versus a (Figure 474).

The intersection point of the two lines can be designated as the θ point or Flory-point. At this point all functions, $[\eta] = K_{[\eta]} \cdot M^a$, $\sqrt{\overline{h^2}} = K_d \cdot M^b$, $s_0 = K_s \cdot M^u$, and $D_0 = K_D \cdot M^{-b}$ have the same exponent $a = b = u = 0.5$. That means that the polymer coils behave as ideal random coils, because under the conditions of the θ point the contracting forces are just compensated by the expanding forces. θ systems will be discussed in more detail on p. 521.

On both sides of the θ point the exponents become different: on the right, in the direction of increasing a values, b becomes greater and u becomes smaller. This is the region of good solvents. Most polymer-solvent systems are found in the range of $a = 0.5$ to $a = 1$. But even systems with a$>$1 are possible and were found with same natural and synthetic polymers, for example, pectic acid, helical proteins, synthetic polyamino acids (compare Figure 515), or polyelectrolytes (compare Figure 540).

$a = 2$ is the limiting case of polymers, whose volume does not increase through expansion in three dimensions, but only in one dimension with increasing molecular weight. This is immediately understandable in the case of maximally extended polymer chains.[41] . But $a = 2$ holds true not only for extended chains, but also for folded or helical chains, if the angle of folding, or the pitch of the helix, remains constant within one $[\eta]$–M-plot. (Compare Figure 476) For systems in which the pitch or the folding angle would increase with increasing molecular weight, we have to expect a$>$2.

---

[41] As shown above [Equations (157) and (127), the diameter of the equivalent sphere $d_{equ[\eta]}$, the end-to-end distance $\sqrt{\overline{h^2}}$, and – with extended chains – the maximum chain length L_{max} are proportional to each other: $d_{equ[\eta]} = F \cdot \sqrt{\overline{h^2}} = F' \cdot L_{max}$. Therefore, we can write for the viscometrically effective sphere volume $v_{equ[\eta]}$:

$$v_{equ[\eta]} = 1/6 \, \pi \, d^3_{equ[\eta]} = 1/6 \, \pi \, (F' \, L_{max})^3 \text{ and } \bar{\rho}_{equ[\eta]} = m_{chain}/v_{equ[\eta]} = \frac{M/N_A}{1/6\pi(F' \cdot L_{max})^3}$$

Therefore:

$$[\eta] = 2.5/\bar{\rho}_{equ[\eta]} = 2.5 \cdot F'' \cdot 1/6 \, \pi \, L^3_{max} \cdot N_A/M. \tag{177}$$

Through variation of the molecular weight M between $M = 10^4$ and $M = 10^6$, one otbains the $[\eta]$ – M relation for maximally stretched (or linearly folded or helical) chains. The experimentally resulting efficiency factor F' is of the order of magnitude of $F' = 0.3$ with extended chain molecules. The maximum chain length L_{max} results from $L_{max} = 2.52 \cdot P$ [Å] (P = degree of polymerisation). For folded chains: $L_{max} = 2.52 \cdot P \cdot \sin \alpha$, if 2α is the folding angle: Because of $L_{max} = const. M$, we can write instead of Equation (177);

$$[\eta] = const. \, M^2. \tag{178}$$

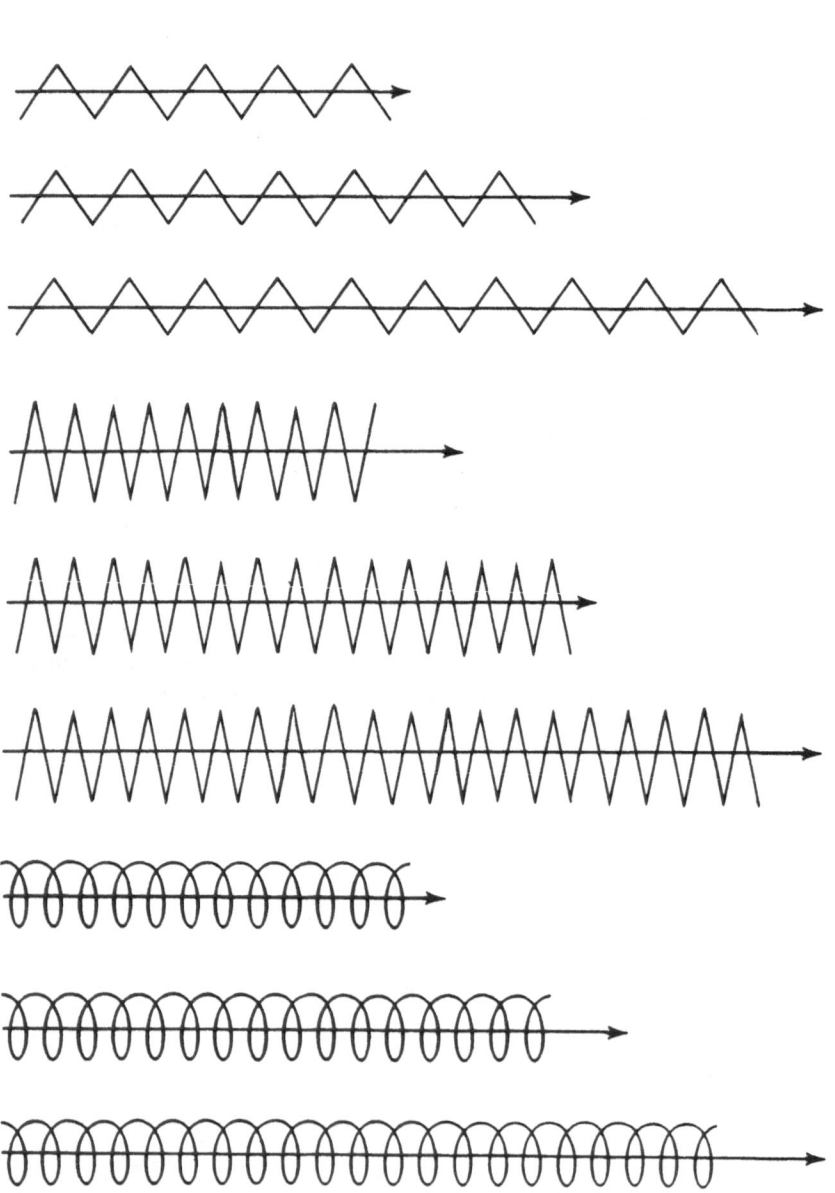

FIG. 476 — Some possibilities of linear chain growth with increasing molecular weight for $a = 2$.

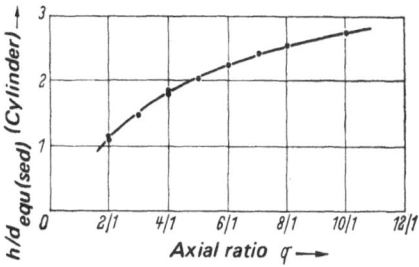

FIG. 477 — Dependence of the ratio $h_1/d_{equ(sed)}$ determined from the sedimentation rate of cylindrical wire models on the axial ratio h_1/h_2 of the wire pieces. (h_1 = length of the wire piece, h_2 = its diameter = 0.1 mm).

In the direction of decreasing a values ($a < 0.5$), the exponent b becomes smaller and u becomes larger. This is the region of poor solvents, in which the polymer chain segments of a coil prefer association with each other over interaction with the solvent. Since in most polymer solvents one has both intramolecular association of the chain segments within the coil and intermolecular association between different coils, not all polymer solutions in this region of a $\leqslant 0.5$ are stable. At the point of intersection of the b and a lines with the ordinate, $a = 0$, i.e., the density and therefore the viscosity of the solutions are independent of the molecular weight, as is the case with ordinary colloidal dispersions (gold sols, and polymer dispersions as in emulsion polymerization). Systems which lie in the region of negative abscissa values, and which would be of great industrial importance, are very difficult to realize because turbidity and gel-like precipitation occurs for the above reasons (compare also p. 518). An example for negative values of a which has been described by Staudinger and Hellfritz is polyisobutylene in benzene at 20°C.

Experimental Results Concerning the Geometrical Shape of the Coils

In the preceding paragraphs we have seen that the phenomena of viscosity and sedimentation, as far as the change of $[\eta]$ and s_0 with the molecular weight is concerned, can be described approximately by the laws of Einstein and Stokes derived for spheres. This should not lead to the conclusion that the coils have a spherical form. At least in the case of sedimentation velocity it was experimentally confirmed that cylindrical models also can fulfill the relationship $u_{sed} = const. \cdot m^{2/3}$, which holds for spheres.

If the coil molecules would have a spherical shape, then the coil diameter as determined by different experimental methods would have to be exactly equal. In reality, however, the optically determined average coil length $\sqrt{h^2}_{opt.}$ is considerably greater than the diameter of the equivalent sphere $d_{equ,[\eta]}$ and $d_{equ,so}$ determined from viscosity and sedimentation measurements, respectively, according to Equations (154) and (155) (Table 466). This result that $\sqrt{\overline{h^2}}_{opt.}$ is greater than d_{equ,s_0} corresponds entirely (as model experiments have shown), to what one would expect from elongated particles. Thus, if one measures the sedimentation velocity u of small model cylinders with increasing axis ratio $h_1/h_2 =$

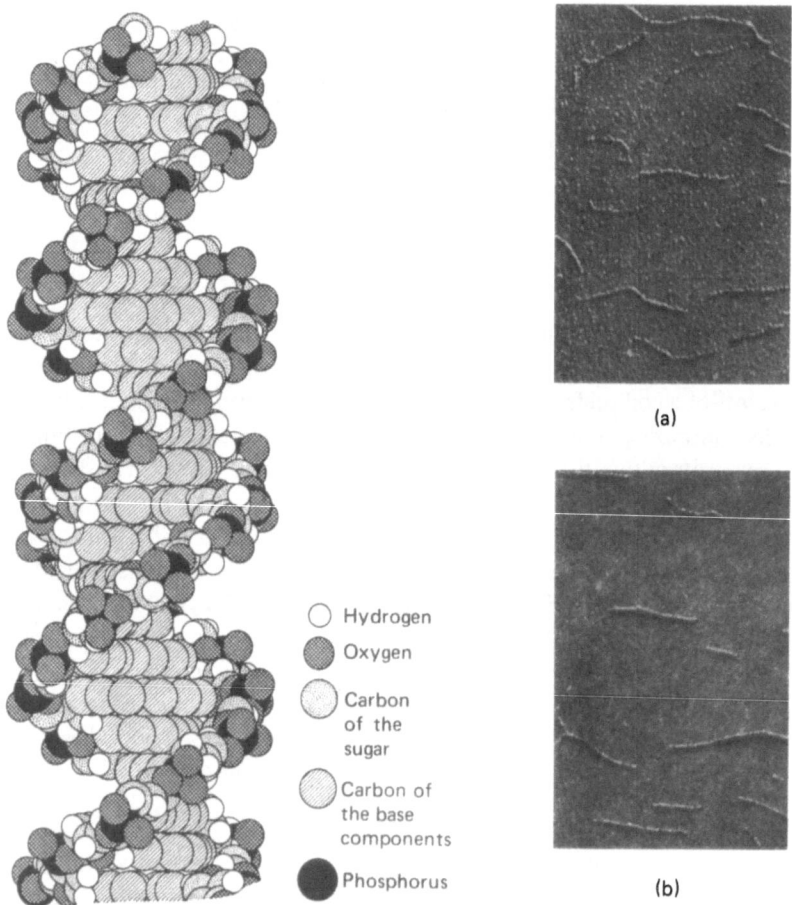

○ Hydrogen

● Oxygen

◉ Carbon of the sugar

○ Carbon of the base components

● Phosphorus

(a)

(b)

FIG. 478-1 — Model of a double-stranded DNA — helix (according to FEUGHELMAN).

FIG. 478-2 — Electron micrograph of rod-like macromolecules a) poly-glutamicacid, $\overline{M}_w = 43,000$ (Magnification 1:120,000) b) collagen, $\overline{M}_w = 360,000$ (Magnification 1:70,000) (from HALL and DOTY).

---

[42] $d_{equ(sed)} = 2mg[1 - (\rho_o/\rho)] /6\pi\eta u$, where u is sedimentation velocity; m, the mass of the sedimenting particle; g, gravity; ρ_o, density of the medium in which the particles are sedimenting; ρ density of the sedimenting particles; and η, viscosity of the medium.

q and calculates according to Stokes' law[42] the corresponding $d_{equ(sed)}$, then one obtains for each axial ratio a quotient $h_1/d_{equ(sed)}$ which corresponds to the $\sqrt{\overline{h^2}}/d_{equ,so}$ value of the coil at least in a qualitative way. If one plots the measured $h_1/d_{equ(sed)}$ values versus the axial ratio $h_1/h_2 = q$, then one obtains a curve such as the one shown in Figure 477.

As one can see, the quotient $h_1/d_{equ(sed)}$ is the larger, the longer the particles are. One can therefore take $h/d_{equ,s}$ as a relative measure for the deviation from spherical shape.

The question whether there are other explanations for the difference between $\sqrt{\overline{h^2}}$ and $d_{equ,so}$ must remain open. A partial draining of the coil, which has been used as an explanation, is improbable because the $[\eta]$ versus $\overline{M}$, and s_o versus $\overline{M}$ relationships on a double logarithmic plot are strongly linear. Furthermore, with a draining coil $d_{equ,so}$ would be larger than $\sqrt{\overline{h^2}}$, as one can easily derive qualitatively from the result of the sedimentation measurements shown in Figure 477: the $1/d_{equ(sed)}$ values approach asymptotically a constant limiting value of about 3 to 3.5. That means that the $d_{equ(sed)}$ values of completely drained particles are (independent of their length) $1/3 \cdot l_{max}$. With polymer chains the ratio q has the order of magnitude of $10^3/1$ to $10^4/1$. As shown in Table 463 with such chains $\sqrt{\overline{h^2}}$ is $1/50$ to $1/100$ of the maximum chain length l_{max}. For a chain of the length $l_{max} = 1,000\text{Å}$, for example, $\sqrt{\overline{h^2}}$ would be about $1,000/50 = 20\text{Å}$ and $d_{equ,s}$ would be about $1,000/3 \cong 300\text{Å}$ for completely drained coils. In this case, therefore, we would have to expect a ratio $\sqrt{\overline{h^2}}/d_{equ,s} \cong 1/15$. Actually, however, one finds $\sqrt{\overline{h^2}}/d_{equ,s} \cong 2/1$ (Table 466).

With massive particles, whose density is known, one can take the ratio of the frictional factors F/F_{Sp} as a measure of the axial ratio of the particles. F is the experimentally determined frictional factor (through diffusion of sedimentation measurements), and F_{Sp} is the theoretical friction factor for example calculated from the Stokes' equation for spherical particles of the same volume ($F_{Sp} = 6\pi r\eta$).

With coils the correct calculation of F_{Sp} is not possible, because one knows only equivalent values for the coil density and not the absolute values. It would formally be possible to calculate absolute values of the axial ratio from the curve in Figure 477. However, one cannot simply compare the average chain end-to-end distance $\sqrt{\overline{h^2}}$ quantitatively with the length l of the rods. Furthermore, one cannot say for certain to what extent the coil form is influenced by the sedimentation motion. It is not impossible that during sedimentation, the particle assumes a shape which gives rise to a lower frictional factor. This means that during sedimentation the coils would become more spherical or drop-like. Nevertheless, Figure 477 permits us to estimate the axial ratio of polymers for which $\sqrt{\overline{h^2}}/d_{equ,s}$ is known. For polymethylmethacrylate, for example, with $\sqrt{\overline{h^2}}/d_{equ,s} \cong 1.95$ (Table 463), we find by means of the calibration curve in Figure 477 an axial ratio of about $4/1$.

The near constancy of the $\sqrt{\overline{h^2}}/d_{equ,s}$ values over a large molecular weight range, as has been observed with many polymers, leads to the conclusion that the geometric form is independent of the molecular weight.

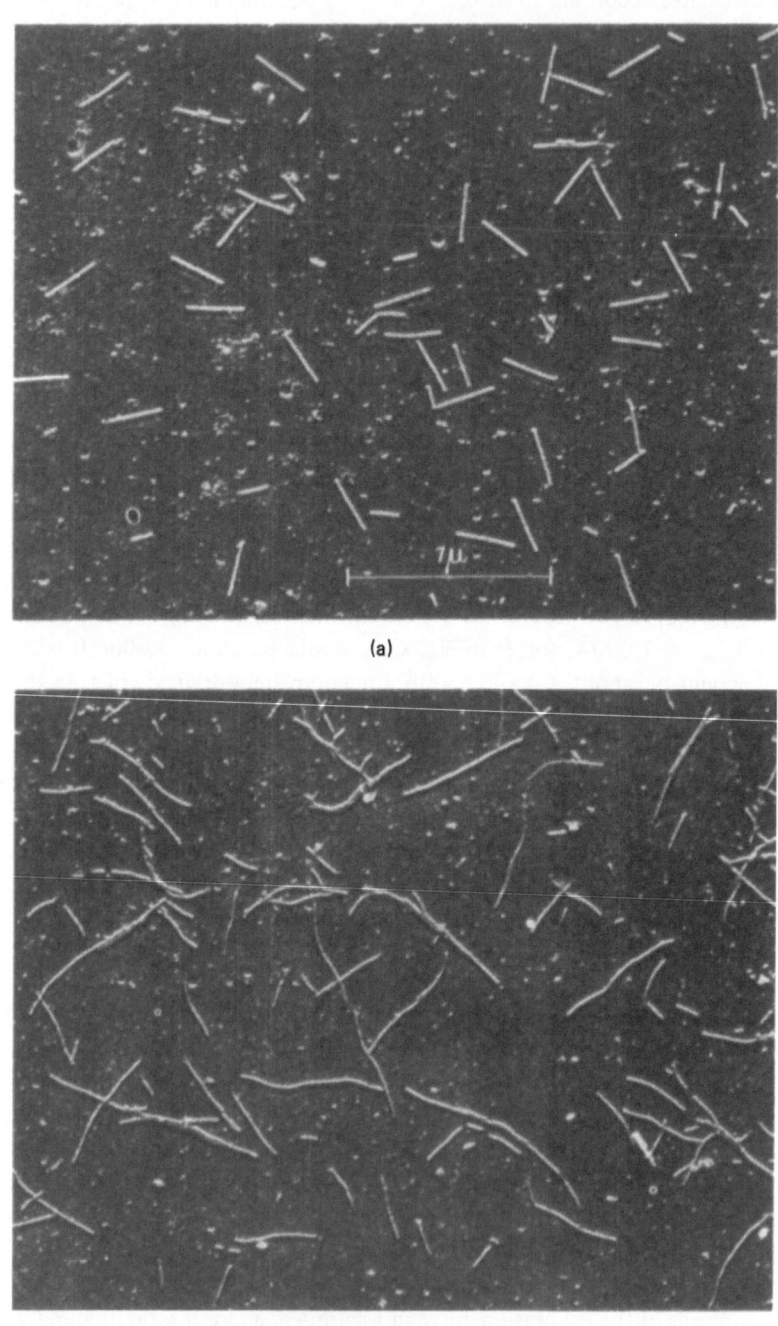

(a)

(b)

FIG. 480 – a) Tobacco mosaic-virus (27,000 x) (SCHRAMM and WIEDEMANN)
b) Potato-Y-virus (27,000 x) (G. SCHRAMM)

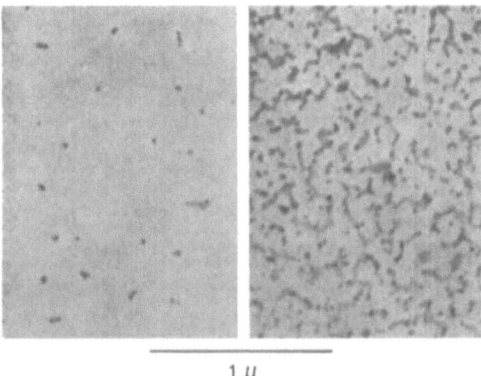

1 μ

FIG. 481 – p-Iodobenzoylglycogen molecules (28,000 x) (E. HUSEMANN and E.
RUSKA)

On the basis of probability considerations, W. Kuhn assumes a bean-shaped form
for the statistical coil with an axial ratio of 6/2.3/1, which corresponds to a
rotational ellipsoid of the same volume with an axial ratio of 4/1.

With the help of statistical wire models, which were built after carrying out
dice-throwing experiments, H. Kuhn determined a most probable axial ratio of
2.72/1.56/1.

Thus even if one cannot give the average dimensions of the coils exactly, one still
finds that the experimental results correspond with the theoretical considerations:
that the coils are in the form of elongated ellipsoids whose axial ratio has the order
of magnitude of 4/1. This does not say that the coil shape with all polymers and in
all solvents is exactly the same. It is possible that the coils in a poor solvent, where
they have a larger density, are more similar to spheres, whereas in good solvents,
where the coil density is less, they are more elongated.

Pictures of Macromolecules with the Electron Microscope

Although the macromolecular coils (depending on the molecular weight) are
somewhere between 100 and 1,000Å large, they cannot be seen in the electron
microscope. This is because they consist of 90% and more of solvent; they therefore
cannot be distinguished from their surroundings. However, with many proteins and
nucleic acids one often finds that the molecular chains organize themselves into
regular spirals or helices which assume the form of more or less stiff or elastic rods
or threads. The tendency to form helices is an effect of the functional groups
situated along the chain at regular intervals. These groups, from one turn of the
helix to the next, can form hydrogen bonds between one another when the chain
assumes a spiral form (compare Figure 597).

Examples of functional groups:

$$
\begin{array}{ccc}
\overset{|}{C}H_2 & \overset{|}{C}=O\cdots & \\
\overset{|}{C}=O\cdots H-\overset{|}{N} & \overset{|}{N}-H\cdots & or \\
\cdots H-\overset{|}{N} & \overset{|}{C}H_2 &
\end{array}
$$

with proteins   with nucleic acids

Because the helix presents a form of minimal potential energy, it is so stable that it is maintained in many solvents [for the transformation: coil ⇌ helix, compare Figures (504-2) and (505)]. Figure (478-1) shows an atomic model of a double-stranded helix of deoxyribonucleic acid, and Figures (478-2) and (480) show several examples of how such helical macromolecules appear under the electron microscope.

In addition to the proteins with rod-like molecules, one also has observed proteins whose molecules are in the form of compact spheres [Figures (482 a and b)].

As far as the inner structure of the spherical macromolecules is concerned, one knows that in the polysaccharide glycogen found in the liver [Figure (481)], one finds a strongly branched molecular structure. We still know very little about the inner structure of the spherical virus particles [Figure (482)].

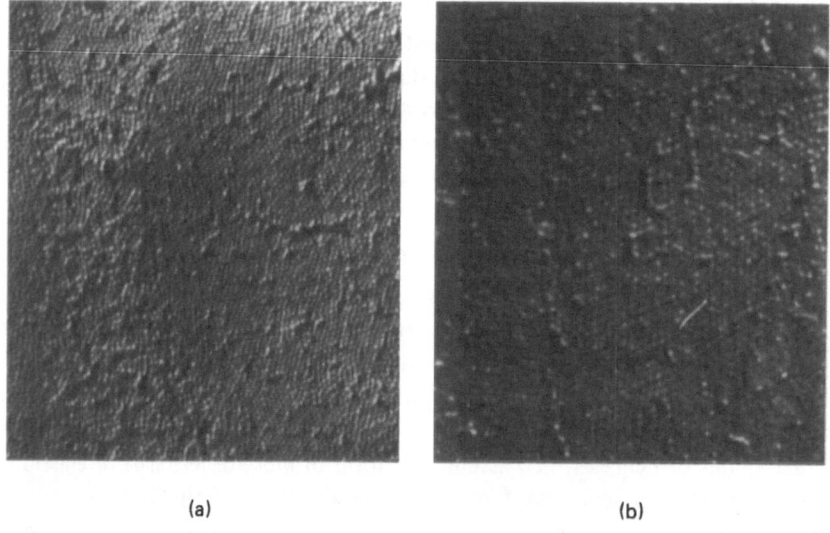

(a)   (b)

FIG. 482 — a) Virus of a tomato disease (39,000 x) b) Rothamsted-Protein (85,000 x) (from WYKOFF).

With proteins, one cannot always assume that particles visible under the electron microscope are equivalent to macromolecules (defined as particles whose atoms are held together by covalent bonds). Thus, in a helix, one often finds two covalent chains wound around each other in the form of a spiral and held together by secondary valence forces [compare Figure (478-1)]. Since the secondary valence forces between the chains often occur through a well-defined number of regularly spaced functional groups, there is a certain justification to considering such particles as individual macromolecules (double-molecules, double-stranded helix).

4. STATES OF MACROMOLECULAR AGGREGATION

41 INTERMOLECULAR FORCES AND AGGREGATION (ASSOCIATION)

The atoms in the molecules of organic compounds are held together by attraction forces, which are generally known as covalent bonds. The attraction is due to a resonance phenomenon between the outer electrons of the atoms which occurs at a distance of the order of magnitude of 1Å and can be regarded as a continuous exchange of electrons (shared electron pair). This resonance has a corresponding sharply defined energy minimum with a maximum energy difference (potential well) of approximately 100 kcal for the hydrogen molecule, which one has to supply in order to split one molecule of hydrogen into two hydrogen atoms.

A similar exchange interaction can also occur between molecules having saturated main valence forces, however at a somewhat larger distance of approximately 4Å. This leads to a potential well which is approximately one or two powers of 10 smaller than that brought about by the electron exchange. Because these attraction forces are brought about by overlap and interchange of vibrating systems with well-defined frequencies and therefore show a dependence on frequency, one characterizes them, in analogy to the frequency dependence of the refractive index, as dispersion forces: When light passes through a material, there results an energetic exchange effect between the high frequency alternating field of the light and the vibrations of the atoms. This is a process which becomes visible in the form of the refraction of light. The dependence of the refractive index on the frequency of the light is called the dispersion.

The fact that the relation between the optical dispersion and the dispersion forces is more than a superficial analogy can be shown by the correlation between heat of evaporation Q and the molar refraction R. The heat of evaporation is work which has to be performed in order to overcome the attractive forces of the molecules and is therefore a measure of ΔE of the dispersion forces, that is, of the interaction between the molecules, or rather of their electron shells with each other. On the other hand, the molar refraction R is a measure for the interaction between the electron shells and the light waves. Both parameters are determined by the polarizability α:

$$Q = f(\Delta E_{Disp}); \quad \Delta E_{Disp} = k \frac{\alpha^2}{r^6} \tag{1}$$

and

$$R = \frac{n^2 - 1}{n^2 + 2} \frac{M}{\varrho} = \frac{4}{3} \pi N_A \alpha. \tag{2}$$

(r, distance between the molecules; n, refractive index).

The polarizability or displaceability α is a measure for the inherent frequency of the electron cloud of a molecule and is defined as the induced dipole moment μ_i caused by the field strength $\epsilon = 1$ volt/cm: $\alpha = \mu_i/\epsilon$.

The fact that ΔE_{Disp} decreases with $1/r^6$ is of fundamental importance for an understanding of the properties and the behavior of polymers in the aggregated states.

The dispersion forces are, however, not the only intermolecular forces (secondary valence forces, van der Waals forces) which bring about the cohesion of liquid and solid materials. With molecules containing polar groups $(-NO_2, {>}C = 0,$

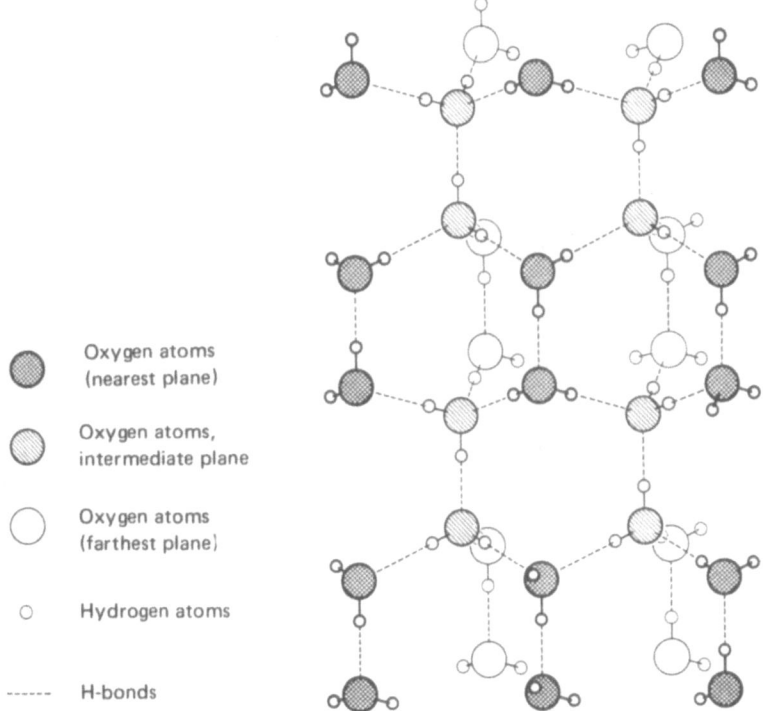

⬤	Oxygen atoms (nearest plane)
◑	Oxygen atoms, intermediate plane
○	Oxygen atoms (farthest plane)
∘	Hydrogen atoms
-----	H-bonds

FIG. 485 — Structure of ice (according to L. PAULING)

$-\overset{|}{\underset{|}{C}}-CN$, $-\overset{|}{\underset{|}{C}}-Cl$) which have permanent dipoles, there is also an electrostatic attraction which in polymers can bring about a considerable increase of their mechanical strength. Insolubility in nonpolar solvents may also be caused by the electrostatic attraction between the polar molecules of a material.

Molecules with a permanent dipole can induce dipoles in nonpolar molecules, especially in systems with aromatic rings where the π electron cloud is easily deformed, so that in this way an electro-static attraction is also brought about between polar and nonpolar molecules. Finally, there is also the possibility of attraction between ions and dipole molecules.

Especially strong intermolecular forces known as hydrogen bonds occur between molecules with active hydrogen atoms and suitable partner-atoms, particularly in water [see Figure (485)], alcohols, carboxylic acids, and carboxylic acid amides. The attractive forces causing the hydrogen bonds are essentially electrostatic forces. In molecular formulas the hydrogen bonds are designated by dotted lines:

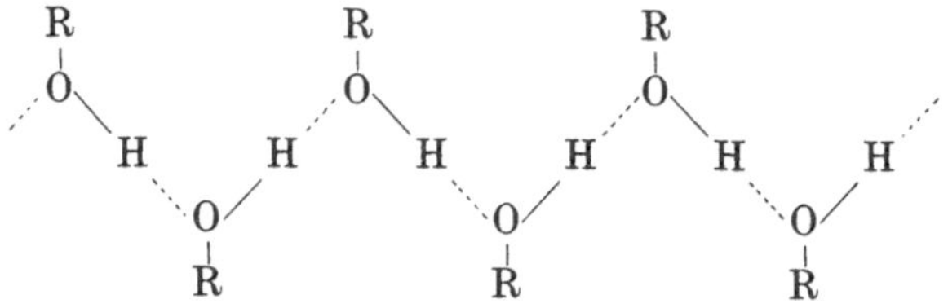

Structure of Alcohols

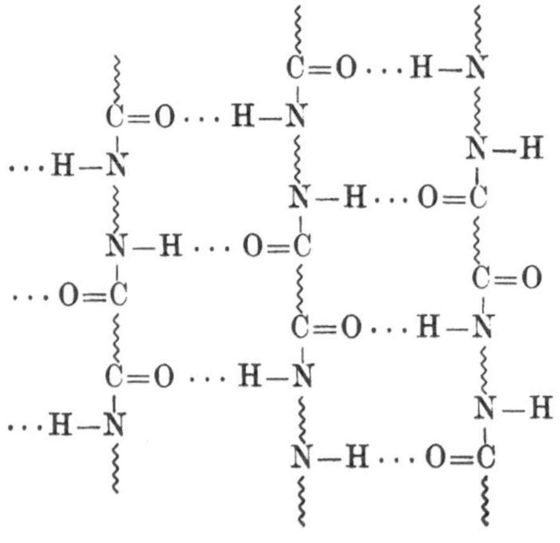

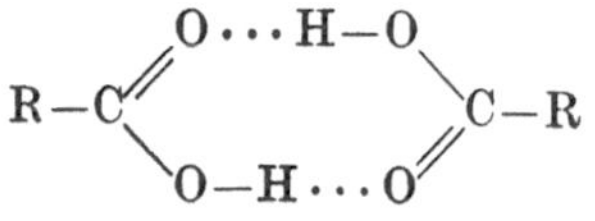

Structure of a Polyamide

Structure of Carboxylic Acids

Hydrogen bonds determine the properties of many macromolecular compounds, especially those of polyamides (both the synthetic polyamides, as well as the natural ones, i.e., the proteins), nucleic acids, and polysaccharides (cellulose).

The gaseous state (the state of minimum, and in the ideal case, completely absent, association), is not possible with macromolecular compounds. Because of the size of the molecules, the intermolecular forces (secondary valence forces) between molecule and molecule, which result in the cohesion of the molecules, are so strong that the necessary energy for evaporation is only available at relatively high temperatures (over 500°C). At these temperatures practically all macromolecular compounds decompose by depolymerization to the monomer, or through some complex processes of carbonization.

Also, the liquid state does not really apply to macromolecular compounds. When macromolecular compounds soften, normally one first obtains a tough rubber-like mass, which at increasing temperatures slowly takes on the properties of a highly viscous plastic material. The viscosity of this melt, like the viscosity of solutions, is dependent on the molecular weight. In the transition region below $\overline{M}$ = 50,000, such plastic melts lose their rubber-elasticity. The technically produced polyamides in particular, such as the 6- and 6, 6- nylons ($\overline{M}$ = 10,000 to 20,000), should be considered viscous liquids at elevated temperatures.

The viscoelastic and rubbery state is characteristic of macrocompounds in their soft state, which they assume above their softening temperature (= glass temperature = solidification temperature). This can be designated with equal justification as a solid state or a liquid state. In the one extreme, for example, with cross-linked rubber, a stretched rubber band returns almost completely to its original state on release of the tension. In the normal case, i.e., with uncross-linked macromolecules above their softening temperature, after release of the stress there remains a more or less strong permanent deformation, the magnitude of which increases with increasing temperature and with decreasing molecular weight. Finally, in the other extreme, after release of the tension there is almost no elastic return. One can therefore define the viscoelastic state as a transition region between viscous liquids and rubbery solids.

In the solid state, macromolecular compounds are usually present in the form of a glass. Normally these glasses are amorphous compounds. Polymers with regular chain structure, especially the isotactic polymers, but also natural polymers, such as proteins, cellulose, and rubber, are able to crystallize. This can be recognized by the increasing turbidity of the glass, which occurs sometimes immediately upon solidification, however, often only after days, or weeks, or even years. With macromolecular compounds crystallization is generally not complete. It can be increased and made more complete by a mechanical orientation—for example, with fibers, by cold or hot drawing. In the crystalline state many macromolecular compounds have a fibrous form. However, fiber structure does not always require crystallinity. Also noncrystallizable compounds, for example, atactic polystyrene and many other noncrystalline polymers can, with sufficiently high molecular weight (over 10^6), assume a fiber structure already upon slight orientation.

The state of least aggregation with macromolecular compounds is the dilute solution, which, therefore, is the only form in which the molecular size and shape of the *individual* macromolecule can be studied. Macromolecular solutions are colloidal solutions, which are especially characterized by their high viscosity.

The viscosity of a solution increases rapidly with increasing concentration. To a corresponding extent one often finds that the liquid character disappears, and the solutions assume the properties of rubber-elastic materials. Then one no longer speaks of a solution, but of a gel. Not all polymers and not all solvents form gels to the same extent. As with rubber, the gel state is observed in its most characteristic form with cross-linked polymers.

These various macromolecular states of aggregation will now be discussed in the following order: the macromolecular solution, the viscoelastic state, the glassy state, and the crystalline state.

42   THE MACROMOLECULAR SOLUTION

421   The Dissolution Process

As all changes of state, the process of dissolution is also determined by the entropy gain and the change in the internal energy of the system:

Every system tends to reduce its internal energy or enthalpy H (i.e., to give off heat) and to increase its entropy, S. The driving force in the system which tries to change its state is the larger, the larger the obtainable increase in entropy (T·ΔS) and the loss in internal energy (−ΔH). Conversely, an increase in internal energy (positive ΔH) and decrease in entropy (negative ΔS) means that no change of state will occur. Both parameters, entropy and enthalpy can change independently of each other. Thus, a process which occurs with an entropy increase, can be coupled either with an energy (enthalpy) decrease (−ΔH), or with an energy (enthalpy) increase (+ΔH), or its ΔH can remain unchanged. Similarly a change in energy can be coupled with both an entropy increase and an entropy decrease. The parameters can change in such a way that both either support or oppose the change in state. It is also possible that a process is favored by the entropy increase, but slowed down by a concurrent enthalpy increase. Whether the process occurs in such a case depends whether the magnitude of ΔH is larger or smaller than the magnitude of T·ΔS:

$$\Delta F = \Delta H - T \cdot \Delta S. \qquad (2a)$$

If one designates a decrease of the internal energy (the system gives off heat, exothermic reaction) by a negative ΔH, then the driving force of the process is the larger, the larger the negative value of the free energy ΔF. Positive values of ΔF mean that a reaction cannot occur by itself. Positive ΔS means an entropy increase, and negative ΔS an entropy decrease during the reaction.

During the dissolution process one always obtains an increase in the mobility of the molecules of the dissolved material, i.e., an increase in the entropy. Thus, dissolution occurs spontaneously when there is no change in internal energy (ΔH = O). If the internal energy decreases (exothermic dissolution process, negative ΔH), then the driving force of the process is increased even further. In the opposite case, with an increase in internal energy (endothermic dissolution process, positive ΔH), the driving force of the dissolution process is diminished, and the more, the larger ΔH. Only when ΔH becomes equal to, or larger than, $T \cdot \Delta S$, will the compound be insoluble.

At a given temperature, the entropy gain will be largest at the beginning of the dissolution process, because then the concentration of the resulting solution is the smallest, and the concentration difference the largest. Gradually the concentration of the dissolved material in the solution increases, and ΔS becomes smaller, until finally $T \cdot \Delta S = \Delta H$, so that $\Delta F = 0$. At this point the system is in equilibrium, and the solution is saturated. Since the entropy increases continously during the dissolution process (i.e., ΔS is always positive), one can achieve saturation only when ΔH is positive, i.e., when the dissolution is an endothermic process. Compounds with positive heats of solution (negative ΔH) are thus miscible with each other in all proportions.

In the last case (an exothermic reaction), the dissolution of macromolecular compounds occurs in principle exactly like that of low molecular weight compounds, i.e., the secondary valence forces between the structural units of the polymer chains are replaced by secondary valence forces between the structural units and the solvent molecules. Always when an entire polymer chain has been completely solvated in this way, it dissolves away from the rest of the material and can then move about in the solvent as a completely free, solvated, solvent-saturated coil. If one looks at the individual phases of this process, the dissolution process of macromolecular compounds is slightly different from that of low molecular weight compounds. Thus, if one compares, for example, the dissolution of glucose in water with the dissolution of cellulose in copper tetraminehydroxide solution, then in both cases glucose molecules i.e., glucose anhydride structural units, are dissolved away from the solid aggregate. However, with glucose each molecule can freely move about in the solvent medium immediately after its combination with water molecules. The glucose molecule can move away as far as it would like to from the crystal from which it has just been separated. This, however, is not the case with the glucose structural unit of the cellulose: after solvation it is still a part of the polymer chain and can only move away to the extent of the length of the solvated portion of the chain. Only when all 500, or 1,000, or 5,000 glucose units which belong to the same chain are also solvated, can the macromolecular coil move away as a unit from the crystallite. Such a dissolution process of macromolecules results in the fact that at the interface, there is always a relatively high concentration of macromolecules, and therefore the dissolution process, even in good solvents, usually occurs quite slowly. Thus solvent molecules can penetrate through the interstices into the interior of the compound and there bring about a partial solvation of the chain segments. Because of the slowness of the dissolution process at the interface, the solvent molecules have sufficient time to penetrate the entire material before it dissolves. Thus, if one does not increase the rate at which the macromolecules diffuse away from the interface by stirring, one finds that first the entire material swells strongly. This means that the diffusion of solvent molecules into the polymer occurs more rapidly than the dissolution of the macromolecules and their diffusion into the solvent. This diffusion into the solvent is slowed even further, as a result of the strong increase in viscosity brought about by the dissolution of the macromolecules in the solvent. This is the greater, the better the solvent. Thus, if one leaves a macromolecular material for a certain time in the solvent, one can observe the gradual transition from swollen gel, to highly viscous solution, to dilute solution.

The above discussion describes the solution process in good solvents, which are miscible in all proportions with the macromolecular compound, and in which the macromolecular compound can be dissolved in principle in the same way as a low molecular weight compound, i.e., each individual molecule is surrounded by solvent molecules, and, depending on the concentration, can move more or less undisturbed in the solution. Particular phenomena of macromolecular solutions caused by the coiling and the intramolecular solvation of the chains will be discussed in the next paragraph 422.

The dissolution of macromolecules differs from that of low molecular weight compounds when ΔH is positive, but its order of magnitude is smaller than $T \cdot \Delta S$, i.e., when the driving force of the process is only the entropy increase, whose magnitude, ΔS, becomes smaller with increasing concentration. (When ΔH is positive and larger than $T \cdot \Delta S$, ΔF becomes positive, i.e., the compound is insoluble). With low molecular weight compounds, one obtains first a dissolution, but the process comes to a halt when the concentration of the resulting solution increases, so that ΔS becomes smaller, and one finally obtains an equilibrium state, when $T \cdot \Delta S = \Delta H$, i.e., $\Delta F = O$. At this stage, just as many molecules combine in unit time to form crystals or aggregate on the existing crystals, as there are molecules that go into solution.

With macromolecular compounds, such a dynamic equilibrium results in the following process: A macromolecule begins to go into solution as described above, such that the individual structural units of the chain are solvated one after the other. However, before the last structural unit is solvated, other structural units of the same polymeric chain, according to the equilibrium, have again linked themselves to other macromolecules. At a later stage these segments of the chain will again become solvated. However, at that point other segments of the chain have again formed secondary valence bonds with neighboring macromolecules. In this way the macromolecules never become completely separated from each other. The solvent penetrates the compounds; however, there always remain so many points where the macromolecules are not solvated (not always the same points, but sometimes here, and sometimes there), so that a secondary valence type of cross-linking remains, and no individual macromolecules separate from the material. The same equilibrium state, which, with low molecular, and difficult to dissolve, compounds, leads to partial dissolution of the material (i.e., a part of it is dissolved, and a part of it remains undissolved) can with macromolecular compounds, when the molecular weight is sufficiently high, lead to a state where the material swells, but none of it goes into solution. One can therefore say that macromolecular compounds are either miscible in all proportions with the solvent, or are completely insoluble (limited swelling). In the strictest sense this only applies to polymers with homogeneous molecular weights. With polymolecular materials, such as one usually deals with, the molecules with lower molecular weight dissolve, while the molecules of high molecular weight remain in the gel state. In the case of insolubility, the polymer swells (depending on the solvation equilibrium and the corresponding density of the secondary valence forces) more or less, i.e., the polymer takes up a limited amount of solvent without releasing the macromolecules into the solution.

Of course, these secondary valence forces do not exist only from one chain to another, i.e., between segments of two different macromolecules, but also within one and the same molecule, from one part of the chain to another. Depending on the chain length of the molecule, (i.e., the degree of polymerization) and the solvation equilibrium, it can happen that the positions where secondary valence links can occur are not close enough together to bring about cross-linking between different macromolecules. In that case the molecules go into solution, and the secondary valence cross-links are only *intra*molecular, i.e., inside the coil. In such solvents the macromolecules are more tightly coiled than in good solvents, and the coil density changes to a smaller extent with the molecular weight (compare pp. 454 and 517).

Whether a macromolecular compound goes into solution in such poor solvents where a solvation equilibrium exists, depends not only on the concentration (frequency) of the nonsolvated segments, but also on the length of the polymer chain, i.e., on the molecular weight of the polymer. Thus, the solvation equilibrium determines for each polymer solvent system a certain concentration of secondary valence cross-link positions. As long as the coil can have this predetermined intramolecular concentration of cross-links predetermined by the solvation equilibrium, one obtains a solution. However, there is a limit to intramolecular cross-linking: the coil cannot have less than a certain density, because any contraction beyond that determined by random flight statistics is a process which leads from a likely to a less likely state and therefore causes an entropy decrease. From a thermodynamic point of view, one always has *inter*molecular secondary valence cross-linking, i.e., a secondary valence link between one macromolecule and another, when the corresponding entropy decrease of the system is matched, or even overtaken, by the entropy decrease caused by the coil contraction which results from *intra*molecular cross-linking. Then the system will choose the way which will give it the ability to achieve the state of cross-linking (determined by the solvation equilibrium) with the lowest entropy decrease. From a certain upper limit of the coil density on, intermolecular cross-linking (i.e., the linking between several coils to a larger aggregate) occurs, because further secondary valence cross-linking inside the coil and the corresponding increase of the coil density force an increase of the internal tension of the coils, the more the coil is away from its statistical density. This internal tension of the coil at forced contraction corresponds to the pressure of compressed gases: in both cases the cause of the pressure is a tendency to go from an improbable, contrived state, to a more probable state. The upper limit of the coil density, from where there is a change-over from intra- to intermolecular association, is the higher, the smaller the molecular weight [i.e., for a given concentration of cross-link points within a coil, the tension resulting from deviation from the statistical state is the larger, the higher the molecular weight, because the statistical (most probable) density of the coil decreases with molecular weight according to the equation: $\bar{\rho}_{coil} = const. \cdot 1/\sqrt{M}$ (compare p. 446)]. A small coil has from the beginning a larger density than a large coil. Therefore, the tension which can occur in a large coil (if because of internal secondary valence cross-linking it is forced to contract to a certain density) is greater than that in a small coil. Conversely, the maximum tension which can be supported before this secondary valence cross-linking becomes intermolecular is reached at a lower concentration of linkage points with large molecules than with small ones. In solvents where a polymer with low molecular weight just dissolves, a fraction of high molecular weight remains insoluble. This is the basis on which the partition of polymolecular polymeric products through fractional precipitation and fractional dissolution to more homogeneous fractions rests (compare p. 389).

422 The Dilute Solution

According to its chemical structure, a macromolecule may be compared to a long chain or a long thin thread. As we have seen, however, macromolecules, are in general, not stretched-out threads, or rods, but occur as random statistical coils. In solution these coils are more or less completely solvated, i.e., they have saturated themselves with the solvent. (Special cases, such as glycogen, strongly cross-linked resins, and some proteins and nucleic acids with helix structures, are not considered here.) The solvent which has been absorbed by the coil is, of course, in constant equilibrium with the surrounding solvent molecules by means of diffusion. However, it is held by the coil through capillary forces to the extent that it follows all the movements of the coil as a whole. This can be seen in the sedimentation behavior in the ultracentrifuge, for example. One designates such solvent-saturated coils as undrained coils. The smallest particles (in the sense of molecules) of the dissolved compound in a macromolecular solution, are therefore strictly speaking not the macromolecules, but the irregularly formed spherical particles which, similar to a sponge full of liquid, consist of a matrix of macromolecular coils and a "filler" of included solvent. The coil and the solvent present in the coil and moving with it form a unit which may be designated a gel molecule because the properties and the behavior of dissolved macromolecules can best be compared to small gel particles.

In order to obtain an idea how strongly the solvent is held in the coil, one can attempt to press out the solvent from a gel (compare Figure 453), or one can place the gel on a filter and suck solvent through the gel. It will be found that it is practically impossible to press a gel dry. Similarly, if one places the gel on a filter, this leads to stopping-up of the filter completely. On the other hand, diffusion processes from a solution into a gel or from a gel into a solution, or also within a gel or from one gel to another (for example, plasticizer migration, compare p. 545) are only relatively slightly hindered by the macromolecular matrix. Thus, the coil is not at all closed off toward the solvent. The solvent molecules can diffuse into the gel and out of it. This has to be taken into consideration when one designates the solvent carried by the coil during a sedimentation run as "included" or "bound" solvent (compare pp. 359 and 468).

Since one does not know the amount of bound solvent, one usually bases the concentration of the solution on the dissolved polymer, i.e., on the pure coil substance. If one uses this concentration in the calculation of the molecular weight, one obtains the molecular weight of the pure coil substance, whereas the actual gel molecules present in solution, i.e., the coil matrix plus the included solvent, have really a considerably higher particle weight.

If one knows the molecular weight, one can determine the coil diameter by means of viscosity measurements, $d_{equ[\eta]}$, or by sedimentation measurements, $d_{equ,so}$ (p. 464). If one knows the coil diameter and the molecular weight of the matrix substance, one also knows the amount of bound solvent and the molecular weight of the gel molecules, so that in this way one can obtain a realistic picture of the state of the macromolecular solution. Figure 493 gives a qualitative, schematic representation of the concepts of chain volume and coil volume, and Table 494 shows quantitative parameters for the chain volume and the coil volume for two polymethylmethacrylate fractions and two cellulose nitrate fractions. From these parameters one can calculate the ratio of the bound and the free solvent at different concentrations.

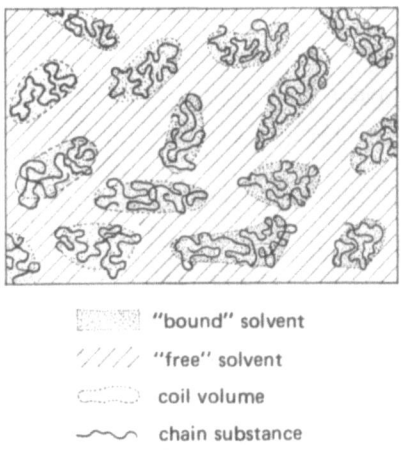

░░░░ "bound" solvent

///// "free" solvent

⌐⌐⌐⌐ coil volume

⌒⌒⌒ chain substance

FIG. 493 — Schematic representation of a dilute macromolecular solution.

As can be readily seen, the coils in these examples carry along between 20 and 700 times their chain volume as solvent, i.e., the gel coils contain, depending on the coil density, approximately 90-99.8% of their volume as solvent. From the values of the coil volume (row 9), and the mass of the chain substance (row 5), one obtains the concentration for which the entire solvent is included in the coils. These values are, therefore, critical concentrations in percent (= g/100 cm³ of solution). They can be found in row 13, and are, of course, except for the factor 100, identical with the density of the empty coils. One sees that the swelling ability of the coils can vary considerably with the molecular size and the type of polymeric compound. For example, with polymethylmethacrylate of $\overline{M}_w$ = 500,000, a 5% solution of polymer is required to take up all free solvent, whereas with cellulose nitrate of $\overline{M}_w$ = 1,000,000, even in a 0.2-0.3% solution, the entire solvent is already held by the coil. In such solutions the coils are present close to each other in dense packing, and the flow of such a solution is really a gliding of the gel particles past each other. Only at very low concentrations can the coils "swim" freely, whereas with concentrations higher than $c_{crit.}$, the coils no longer have the possibility of extending themselves to a size corresponding to a statistical coil.

Recent investigations (Vollmert and Stutz) have shown that even in concentrated polymer solutions ($\cong$ 30%)—in which the coils are closely pressed together—no mutual penetration and entanglement occur. There is only a certain mutual attachment (hook-up) in relatively small contact zones (compare p. 550).

TABLE 494. Chain volume and coil volume for polymethylmethacrylate and cellulose nitrate in acetone.

	Polymethylmethacrylate in acetone		Cellulose nitrate in acetone	
Experimental values:				
1. Molecular weight $\overline{M}_w$	500000	5000000	100000	1000000
2. Sedimentation rate s_0	$32 \cdot 10^{-13}$ sec	$86 \cdot 10^{-13}$ sec	$13 \cdot 10^{-13}$ sec	$27 \cdot 10^{-13}$ sec
3. Viscosity-number $[\eta]$	70 cm³/g	405 cm³/g	300 cm³/g	2600 cm³/g
4. Density of the dissolved polymer $\rho_{Pol} = 1/V_{spec.}$	1.25 g/cm³	1.25 g/cm³	1.75 g/cm³	1.75 g/cm³
Calculated values (from the above experimental values):				
5. Mass of the chain substance $m_{Pol} = M/N_A$	$0.83 \cdot 10^{-18}$ g	$8.3 \cdot 10^{-18}$ g	$0.167 \cdot 10^{-18}$ g	$1.67 \cdot 10^{-18}$ g

TABLE 494. (Cont'd).

	Polymethylmethacrylate in acetone		Cellulose nitrate in acetone	
6. Volume of the chain $v_{chain} = m/\rho_{Pol}$	$0.66\cdot10^{-18}$ cm^3	$6.6\cdot10^{-18}$ cm^3	$0.095\cdot10^{-18}$ cm^3	$0.95\cdot10^{-18}$ cm^3
7. Viscometrically effective coil density* of an equivalent sphere $\overline{\rho}_{equ[\eta]} = 2.5 \cdot 1/[\eta]$	0.036 g/cm^3	0.0062 g/cm^3	0.00835 g/cm^3	0.00096 g/cm^3
8. Coil density* of an equivalent sphere measured by sedimentation $\overline{\rho}_{equ\ s_0} = \dfrac{M/N_A}{1/6\pi d^3_{equ\ s_0}}$	0.05 g/cm^3	0.01 g/cm^3	0.025 g/cm^3	0.0023 g/cm^3
9. Coil volume $v_{equ\ s_0} = 1/6\ \pi d^3_{equ\ s_0}$	$16.5\cdot10^{-18}$ cm^3	$840\cdot10^{-18}$ cm^3	$6.4\cdot10^{-18}$ cm^3	$735\cdot10^{-18}$ cm^3
10. Coil diameter** $d_{equ\ s_0}$	315 Å	1170 Å	230 Å	1120 Å
11. Mass of the gel-coil***	$13.3\cdot10^{-18}$ g	$660\cdot10^{-18}$ g	$5.2\cdot10^{-18}$ g	$570\cdot10^{-18}$ g
12. $\dfrac{\text{Coil volume}}{\text{Chain volume}}$	25/1	125/1	70/1	760/1
13. Critical concentration in % $c_{crit} = \overline{\rho}_{coil} \cdot 100$	5%	1%	2.5%	0.23%

*The density of the equivalent spheres can be obtained from $[\eta]$ and s_0 according to EINSTEIN'S, resp. STOKES', Law. Due to the fact that the form of the coils is not spherically symmetrical, the real coil density is always larger than the coil density of the equivalent sphere. Since the equivalent value measured by sedimentation comes closest to the real coil density, this was used for the calculation of the values listed under 9) to 13) in Table 494. Compare p. 505-510).

**Since the coils do not have a geometrically regular shape, they do not have a real diameter. The diameter-values obtained from sedimentation and viscosity measurements, are those of equivalent spheres. (For the meaning of the equivalent sphere, see p. 463-465 and 505-509).

***Mass of the included solvent = (coil volume-chain volume). $\rho_{acetone}$; mass of the gel-coil = mass of the included solvent + mass of the chain substance. With high viscosity-numbers, the chain volume is much smaller than the coil volume, so that it can be neglected. Then the mass of the gel-coil = coil volume · density of the solvent.

TABLE 496. Polymer concentration and gel-coil concentration for polymethylmethacrylate and cellulose nitrate in acetone (calculated from the data in Table 494)

Polymer concentration (g in 100 cm^3 soln.) (from $\bar{\rho}_{equ\ s_0}$)]	Gel-coil concentration (g in 100 cm^3 - soln.)	Polymer concentration (g in 100 cm^3 soln.) (from $\bar{\rho}_{equ[\eta]}$)
Polymethylmethacrylate		
$M = 500\,000$		
5.0% (Crit. Conc.)*	81% (Crit. Conc.)**	3.6% (Crit. Conc.)*
3.1%	50%	2.3%
0.62%	10%	0.4%
0.06%	1%	0.03%
$M = 5\,000\,000$		
1% (Crit. Conc.)	79% (Crit. Conc.)	0.62% (Crit. Conc.)
0.64%	50%	0.38%
0.13%	10%	0.077%
0.013%	1%	0.008%
Cellulosenitrate $M = 1\,000\,000$		
0.23% (Crit. Conc.)	79% (Crit. Conc.)	0.096% (Crit. Conc.)
0.15%	50%	0.06%
0.075%	25%	0.0385%
0.03%	10%	0.012%
0.003%	1%	0.0012%

$$* \ c_{crit} = \frac{\text{mass of the chain substance} \cdot 100}{\text{coil volume}} = \bar{\rho}_{equ} \cdot 100$$

** critical gel-coil concentration in %

$$= \frac{\text{mass of the gel-coil}}{\text{coil volume}} \cdot 100 \approx \rho_{solv.} \cdot 100$$

Since with high molecular weights the chain substance may be neglected in comparison with the absorbed solvent, the critical gel-coil concentration is 100 times the solvent density.

As a result of this close contact between the coils in concentrated polymer solutions, one observes a steep increase in the viscosity with increasing concentration.

The gel character of solutions above the critical concentration can be observed in many cases by the appearance of a structural viscosity, i.e., a viscosity which is dependent on the velocity gradient.

Table 496 gives an overview over the gel-coil concentration for different polymer concentrations, which also represents the percent of bound solvent present in the total solvent. The concentration range for each polymer begins with the critical concentration, i.e., the highest possible concentration where the coils may still reach their maximum extension in volume.

Table 496 gives two values for each polymer concentration: one is the concentration obtained from the viscometrically active coil density $\bar{\rho}_{equ,[\eta]}$, and the other is the concentration obtained from the coil density $\bar{\rho}_{equ,so}$ determined from the sedimentation rate in the ultracentrifuge. The difference is a result of the fact that the coils are not spherical particles, as will be discussed further in later sections (p. 509, compare also p. 428).

One sees from Table 496, that the concentration of the colloidal particles in macromolecular solutions is always much higher than the concentration of the polymer itself (= concentration of the chain substance), and the difference between the two concentrations is the larger, the higher the molecular weight. This is a result of the decreasing coil density, which decreases according to the relation $\bar{\rho}_{coil} = k \cdot M^{-a}$ with increasing molecular weight. Accordingly, the critical concentration, above which no free solvent is available, is the smaller, the higher the molecular weight.

According to the dimensions of the gel-coils, which, depending on molecular weight and solvent, have diameters between 100 and 1,000 Å, macromolecular solutions belong to the group of colloidal dispersions. In contrast to the colloidal particles of ordinary dispersions, with macromolecular solutions the colloidal particles are identical with the macromolecules, and therefore one speaks of molecular colloids.

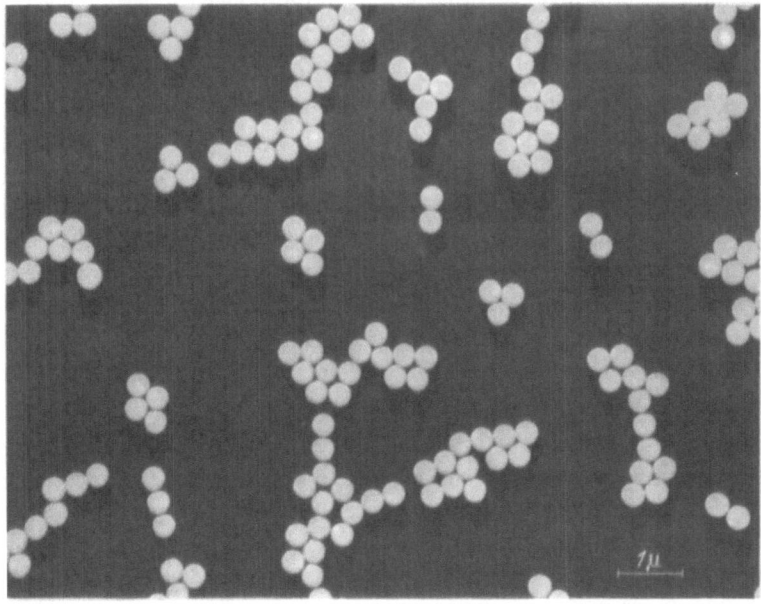

FIG. 497 — Electron micrograph of an aqueous polystyrene dispersion (prepared by emulsion polymerization) as an example of a dispersion-colloid (from K. SCHÄFER, courtesy BASF).

Figure 497 shows an example of a normal dispersion of solid particles in water, polystyrene latex, under the electron microscope. The spherical particles have a dimension of approximately 0.5μ (= 5,000Å). They are thus comparable to the size of larger gel-coils in a macromolecular solution. However, they are easily visible, whereas the macromolecular gel-coils in general are not, because a macromolecular gel-coil does not sufficiently contrast with its environment since it consists of more than 90% of solvent. Only in some special cases of intramolecular cross-linked macromolecules with extremely high molecular weight (microgels) has it been possible to get micrographs of dissolved polymer molecules by means of the phase contrast technique (compare Figure 621-2).

In spite of their size, the colloidal particles undergo a Brownian motion, which can easily be seen by observing a polymer dispersion, similar to that in Figure 497, under the light microscope.

With gel-coils, the molecular motion is not (as with latex particles) simply a three dimensional movement of the entire particle (*inter*molecular Brownian motion), but also the chain segments of the coil matrix are in continuous motion (*intra*molecular Brownian motion). This leads to the result that also the shape of the gel-coil undergoes continuous changes. This must be taken into consideration when one speaks of particle shape with such molecular colloids. As was discussed in the chapter on molecular shape, the most probable form of the coils is that which is similar to a bean-shaped ellipsoid. This means that a large fraction of the coils, at a given time t, is present most often in such a form. One can also express this, by saying that a particular coil, observed over a sufficiently large time Δt, is present as a bean-shaped ellipsoid during most of this time.

The absence of a definite separation boundary and the deviation of the average coil shape from that of the spherical particle make all statements about the coil diameter somewhat ambiguous. One obtains different numerical values depending on which physical phenomenom one considers. Thus by light-scattering, one obtains the parameter $\sqrt{\bar{h}^2}$, i.e., the average end-to-end distance of the chains. By means of viscometric measurements, one obtains the diameter of an equivalent sphere $d_{equ[\eta]}$, and from sedimentation measurements in the ultracentrifuge, one obtains the value of $d_{equ,so}$, which is somewhat smaller than $d_{equ[\eta]}$. As model experiments show (compare p. 508), the volume of the equivalent spheres is somewhat larger than the actual coil volume.

Through the intramolecular Brownian motion the gel-coil undergoes a constant intensive blending action, so that the reactivity of the chain molecule is not considerably reduced in any way by the gel character of the coil. This explains why cross-linked gels or those obtained by precipitation (in which there is tight coiling of the molecules) are still highly active reaction media. (For example, the methylation of carboxyl group-containing polymer gels with diazomethane still occurs quantitatively in a few minutes at $-20°C$ to $-40°C$). This is particularly important for the reaction of living organisms, where nearly all reactions in the cell occur in a gel state.

There is considerable difference between colloids of macromolecular compounds and dispersed colloids from compact materials: the density of the coils and therefore their swelling ability (i.e., their ability to take up solvent and to hold it) is not constant for one and the same materials, but changes with the particle size. Thus, one finds that the coil density increases with increasing molecular weight, and therefore the swelling ability increases. This is a property which can be derived from the statistical laws for coils, and which has to be taken into account in the interpretation of such physical processes as sedimentation, diffusion, light-scattering, and viscosity. It should also be taken into consideration in interpreting such natural processes as the swelling or the contraction of tissue.

423   The Viscosity of Dilute Solutions

The viscosity of liquids is caused by the frictional resistance which occurs when molecules of the liquid are undergoing motion with respect to each other (Figure 499). If P_1 and P_2 are two glass plates with a liquid between them, then the force one has to apply in order to move the plate P_2 with the velocity u in a direction opposite to that of plate P_1, is the larger, the larger the frictional resistance of the molecules against each other. This frictional resistance, in turn, is the larger, the larger the secondary valence forces between the molecules. The secondary valence forces per molecule are in general the larger, the larger the molecules, and they can become especially large if groups which can form hydrogen bonds are present (such as OH groups, CO–NH groups, etc.). For example, glycerol has an especially high viscosity because of its OH groups; on the other hand, paraffin oils have a high viscosity because of their large molecules.

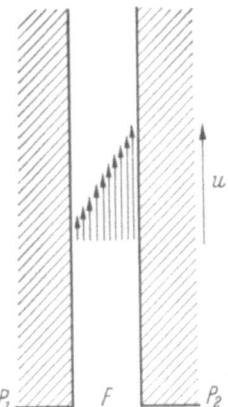

FIG. 499 – Increase in the rate of flow of a liquid layer F between two plates P_1 and P_2.

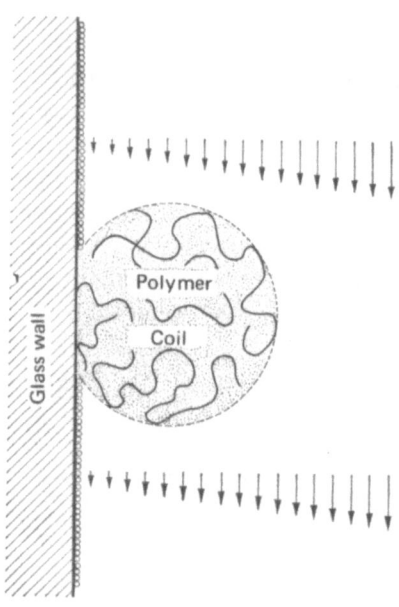

FIG. 500 — Wall-layer of a streaming polymer solution in a capillary.

This effect is increased to the extreme with melts of macromolecular compounds, which hardly flow without the mechanical application of force. On the removal of this force (pressure or tension), they show elastic recovery.

In a colloidal solution, the relatively large colloid particles and the relatively small solvent molecules are combined to a homogeneous system. If one assumes that in a macromolecular solution both components of the system have spherical particles, then the coils have a diameter of the order of magnitude of 500-1,000Å, whereas the solvent molecules have a diameter of 5-10Å. A model of such a system on a proportional scale would therefore consist of a mixture of small pinhead-sized glass spheres (d = 1 mm) and large glass spheres of 10 cm in diameter.

In a flow gradient (Figure 500), the large particles of the colloidal system act like breaking points and, therefore, increase the frictional resistance between the individual layers. This can immediately be seen if one looks first at the stationary glass wall, where at least one monomolecular layer of the solution is held by adsorption. As Figure 500 shows, the adsorbed colloidal particles penetrate and interrupt the streaming solution, thus creating braking points which slow down the flow of the solvent molecules, which is the faster, the greater their distance from the wall. In the interior of the solution, the large spheres flow with the streaming solvent, but since in the entire space F between the two plates there is a flow gradient, they are always partly present in a faster layer and partly in a slower flowing layer, which leads to a slowing down of the faster flowing layer. The flowing layers are thus linked to each other by means of the polymer molecules, which increases the frictional resistance between them. This leads to an increase in the viscosity, which becomes larger, the larger the concentration of the colloidal particles in the solution, as described by the viscosity law derived theoretically by Einstein for spherical colloids:

$$\eta_{rel} = 2.5\varphi + 1. \tag{3}$$

In this equation $\eta_{rel} = \eta/\eta_0$ is the relative viscosity, which describes how many times greater the viscosity of the solution is, than that of the solvent, and, where φ is the volume fraction of the dissolved phase, i.e., the volume of the colloidal particles in 1 cm^3 of the solution.

4231   The Einstein Viscosity Law for Macromolecular Solutions

As has been shown previously (compare pp. 375 and 450), one can also write the Einstein law in the following form:

$$[\eta] = 2.5/\bar{\rho}. \tag{4}$$

In this equation the viscosity number $[\eta]$ is defined as $[\eta] = \eta_{sp}/c$, where c tends to 0 (compare p. 376), and $\eta_{sp} = (\eta - \eta_0)/\eta_0$, which is the relative viscosity increase.

$\bar{\rho}$ is the density of the colloidal particles. Since the colloidal particles of macromolecular solutions consist of the coil matrix and the included solvent, one can calculate either with the density of the gel-coil (matrix + solvent) or with the density of the empty coil. Which density one has to use (i.e., which density is used in the Einstein equation) depends on what is used as the basis of the concentration in the expression η_{sp}/c. Thus $\bar{\rho}$ is the density of the gel-coil if one calculates the concentration c (g/cm^3) on the basis of the gel-coil (as in Table 496 for polymethylmethacrylate) [One uses the density of the equivalent sphere which corresponds exactly to the coil. See also pp. 451 and 506-509]. On the other hand, usually one does not know the amount of included solvent and therefore does not know the mass of the gel-coil. One therefore assumes that c is the concentration of the polymer itself, so that $\bar{\rho}$ is the much smaller density of the empty coil (mass of the chain substance divided by coil volume). As can be seen from Table 494, for example, for polymethylmethacrylate of $\overline{M}_w = 500,000$, the density of the gel-coil is approximately 0.8 g/cm^3, whereas the density $\bar{\rho}_{equ,[\eta]}$ of the empty coil is only 0.036 g/cm^3. In practice, one can only count on the basis of the polymer concentration, so that $\bar{\rho}$ is the density of the empty coil. Since the coils are not spherical particles, but have an irregular shape, $\bar{\rho}$ is not completely identical with the coil density, but corresponds to the density of spheres which are equivalent to the coils in terms of their viscosity-increasing efficiency. One therefore has to use in the Einstein equation for the density $\bar{\rho}$, the density of the equivalent sphere $\bar{\rho}_{equ,[\eta]}$, to describe the viscosity increase in macromolecular solutions;

$$[\eta] = 2.5/\bar{\rho}_{equ[\eta]}. \tag{5}$$

Since the density differs from molecule to molecule and changes continually with each individual coil, the coil density is always an average.

In this form the Einstein viscosity law tells us that the viscosity number of the coil is inversely proportional to the coil density. The validity of the law has been shown to hold for normally dispersed colloids, as well as for the molecular colloids of macromolecular compounds (chain-molecule coils).

Density changes of coils can also be measured (in addition to viscosity measurements), by sedimentation measurements, diffusion measurements, and light-scattering measurements. As has been shown earlier (compare pp. 468 and 465), the following relationships hold:

for sedimentation:

$$s_0 = \text{const.} M^{2/3} \rho_{coil}^{1/3} \text{ (Stokes)}, \tag{6}$$

for diffusion:

$$D_0 = \text{const.} M^{-1/3} \rho_{coil}^{1/3} \text{ (Nernst)}, \tag{7}$$

for light scattering:

$$\sqrt{\overline{h^2}}_{opt} = \text{const.} M^{1/3} \rho_{coil}^{-1/3} \text{ (Sphere volume)}, \tag{8}$$

for viscosity:

$$[\eta] = \text{const.}/\bar{\rho}_{coil} \text{ (Einstein)}. \tag{9}$$

Since with coils the density always decreases with increasing molecular weight according to Equation (10) (for derivation see pp. 445 and 455):

$$\bar{\rho}_{coil} = \text{const.} \cdot \bar{M}^{-a}. \tag{10}$$

One can replace the coil density in Equations (6) to (9), according to Equation (10) by (const.$\cdot \bar{M}^{-a}$):

$$s_0 = \text{const.} M^{(2-a)/3} = \text{const.} M^u, \tag{11}$$

$$D_0 = \text{const.} M^{-(1+a)/3} = \text{const.} M^d, \tag{12}$$

$$\sqrt{\overline{h^2}} = \text{const.} M^{(1+a)/3} = \text{const.} M^b, \tag{13}$$

$$[\eta] = \text{const.} M^a. \tag{14}$$

Thus, for a particular polymer, if one finds from viscosity measurements a certain value for the exponent a of the coil density function, the other exponents, calculated according to $u = (2-a)/3$, $d = -(1+a)/3$, and $b = (1+a)/3$, can be compared with those obtained experimentally from sedimentation, diffusion, and light-scattering measurements, if the proportionality between $[\eta]$ and $\bar{\rho}$ demanded by the Einstein viscosity law holds true. The relations (11) to (14) have not been checked carefully with all polymers. However, in cases where they have been determined exactly for a sufficiently large molecular weight range (for example, with polymethylmethacrylate, isobutylene and cellulose, and partly also with other vinyl polymers), good correspondence between experimental and calculated exponents has been obtained.

Figures 503 and 504-1 show the relationship described by Equation (11) between the exponent a of the $[\eta]$ versus $\overline{M}$ function, and the exponent u of the s_0 versus $\overline{M}$ function, for an example particularly interesting to biologists. (Further examples for the validity of the exponent relationships (11) to (14) can be found on pp. 466, 471, and 472.) This example deals with solutions of a polynucleotide, prepared by one and the same enzymatic system (adenyl-ribose-polyphosphate), at different pH values. In neutral aqueous solutions, the chains are statistically coiled. During the transition to an acid medium, association takes place, probably under the formation of a spiral structure (double helix). This leads to an increase in the molecular weight and a stiffening of the chain.

If one calculates the value of the exponent u of the s_n versus $\overline{M}$ function according to Equation (11), one obtains for case I: $u = [2-a]/3 = 0.46$ and for case II: $u = [2-a]/3 = 0.357$. One can see that in both cases (coil and helix, respectively) the calculated and the experimental exponents correspond.

This example is also interesting because here the increase of the viscosity with molecular weight occurs much more strongly because of the helix structure. However, the absolute values of the viscosity number are higher for the coil structure because the helix contains a crystalline spiral structure, which leads to a

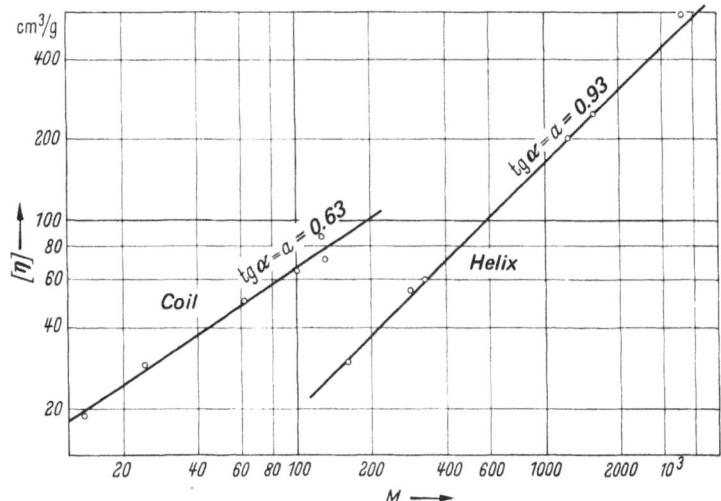

FIG. 503 — $[\eta]$ vs. M diagram of adenylribose polyphosphate at different pH-values (FRESCO and DOTY).

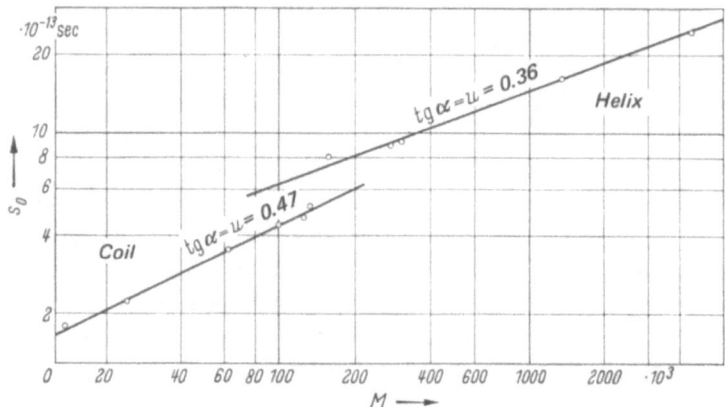

FIG. 504-1 — s_0 vs. M diagram for adenylribosepolyphosphate at different pH-values (FRESCO and DOTY) (Formula of adenylribosepolyphosphate on p. 18).

relatively high particle density. Short helix segments can be regarded as regularly wound coils with a high density. With longer segments, the entire helix again forms a statistical coil. This is possible in this special case, because the helix does not consist of a single element, but of a large number of smaller polynucleotide chain segments (see Figure 504-2). An undisturbed helix forms straight, or only slightly bent, rods (compare Figure 478).

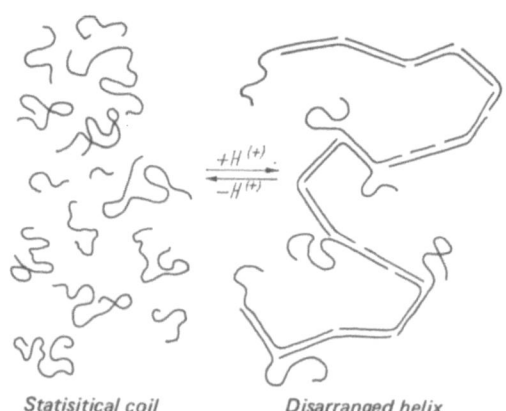

Statisitical coil   Disarranged helix

FIG. 504-2 — Schematic representation of the helix-coil transition with adenylribosepolyphosphate (according to DOTY) (Compare also Fig. 505).

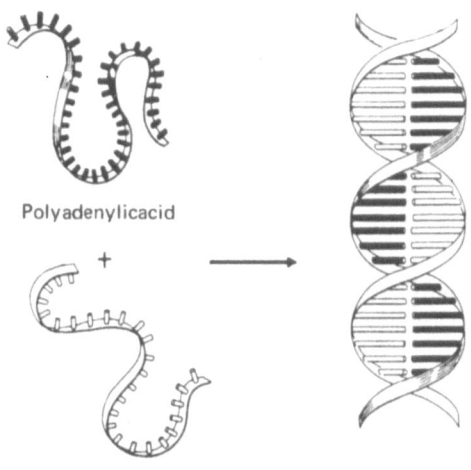

Polyadenylicacid

+

Polyribothymidylicacid   Double-stranded helix

FIG. 505 — Diagram showing the formation of a double-stranded helix. The rods represent complementary functional groups between which hydrogen-bonds can form. (After RIECH)

Insofar as the exponential relationships in Equations (11) to (14) are valid (i.e., the laws of Stokes, Nernst and Einstein), one can determine changes in the coil density by sedimentation, diffusion, or viscosity measurements. The case where one would like to obtain absolute values of the coil density is more complicated. In principle, it is of course possible, with the aid of the Einstein viscosity equation or the sedimentation equation, to obtain absolute values for the coil density from the easily determined values of $[\eta]$ or s_o. However, these equations have been derived for spherical particles, and one therefore also obtains density values for spherical particles which in solution give rise to the same viscosity increase, or the same sedimentation velocity, as the actual coils with their irregular shape (on the average elongated particles of axial ratios between 3/1 to 5/1; compare pp. 477-481). Since one cannot expect that the deviation from spherical shape shows up in the same way for different physical processes, one can also not expect that the viscometrically determined coil density is completely identical with that obtained from sedimentation measurements. In Table 506 are shown the coil densities calculated from experimental values of $[\eta]$ and s_o according to the Einstein and Stokes equations, respectively, for a series of polymethylmethacrylate fractions.

As the data in the table show, the viscometric coil densities $\bar{\rho}_{equ,[\eta]}$ are smaller than the corresponding sedimentation values. This means that with viscosity measurements the deviation of the coils from a spherical shape is more pronounced than in sedimentation. In addition to the deviation from spherical shape, one can also assume that certain other properties, in which the coils are different from "Einstein" or "Stokes" spheres, might explain this difference: for example, internal solvation. However, this would only cause difference from closed spheres, if it would lead to a draining of the coil. As has been discussed in the chapter on coil shape (compare pp. 468-473), the coils are not freely drained, and most probably the elongated form of the coil is the main reason for the difference observed in the coil density determined by different methods.

TABLE 506. Comparison of viscometric coil-densities and coil-diameters of polymethylmethacrylate in acetone with those obtained from sedimentation.

1. M	2. s_0^* in Svedberg	3. $d_{equ\,s0}$ in Å	4. $\bar{\rho}_{equ\,s0}$ g/cm^3	5. $[\eta]^*$ cm^3/g	6. $\bar{\rho}_{equ[\eta]}$ g/cm^3	7. $d_{equ[\eta]}$ Å	8. $\dfrac{\bar{\rho}_{equ\,s0}}{\rho_{equ[\eta]}}$	9. $\dfrac{d_{equ\,s0}}{d_{equ\,[\eta]}}$
$2 \cdot 10^4$	8.0	50.5	0.495	7.2	0.348	56.7	1.42	0.89
$0.5 \cdot 10^6$	32.0	315	0.0505	70	0.0357	353	1.42	0.89
$1 \cdot 10^6$	43.4	466	0.0315	120	0.0208	533	1.52	0.87
$2 \cdot 10^6$	58.5	691	0.0192	202	0.0124	798	1.55	0.87
$3 \cdot 10^6$	70.0	867	0.0147	275	0.0091	1012	1.62	0.86
$5 \cdot 10^6$	86.5	1170	0.010	405	0.0062	1367	1.61	0.86
$7 \cdot 10^6$	100.0	1415	0.0080	520	0.0048	1660	1.65	0.85

*Measurements of G. MEYERHOFF and G.V. SCHULZ.

$$3. \quad d_{equ\,s0} = \frac{2 M (1 - V\rho_{solv})}{6 \pi \eta s_0 N_A} \quad \text{(Derivation on p. 464).} \tag{15}$$

$$4. \quad \bar{\rho}_{equ\,s0} = \frac{M/N_A}{\pi/6 \, d_{equ\,s0}}, \tag{16}$$

$$6. \quad \bar{\rho}_{equ\,[\eta]} = 2.5 \cdot 1/[\eta] = \frac{m}{v_{coil}} = \frac{M/N_A}{1/6 \, \pi \, d_{equ[\eta]}^3}, \tag{17}$$

From which one obtains $d_{equ[\eta]}$:

$$7. \quad d_{equ\,[\eta]} = \left(\frac{6 M [\eta]}{2.5 \, \pi N_A} \right)^{1/3} \tag{18}$$

Therefore, on the basis of viscosity and sedimentation measurements, one can only say that the coils *behave*, as far as their viscosity is concerned, as if they were closed spheres with a density $\bar{\rho}_{equ,[\eta]}$ and a diameter $d_{equ,[\eta]}$, and in sedimentation like spheres with a density $\bar{\rho}_{equ,so}$ and a diameter $d_{equ,so}$. Thus in speaking about the diameter, or respectively, the density of equivalent spheres, one has to realize that this does not tell us anything about the actual shape of the coil. $d_{equ,[\eta]} = 353$ Å is not the coil diameter, but simply one of several experimentally measurable values for the coil size, and $\bar{\rho}_{equ,[\eta]} = 0.0357$ g/cm^3 is not the real coil density, but the density of theoretical spheres which in a solution bring about the same viscosity increase as a coil of polymethylmethacrylate of molecular weight 500,000 in acetone at 25°C (compare Table 506).

It is possible to estimate the direction and the magnitude by which the real coil density deviates from the equivalent value $\bar{\rho}_{equ,so}$ by the following simple experiment: If one deforms a sphere, for example, a sphere made of wax, to an ellipsoid or to a rod and compares the sedimentation velocity of the sphere and the rod, one observes that the average sedimentation velocity u of the rod (u depends on the position of the rod) is smaller than that of the sphere. Now one can calculate the diameter of the sphere which would have the same sedimentation velocity as the rod, using the Stoke's equation:

$$d_{equ(sed)} = \frac{2\,m \cdot g\,(1 - \varrho_0/\varrho)}{6\,\pi\,\eta\,u_{(sed)}}, \tag{19}$$

where m, the mass of the ellipsoid, equals the mass of the original sphere; ρ, the density of the ellipsoid, equals the density of the original sphere; $u_{(sed)}$, the sedimentation velocity of the ellipsoid, equals the sedimentation velocity of the equivalent sphere < sedimentation velocity of the original sphere; and where g is the acceleration due to gravity; ρ_0, the density of the solvent (dispersion medium); and η, the viscosity of the solvent (dispersion medium).

As can be seen, d must become larger as u becomes smaller, i.e., the diameter of a sphere which is equivalent in terms of the sedimentation velocity to the rod is larger than the diameter of the original sphere equivalent to the rod in volume, because u_{sed} of the rod is smaller than u_{sed} of the sphere from which is was formed. Corresponding to the larger d_{equ}, the density $\rho_{equ(sed)}$ (for the same mass) of the equivalent sphere is smaller than the actual density ρ of the sedimenting rod or ellipsoid. All sedimentation experiments with cylindrical models have shown that the effective density of rods with an axial ratio between 2/1 and 7/1 is approximately twice as large as the density $\rho_{equ(sed)}$ of the corresponding equivalent spheres. As Table 508 shows, the density $\rho_{equ(sed)}$ (in the region under examination) is only relatively little dependent on the axial ratio of the rods, so that the result (even though one knows the axial ratio of the coils only as an order of magnitude) can also be applied to the conditions existing in a macromolecular solution. Thus, the actual density of the coils with their irregular ellipsoidal shape is larger than the equivalent density $\bar{\rho}_{equ,so}$ which is calculated from the sedimentation constant. Since $d_{equ,[\eta]}$ is larger than $d_{equ,so}$, the difference between the actual coil density and the equivalent coil density calculated from the viscosity number according to the Einstein equation $\bar{\rho}_{equ,[\eta]} = 2.5/[\eta]$ must be even greater.

The actual value by which the equivalent density $\rho_{equ,so}$ is smaller than the actual coil density, can be obtained from the model experiments only in an approximate fashion, since it is not possible to compare the coil exactly with cylindrical wire models. With ellipsoids, however, the deviation of the sedimentation velocity from that for a sphere, should not be as large as with cylinders, so that the value of the quotient $\rho_{rod}/\rho_{equ(sed)} \cong 2$ (resulting from the values in Table 508) may be regarded as the maximum deviation. Thus, the actual density of the coils can be at most twice as great as the equivalent density $\bar{\rho}_{equ,so}$, and at most three times as great as the equivalent density $\bar{\rho}_{equ,[\eta]}$.

TABLE 508. Density of equivalent spheres, $\rho_{equ(sed)}$, as a function of the axial ratio, q, for the sedimentation of wire model pieces (diam. = 0.01cm) in paraffin oil.

q	$\rho_{equ(sed)}$ (experimentally determined from u_{sed})	ρ_{rod}	$\dfrac{\rho_{rod}}{\rho_{equ(sed)}}$	$\dfrac{d_{equ\,(vol)}}{d_{equ\,(sed)}}$
2/1	4.4	7.8	1.8	0.82
3/1	4.2	7.8	1.9	0.81
4/1	4.6(?)	7.8	1.7	0.84
5/1	4.0	7.8	1.9	0.81
6/1	3.7	7.8	2.1	0.78
7/1	3.4	7.8	2.3	0.76

With cross-linked gels, the coil density can be measured directly by determining the amount of solvent taken up during swelling (compare Figures 453 and 544). If one decreases the degree of cross-linking more and more, the amount of solvent taken up at maximum swelling becomes larger and larger, until at the critical degree of cross-linking the gel goes over into a solution, where the coil density can be determined viscometrically. In this way the coil density was determined for a polystyrene of molecular weight 10^6 and was found to be larger by a factor of 2 than the corresponding viscometric coil density $\bar{\rho}_{equ,[\eta]}$. Therefore, we assume that the equivalent value for the coil density from sedimentation $\bar{\rho}_{equ,so}$ will be only slightly different from the actual average coil density, in any case less different than that obtained with the cylindrical models.

That the equivalent value $\bar{\rho}_{equ,so}$ is smaller than $\bar{\rho}_{coil}$ can be assumed because the ellipsoidal particles with elongated forms sediment more slowly than spheres of the same mass and the same volume. That the equivalent value $\bar{\rho}_{equ,[\eta]}$ is smaller than $\bar{\rho}_{coil}$ depends on the volume demands on the coil: because of temperature fluctuations, the coils undergo an irregular rotational movement; thus, the volume they require in the sense of the Einstein equation (that is, the volume fraction φ) is larger than the actual volume. Correspondingly, the viscosity increase brought about by elongated ellipsoids is larger than that brought about by spheres of the same mass and the same volume. Therefore, the density $\bar{\rho}_{equ,[\eta]} = 2.5 \cdot [\eta]$ will be smaller, and the corresponding diameter $d_{equ,[\eta]}$ larger than the values ρ and d of the coils, which one would find if the coils would be transformed from the ellipsoid form into spheres of the same mass and volume (compare Figure 509).

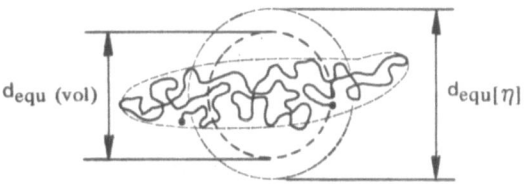

FIG. 509 — Diagram illustrating the concept $d_{equ[\eta]}$.

The last column in Table 506 shows the ratios $\bar{\rho}_{equ,s_0}/\bar{\rho}_{equ,[\eta]}$ and $d_{equ,s_0}/d_{equ,[\eta]}$. These are found to be constant within the limits of experimental accuracy in the determination of the relations $[\eta] = f(M)$ and $s_0 = f(M)$ and the slopes of the resulting lines. This again confirms the validity of the exponential relationships [Equations (6) to (9)]. Since the proportionality of d_{equ,s_0} and $d_{equ,[\eta]}$ has been confirmed for a number of polymers, and since viscosity measurements are always more easily carried out than diffusion measurements, it is possible to determine the coil diameter $d_{equ(diff)} = d_{equ,s_0}$, which is needed for the calculation of the molecular weight from the sedimentation velocity according to the Svedberg equation (compare p. 362), also from viscosity measurements.

As far as the magnitude of the viscosity increase, which one observes on the dissolution of macromolecular compounds, is concerned, the colloidal particles in macromolecular solution are not different from ordinary dispersed colloids, as long as the concentration of the solution does not refer to the actual polymer coils (i.e., the weight of the polymer), but refers to the gel-coils. It is a striking and at first surprising result that as little as 0.04 g of cellulose nitrate ($\bar{M} = 10^6$) doubles the viscosity of $100 \, cm^3$ of acetone[1], whereas one needs approximately 1,000 times that amount of a normal dispersion colloid in an aqueous dispersion to achieve the same effect.

The impression of the peculiarity of this phenomenon disappears, if one considers that the colloid concentration (= concentration of the gel-coils) of the 0.04% cellulose nitrate-containing solution is actually about 31%. By comparison, for an aqueous polystyrene dispersion (as is obtained in emulsion polymerization), the concentration at which the viscosity of the dispersion is twice the viscosity of water is approximately 42% ($\rho_{styrene} \cong 1.05$ g/cm^3), assuming that with such high concentrations there is no deviation from the Einstein law.

If one observes such deviations even with polystyrene dispersions with their hard, exactly spherical, colloidal particles with sharp boundaries against the closed phase (compare Figure 497), one can understand much more easily that for macromolecular solutions with concentrations below 0.1% such deviations will occur. They are eliminated by extrapolating a series of η_{sp}/c values measured at different concentrations to the limiting value $c = 0$.

Colloidal solutions of macromolecular compounds and colloidal dispersions of ordinary solid particles are different in that the colloidal particles in the macromolecular solution are only formed on dissolution as the coiled chain molecules take up solvent. Also, the amount of solvent bound by the statistical coils is not constant, but depends on the molecular size, the solvent, and the temperature, and can undergo considerable changes. The statistical coil can expand and contract, so that also the volume which the macromolecules occupy in the form of colloidal gel particles (the value φ in the Einstein law) can be completely different. Thus, if the coils contract, they squeeze out solvent and the solution dilutes itself. On the other hand, if they expand, solvent is absorbed, the free solvent volume becomes smaller, and the concentration (with regard to the volume of the coils!) increases. It is easiest to describe these phenomena as changes in the coil density.

4232 Viscosity Changes Resulting from Density Changes of the Coils

The coil density is subject to different influences: it changes from compound to compound and from solvent to solvent; it changes with temperature; and it changes with the particle size. But with such macromolecular electrolytes as polyacrylic acid or pectin, one also finds that the coil changes with the degree of dissociation. All these changes in the coil density are accompanied by an inversely proportional change in the viscosity.

The Increase in the Viscosity Number with the Molecular Weight

The dependence of viscosity on molecular weight is a characteristic property of macromolecular solutions, which distinguishes them from all other colloidal systems. W. Kuhn first realized that this characteristic results from the random flight character of the coil and the related dependence of the coil density on the

---

[1] Measured: $[\eta] = 2,600$ cm^3/g (compare Table 494). On doubling of the viscosity of $\eta = 2\eta_0$ or $\eta/\eta_0 = \eta_{rel} = 2$ and $\eta_{rel} - 1 = \eta_{sp} = 1$. If one sets $\eta_{sp} = 1$, in $[\eta] = \eta_{sp}/c = 2,600$, then one obtains l/c $= 2,600$ and c $= 1/2,600 = 0.000385$ g/cm^3 = approx. 0.04%. (In reality, the viscosity increase at this concentration is even greater than $2\eta_0$, since η_{sp}/c increases with concentration. $[\eta] = \eta_{sp}/c$ thus holds in a strict sense only at c = 0.) The corresponding gel particle concentration is calculated from the value for the critical concentration (Figure 544) and Tables 494 and 496). With a critical polymer concentration of 0.096% (= 2.5/$[\eta]$ •100), the gel-coil concentration is 78%: (mass of the gel-coils•100/coil volume = $570 \cdot 10^{-18}$g•100/735.10^{-18}cm^3; compare Table (494): Therefore, at a polymer concentration of 0.0385%, the gel-coil concentration is 31%.

molecular weight. As has been derived on page 445, for statistical coils the square-root law holds:

$$\bar{\rho}_{coil} = const.\cdot\overline{M}^{-0.5}. \qquad (20)$$

This equation is valid for the statistical coil in the absence of external forces. With real coils, which are subject to the chain-stiffening action of the solvent, the exponent is only equal to 0.5 in so called θ solvents, but in general it lies between 0.6 and 0.8. If one designates the exponent in general with a, one can write:

$$\bar{\rho}_{coil} = const.\cdot\overline{M}^{-a} \qquad (21)$$

If one now substitutes in the Einstein equation $[\,[\eta] = 2.5/\bar{\rho}_{coil}]$ the density ρ [according to Equation (21)] by $const.\cdot\overline{M}^{-a}$, one obtains for the dependence of the viscosity number $[\eta]$ on the molecular weight:

$$[\eta] = K_{[\eta]}\cdot M^{a}. \qquad (22)$$

This relationship, with $a = 1$, was found by H. Staudinger empirically with solutions of cellulose derivatives, and was reformulated by H. Mark as an exponential function long before it was theoretically explained by W. Kuhn. It is considered as a general viscosity law for macromolecules. [In the literature, one often finds the designations "Staudinger–Kuhn equation" and "Mark–Houwink equation."] This relationship is obtained directly from the Einstein viscosity law as long as one takes into account the fact that the coils change their density according to $\bar{\rho} \propto \overline{M}^{-a}$.

Equation (22) has been found to be true for all carefully investigated macromolecular compounds. The value of the exponent a with most compounds with statistical chain molecules lies close to the ideal value 0.5, usually between 0.6 and 0.75. For compounds with especially stiff chains, such as derivatives of polysaccharides, one finds values around $a = 1$. The exponent $a = 1$ is, however, not an upper limit; nor is $a = 0.5$ a lower limit. For example, pectin, a natural polygalacturonic acid ester, has a value of $a = 1.2$. Yet one can prepare solutions of polyisobutylene in benzene with a smaller than 0.5 (below the θ temperature). In many cases, however, such solutions with small a values lead to a partial precipitation of the polymer, and they become turbid or gel. Finally, there is also a series of macromolecular compounds whose solutions behave similar to normal colloidal dispersions of solids as a result of their special structure. With such solutions $a = 0$, and $[\eta]$ is independent of the molecular size. The strongly cross-linked polysaccharide, glycogen, and a group of proteins behave in this way. The ideal value $a = 0.5$ can be achieved for many polymers by suitable choice of a solvent where the chain-stiffening (and therefore coil-expanding) secondary valence forces between the chains and the solvent are exactly compensated for by the forces between the chain segments themselves (which lead to an increase in the density of the coil). Usually one plots Equation (22) in the form of an $[\eta]$ versus $\overline{M}$ double logarithmic plot and obtains a line with the slope $\tan \alpha = a$ and the intercept at the ordinate $K_{[\eta]}$. Since such curves are of great utility for the viscometric determination of the molecular weight, a number of examples are shown there (compare Figures 379, 454, and 455, and also Table 380).

As the derivation of Equation (22) from the Einstein viscosity law shows, the increase of the viscosity with the molecular weight is brought about by the fact that the coils, because of their random flight character, are the more expanded, the higher the molecular weight. The smaller the coil density, the larger is the volume which is taken up by a given amount of the macromolecular compound, and therefore the larger the volume fraction which a colloidal particle in a colloidal solution takes up, and the larger the viscosity η of the solution compared to the viscosity η_o of the solvent, according to the Einstein relation: $\eta/\eta_o = 2.5\,\varphi + 1$.

The constant $K_{[\eta]}$ of the viscosity Equation (22) can be examined more closely if one substitutes in the Einstein equation for the coil density $\bar{\rho}_{equ,[\eta]}$ the mass of the polymer chain m and the coil volume $v_{equ,[\eta]}$:

$$[\eta] = 2.5/\bar{\rho}_{equ,[\eta]} = 2.5\,v_{equ,[\eta]}/m = \left(2.5 \cdot 1/6\pi d^3{}_{equ[\eta]}\right)/M/N_A \qquad (23)$$

The diameter $d_{equ,[\eta]}$ is proportional to the average chain end-to-end distance (compare Table 466):

$$d_{equ\,[\eta]} = F \cdot \sqrt{\bar{h^2}} \qquad (24)$$

The form factor, F, corresponds to the deviation of the coil shape from that of the sphere. As can be derived from random flight statistics (compare p. 443) one can write for $\sqrt{\bar{h^2}}$:

$$\sqrt{\bar{h^2}} = \sqrt{A_m \cdot L_{max}} = \sqrt{A_m \frac{l_{Mon}}{M_{Mon}} M_P} \cdot \qquad (25)$$

A_m is the length of the segment in the random flight coil (the so called statistical chain element), and L_{max} is the maximum chain length of the polymer molecule which is obtained from the length l_M of a monomer unit and the degree of polymerization P: $L_{max} = l_M P = M_p \cdot l_M/M_M$, whereby M_P is the molecular weight of the polymer and M_M is the molecular weight of the structural unit.

As has been shown in the discussion of the coil properties, the square-root law holds for ideal coils. With real coils, the diameter changes not with $M^{0.5}$ but with M^b where $b = (1+a)/3$ (compare p. 466). Then Equation (25), i.e., Equation (24) changes as follows:

$$d_{equ[\eta]} = F\left(A_m \cdot \frac{l_M}{M_M}\right)^{(1+a)/3} \cdot M_P^{(1+a)/3}. \qquad (26)$$

If one substitutes this equation into the Einstein relation (23) one obtains:

$$[\eta] = 2{,}5 \; \frac{\frac{1}{6}\pi\,[F(A_m \cdot l_M/M_M)^b \cdot M_P^b]^3}{M_P/N_A}.$$

If one replaces b again with $(1+a)/3$ one obtains:

$$[\eta] = 2{,}5 \tfrac{1}{6} \pi \; N_A \cdot F^3 \cdot \left(A_m \cdot \frac{l_M}{M_M} \right)^{1+a} \cdot M_P^a \qquad (27)$$

and $K_{[\eta]}$ is:

$$K_{[\eta]} = 2.5 \; 1/6\pi N_A F^3 (A_m \cdot l_M/M_M)^{1+a}. \qquad (28)$$

This equation shows that $K_{[\eta]}$ (and consequently also the viscosity number $[\eta]$) depends on two characteristic parameters, F and A_m. The factor F can be determined by viscosity and light-scattering measurements as the ratio $d_{equ,[\eta]}/\sqrt{\overline{h^2}}_{opt}$ and is for most polymers in the neighborhood of 0.6 (compare Table 466).

The statistical chain element A_m can be calculated from $K_{[\eta]}$ by means of Equation (28). According to its meaning in the random flight statistic, A_m should be a measure for the flexibility of polymer chains. As we will see in the next section but one, this holds true only with θ systems.

Dependence of the Viscosity Number on the Structure of the Polymer Chain

The formation of a statistical coil depends on the condition that the statistically coiled chain and the fully stretched-out chain are not different, or at least not too different, energetically from each other. If this is the case, however, (i.e., if both states are different), then the thermodynamically more stable state, i.e., the state with the lower energy, is preferred. This can lead to the possibility that instead of statistical coiling one obtains regular chain-folding (helix formation), as can be observed with many proteins, for example. One can consider this state as an intramolecular crystalline state, which remains unchanged in solution provided the forces which bring it about are strong enough. The regular chain-folding with proteins is brought about by the hydrogen bonds between CO and NH groups (compare Figure 597).

In addition to the energetic preference for a particular arrangement of the atoms in the chain, a more stretched-out state can also be preferred as a result of steric hindrance. In every case an expansion of the coils above the ideal statistical state takes place because of the volume demands of the chain ("excluded volume effect"): in random flight, a molecule (for example, a gas molecule undergoing thermal motion) can, as often as it wants to, cross its own path, so that the line of flight can lead several times through one and the same point in space. This is impossible with real macromolecular coils, because the space which at a given time is occupied by a part of the molecular chain is not available for a different part of the chain. This effect is increased even further if the polymer chain has bulky or voluminous side groups, for example, with polyvinyl carbazole:

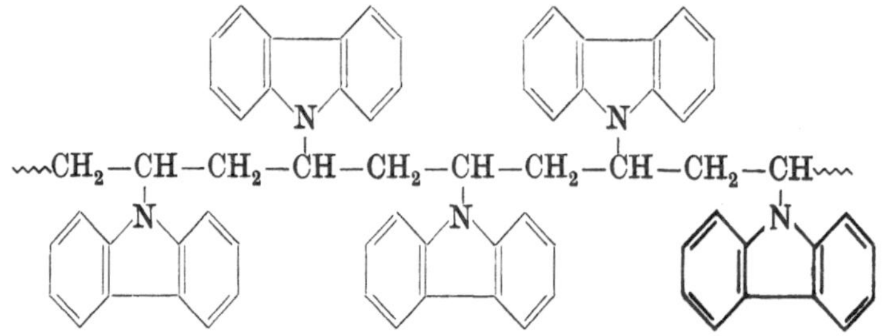

With such macromolecules, a tightly coiled state is impossible for steric reasons. The same holds true when the chain itself contains bulky structural units (for example, cellulose and cellulose derivatives or other polymers containing ring units in the chain, such as aromatic polyimides).

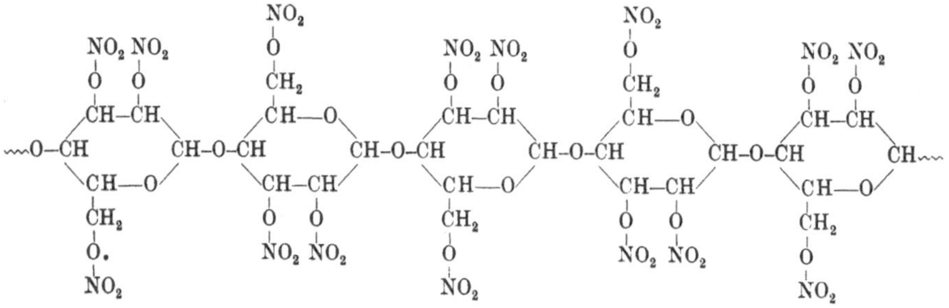

Cellulose nitrate (disregarding its steric configuration)

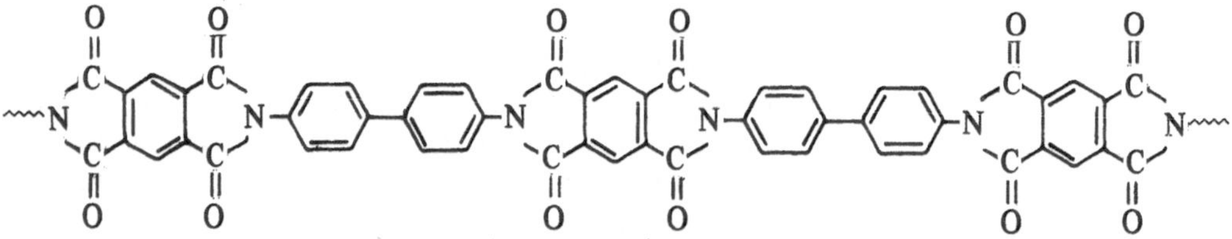

Aromatic polyimide

It is easy to see that the chains here are stiffer, and the coils therefore must be more extended, than for example, with polyethylene or polyisobutylene. Therefore, the viscosity increase, which such compounds with bulky molecules and low coil density bring about, is especially high. For example, cellulose nitrate with a molecular weight of 1 million has a viscosity number of $[\eta] = 2,600$ cm^3/g, whereas polymethylmethacrylate of the same molecular weight and in the same solvent has only a viscosity number of $[\eta] = 120$ cm^3/g. Another instructive example for the influence of the primary structure of the polymer chain on the viscosity number $[\eta]$ is shown in Figure 526. The $[\eta]-M$ relation for polymethylmethacrylate is compared with the $[\eta]-M$ curves for polystyrene and for two polyamide carboxylic acids, which differ only in the diamine component: 4,4′ –diaminodiphenylether in the one and benzidine in the other one. The comparison shows that the polyamide carboxylic acids with their big, stiff structrual units have considerably higher [η] values than the polyvinyl compounds with their more flexible chains.

Not only the absolute value of the viscosity number, but also its change with the molecular weight is strongly influenced by the stiffness of the chain. Usually, the stiffer the chains, the more rapidly the average coil density decreases with increasing molecular weight, and thus the viscosity number increases. Macromolecular compounds with stiff chains often have high exponents a in the viscosity equation and vice versa. Thus, with cellulose nitrate, a = 1, whereas with polymethylmethacrylate in acetone, a = 0.73, and with glycogen a = 0 (compare Figures 454 and 455).

Especially strong differences in the molecular form are found with several polypeptides (synthetic and natural), in which the polymer chains, because of the regular arrangement of the CONH groups and their tendency to form hydrogen bonds, can form spirals under certain conditions in a solution. Such macromolecules with helix structure have the form of rods, as can be seen under the electron microscope (compare Figures 478 and 480). With polybenzylglutamates (prepared by polymerization of the N-carboxyanhydride of benzylglutamate, see p. 248), it was observed that the polymer is present as a helix in chloroform, and in

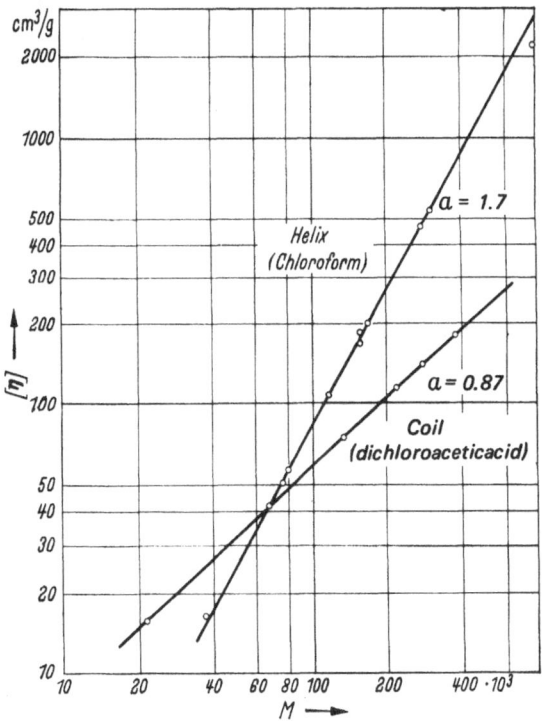

FIG. 515 — [η] vs. M plot for polybenzylglutamate in chloroform and in dichloroaceticacid (DOTY).

the form of a statistical coil in dichloroacetic acid. **Figure 515** shows that the viscosity number for the relatively stiff-chained helix increases much more rapidly with the molecular weight ($a = 1.7$), than the coil form ($a = 0.87$) (compare also Figure 504-2).

A completely different influence on the viscosity of macromolecular solutions is exercised by branched structures, such as the ones shown schematically in Figure 516.

With such structures, the molecular weight is the sum of the molecular weights of the individual branches. The coil density, however, corresponds to the molecular weight of a single branch, because the coil statistics apply to linear chains. Therefore, the entire coil has the density which would correspond to a single branch, if it could move about freely. The density of branched molecules is larger, and the viscosity smaller, than that of compounds with linear chains of the same molecular weight under the same conditions (solvent, temperature, etc.). This is because the average coil density decreases with increasing molecular weight (according to M^{-a}), and, conversely, increases with decreasing molecular weight, and because the partial molecular weight of the individual branches is considerably smaller than the molecular weight of the entire molecule. A systematic quantitative examination of these relationships has not been possible as yet, since it is only in recent years that we can attach branches or side chains to back-bone polymer chains at will. Thus with most graft-polymerization techniques, up to now one has usually also observed cross-linking, and furthermore it was not possible to determine exactly the length of the side chains and their number.

If one has fractions of a polymer, in which the higher fractions are more strongly branched than the lower ones, then the viscosity number with the higher fractions no longer increases, or only slowly, so that there will be deviations from the viscosity equation $[\eta] = K_{[\eta]} \cdot \overline{M}^a$. Conversely, a curvature of the $[\eta] - \overline{M}$ curves on the double logarithmic plot may often be an indication that there is increasing chain branching with increasing molecular weights.

The stronger the degree of branching, i.e., the smaller the linear chain segments between branching points, the smaller the increase of the viscosity will be, until finally the net structure within the individual macromolecule is so tight that the typical internal elasticity of the coils disappears completely, and the macro-

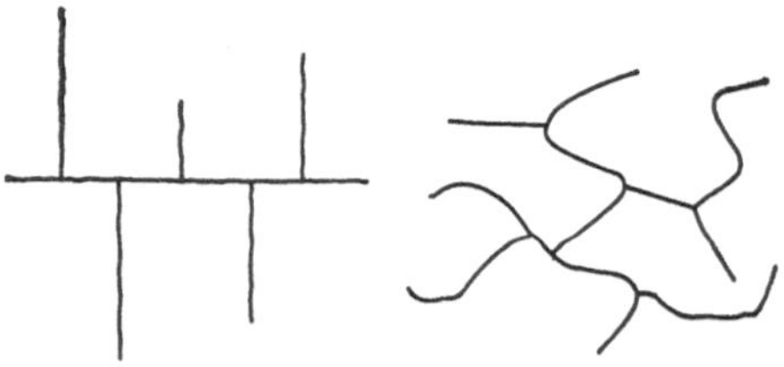

FIG. 516 — Schematic representation of branched macromolecules.

molecules will not be distinguishable from ordinary compact colloidal particles. This is the case, for example, with the branched liver polysaccharide glycogen, and certain proteins. On dissolution of such compounds, there is only a very slight increase in viscosity, and the viscosity number $[\eta]$ is independent of the molecular weight (Figure 455).

Dependence of the Viscosity Number on Solvent and Temperature

There is a dynamic equilibrium between the dissolved molecules of a saturated solution and the original solid body, such that in unit time just as many molecules go into solution as are precipitated from the solution. During solvation, molecules of the dissolved material form secondary valence bonds with the molecules of the solvent, and on precipitation from the solution secondary valence bonds are reformed between the molecules themselves. (crystals or droplets). In this respect, there is no major difference between low molecular and macromolecular compounds. However, a macromolecular solution is more "elastic" in its behavior, because aggregation exists not only between macromolecule and macromolecule, but also within a single coil. Then, within a macromolecule or coil, there are chain segments which are solvated and others which are associated with segments of the same chain (Figure 517). This is not necessarily a reason for precipitation of the dissolved polymer to take place. The solvated portions of the coil keep the entire material in solution; however, the degree of coiling increases as a result of the intramolecular association: the coil contracts. The macromolecules have the chance to loose their freedom of movement in a stepwise manner: they first decrease the freedom of movement of the polymer chain within the coil, and only then, when the solvent becomes even poorer, is the movement of the entire coil restricted.

Like the solubility equilibrium between dissolved and undissolved molecules in low molecular solutions, there is an equilibrium between solvated and associated chain segments in macromolecular solutions: the better the solvent, the fewer the secondary valence links which form within a coil, and the lower the coil density and, therefore, the higher the viscosity. The poorer the solvent, the more undissolved associated segments can be found within a coil, and, therefore, the coil is tighter, and the viscosity is lower.

Thus by choosing suitable solvents, one has the ability to vary the coil density of a dissolved compound within certain limits. How wide these limits are depends

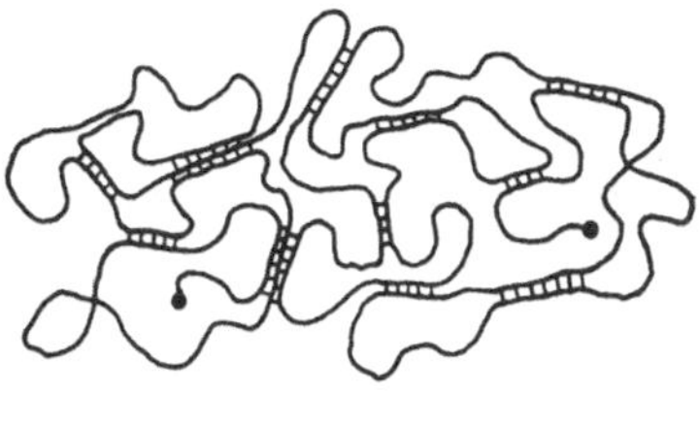

FIG. 517 – Coil showing *intra*molecular associations.

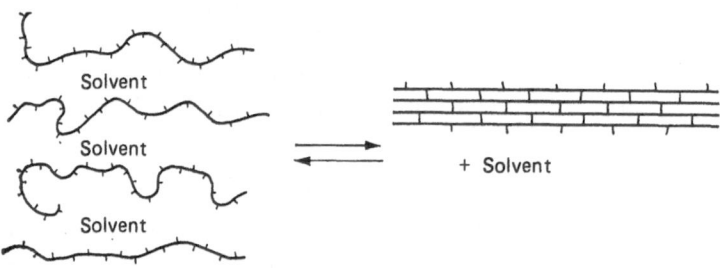

FIG. 518 – Solution equilibrium.

mostly on the constitution of the polymer chain, which determines the strength of the secondary valence forces between the chains. With polymers having strong secondary valence forces, for example, hydrogen bonds (cellulose, polyamides), or with polar groups (polyacrylonitrile), as well as with compounds which, because of their regular chain structure, tend to crystallize (polyvinylidene chloride, poly-ethylene, and isotactic polymers), the choice of a solvent is limited. For such polymers the schematical reaction represented in Figure 518 is favored thermo-dynamically from left to right, unless there is a solvent present, which attaches itself as strongly, or stronger, to the structural units of the polymer (through secondary valence forces), as the bonds formed between one chain and another. In other words, if there is any solvent at all which can dissolve these polymers, it has to be a good solvent. Therefore, in such solutions the coil density is relatively low and the viscosity correspondingly high.

This is different with polymers which have relatively weak intermolecular forces, such as polystyrene, polyvinylether, polyisobutylene, polybutadiene, and several copolymers: for these polymers there is usually a series of solvents with increasing solvent power, such that there are some, where the viscosity is relatively low and the coil density relatively high, and others where the coil density is low and the viscosity correspondingly high. Good solvents are in general compounds whose molecules have a constitution similar to the structural units of the polymer. Thus such polymeric hydrocarbons as polyisobutylene and polybutadiene are readily soluble in hydrocarbons such as hexane, heptane, or cyclohexane, and in those aromatic liquids with low viscosity number, such as benzene or toluene. Conversely, polystyrene with its benzene rings along the chain, is readily (i.e., with high viscosity number) soluble in benzene and ethylbenzene, and only poorly soluble in parafinic hydrocarbons, such as a cyclohexane. Similarly, polystyrene (as a hydrocarbon) is not soluble in acetone or methylacetate and only poorly in methylethylketone (low viscosity number), whereas polyacrylates are readily soluble in acetone and in general in ketones and esters.

Conversely, it is possible to "construct" a polymer for a given solvent by copolymerization of monomers which, if polymerized alone, would give polymers with different solubility, such that one can obtain a maximum or minimum viscosity increase in that solvent. This process is used to a great extent in industry, for example, in the production of lube oil additives (V.I.-improvers), which decrease the temperature dependence of the viscosity for such oils. These additives (mostly complex methacrylate or acrylate copolymers), when they are dissolved in mineral oils, make it possible that at low temperatures (20°C), there is only a small viscosity increase (poor solvation and relative large density of the coil). With higher temperatures, however, the solvation becomes better, the coil density smaller, and the viscosity-increasing function of the additives considerably larger. In this manner it is possible to counteract the normally rather steep viscosity decrease of mineral oils on heating, and the viscosity-temperature curve becomes flatter.

It is a general rule that the increase in the viscosity number is the larger with increasing temperature, the larger the coil density at ordinary temperatures. The viscosity number is the relative viscosity increase. The viscosity of the solution itself decreases with increasing temperature, although $[\eta]$ increases. With solutions of polymers in good solvents, however, the temperature dependence of the viscosity number is only small. For example, the viscosity number of polyisobutylene in the good solvent cyclohexane, $[\eta] = 1{,}250$ cm^3/g, (according to measurements of Staudinger and Hellfritz) does not change at all on warming from 20°C to 60°C, whereas the viscosity number of the same polyisobutylene with benzene as the solvent increases from about 40 cm^3/g at 20°C to 700 cm^3/g at 60°C.

The opposite case, where solvation decreases with increasing temperature, is restricted to exceptions such as methylcellulose and polymethylvinylethers, which precipitate on heating of the aqueous solution. Most probably such polymers are only water-soluble because the ether group can, at low temperature, build oxonium complexes with water, and these are destroyed at higher temperatures, so that the polymer again becomes insoluble.

The strong viscosity increase, which is brought about in good solvents by polymer molecules, is not only the result of the shift in the solubility equilibrium, i.e., the breaking of intramolecular association (secondary valence links) according to Figure 517. In addition to this, a good solvent also causes chain-stiffening, probably because with a linear chain a larger number of solvent molecules can link themselves to the monomer segments in the chain, than with a strongly folded chain, and, therefore, the stretched-out state is thermodynamically favored in good solvents. This can readily be seen from the fact that ring-closure reactions with small molecules with a chain length of five to six atoms occur with better yield in poorly solvating solvents than in good solvents (Lüttringhaus).

The tendency for coil expansion is also increased by the excluded volume effect: whereas the ideal statistical coil (random flight coil) can occupy one and the same place in space several times with a single chain segment, this is not possible with an actual coil. In compensation for this excluded volume in the interior, the coil occupies a larger total volume. Therefore the coil volume, even without the influence of a strongly solvating solvent, is already larger, and the density smaller, than in the ideal case. In order to reduce the coil density to its ideal value, one

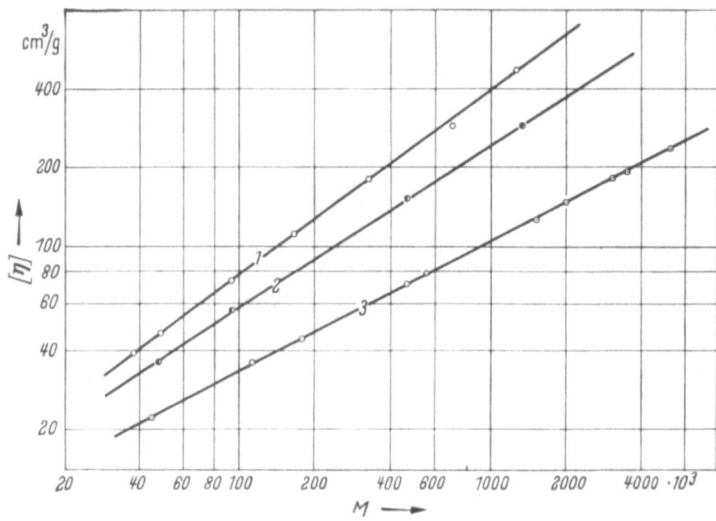

FIG. 520 – $[\eta] = f(\overline{M})$ for polyisobutylene in different solvents at 20°C 1) in cyclohexane; 2) in diisobutylene; 3) in benzene (Measurements of FLORY).

therefore needs, working in opposition to coil expansion, an intramolecular association, brought about by secondary valence forces between different parts of a polymer chain. Since these forces can only make themselves felt when they are not completely neutralized by the secondary valence forces between the polymer chain and the solvent molecules, one has to look for the ideal coil in poor solvents rather than in good solvents.

The influence of the solvent not only makes itself felt in terms of the viscosity number, but also affects the change in viscosity number with molecular weight. Good solvating solvents stiffen the chain and expand the coil, and therefore the normal decrease of the coil density (due to random flight statistics) according to $\bar{\rho}_{equ} \propto M^{-0.5}$ is increased, so that the exponent a in good solvents is not 0.5 but 0.6 to 0.8, and with polymers with bulky chains, even 1. (In the extreme case of the maximally extended chain, or with helix rods, one finds $a = 2$; compare Figures 515 and 526). In poor solvents, on the other hand, the expanding action of the excluded volume effect, and of solvation is compensated for by intramolecular secondary valence links (association), so that for a series of polymers one can find solvents where the decrease in the coil density and the corresponding increase in the viscosity occurs according to the square-root law for ideal coils ($a = 0.5$). Since the viscosity number in systems with little solvation, (as has been described above), is strongly dependent on temperature, the exponent a has the value 0.5 only at a certain temperature. Figures 520, 521-1, and 521-2 show the solvent- and temperature-dependence of the $[\eta]$ versus $\overline{M}$ relation for a number of examples.

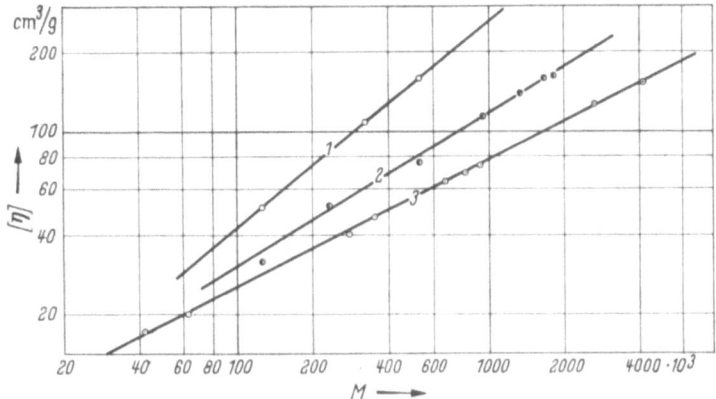

1) in benzene (MEYERHOFF)
2) in methylethylketone (OUTER, CARR, and ZIMM)
3) in cyclohexane (34°C) (CANTOW)

FIG. 521-1 $-[\eta] = f(M)$ for polystyrene in different solvents at 20°C

Solvents, in which at normal temperatures (0°C-40°C) the exponent a has the value of 0.5, are called θ solvents, and the temperature at which $a = 0.5$ is called the θ or Flory temperature (it is measured in degrees K). At this temperature, not only is the exponent of the $[\eta]$ versus $\overline{M}$ function = 0.5, but also the exponent of the $\sqrt{\overline{h^2}}_{opt}-\overline{M}$, the $s_o-\overline{M}$, and the $D_o-\overline{M}$ function have the ideal value 0.5, so that in the $a-b-u-$ diagram see (Figure 474), the two straight lines intersect at a = 0.5 (θ point).

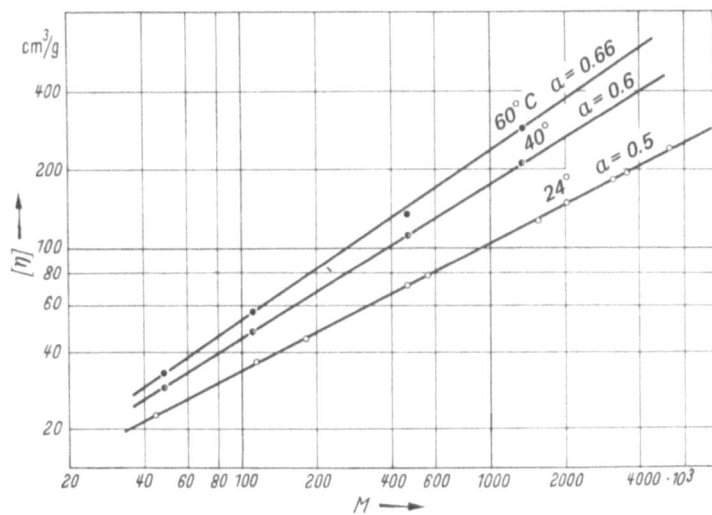

FIG. 521-2 $- [\eta] = f(M)$ for polyisobutylene in benzene at different temperatures (P.J. FLORY).

TABLE 522. Comparison of the K_θ – values of some polymers with different ratio $1_{mon}/M_{mon}$
(a = 0.5 for all polymers listed)

Polymer	Structural unit	Length of the structural unit 1_{mon}	Molecular weight of the structural unit M_{mon}	$\dfrac{1_{mon}}{M_{mon}}$	K_θ
polymethyl-methacrylate	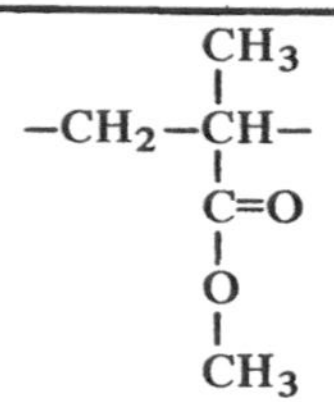 $-CH_2-CH-$, $C=O$, $-O-CH_3$, CH_3 (side)	2.5 Å	100	0.025	0.056
poly-styrene	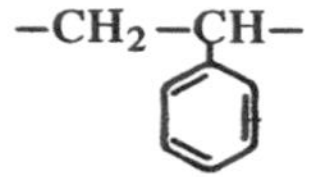 $-CH_2-CH-$ (phenyl)	2.52 Å	104	0.024	0.082
polyvinyl-pyrrolidone	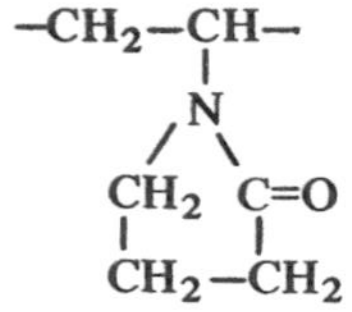 $-CH_2-CH-$, $-N-$, CH_2 $C=O$, CH_2-CH_2	2.52 Å	111	0.023	0.074
polyvinyl acetate	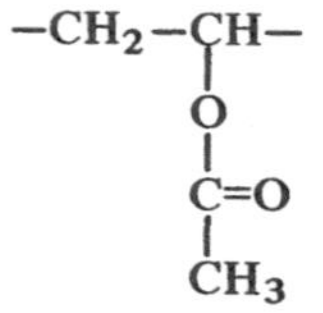 $-CH_2-CH-$, $O-C=O$, CH_3	2.52 Å	86	0.029	0.090

	Structure				
polyiso-butylene	$-CH_2-CH(-CH_3)-CH_3$	2.52 Å	57	0.044	0.10
poly-ethylene-oxide	$-CH_2-CH_2-O-$	~3.75 Å	44	~0.085	0.13
poly-propylene	$-CH_2-CH(-CH_3)-$	2.52 Å	42	0.060	0.13
poly-1,4-butadiene	$-CH_2-CH=CH-CH_2-$	~5 Å	54	~0.093	0.18
Nylon	$-NH-CH_2-CH_2-CH_2-CH_2-CH_2-C(=O)-$	~9 Å	113	~0.080	0.25
cellulose tricaprylate	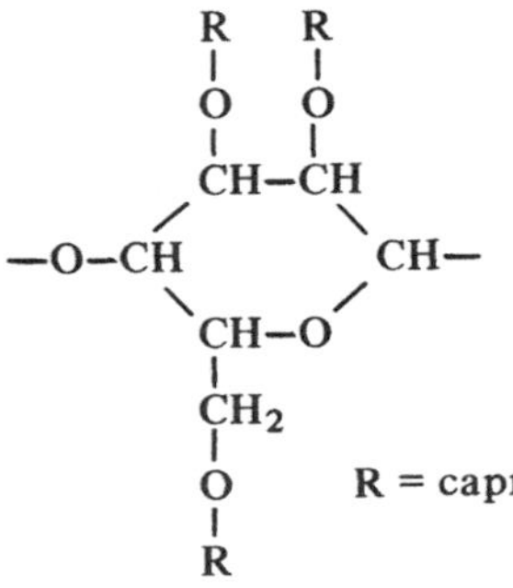 (structure, R = caprylate)	~6 Å	540	~0.011	0.11

The ideal temperature θ was introduced by P. Flory in a quantitative theoretical treatment of the interaction between solvent and polymer chain. According to Flory one can write for the chemical potential $(\mu_1 - \mu_o)_E$ of the chain-solvent reaction in the volume element $(\mu_1 - \mu_1^o)$:

$$(\mu_1 - \mu_1^o)_E = -RT\,\Psi_1(1 - \theta/T)v_2^2 \qquad (29)$$

At $T = \theta$ the interaction potential becomes zero, so that one can also call θ the ideal temperature. With real systems this potential can not actually become zero. It only seemingly becomes zero, i.e., in its effect on the coil density, when the potential of the intramolecular chain association and the potential of the association between the chain and the solvent (solvation potential) become equal in their actual value. Since the coils behave as ideal coils at the θ temperature, the coil density $\bar{\rho}_\theta$ is described by random flight statistics (see Equations 124-129, p. 441 ff). According to Equation (128), p. 445, we can write:

$$\bar{\rho}_{equ,[\eta]} = \overline{M}_{[\eta]}/1/6\pi d^3{}_{equ,[\eta]}\cdot N_A. \qquad (30)$$

$d_{equ,[\eta]}$ can be replaced by $F\cdot\sqrt{\overline{h^2}}$ ($F = d_{equ,[\eta]}/\sqrt{\overline{h^2}}_{opt}$, compare p. 512 and Table 466):

$$\bar{\rho}_{equ,[\eta]} = \overline{M}_{[\eta]}/1/6\pi(\sqrt{\overline{h^2}})^3\cdot F^3\cdot N_A. \qquad (31)$$

According to random flight statistics $\sqrt{\overline{h^2}} = \sqrt{A_m}\cdot\sqrt{L_{max}}$ (Equation 126, p. 443). Since the maximum chain length $L_{max} = l_{mon}\cdot P$, and $M_{pol} = M_{mon}\cdot P$, we obtain:

$$\bar{\rho}_{equ,[\eta]} = M_{mon}/1/6\pi\cdot N_A\cdot F^3\cdot A_m^{1.5}\cdot l_{mon}^{1.5}\cdot \overline{P}_w^{0.5} \; [g/cm^3]. \qquad (32)^2$$

And, since $[\eta] = K_\theta\cdot M^{0.5} = 2.5/\bar{\rho}_{equ,[\eta]}$, we find for K_θ:

$$K_\theta = 2.5\cdot 1/6\pi\cdot N_A\cdot F^3\cdot A_m^{1.5}\cdot l_{mon}^{1.5}/M_{mon}^{1.5} \; [cm^3\cdot mol^{0.5}/g^{1.5}]. \qquad (33).$$

The same expression will be found of course by replacing in Equation (27) the general exponent $(1 + a)/3$ by 0.5.

Equation (33) shows that K_θ is a measure for the length of the statistical chain element A_m, if the corresponding polymers have the same values for the ratio l_{mon}/M_{mon}, as it is the case with a series of polyvinyl compounds. In good solvents the form-factor $F = d_{equ,[\eta]}/\sqrt{\overline{h^2}}$ has values of about 0.5-0.6 (compare Table 466): In θ solvents, F is found to be more close to 1.

Figure 526 shows some logarithmic $[\eta] - M$ curves extrapolated to $\lg M = 0$. It can be seen that the range of the curves usually determined by viscosity measurements is relatively small, and that the physical meaning of $K_{[\eta]}$ can not be interpreted at the same time for several curves with different slopes.

The validity of the viscosity law $[\eta] = K_{[\eta]}\cdot M^a$ is limited to random coil molecules. Macromolecules have the character of random coils only above a certain molecular weight. There is a transition stage in which the macromolecules cease to

---

[2] The molecular weight M is used here in the sense of mole: $M = m\cdot N_A\,[g/mol]$.

be undrained coils and begin to show the character of open chains with a fixed valence angle and $L_{max} \leqslant A_m$. This transition range between M = 500 and M = 5,000 (dependent on the primary chain structure) has been investigated much less carefully than the coil zone. In some cases experimental investigations have shown, that the $[\eta]-M$ curves run nearly parallel with the abscissa in the oligomer range. (W. Burchard).

The discussion of Figure 526 has shown that K $[\eta]$ cannot be interpreted as an intercept (in the sense of $K[\eta] = [\eta]_{M=0}$), because the actual curves do not intersect the ordinate at lgM = 0. Therefore, $K_{[\eta]}$ is suitable only in order to compare the $[\eta]-M$ functions with one another in the sense of $K_{[\eta]} = [\eta]/M^a$ within the range of M = 10^4 to M = 10^7: $K_{[\eta]}$ is a measure for the degree of increase of viscosity caused in a solvent by a certain polymer at a given molecular weight M and a given slope a. The viscosity number is not only influenced by the structural parameters, which are contained in $K_{[\eta]}$ [compare Equations (28) or (33)], but also through the mutual energetic interaction between the polymer chain and the solvent molecules.

If we want to exclude this influence of interaction, we have to compare the K_θ values, i.e., the $K_{[\eta]}$ values at the θ temperature. This has been done in Figure 527 and Table 522: Now the K_θ values can be correlated (at least qualitatively) with the structural parameters: A_m (the length of the statistical chain element, compare Figure 443). l_{mon}/M_{mon} (the ratio of the length and the molecular weight of the structural unit) and F (the form-factor $d_{equ,[\eta]}/\sqrt{\overline{h^2}}$).

The influence of l_{mon}/M_{mon} can be demonstrated by comparison of polyvinyl compounds such as polystyrene, polymethylmethacrylate, polyvinylacetate with 1,4-polybutadienes, or nylon (see Table 522): polybutadiene, polyethyleneoxide, and nylon have higher K_θ values than polyvinyl compounds, as is to be expected according to Equation (33). The influence of l_{mon}/M_{mon} can become evident only if the other parameters A_m and F do not change considerably in the same time. As one can see with the example of cellulose tricaprylate, the influence of a smaller l_{mon}/M_{mon} can be compensated by a higher A_m. The higher A_m results from the inflexibility of the relatively big structural unit of cellulose tricaprylate. The comparison of cellulose nitrate and polymethylmethacrylate (Table 494) shows that the influence of A_m becomes especially effective in good solvents, i.e., in combination with the coil-expanding efficiency of complete chain solvation.

The influence of l_{mon}/M_{mon} can be understood if one considers that this ratio is a measure for the average thickness of a polymer chain: The greater the ratio l_{mon}/M_{mon}, the shorter and thicker the chain at a given molecular weight of the polymer and therefore, the higher the average coil density and the lower $[\eta]$ (and vice versa). In this connection it is useful to compare a model of a polystyrene, or a polymethylmethacrylate chain with a model of a polybutadiene, or a nylon chain (see Figure 528).

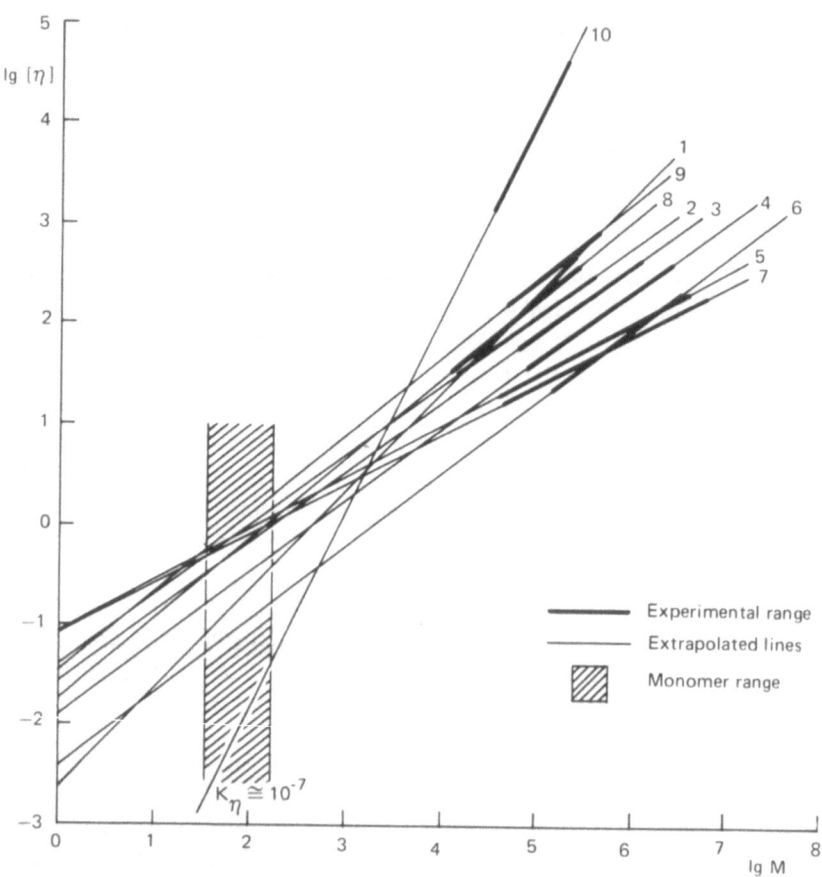

1   Cellulosenitrate (acetone, 20°C)
     $K_{[\eta]}$ = 0.0028 [cm³ · molᵃ/g¹⁺ᵃ]
     a = 0.98
2   Bisphenol-A-polycarbonate (tetra-
     hydrofuran 20°C)
     $K_{[\eta]}$ = 0.040 [''] a = 0.70
3.   Polyisobutylene (cyclohexane,
     30°C)
     $K_{[\eta]}$ = 0.026 [''] a = 0.70
4   Polystyrene (toluene, 22°C)
     $K_{[\eta]}$ = 0.017 [''] a = 0.72
5   Polyisobutylene (benzene, 25°C)
     $K_{[\eta]}$ = 0.083 [''] a = 0.52
6   Polymethylmethacrylate (acetone,
     20°C)
     $K_{[\eta]}$ = 0.004 [''] a = 0.73

7   Polystyrene (cyclohexane, 34°C)
     $K_{[\eta]}$ = 0.082 [''] a = 0.50
8   Polyamide-carboxylic acid from
     PMDA and 4.4' -diamino-diphenyl-
     ether (DMA/LiBr, 25°C)
     $K_{[\eta]}$ = 0.018 [''] a = 0.80
9   Polyamide-carboxylic acid from
     PMDA and benzidine (DMA/LiBr,
     25°C)
     $K_{[\eta]}$ = 0.034 [''] a = 0.78
10   Polyamide-carboxylic acid from
     PMDA and benzidine (DMA/TEA,
     25°C)
     $K_{[\eta]}$ ≈ 10⁻⁷ [''] a ≈ 2

FIG. 526 − [η]-M-curves of some polymers extrapolated to lgM = 0.

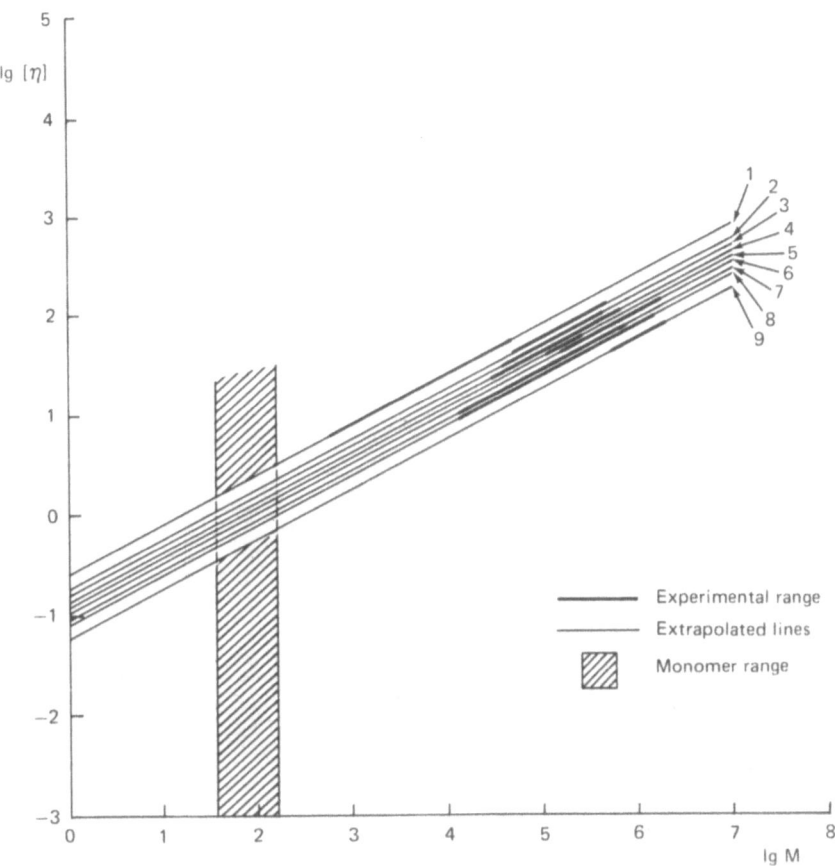

1	Nylon 6.6 (HCOOH/H_2O, 25°C) $K_\theta = 0.25$ [$cm^3 \cdot mol^{0.5}/g^{1.5}$]	6	Polyisobutylene (benzene, 24°C) $K_\theta = 0.10$ ["]
2	cis-1.4-Polybutadiene (isobutyl-acetate, 20.5°C) $K_\theta = 0.18$ ["]	7	Polystyrene (cyclohexane, 34°C) $K_\theta = 0.082$ ["]
3	Polyvinylchloride (benzylalcohol, 155°C) $K_\theta = 0.15$ ["]	8	Polyvinylpyrrolidone (H_2O/acetone, 25°C) $K_\theta = 0.074$ ["]
4	Polyethyleneoxide $K_\theta = 0.13$ (H_2O/K_2SO$_4$, 35°C)	9	Polymethylmethacrylate (toluene/methanol, 26°C) $K_\theta = 0.056$ ["]
5	Cellulosetricaprylate (dimethylacetamide, 140°C) $K_\theta = 0.11$ ["]		

FIG. 527 — [η]-M-curves of some polymers under θ-conditions (a = 0.5), extrapolated to lgM = 0.

It is interesting to see, but difficult to understand, that the $[\eta]$ –M straight line
of polymethylmethacrylate in acetone intersects the $[\eta]$ –M line of the same
polymer in toluene/methanol or other θ curves (compare Figures 526 and
527). This means that with molecular weights $>10^5$, the coil density of
polymethylmethacrylate in acetone is lower than in toluene/methanol, and with
molecular weights $<10^5$, the coil density in toluene/methanol is lower than in
acetone.

Stockmayer and Fixman have derived the following relation:

$$[\eta]/M^{0.5} = K_\theta + 0.51 \cdot \phi \cdot B \cdot M^{0.5}, \qquad (34)$$

where ϕ is the Flory constant; B, the polymer-solvent interaction constant; and $[\eta]$,
viscosity number in a particular solvent.

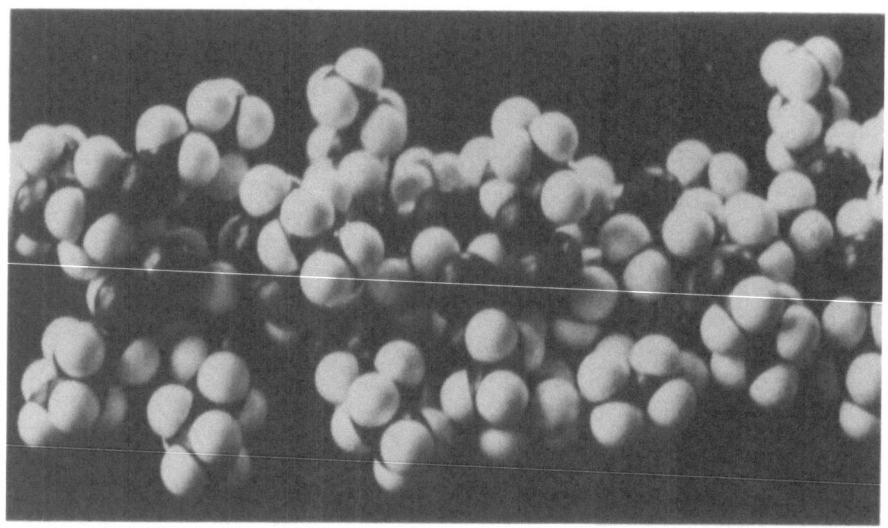

a) polybutylacrylate

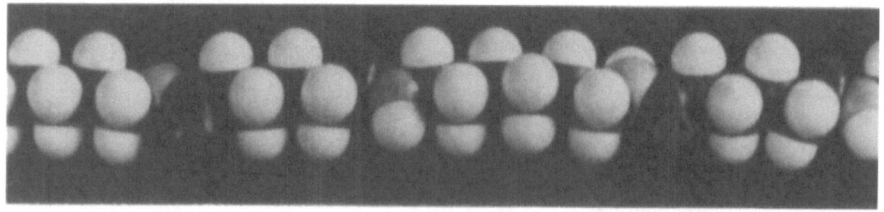

b) nylon-6, 6

FIG. 528 – Molecular models of extended chains

We can therefore determine K_θ by plotting $[\eta]/M^{0.5}$ against $M^{0.5}$ provided the $[\eta]$ versus $\overline{M}$ relationship is known in any particular solvent. The expression $\sqrt[3]{K_\theta/\phi} = \sqrt{\overline{h^2}/M}$ is called the "unperturbed dimension."

Dependence of the Specific Viscosity on Concentration

According to the Einstein viscosity law, the specific viscosity η_{sp} should be directly proportional to the concentration (compare also pp. 375 and 450):

$$\left.\begin{array}{l} \eta/\eta_0 = 2.5\,\varphi + 1\,, \\ \eta_{\mathrm{rel}} - 1 = 2.5\,\varphi\,, \\ \eta_{sp} = 2.5\,\dfrac{c}{\overline{c}_{\mathrm{equ}[\eta]}}\,. \end{array}\right\} \tag{35}$$

The Einstein viscosity law holds only for dilute solutions where the colloidal particles do not interfere with each other's freedom of movement, and thus the proportionality between η_{sp} and c holds only under the same conditions. If one considers that in macromolecular solutions, the concentration of the colloidal particles in 0.05-1% solutions is already 50% (compare Table 496), and that therefore a seemingly quite dilute solution in reality is rather concentrated, then one will not be surprised that the η_{sp}/c versus c curve for polymers with high molecular weight, that is, lower coil density, does not yield a straight line parallel to the x-axis, as it should be expected according to the Einstein law. Instead it rises more or less steeply. We have to remember that high molecular weight and low coil density do not always have to go hand in hand: in poor solvents compounds with high molecular weight can sometimes have a larger coil density, and, therefore, a lower volume, than lower molecular weight compounds in good solvents.

As has already been discussed in the treatment of molecular weight determination by viscometry, one can describe the η_{sp}/c versus c curve, similar to the $p/c-c$ curve, as a power series (the Huggins equation):

$$\eta_{sp}/c = [\eta] + k_H \cdot [\eta]^2 \cdot c. \tag{36}$$

Equation (36) is the equation for a straight line with the slope $\tan \alpha = k_H \cdot [\eta]^2$ and shows that η_{sp}/c becomes the steeper with increasing concentration, the higher the viscosity number of the system. Actually, the slope is proportional to the square of the viscosity number. The qualitative correctness of this conclusion can be seen directly from the η_{sp}/c versus c curves (Figures 530-532).

Quantitatively, the Huggins equation can describe the η_{sp}/c versus c curves often only in a limited concentration range, which is the larger, the lower the viscosity number. With high viscosity numbers, i.e., with compounds with low coil density (for example, cellulose derivatives) or, in general with high molecular weights, the increase of η_{sp}/c is found to be steeper than linear even at relatively low concentrations. We have previously seen (p. 382), that in such cases one can still obtain straight lines, if one plots η_{sp}/c versus η_{sp} according to the Schulz-Blaschke equation.

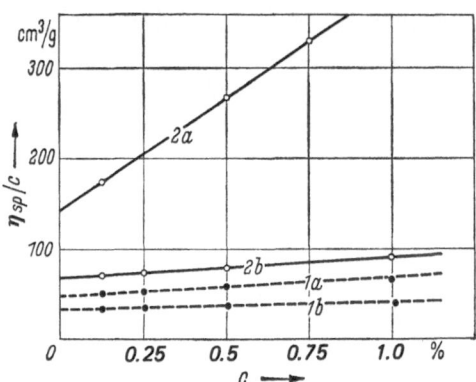

1a) polystyrene ($\overline{M}$ = 100,000) in benzene at 20°C

1b) polystyrene ($\overline{M}$ = 100,000) in methylethylketone at 20°C

2a) polystyrene ($\overline{M}$ = 400,000) in benzene at 20°C

2b) polystyrene ($\overline{M}$ = 400,000) in methylethylketone at 20°C

FIG. 530 — Influence of the solvent on the η_{sp}/c vs. c relationship.

The slope $k_H \cdot [\eta]^2$ is taken as a measure of the energetic interaction between the macromolecule and the solvent, that is, the interchange reaction between the solvent molecules and the structural units of the polymer chain, which leads to the formation of secondary valence links. The formation of such secondary valence links between a macromolecule and the solvent, by destruction of the secondary valence links between macromolecule and macromolecule (which cause the cohesion found in the pure, solvent-free material) is what happens during the dissolution process of the material. Depending on the type of solvent and the type of the dissolved polymer, the link between solvent and polymer is more or less strong depending on the position of the solvation equilibrium. Also, in the dissolved state, a smaller or larger portion of polymer-polymer secondary valence links can still be present: the larger this portion is, the poorer the solvent is; the smaller the amount of these polymer-polymer secondary valence links, the better the solvent, i.e., the more thermodynamically favored the link between polymer and solvent is in relation to the link between polymer and polymer.

To take $k_H \cdot [\eta]^2$ as a measure of the interaction seems logical because $k_H \cdot [\eta]^2$ is larger, the better the solvent, i.e., the more complete the solvation of the macromolecules. This is equivalent to the dependence of the reduced osmotic pressure p/c on the concentration c (compare p. 343):

$$p/c = \frac{RT}{M} + Bc. \tag{37}$$

The analogy between Equations (36) and (37) becomes even more striking if one remembers that $[\eta] = \lim_{c \to 0} \eta_{sp}/c$ and $RT/M = \lim_{c \to 0} p/c$:

$$\eta_{sp}/c = \lim_{c \to 0} \eta_{sp}/c + k_H [\eta]^2 \cdot c \quad \text{(Huggins)} \tag{38}$$

$$p/c = \lim_{c \to 0} p/c + Bc \quad \text{(osmotic pressure)} \tag{39}$$

The Huggins expression $k_H \cdot [\eta]^2$ plays the same role in the dependence of η_{sp}/c on c, as the second virial coefficient, B, in the dependence of p/c on c. Thus one may assume that there is a certain relationship between these parameters, even if the deviation from ideal behavior is not always due to the same reason in both cases. A comparison of the p/c versus c curves (Figure 344) with the η_{sp}/c versus c

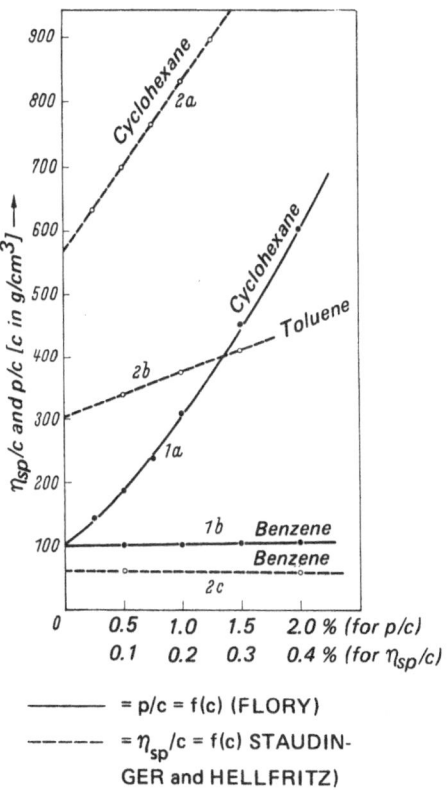

= p/c = f(c) (FLORY)

- - - - = η_{sp}/c = f(c) STAUDIN-
GER and HELLFRITZ)

FIG. 531 — Curves showing the influence of the solvent on the p/c vs. c and the η_{sp}/c vs. c relationships for polyisobutylene.

curves (Figures 530-532) shows that there is indeed a striking parallel. One cannot expect a simple relationship between the second virial coefficient, B, and the Huggins constant, k_H, since B differs from system to system. However k_H is a parameter which is (should be) independent of the system and which for many polymers has values of 0.3 to 0.5 (for η_{sp}/c in cm^3/g).

If there is a common reason for the increase of η_{sp}/c, and p/c with the concentration c, then it must not be looked for immediately in the interaction between macromolecules and solvent molecules, but rather in the influence of the macromolecules over one another, which increases more or less rapidly with increasing concentration. The Einstein viscosity law, just as the van't Hoff equation for osmotic pressure, holds only for extremely dilute solutions, where the dissolved colloidal particles do not interfere with each other with regard to their rotational and vibrational movement, and, therefore, where the collisions can occur elastically

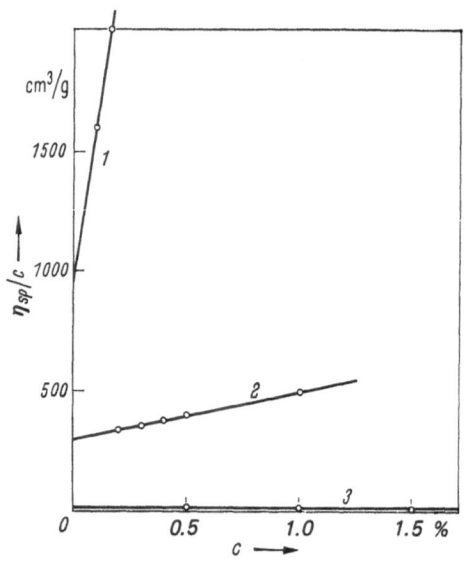

1) cellulose in $[Cu(NH_3)_4] (OH)_2$-solution;
   $\overline{M} = 360,000$; (HUSEMANN and SCHULZ)

2) polyisobutylene in cyclohexane;
   $\overline{M} = 500,000$; (STAUDINGER and HELLFRITZ)

3) glycogenacetate in chloroform;
   $\overline{M} = 860,000$; (STAUDINGER and HUSEMANN)

FIG. 532. – Curves illustrating the influence of the chain structure on the η_{sp}/c vs. c relationship.

without energy loss. If this condition is fulfilled, the second virial coefficient is zero, and, therefore, η_{sp}/c does not change with increasing concentration. The interaction between solvent and polymer chains plays a role insofar as the expansion of the coil (over and above the density corresponding to the ideal coil) is concerned, therefore, the mutual interference between the coils is the greater, the better the solvent.

If one is concerned with the concentration of a macromolecular solution, one always has to keep in mind that there are two concentrations: the concentration based on the weight of the actual polymer, and the gel-coil concentration in the solution brought about by the absorption of solvent. For the determination of the mutual interaction of the macromolecules in solution, it must be the gel-coil concentration which is important and this depends on the coil density. In many systems, even for polymer below 0.01 g/100 cm³ this is still so high that the solutions are considered concentrated in the sense of the Einstein law. However, if the solutions are more dilute, the accuracy of measurement becomes too small, and often anomalies are observed such as a steep increase of the η_{sp}/c values with decreasing concentrations. Thus, with many systems, it is not possible to continue the viscosity measurements to the dilution required to obtain η_{sp}/c versus c curves parallel to the abscissa.

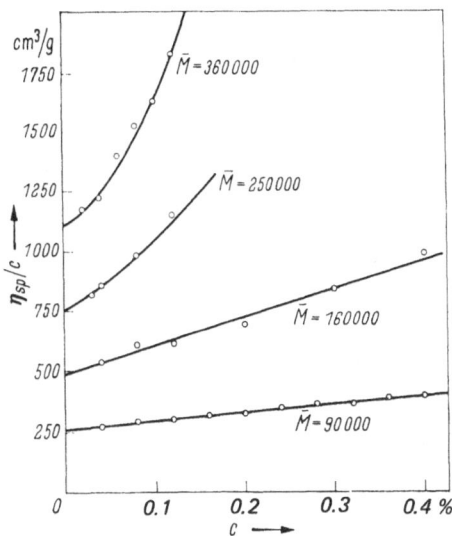

FIG. 533 — Influence of the molecular weight on the η_{sp}/c vs. c — curves for cellulose in $[Cu(NH_3)_4]$ $(OH)_2$ — solutions. (HUSEMANN and SCHULZ)

The solvation, i.e., the reaction between solvent and polymer chain, influences the steepness of the η_{sp}/c versus c-and the p/c versus c curves, because it strongly influences the coil density. For example, one and the same polyisobutylene has in the good solvent, cyclohexane, a viscosity number of 1250 cm^3/g, corresponding to a coil density of $\bar{\rho}_{equ,[\eta]} = 2.5/1250 = 0.002$ g/cm^3, and in the poor solvent, benzene, a viscosity number of 40 cm^3/g, corresponding to a coil density of $2.5/40 = 0.062$ g/cm^3. Thus, in benzene, the critical polymer concentration, where the total volume of the solution is taken up by the gel-coils, is 6.2%, and in cyclohexane, only 0.2%. The corresponding critical gel-coil concentration is 88% in benzene and 78% in cyclohexane.[3] Therefore, the 1% polyisobutylene solution of this compound has in benzene a gel-coil or colloidal particle concentration of 14%, whereas a corresponding 1% solution of the same polyisobutylene in cyclohexane has a colloidal particle concentration of approximately 400% (c always in g/100 cm^3).

Thus, if one wants to measure at a concentration of 0.5% based on the colloidal particles, in the case of benzene it is sufficient to have a polyisobutylene solution of approximately 0.04%, whereas in the case of cyclohexane, one has to reduce the polymer concentration to 0.00125% in order to achieve the same degree of dilution. Table 534 presents a concentration series for different polymers, which shows how large the differences between the colloidal particle concentration are for the same polymer concentration. One readily sees the influence of the solvent, of the molecular weight, and of the chain structure:

TABLE 534. Colloid-particle concentration of different polymer systems with polymer concentrations between 0.05% and 0.8% at 20°C

| Polymer-concentration in % | Colloid particle concentration = gel-coil concentration in %* | | | | | | | |
| | Polyisobutylene $M = 9 \cdot 10^6$ | | Polystyrene $M = 5 \cdot 10^5$ | | Polymethyl-methacrylate in Acetone | | Cellulosenitrate in Acetone | |
	Benzene	Cyclo-hexane	Methyl-ethyl-ketone	Benzene	$M = 10^5$	$M = 10^6$	$M = 10^5$	$M = 10^6$
0.01	0.14	4	0.2	0.4	0.07	0.4	1	8.2
0.05	0.7	20	1.1	2.2	0.37	2	5	41
0.1	1.4	40	2.2	4.5	0.74	4	10	82
0.2	2.8	80	4.5	9	1.5	8	20	164
0.4	5.6	160	9	18	3	16	40	330
0.8	11	320	18	36	6	32	80	660
1.0	14	400	22	45	7.4	40	100	820

*Critical gel-coil concentration $\approx \rho_{solv.} \cdot 100\%$, compare Table 494, note***.

---

[3]Critical gel-coil concentration $= \rho_{solv.} \cdot 100\%$, compare Table 496, note**.

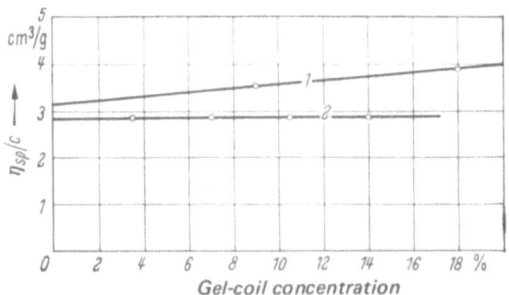

1) polyisobutylene in cyclohexane

2) polyisobutylene in benzene

FIG. 535 $-\eta_{sp}/c$ vs. c plot for polyisobuty-lene in cyclohexane and in benzene, using the gel-coil concentration (instead of the polymer concentration). This is the same system as in Fig. 531, where the polymer concentration was used.

Whereas in benzene the colloidal particle concentration hardly exceeds 10%, for the same polyisobutylene in cyclohexane, within the same concentration range, it increases from 4% to 400%. The influence of the molecular weight makes itself felt for example with polymethylmethacrylate: the gel-coil concentration increases from 0.7 to 7% for $\overline{M} = 10^5$, and from 0.4 to 40% for $\overline{M} = 10^6$. Comparison of cellulose nitrate with polymethylmethacrylate or polyisobutylene also shows the influence of the chain structure. A concentration increase from 0.4 to 40% for polymethylmethacrylate corresponds to the very high values of from 8 to 800% for cellulose nitrate. Figures 530 to 532 show how the differences in the gel-coil concentration affects the shape of the η_{sp}/c versus c, and p/c versus c curves.

The Huggins coefficient $k_H \cdot [\eta]^2$ is always very small when the colloidal particle concentration in the measured (and that means also the measurable) concentration range remains so low that a really dilute solution exists. This may be because a low concentration results from a poor solvent or from a low molecular weight: in both cases the coil density is relatively high, and therefore the amount of absorbed solvent is relatively small, so that for a given polymer concentration, much free solvent is available.

One sometimes states that the second virial coefficient in θ solvents, or at the θ temperature, is equal to zero. This is correct as long as one does not state the converse, i.e., that when the second virial coefficient is zero, one has a θ solvent, or a solvent at a θ temperature. Under θ conditions, the solvation and the aggregation within the coil, are in equilibrium, so that coiling of the chain is only determined by random flight statistics ($a = 0.5$). As was previously shown, (p. 520), this condition can only be realized in poor solvents, where the coil density is relatively large, so that the condition for B→O and $k_H \cdot [\eta]^2 \to O$, i.e., very large dilution, is realized. However, this condition is not realized only in θ solvents. With low molecular weights (for example, below 200,000), the coil density is also relatively high, so that dilute solutions can be prepared even when a is greater than 0.5.

High values for the Huggins coefficient, $k_H \cdot [\eta]^2$, are found, if within the measurable concentration range, the gel-coil concentration increases to high values (50-100%). This may occur because of good solvation, because of a high molecular weight, or because of stiff chain structure. As the values for cellulose nitrate ($\overline{M} = 10^6$) show, the concentration of the coils can reach especially high values with stiff chains. Such solutions not only have no free solvent, but they even show a lack of solvent. Because of this lack of solvent, the coils are unable to reach the volume corresponding to their statistical equilibrium. This leads to a strained system, where the coils are pressed closely together and become hooked up with each other to a certain extent. The increase in inner tension of such a system with increasing concentration explains the steep slope of the η_{sp}/c values for solutions of cellulose and cellulose derivatives.

From a colloid-chemical point of view, it would be better to use the gel-coil concentration rather than the polymer concentration in the η_{sp}/c versus c diagram. Then one would obtain curves which would be similar to those obtained with normal solid dispersions. Figure 535 shows this for polyisobutylene in cyclohexane. The curves, which in Figure 531 are far apart, in this diagram lie very close together, and the steeply rising cyclohexane curve (2a) in Figure 531 now has almost no slope. In practice, such a method of plotting is of little value because one doesn't know the gel-coil concentration and only calculates it from the $[\eta]$ value obtained by extrapolation according to the Einstein law $\overline{\rho}_{equ,[\eta]} = 2.5/[\eta]$.

What applies to the slope of the η_{sp}/c versus c curves also applies roughly to the slope of the p/c versus c curves ($\tan \alpha = B$). If one looks into this more carefully, however, one notices that the second virial coefficient increases strongly during the transition from poor to good solvents, however, not with increasing molecular weight, whereas $k_H \cdot [\eta]^2$ becomes larger in both cases.

Viscosity, i.e., the property of fluids by virtue of which they offer resistance to flow or to any change in the arrangement of their molecules, is a rheological phenomenon and relates basically to the mutual mechanical influence of the macromolecules on one another. Osmotic pressure, caused by the tendency of a solution to increase its entropy by dilution, is a phenomenon which relates (with macromolecular solutions) primarily to the interaction of the macromolecules and the solvent molecules.

The experimental investigation of the dependence of p/c and η_{sp}/c on the concentration has shown that both phenomena are influenced by hindrance of the mobility of the coil molecules in the solution. However, the mechanism which causes the hindrance is not the same for both phenomena (for example, coil expansion by change of the solvent or by change of the molecular weight).

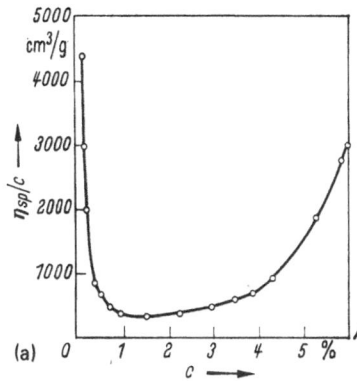 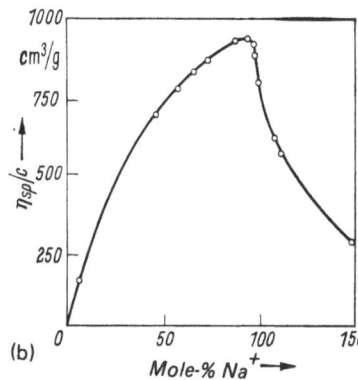

a) η_{sp}/c vs. c -curve of polyacrylicacid in water

b) change in the relative viscosity increase, η_{sp}/c, on neutralization of a 0.5% poly-acrylicacid solution in water

FIG. 537(a and b) — Viscosity effects with polyelectrolytes (H. STAUDINGER).

If the slope of the η_{sp}/c versus c curves is based on the mutual interference of the coils caused by high concentration, then one should expect that the slope below a certain very low concentration c becomes smaller and finally becomes zero. The decrease of the slope of the η_{sp}/c versus c curves at concentrations below 0.1% can actually be observed quite often. However, it does not stay at the zero slope, but instead, the η_{sp}/c curves rise again in the vicinity of the ordinate. The reason for this behavior could not be definitely cleared up until now.

Viscosity Changes with Polyelectrolytes

Polyelectrolytes are macromolecular acids and bases. The acids are often found in nature, for example, the polyuronic acids, (pectins, alginic acids, plant mucilages, etc.), and they are also easily prepared by polymerization of their monomers: for example, the polycarboxylic acids (polyacrylic acid, polymethacrylic acid, poly-itaconic acid, polyamide carboxylic acids, polyamide sulfonic acids and the maleic and fumaric acid copolymers). Polyelectrolytes can also be polymeric bases, such as polyvinylpyridine, polyaminoalkylacrylates, and polymeric quaternary ammonium-hydroxides.

Along with the factors which usually influence the coil density (and, therefore, the viscosity) of polyelectrolytes, such as molecular weight, structure of the polymer chains, extent of solvation, and temperature, one also has to consider the degree of dissociation:

$$\text{wwCH}_2\text{—CH—CH}_2\text{—CH—CH}_2\text{—CH\text{ww}} \rightleftharpoons \text{wwCH}_2\text{—CH—CH}_2\text{—CH—CH}_2\text{—CH\text{ww}}$$
$$\begin{array}{ccc} | & | & | \\ \text{COOH} & \text{COOH} & \text{COOH} \end{array} \qquad \begin{array}{ccc} | & | & | \\ \text{COO}^- & \text{COO}^- & \text{COO}^- \end{array}$$
$$+ \, n\,\text{H}^+$$

In the dissociated state, there is a concentration of negative (or with polymeric bases, positive), groups along the polymer chain which repulse each other electrostatically. This results in an expansion of the coil over and above the normal state, leading even to a stretching of the polymer chain if the degree of dissociation is high enough. In every case, the transition between nondissociated and dissociated polyelectrolytes is always coupled with a strong increase in the viscosity number, and conversely, in the transition from the dissociated to the nondissociated state, the viscosity drops.

In this way, one can explain the peculiar maximum of the specific viscosity obtained in the neutralization of aqueous polyacrylic acid solutions, or pectinic acid solutions, where dissociation is only relatively small. During neutralization, the quantitatively dissociated sodium salts are gradually formed, and the specific viscosity increases until at the neutralization point, all carboxylic acid groups are present as negatively charged carboxylate anions, and the coil density has reached its minimum corresponding to maximum stretching of the polymer chain. With further addition of sodium hydroxide, the dissociation is again repressed (addition of an excess of ions of the same type), the coils contract again, and the specific viscosity drops off. The same decrease of specific viscosity is observed on the addition of sodium ions in another combination, for example, as NaCl, to sodium acrylate or sodium pectate solutions. Similarly, one can observe a viscosity decrease in the esterification of the carboxyl groups of pectinic acid. Changes of the viscosity itself are often complicated by a dilution effect, so that under unsuitable circumstances, it is not always possible to observe them. One has to eliminate the concentration change in such cases, by plotting η_{sp}/c or the viscosity number (Figure 537b).

On simple dilution of an aqueous polyacrylic acid solution, the specific viscosity goes through a minimum (Figure 537a). First there is the effect of dilution and the specific viscosity decreases. Then with further dilution, one observes increasing dissociation, and the increase in η_{sp}/c corresponding to this dissociation becomes more important than the viscosity decrease that results from dilution.

Strong increase of viscosity has been found not only in aqueous solutions of polyelectrolytes but also with solutions in organic solvents. Figure 539 shows a comparison of the increase of the η_{sp}/c values (polymer concentration: 0.5%) of a solution of a polyamide carboxylic acid and a polymethacrylic acid in dimethyl-acetamide (DMA) as a solvent with increasing addition of triethylamine (TEA). Contrary to the behavior in aqueous solutions with addition of sodium hydroxide here we find no maximum at the neutralization point (1 mole TEA per mole carboxylic group). Instead, the curves tend toward a saturation value which is reached only at an excess of from 10 to 15 moles of TEA. The reason for this behavior becomes clear if we consider the dissociation equation.

$$\vdash COOH + TEA \rightleftharpoons \ \vdash COO^- + TEA + H^+.$$

TEA can not dissociate itself so that the relatively small degree of dissociation is increased by an excess of TEA according to the mass action law.

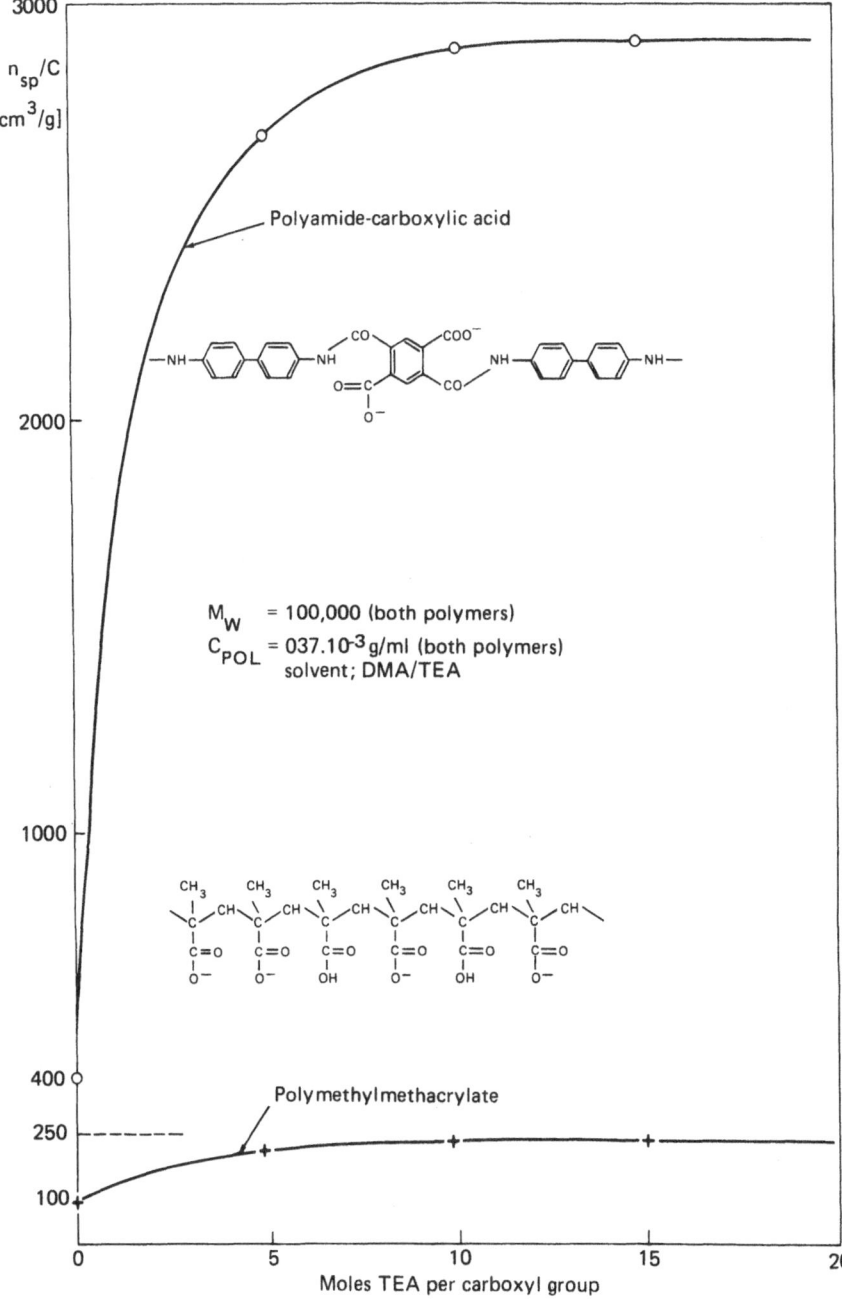

FIG. 539 – Comparison of the increase of η_{sp}/c for two polymers with different primary chain structure as a demonstration of the influence of stiffening structural units the viscosity increase of polymer solutions.

+ : polymethylmethacrylate in dimethylacetamide (DMA) with increasing additions of triethylamine (TEA).

O : the same for the polyamide-carboxylic acid from PMDA and benzidine.

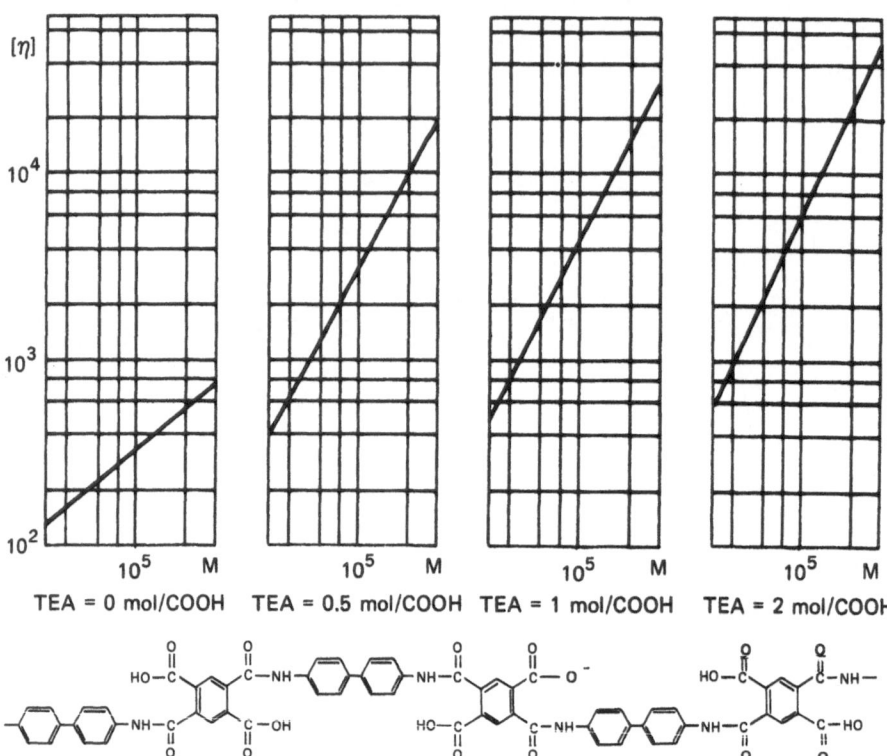

TEA = 0 mol/COOH   TEA = 0.5 mol/COOH   TEA = 1 mol/COOH   TEA = 2 mol/COOH

FIG. 540 — Increase of the exponent a in the $[\eta]$-M-relation through addition of triethylamine (TEA) to a solution of polyamide-carboxylic acid in dimethylacetamide (DMA) according to VOLLMERT and HORVATH.

The comparison of the two polymers shows: with polymethacrylic acid an increase of the specific viscosity from 100 to 250 cm³/g and with polyamide carboxylic acid (PMDA — benzidine) from 400 to almost 3,000 cm³/g. These results demonstrate the strong influence of the primary structure of the polymer chain, in this case, the stiffening influence of the rigid and inflexible aromatic and heterocyclic structural units, which "amplifies" the influence of the ionic groups along the polymer chain. In the nondissociated state the η_{sp}/c values of both polymer solutions have the ratio 4:1, whereas in the partially dissociated state the ratio is 12:1. This confirms the fact that the influence of the primary structure is especially strong in good solvents.

Dissociation influences not only the increase of the η_{sp}/c values, but also the slope of the logarithmic $[\eta]$—M plot is strongly increased: with increasing degree of dissociation, the exponent a of the $[\eta]$—M law reaches the value $a = 2$, which is the theoretically derived exponent for rod-like polymer molecules. Figure 540 shows as an example for the change of the $[\eta]$—M slope polyamide carboxylic acid in DMA with increasing amounts of TEA.

424   Concentrated Solutions and Gels

The state of dilute solutions is characterized by the fact that the dissolved molecules do not interact with each other, i.e., that they do not exert any forces on each other, do not form any secondary valence links between each other, and do not interfere with each other in any way. As has been shown in the last section, the properties corresponding to this state (because of the low density of the coil and the correspondingly large volume requirements) can be determined only by means of extrapolation. With relatively low polymer concentrations, the entire available solvent is taken up by the coil (critical concentration). The system no longer has the character of a solution, but that of a liquid where similar molecules, the gel-coils, are in direct contact with each other and in their movements rub against each other. According to their character, the gel-coils are (in contrast to the spheres of a polystyrene dispersion) not hard, but soft, and are easily deformable. They are deformable to a higher degree, the higher the solvent content of the gel-coil, and the smaller, therefore, the coil density.

The viscosity increases more rapidly with increasing concentration in the region of critical and supercritical concentrations, than is the case with real solutions. Furthermore, one observes more and more a clear dependence of the viscosity on the velocity gradient. The viscosity can decrease or increase with increasing gradient, depending on the polymer-solvent system. However, in most cases the viscosity decreases with increasing gradient.

4241   Dependence of the Viscosity on the Velocity (Shear) Gradient

The decrease in viscosity with increasing velocity gradient can be explained as follows: in a stationary macromolecular solution, whose concentration is above the critical concentration, the gel-coils are present in a type of very close packing, and in the outer regions of the coils (in the contact zones) a more or less intensive touching may occur, which can be correlated to more or less strong intramolecular secondary valence cross-linking. This means that chain segments, which belong to different coils, can aggregate in bundles with more or less pronounced crystalline order, depending on the primary structure of the polymer. This leads to a certain amount of reinforcement of the solution, which in favorable cases can lead to real gel formation. This phenomenon is called thixotropy. Thixotropy is a phenomenon which one observes not only with macromolecular solutions, but also in general with colloidal systems with anisotropic particles, for example, with small particle clay suspensions. Since the chain bundles which bring about this reinforcement usually contain only a few, perhaps only two, chains, the thixotropic ordering occuring with stationary macromolecular solutions is usually not observable by means of X-ray diffraction. The linkage points are usually not permanent, but are again disrupted by the solvent, and new associations occur at different points. If in such solutions one creates a strong velocity gradient, for example, by strong agitation, stirring, or flow through a capillary, this gives rise to shearing forces, which break apart the coil aggregates held together only by weak intermolecular forces, and the viscosity will decrease the more, the stronger the flow gradient, and the corresponding shear forces.

The extent of the magnitude of the shear forces created by flow gradients can be seen from the fact that they are even able to break main valence bonds. For example, if one presses macromolecular solutions at high pressures through a narrow opening, one observes considerable chain degradation (decrease of the molecular weight), which is the greater, the greater the molecular weight and the more voluminous the coils. Also the shear gradients which occur in a mill or in certain other mixing equipment (Waring Blendor, etc.) are able mechanically to degrade macromolecules in solution.

The reinforcement of a solution brought about by association leads only in certain cases to the formation of a solid gel. In general, the character of the liquid (solution) is maintained, and the reinforcement is observed only in the decrease in the viscosity with increasing shear gradient, and one says: "the solution shows structural viscosity (non-Newtonian viscosity)."

As can be shown by measurements of streaming birefringence, strong flow gradients always bring about straightening (i.e., parallel orientation) of the elon-

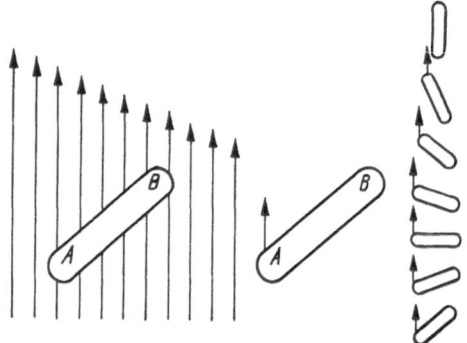

FIG. 542 — Illustration describing the parallel orientation of coils in a flow gradient. The force acting on point A is larger than that acting on point B. The coil therefore undergoes an acceleration at point A, resulting in a clock-wise rotation. Only after the coil has become oriented with its long axis parallel to the direction of flow, will the forces acting on points A and B be equal. This equilibrium state is, however, only a temporary one, because any small deviation from the parallel position is sufficient to start another rotation of the particle.

gated coil-ellipsoids in the direction of the flow. This, too, can bring about a decrease in the viscosity, because the interlocking between the flowing layers, brought about by the dissolved coils, is less when these are parallel to the direction of flow than in other positions. This effect makes itself felt especially with dilute solutions.

The opposite effect, an increase in the viscosity because of a shear gradient, is observed only in very few cases. This phenomenon (rheopexy) is similar in its molecular mechanism to the strengthening of films and fibers by means of cold drawing. Thus, with the appearance of a flow gradient, not only are the coils turned and oriented in the direction of flow, but they are also deformed (if they are large enough): the coils are made longer by the flow gradient. Thus, chain segments of different coils are forced into closer contact, they are pulled past each other, and they have to flow by each other. If now the structure of the polymer has a certain regularity, the parallel orientation of the chain segments gives them an opportunity to occupy a position of minimal internal energy. If the resulting secondary valence bonds are so strong that they can resist the shear-stress (and this is obviously a prerequisite for the occurrence of this effect), then the shear action is increased by the increase in association between the coils, which again leads to a further strengthening, etc. The viscosity increase in the flow gradient (if the prerequisites for its occurrence exist) increases according to a "dynamo" principle, so that a rubber-like gel is rapidly formed, which no longer flows, and finally, if one forces it apart, it tears. If the external force, i.e., the stress on the polymer chains due to the flow gradient, is removed, then the tendency to occupy the state of maximum possible entropy gains the upper hand, and the chains return to the more probable state of statistical coiling. The occurrence of rheopexy is favored by the possibility of the formation of hydrogen bonds between the structural units of the polymer, by a regular chain structure, by low coil density caused by stiff chains, and by a high molecular weight.

43 THE GEL STATE

431 Character, Properties, and Significance of Gels

The gel state can be distinguished from the state of a macromolecular solution with critical or supercritical concentration by the fact that the coils no longer move as units and can no longer interchange their places. From this point of view the transition from a concentrated solution to a gel corresponds exactly to the transition of a liquid to a solid. But, whereas in the solid body only the atoms are able to carry out small vibrations, with a gel it is possible for quite large segments of the polymer chain to move about relatively freely in the included solvent. How great this freedom of movement is depends on the coil density (at a given cross-link density).

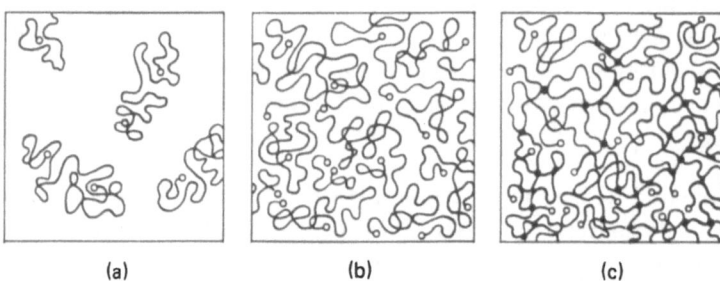

(a)　　　　　　(b)　　　　　　(c)

FIG. 544 — Schematic representation of the transition from a dilute solution (a), to a concentrated solution (b), and finally to a gel (c).

Figure 544 shows schematic diagrams respresenting a dilute solution, a concentrated solution (with critical or supercritical concentration), and a gel. This is meant to show that the structure and the character of the gel is already present in the individual coils. The coils only have to be pressed close enough together (concentrated solution) and to become cross-linked with each other at a few points. Then from the $6 \cdot 10^{18}$ individual coils, which for example are present in 10 g of a polymer of $\overline{M} = 10^6$, a single giant coil, i.e., the macroscopic gel, is formed.

The linkages between the coils can occur through secondary valences as well as covalent links (secondary valence gels, covalent gels). The formation of secondary valence gels occurs frequently in poor solvents, where solvation is incomplete. Since the linkages caused by secondary valences are not of long duration and form sometimes here, sometimes there, and then open up again, it requires a certain minimum size of the molecule to ensure that a gel can form for a given solvation equilibrium. Corresponding to great differences in the strength of intermolecular forces, there are polymer systems which, because of their constitution, have either little tendency or a strong tendency to form secondary valence gels. Polymers which easily form secondary valence gels are, for example, cellulose nitrate (which, with approximately 90% glycerine nitrate forms explosive gelatins); or polyvinylchloride, which in the form of concentrated solutions with a large number of organic liquids (the so-called plasticizers) can be milled to give, depending on the plasticizer, tough, rubber-like or leather-like, secondary valence gels; or natural polymers such as gelatins, pectin, and starch, whose ability even in low concentration to bring about gelation of a solution finds much use in the home and in industry. The secondary valence gels are usually thermo-reversible, i.e., on warming they again become solutions.

Gels in which the coils are bound to each other by covalent bonds are obtained either by swelling of cross-linked polymers or by cross-linking of dissolved polymers with functional groups in the chain by the addition of bi- or multifunctional compounds which react with functional groups built into the polymer chain. The methods for the formation of cross-linked polymers by copolymerization with divinyl compounds (divinylbenzene, diacrylates, divinyl ether), or with the help of functional groups, has already been discussed in Chapters 1 and 2 (compare pp. 24-30, 296-301, and 322-326). For the structure of gels see p. 548.

4311 Plasticized Polyvinylchloride as a Thermo-reversible Secondary Valence Gel

All solvents of a polymer are suitable for the formation of covalent gels, because the covalent cross-links cannot be broken by any solvent and the swelling can be selected at will by the degree of cross-linking. On the other hand, for the formation of secondary valence gels, one can only use poor solvents which will not prevent all secondary valence bonds between the polymer coils by solvation. The solvation equilibrium (compare Figure 518) must be of such a nature that both equilibrium partners (solvated chains $\rightleftharpoons$ associated chains + solvent) are present in comparable concentrations. This sets the prerequisites for the most commonly used solvents for the plasticization of polymeric materials, such as polyvinylchloride.

The plasticizers must also obey additional requirements. Thus the solvation equilibrium of the system (for example, polyvinylchloride + plasticizer = plasticized PVC) should be temperature dependent, because at higher temperatures the solvation has to increase, because on deformation (for example, on a mill or in the extruder) one would like to have a solution which is capable of flowing. This is necessary for the fabrication of plasticized PVC into films, tubes, and leather-like materials. On the other hand, the temperature dependence of the solvation equilibrium should not be too great, otherwise with decreasing temperature the concentration of the cross-linking points increases too rapidly, and this brings about an undesirable hardening and embrittlement of the products prepared from plasticized PVC, especially at temperatures below $0°C$. In addition to these properties required of technical plasticizers for the formation of gels and the retention of the gel state, such plasticizers must also be, as far as possible, colorless, odorless, and nontoxic and should have only very low vapor pressure at $20°C$. Furthermore, they should not "migrate," so that in contact between plasticized PVC and other plastics, they do not diffuse into these other materials.

The following compounds are commonly used in industry in the production of plasticized PVC: phthalic acid esters, (dimethylphthalate, dibutylphthalate, diethyl-hexylphthalate), adipic acid esters, sebacic acid esters (which are particularly low temperature resistant), and polyesters with a relatively low degree of polymerization. Such polymeric plasticizers show especially low migration tendency.

4312 The Mechanical Properties of Gels

In gels, the movement of the coil in the sense of a continuing place exchange, which is characteristic of concentrated solutions, is more or less completely impossible. The coils are no longer able as units to move about, and their position in space is fixed. Thus the gels no longer flow, but can only react with an elastic retractive force to the influence of any deforming force. In contrast to normal solid bodies (for example, crystals), on deformation of gels there is no change in the distances between atoms or in the valence angles, and therefore there is no change in the potential energy of the atoms. On the other hand, parallelization of the chain segments occurs, i.e., the coils on deformation are forced to a less probable state, and there is a decrease in the entropy of the system, such as is found on the compression of gases. Thus, the stress of a deformed gel corresponds entirely to the pressure of the compressed gas and the elastic retractive force of the gel results from the desire of the system to return to the original state of higher entropy. This is in contrast to a stretched steel spring, which tends to return to a state of lower internal energy. Thus one considers the elasticity of gels, just as the elasticity of rubber and other elastomers, as an entropy elasticity.

The gels behave like solid bodies in respect to their elasticity. The characteristic difference between gels and solid bodies is that in spite of the fact that the macromolecules are fixed in space, each gel molecule (gel-coil), considered on its own, is a relatively dilute solution containing in the included solvent the polymer chain as the dissolved substance, which, at low coil density, is dissolved only to 1%, or even less. Accordingly, the segments of the gel coil have a relatively large freedom of movement. The extent of the actual movement of the chains, which can be called micro-Brownian motion, is dependent not only on the polymer concentration in the coil, i.e., the coil density, but also on the constitution of the chains. Since the movement of the chain in its own solvent can consist only of continuous changes in the form of the chain, it requires more or less unhindered free rotation around the chain axis, which in turn is again considerably influenced by the constitution of the chain. Thus, for example, the chains of macromolecular hydrocarbons (such as polyisoprene, polybutadiene, polyisobutylene, polyethylene, and polypropylene) and vinyl polymers (such as polyacrylic acid, and polyvinyl ethers) are relatively freely mobile, whereas polymers with large substituents along the chains (such as polystyrene and polyvinylcarbazole) have a restrained mobility. Polysaccharides, such as cellulose or pectin, have stiffer chains.

The micro-Brownian motion of gels is influenced by the degree of cross-linking even more strongly than by the chain structure. Cross-linking in general is not restricted to the linking of macromolecules with each other. It also covers the linking of chain segments within a single coil, which in certain cases is even the preferred or only type of cross-linking (for example, in dilute solutions of polymers in poor solvents, where the coils are contracted more strongly through intra-molecular secondary valence cross-links than would be the case with complete solvation). The higher the number of cross-links in a given volume at a given polymer concentration, the shorter are the chain segments between the cross-linking points and the smaller the possibility for movement. With increasing cross-linking, the gels become harder and tougher, or more brittle, depending on the type of the polymer and the swelling agent.

In general, the character of the solvent-free polymer is also maintained in the gel state. Brittle or glassy polymers, such as polystyrene, polymethylmethacrylate, polyvinylcarbazole, or polyvinylacetate, form brittle gels on cross-linking, which, in spite of their gel-elasticity, tear or break even under weak mechanical tension. On the other hand, rubber-like polymers, such as natural rubber, butadiene copolymers, polyacrylates, polyvinyl ethers, and polyisobutylene, maintain a certain toughness even in the gel state. Particularly strong mechanical gels, are the secondary valence gels of polyvinylchloride, which find extensive practical applications as rubber-like or leather-like materials.

With these exceptions, the gels have much smaller mechanical strength than the solvent-free polymers. This also applies to covalent gels, although here in every case of fracture of the gel covalent bonds must be destroyed. The low strength of gels is obviously related to the fact that the polymer chains are isolated from each other by the solvent, and are therefore exposed as single units to the tension (stress) and shear forces occuring during deformation.

4313 Characterization of Gels

Although gels differ from solutions because of the fixed spatial positions of the individual macromolecules, this difference is not very sharp. Just as the rubber-elastic state, the gel state represents a transition region between the liquid and the solid state. Elasticity, which is a characteristic property of the solid state, is observed, to a more and more pronounced extent, only with increasing degree of cross-linking. The lower the concentration of the cross-links, the less pronounced is the elasticity, i.e., after removal of the deforming forces, the gel does not completely return to the original state before deformation. There remains a more or less permanent deformation, which for a certain gel, is the larger, the longer the deformation has lasted.

[With very short deformations, even a liquid can be elastic. This is always the case when the time of deformation is shorter than the time the liquid needs in order to escape from the deforming force by changing the spatial arrangement of the molecules. This time is described as the relaxation time] (compare p. 578).

If the degree of cross-linking is low enough, the force of gravity is sufficient to bring about a permanent gel deformation, i.e., such gels no longer keep their shape but slowly flow apart. Therefore it is difficult to determine the mechanical strength quantitatively, if the gels are soft. With increasing degree of cross-linking gels become elastic and one can determine the modulus of elasticity, the tensile strength, and the elongation at break (as with solid bodies).

A general characterization of gels consists in the determination of the swelling power, which tells us how many cubic centimeters of solvent are taken up at a maximum swelling by 1 g of a cross-linked polymer. The swelling power, just as the viscosity, is a function of the coil density and is therefore influenced (like the viscosity) by the same factors which influence the coil density: i.e., solvent, the constitution of the polymer chain (chain mobility and chain branching), and (with polyelectrolytes) the degree of dissociation, which depends mainly on the degree of neutralization (pH) and the solvent. The dependence of the viscosity on the molecular weight corresponds with gels to the dependence of the swelling power on the degree of cross-linking, because this determines the length of the linear chain segments between two cross-links: the fewer the cross-links, the longer the chain segments and vice versa.

The swelling power as defined above is the reciprocal value of the coil density (density of the empty coil), which can be directly measured here as the "gel density." From Figure 544 this relationship between swelling ability and the coil density can be seen directly (compare also p. 454). One only has to take into consideration that the cross-linking, which follows the transition of the material to a gel, in general causes an increasing of the coil density so that, quantitatively speaking, the gel density is not identical with the density of the uncross-linked coil.

The swelling ability does not vary parallel to the softness of the gel. Gels with low coil density and correspondingly large swelling ability do not at the same time have to be softer than a gel with a small swelling ability. Small swelling ability can also be caused by a low molecular weight of the polymer chain. Furthermore, the chain stiffness, which depends on the chemical structure of the chain, influences the strength of the resulting gel.

4314   The Significance of Gels in Nature

The gel state occupies a very important place in nature. Especially the animal organism is a demonstration of the infinite variety of the aggregate state, because all soft parts of the animal body (muscles and internal organs) are protein gels, with water as the swelling agent, where the degree of cross-linking, the molecular size, and the coil density are undergoing continuous change.

Movement and work by muscle contraction depend on very rapid swelling and deswelling processes, which are probably brought about by the increase and decrease in the coil density [which, with polyelectrolytes, is caused by small changes of pH (compare p. 537)]. The expansion and contraction brought about by swelling is normally a three-dimensional process. However, it can occur linearly, if one arranges alternating thin layers (lamellae) consisting of contractile layers with pH-variable coil density, and noncontractile, strongly cross-linked layers. These layers are stuck together in large numbers by cross-links. Then the swellable layers are prevented by the nonswelling layers from expansion in the plane of the lamella, but not perpendicular to it. Such a model muscle can then contract only in a single direction (Figure 549). If the swelling layer is a polyacrylic acid gel, then in acid medium there is a contraction and in alkaline medium an expansion.

432   Structure of Supercritical Solutions and Gels

4321   The Individual Coil Structure Model (ICS Model)

If one increases the concentration of a dilute solution by removing the solvent, the coils are forced to come closer together until finally, at a critical concentration, the total available volume of the solution is filled by the coils. The question then arises: how will the coils interact if even more solvent is removed, until finally the solvent free polymer is left as a rubber or glass.

Figure 550 shows schematically the two possibilities for the situation that might occur. According to the upper scheme (a), the chain segments of the coils move gradually into each other and the coils penetrate one another, or as one says, the polymer chains become "entangled." According to the lower scheme (b),the, coils remain as individuals by contracting within themselves and forming tighter and tighter particles.

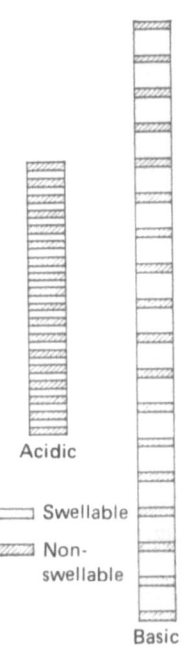

Acidic

☐ Swellable

▨ Non-
swellable

Basic

FIG. 549 — Model of a muscle with striations perpendicular to the fiber axis (long axis) (after W. KUHN). White areas are contractile layers consisting of a lightly crosslinked mixture of polyacrylic- acid and polyvinylalcohol; shaded areas are non-contractile layers consisting of strongly crosslinked poly- vinylalcohol.

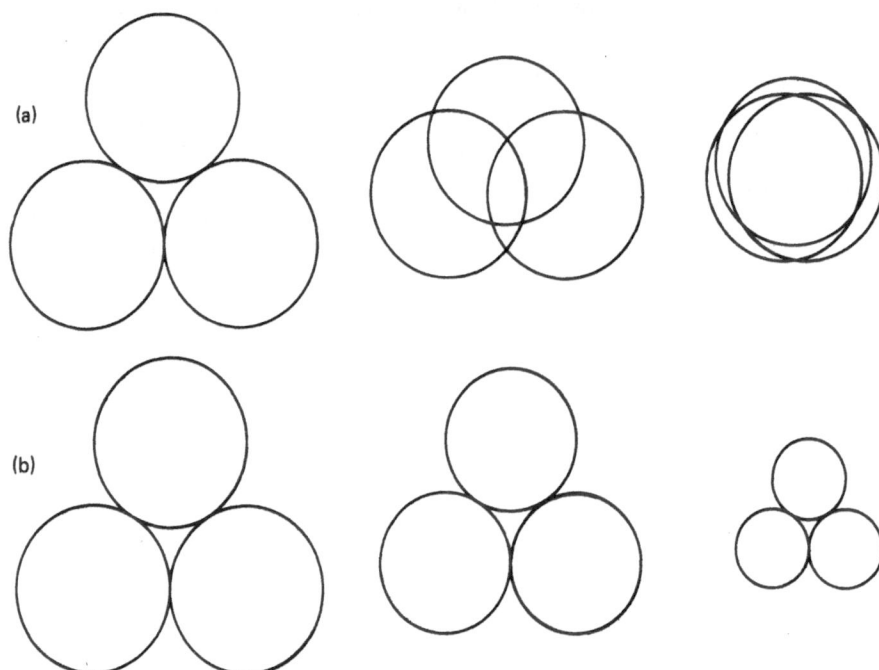

FIG. 550 — Schematic representation of the behavior of polymer coils when their
concentration becomes $>c_{critical}$ (by stepwise evaporation of a part of the solvent.)

a = interpenetration of random coils (P.J. FLORY)
b = shrinking of random coils (S.H. MARON)

Only in the case of mutual coil penetration (interpenetration) would it be
correct to represent the structure of concentrated solutions and gels (as it has
usually been taken for granted up to now) by a complete disarray of lines (see
Figure 551a).

In the other case, however, where the coils remain individuals, the structure of
the gels can be represented by the scheme in Figure 551b.

Through the experimental work of S. H. Maron et al., J.C.H. Hwa, H. Wesslau,
and B. Vollmert and H. Stutz, it has been confirmed that for such vinyl polmers as
polystyrene or polyacrylates, the scheme pictured in Figure 551b is the correct
one, i.e., the coils do not penetrate into each other, but form an aggregated state
consisting of a lattice or cell-structure of tightly packed polyhedra which are only
interlocked at their touching surfaces.

The correctness of the cell structure of polymer coils in the aggregated state can
be demonstrated convincingly by the following simple and clear experiment:
polybutylacrylates with OH groups (X) on the chain are reacted with polybutylacrylates with acid chloride groups (Y) on the chain in solution at different
polymer concentrations, and at different X and Y concentrations in such a
concentration ratio that the molar ratio $[X]_m/[Y]_m = 1$. If the polymer coils

could penetrate each other and become entangled at will, according to the model in Figure 551a, then one would expect after a certain reaction time a complete conversion, as it is obtained if adipoyldichloride is reacted with the OH group-containing polyacrylate. In reality the maximum obtainable conversion (depending on the polymer concentration) is only between 10% and 20%, even if the reaction begins only several hours after mixing the two polymer solutions, so that the chain segments would have time to diffuse as far as is sterically possible for them. This means that only the functional groups in the border zones of the coils are capable of reacting with other, however not those contained in the interior of the coils. This reaction therefore almost presents a picture of the coil structure at any given time, because (by an analytical determination of the unconverted groups) one knows exactly how far the coils have reacted with each other.

Figure 552 shows as the result of such an experiment the width of the touching or contact zone of coils of a polyacrylate gel expressed in percent of the total volume as a function of the polymer concentration.

Since the polymer coils are rather loose and porous structures (compare p. 493), it is somewhat surprising that even in concentrated solutions or gels, whose volume is much too small for a normal side by side situation of all the coils, they do not penetrate each other and become entangled, but instead, even on continuing the removal of the solvent, they remain as individual coils which only contact one another in relatively small and narrow boundary zones without entanglement.

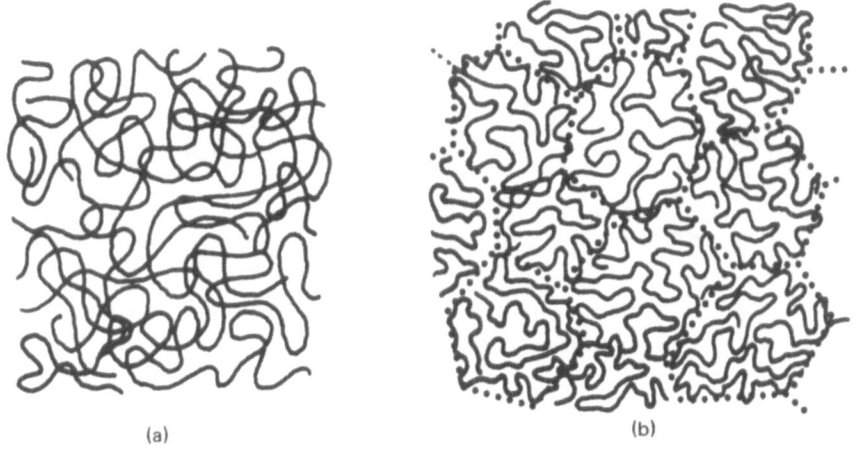

(a)   (b)

a) Solution with matted and entangled chains ("Spaghetti" structure)

b) Solution with individual coils ("Cell" structure)

FIG. 551 — Diagrams illustrating the structure of concentrated solutions and gels.

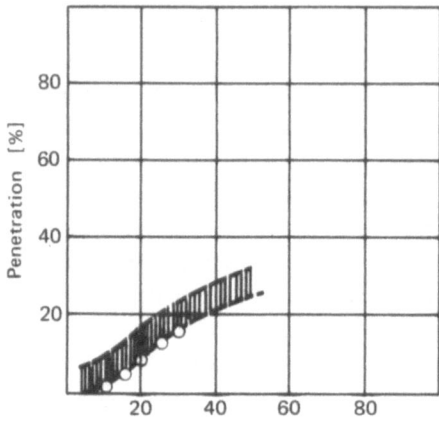

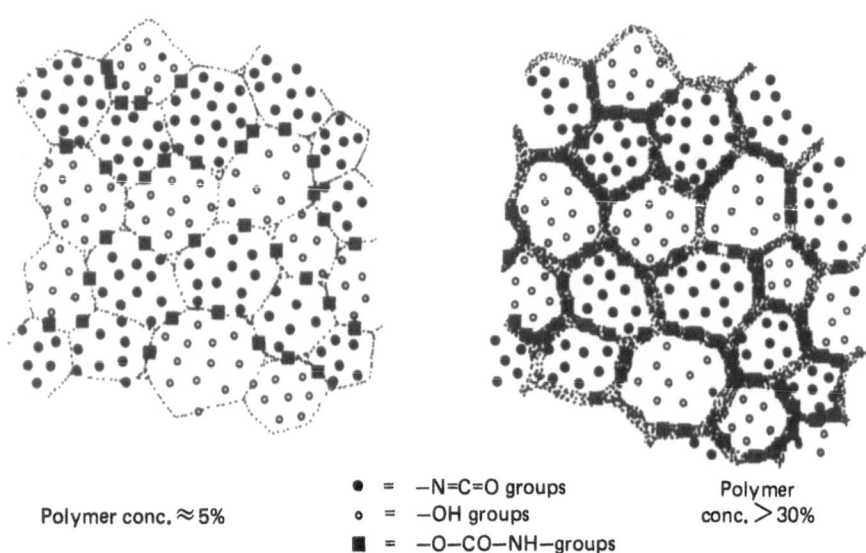

Polymer conc. ≈ 5%

● = —N=C=O groups
o = —OH groups
■ = —O—CO—NH—groups

Polymer
conc. > 30%

FIG. 552 — Depth of interpenetration of the coils in a polyacrylate gel as a function of the polymer concentration. (VOLLMERT and STUTZ)

The inability of coils for interpenetration is probably a steric phenomenon. Random coils may be considered unoriented spatial lattices, which are stabilized by the tendency to maintain a maximum level of entropy. Two spatial lattices can move into each other only if all the bars of both of the lattices are parallel. This state, however, is extremely improbable in the case of random coils. From this point of view one will find it not so surprising that random coils do not interpenetrate.

Structure and Representation of Cross-linked and Branched Polymer Systems

Usually a cross-linked system of polymer chains is described by a diagram such as the one in Figure 553. This type of representation goes back to the old Staudinger concept of the thread-like form of the macromolecules.

One has to use this scheme in order to represent cross-linking reactions (compare pp. 24-30). However, it is not sufficient as a model for the configuration of a cross-linked polymer system since a polymer network is a densely packed system of polymer coils according to Figure 552, which are tied to each other via co-valent bonds in the boundary zones. In addition to the *inter*molecular cross-links, which tie the coils to each other, there are usually also cross-links in the interior of the coils whereby different chain segments in one and the same coil are linked to each other and these internal cross-links are known as *intra*molecular or intra-catenary cross-links. In Figure 554 the configurational structures of cross-linked polymer systems of both types are represented schematically.

It can readily be seen that with the polymers which are both *inter*molecularly and *intra*molecularly cross-linked, we are dealing with a homogeneously cross-linked system in which the cross-linking points are distributed in a homogeneous statistical manner, wheras in the exclusively *inter*molecularly cross-linked polymers, the cross-links are only to be found in the boundary layers of the coils (compare also Figure 552).

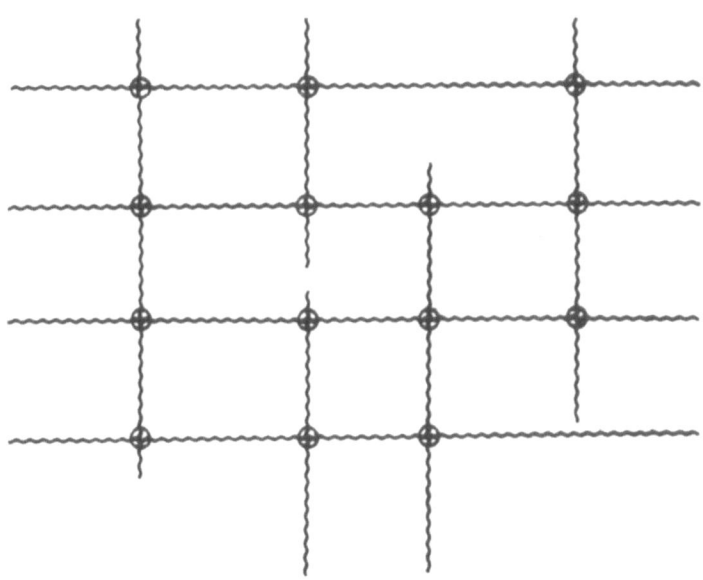

FIG. 553 – Network-structure (schematic), without consideration of chain-coiling.

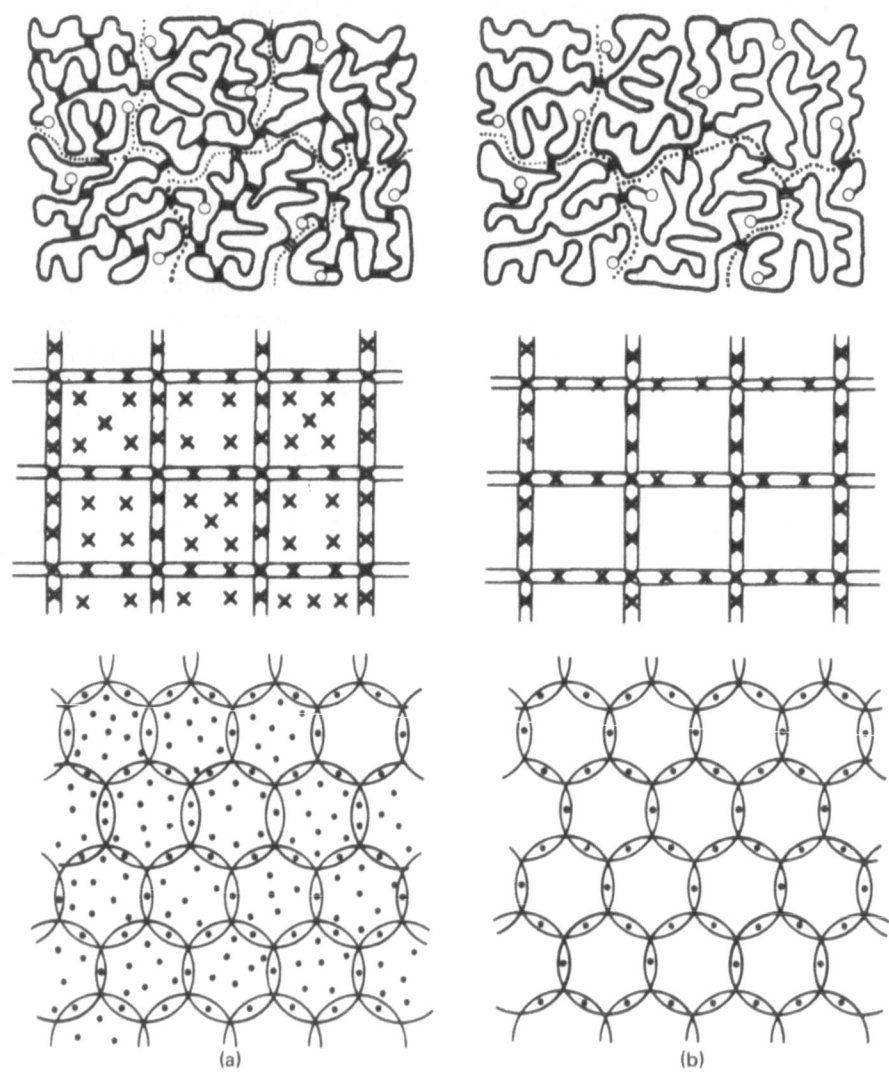

a) Homogeneously crosslinked system (inter- and intra-molecular crosslinks)

b) Only inter-molecularly crosslinked system

$\bowtie$ = intra-molecular crosslink

$\equiv\mathsf{(}$ = inter-molecular crosslink

x
• = crosslinking points

FIG. 554 — Schematic representations of network-structures according to the ICS-model.

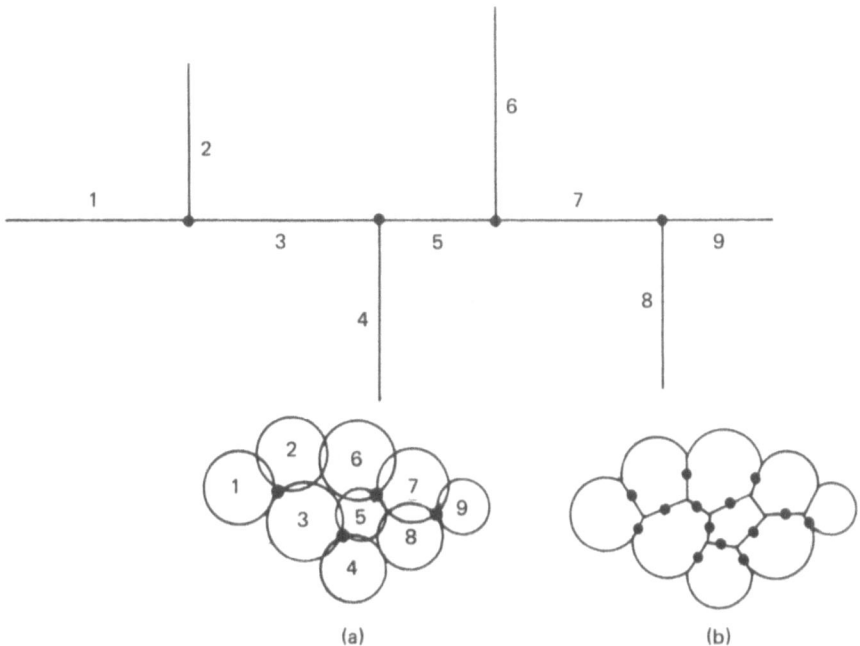

Top: straight-line model (gives no information about the chain conformation)

Bottom: coil model of a branched polymer (a), compared with an inter-molecularly "crosslinked" aggregate consisting of the same number of individual coils (b).

FIG. 555 – Diagrams illustrating branched polymers.

It was already shown on page 25, that there is no principal difference between cross-links and branching and that a branched macromolecule is nothing other than a small volume element of a cross-linked system, which in general, depending on the degree of cross-linking, contains from 10 to 100 coils. However, in the case of microgels, it may even contain from 10^4 to 10^6 coils (for example, with the rubber particles of impact-resistant polystyrene, compare p. 618 ff.). Just as the old network model leads to incorrect conclusions, if one uses it for the description of chain conformation, the often used "Christmas tree"-structure, or the "Fishbone Diagram" for branched polymers and graft-copolymers, are equally misleading.

Figure 555 shows the diagram for the constitution of a branched polymer chain (long chain branching) compared with the corresponding coil scheme. It can be seen how little the branched macromolecule differs from a system consisting of nine *inter*molecularly cross-linked coils. It is, of course, justified to consider the system as *one* large *intra*molecularly cross-linked coil of a branched polymer chain. Similarly, with a graft-copolymer one should always distinguish between the diagram of its constitution and the diagram of its configuration (compare Figure 290).

4322 Deformation of Gel and Rubber

The retractive force, which occurs on deformation of a rubber-elastic system (gel or rubber), is caused by the partial or complete stretching of chain segments by the deformation. The larger the entropy decrease coupled with this deformation, the larger will be the tendency of the system to return to its original state (state of maximum entropy) (compare also pp. 567-570). Naturally the entropy difference, and therefore the retractive force, will become larger, the greater the deformation. The magnitude of the retractive force for a given deformation, or, which comes to the same thing, how rapidly it increases with increasing deformation (E-modulus, modulus of elasticity), depends on the structure of the polymer sytem. It is easily seen that the entropy difference increases more rapidly, i.e., the modulus of elasticity is larger, the more chain segments reach a state of maximum stretching because of a certain increase in the deformation. This state of maximum stretching is reached for a given deformation by a greater number of chain segments, the shorter the chain segments between two cross-linking points, i.e., the larger the number ν_e of the chain segments per cm^3. The E-modulus will therefore increase with increasing degree of cross-linking as a linear function of ν_e, as was first derived by P. Flory

$$E = \text{constant} \cdot \nu_e \cdot (\alpha - \alpha^{-2}) \tag{40}$$

According to Flory one can write:

$$f = \nu_e \, RT \cdot A_o v_2^{-1/3} (v_2^o)^{2/3} \, (\alpha - \alpha^{-2}) \tag{41}$$

where f is the retractive force; ν_e, the number of moles of chain segments between two cross-links in 1 cm^3; $N_A \cdot \nu_e/2$, the number of cross-link points in 1 cm^3 of gel at equilibrium swelling (N_A = Avogadro's number); $v_2 = v_{polymer}/v_{gel}$ in the swelling equilibrium (= degree of swelling); A_o, the surface area of the sample in cm^2 before swelling; v_2^o, the volume fraction of the polymer before swelling; and α, relative deformation.

According to the classical network model (with spaghetti structure) only *inter*molecular cross-linking points have to be considered. Although one has always distinguished between *inter*molecular and *intra*molecular cross-linking, one has interpreted *intra*molecular cross-linking only in terms of loops (see Figure 557) which according to theoretical considerations of Flory, can be disregarded as ineffective (as far as elasticity is concerned) since they are not stretched on deformation and also since there are relatively few of them. With the ICS model, there are not only loops as *intra*molecular cross-links but one also has far more *intra*molecular cross-links whose chain segments are stretched or oriented on deformation (compare Figure 557) and which therefore are more or less equivalent in their elastic effectiveness to the *inter*molecular cross-links. According to the experiments of H. Wesslau and those of B. Vollmert and H. Stutz, approximately 70 to 80% of the cross-links in a homogeneously cross-linked polymer system are *intra*molecular cross-links.

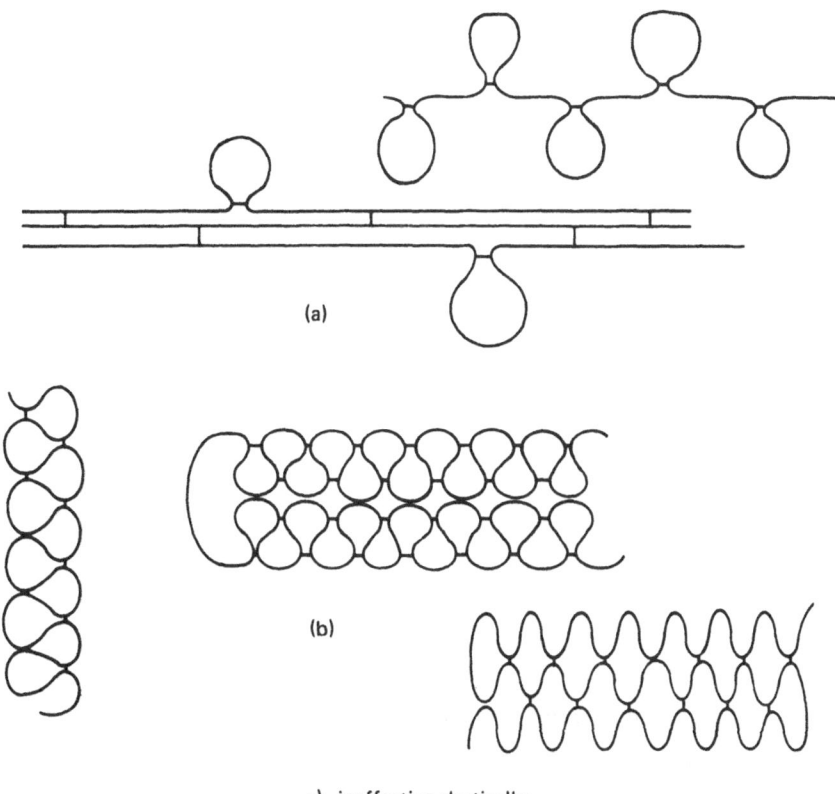

a) ineffective elastically

b) effective elastically

FIG. 557 — Diagram illustrating intra-molecular crosslinking.

That the *intra*molecular cross-links of the homogeneously cross-linked systems are really elastically effective, can easily be shown by comparison of the moduli of elasticity of a gel which is only *inter*molecularly cross-linked and a gel which is cross-linked both *inter-* and *intra*molecularly.

Figure 558 shows the result of such an experiment. The gels which are only *inter*molecularly cross-linked were prepared by reaction of a 3 mole percent COCl group-containing polyacrylate with a 3 mole percent OH group-containing poly-acrylate. In order to prepare gels which were also *intra*molecularly cross-linked, polyacrylate copolymers with 3 mole percent OH groups had been cross-linked with adipoylchloride. It can be seen that the additional *intra*molecular cross-links increase the E-modulus considerably.

For the theoretical treatment of the rubber-elastic deformation of a homo-genously cross-linked system, the distinction between *inter-* and *intra*molecular cross-links can be disregarded. Although, as can be seen from the different slope of the two curves in Figure 558, the two types of cross-linking are not completely

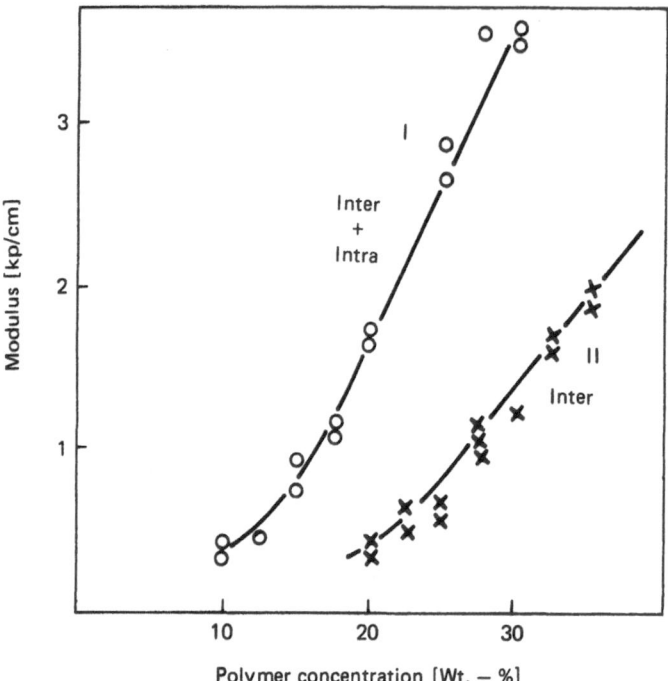

FIG. 558 — Influence of the type of cross-linking
on the E-Modulus of gels (VOLLMERT, STUTZ).

equivalent in their effect on the E-modulus, one is unable with normal homogeneously cross-linked systems to distinguish the effectiveness of the different types
of cross-linking and therefore has to be satisfied with an experimentally deter-
minable constant which contains a general structural parameter (see p. 556).

For a better understanding of the two curves in Figure 558, one has to realize
that on plotting $E = f(\nu_{e(inter)})$ for the compounds in the upper curve, the total
number of cross-link points present remains constant, because all products whose
E-modulus was measured have the same number of functional groups. The only
variable is the concentration of the polymer during the cross-linking reaction and
therefore the depth of penetration of the coil surfaces into each other, and therefore also the number of *inter*molecular cross-links. This means that for the compounds in the upper curve only the ratio between *inter-* and *intra*molecular cross-
links is displaced with increasing $\nu_{e(inter)}$, or E-values, more and more in favor of
*inter*molecular cross-links, i.e., the "cell-wall" of the system becomes somewhat
stiffer and thicker.

4323 Chain Entanglement

A further correction of the classical picture of the structure of cross-linked
polymers systems obtained by abandoning the network model in favor of the
individual coil structure model (ICS-model), concerns the question of the mechanical entanglements (Figure 557) which have to result if one assumes coil in-

terpenetration and chain entanglement as pictured in Figure 551a. Since coil interpenetration (at least in the normal case) does not exist, one therefore does not have to postulate the mechanical entanglement of the chains. As evidence for the entanglement of the chains, one formerly used the argument that the physical degree of cross-linking (as it is determined by measurement of the degree of swelling from the Flory-Rehner equation,[4] or by measurement of the E-modulus in the Flory equation), is larger than the chemical degree of cross-linking determined by chemical analysis. More recent experiments which investigated the ratio between the so-called physical degree of cross-linking and the analytically determined chemical degree of cross-linking (B. Vollmert and H. Stutz) have shown that with cross-linked systems which have only *inter*molecular cross-links the situation is exactly the reverse: the number of cross-links determined according to the Flory equation is clearly lower than the effectively present number of cross-links determined by chemical analysis (Figure 560). Only after the addition of *intra*molecular cross-linking points, the ratio is reversed and the number of cross-linking points (determined as $\nu_e/2$) is larger than the number of cross-links determined by chemical analysis. In Figure 560 the mechanically determined degree of cross-linking $\nu_e/2$ is plotted against the analytically determined degree of cross-linking.

If both the mechanically and the analytically determined degree of cross-linking coincide, all the points would be on the 45° diagonal. Since the gel products (o and x) shown in Figure 560 differ from each other only in that in those represented by o there is a certain number of *intra*molecular cross-linking points over and above the other cross-links, the experimentally determined E-modulus of the homogeneously cross-linked o-compounds, which is higher than the theoretical value, must be influenced by the additional *intra*molecular cross-links. Entanglements, if they exist, would be present in both systems to the same extent, and could therefore not explain the discrepancy between the experimental and the theoretical E-modulus. One therefore can disregard mechanical entanglements, because there is no evidence that they exist, in fact the inability of the coils to interpenetrate seems to demonstrate exactly the opposite.

4324 Incompatibility of Polymer Mixtures

It is a very striking phenomenon that different polymers, with very few exceptions, are incompatible with each other. This means that if one mixes two polymers A and B, one does not obtain a homogeneous mixture of single macromolecules of polymers A and B, but the macromolecules of these two polymers remain grouped in larger aggregates. The size of the aggregates of polymers A and B in the mixture depends first on the intensity of mixing, but also on the chemical constitution of the chains of polymers A and B. There will always remain a strong tendency for phase separation.

---

[4] $V_1 \cdot \nu_e = - [\ln(1 - V_2) + V_2 + \chi_1 V^2_2]/(V_2{}^0)^{2/3} \cdot V_2{}^{1/3} - V_2/2$, where χ_1 is the Huggins interaction parameter; V_1, the molar volume of the solvent, i.e., swelling agent; and ν_e, V_2, $V_2{}^0$, same meaning as in the Flory-Rehner equation.

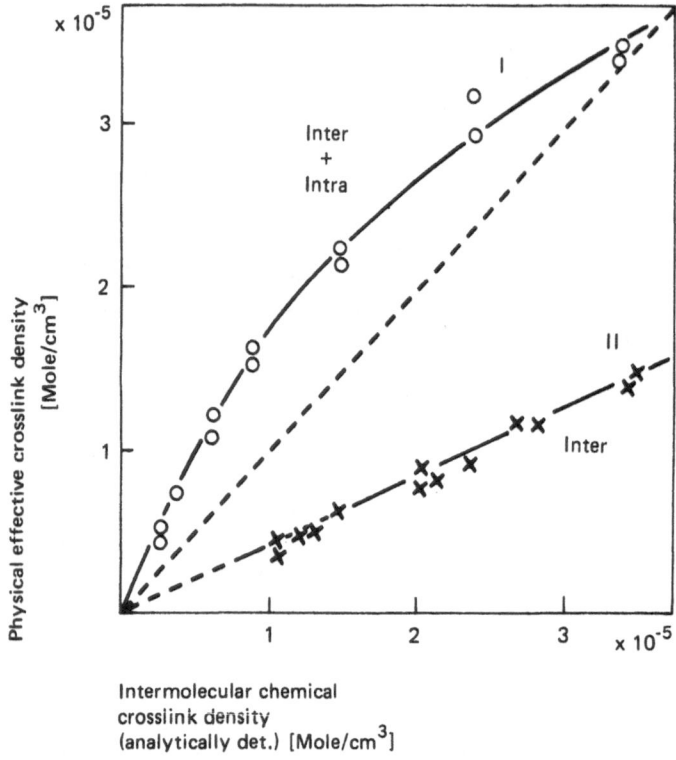

FIG. 560 — Comparison between the mechanically determined and the analytically determined degree of crosslinking (VOLLMERT, STUTZ).

The reason for this incompatability is the inability of the polymer coils (not only of polymer coils in general, but also of coils of one and the same polymer) to penetrate into each other. A result of this sterically prevented interpenetration is that the entropy difference between the two-phase state and the state of molecular mixing is orders of magnitude smaller than is the case with low molecular compounds (for example, the monomers). Thus, the entropy increase on mixing is a function of the number of individual particles being mixed. Since the degree of polymerization of most industrially prepared polymers is of the order of from 3,000 to 5,000, the number of molecules being mixed is smaller by a factor of from 10^{-3} to 10^{-4} than the corresponding monomers. However, this is only the case if the coils do not interpenetrate. With complete interpenetration of the coils, the state of the polymer mixture, as far as the entropy is concerned, would be less different from that of a monomer mixture.

The entropy difference is the driving force for the mixing of different compounds. Opposed to the positive value of the entropy difference ($T \cdot \Delta S$) as the driving force in a mixing process, (if one deals with molecularly homogeneous mixtures considered in the sense of an endothermic solution process), there is always a positive enthalpy value ΔH, i.e.,

$$\Delta F = \Delta H - T\Delta S. \tag{42}$$

Therefore, the mixing process can only bring about complete compatibility (molecularly dispersed distribution) if $T \cdot \Delta S$ is larger than ΔH. This is the case with many low molecular weight organic compounds, because obviously they dissolve in each other. However, it is not true for macromolecular compounds. This can be understood because the entropy of mixing ΔS is several orders of magnitude smaller with polymers than with monomers. Molecularly homogeneous mixtures of polymers should only be expected when ΔH is negative, that is, if the attraction forces between different polymer molecules are larger than, or at least equal to, those between the same type of polymer molecules.

44 THE RUBBER-ELASTIC STATE (VISCOELASTICITY)

441 Rubber-elasticity (Viscoelasticity) of Melts of Macromolecular Compounds

When a solid (crystalline or glassy) macromolecular compound softens, this does not occur in such a way that at a given temperature the intermolecular forces between the molecules are overcome by the molecular motion, so that when a deforming force is applied, the molecules are moved with respect to each other and the material becomes a liquid. Rather, because of the coil structure of the macromolecules, one has to distinguish between a movement of chain segments within one coil, the micro-Brownian motion, and the movement of entire coils with respect to each other, the macro-Brownian motion. In the glassy state, both types of molecular motion are arrested, i.e., "frozen-in." With increasing temperature, one after the other, the segments of the chain begin to melt ("thaw out"). In the neighborhood of the softening point the movable chain segments become larger and larger, and under the influence of a deforming force the sample slowly changes its shape. With polystyrene, for example, deformation when a force is applied begins between 70°C and 90°C.

Naturally, the temperature at which deformation occurs, depends on the magnitude of the deforming force. Since the deformation occurs at first only slowly, the temperature at which the deformation has reached a certain magnitude, must also be dependent on the rate of the temperature increase. In practice, therefore, the conditions under which the softening temperature has been determined must be clearly stated.

With further increase in temperature the thawed-out chain segments become more and more mobile, so that finally for each chain there remain only a few linkage points. The energy which is required to "melt" a structural unit or a chain segment is always the same, regardless of whether secondary valence links within a coil, or between chain segments of different coils, are removed. Therefore, the links which still remain at a certain temperature must be distributed at random over the entire material (intramolecular and intermolecular cross-links), and there will be a temperature at which each macromolecule is linked to its neighbors only by relatively few secondary valence forces. In this state a considerable deformation, for example, stretching of the material, is possible even with a relatively small deforming force. However, if the deformation occurs rapidly and does not continue for very long, it will not give rise to a movement of the macromolecular coils with respect to each other (this is prevented by the secondary valence cross-links) but only to a deformation of the coils (Figures 562 and 568). In this process the coils are forced into an improbable shape and the entropy of the system decreases. Therefore, as soon as the deforming force is removed, the coils again assume their most probable form, which results in the fact that the stretched sample also returns to its original length.

This behavior of a material, i.e., strong deformation under small deforming forces and elastic recovery, was first observed with natural rubber. The phenomenon is therefore known as rubber-elasticity. Conversely, a material which exhibits this behavior is characterized as "rubbery."

4411   Rubber-Elasticity and Relaxation

Rubber-elasticity is not a special behavior of a few macromolecular compounds, such as rubber or polyisobutylene, but it is a property which all macromolecular compounds have above their softening temperature, as long as they have a softening temperature and do not first decompose. [Softening temperature = freezing point = glass temperature (compare p. 573)]. However, the elasticity of polymers above their softening point is not a pure elasticity, i.e., the recovery is not complete. On stretching rubber-elastic polymers, for example, the polymer is always somewhat

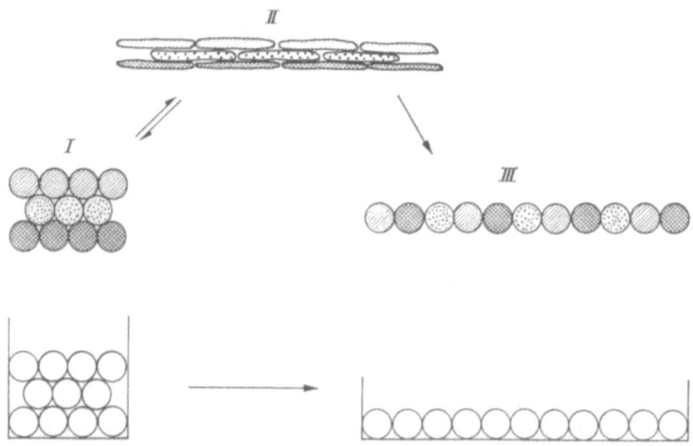

FIG. 562 — Diagram illustrating the deformation of polymers in the rubber-elastic state.

longer after stretching and release than before stretching, and the remaining deformation remains the larger, the larger the degree of stretching, and the longer the material was held in the deformed state. Thus, if the time of stretching is longer, the stretched molecules have time to change their position with respect to each other, so that in their new position they can again attain their most probable shape. In Figure 562 this is schematically shown in the transition from II to III: the change brought about in the macroscopic proportions by stretching remains unchanged. The inner tension of the system is relieved by changes in the position of the molecules, a process which is called relaxation, and which occurs in every liquid when it flows. However, with normal low viscosity liquids, the relaxation occurs so rapidly, that no elastic recovery can be observed. Thus, one could consider rubber-elastic melts of macromolecular compounds as tough, viscous liquids, or liquids with a extremely long relaxation time. With rapid, short deformations they are elastic, and with longer, continuing deformations, they flow. With increasing temperature, the character of a liquid becomes more and more pronounced.

4412 Thermoplastic Deformation

Since flow under the deforming action of gravity usually occurs only very slowly, the deformation of polymers which are processed as raw materials for plastics is carried out in industry under the application of high pressures (hydraulic presses, extruders, and injection molding machines). The melts of such polymers are not characterized as liquid or elastic, but only as "plastic," and the polymers themselves are called "thermoplastic" if their softening temperature lies above room temperature. Polymers whose softening temperature lies below room tem-

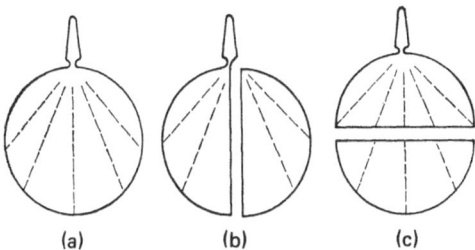

(a)   (b)   (c)

a) flow-lines with an injection-molded plastic disc

b) fracture along the flow-lines occurs easily even with small deformations

c) fracture across the flow-lines is only possible at larger deformations and using a considerably greater force

FIG. 563(a-c) — Diagram illustrating the effect of molecular orientation in injection-molded materials (e.g. polystyrene, polymethylmethacrylate).

perature, and which therefore at ordinary temperatures are rubber-elastic, are known as elastomers. Elastomers, as long as they are not cross-linked, are also always plastomers. Similarly, the thermoplastic polymers are always thermoelastic. Plastic deformation of elastomers at ordinary temperatures is known as cold flow. In the processing of thermoplastic materials by injection molding, the grainy mass is first heated to about 100°C above its softening temperature and then (with a piston or a screw) pressed through a narrow opening into a cold or slightly warm form. This means that a strong deformation of the plastic-elastic mass occurs in the flow gradient and the macromolecules undergo orientation according to Figure 562-II. Since cooling in the form occurs rapidly, the mass does not have the time to release the inner tension through changes in the place of the molecules. Relaxation (see Figure 562-II→III) is not possible and the macromolecules in their deformed state are frozen in the molded product. If, therefore, the molded material is at some time heated above its softening point, it contracts corresponding to the rubber-like character of the melt (see Figure 562-II→I, recovery). A technical application of this effect are the so-called shrinkage films: cross-linked films, which are cooled in the stretched state.

If heating (tempering, annealing) is carried out in the form (in the mold), then recovery by means of shrinkage is impossible; and if heating is carried out long enough, there is a slow release of the tension (relaxation) according to Figure 562, II→III.

The inner tension (stress, brought about by the rubber-like character of the polymers above their softening temperature), which is found in nontempered molded objects, is very undesirable because it results in a lower tensile strength parallel to the direction of flow, than perpendicular to the direction of flow (Figure 563).

442 Rubber-elasticity of Cross-linked Polymers

With the melts of macromolecular compounds, the property of rubber-elasticity is clearly observed only in a certain temperature interval. In the neighborhood of the softening temperature, the polymers are still relatively stiff, deformation requires a considerable force, and recovery occurs only slowly. Much above the softening temperature, the melt is easily deformable; however, the elastic recovery is overshadowed by the tendency of the material to flow due to macro-Brownian motion. The width of the clearly defineable rubber-elastic behavior region depends on the chain structure of the polymers and also on their molecular weight. The higher the molecular weight, the broader the rubber-elastic temperature range and vice versa. Polymers with molecular weights under 50,000 (for example, technical polyamides) are viscous liquids, not rubbers, above their softening, i.e., melting temperature.

Covalent links, in contrast to secondary valence links, are never in a state of reversible, temperature-dependent dynamic equilibrium. This has two important consequences for covalently cross-linked elastomers: (1) the number of cross-linking points is independent of temperature, and (2) the position of the cross-linking points is not changeable. Thus, it is possible to fix the concentration of the cross-links (degree of cross-linking) by choice of the reaction conditions, without having to worry that at higher temperatures the degree of cross-linking might become smaller and elasticity might disappear.

TABLE 565. Industrial elastomeric materials.[1]

Trade name	Chemical Composition	Vulcanization	Glass temp. °C
Natural rubber	cis-1,4-Polyisoprene	Sulfur	−70 to −75
Synthetic rubber SBR, Buna S,	Butadiene-Styrene (25%)-copolymer	Sulfur	−50 to −60
NBR, Buna N, Perbunan	Butadiene-Acrylonitrile (25 − 30%) - copolymer	Sulfur	∼ −40
C 23, Nordel	Propylene-ethylene-copolymer	Sulfur	∼ −40
Butyl rubber	Isobutylene-Isoprene (5%)-copolymer	Sulfur	∼ −70
Tufsyn, Ameri-pol CB Cis 4 (-Rubber)	cis-1,4-Polybutadiene	Sulfur	∼ −100
Cariflex I 300	cis-1,4-Polyisoprene	Sulfur	∼ −70
Vulcollan[2]	Polyester-polyurethane	Diisocyanate	−30 to −60
Lactoprene	Ethylacrylate-Chloroethyl-vinyl-ether-copolymer	Diamine	∼ −25
PVC	Polyvinylchloride (with plasticizer)	Secondary valence forces	depends on the plasticizer
Hypalon	Sulfochlorinated Polyethylene	Diamine	∼ −30
Viton A	Vinylidenefluoride-Hexafluoro-propylene-copolymer		
Thiokol	Polyethylene-tetrasulfide	ZnO+Sulfur	
Silicone rubber	Polysiloxane with $\overset{\displaystyle R}{\underset{\displaystyle Cl}{O-Si-O-}}$ units	H_2O	

[1]Polymers, which at room temperature (and below) exhibit rubberelastic behavior, are called elastomers.

[2]The materials listed below the dashed line are specialty rubbers which are mainly used as insulating materials, caulking compounds, tubing, and for other special applications.

By this decoupling temperature and degree of cross-linking through covalent cross-linking, it is possible to produce soft rubber with high elasticity, i.e., compounds which undergo strong deformation even at low deforming forces, but remain almost free of permanent deformation.

The low resistance against elastic deformation is the result of a relatively large temperature interval between the freezing point and use temperature.

Thus, the secondary valence forces are more or less eliminated and the chains, i.e., the chain segments, have a high degree of mobility which is only slightly restricted by the cross-links, because there are only relatively few cross-linking points. The reason that even with this low degree of cross-linking there is no flow under deformation and that these elastomers are therefore highly elastic is that the number and position of the cross-linking points are fixed with elastomers cross-linked by main valence links. This prevents a spatial rearrangement of the macromolecules with respect to each other. The coils are not able to glide past each other, so that a relaxation of the stress caused by the deformation (stage II, Figure 562) can no longer occur by place exchanges among the molecules (Figure 562, II→III). The coils must therefore remain in their stretched form, even if the stress continues for a longer time. The great technical importance of rubber was achieved only through such covalent cross-linking by a process called vulcanization (invented by C. Goodyear).

Depending on the constitution of the polymer chain, there are several possibilities for producing this type of cross-linking: natural rubber (cis-1,4-polyisoprene) and synthetic rubber (butadiene-styrene copolymers) have an olefinic unsaturated chain, and vulcanization occurs by reaction of the double bonds with sulfur. In addition to the butadiene and isoprene copolymers, there is also a series of saturated polymers (for example, ethylene-propylene copolymers or polyisobutylene), which are rubber-elastic at room temperature. Since they are saturated, they cannot be vulcanized with sulfur (see Table 565), and one has to (by means of copolymerization or by condensation) build certain suitable reactive groups (double bonds or functional groups) into the polymer chain by means of which cross-linking then becomes possible (compare p. 322).

In spite of the fact that the different rubber-elastic polymers have qualitatively the same behavior, their mechanical properties, in addition to their very different chemical properties, also show considerable quantitative differences, which result from the different constitution of their polymer chains. In addition, with all elastomers it is possible to influence the modulus of elasticity and the maximum deformation by means of the degree of cross-linking: the higher the cross-linking, the larger would be the modulus of elasticity and the smaller the maximum elongation (elongation at break). According to its definition, the modulus of elasticity with elastomers is paradoxically much smaller than with all other materials, and it is the smaller, the more "elastic" the rubber. With strong cross-linking one finally obtains materials which have completely lost their elasticity, and which are hard and rigid (hard rubber). Conversely, with decreasing degree of cross-linking, the ability to stretch becomes larger; however, especially with large deformations applied over long times, it becomes less and less completely recoverable.

4421 The Elongation of Rubber-elastic Materials

As an example for the deformation of a rubber-elastic polymer there is shown in Figure 567 the elongation of a rubber band as a function of the deforming force [tension (stress), elongation (strain) diagram]. Figure 568 shows in a schematic manner the coil deformation occurring during stretching.

It is easy to understand the nature of the curve after the previous discussion of rubber-elasticity as a macromolecular state of aggregation. Under low tension (1-10 g) there is no visible deformation. This is because the polymer chains in their statistically coiled form are in a state of maximum entropy, so that a certain minimum force is required (corresponding to the activation energy of a reaction) to remove them from this stable state and to force them into a more elongated state. When this limit of stress has been reached, first the coils or coil segments where the cross-linking is least strong, are deformed, i.e., where the length of the chain between two cross-linking points is largest. For this chain length there will be a similar distribution, as for the molecular size in a polymeric material: very large uncross-linked segments and especially short ones will be relatively few, compared to the chain segments with average length. At low stress at first only the slightly cross-linked parts of the material take part in the deformation; as the stress becomes larger, more macromolecules, i.e., molecular regions, are deformed. At this early stage (up to 100 g of stress), the slope of the curve increases, i.e., the elongation per gram of increasing stress becomes larger with increasing stress. Finally, the deforming force has become so large that with increasing stress practically all the chain segments of middle length have been forced to abandon their statistically coiled state and take part in the elongation process. In this stage the stress-strain diagram has reached its maximum slope, i.e., the deformation per

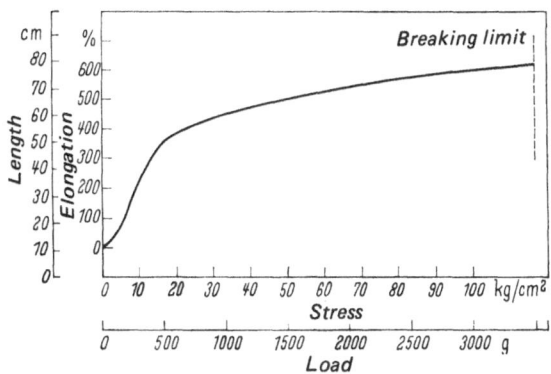

FIG. 567 – Elongation of a rubber band as a function of the load (Cross-section: 3mm^2, length: 11cm)

gram of increasing stress remains now constant (nearly linear portion of the curve
between 10 and 500 g). The proportionality of length and stress is maintained as
long as the number of chain segments taking part in the orientation process does
not change. When the stress has reached a given magnitude, the slope of the curve
decreases again, because now more and more chain segments (chain between two
cross-linking points) come closer to a state of maximum possible deformation and
therefore no longer take part in the elongation. The result, shown in the above
example (Figure 567), is that at 50 cm length there is a rapidly increasing stiffening
of the rubber. For the elongation from 10 to 50 cm it is sufficient to have a force
of 500 g, but then, in order to stretch the rubber band 30 cm more (up to its
breaking point), one has to use 10 times that force. Through this great force the
molecular coils are stretched to the breaking point. Because of the uneven
distribution of the cross-linking points, not all chains are completely stretched at
lower stress. In order to stretch even the last few chains, one hs to deform the
entire molecular material which can only be brought about through very high
tension and only with elastomers whose molecules have great chain mobility, such
as polybutadiene, polyisoprene, and their copolymers. In other cases, for example,
with the polyacrylic ester elastomers, even at rather low tension, chain rupture
occurs and, therefore the stretched band breaks.

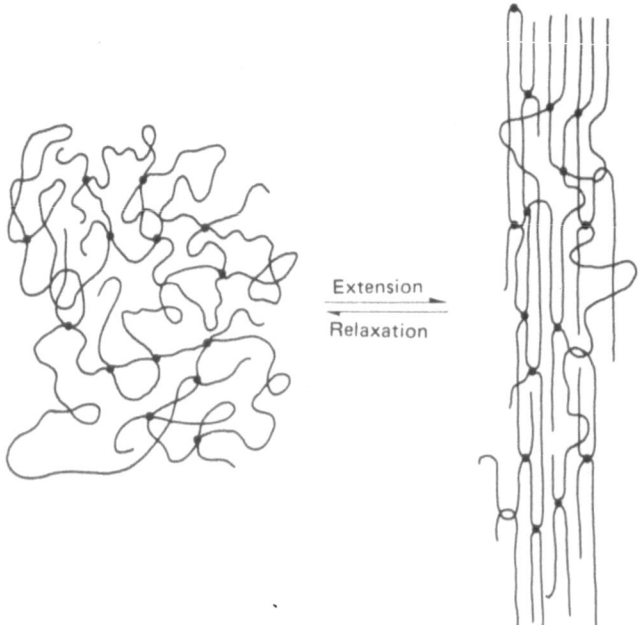

Extension

Relaxation

FIG. 568 — Schematic representation of the transition from the randomly coiled
to the oriented state on elongation of a crosslinked rubber (compare the effect on
the x-ray diagram shown in Fig. 571).

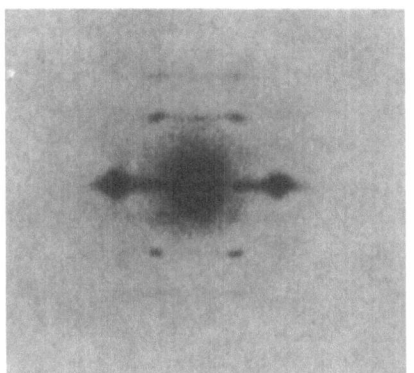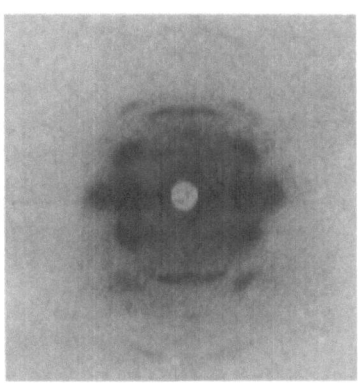

FIG. 569 — X-ray diagrams of cellulose (left) and silk (right). (Photographs of ANDRESS and of TROGUS and HESS).

The disorientation of the entire molecular material brought about by the strong tension is not completely reversible because of the relaxation phenomenon discussed above. Because of the cross-linking, one therefore observes a rather limited flow. This is the reason why in the stress-strain diagram the recovery curve (decreasing elongation under decreasing tension) does not coincide completely with the tension curve (increasing elongation under increasing tension). One obtains a hysteresis loop, and the width of the loop is a measure of the permanent deformation. In addition to the rubber-elastic deformation, which is made possible by the orientation of the coiled macromolecules, at high tensions one also obtains a small normal elastic deformation, which is related to changes in the distances between the atoms and in the valence angles.

The end point of the rubber-elastic deformation is especially sharp when the rubber crystallizes in the stretched state, which happens, for example, with natural rubber, (cis-1,4-polyisoprene), as well as with synthetic cis-1,4-polyisoprene or cis-1,4-polybutadiene and polyisobutylene, all of which are characterized by their regular steric structure. Through crystallization, the parallel-oriented chains are tied more closely together over larger regions, so that even at high stress a short gliding of the chains is no longer possible. Through this, the rubber-elastic deformation comes to a sudden end and also the tensile strength of the stretched rubber is increased. If one cools a stretched rubber with stereo regular chains to below its glass temperature (freezing point), that is, to approximately $-70°C$ to $-100°C$, then it shows on mechanical fracture a fiber structure. Also in the X-ray diagram one obtains a typical fiber diagram. In Figures 569 and 571, the X-ray diagrams of cellulose and silk are compared with that of stretched and unstretched natural rubber. (For the formation of a fiber diagram through diffraction of X-rays, compare p. 606 ff.). At the same time, Figure 571 presents a confirmation of the transition from the statistically coiled (amorphous) state into the oriented (crystalline) state on stretching.

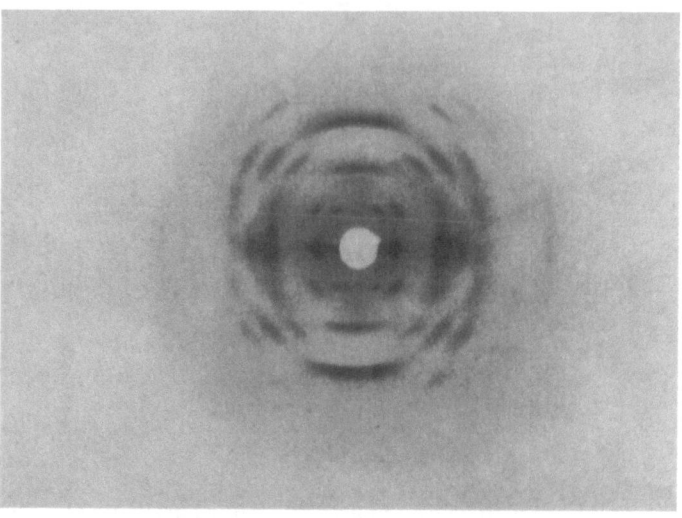

FIG. 570 — X-ray diffraction diagram of a lobster tendon (chitin) (K.H. MEYER and PANKOW).

The crystal lattice forces (heat of crystallization ΔH) are so small with rubber that at room temperature the crystal lattice is stable only in the stretched state. As soon as the stress is removed, the system can increase its entropy. Depending on the temperature, the magnitude of $T \cdot \Delta S$, the entropy term, is larger than the amount of the crystallization enthalpy ΔH, so that ΔF for the transition from the stretched to the coiled state, according to the relation $\Delta F = \Delta H - T \cdot \Delta S$ is negative. Only with temperatures near the glass temperature, is the magnitude of $T \cdot \Delta S$ smaller than the value of ΔH, so that the crystallized rubber at low temperature remains in the crystalline state, or is transformed into this state, even without applied stress.

4422 Rubber-elasticity–Normal elasticity

Rubber-elasticity was for many years a mysterious phenomenon, and before one knew about the macromolecular coil structure of rubber, many, today fantastic-al-sounding theories for rubber-elasticity had been proposed. These were all derived from mechanical models (Christmas-tree structure, spiral-springs, and similar models). The first step for the correct interpretation of rubber-elasticity was made by H. Wöhlisch when, on the basis of some of his experimental observations, he compared rubber-elastic deformation to the compression of gases.

Thus, if one heats a stretched rubber band, the tension increases, which is quite the opposite to what is found with a stretched wire or a spiral spring of steel, where the stress is reduced by heating. However, it is completely analogous to the behavior of a compressed gas whose pressure increases on heating, i.e., the recovery force of a stretched rubber becomes larger on heating, just as the recovery force of a compressed gas becomes higher on heating. This means that the driving force in the contraction of stretched rubber is not the tendency for the reduction of the internal energy (negative ΔH), as would be the case with the contraction of a spring, but that rubber contraction is the result of the tendency to increase the entropy (positive value of $T \cdot \Delta S$), just as with the expansion of a compressed gas.

The molecular mechanism of the entropy change on stretching and contraction of rubber could be explained readily after the coil structure of polymers became known and the X-ray diagrams showed the transition from the amorphous to the crystalline state on stretching of natural rubber. (K. H. Meyer). The entropy decrease on stretching corresponds to the transition from the most probable coil

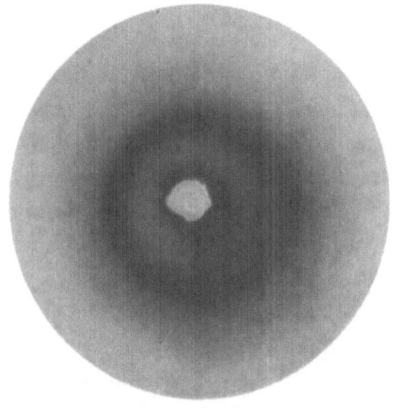

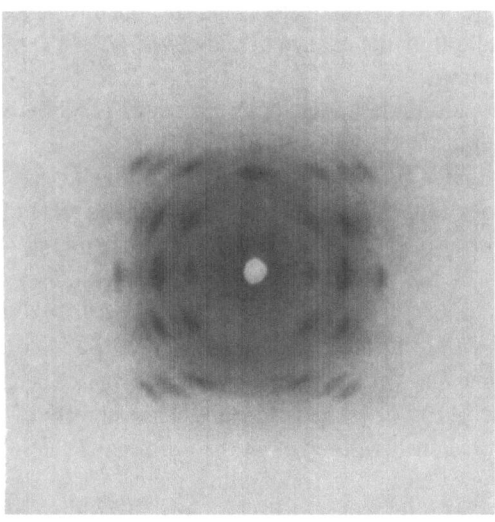

FIG. 571 – X-ray diagram of un-stretched (top) and stretched (bottom) natural rubber (P. SCHERRER, K.H. MEYER and H. MARK)

structure of the chain to the improbable state of the stretched chains, and the entropy increase on recovery (retraction) corresponds to the return from the improbable oriented chain structure to the more probable statistical coiling after release of the tension. The higher the temperature, the greater the chain mobility (intramolecular or micro-Brownian motion), and the larger the entropy loss in the transition to the stretched state where the mobility of the chains is strongly reduced. The greater $T \cdot \Delta S$, the greater will be the driving force for the return to the statistically coiled state, and therefore, the tension (stress) must increase on heating. Crystallization in the stretched state, which is coupled with heat being given off, corresponds to the liquefaction of a gas on compression.

45   THE SOLID STATE

If one cools liquids, the energy of motion of the molecules and atoms become smaller and smaller, and finally, a temperature is reached where the thermal motions of the molecules are so small, that they can no longer free themselves from their secondary valence force fields. The kinetic energy of the thermal motion at that stage is then sufficient only for rotational and vibrational motions of atoms and atomic groups about a position of minimal potential energy.

Thermodynamically considered, this is a temperature where the entropy term $T \cdot \Delta S$ of the thermodynamic equation of state is equal in magnitude to the heat of crystallization ΔH, i.e., the liquid is in equilibrium with the crystals: $\Delta F = \Delta H - T \cdot \Delta S$; where ΔH is the difference in internal energy, i.e., the enthalpy of the system, before and after the change in state; ΔS, the difference in entropy of a system before and after the change of state; and ΔF, the difference in free energy of the system before and after the change of state. When $\Delta H = T \cdot \Delta S$, $\Delta F = 0$, and the system is in equilibrium.

With the smallest additional temperature decrease, $T \cdot \Delta S$ becomes smaller than ΔH, and precipitation of the crystals begins.

If a molecule of the liquid phase is trapped by a growing crystal then its spatial position becomes fixed, and it loses its free mobility. The value of the translational component of its kinetic energy remains, however, in the system as vibrational energy. The process of the incorporation of the molecule into the crystal lattice represents a free fall in the secondary valence force field. Thus, the trapped molecule goes from a state of higher potential energy to a state of lower potential energy, just as a falling body does in a force field. Its kinetic energy increases to the extent to which the potential energy decreases. This increase of kinetic energy at the expense of potential energy shows up in the warming of the system. The heat of crystallization can only lead to a temperature increase if the system was super-cooled, because otherwise the released heat of crystallization must be used again as heat of melting. If the heat is not removed, then $T \cdot \Delta S$ becomes larger again and the crystallization process stands still. If on the other hand, the heat is removed, additional molecules can be transferred from the liquid phase to the crystalline material. This again releases additional heat, so that in spite of the cooling from the outside, the inner temperature remains constant as long as any liquid is left.

For all the molecules of a compound, in general there is only a single spatial arrangement of minimal potential energy, which therefore all molecules of the compound try to atfain. The necessary result is that in the transition from the liquid to the solid state, one always obtains the same spatial arrangement of the molecules: the crystal lattice.

This is in principle not different with macromolecular compounds than with low molecular ones: a macromolecular compound will also crystallize on cooling from solutions and melts; however, the ability to crystallize assumes a sterically regular chain structure. This assumption is always true for polymers which do not have any asymmetric carbon atoms in the chain, for example, unbranched polyethylene, polytetrafluoroethylene, polyvinylidenechloride, polyisobutylene, and linear polyamides and polyesters, but it does not hold for vinyl polymers such as polystyrene, polyvinylchloride, the polyacryates, polyacrylonitrile and polyvinyl ethers. Therefore, the former type of polymer always crystallizes, no matter how it is prepared, whereas the vinyl polymers crystallize only when they are polymerized in such a way that they have an isotactic or syndiotactic structure, i.e. a stereoregular structure.

451 The Glassy State

If one cools the melt of a polymer with irregular steric structure below the melting point which the same polymer would have with an isotactic structure, then in general one does not observe any change. Only if one cools further, does one finally reach a temperature range where the melt solidifies to a glass. This temperature is called the glass temperature, or freezing point.

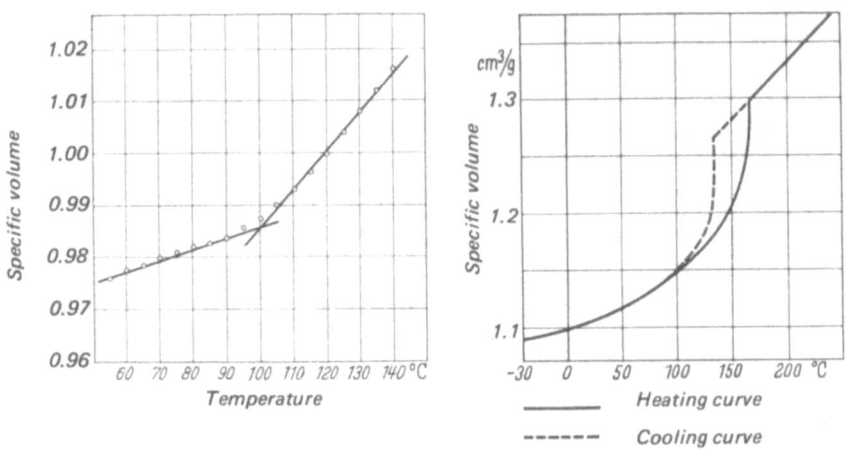

FIG. 573 — Diagrams illustrating the glass-temperature. Left: Change of the specific volume with temperature for polystyrene (JENCKEL and ÜBERREITER) Right: the corresponding curve for polypropylene (H. WILSKI)

The glassy solidification of a polymer is not a result of a continuous inclusion of the molecules into an arrangement of minimal potential energy (as with crystallization), but concerns a gradual slowing down of molecular motion. The motions of the macromolecules gradually freeze without giving rise to any changes in the structure, or rather the absence of structure, of the polymer as it exists above the glass temperature, i.e., in the rubber-elastic region. Therefore, it is quite reasonable to characterize such a glass as a supercooled or frozen melt.

The glassy state is in a general sense the solid state of a macromolecular compound into which it transfers from the rubber-elastic state on cooling. This also applies to those polymers which are able to crystallize. The form of the solid state (glassy or crystalline) these polymers assume on cooling depends on the rate of cooling. By sudden quenching, one can obtain a glass in every case. Depending on their crystallization tendency, the crystallizable polymers enter more or less quickly into the crystalline state; especially fast, if one anneals the sample, i.e., by keeping it at temperatures somewhat below the softening point. Also by stretching and drawing (for example, with fibers and films), one can increase the rate of the transition from the glassy to the crystalline state. The glassy state is unknown only with crystalline polymers which decompose before they can melt (for example, polysaccharides, proteins, and nucleic acids).

Since the macromolecules in the glass do not take up a position of minimum potential energy (which would be the same for all chains and chain segments and is prevented by the irregularity of the chain structure), the freezing or solidification to a glass is not related to a stepwise change in the internal energy at a certain given temperature. Therefore on cooling there is no discontinuity in temperature. A glass, in this connection, is quite comparable to a mixture which consists of an infinitely large number of compounds, where from one to the next, the melting point is a few degrees higher. The melting point of such a mixture is lower than the melting point of the lowest-melting component (melting point depression). Furthermore, the melting point will no longer be sharp. Similarly, with a glass, the softening temperature lies below the melting point of a crystalline polymer corresponding in its chemical constitution to that of the glass. For example, crystalline isotactic polystyrene melts at 232°C, whereas glassy, atactic polystyrene softens at 100°C. The softening of a glass does not occur at a sharply limited temperature, such as the melting of a crystal, but covers a temperature range of 5–20°C.

In spite of this, the freezing temperature may be exactly defined. Thus, if one measures the change in density (or, the specific volume) as a function of temperature, one notices that in the glass, as well as in the melt, there is a linear dependence; however, the slope of the curve is smaller with the glass than with the melt, i.e., the overall curve has a break within the freezing range, which can be determined quite accurately by extrapolation of the two branches of the curve. This point can be defined as the freezing or glass temperature. Figure 573 shows the curve for polystyrene, and in comparison the corresponding curve for crystalline polypropylene.

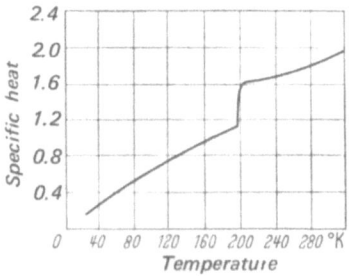

FIG. 575 — Change of the specific heat of natural rubber with temperature (BEKKEDAHL and MATHESON).

In addition to the density, other physical properties, such as thermal conductivity, the dielectric constant, or the refractive index, show anomalies in their dependence on temperature in the softening region. Figure 575 shows the temperature dependence of the specific heat, and Figure 576 as a further example, the change of the modulus of elasticity and the logarithmic damping decrement (which is determined by torsional vibration measurements).

By torsion measurements, it is also possible to determine the freezing point of polymers, or polymeric phases, which are present as the amorphous component in a crystalline polymer such as polypropylene or nylon. Similarly, one can determine the freezing temperature of components present in mixtures (blends) of polymers or graft-copolymers, for example, the rubber component in high impact polystyrene (for the structure see pp. 291-293 and 618). As the examples in Figure 576 show, in such systems the freezing range of the amorphous, i.e., the rubber-elastic phase, does not coincide with the much higher softening range, or melting point, of the hard phase. The freezing phenomenon shows up here only in the jump-like increase of the brittleness.

The irregularity in the property-temperature curve does not lie at exactly the same temperature for all physical properties. Correspondingly, one also obtains, depending on which property one considers, slightly different values for the glass temperature T_g.

In daily practice, for the characterization of the softening point, one usually does not employ the glass temperature, but the Vicat-number, the Martens-number, or the F.P. (freezing point). These are conventional values which are defined by the temperature, where under well-defined conditions a certain deformation of a molded sample takes place. Vicat-number and F.P. are usually only slightly different from the glass temperature, whereas the Martens-number lies 20-30°C lower.

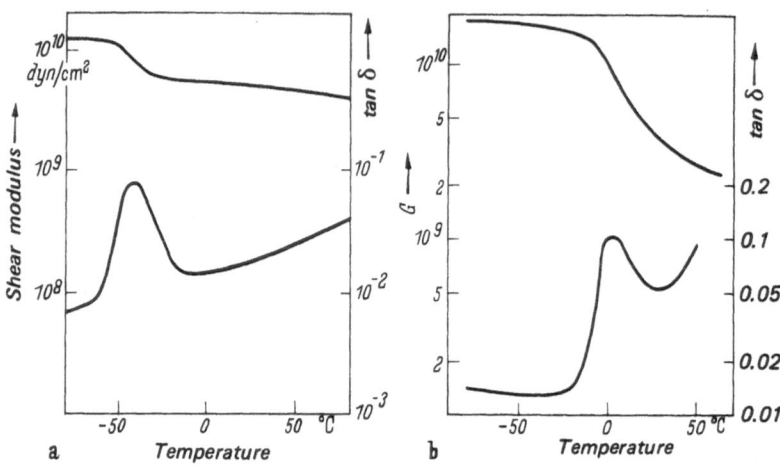

FIG. 576(a and b) — Effect of temperature on the shear modulus G and the loss factor tan δ of mechanical damping. a) impact-resistant polystyrene b) polypropylene (measurements of STAVERMAN and HEIJBOER)

For the position of the glass temperature, the freedom of motion, not the molecular weight or length of the polymer chain, is the determining factor. This mobility depends first, on the internal stiffness of the chain, and second on the strength with which the neighboring chains or chain segments hold together, i.e., the strength of the secondary valence forces between the chains. Both factors are given by the chemical structure of the chain (compare the section "Structure and Properties in the Solid State," p. 607 ff.).

In addition to these influences which depend on the constitution of a given polymer, and therefore cannot be varied at will, there is a further influence on the chain mobility which is an external influence: the influence of solvents and plasticizers. A small concentration (1-3% of solvent; this includes also its own monomer) usually does not change the outer appearance of the polymer. However, it loosens up the cohesion between the polymer chains to the extent that the softening point is lowered by 5-10°C. This is the reason why with plastics one usually has to remove any residues of monomer or solvent which are still present from the polymerization by a vacuum-degasification process, if one would like to have a product with the highest possible softening temperature. On the other hand, one sometimes adds to plastics a small amount (2-3%) of a high-boiling liquid in order to reduce the viscosity of the melt and thus to facilitate the shaping of the products in extruders or injection molding machines (= internal lubrication); but, as a consequence, one has to accept a certain lowering of the softening point. With low molecular weight polymers the greater mobility of the chain ends has the same effect as a plasticizer: in the molecular weight range below 100,000, one observes a slow decrease in the glass temperature with decreasing molecular weight (compare also p. 5 and Figure 6).

By plasticization one usually understands the addition of larger amounts (20-40%) of a poor solvent (in a thermodynamic sense) to a hard plastic, so that at normal temperatures it will become rubbery or leather-like (compare p. 544).

The melting points of crystalline polymers are considerably higher than the softening points of the corresponding glassy polymers with the same chemical structure. For example, the glass temperature of atactic polypropylene is about 0°C, whereas the melting point of the crystalline isotactic polymer is 175°C. Similarly, atactic polystyrene softens at 100°C, while the isotactic polystyrene melts only at 232°C. This is because a considerably higher amount of kinetic energy of the molecules i.e., the chain atoms) is necessary in order to shake these out of their stable position in the crystal lattice, similar to a ball which lies in a hole and which can roll away only if it is given a push which is strong enough to push it over the edge of the hole. A sphere which lies on a plane can be made to roll with a considerably less energetic push.

4511   The Mechanics of Elastomers and Glasses

Glasses as well as elastomers are usually quite far removed from the state of an ideal elastic solid. Although they are elastic on short-term deformations, they flow when the deformation is maintained over a longer period (relaxation of viscoelastic materials). This statement also holds for many metals and is therefore not at all typical for polymeric materials. Only the very large difference of the E-moduli between the glassy state and the rubbery state of one and the same polymer is typical. This brings about a very steep damping maximum at the glass temperature and a very strong dependence of the relaxation time on structural parameters.

In order to describe the mechanical behavior of solid materials, one must investigate the dependence of the stress on the deformation and their time dependence. For ideal solids, Hooke's law holds, which states that the stress of a deformed body is proportional to the elongation, and vice versa:

$$\epsilon = (1/E) \cdot \sigma, \qquad (43)$$

where ϵ (elongation) is relative deformation; σ (stress), forces per $cm^2 = k/F$; and E, modulus of elasticity.

The ideal (nonflowing) solid represents one extreme, and the ideal (only flowing, i.e., without the appearance of an elastic retractive force) liquid represents the other extreme. The behavior of the ideal liquid on deformation can be described by Newton's law. One considers a layer of liquid between two glass plates (1 cm. apart) with area F, displaced with respect to each other with a velocity $d\epsilon/dt$ by a force k:

$$\dot{\epsilon} = (1/\eta) \cdot \sigma, \qquad (44)$$

where $\dot{\epsilon}$ is $d\epsilon/dt$ and η, viscosity.

If one represents a Hookean solid by the symbol I (Figure 579), and a Newtonian liquid by the symbol II, then one can represent real solids which at short deformation times and small loads are elastic, but on longer deformation times start to flow, by the symbols III (Maxwell model), IV (Voigt-Kelvin model) and their combination by V (Relaxation model) and VI (Retardation model). Models V and VI are equivalent as far as the description of mechanical behavior is concerned.

The deformation-time curves corresponding to the models in Figure 579 are idealized insofar as the deformation process in the Hookean region is assigned an infinitely short time in order to better show the contrast with the time required for deformation with superimposed flow (Voigt-Kelvin model).

If one differentiates Equation 43 with respect to time t, one can put together the two equations and one obtains:

$$d\epsilon/dt = (1/\eta) \cdot \sigma + (1/E) \cdot d\sigma/dt \tag{45}$$

or on integrating

$$\sigma = \epsilon_0 \cdot E \cdot e^{-(E/\eta)t} = \epsilon_0 \cdot E \cdot e^{-t/\tau}. \tag{46}$$

This gives an equation which describes the Maxwell model. The quotient η/E is called the relaxation time τ.

The corresponding equation for Model VI (Figure 579) is accordingly:

$$\sigma/\epsilon_0 = E_1 \cdot e^{-t/\tau_1} + E_2 \cdot e^{-t/\tau_2}. \tag{47}$$

If one considers that one stretches a viscous system, such as an unvulcanized elastomer (for example, polyisobutylene), and then holds its length constant at a certain elongation and from this time on measures the stress at different times, i.e., the stress which is necessary and sufficient in order to maintain the elongation at this value, then the time in which the stress decreases to $1/e = 1/2.7$ of its original value is identical with the relaxation time τ. The relaxation time τ is the smaller, the larger the modulus of elasticity at a given viscosity, and it is the larger, for a given E-modulus, the higher the viscosity. (The molecular mechanism of stress equilibration by flow is described in Figure 562.)

Stress equilibration through flow in solid bodies is a structural reason for the damping of mechanical vibrations (periodically occurring deformations). Damping is the stronger, the higher the viscosity η (internal friction). The pronounced damping maximum at the glass temperature is characteristic for polymeric materials. Whereas low molecular weight compounds on melting almost completely lose their elasticity (i.e., the damping curve of a vibration becomes infinite at the melting point) with a polymer, elasticity does not disappear but is covered by an extremely high viscosity in the region of softening (i.e., in the temperature range of the glass temperature). On further heating, the internal friction becomes smaller and the rubber-elasticity of the macromolecular compound appears (assuming high molecular weight). On still further heating, usually the elasticity of polymer melts disappears, unless there is some intermolecular covalent cross-linking.

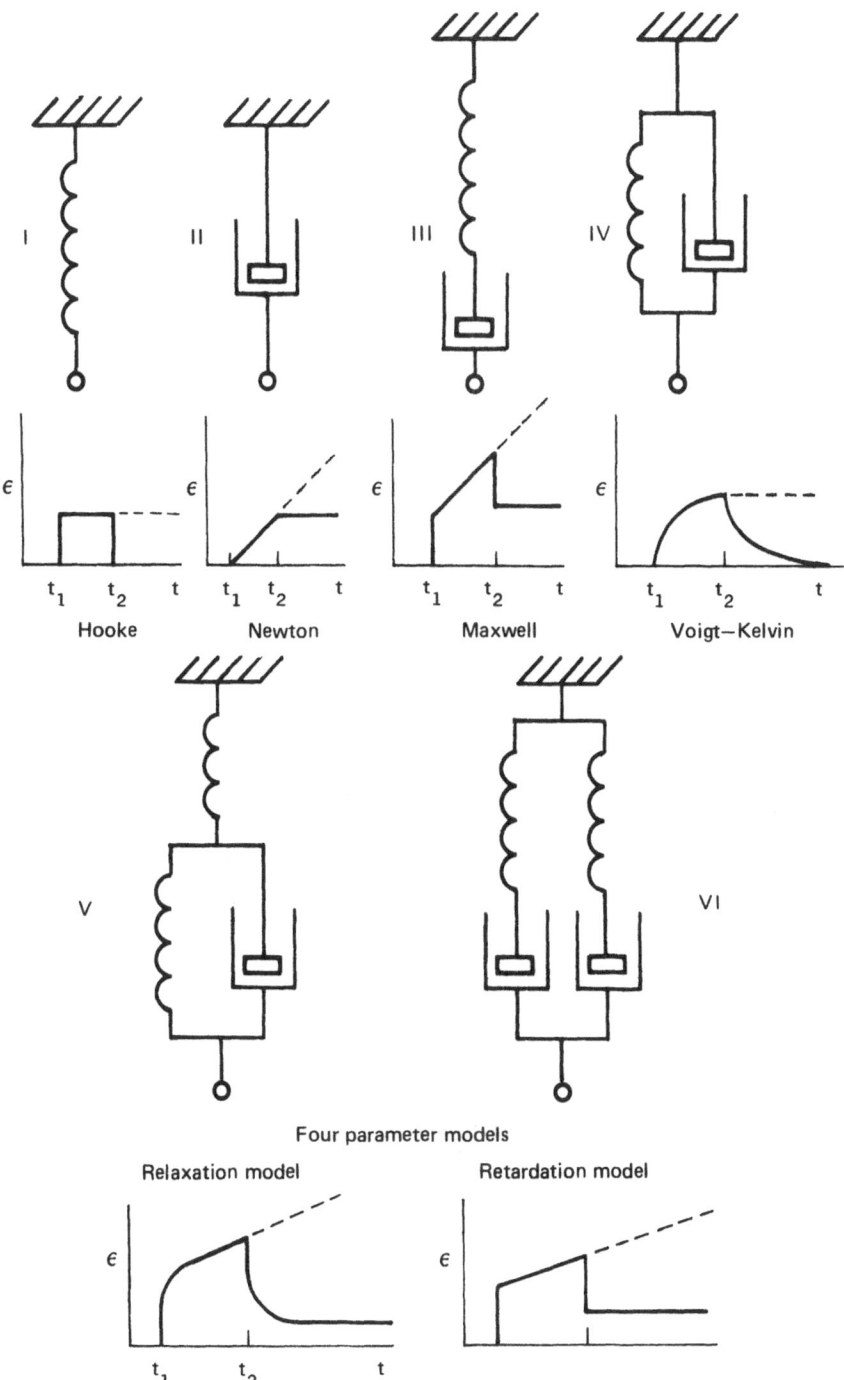

FIG. 579 — Models for elastic, liquid, and viscoelastic systems with the corresponding $\epsilon = f(t)$ – functions.

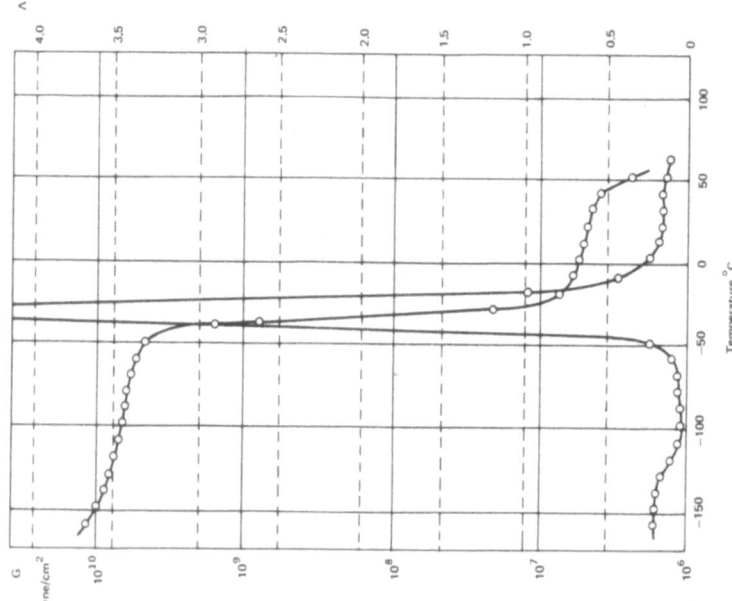

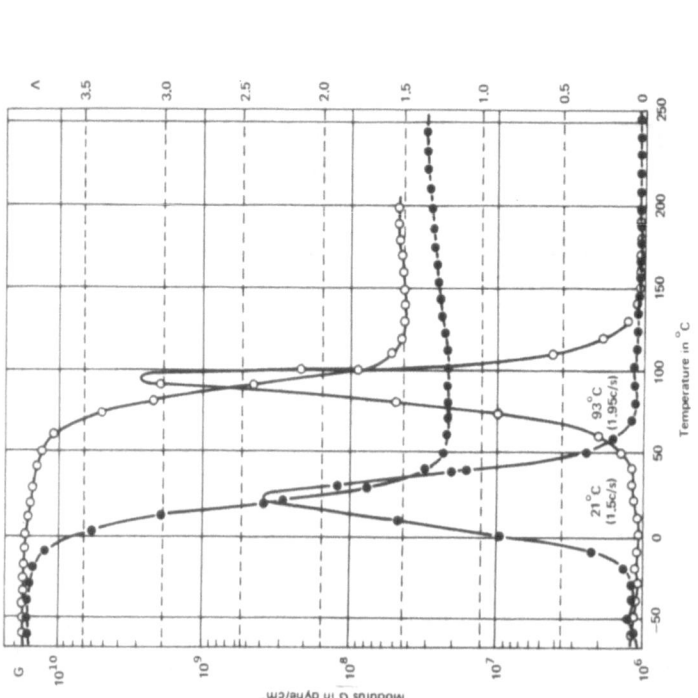

G = torsion modulus

Λ = logarithmic damping decrement

FIG. 580 — The effect of temperature on the shear-modulus and the logarithmic decrement for two polyester resins (left), and for poly-n-butylvinylether (right) (SCHMIEDER and WOLF)

Figure 580 shows the typical changes in the mechanical behavior as a function of temperature, as they would be observed with the torsion pendulum of Schmieder and Wolf. Figure 581 shows a schematic presentation of the experimental apparatus. The polymer sample (in the form of a thin plate) (40 x 10 x 1 mm) is joined to a light disc and made to oscillate. The resulting sine-curve is photographically registered by means of a light beam reflected from a mirror fastened to the torsion axis. The entire apparatus is enclosed in a thermostat which permits variation of the temperature from −150°C to +300°C. The change in the E-modulus, respectively the shear modulus G, is measured by following the change in the frequency. The diminution of the amplitude of the sine-curve yields the damping.

The appearance of the damping maximum at the glass temperature can be explained by means of the mechanical model shown in Figure 582.

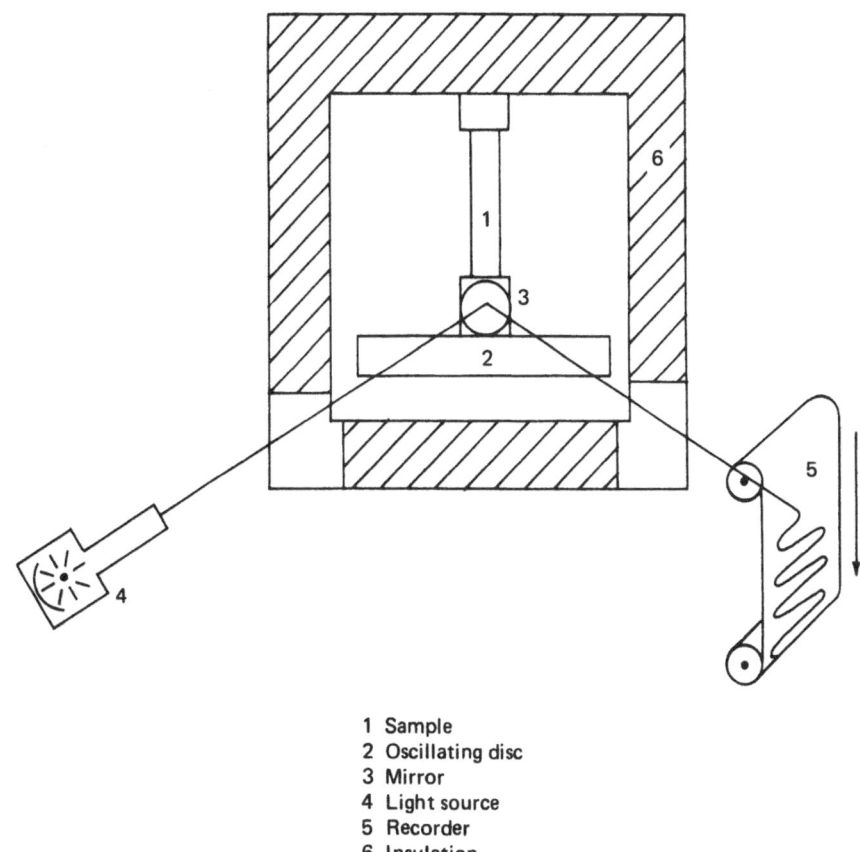

1 Sample
2 Oscillating disc
3 Mirror
4 Light source
5 Recorder
6 Insulation

FIG. 581 − Apparatus for determining the torsion-modulus, G, and the logarithmic damping decrement, Λ (SCHMIEDER and WOLF)

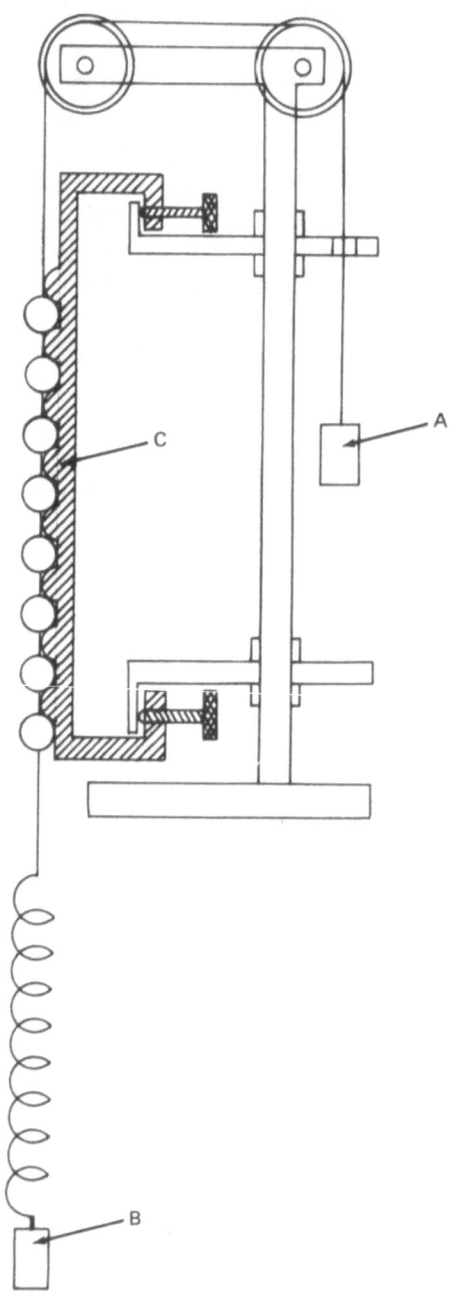

FIG. 582 — Model used to explain the damping maximum at the glass-temperature.

A chain of spheres is suspended by means of two weights A and B in such a way that in the state of rest, even if it hangs free, no movement occurs. If one now moves (in a horizontal fashion) the fixed hemisphere-wall C, so that each sphere occupies a gap between two hemispheres then one can pull down the weight B a certain amount or lift it without changing the position of the spheres. If one now releases B, an up and down oscillation begins. The oscillation is dampened only slightly as long as the series of spheres does not change its positions. If one now diminishes the friction, i.e., cohesion between the spheres and the hemispheres by moving C to the right, then a moment is reached when the spheres glide away to the top or to the bottom. The oscillation is strongly damped immediately and stops completely after a short time, i.e., the energy of oscillation has been used up by the friction of the spheres. In principally the same way, on heating of a glassy polymer, at the glass temperature the cohesion between the chains and chain segments is loosened to the extent that a movement of the segments becomes possible. In this stage, the stress caused by the torsional deformation is rapidly decreased by displacement of the molecules or molecular groups with respect to each other. If one uses the terminology of Equation 46, this means that the relaxation time becomes shorter, shorter, in fact, by several orders of magnitude.

If with an undamped oscillation the stress curve and the deformation curve are in phase (the curves coincide), the stress curve gets ahead of the deformation curve on appearance of damping, i.e., the stress curve has already passed its maximum when the deformation just reaches its maximum (Figure 583).

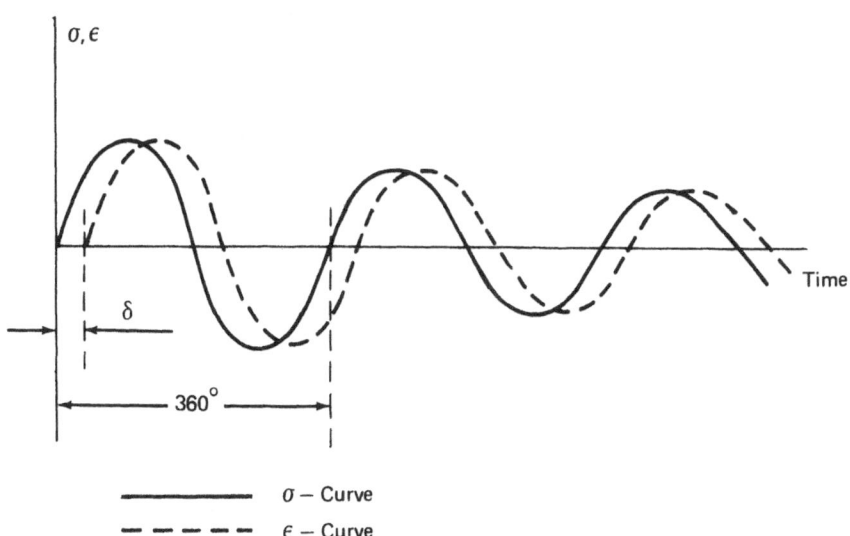

FIG. 583 — Phase displacement δ between ϵ and σ with a damped sine oscillation curve.

Damping is measured through the tangent of the phase difference between stress and deformation (tan δ). This value of tan δ is identical with the relative loss in energy, i.e., the quotient of the imaginary part E'' and the real part E' of the oscillation energy.[5] This is called the loss factor d.

$$d = \tan \delta = E''/E'. \tag{48}$$

The static Equation (47) for Model VI changes with the oscillation model into the corresponding dynamic equation:

$$\frac{\sigma_\omega}{\epsilon_o} = E'_\omega = E_1 \left(1 - \frac{1}{1+(\omega\tau_1)^2}\right) + E_2 \left(1 - \frac{1}{1+(\omega\tau_2)^2}\right) \tag{49a}$$

$$E''_\omega = E_1 \cdot \frac{\omega\tau_1}{1+(\omega\tau_2)^2} + E_2 \cdot \frac{\omega\tau_2}{1+(\omega\tau_2)^2} \tag{49b}$$

Correspondingly, for the interpretation of oscillation experiments, the loss modulus E'', or (as in Figure 580) the loss factor d, or tan δ, (or also the logarithmic damping decrease Λ) is plotted.

$$\Lambda = \pi \cdot \tan \delta. \tag{50}$$

That the damping again becomes smaller on increasing the temperature happens because the normal elasticity present below T_g goes over into an entropy elasticity (rubber-elasticity) above T_g. The retractive force necessary for this arises because the molecular motion under the influence of stress in the polymer melt (especially if the polymers are covalently cross-linked, but also if the cross-links are only brought about by intermolecular forces) corresponds to a coil deformation from a statistically probable state into a long stretched-out, improbable state. The desire of the system to return to its state of maximum entropy brings about the creation of a retractive force. The latter, however, becomes noticeable through effective return motion only if the displacement of the molecules can occur without too large an internal friction. Only the fact that the transition from the glass to the rubber goes through a state of extremely high viscosity causes the steep increase in damping. To the degree by which the viscosity decreases at higher temperature, the energy-driven oscillation of the glass is replaced by the entropy-driven oscillation of the elastomer.

---

[5] The total energy of an oscillation is $E^* = E' + iE'' = E' = (1 + id)$, where $d = E''/E'$. E' is the reversible part (which is always converted from potential into kinetic energy) of the oscillation energy and is called the "real" part, or the "dynamic modulus." E'' is the part of the oscillation energy which is converted to heat by the corresponding friction, and it is called the "imaginary" part, or the "loss modulus."

The driving force for the rubbery return motion disappears only if the coil deformation is reversed through place exchange of the macromolecules (*inter*molecular displacement) as described in Figure 562.

The rubber-elastic deformation is a relaxation in the sense of solid state physics, because it concerns a flow process. One has to consider, however, that this flow process involves only molecular groups and molecular segments within one macromolecule, not the coils themselves. The *intra*molecular flow process stops by itself through the appearance of a counterforce determined by the decrease in entropy which is connected with the flow process. Since this entropy decrease is brought about by the deformation of the statistical coil, rubber-elasticity (at least in a practical domain) is tied to the presence of macromolecules, i.e., the chain segments between the cross-link points have to be long enough in order to form coils themselves in the sense of random flight statistics (compare pp. 441 and 444). Through the *intra*molecular flow process during rubber-elastic deformation, the sharp decrease in the E-modulus by three or four powers of ten at the glass temperature is explained. In a liquid system even very small forces bring about large deformations and according to Hooke's law Equation (43) the E-modulus is then correspondingly small. In contrast to normally elastic bodies where the E-modulus decreases with increasing temperature, the E-modulus with rubber-elastic bodies increases with increasing temperature (this is of course easily observed only with cross-linked elastomers or with higher frequency deformations, because with uncross-linked systems the *intra*molecular bonds are loosened more and more with increasing temperature, so that the elasticity disappears through *inter*molecular relaxation). The positive temperature coefficient of the E-modulus with rubbers is caused by the fact that with increasing temperature the entropy difference between the elongated state of the coils brought about by deformation and their statistically most probable form becomes larger and larger. Since this entropy difference is the driving force for the rubber-elastic return motion, the stress of a stretched elastomer increases on heating just as the pressure of a gas increases on heating. This effect is large enough so that it can be confirmed by a simple experiment (Figure 586). If one heats the rubber band with an infrared lamp or with a heated tube, the weight will be lifted by a few millimeters to a few centimeters (depending on ΔT and the nature of the elastomer). In order to maintain the deformation constant, one therefore has to increase the weight.

Since the changes in relaxation phenomena with macromolecular compounds are determined by their structure, torsion pendulum measurements as a function of temperature are useful in order to relate the appearance of motion with the chain structure. There are polymers which are completely frozen below their glass temperature; however, there are others which in addition to the damping maximums at the glass temperature (at the melting point, with crystalline polymers) have further maximums at lower temperatures. Polystyrene, for example, has only one damping maximum. However, by blending in rubber, one can form a polymer mixture which at $-50°C$ to $-60°C$ has a second maximum (compare Figure

576a). If the addition of the rubber is carried out in such a way that the phase boundaries are tied to each other by grafting, one obtains the industrially important high-impact polystyrenes (ABS polymers, if instead of the homopolymers one uses copolymers with acrylonitrile). However, one should not assume that with all polymeric materials with high impact resistance this property depends on the presence of a second damping maximum. Damping maximums at low temperatures are not in all cases an explanation for impact resistance. Plexiglas, for example, has a relatively pronounced damping maximum at $-70°C$. However, its impact resistance is in no way higher than that of polystyrene.

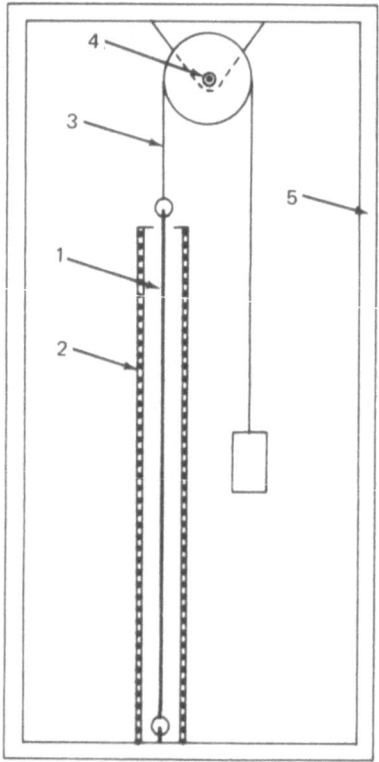

1   stretched rubber-band
2   heating tube
3   nylon thread
4   ball-bearing
5   metal frame

FIG. 586 — Experimental set-up to demonstrate the contraction of stretched rubber on heating.

The additional maximums at lower temperatures show that at these temperatures some sort of molecular motion (vibration, rotation of groups, etc.) is frozen in or begins to appear. How far this influences the mechanical strength of the material at normal temperatures has to be investigated from case to case.

TABLE 587. E-moduli and impact strength of polymeric materials.

Polymeric Materials	E – Modulus [kp/cm^2]	Impact Strength (DIN) [$cm\ kp/cm^2$]
Polyethylene, High pressure	2 000 – 4 000	(> 100)
Polyethylene, Low pressure	5 000 – 10 000	> 100
Polypropylene, isotactic	≈ 15 000	> 80
Polystyrene	≈ 33 000	18 – 30
Styrene-Acrylonitrile	≈ 36 000	25 – 30
Polystyrene, High impact	22 000 – 26 000	60 – 80
ABS – Polymers	22 000 – 26 000	80 – 100
Plexiglass	≈ 30 000	≈ 20
Polyvinylchloride	≈ 30 000	≈ 100
Polyvinylcarbazole	≈ 45 000	≈ 10
Polyamide (Nylon 6,6)	13 000 – 17 000	> 100
Polycarbonate	≈ 22 000	> 100
Celluloseacetate	≈ 20 000	≈ 50
Cellulosenitrate	≈ 25 000	≈ 100
Melamine resin (molded)	80 000 – 100 000	≈ 7
Phenolic resin (molded)	70 000 – 150 000	2 – 12
Glas-fiber reinforced Polyester resins Epoxy resins	100 000 – 420 000	> 100
Aromat. Polyimides	≈ 27 000	> 100
Aromat. Polyether (PPO)	≈ 25 000	> 100
Silicateglasses	400 000 – 900 000	5
Aluminum alloys	≈700 000	40
Cast iron	400 000 – 750 000	10 – 15
Construction steel	2 100 000	100

Table 587 presents a summary of the E-moduli of polymeric materials. In order to be able to compare the E-moduli of polymeric materials with other materials, the table includes some nonpolymeric materials. One sees that the stiffness (hardness) of polymeric materials is lower than that of silicate glasses and metals by an order of magnitude. In general, with increasing hardness there occurs increasing brittleness, but not always. Steel is a material in which both properties, high impact resistance and high hardness, are combined. That such a combination is also possible with glasses can be seen from the example of the polycarbonates and PPO.

452   The Crystalline State

4521   Crystallization, Chain-folding

A prerequisite for the ability to crystallize (crystallizability), is not only that the structure of the polymer chain is regular as far as composition is concerned, but also that it is regular from a steric point of view (compare p. 135). Only in exceptional cases, with an especially great tendency to crystallize, can one find crystalline components even with atactic vinyl polymers (the vinylidene polymers are always sterically regular), for example, with polyvinylalcohol, and to a very small extent also with polyvinylchloride. This is a proof for the fact that atactic chains also have segments with sterically regular structure.

If this prerequisite is fulfilled, this does not mean that the polymer will now crystallize under all circumstances. In fact on cooling crystallizable polymers, the chains are often so strongly hindered in their mobility near the melting point (recognizable by the high viscosity of the melt), that on further cooling they solidify to a glass because it is simply no longer possible for them to assume the regular form needed for crystallization. This is often the case especially with polymers with bulky substituents on the chain. For example, if isotactic polystyrene [which can be obtained in the form of an X-ray crystalline powder by polymerization of styrene with $TiCl_3 - Al(C_2H_5)_3$ or with $TiCl_3 - Al(isobutyl)_3$ in hexane or heptane] has once been melted, it can be brought to crystallize again only by long annealing, whereas polyethylene is always easy to obtain in crystallized form even from the melt.

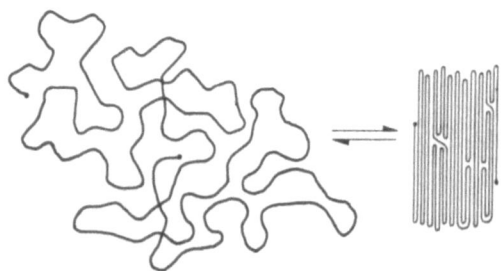

FIG. 588 — Diagram illustrating the equilibrium between solvated statistical coils and crystalline, folded coils.

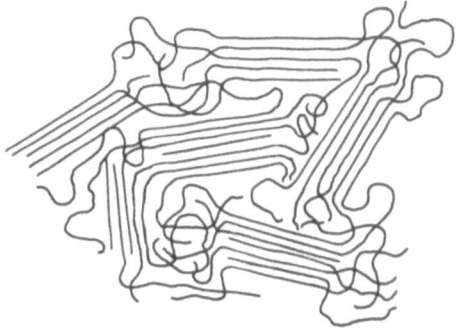

FIG. 589 – The fringed micelle model.

Macromolecular compounds do not become completely crystalline. Even the so-called single crystals, which one can observe under the electron microscope (Figure 590), are not completely crystallized. This is because of the special manner in which polymer chains crystallize, which always leads to the existence of some, more or less large, noncrystalline areas in addition to the crystalline domains. It was formerly thought that in the crystalline regions of macromolecular compounds the individual polymer molecules lie next to each other in a parallel fashion. The crystalline regions were assumed to be separated from each other by regions with an irregular arrangement of the molecular chains, with the individual polymer chains passing through several crystalline and amorphous regions. This type of structure is called a "fringed micelle." This picture cannot be correlated with the coil character of the molecules and can therefore not apply to polymers that crystallize from solution or from the melt. The results of electron microscope investigations of polyethylene single crystals are also not compatible with the idea of a fringed micelle. On the other hand, it is possible that in the fibrils of cellulosic fibers, and in other natural polymers, there are structures such as, or similar to, the fringed micelles (compare p. 600).

As was previously discussed in detail, the polymer molecules in solution, as well as in the melt, are not in the form of stretched-out threads or rods, but in the form of irregular statistical coils. It is not easy to see how such a coil can become stretched out for formation of crystal seeds to take place. Since, on the other hand, there is no longer any doubt about the coil form of the macromolecules, one has to assume that even in the crystalline state the coil is the smallest unit—of course, not in a state of statistical disorder, but in a folded form (see Figure 588).

At the points where the chain folds or forms loops, the crystalline order has been disturbed even within the coil. Other noncrystalline regions occur in the boundary areas between different coils. The idea of the folded chain in crystalline polymers has been confirmed by X-ray investigations of single crystals of polyethylene, polypropylene, and other polymers. Such crystals, which one can prepare

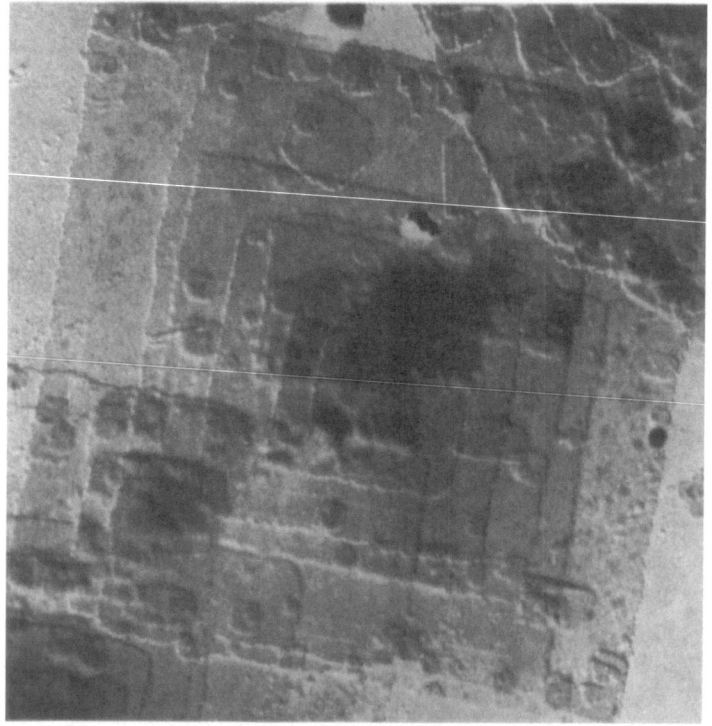

Top: polyethylene (10,000 x)

Bottom: poly-4-methylpentene-1 (8300 x)

FIG. 590 — Electron micrographs of single crystals of polyethylene and poly-4-methylpentene-1 (KELLER, FRANK and O'CONNOR).

by careful cooling of solutions and melts, grow as very thin lamellae, one above the other, as shown in Figure 590. Since the thickness of these lamellae is usually of the order of 100Å, and the polymer chains (as has been shown in electron diffraction experiments with polyethylene, polypropylene, and other single crystals) are arranged perpendicular to the plane of the crystal lamellae, the polymers must be folded, because the fully stretched chains have, depending on molecular weight, a length of 1,000-10,000Å. With low-molecular n-paraffins the chain arrangement is also perpendicular to the crystal lamellae. However, the thickness of the lamellae is proportional to the chain length (experimentally investigated up to $C_{82}H_{166}$), which means that these chains are not folded.

Most probably the linear chain segments in the crystalline lamellae are not all equally long, so that the surface of the lamella (on an ångström-scale) must be described as rough, and the degree of folding has to be considered as an average. On the other hand, longer chain ends and loops which protrude from the lamella surface lead to a higher potential energy (surface tension), so that such imperfections (dislocations) can not be very numerous.

The process of crystalline chain-folding will not be restricted to the growing lamellae but also will take place in single coils corresponding to short-range orientation in low molecular weight solutions and melts. Such preordered or precrystallized coils, which, in the form of more or less completely desolvated particles, are present in the solution equilibrium and, therefore, have only a limited life-time, will either become resolvated or attach themselves as a unit to the growing crystal surface, or, and this is much rarer, they will become seeds for new crystals. The chain shown in Figure 591 can, therefore, be the end of a growing lamella, but it can also represent a single coil (nucleus).

The question why the thickness of the individual lamellae is generally independent of molecular weight and usually lies around 100Å is still not definitely and satisfactorily resolved in detail.

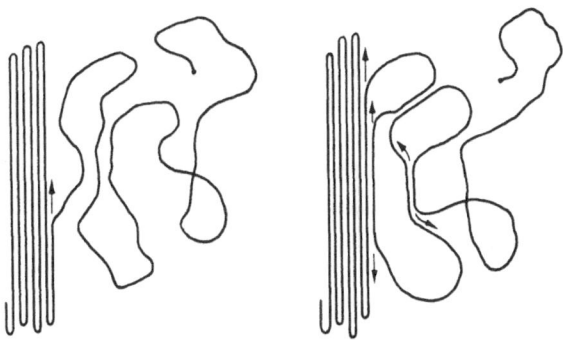

FIG. 591 — Diagram illustrating the hypothetical growth of a single crystal lamella in a polymer solution.

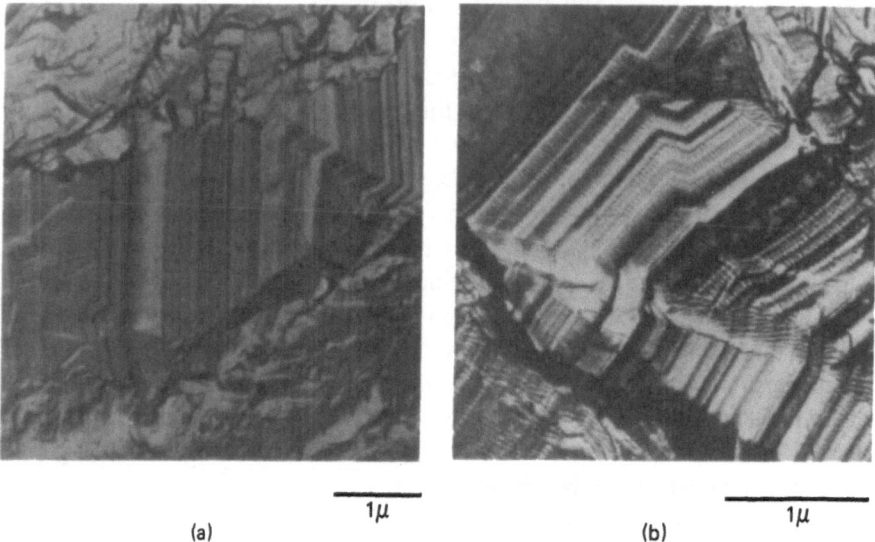

(a)   1μ   (b)   1μ

FIG. 592 — a) Fracture surface of an extended chain lamella with kink bonds grown in a polyethylene melt at 227°C and 4.5 kbar. b) Fracture surface of an extended chain crystal after annealing for 10 hours at 110°C. The ripples show the folded chain lamellae formed by recrystallization during the annealing process. (B. WUNDERLICH)

Rånby has found that the thickness of the lamellae with polyethylene single crystals and, therefore, the length of folding of the chains, is dependent on the crystallization temperature in different solvents.

According to the experiments of B. Wunderlich, the length of chain-folding seems to be strongly increased through crystallization of polyethylene at 227°C under high pressure (4.5 kbar). The formation of extended chain crystals (compare Figure 592) can be explained by a two step mechanism: (1) addition of a folded coil (in the sense of Figures 588 and 591) to a growing crystallite surface and (2) chain extension on the front of the crystallite (lamella) to the thermodynamically stable folding length. By annealing of extended chain crystals of polyethylene at temperatures close to the melting point under normal pressure, a lamella structure with a folding length between 150 and 250Å appears (ripples in Figure 592b).

Polymers with extended chain crystals arise also by crystallization polymerization, that is, by direct addition of monomers in a polymer crystal lattice under simultaneous polymerization, as it occurs, for instance, in the polymerization of di-p-xylene at 600°C by vacuum vapor phase pyrolysis (see section 214).

Pechold has proposed his "kink-model" of polymer crystals. Kinks are (in connection with the crystalline state) irregularities in the polymer chains caused by repeated changing of the chain direction in the sense of Figure 593. Kinks are able to carry out rotatory motion similar to the movement of a crankshaft, if in the neighboring chains there are corresponding kinks. Such movements can run along the whole chain and represent residual Brownian motion even in the crystalline state. They are able to dissipate mechanical energy transferred to a polymeric material, through a blow, for instance.

4522 Morphological Structures

Usually macromolecular compounds do not crystallize as single crystals. More often one observes spherulites as the larger morphological units (Figure 594). On stronger magnification these turn out to be bundles of very fine crystalline fibers (fibrils) (Figures 594b and c) or lamellae (Figure 592). The thickness of the individual fibrils is again of the order of 100Å. From the results of birefringence and micro X-ray diffraction experiments, one assumes that the molecular chains in the fibrils are oriented perpendicular to the fibril axis. One can therefore consider the fibrils to be twisted lamellae. From the measurements available, it is not yet possible to present universal models for the arrangement of the chain molecules within the different morphological structures.

4523 Spiral Form of the Chains (Helix Structure)

In general, polymer chains with side groups, such as polyvinyl compounds or proteins, do not have a stretched-out zig-zag form in the crystalline state. Isotactic arrangement of the side groups is not possible for sterical reasons. Instead, they usually prefer the form of a helix such that the third, fourth, or fifth substituent, respectively, assumes the same position in space as the first. One speaks of a three, four-, or five-type helix (compare also Figure 188). Figure 416 shows such spiral chains as have been found for isotactic polymers. The isotactic polymers of propylene, 1-butene, styrene, and vinylmethyl ether have a three-type helix or 3_1-helix (Figure 595 I). For a more complete description, one can also give the pitch of the helix.

The spiral structure of polymer chains was first observed with protein molecules, where the helix is built up in such a way that there is always a CO group lying over an NH group, so that hydrogen bonds form between the turns of the helix (Figure 597). The polypeptide helix, therefore, has a high stability so that the protein-molecules often maintain the helix form even in solution. Considered as a unit, such molecules have the form of straight or slightly bent rods. One might consider such rods as molecular crystals, because helix formation is, of course, nothing other than an intramolecular crystallization process (compare also the electron microscope

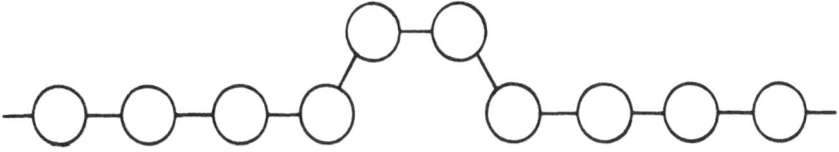

FIG. 593 – Schematic representation of a kink in a polymer chain.

(a)

(b)

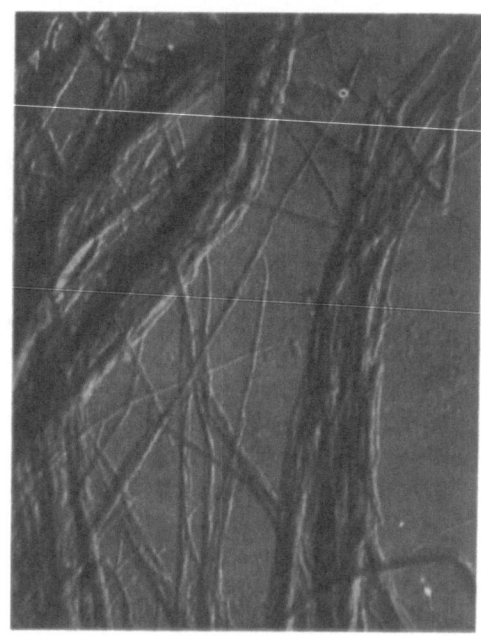

(c)

a) polyethylene spherulites (polarizing microscope, 1000 x) (H. EMMETT)

b) fibrillar spherulites of Nylon-6 (electron microscope, 2600 x) (E.W. FISCHER and H.A. STUART)

c) fibrils of a Nylon-6, 10 spherulite (electron microscope, 21,000 x) (A. KELLER)

FIG. 594(a-c) — Morphological structures of synthetic polymers.

pictures on p. 478). Very long helices, for example, the double helix of DNA (compare Figure 478-1), with molecular weights around 10^9 form random coils (in solution) or superspiraled tertiary and quarternary structures (compare Figure 602) in spite of their relatively stiff helix chain.

4524   Fiber Structure

The crystallization of macromolecular compounds can be greatly speeded up by exposure to a shear gradient, for example, by stretching or drawing the solution or the melt of the polymer during, or even after, cooling. This is the case in the different spinning processes for the production of synthetic fibers or thin films. The melts or solutions are first pressed through thin spinerets or slits, and the resulting fibers or films are stretched while they are cooling, or after they have cooled, or the solvent has evaporated. Crystallization occurs during this process, provided the

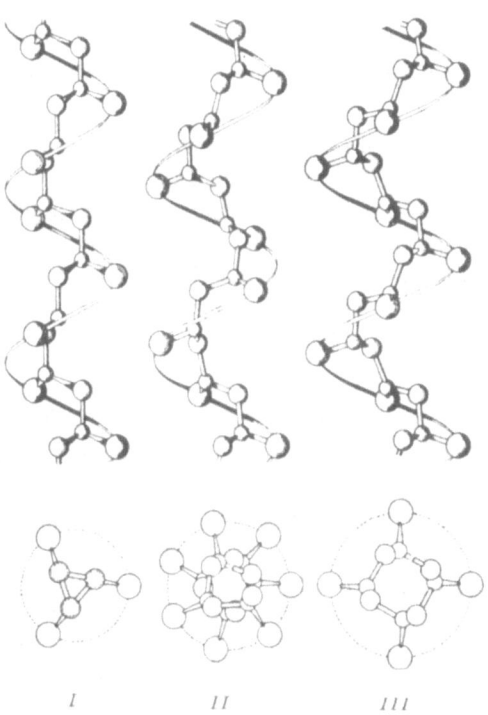

$I = 3_1$-helix; $II = 7_2$-helix; $III = 4_1$-helix

FIG. 595 − Chain configurations of isotactic polymers (according to NATTA)

structural prerequisites are present, and the threads take up a typical fiber structure. Especially striking is the crystallization increase (caused by a forced orientation of the polymer chains) on stretching of natural rubber, which in the stretched state shows a clear fiber diagram, i.e., is crystalline, whereas on release of the tension, it returns to its amorphous state (compare Figure 571). The same is true also for other elastomers with sterically regular structure, for example, polyisobutylene, or synthetic poly-*cis*-1,4-isoprene or -butadiene. These rubbers, even though of sterically regular structure (*cis*-1,4-) and therefore able to crystallize, do not (at normal temperature) go into the crystalline state on their own, but only if they are brought first into a stretched state. The thermodynamic reason for this is that the polymer chains in the stretched state have less mobility than in the unstretched state, so that the magnitude of $T \cdot \Delta S$ in the transition from amorphous to crystalline for the stretched state is lower than for the unstretched state, so that $T \cdot \Delta S$ (positive) is smaller than ΔH (negative), and therefore the crystallization becomes thermodynamically possible (negative ΔF). The lower mobility of the chains in the stretched rubber is the result of the deformation of the coils on stretching, from which a parallel orientation of the chain segments results (compare Figure 568). A similarly stretched and deformed coil is probably present if the polymer chains are oriented by force in a flow gradient.

In cases where the crystallization is preceded by a deformation of the coils (in the sense of orientation of the chain segments), one has to expect, and this is confirmed by X-ray diffraction experiments, that the polymer chains in the crystalline material are arranged in a way similar to the way those in stretched rubber are arranged (Figure 568), i.e., parallel to the fiber axis, or in general parallel to the direction of flow or the direction of stretching (and not as in the fibrils shown in Figure 594, where one assumes that the chains lie perpendicular to the fibril axis).

On cooling the melts of crystallizable polymers (polyethylene, polypropylene, nylon, and Dacron) at first a crystalline structure is formed, in which the molecular chains are perpendicular to the fibril axis (as the results of X-ray diffraction have shown); however, on stretching of the cooled melt, there must be a reorientation of the polymer chains parallel to the fibril or fiber axis.

That polymer chains are oriented in the direction of the fiber is also supported by the fact that one can obtain products with quite noticeable fiber structure from noncrystallizable polymers if one lets their rubber-elastic melts solidify in a flow gradient. For example, if polystyrene of high molecular weight ($M \geqslant 10^6$) is injection molded into plates or rods, these can be defibrillated by mechanical working similar to wood. This shows at least that the fiber-like texture observed with many crystalline polymers is not tied to any special structures (for example, the crystalline fibrils), and that a structure with the polymer chains oriented parallel to the fiber axis is sufficient for the appearance of fibrous texture.

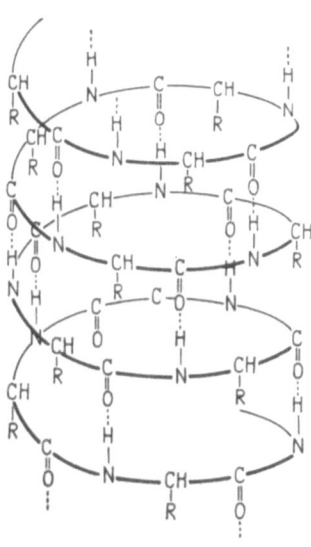

FIG. 597 – Molecular model (left) and schematic representation (above) of a polypeptide helix (PAULING, COREY and BRANSON and DOTY).

4525 Cold Drawing

Some fiber-forming polymers, especially polyamides, such as nylon and perlon, can be cold-drawn in the form of ribbons or fibers to many times their original length. This process is quite similar to the stretching of a rubber band (Figure 567): up to a certain tension (which for a nylon band is considerably greater than for a rubber band of the same thickness), there is practically no deformation. When this stress limit has been reached, the band begins to give and to elongate within a narrow tension range by a large, but strictly limited amount (with nylon, for example, about 400%). From this point on, a continuing increase of tension, even by several times the previous tension, produces no considerable further extension of the band, till finally, at very high stresses (over 600 kg per square millimeter of cross-section), the nylon band breaks.

Only after the considerable increase of the tensile strength produced by the drawing process do fibers attain the valuable properties which make them suitable for textile and industrial applications.

In a molecular sense, drawing is a flow process, similar to rubber-elastic stretching, in which a parallel orientation of the polymer chains and the chain segments occurs until the chains are so strongly interlocked because of cystallization that a further gliding past each other is impossible. However, even before drawing there is some crystallization. One therefore has to assume that in the flow process due to the drawing, there is also a parallel orientation of the crystallites present before the drawing.

FIG. 598 — Bacterial cellulose fibrils (Acetobacter xylinum, 14,000 x) (from WYCKOFF).

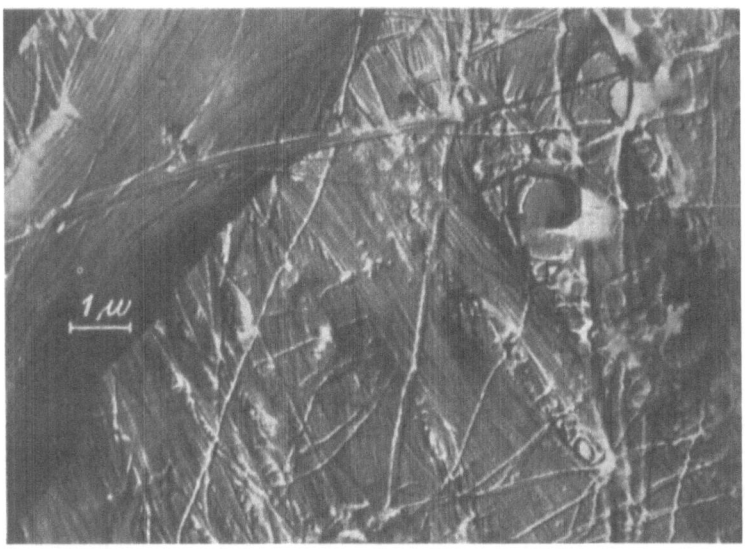

FIG. 599-1 — Collagen fibers from the appendix of an ox (BRÜCHE and SCHULZE).

FIG. 599-2 — Cellulose fibrils from the cell-wall of algae (Valonia ventricosa) (ELÖD and ZAHN).

In the X-ray diffraction diagram, the transition from the crystalline state with random spatial arrangement of the crystallites (i.e., before drawing) to the crystalline state with the polymer chains and the crystallites oriented parallel to the stretching direction (after drawing), one observes a transition from a Debye-Scherrer diagram to a fiber diagram (layer-line diagram) (compare Figure 611).

4526 Morphological Structures of Natural Polymers

Nature uses the ability of macromolecular compounds to form fibers as a means of building up cell-wall materials which have an extraordinary strength (cellulose fibers, sinews, and leather).

Under the electron microscope, one observes a felt of fine fibrils (Figures 598-601) with polysaccharide fibers (cellulose) and with protein fibers (collagen and silk). With collagen the individual fibrils also show an arrangement of stripes perpendicular to the fiber axis.

The arrangement of the macromolecular chains differs from substance to substance and is not completely understood in all details. In general one has to consider the following: since the macromolecules of the cell wall are formed by enzymatic synthesis directly wherever they are needed, one can imagine that the monomers are incorporated, similar to bricks, into a wall of an existing crystalline structure, or they are added to its surface. It is therefore possible that here the polymer chains are not folded, but are present in maximum length, or as a helix parallel to the fibril axis. The fibrils consist, in such cases, of bundles of more or less completely parallel polymer chains, so that the picture of the fringed micelle (compare Figure 589) still has justification with natural fibers.

Also the fibrils or larger morphological units can be arranged in a regular fashion in the fiber. Thus one sees in Figure 599 with collagen an aggregation of parallel fibrils to larger bundles. With Valonia, fibrils run parallel within lamellae, and the fibrils of adjacent lamellae are arranged perpendicular to each other. With cotton and other cellulosic fibers (for example, palm fibers), one has observed a spiral-like arrangement of lamellae, which in their turn consist of bundles of fibrils (Figures 600 and 601).

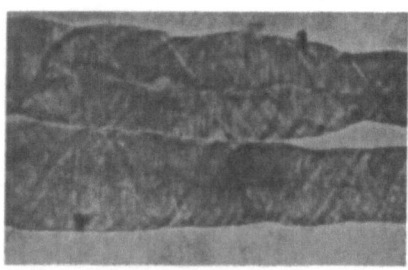

FIG. 600 — Nitrated cotton (400 x) swollen in a mixture of alcohol and acetone (Magda STAUDINGER).

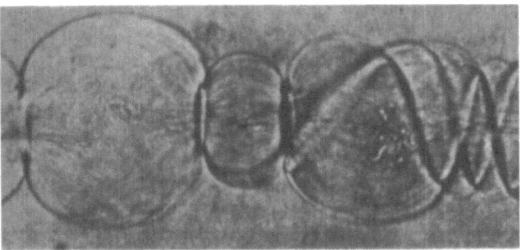

FIG. 601-1 — Native cotton fiber swollen in Schweizer's Solution. (Magda STAUDINGER).

One gains the impression that in nature the morphological structures are built in each case to fit a specific purpose, rather than in a complete random fashion.

This becomes evident especially if we consider the tertiary structures of enzymes (see p. 255) or DNA, which is shown in a schematical representation in Figure 602a. From the double helix A the D form results by repeated superspiraling.

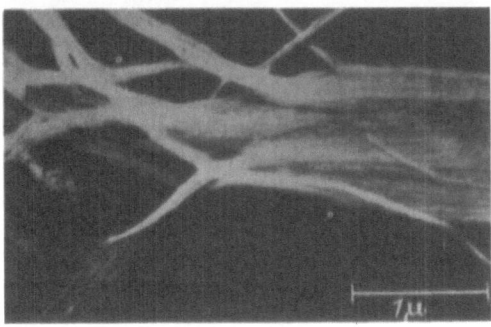

FIG. 601-2 — Silk fibroin (30,000 x) (H. ZUBER)

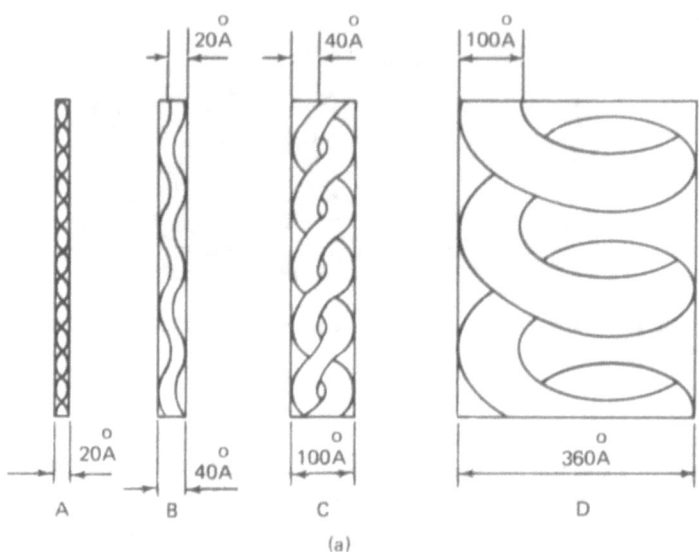

(a)

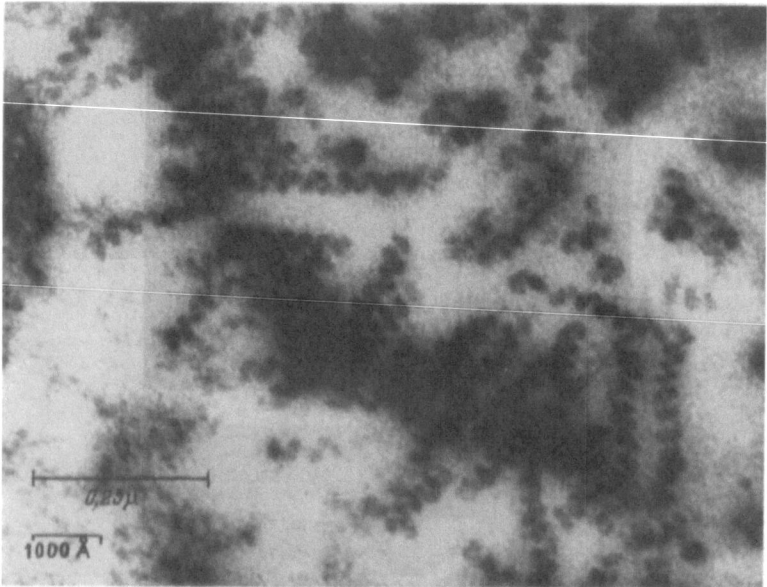

(b)

a) hypothetical representation of possible intermediate steps.
   (C. BRESCH).

b) morphological structure of a chromosome. (G.D. PAPPAS
   and P.W. BRANDT).

FIG. 602 — Tertiary and quarternary structures as steps between molecular
structure (primary structure) and morphological structure.

The C strands with a thickness of 100Å seem to be identical with the fibrils, which have often been found in electron micrographs of chromosomes as "side chains." In some cases the DNA-fibrils of chromosomes are spiraled again as shown in Figure 602) corresponding to scheme D.

Superspiraling seems to be a suitable way for giving the necessary mechanical stability to the DNA macromolecules. Single macromolecules are sensitive to mechanical deformation by shearing forces. This sensitivity is higher, the higher the molecular weight (compare p. 542), and DNA has extremely high molecular weights of (about 10^8-10^{10}).

4527 The Fiber Diagram

An absolute necessity for the investigation of the crystalline state are the methods of X-ray and electron diffraction. If a monochromatic light beam falls on a grating with regularly spaced lines, where the space between the lines is of the order of magnitude of the wavelength of the light, then, because of the resulting interferences, the light is eliminated in certain directions and amplified in others (Figure 603-1). The maximum reinforcement of the light occurs at angles φ, where the path difference r is a multiple of the wavelength λ:

$$r = n \cdot \lambda = d \cdot \sin \varphi. \qquad (51)$$

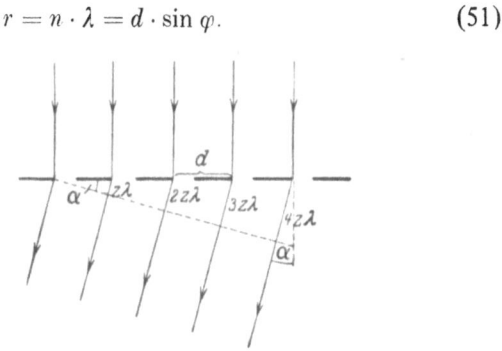

FIG. 603-1 — Diffraction at a line-lattice.

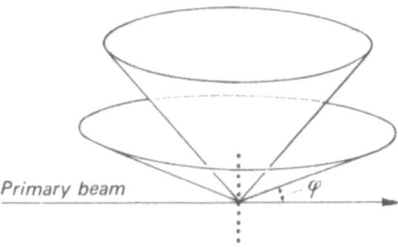

FIG. 603-2 — Diffraction at a point-lattice.

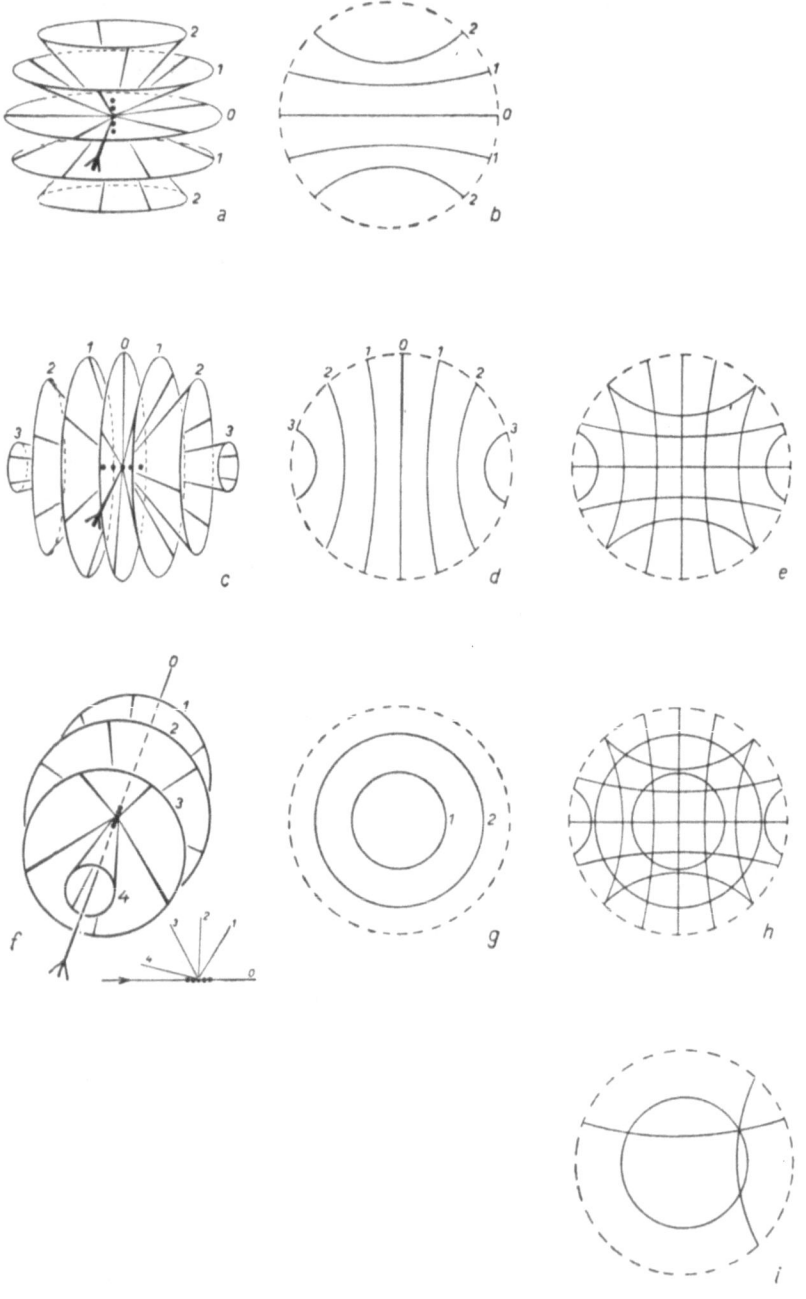

FIG. 604(a-i) — Diagrams illustrating diffraction at a 3-dimensional point-lattice (i.e. a crystal lattice). The three rows of points are non-coplanar. (After A.J.A. VAN DER WIJK).

In principally the same way, X-rays are diffracted by the crystal lattice. Only one has to take into consideration that a crystal lattice is not a two-dimensional grating, but a three-dimensional point-lattice. With such a spatial point-lattice, the directions of maximum reinforcement form concentric cones, and on a photographic plate one observes a series of hyperbolas, shown in Figure 603-2 for the (impossible) case where only *one* series of points exists. The relationship between the distance d between the lattice points and the angle φ is again given by Equation (51). However, in the crystal there is not only one series of lattice points which causes scattering, but an infinity. They can be ordered in three directions, perpendicular to each other, and one only obtains interference scattering, where the relation $n \cdot \lambda = d \cdot \sin\varphi$ is realized for three different noncoplanar point series. The diffraction of the X-rays may be regarded as a reflection from the so-called lattice planes which lie in such a way that they bisect the angle φ. As becomes clear from Figure 604, this does not result in lines (on a flat film), but instead this lattice gives rise to a series of points, which lie at the intersection of any three lines for which Equation (51) holds. If one now compares with this the fiber diagrams of chitin and rubber, shown in Figures 570 and 571, one recognizes immediately this line diagram, if one considers that the corresponding points of first and second order are connected by lines.

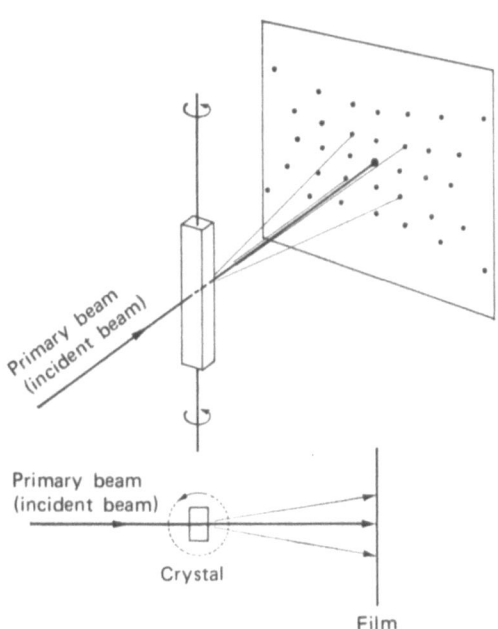

FIG. 605 − Arrangements for a rotating crystal diffraction picture.

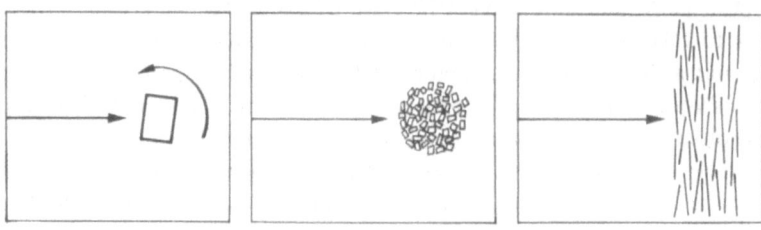

a) single cystal

b) fiber, cross-section

c) fiber, longitudinal section

FIG. 606(a-c) – Diagrams illustrating the origin of the layer-line x-ray diagram. obtained with fibers.

With single crystals one obtains such layer-line diagrams if one rotates the crystal during the irradiation about one axis perpendicular to the direction of the X-rays (Figure 605). In this manner the crystal passes all positions where the relation $n \cdot \lambda = d \cdot \sin\varphi$ is fulfilled for three points at the same time, and the layer lines of the X-ray diagram are formed one after the other. As we have seen previously, fibers are not single crystals, but bundles of many more or less well-oriented (parallel or helical) crystallites or fibrils. As the appearance of a layer-line diagram with fibers indicates, the crystallites in the fibers are parallel to the fiber axis, but are not oriented within the fiber cross-section (Figure 606b), so that the different spatial arrangements, which with a single crystal pass through the beam one after the other (on rotation perpendicular to the primary beam), are already realized in the fiber because of the presence, next to each other, of the many crystallites. Therefore, a fiber with the X-ray beam perpendicular to the axis gives a more or less well-defined layer-line diagram (fiber diagram) without any rotation about the fiber axis. The type of fiber diagram obtained will depend on the extent of parallelization of the fiber crystallites.

As can be seen with the aid of Figures 604 and 605, the distance between the layer-lines is determined by the angle φ, and therefore according to Equation (51), by the distance between the points in the perpendicular point series. With sodium chloride crystals, this is simply the distance between the sodium and chloride ions in the lattice. With the complicated lattices of macromolecules, and in general with molecular lattices, the atomic distances are not equal, and d is therefore the distance after which certain atomic groupings repeat themselves. This is called the identity period. It is not possible to determine a certain chain configuration from the identity period alone.

If one imagines that the crystal in Figure 605 is maintained in a certain position, where interference conditions for maximum amplification ($n \cdot \lambda = d \cdot \sin\varphi$) are realized according to Figure 604 for a certain point series, and that the crystal is now rotated in this position about the primary beam as the axis, then the interference pattern remains unchanged, i.e., each interference beam describes a cone during the rotation about the primary beam and therefore describes a circle on the photographic plate. These circles are identical with the so-called Debye-Scherrer circles, which are always obtained when a crystalline powder is exposed to X-rays. In that case the individual crystals take up all possible orientations in space, so that the circles which are obtained with a single crystal by rotation about the primary beam through gradual smearing-out of the point diagram are obtained immediately in a Debye-Scherrer powder diagram.

More or less well-defined spherical smearing-out of points in a fiber diagram (arcs) signifies that there is poor parallel orientation of the crystallites or fibrils with respect to the fiber axis. This may also be the result of a helical-arrangement (Figures 600 and 601). Conversely, the absence of arcs indicates that the crystallites must be especially well-oriented in the fiber, or that there are no crystallites present at all, as with single crystals (Figure 610).

Figure 611 shows X-ray diffraction diagrams of a nylon filament before and after drawing. Polyamide fibers and ribbons can be cold-drawn to several times their length, as has been discussed previously. Even though the material is already crystallized before drawing, the crystallites (just as with a crystalline powder) are not ordered, so that the result of X-ray diffraction is a Debye-Scherrer diagram with concentric circles. On drawing, the polymer chains and the crystallites are oriented, and this can be recognized by the appearance of a fiber diagram.

453 Structure and Properties in the Solid State

4531 Structure and Softening Temperature with Glassy Polymers

If one speaks about macromolecular compounds in the solid state, this is not a completely unambiguous description, because rubber-elastic materials can be regarded as solids as well as liquids. Nothing is gained by considering them strictly

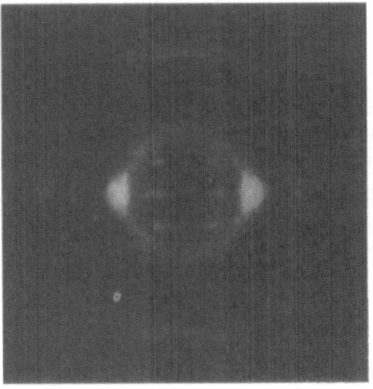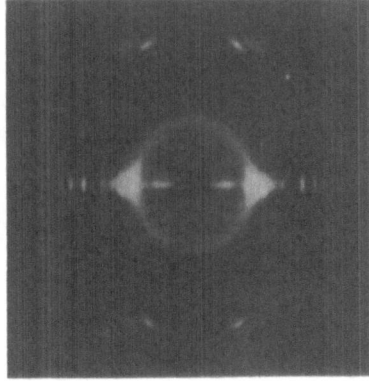

FIG. 607 − Left: fiber-diagram of drawn 6-Nylon (perlon) Right: fiber-diagram of drawn polyethylene (R.P. PALMER).

as solid materials, because this would not give proper recognition to the characteristic properties of an elastomer: the fact that it is both a liquid and a solid. Liquid, because large segments of the molecules can move about freely, and solid, because it is elastic. But it is really neither a liquid nor solid in the usual sense. One therefore tries to describe the solid state of macromolecular compounds by using instead the term "hard." This would be entirely correct as long as we are considering glasses. However, there is also a series of solid macromolecular materials, such as polyethylene, polypropylene, polyamides, polycarbonates, polyformaldehyde, impact-resistant polystyrene, and polyvinylchloride, which cannot be characterized as hard (in the sense of a glass). One therefore subdivides solid macromolecular materials into "hard-tough" materials and "hard-brittle" materials. The rubber-elastic polymers according to this classification should be characterized as "soft-tough." Whether a macromolecular compound is rubber-elastic or hard (glassy or crystalline) at normal temperature (20°C) depends on the mobility of its molecular chains. The mobility of the molecular chains at a certain temperature is determined by three important factors: first, the degree of steric interference between atomic groups appearing as substituents along the chain; second, the strength of the secondary valence forces, which are active between adjacent chains, or from one chain segment to another chain segment, and which are also influenced greatly by the type of substituent on the polymer chain; and, finally, by the regularity of the structure. Naturally, the energy contribution which is required to bring about rotation about the carbon-carbon bond (or the C–O or the C–N bond, for polymers with heteroatoms), has an influence on the mobility of the chains. This energy contribution for most polymers is of the same order of magnitude, so that one can neglect it in comparing polymers with different constitution. Only with diene polymers (polybutadiene, polyisoprene) is the energy contribution of rotation about the carbon-carbon single bond (next to the double bond) smaller, so that with such unsaturated polymers the chain mobility is probably greater than with polymers with saturated carbon-carbon chains (compare Table 438).

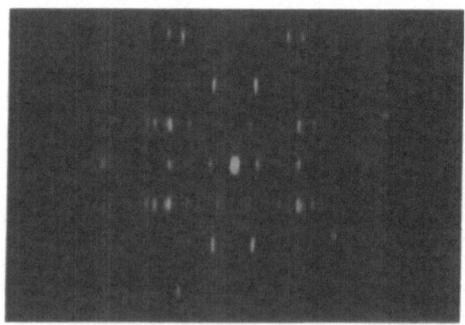

FIG. 608 – Fiber-diagram of chrysotile
(an asbestos form) (WARREN and BRAGG).

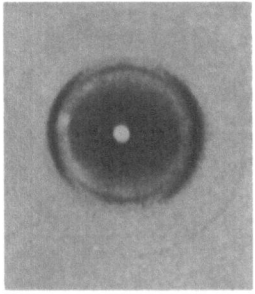

FIG. 609 – Fiber-dia-
gram of cotton (MEYER
and MARK).

When the free rotation of the chains is sterically hindered, the degree of mobility which the chains gain on softening of the polymer is not very great. Conversely, the entropy loss ΔS is not large when such polymers go from the rubber-elastic to the glassy state, i.e., the magnitude $T \cdot \Delta S$, which in its sign opposes the glassy solidification (freezing), is already so small at a relatively high temperature T, that it is overshadowed by the magnitude of ΔH, the loss of inner potential energy on freezing of the molecular chains. This means that the solidification process begins already at relatively high temperatures. Typical examples for such vinyl polymers with large and stiff substituents and high glass temperature, are poly-α-methylstyrene and polyvinylcarbazole. But also polyvinylidene chloride and poly-methylmethacrylate are, as models have shown, sterically hindered in their chain mobility. Even when the rotation about the carbon-carbon axis of the chain does not suffer any direct hindrance, bulky and stiff substituents hinder the micro-Brownian motion because the molecule is not alone in space, but is restricted by its own chain segments or those of neighboring molecules. In this sense polystyrene and polyvinylcyclohexane also could be regarded as sterically hindered.

It is, of course, not possible to clearly separate the influences of steric hindrance and secondary valence forces. The influence of the secondary valence forces is always present, and one has to subtract their contribution, if one wants to obtain a measure of the effect of the steric hindrance on the freezing point, (i.e., the softening point), of the polymer. With polymers whose glass temperature lies above room temperature, one usually speaks about the softening point, or softening range, and with polymers which at normal temperature are rubber-elastic, one speaks, instead of the glass temperature, often of the freezing temperature.

Table 623 shows for the individual polymers, in addition to the structural formula, the heat of evaporation of the monomer, which presents a measure for the magnitude of the intermolecular forces which one has to expect in the corresponding polymers.

FIG. 610 – Electron diffraction diagram of a polyethylene single crystal (A. KELLER).

The influence of steric hindrance is most clearly seen, if one compares polymers where the secondary valence forces are approximately equal. This is the case with polystyrene (#4) and poly-α-methylstyrene (#2), and with polymethylacrylate and polymethylmethacrylate (#8 and #5). As can be seen from the Stuart-Hirschfelder models, the α-methyl group brings about a restriction of the rotation about the carbon-carbon bond and therefore a lowering of the chain-mobility. There is an increase in the softening point by 80°, respectively 100°C, which must be a result of the increased chain stiffness caused by the α-substituents.

The smaller the chain substituents, the weaker the van der Waal's forces, the larger the mobility of the chains, and the lower the temperature where the polymer solidifies to a glass.

This is confirmed if one compares in Table 623 the series of polymers from polyvinylcarbazole through polystyrene to polybutadiene and the amorphous copolymer of ethylene and propylene (#15). For the freezing process, as for all physical-chemical state changes, the relation $\Delta F = \Delta H - T \cdot \Delta S$ holds. The process occurs only when the heat given off by the system on solidification, (measured by ΔH), is larger than the entropy term $T \cdot \Delta S$. Since with polymers #9 to #16 in Table 623 the value of ΔH (due to the low van der Waal's forces) is only small, T must lie considerably below 273°K, in order that $T \cdot \Delta S$ should be smaller than ΔH.

ΔH on softening of a glass does not occur in the form of a heat of melting, because the value of ΔH, because of the irregular arrangement of the chains and chain segments, is not the same for all parts of the material and freezing therefore occurs in different molecular regions of the material at different temperatures. Therefore, with glasses one speaks of a second order transition point.

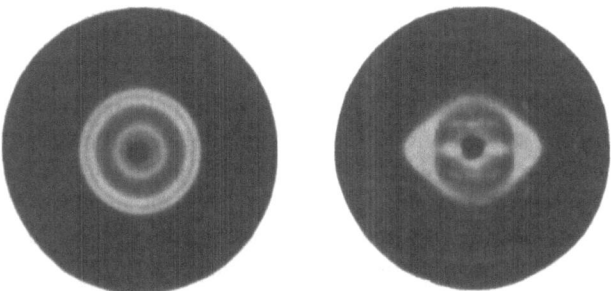

FIG. 611 – Fiber-diagrams of Nylon-6, 10 before and after drawing (FULLER, BAKER and PAPE).

A striking exception is polybutylacrylate (#9), with its long side chains and relatively strong secondary valence forces, which may be compared to those of polystyrene. In spite of this, polybutylacrylate (and the same holds in increasing degree for the polymers with longer substituents, such as poly n-hexylacrylate and n-octylacrylate) is not a solid resin, but a soft elastomer, which only solidifies to a glass at −50°C to −60°C. In this case one speaks of internal plasticization. The long paraffin side chains, because of their high mobility, act as a swelling agent which surrounds each individual polymer chain and shields the polar ester groups, so that their contact with other chains or chain segments is made impossible, or at least more difficult. Thus, the polymer chains are surrounded by their side chains just as by a solvent, and this, as the low glass temperature shows, leads to considerable chain mobility.

Just as the low molecular paraffins and paraffinic carboxylic acids, the polyacrylates with side chains of ten or more carbon atoms show a certain tendency to crystallize, which can be confirmed by X-ray diffraction. Such polyacrylates have a wax-like, soft consistency, which results from the fact that only the paraffin side chains crystallize, but not the entire polymer chain as with polyethylene.

One sees with this example that the chain-stiffening action of substituents is not so much a function of their size as of their mobility. A confirmation of this is polycyclohexylacrylate, where the hexyl residue, because of its cyclic structure, no longer has great mobility, and the polymer is (as one must expect) a hard and brittle glass at normal temperature, which only softens at 40°C to 50°C, whereas poly-n-hexylacrylate is rubber-elastic down to −60°C.

From the same example one also sees that the substituents exert an influence, not only as a function of their size on the chain mobility, but also indirectly over the secondary valence forces, because the spatial arrangement of the polymer chains and chain segments and their distance from each other are determined to a high degree by the side groups. The high strength of macromolecular compounds is especially influenced by the fact that the polymer chains can interact energetically with each other over longer chain regions. The attractive forces between the chains will be the larger, the closer the chains are able to come to each other and approach the optimal distance for the secondary valence forces to be operative. This depends on the type and number of substituents, but also on the chain regularity (or irregularity) as far as the steric structure of the chain is concerned. The great influence of the chain distance r on the properties of polymers is easily understood if one considers that secondary valence forces decrease according to $1/r^6$

An example of the influence of different types of substituents, is the comparison between polymethylacrylate and polybutylacrylate (#8 and #9). The larger the substituent, the smaller the contribution of the CO groups for holding together the polymer chains. With polybutylacrylate, -hexylacrylate, or -octylacrylate, the influence of the ester group on the intermolecular cohesion is more or less eliminated, and the material character of such polymers is therefore determined essentially by the alkyl residues of the alcoholic components. Thus it happens that polymethylacrylate, in spite of its low molar cohesion, has the highest softening point in this series of homologues: because here the shielding action of the alcohol component is the lowest.

A further example is provided by the comparison of polyisobutylene and polyvinylchloride (#11 and #17). The magnitude of the secondary valence forces is of the same order with the two monomers. However, with the polymers, the forces show up in a completely different manner. Polyisobutylene down to $-65°C$ is a soft elastomer; polyvinylchloride, on the other hand, is a hard polymer which only becomes rubber-elastic at 80°C. The secondary valence forces of the C-C chain itself cannot play an important role with polyisobutylene because of the shielding action of the CH_3 groups.

In other terms this means that the secondary valence forces of a polymer are not simply additive from those of the monomer, and therefore if one wants to compare polyisobutylene and polyvinylchloride, one should not compare the heat of evaporation of isobutylene and vinylchloride, but better that of methane (2,300 cal/mole) and methylchloride (5,200 cal/mole), or ethane (3,900 cal/mole) and ethylchloride (6,000 cal/mole).

Apparently with the polyvinylchloride chain the carbon-chlorine dipole is quite active, which results in the relatively high softening point of polyvinylchloride. The rather regular arrangement of the polyvinylchloride chain also plays a role, which shows up as a slight, but definite tendency to crystallize. One can increase the crystallizability of polyvinylchloride by polymerizing vinylchloride at low temperature ($-20°C$ to $-40°C$). With lower polymerization temperature the syndioactic components increase, and the softening point correspondingly increases also.

4532 Structure and Properties of Crystalline Polymers

The third factor which influences the chain mobility and, therefore, the physical properties is the steric regularity of the chain. With the example of polyvinylchloride it could be seen that this influence makes itself felt because with a sterically regular chain structure, the secondary valence forces between chain and chain, or between chain segment and chain segment, become fully active.

The most impressive examples for this, are low pressure polyethylene, isotactic polypropylene, and polyformaldehyde (#22, 23, and 24). Low pressure polyethylene has a continuous $-CH_2-CH_2-CH_2$ paraffin chain, esentially without side groups, and because of this regular structure has a strong tendency for crystallization. In spite of the low secondary valence forces (polyethylene has the lowest molar cohesion of all polymers) its melting point is 136°C, i.e., higher than the softening point of polystyrene. If for some reason it could not crystallize, then its freezing temperature would lie at −77°C, as can be found by extrapolation of the glass temperatures of ethylene copolymers of increasing ethylene content. The same phenomenon can be observed with polytetrafluoroethylene, polypropylene, and polyformaldehyde: in spite of relatively low attractive forces between the molecules, they have high softening i.e., melting points. The available forces give rise to such a high cohesion because the molecules have no bulky side groups to prevent the chains from approaching each other to a distance of optimal influence of the secondary valence forces.

How strong the influence of the side groups is, can be seen clearly from the fact that high pressure polyethylene,[6] which has only relatively few side groups and side chains in the macromolecule (one per 100 structural units), has a softening point which lies 10°C to 20°C lower than that of the more linear low pressure polyethylene (one side group per 1,000 structural units). If the CH_3 side groups become numerous, then the softening point continually decreases until the copolymer of ethylene and propylene finally is an elastomer, which only solidifies to a glass at −40°C to −60°C. The freezing temperature of this elastomer is so low only because the CH_3 groups are distributed in a random fashion along the polymer chain. Thus, isotactic polypropylene, where the CH_3 groups occur in regular intervals along the polymer chain and are always in the same steric position, is crystalline and melts only at +170°C. The same is observed with isotactic polystyrene, which melts about 130°C higher than atactic polystyrene softens, because the phenyl residues along the isotactic chains are arranged in a sterically regular manner so that a closer approach between the chains (in helix form) is possible over longer ranges, which leads to the formation of a crystal lattice. This is the same with polyformaldehyde, which melts at +170°C, whereas the non-crystallizable copolymer with acetaldhyde only solidifies at temperatures below 0°C (depending on the composition).

---

6 There are two polyethylene types: the relatively soft, high pressure polyethylene, with branched polymer chains and a higher content of noncrystalline polymer, and the more hard and stiff, low pressure polyethylene, with almost unbranched, linear chains and a lower content of noncrystalline material. High pressure polyethylene is prepared commercially by polymerization of ethylene at a pressure of 1,500-3,000 atm. and 180°C-200°C with oxygen as initiator. Low pressure polyethylene is obtained either with Ziegler catalysts ($AlR_3/TiCl_4$) or with Phillips' catalyst (chromiumoxide on Al_2O_3), in the presence of hydrocarbons at 80°C-100°C.

A sterically regular arrangement of the structural units does not give rise to a specially strong tendency to crystallization with all polymers. Polyisobutylene, for example, (as a vinylidene polymer) has a definite sterically regular structure. While it is crystallizable in the stretched state (forced crystallization), it normally does not crystallize, but remains rubber-elastic until $-70°C$, and then solidifies on further cooling to a glass. This is in contrast to polypropylene, (from which it differs only by an additional CH_3 group in the structural unit), which melts at $+170°C$. Obviously there is a shielding effect of the additional CH_3 group, which prevents an optimum approach of the chains. One sees that a CH_3 group in the structural unit, because of its bulkiness and the resulting steric hindrance, can raise or lower the softening range of a polymer: The glass temperature is increased because of a reduction in the chain mobility (compare polymethylmethacrylate and polymethylacrylate, or polystyrene and poly-α-methylstyrene), and the melting point is moved to lower temperatures because of reduction of crystallization (polypropylene and polyisobutylene). It is not possible to predict the effect of restriction of crystallization by side groups (i.e., when it will occur and when it will not), as is shown by the comparison of polyethylene and polypropylene, and of polypropylene and polyisobutylene. One CH_3 group per structural unit (polypropylene) does not disturb crystallization in the least, whereas the additional CH_3 group per structural unit restricts the approach of the chains so strongly that no crystallization occurs until the beginning of solidification to the glass at $-60°C$ to $-70°C$.

Additional examples of polymers, where sterically regular structure is not coupled to a high melting point, are the cis-1,4-diene polymers: cis-1,4-polybutadiene with a glass temperature of approximately $-100°C$, and cis-1,4-polyisoprene (natural rubber) and also synthetic cis-1,4-polyisoprene) with a glass temperature of approximately $-70°C$. In this case, it is the special steric conformation of the polymer chain caused by the double bond (cis-structure), which results in the low crystallization tendency. The corresponding trans-polymers have considerably higher melting points: trans-1,4-polyisoprene (guttapercha) is a hard resin at normal temperature (structural formulas: compare pp. 57, 178 and 190) and trans-1,4-polybutadiene melts at $+145°C$.

4533 The Property Combination: "Hard and Tough"

In contrast to the hard polymers (1-7 in Table 623), there are other hard polymers (17-28 in this table), that one would expect to be rubber elastic at normal temperatures (similar to polymers 8-16) because their molecules neither have stiff bulky groups, nor are they restricted in their mobility by especially high van der Waal's forces (except for the polyamides). These polymers have such high melting points only because they crystallize as a result of their regular steric structure. Polyvinylchloride and polyacrylonitrile do not show crystallinity on the X-ray diagram. However, one can assume with certainty that in these polymers there are regions with oriented polymer chains, which play the same role as the crystallized portions in polyethylene and polypropylene. For example, the glass temperature of polyacrylonitrile is approximately 100°C, as can be measured by determination of the damping maximum (compare p. 575). However, in spite of this, polyacrylonitrile remains a solid above 100°C until it decomposes between 200°C and 300°C (compare p. 316).

As has been discussed already in a previous section, the crystallization of macromolecular compounds is never 100%, so that with the polymers of the type of polyethylene, polypropylene, and polyformaldehyde, one does not have a uniform material, but an internal mixture of a hard, crystalline component and an amorphous, rubber-elastic, soft part.

This mixture of a hard-brittle and a tough-soft component is responsible for the material properties of a crystalline polymer of the type of polyethylene. In contrast to the glasses (polystyrene, Plexiglass, and the silicate glasses, whose mechanical properties are characterized by the term "hard and brittle"), one has to describe the mechanical properties of polyformaldehyde, polyethylene, and polypropylene as "hard and tough." These polymers maintain their shape, and even though they are not rubber elastic, they are relatively unbreakable and in this way resemble the metals, which are actually being replaced to a larger extent by such polymeric materials.

The fact that the impact resistance of crystalline plastics of the polyethylene type is related to their content of noncrystallized rubber-elastic polymer, can be seen readily from the fact that such plastics become immediately brittle if one cools them below, or even close to the freezing point of the amorphous component. This is, for example, the case with polypropylene below 0°C, and with polyethylene below −70°C. One only has to place dry ice into a bucket of polyethylene or polypropylene and drop it, to see that it will break into splinters just like glass.

The resistance of such plastic materials against sudden deformation, as it occurs under a blow, can be explained by the fact that the energy of the blow is taken up by the rubber-elastic components, which are able to convert this energy into energy of Brownian motion of the chain segments. They are able to do this because of the high mobility of the chain segments, and therefore no break in the chain results. In this way the energy of the blow is made harmless. With metals that process is quite similar, but there the electrons of the electron gas distribute the energy of the blow in a short time over larger regions of the material. It is different with glasses, where the molecular motion is more or less frozen in and where a blow, if it is strong enough, leads to a high local deformation energy, which results in the breaking of secondary valence and covalent bonds, and the material disintegrates.

One can not completely explain in this way the high impact resistance of cellulose esters such as cellulose-acetobutyrate (mixed acetic acid/butyric acid ester), or polyamides such as nylon or perlon, and certain polyesters and polycarbonates such as polyethyleneglycolterephthalate (Terylene, Dacron, Diolen, Trevira), poly-bisphenol-A-carbonate (Macrolon, Lexan) and poly-2,6-dimethyl-phenylene ether (PPO, Noryl). Of these polymers, only the polyamides and the polyesters are crystalline, so that one may assume the existence of two phases, one crystalline and one amorphous.

The toughness of cellulose esters is caused by the plasticization which results from the addition of suitable low molecular compounds (such as camphor for cellulose nitrate), or by the introduction of longer ester side chains in the molecule. For example, with acetobutyrate, the butyl side chains (similarly as with polybutylacrylate) bring about internal plasticization. By means of the length and the concentration of the plasticizing side chains, one can arbitrarily influence the degree of softening within certain limits.

Normally by the addition of small amounts of a plasticizer, the physical character of a polymer (at room temperature) is not influenced. For example, polystyrene remains hard and brittle if one adds 2-3% paraffin oil or butylstearate as a lubricant for lowering the melt viscosity. However, if one adds larger amounts, for example, 20-40% of a phthalic acid ester to polyvinylchloride, then one obtains soft rubber-like gels. There are no concentration ranges in between where the gels are hard and tough, because the plasticizer addition always lowers the softening point considerably. But with cellulose esters, there is such a region of hard gels, which probably results from the relatively high stiffness of the cellulose molecules. In this region the plasticizer concentration is already high enough to take up mechanical energy, but not yet large enough to make the entire gel rubber-elastic.

The polyamides, polyterephthalates, polycarbonates and poly2,6-dimethylphenyleneoxide (PPO) occupy a special position among plastics. Their molecular weight lies around 20,000 (40,000-80,000 with PPO and polycarbonates) and is therefore by an order of magnitude lower than that of other plastic materials. Polystyrene, polymethylmethacrylate, polyvinylchloride, polyethylene, and polyacrylonitrile, with such low molecular weights, have such a low mechanical strength that they are not usable as plastic materials. This shows clearly that the excellent mechanical strength of the polyamides and polycarbonates is to a large extent a function of the strong secondary valence forces, which are active between the polymer chains because of the regularly ordered polar groups.

Secondary valence links are by their very nature more "elastic" than covalent forces: they are easier to break; however, after a change of orientation of the molecular chains, coils, and crystallites, they readily reform, so that the disturbances brought about by mechanical deformation can repair themselves under suitable circumstances. The polymer chains hang together very tightly and if they are torn apart in a locally limited zone, they hold together even more strongly in the immediate neighborhood, so that the defect brought about by mechanical deformation has no chance to spread. In this manner, mechanical energy can also be transformed into heat: the deformation energy is first taken up by the breaking of hydrogen bonds. As a result of the same process, new hydrogen bonds are simultaneously formed at other positions, and heat is liberated (heat of crystallization).

A prerequisite for the functioning of this mechanism is a certain chain mobility, which permits deformation of a molecule or a coil. This mobility does not have to be as large as with elastomers, where the chain segments between cross-links move by themselves as in a liquid. However, it must be possible to achieve a change in their position under the externally applied deforming force without breaking covalent bonds. With the polyamides, this prerequisite is fulfilled by the noncrystalline regions, which can take up water as a plasticizer and swell slightly. If one removes this water by longer heating in vacuum, polyamides become brittle and regain their toughness only if they take up new water by coming in contact with moist air or with water. The polycarbonates are not crystallized, and because of the lower packing density of the polymer chains, a certain minimum of chain mobility is maintained. The polycarbonates become brittle on crystallization.

The property combination "hard and tough," which includes the polymers 17-28 in Table 623, allows only a rough classification. According to their structure and the resulting different reactions to suddenly applied mechanical forces, one can distinguish three groups: polymers of the type of polyethylene, the cellulose esters, and polymers like the polycarbonates. These have all been discussed separately above. To the first, in addition to polyethylene, belong polypropylene, polytetrafluoroethylene (Teflon), and polyformaldehyde (Delrin), and to the last type, in addition to the polyamides, belong also polyethyleneglycolterephthalate, and the polycarbonates of bisphenol-A.

In addition to this classification, which is based on chemical structure, one can also classify these polymers by their properties. However, neither the hardness nor the toughness is the same with all "hard" and "tough" polymers. With polyethylene, especially high pressure polyethylene, the property "hard" is not very apparent (compared to a soft rubber, or a wax, it is hard, but compared to iron or steel, it is soft) whereas with polycarbonates, their hardness may already be compared to steel, and one therefore has to state the properties "hard" and "tough" by actual numerical figures.

The hardness can be quantitatively expressed in different ways, for example, by the modulus of elasticity (resistance against torsion or bending) or by the surface hardness, for which there are a number of conventional methods of measurement and ways of expression (ball hardness, Brinell hardness). The toughness can be represented by the impact resistance, or by the tear resistance, which can be determined in different ways, for example, by measuring the energy loss which a metallic pendulum suffers in breaking a notched plastic sample, or by the height of fall of a pendulum, or some other object, at which a plastic sample still doesn't break. It must be taken into consideration, however, that each method yields different numerical values.

In addition, one also measures the bending strength, the tensile strength, and the elongation (the percent elongation which a sample undergoes before it breaks), in order to quantitatively describe the mechanical behavior of a polymer.

Since most synthetic macromolecular compounds are used as plastic materials, the relationship between constitution and mechanical properties is of great interest for the synthesis of new polymeric materials and for the improvement of existing materials.

4534   Impact-Resistant Polymer Blends

It is possible to achieve the "hard and tough" combination of properties possessed by the plastics of the polyethylene type, such as crystalline polyethylene, polypropylene, and polyformaldehyde, by combination of a hard-brittle component with a tough-soft (rubber-elastic) component, in order to produce impact-resistant (hard-tough) materials from brittle plastics. This approach is used on a large technical scale with polystyrene, styrene copolymers, and polyvinylchloride.

With polyethylene, polypropylene, etc., both components of the material, i.e., the crystalline hard component and the noncrystalline rubber-elastic phase, are chemically the same material. This is not the case with blends of polystyrene with rubber-elastic polymers, because there is no rubber-elastic polystyrene or polyvinylchloride at normal temperatures. Different polymers are, however, usually not miscible with each other and in order to overcome this, the rubbery component has to be combined by grafting at the phase boundary areas with the macromolecules of the hard-brittle component.

By far the largest amount of impact-resistant polystyrene (approximately half of the total polystyrene production goes into impact-resistant polystyrene) is prepared nowadays in the following way: one dissolves a gel-free synthetic elastomer (usually polybutadiene or butadiene copolymers) in monomeric styrene, so that one obtains a solution of approximately 10% elastomer. This is polymerized by heating to 80°C-140°C.

Soon after the start of the polymerization the polystyrene precipitates in the form of fine gel-particles (because of its incompatibility with increasing conversion). This is clearly seen under the phase contrast microscope (Figure 619). If one continues the polymerization to complete conversion without stirring, i.e., to a conversion of 97-99%, then the polystyrene particles become larger and larger and finally are separated from each other only by a relatively thin elastomer layer. Because of the cross-linked structure of the elastomer, the entire product is insoluble. Its morphological structure can be examined by looking at a microtome section under the phase contrast microscope (Figure 620-1) or, even better, in an electron microscope (Figure 620-2). One observes a type of cell structure in which the polystyrene occupies the inner cell volume, whereas the rubber-rich phase, which is only present in small amounts (about 10%), forms the cell walls.

The structure of impact-resistant polystyrene is quite different if one stirs the mixture (at least in the initial stages of polymerization). As has been found by G. Molau, the influence of stirring at conversions of 8-10% causes a phase inversion. The rubber phase coagulates to small spherical particles with diameters between 0.1 and 0.5 μ which are embedded in the polystyrene phase. This process of phase inversion can be seen clearly in the series of microphotographs 1 to 4 in Figure 619. First, the polystyrene-rich phase (dark) is shown in the dispersed form. In the stage of the phase reversal one recognizes both domains: those in which polystyrene is dispersed and those in which rubber-rich particles are dispersed in the polystyrene solution. Apparently with higher polystyrene concentrations the system with the rubber as the dispersed phase is the stable one. The necessary inversion of the phases to reach the stable system occurs, however, only then when the inertia resulting from the high viscosity is overcome by addition of an activation energy. As experience has shown, relatively small shearing forces brought about by slow stirring are sufficient.

Figure 621-1a shows the structure of a product prepared under stirring by means of an electron microscope picture. It can be seen that the rubber particles are not homogeneous, but contain a large number of smaller polystyrene beads.

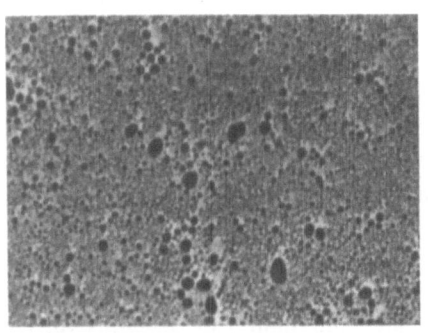

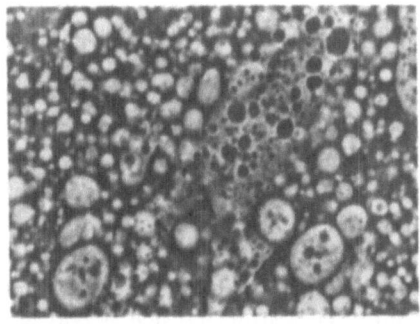

a) 6 hours after the start of the polymerization (conversion = about 5%)

b) after 10 hours (conversion = about 9%)

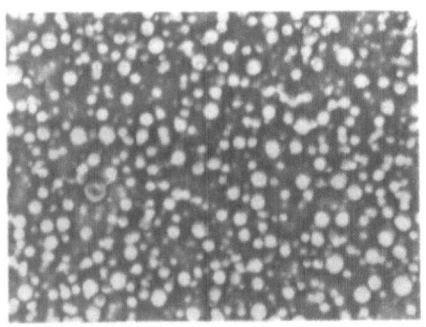

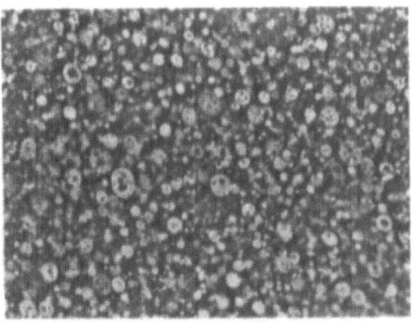

c) after 48 hours (conversion = about 98%), evaporated to form a film on the sample stage

d) like c), but as a microtome section

Scale: 20μ
light = rubber-rich phase
dark = polystyrene-rich phase

FIG. 619 — Morphological structure of rubber solutions during polymerization of the styrene monomer. (Temperature of polymerization: 80°C; pictures taken with phase-contrast microsope)

High-impact polystyrene of this type (as it is to be expected judging from its structure) is soluble. The solutions contain, in addition to polystyrene macromolecules, the rubber-rich graft-copolymer particles which are present as microgels in a maximum swollen state in the solution (Figure 621-2). Judging from its structure (Figure 621-1b), these particles are large branched macromolecules (M $\cong$ 10^8 to 10^{10}) which are held together through graft and network bonds at their phase boundaries, and which are similar in their form to the cell structure of certain living organisms.

It would, however, be misleading to compare such cell structures with those present in nature and in living cells. It is not the form which is characteristic and mysterious in the living cell, but the organization. In order to elucidate organization in the living cell, experiments such as the addition of heated amino acid mixtures to water (S. W. Fox), whereby colloidal systems with spherical particles (d $\cong$ 1 to 5μ) are formed, cannot be considered as realistic. The appearance of such colloidal structures, which have been characterized in a rash manner as prebiotic structures, also in industrially prepared polymer systems (which do not contain water, or proteins, or nucleic acids) clearly shows the completely unspecific character of such structures.

Polymer mixtures with morphological structures similar to those shown in Figure 619(d) can be prepared not only with polybutadiene elastomers, but also with other elastomers such as polyvinyl ethers, polyisobutylene, or polyacrylates. Such mixtures, however, are not impact-resistant materials. With polyacrylates it could be shown that the property of impact resistance appears only if the elastomer is chemically bound to the polystyrene phase at the phase boundaries, i.e., if one makes sure that there is a graft-copolymer zone (B. Vollmert).

With polybutadiene elastomers such graft-copolymers are formed very easily during the polymerization of rubber-monomer solutions by chain transfer (compare

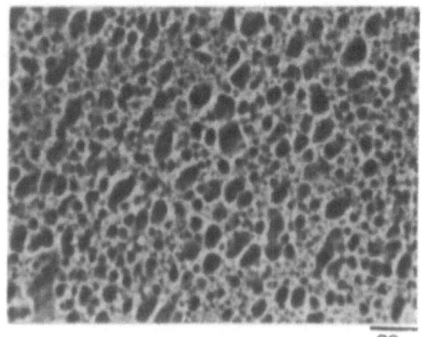 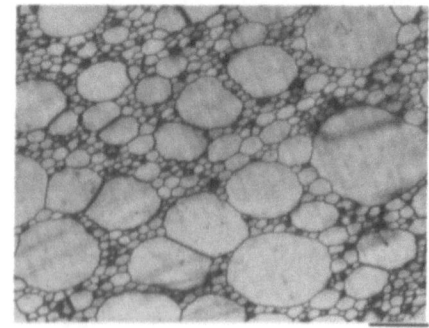

1) phase contrast micrograph: dark = $\overline{20\mu}$ 2) electron micrograph of the same sample;
   polystyrene, light = rubber   light = polystyrene $\overline{4\mu}$

FIG. 620-(1 and 2) — Impact-resistant polystyrene, produced by polymerization of a 5% rubber solution in styrene without stirring (microtome sections).

p. 293). In addition, because of the double bonds in or on the polybutadiene chain, there is also grafting and cross-linking through copolymerization of a small part of the double bonds (especially the vinyl groups on the chains) with the growing polystyrene chains, or through chain transfer reactions (see p. 299). With poly-acrylate elastomers, which have the advantage of better age resistance, grafting occurs over ester linkages (compare pp. 296-301).

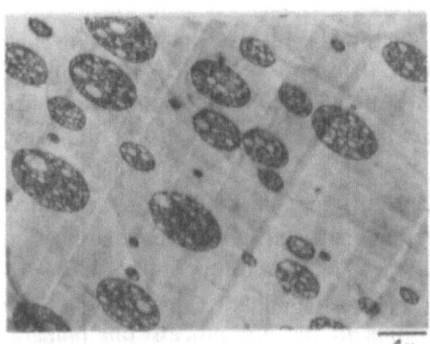

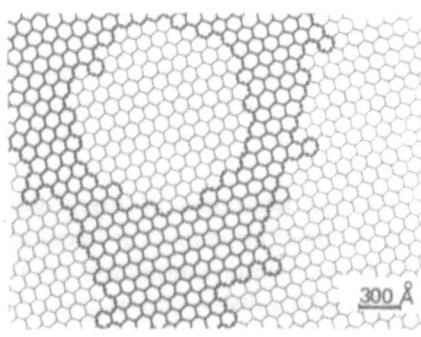

a) microtome section, electron micrograph;
    (light = polystyrene)

b) schematic representation of the molecular
    structure of a small section of an edge of
    one oı the rubber-graftcopolymer particles
    shown in a).
    (light lines = contact zones between
    polystyrene coils
    dark lines = contact zones between
    rubber coils
    〰〰〰 = graft-copolymer zones)

FIG. 621-1 — Impact-resistant polystyrene, produced by polymerization of a 5% rubber solution in styrene (with stirring).

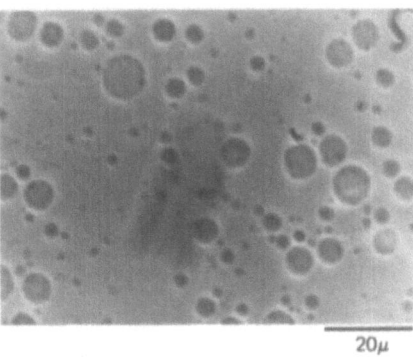

FIG. 621-2 — Phase-contrast photograph of a solution of impact-resistant poly-styrene in toluene (Sample from Fig. 621-1a.)

Polymer mixtures produced in a similar manner from butadiene-acrylonitrile copolymers as the rubbery component and styrene-acrylonitrile copolymers as the hard component are called ABS polymers. Impact-resistant variations are possible not only with polystyrene, but also with Plexiglas and polyvinyl chloride, where one uses chlorinated polyethylene as the rubbery component.

4535 Foams

By "foam" one understands a highly dispersed two-phase system, where a gas is dispersed in a liquid or a solid phase. Many hard and rubber-elastic macromolecular compounds have become very important in the form of foams.: foam rubber, polyurethane foams, foamed polystyrene (Styrofoam and Styropor), urea-formaldehyde foams, etc. The gas phase can be present in the foam in the form of open or closed pores.

For the fabrication of foams there are different methods: (1) Mechanical distribution of air in melts or latices or prepolymers, and subsequent solidification by vulcanization or hardening of the foamed mass. (2) Addition of blowing agents which under suitable circumstances form small gas bubbles. One prepares foam rubber and urea-formaldehyde foams according to the first process; one prepares foamed polystyrene, polyurethane foams, and many other foams according to the second process. As the blowing agent, one can use such low boiling solvents as pentane (with styrofoam), or fluorocarbons (Freon), or nitrogen and CO_2-yielding compounds, such as azobisisobutyronitrile and ethylene carbonate. The blowing gas can also be formed directly in the synthesis reaction (polyurethane foam, compare: p. 239). Among the low boiling solvents, those which are of special use as blowing agents are those which do not dissolve the polymer, but which are soluble to a limited extent in the polymer without changing its external appearance. On heating, the polymer (for example, polystyrene) becomes plastic soft, and the overheated blowing agent boils under the formation of a foam with fine pores.

Foams have extremely low densities (0.1-0.01 g/cm^3) and are used as cushioning, insulation, and packing materials.

TABLE 623. Chain structure, intermolecular forces and physical properties of polymers.

	Chemical composition of the polymer chain	Heat of * evaporation of the monomer	Softening** range of the polymer	Physical properties at room temp.
1	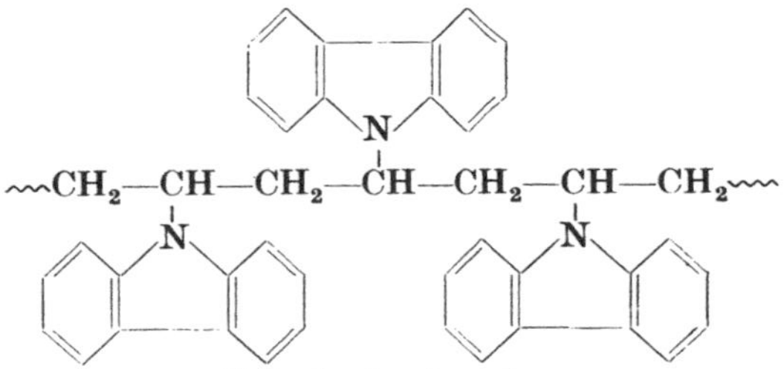 Polyvinylcarbazole	9500	200°C	hard-brittle
2	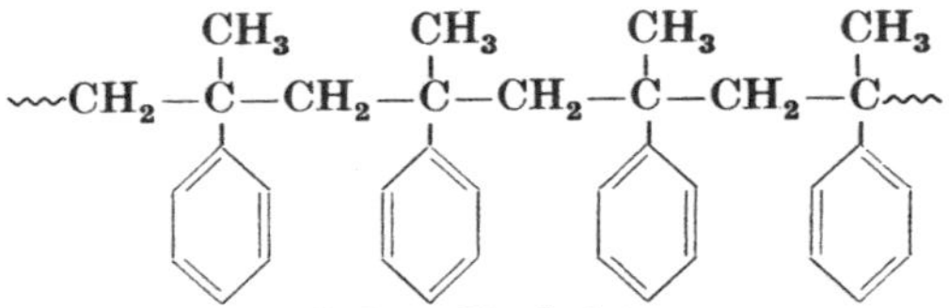 Poly-α-Methylstyrene	8900	175°C	hard-brittle
3	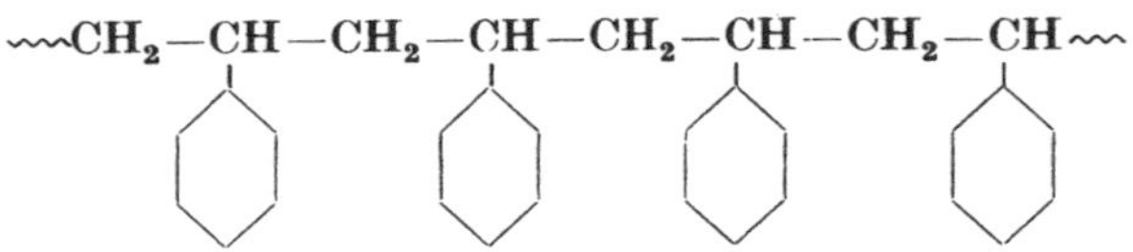 Polyvinylcyclohexane	atactic amorphous 8700 isotactic, cristalline	140°C 325°C	hard-brittle hard-brittle

623

TABLE 623. (Cont'd.)

	Chemical composition of the polymer chain	Heat of * evaporation of the monomer	Softening** range of the polymer	Physical properties at room temp.
4	Polystyrene	atactic, amorphous 8700	100°	hard-brittle
		isotactic, cristalline	230°C	hard-brittle
5	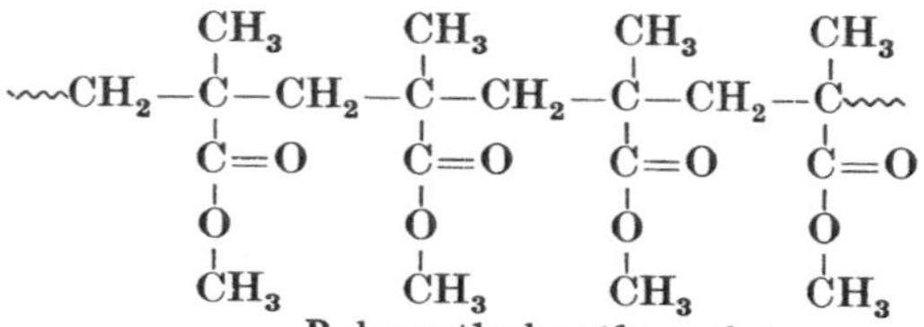 Polymethylmethacrylate	7800	100°C	hard-brittle
6	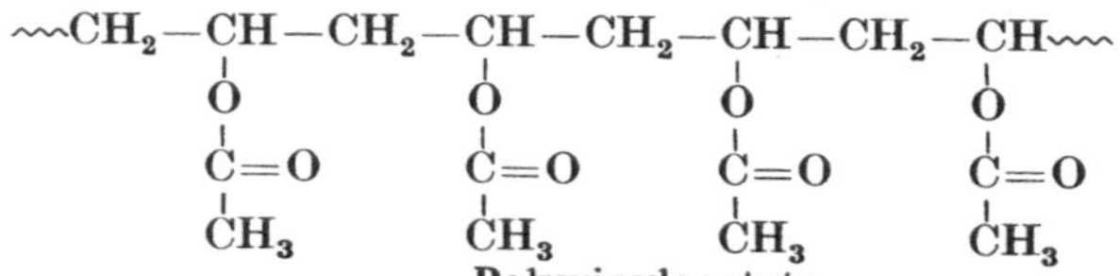 Polyvinylacetate	7200	40°	hard-brittle
7	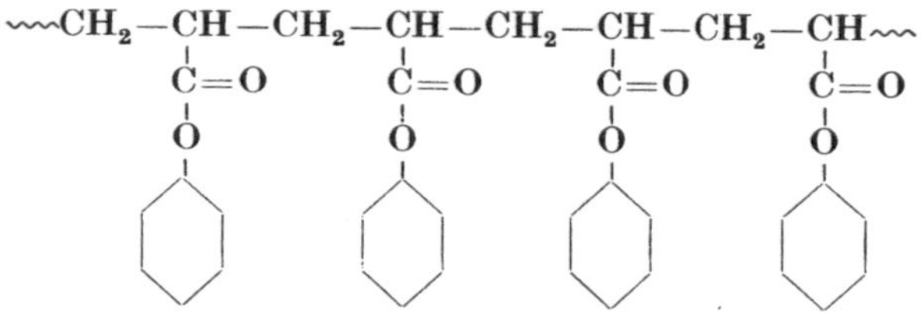 Polycyclohexylacrylate		40°C	hard-brittle

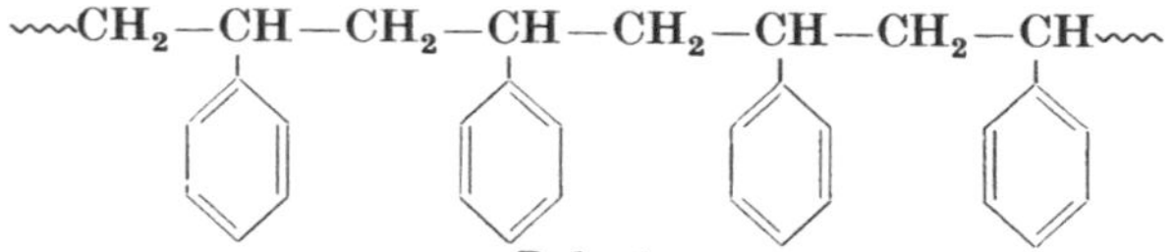

624

No.	Structure		Mol. wt.	Temp.	Property
8	$\sim$CH$_2$—CH—CH$_2$—CH—CH$_2$—CH—CH$_2$—CH$\sim$ with C=O—O—CH$_3$ side groups	**Polymethylacrylate**	7400	0° C	soft-tough
9	$\sim$CH$_2$—CH—CH$_2$—CH—CH$_2$—CH—CH$_2$—CH$\sim$ with C=O—O—CH$_2$—CH$_2$—CH$_2$—CH$_3$ side groups	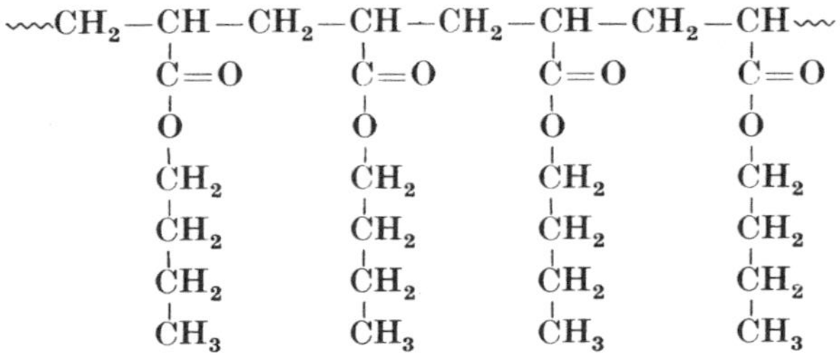 **Polybutylacrylate**	8800	$-$50° C	soft-tough
10	$\sim$CH$_2$—CH—CH$_2$—CH—CH$_2$—CH—CH$_2$—CH$\sim$ with O—CH$_2$—CH(CH$_3$)CH$_3$ side groups	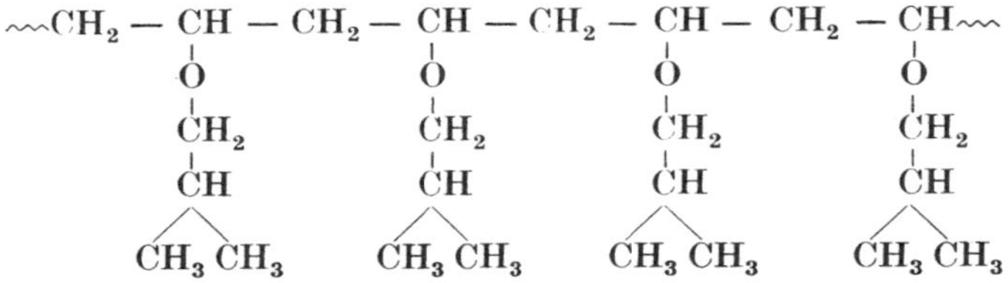 **Polyvinylisobutylether**	7500	$-$60° C	soft-tough
11	$\sim\sim$CH$_2$—C(CH$_3$)(CH$_3$)—CH$_2$—C(CH$_3$)(CH$_3$)—CH$_2$—C(CH$_3$)(CH$_3$)$\sim$	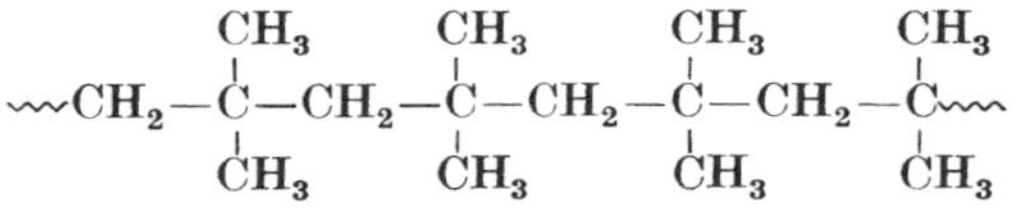 **Polyisobutylene**	5600	$-$60° C	soft-tough

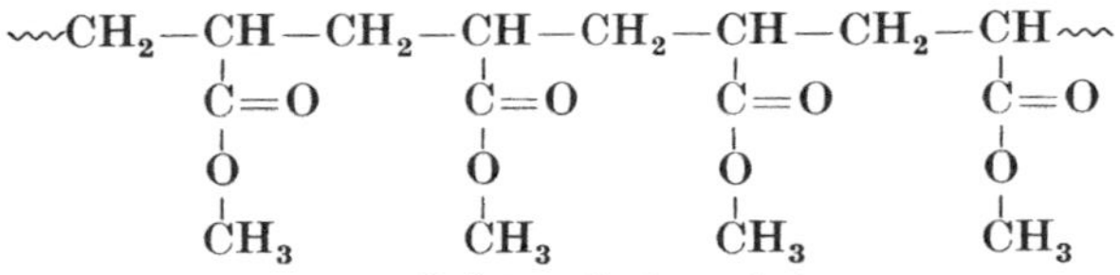

TABLE 623. (Cont'd.)

	Chemical composition of the polymer chain	Heat of evaporation of the monomer *	Softening** range of the polymer	Physical properties at room temp.
12	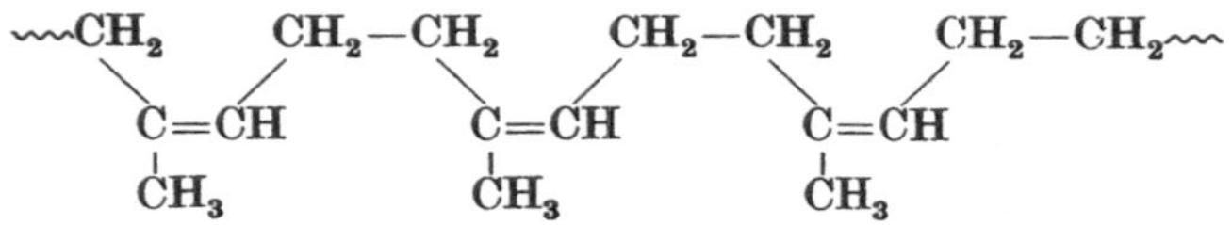 $\sim\sim CH_2-\underset{\underset{CH_3}{\mid}}{C}=CH-CH_2-CH_2-\underset{\underset{CH_3}{\mid}}{C}=CH-CH_2-CH_2-\underset{\underset{CH_3}{\mid}}{C}=CH-CH_2-CH_2\sim\sim$ Cis-1,4-polyisoprene (Natural Rubber)	6500	-70^0 C	soft-tough
13	$\sim\sim CH_2-\underset{\underset{Cl}{\mid}}{C}=CH-CH_2-CH_2-\underset{\underset{Cl}{\mid}}{C}=CH-CH_2-CH_2-\underset{\underset{Cl}{\mid}}{C}=CH-CH_2\sim\sim$ Polychlorobutadiene (Neoprene)		-60^0 C	soft-tough
14	$\sim\sim CH_2-CH=CH-CH_2-CH_2-CH=CH-CH_2-CH_2-CH=CH-CH_2\sim\sim$ 1,4-Polybutadiene	cis-1.4: 5700 trans-1.4:	-100^0 C $+145^\circ C$	soft-tough hard-brittle
15	$\sim\sim CH_2-\underset{\underset{CH_3}{\mid}}{CH}-CH_2-CH_2-CH_2-\underset{\underset{CH_3}{\mid}}{CH}-CH_2-CH_2-\underset{\underset{CH_3}{\mid}}{CH}\sim\sim$ Copolymer of Propylene and Ethylene	4000	-40°C to -60°C	soft-tough
16	$\sim\sim CH_2-O-\underset{\underset{CH_3}{\mid}}{CH}-O-CH_2-O-CH_2-O-\underset{\underset{CH_3}{\mid}}{CH}-O-CH-O-CH_2\sim\sim$ Copolymer of Formaldehyde and Acetaldehyde		below 0°C	soft-tough

#	Structure			
17	$\sim\sim CH_2-CH-CH_2-CH-CH_2-CH\sim$ $\overset{|}{Cl}$ $\overset{|}{Cl}$ $\overset{|}{Cl}$ Polyvinylchloride	5800	80° C	hard-tough
18	$\sim\sim CH_2-CH-CH_2-CH-CH_2-CH\sim$ $\overset{|}{OH}$ $\overset{|}{OH}$ $\overset{|}{OH}$ Polyvinylalcohol		over 250°C	hard-tough
19	$\sim\sim CH_2-CH-CH_2-CH-CH_2-CH\sim$ $\overset{|}{CN}$ $\overset{|}{CN}$ $\overset{|}{CN}$ Polyacrylonitrile	7800	over 250°C	hard-tough
20	$\sim\sim CH_2-\overset{\overset{\displaystyle Cl}{|}}{\underset{\underset{\displaystyle Cl}{|}}{C}}-CH_2-\overset{\overset{\displaystyle Cl}{|}}{\underset{\underset{\displaystyle Cl}{|}}{C}}-CH_2-\overset{\overset{\displaystyle Cl}{|}}{\underset{\underset{\displaystyle Cl}{|}}{C}}$ Polyvinylidenechloride	6500	over 170°C	hard-tough
21	$\sim\sim\overset{\overset{\displaystyle F}{|}}{\underset{\underset{\displaystyle F}{|}}{C}}-\overset{\overset{\displaystyle F}{|}}{\underset{\underset{\displaystyle F}{|}}{C}}-\overset{\overset{\displaystyle F}{|}}{\underset{\underset{\displaystyle F}{|}}{C}}-\overset{\overset{\displaystyle F}{|}}{\underset{\underset{\displaystyle F}{|}}{C}}-\overset{\overset{\displaystyle F}{|}}{\underset{\underset{\displaystyle F}{|}}{C}}-\overset{\overset{\displaystyle F}{|}}{\underset{\underset{\displaystyle F}{|}}{C}}$ Polytetrafluoroethylene (Teflon)	4200	327° C	hard-tough
22	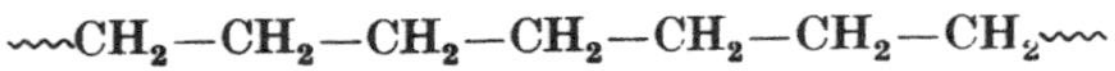 $\sim\sim CH_2-CH_2-CH_2-CH_2-CH_2-CH_2\sim\sim$ Polyethylene, linear	3600	*136°C*	hard-tough
23	$\sim\sim CH_2-\overset{\overset{\displaystyle H}{|}}{\underset{\underset{\displaystyle CH_3}{|}}{C}}-CH_2-\overset{\overset{\displaystyle H}{|}}{\underset{\underset{\displaystyle CH_3}{|}}{C}}-CH_2-\overset{\overset{\displaystyle H}{|}}{\underset{\underset{\displaystyle CH_3}{|}}{C}}-CH_2\sim$ Polypropylene, isotactic	4700	170° C	hard-tough

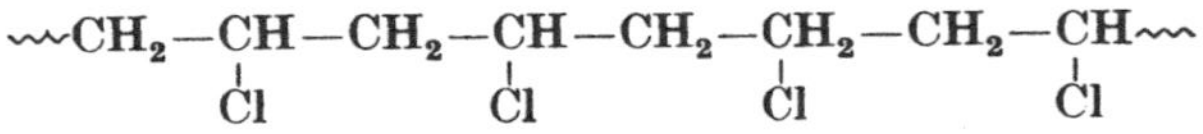

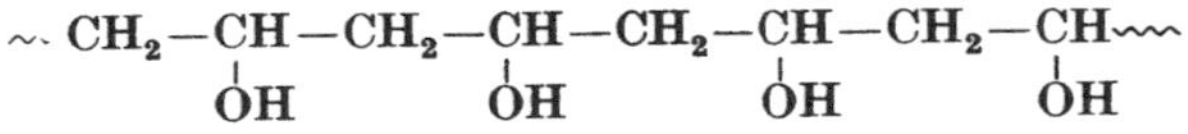

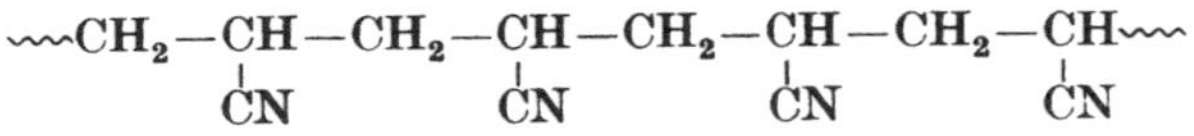

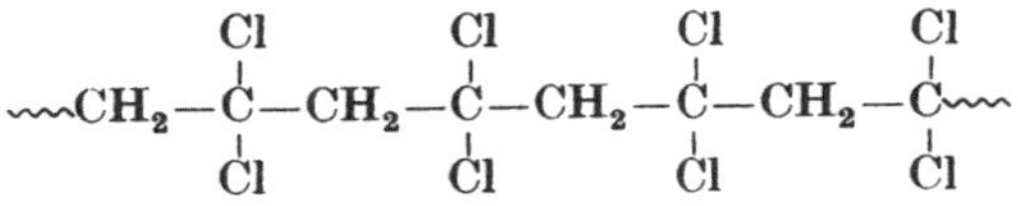

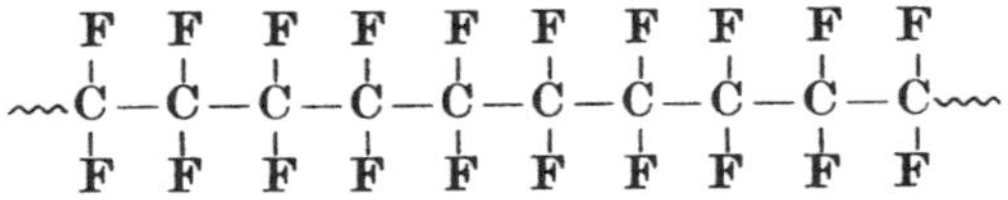

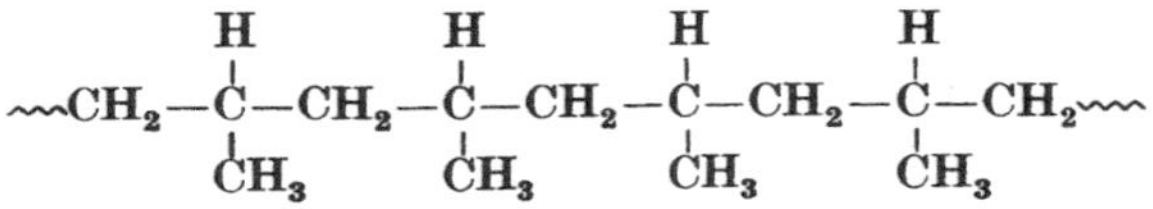

TABLE 623. (Cont'd.)

	Chemical composition of the polymer chain	Heat of evaporation of the monomer [*]	Softening range of the polymer [**]	Physical properties at room temp.
24	$\sim CH_2-O-CH_2-O-CH_2-O-CH_2-O-CH_2-O\sim$ Polyformaldehyde (Delrin, Celcon)	5200	170° C	hard-tough
25	$\sim C-(CH_2)_4-C-NH-(CH_2)_6-NH-C-(CH_2)_4-C-NH-(CH_2)_6-NH\sim$ (with $=O$ at each C) Nylon 66 (Adipic acid-Hexamethylenediamine-Polycondensate)		*262°C*	hard-tough
26	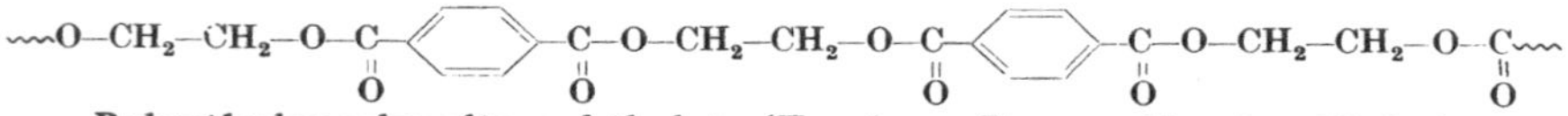 Polyethyleneglycolterephthalate (Terylene, Dacron, Trevira, Mylar)		260° C	hard-tough
27	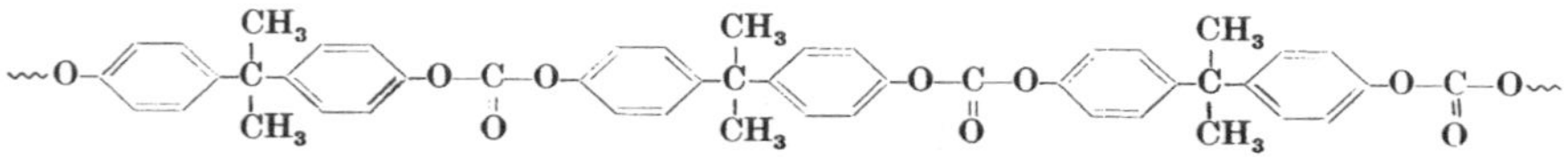 Polycarbonate (from Bisphenol-A and Phosgene (Makrolon, Lexan)		*165°C* *222°C*	hard-tough
28	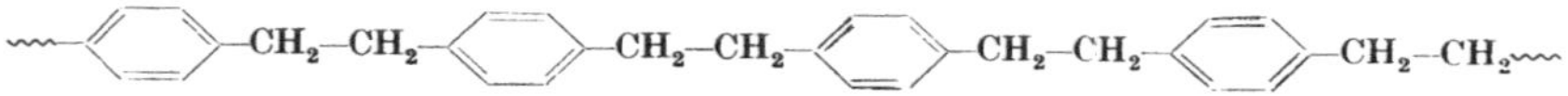 $\sim CH_2-CH_2-[\text{ring}]-CH_2-CH_2-[\text{ring}]-CH_2-CH_2\sim$ Poly-p-xylene (Parylene)		over 200°C	hard-tough

*Approximate values according to Trouton's Rule.

**Only one temperature is listed in this column, and it should be read as "about 200°C", etc. Where italics are used, however, the temperature is the crystalline melting point.

628

APPENDIX

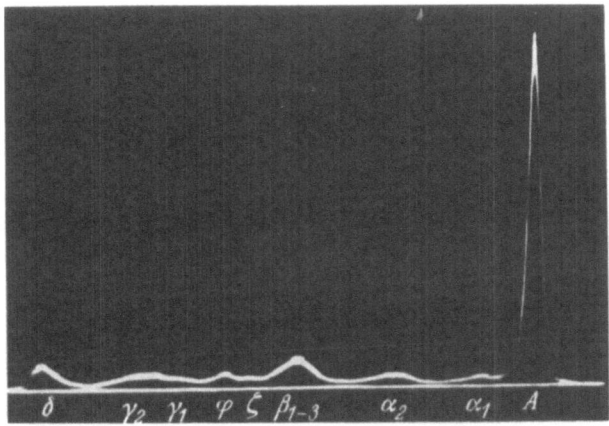

FIG. 629-1 – Electrophoresis diagram of human plasma (according to E. WIEDEMANN)   A = albumin γ = fibrinogen α and β globulins.

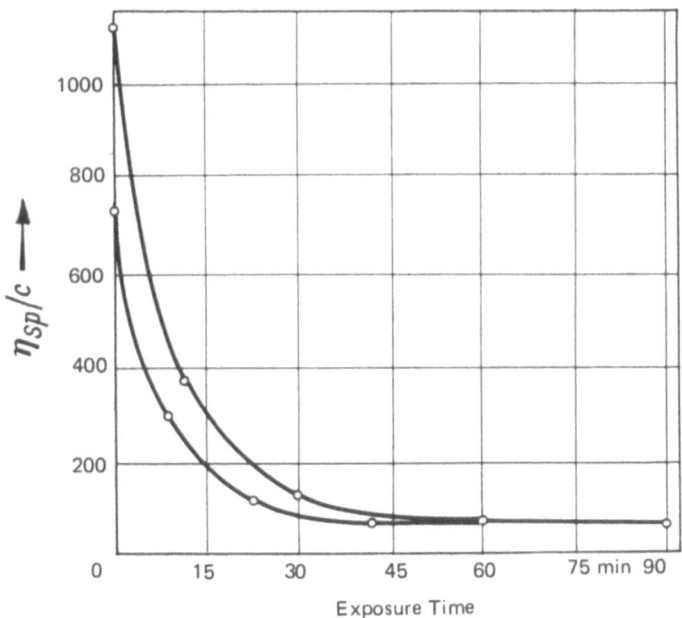

FIG. 629-2 – Degradation of polyacrylamides of different molecular weight with ultrasound in dilute aqueous solutions (according to A. HENGLEIN) (η_{sp}/c measured at c = 0.1%)

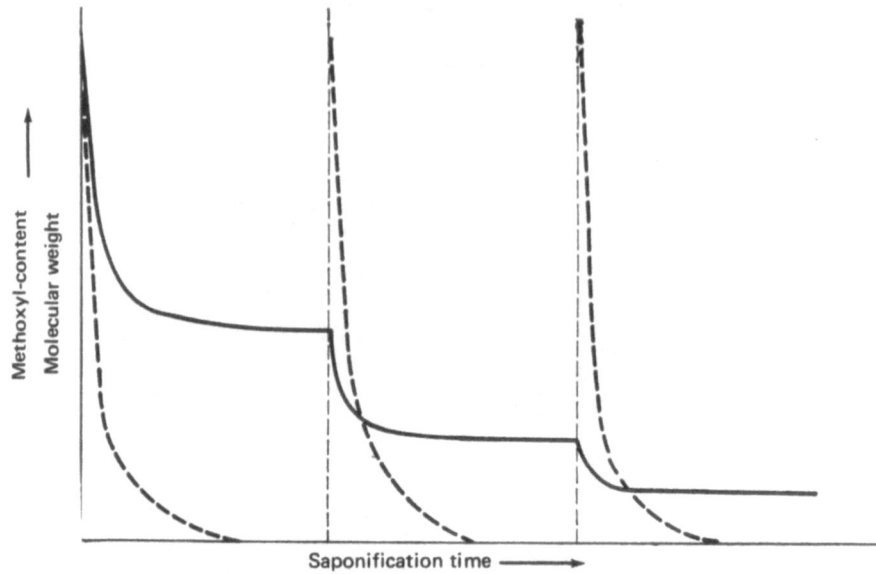

FIG. 630 – Stepwise alkaline degradation of pectin during saponification of the methylester.

————— decrease of the molecular weight
------------- decrease of the methoxyl-content

TABLE 630. Chain-splitting enzymes

Enzyme	Polymer	Occurrence
Cellulases	Cellulose	Bacteria and fungi
Pectinases	Pectin	Penicillium- and Aspergillus-Species
Amylase	Amylose	Malt, saliva, pancreas, liver, bacteria (Bacillus subtilis)
Phosphorylase	Amylose Amylopectin Glycogen	Yeast, muscle, potatoes
Pepsin	Proteins	Stomach lining
Trypsin	Proteins	Pancreas
Carboxypeptidase	Proteins with free terminal carboxylgroup	Pancreas, bacteria
Ribonuclease	Ribonucleic acid (RNA)	Pancreas, bacteria (Azotobacter vinelandii)
Deoxyribonuclease	Deoxyribonucleic acid (DNA)	Pancreas, bacteria (e.g. Escherichia coli)
Phosphodiesterase	Nucleic acids (RNA, DNA)	Snake poison, sweetbreads

INDEX

631

Peter Corke and James Trevelyan (Eds)

Experimental Robotics VI

With 358 Figures

 Springer

Series Advisory Board

A. Bensoussan · M.J. Grimble · P. Kokotovic · A.B. Kurzhanski ·
H. Kwakernaak · J.L. Massey · M. Morari

Editors

Peter Corke, PhD
CSIRO Manufacturing Science and Technology, PO Box 883,
Kenmore 4069, Australia

James Trevelyan, M.Eng.Sc.
Department of Mechanical and Materials Engineering,
The University of Western Australia, Nedlands 6907, Australia

ISBN 978-1-85233-210-5   ISBN 978-1-84628-541-7 (eBook)
DOI 10.1007/978-1-84628-541-7

British Library Cataloguing in Publication Data
Experimental robotics VI. - (Lecture notes in control and
   Information sciences ; 250)
   1.Robotics - Congresses
   I.Corke, Peter I., 1959- II.Trevelyan, James P.
   629.8'92

Library of Congress Cataloging-in-Publication Data
A catalog record for this book is available from the Library of Congress

Typesetting: Camera ready by contributors

69/3830-543210. Printed on acid-free paper SPIN 10718605

Preface

Robotics experiments are complex, require multiple skills and frequently large teams of people, and are often expensive. These factors combined means that relatively little experimentation occurs in the field of robotics. To redress this issue the first International Symposium on Experimental Robotics was held in Montreal in 1989. Subsequent meetings have been held every two years in a rotating sequence through North America, Europe and Asia. The symposia seek to extend the field of robotics by experimentally validating theoretical results, and conversely by experimentation determining the limitation of current theory.

The symposia are small but very special conferences and this one was no exception. It brought together many of the world's leading researchers in a small and intimate conference, small enough for everyone to meet each other. The papers in this volume represent the best of the world's contemporary robotics research in a small and compact volume, and form an essential part of any collection for research institutions, even individual researchers. The papers are presented in a highly readable and accessible form, with plenty of references and contact information to obtain further details.

After the delightful banquet cruise on Sydney harbour most of the delegates were gathered together near the famous Opera House, waiting for a bus to take them back to their hotels. A passer-by asked them "what are you people trying to achieve?" The response was remarkable: all the delegates were lost for words, even though many had taken various drinks which significantly eased their usual inhibitions on speaking out of turn.

Robotics research is in trouble. The era of 'intelligent machines' proclaimed three decades ago has not materialized, and robotics researchers are anxiously looking behind and around themselves, worried that robots are not becoming the ubiquitous machines some people thought they might become by now. We still do not know how to make the 'intelligence' and the machines may need to become more reliable and powerful to achieve the independence needed for full autonomy.

At the 1997 conference on Field and Service Robots held in Canberra Hugh Durrant-Whyte commented "In the past 15 years remarkably little progress has been achieved by the robotics research community. We end up developing systems in which the original theory, techniques and technology are often too fragile, too expensive and inappropriate for industrially-hard application".

One of us (JPT) has made similar comments to students for the last 10 years, posing the question "Where are all the robots which forecasts predicted 20 years ago? Why are there so (relatively) few robots being used?"

As robotics researchers look back on the last two decades, it is often difficult to see where the results of their work have led. Funding agencies have had similar reservations, except perhaps in Japan where the government has embarked on a seven year programme to develop "humanoid robots".

However, this is merely a problem of perceptions and definitions. While few of the innovations which emerge from our work ever appear in the form of robots, or even parts of robots, our results are widely applied in industrial machines which we choose not to define as robots: a common example is the use of computer vision for industrial measurement and inspection. We often find that our robotics research leads not to robots, but better tools which extend the abilities of human workers to the point where they surpass the performance of our robots! This makes it difficult for researchers and other people to understand and appreciate the very significant contributions which emerge from this work.

Many of the papers in this conference describe machines which will never become commonplace. However, many of the solutions needed to make these machines work will become a commonplace, yet largely invisible testament to the efforts reported here.

The program committee for the 1999 meeting comprised:

Alicia Casals	Spain	Raja Chatila	France
Peter Corke	Australia	John Craig	U.S.A.
Paolo Dario	Italy	Vincent Hayward	Canada
Gerd Hirzinger	Germany	Oussama Khatib	U.S.A.
Jean-Pierre Merlet	France	Yoshihiko Nakamura	Japan
Daniela Rus	U.S.A.	Kenneth Salisbury	U.S.A.
Joris De Schutter	Belgium	James Trevelyan	Australia
Tsuneo Yoshikawa	Japan		

We are grateful to the University of Sydney for providing the wonderful environment in which the meeting was held. On behalf of all participants we would like to thank the local arrangements team of Hugh Durrant-Whyte, Lyn Kennedy, David Rye, Eduardo Nebot, Celia Bloomfield and Anna McCallan. Special thanks also to Ms Cuc Tran of CSIRO for assistance with manuscripts, and to our sponsors CSIRO Manufacturing Science and Technology and the Australian Centre for Field Robotics.

Peter Corke (Urbana-Champaign, USA)
James Trevelyan (Perth, Australia)
July 1999.

List of Participants

Prof Dr Helder Araujo
Institute of Systems and Robotics
Department of Electrical Engineering - POL02
University of Coimbra
3030 Coimbra PORTUGAL

Ms Jasmine Banks
C S I R O - Q C A T
2634 Moggill Road
Pinjarra Hills QLD 4069

Mr Oliver Brock
CS-Department, Robotics Laboratory
Stanford University
CA 94305
U S A

Dr Joel Burdick
California Institute of Technology
1200 E California Blvd M/C 104-44
Pasadena CA 91125
U S A

Dr Peter Corke
C S I R O
P O Box 883
Kenmore Qld 4069

Mr Barney Dalton
Department of Mechanical Engineering
University of Western Australia
Nedlands WA 6907

Prof Minoru Asada
Department of Adaptive Machine Systems
Osaka University
Suita
Osaka 565-0871 JAPAN

Mr Adrian Bonchis
C M T E
PO Box 883
Kenmore Qld 4069

Prof Rodney A Brooks
MIT Artificial Intelligence Laboratory
545 Tech Square - Rm 940
Cambridge MA 02139
U S A

Mr Howard Cannon
Caterpillar Inc
Technical Center E855
P O Box 1875
Peoria IL 61656 U S A

Dr Eve Coste-Maniere
INRIA Sophia Antipolis
BP93 06902 Sophia Antipolis
Cedex
FRANCE

Prof Joris De Schutter
Katholieke Universiteit Leuven
Celestijnenlaan 300B
B-3001 Heverlee
BELGIUM

Dr "Dissa" Dissanayake
ACFR Department of Mechanical &
Mechatronic Engineering
University of Sydney
NSW 2006

Mr David Downing
University of Melbourne
Gratton Street
Parkville Vic 3052

Prof Steve Dubowsky
Massachusetts Institute of Technology
77 Massachusetts Avenue Room 3-469
Cambridge MA 02139
U S A

Prof Hugh Durrant-Whyte
ACFR Department of Mechanical &
Mechatronic Engineering
University of Sydney
NSW 2006

Mr Tim Edwards
Department of Systems Engineering
Australian National University
Canberra ACT 2600

Mr Christoforou Eftychios
Department of Mechanical Engineering
University of Canterbury
Private Bag 4800
CHRISTCHURCH NEW ZEALAND

Prof Michael Erdmann
School of Computer Science
Carnegie Mellon University
5000 Forbes Avenue
Pittsburgh PA 15213-3891 U S A

Prof Bernard Espiau
INRIA R.A.
Zirst 655 av de'l Europe
38330 Montbonnot Saint Martin
FRANCE

Mr Chris Gaskett
Robotic Systems Laboratory
Department of Systems Engineering
RSISE
Australian National University
Canberra ACT 0200

Dr Peter Gibbens
Department of Aeronautical Engineering
University of Sydney
NSW 2006

Dr Quang Ha
ACFR Department of Mechanical &
Mechatronic Engineering
University of Sydney
NSW 2006

Prof Vincent Hayward
Center for Intelligent Machines
McGill University
3480 University Street
Montreal QC H3A 2A7 CANADA

Mr Andreas Hein
Humboldt-University Berlin SRL
MUG Virchow-uliniuum
Augustenburger Plate 1
13353 Berlin GERMANY

Mr Jochen Heinzmann
Robotic Systems Laboratory
Department of Systems Engineering
RSISE
Australian National University
Canberra ACT 0200

Prof John Hollerbach
Department of Computer Science
University of Utah
50 S Central Campus Drive
Salt Lake City UT 84112 U S A

Dr Geir Hovland
ABB Corporate Research
Bergerveien 12 P O Box 90
n-1361 Billingstad
NORWAY

Mr H Ikegami
Manufacturing Science & Technical Centre
7th Floor Mori Bldg No 9 1-2-2 Atago
Minato-Ku
Tokyo JAPAN

Prof Ray Jarvis
Depart of Electrical & Comp. Sys. Engineering
Monash University
Wellington road
Clayton Vic 3168

Prof Oussama Khatib
Computer Science
Stanford University
Stanford CA 94305
U S A

Dr Atsushi Konno
Department of Aeronautics & Space Engineering
Tohoku University
Aramaki-aza-Aoba 01, Aoba-Ku
Sendai 980-8579 JAPAN

Prof John Leonard
Massachusetts Institute of Technology
5-422 77 Massachusetts Avenue
Cambridge MA 02139
U S A

Mr Gareth Loy
Robotic Systems Laboratory
Department of Systems Engineering
RSISE
Australian National University
Canberra ACT 0200

Mr Richard Mason
California Institute of Technology
1200 E California Blvd M/C 104-44
Pasadena CA 91125
U S A

Dr Guillaume Morel
University of Strasbourg
ENSPS Blvd S Brant
F67400 Alkrich
FRANCE

Prof Hirochika Inoue
Department of Mechano-Informatics
University of Tokyo
7-3-1 Hongo, Bunkyo-Ky
Tokyo JAPAN

Dr M Kaneko
Hiroshima University
1-4-1 Kagamiyama
Higashi-Hiroshima 739-8527
JAPAN

Dr Pradeep Khosla
HbH1201
Carnegie Mellon University
Pittsburgh PA 15213

Ms Sharon Laubach
California Institute of Technology
1200 E California Blvd M/C 104-44
Pasadena CA 91125
U S A

Prof Pedro Lourtie
DGESup
Av Duque D'Avila 137
P-1050 Lisboa
PORTUGAL

Dr Stuart Lucas
University of Melbourne
Gratton Street
Parkville Vic 3052

Dr Matt Mason
Computer Science Department
Carnegie Mellon University
5000 Forbes Avenue
Pittsburg PA 15213-3891 U S A

Prof Yoshi Nakamura
Department of Mechano Informatics
University of Tokyo
Hongo, Bunkyo-ku
Tokyo 113 JAPAN

Dr Eduardo Nebot
ACFR Department of Mechanical &
Mechatronic Engineering
University of Sydney
NSW 2006

Dr Rhys Newman
Robotic Systems Laboratory
Department of Systems Engineering
RSISE
Australian National University
Canberra ACT 0200

Ms Allison Okamura
Stanford University
560 Panama Street
Stanford CA 94305
U S A

Mr Arthur Quaid
Robotics Institute Smith Hall
Carnegie Mellon University
5000 Forbes Avenue
Pittsburgh, PA 15213 U S A

Prof M Renaud
L A A S - C N R S
7 Avenue du Colonel Roche
31077 Toulouse
Cedex 4 FRANCE

Dr Jonathan Roberts
CSIRO Manufacturing Science and Tech-
nology
P O Box 883
Kenmore Qld 4069

Prof Daniela Rus
HB6211 Sudikoff
Dartmouth
Hanover NH 03755
U S A

Dr David Rye
ACFR Department of Mechanical &
Mechatronic Engineering
University of Sydney
NSW 2006

Prof Brad Nelson
Department of Mechanical Engineering
University of Minnesota
111 Church Street SE
Minneapolis MN 55455 U S A

Prof Yoshi Ohkami
Tokyo Institute of Technology
2-12-1 Ohokayama
Ohta-ku Tokyo
JAPAN

Prof Dinesh Pai
Computer Science, 2366 Main Hall
University of British Columbia
Vancouver BC V6T 1Z4
CANADA

Mr Danny Ratner
School of Information Technology & Com-
puter Sci
University of Wollongong
Northfields Avenue
Wollongong NSW 2522

Dr Alfred Rizzi
Robotics Institute
Carnegie Mellon University
5000 Forbes Avenue
Pittsburgh, PA 15213 U S A

Dr Sebastien Rougeaux
Department of Systems Engineering
RSISES
Australian National University
Canberra ACT 0200

Mr Hendrik Rust
Fraunhofer I P A
Nobelstr 12
70453 Stuttgart
GERMANY

Dr Kenneth Salisbury
M I T Artificial Intelligence Lab
545 Technology Square NE43-837
Cambridge MA 02139
U S A

Dr Andrew Samuel
University of Melbourne
Gratton Street
Parkville Vic 3052

Dr Pavan Sikka
C S I R O
P O Box 883
Kenmore Qld 4069

Dr Sanjiv Singh
Field Robotics Centre
Carnegie Mellon University
Pittsburgh PA 15213
U S A

Mr Mike Stevens
ACFR Department of Mechanical &
Mechatronic Engineering
University of Sydney
NSW 2006

Mr T Sunaoshi
Tokyo Institute of Technology
2-12-1 Ohokayama
Ohta-ku Tokyo
JAPAN

Prof Atsuo Takanishi
Waseda University
3-4-1 Okubo, Shinjuku-ku
Tokyo 169-8555
JAPAN

Mr Ken Taylor
Robotic Systems Laboratory
Department of Systems Engineering
RSISE
Australian National University
Canberra ACT 0200

Dr Craig Tischler
University of Melbourne
Gratton Street
Parkville Vic 3052

Mr Samuel A Setiawan
Waseda University
3-4-1 Okubo, Shinjuku-ku
Tokyo 169-8555
JAPAN

Mr Chanop Silpa-Anan
Robotic Systems Laboratory
Department of Systems Engineering
RSISE
Australian National University
Canberra ACT 0200

Prof Mandyam Srinivasan
Research School of Biological Sciences
Australian National University
Canberra ACT 2601

Mr Tom Sugar
GRASP Laboratory
University of Pennsylvania
34011 Walnut Street Suite 300c
Philadelphia PA 19104-6228 U S A

Dr Jerome Szewczyk
Laboratoire de Robotique de Paris
10/12 avenue de l'Europe
78140 Velizy
FRANCE

Dr Juan Domingo Tardos
Universidad de Zaragoza
Maria de Luna 3
E-50015 Zaragoza
SPAIN

Mr Simon Thompson
Robotic Systems Laboratory
Department of Systems Engineering
RSISE
Australian National University
Canberra ACT 0200

A/Prof James Trevelyan
University of Western Australia
Nedlands WA 6907

Prof Takashi Tsubouchi
Intelligent Robot Laboratory
University of Tsukuba
1-1-1 Tennoudai, Tsukuba 305 8573
JAPAN

Prof Luc Van Gool
University of Leuven
Kardinaal Mercierlaan 94
B-3001 Leuven
BELGIUM

Ms Koren Ward
University of Wollongong
Ohio State University

Prof YangSheng Xu
Department of MAE
The Chinese University of Hong Kong
Shatin
HONG KONG

Dr Alex Zelinsky
Robotic Systems Laboratory
Department of Systems Engineering
RSISE
Australian National University
Canberra ACT 0200

Prof Masaru Uchiyama
Department of Aeronautics & Space Engineering
Tohoku University
Aoba-yama 01, Sendai 980-8579
JAPAN

Dr Ken Waldron
Department of Mechanical Engineering
Ohio State University
Columbus Ohio 43210-1154
U S A

Dr David Wettergreen
Robotic Systems Laboratory
Department of Systems Engineering
RSISE
Australian National University
Canberra ACT 0200

Prof T Yoshikawa
Department of Mechanical Engineering
Kyoto University
Kyoto 606-8501
JAPAN

Contents